Particle Technology Forum

AA003013

Presentations at the 2008 AIChE Annual Meeting

Philadelphia, Pennsylvania
16 – 21 November 2008

Volume 1 of 2

ISBN: 978-1-61567-218-9

Printed from e-media with permission by:

Curran Associates, Inc.
57 Morehouse Lane
Red Hook, NY 12571

Some format issues inherent in the e-media version may also appear in this print version.

Printed by Curran Associates, Inc. (2009)

For permission requests, please contact AIChE
at the address below.

AIChE
3 Park Avenue
New York, NY 10016-5991

Phone: (203) 702-7660
Fax: (203) 775-5177

www.aiche.org

Additional copies of this publication are available from:

Curran Associates, Inc.
57 Morehouse Lane
Red Hook, NY 12571 USA
Phone: 845-758-0400
Fax: 845-758-2634
Email: curran@proceedings.com
Web: www.proceedings.com

TABLE OF CONTENTS

VOLUME 1

Characterization of Crystal Size and Shape 1
Michel Kempkes, Jan Eggers, Thomas Vetter, Marco Mazzotti

Modeling the Manipulation of Crystal Morphology Distributions 3
Christian Borchert, Nandkishor Nere, Andreas Voigt, Kai Sundmacher, Doraiswami Ramkrishna

Parameter Estimation of Preferential Crystallization Using a Detailed Mathematical Model 5
Yoshiaki Kawajiri, Ivan Angelov, Grzegorz Ziomek, Martin Peter Elsner, Andreas Seidel-Morgenstern

Dynamic Modelling of Particle-Droplet Collisions 6
M.J. van der Hoeven, Tony Howes, James D. Litster, W.J. Wildeboer, G.M.H. Meesters, I.T. Cameron

Shear-Rate Dependence of the Initial Cluster Aggregation Kinetics of Polystyrene Latices Under Turbulent Conditions in Stirred Tank 7
Lyonel Ehrl, Miroslav Soos, Massimo Morbidelli

Shear Induced Aggregation Rate of Colloidal Nanoparticles and Clusters In the Presence of Repulsive Interactions 8
Marco Lattuada, Massimo Morbidelli

Insights in Scaling and Optimization of Ball Mill Operations - A Discrete Element Method Based Approach 9
Joerg Theuerkauf, Kaustubh S. Mujumdar

Characterization of Granular Flow In a Bladed Mixer Using the Discrete Element Method 10
Brenda Remy, Johannes G. Khinast, Benjamin J. Glasser

Mixing Analysis of Binary System of Particles in An Agitated Bed by Discrete Element Method 11
Ajit Mujumdar, Masayuki Horio, Toshio Maetani, Naomichi Nakamura

Rheological Behavior of Quasi-Static Granular Flows Under Unsteady Simple Shear 14
Jin Sun, Lee R. Aarons, Sankaran Sundaresan

Using Janus Particles to Control Mixing/Segregating of Adhesive Granular Materials in Rotating Tumblers 16
Isabel Figueroa, Joseph J. McCarthy

Granular Mixing and Segregation In Zig-Zag Chute Flow 24
Suman K. Hajra, Deliang Shi, Watson L. Vargas, J. J. McCarthy

Discrete Element Method Simulations of Dry Impregnation Processes in Catalysts 25
M. Silvlna Tomassone, Charles Radeke, Arthur W. ChEster

A DE Numerical Study upon Particulate Solids Flow from Model Silos 26
Guoxian Xiao, Delong Xu, S. Ding

Measuring Particle Attrition In a Jet Cup 33
Ray A. Cocco, S.B. Reddy Karri, Yeook Arrington, Ted Knowlton

Shear and Composition Effects on the Flow and Electrostatic Properties of Pharmaceutical Powders 34
Alisa Vasilenko, Kalyana Pingali, Fernando J. Muzzio

Experimental and Numerical Analysis of Stress Generation In Non-Circular Pharmaceutical Tablets - A Manufacturing Case Study 35
Stephen L. Conway, Adam Procopio, Marty J. Smith, Michael Gentzler

Scale up and Optimization of Fluid Bed Coating and Layering Processes Using In-Situ Particle Characterization 36
Andrew Birkmire, Kim T. Walter, Eric Dycus, Benjamin Smith, Frederic Prulliere

Quantitative Large-Scale Validation of Image-Based Sensors for Online Particle Size Characterization during Crystallization 37
Ying Zhou, Rajagopalan Srinivasan, L Samavedham

Characterization of Density of Cohesive Granular Materials Using An X-Ray CT Scanner 39
M. Silvina Tomassone, Charles Radeke, Divya Jonnavittula

Application of Helmholz Resonance to Determination of Bulk Density and Fill Fraction of Dry Particulate Solids 40
Emile S. Webster, Clive E. Davies

Newton's Cradle Undone: Experiments and Collision Models for the Normal Collision of Three Solid Spheres 44
Carly M. Donahue, Christine M. Hrenya, Kenshiro Nakagawa, Alexandra Zelinskaya

Stokesian Dynamics for Rigid Body Motion of Particle Aggregates 45
Yogesh M. Harshe, Marco Lattuada, Massimo Morbidelli

Silica Precipitation during Analcime Dissolution 46
Elizabeth A. Gorrepati, H. Scott Fogler

Comparison of the Use of Different Solubility Models for the Cooling Crystallization of Acetaminophen 47
David Widenski, Ali Abbas, Jose A. Romagnoli

Development and Validation of History-Dependent Inter-Particle Bonding Terms In Compacted Powder Beds 55
Athanas A. Koynov, Alberto Cuitino

Discrete Element Method Simulation of a Dense Phase Conveying System 56
Bing Du, Ting(Tim) Han, Nestor A. Vasquez, Joerg Theuerkauf

Discrete Element Method Simulation of a Dilute Conveying System 57
Bing Du, Ting(Tim) Han, Nestor A. Vasquez, Joerg Theuerkauf

Computation of Turbulence and Low Sherwood Numbers in Fluidized Beds 58
Benjapon Chalermsinsuwan, Pornpote Piumsomboon, Dimitri Gidaspow, Mayank Kashyap, Ronald W. Breault

Influence from Particle Size Distributions on the Cfd Simulations and Experiments of Bubbling Fluidized Beds 66
Sanoja A. Jayarathna, WenJun Wu, Chameera K. Jayarathna, Britt M. Halvorsen

Jetting Observations In a Cold-Flow Fluidized Bed 75
Nathan P. Franka, Theodore J. Heindel

Imaging Gas/Solid Jet Penetration in a Gas-Solid Fluidized Bed Using Electrical Capacitance Volume Tomography 76
Fei Wang, Qussai Marashdeh, W. Wasito, L. -S. Fan

A Rotating Chimney to Increase the Flexibility of Rotating Fluidized Beds 83
Juray De Wilde

The Effect of Bed Depth on the Hydrodynamics of a Fluidized Bed of Fcc Catalyst 87
Todd S. Pugsley, Shayan Karimipour

New Nanopowders and Nanostructured Materials Derived from Liquid-Feed Flame Spray Pyrolysis (LF-FSP) 88
Richard M. Laine, Min Kim, Jose Azurdia, Bingyuan Huang

Generating Nanoparticle Test Aerosol Using Liquid Flame Spray 89
Mikko Aromaa, Jyrki M. Mäkelä, Antti Rostedt, Thomas J. Krinke, Kauko Janka, Marko Marjamäki, Jorma Keskinen

Flame Synthesis of Nano-Scale Nio/ysz and NiO/CeO2/YSZ Powders for Solid Oxide Fuel Cell Anode Catalyst Applications 90
Ranjan Pati, Osifo Akhuemonkhan, Michael B. Pomfret, Hillary Sadoff, Robert A. Walker, Sheryl Ehrman

Synthesis of ZnS Nanoparticles by Spray Pyrolysis 91
Sha Liu, Hongwang Zhang, Mark T. Swihart

Catalysis by Flame Synthesized Supported Gold Nano-Particles 92
Sachit Chopra, Jeroen van Bokhoven, Jagdeep Singh, Gregory Beaucage

Flame Aerosol Synthesis of Functional Composite Nanoparticles 93
Bing Guo, Hoon Yim

Investigation of Ligand Solvation Effects for Alkyl Thiol Stabilized Metal Nanoparticles In CO2 Expanded Liquids Using Small Angle Neutron Scattering 94
Gregory Von White II, Christopher L. Kitchens

A New Ion-Mobility Method for Probing Reaction Kinetics at the Nanoscale 95
Lei Zhou, X. Ma, Anshuman Lall, G. M. Mulholland, Michael Zachariah

Functionalization and Characterization of Gold Nanoparticles 96
Sirnegeda D. Techane, Lara J. Gamble, David G. Castner

The Quality of SiO2 Coatings on Flame-Made TiO2-Based Nanoparticles 97
Alexandra Teleki, Kamal M. Akhtar, Sotiris E. Pratsinis

Properties and Characterization of Pharmaceutical Powders Modified by Dry Coating 98
Lauren Beach, Yuhua ChEn, Laila Jallo, Ajit Mujumdar, Rajesh Dave

Prediction of Inter-Particle Adhesion Force from Surface Energy and Roughness 99
Laila Jai Jallo, Yuhua ChEn, Lauren Beach, Ajit Mujumdar, Dave Rajesh

Axial Particle Concentration Distribution In Two-Phase and Three-Phase Fluidized Beds with Binary Solid Mixtures 100
Dong Hyun Lee, Byung Soon Chun, Young Lee, Ah-Hyung Alissa Park, Sang Done Kim, John R. Grace, Norman Epstein

Phase Transfer in a Droplet-Solid Sphere Collision 101
Dawei Wang, Zheng Shen, Chao Zhu

Enhanced Fluidization of Agglomerates of Nanoparticles 110
Jose A. Quevedo, Robert Pfeffer

Emulsion Phase Density In Deep Fluidized Beds of Geldart Group a Materials 111
Allan Issangya, Reddy S. B. Karri, Ted Knowlton

Investigation of Fluidised Bed Dryer by Frequency-Depend Measurement and Different Normalisation Methods with Electrical Capacitance Tomography 112
H. G. Wang, W. Q. Yang

Tomographic Imaging during Semi-Batch Reactive Precipitation of Barium Sulphate In a Stirred Vessel 124
Thomas L. Rodgers, David Stephenson, Michael Cooke, Trevor A. York, Reginald Mann

Time-Resolved X-Ray Tomography of a Fluidized Bed 125
Robert F. Mudde

Improvement of Granular Mixing In Rotating Cylinders 134
Suman K. Hajra, Tathagata Bhattacharya, J. J. McCarthy

Using Dem to Identify Mixing Mechanisms In Bladed Mixers 135
Stefan Radl, Benjamin J. Glasser, Johannes G. Khinast

Understanding Mixing to Enhance Coating Performance In a Pancoater 136
Ekneet Sahni, Robert Yau, Bodhisattwa Chaudhuri

Using Dem Models to Understand Segregation In Mixtures with Continuous Grain Size Distribution In Long Cylinders 137
Atul Dubey, Fernando J. Muzzio, Maria. S Tomassone

Mixing of Nanosized Particles by Magnetically Assisted Impaction Mixing in Dry and Fluid Suspension 138
James V. Scicolone, Ajit Mujumdar, Rajesh Dave

Simulation of a Continuous Granular Mixer: Effect of Operating Conditions on Flow and Mixing 143
Avik Sarkar, Carl R. Wassgren

Nano-Mixing of Dipyridamole Drug and Excipient Nanoparticles by Sonication in Liquid CO2 144
Ganesh Sanganwar, Ram B. Gupta

Forces Between Functionalized Silica Nanoparticles 145
Ahmed E. Ismail, J. Matthew D. Lane, Michael E. Chandross, Kelly Anderson, Chris Lorenz, Gary S. Grest

Molecular Modeling of Particle Nucleation In Emulsion Droplets: Towards An Improved Understanding of Nanoparticle Production Processes 146
Andreas Voigt, Michael Fricke, K. Sundmacher

Computational Modeling of Silicon Nanoparticle Synthesis In a Laser-Driven Reactor 148
Hongyi Dang, Mark T. Swihart

Scale-up of a Nanoparticle Size-Selective Fractionation Process Using CO2-Expanded Liquids 149
Steven R. Saunders, Christopher Roberts

Kinetic Study and Modeling of Metal Particle Formation 150
Roberto Irizarry, Madeline Leon, Miguel Castro

Efficient Optimization of Particulate Systems Described by Monte Carlo Method: Determination of Nucleation Parameters from Kinetic Data 151
Roberto Irizarry

Large Scale Synthesis of Luminescent Rare-Earth Doped Oxide Nanoparticles 152
Yuanbing Mao, Jane P. Chang

Solids Drying Process Simulation by Solidsim - New Tool for Flow Sheeting of Solids Processes 153
Ting(Tim) Han, Joerg Theuerkauf

Characterization of Nano-Sized Ni/ce(x)Zr(1-x)O2 Catalyst for the Auto Thermal Reforming 154
No Author Available

Characterization of Ni/Ce(x)Zr(1-x)O2 Catalyst for Autothermal Reforming 155
Jin-Hwa Kong, Dong-Ju Moon, Jong-Ho Kim, Nam-Cook Park, Jae-Soon Shin, Young-Chul Kim

Generation and Geometrical Analysis of Dense Clusters with Variable Fractal Dimension Providing a Basis for the Interpretation of Data from Optical Particle Characterication Techniques 156
Lyonel Ehrl, Marco Lattuada, Miroslav Soos, Massimo Morbidelli

The Breakage Matrix Approach In Evaluating Inadvertent Breakage of Dense Particulate Mixtures 157
A. Abu-Nahar, U Tüzün

Modeling of An Industrial Vibrating Double-Deck Screen of a Urea Granulation Circuit 159
Ivana Cotabarren, José Rossit, Juliana Piña, Verónica Bucalá

Simulation of Continuous Granular Mixing 169
Avik Sarkar, Carl R. Wassgren

Coagulation of Highly Concentrated Aerosols 170
Beat Buesser, Sotiris E. Pratsinis

Preparation of Dextran Particles by Using Sas Process 171
Dong-Yuk Kang, Dong-Lyun Cho, Choon-Hyoung Kang

Effects for TiO2 Structures In Dye Sensitized Solar Cell 172
Byoung-Jun Min, Bok-Min Kim, Seon-Gyun Rho, Sung-Yong Cho, Seung-Jai Kim, Choon-Hyoung Kang

The Role of Gas-Aerosol Mixing during In-Situ Coating of Flame-Made Titania Particles 173
Alexandra Teleki, Martin C. Heine, Frank Krumeich, Kamal M. Akhtar, Sotiris E. Pratsinis

New Flame Pyrolysis Reactor for Carbon Black Synthesis: Soot Surface Growth and Oxidation 174
X. Ma, Anshuman Lall, Dale Huang, Michael Zachariah

Preparation of CeO2/SiO2 Nano-Abrasive Particles 175
Sang-Ho Chung, In-Seon Park, Young-Moo Park, Dae-Won Lee, Kwan-Young Lee

Shear-Induced and Brownian Coagulation 176
Max L. Eggersdorfer, Beat Buesser, Sotiris E. Pratsinis

Three-Dimensional Gas-Solid Fluidized Bed Cfd Simulation Using a Lattice Boltzmann Based Drag Correlation 178
Milton Mori, Maximilian J. Hodapp, Maria G. E. Silva

Optimization of Particle Agglomeration and Breakage Operations with the Attainable Region (AR) Theory 187
Matthew J. Metzger, Benjamin J. Glasser, Ngangezwe Khumalo, Brendon Hausberger, Diane Hildebrandt, David Glasser

Force Model Concerns for Glued Sphere Dem Simulations 188
Madhusudhan Kodam, Rahul Bharadwaj, Jennifer Sinclair Curtis, Bruno C. Hancock, Carl R. Wassgren

Effect of Dem Parameters on the Collision Distribution of a Small Shearing Region below the Surface of a Rotating Drum 189
Ben Freireich, James D. Litster, Carl R. Wassgren

Behavior of the Elastomer Particles with Different Surface Charge Groups during Aggregation Under Static and Flow Conditions 190
Miroslav Soos, Cornelius Gauer, Hua Wu, Massimo Morbidelli

Simulation of a High-Flux Riser-Reactor Flow Using Cfd Techniques 191
Leonardo Machado da Rosa, Jaci Carlo Schramm Camara Bastos, Milton Mori, Waldir Pedro Martignoni

Synthesis of Doped TiO2 Metal Oxide Nanoparticles in Supercritical CO2 202
Gabrielė Aksomaitytė, Esther Alonso, Maria José Cocero

Simulation of a Confined Turbulent Two Phase Jet Flow 204
Milton Mori, Jaci C. S. C. Bastos, Rodrigo K. Decker, Udo Fritsching

Study of Different Approaches for Modeling Cyclones Using Cfd 218
Daniel de Brito Dias, Milton Mori, Waldir Pedro Martignoni

Investigating Size-Resolved Nanoparticle Reaction Kinetics: A Novel Tandem Ion-Mobility Method 228
Lei Zhou, X. Ma, Anshuman Lall, G. M. Mulholland, Michael Zachariah

Nanofluidization as Affected by Vibration and Electrostatic Fields 229
Daniel Lepek, Robert Pfeffer, Rajesh Dave, Miguel Angel Sanchez Quintanilla, Jose Manuel Valverde, Antonio Castellanos

Preparation and Characterization of Nanodispersed Ag-Based Sorbent for Removal of Sulfur from Logistic Fuels 230
Alexander Samokhvalov, Sachin Nair, Zenda Davis, Bruce Tatarchuk

Synthesis of Nanoscale Cobalt-Iron Spinel Oxides Via Atomic Layer Deposition and Their Applications for Solar Thermochemical Water Splitting 232
Jonathan R. Scheffe, Nathan P. Siegel, Mark D. Allendorf, Alan W. Weimer

Effect of Interactions on Nanoparticle Dispersion In a Polymer Solution 233
Deepika R. Gollamandala, Ileana C. Carpen

Numerical Simulation of Granular Flow In a Bladed Mixer 234
Brenda Remy, Johannes G. Khinast, Benjamin J. Glasser

Verification and Validation Study of Continuum Granular Frictional Flow Theories Using a Discrete Technique and Experimental Data 235
Sofiane Benyahia

Development of An Automatic Fluidization Rig 244
S. Ding, Are Dyrøy, Morten Karlsen

Effect of Process Variables in the Preparation of Bohemite Sol on the Microstructure of Gamma-Alumina Particles 245
Chul-Joo Kim, Ho-Sung Yoon, Sung-Don Kim, Hyungkyu Park, Jung-Soo Sohn

Preparation and Growth of Silica Sol from Sodium Silicate Solution 246
Ho-Sung Yoon, Chul-Joo Kim, Sung-Don Kim, Hee Dong Jang

Preparation of Anhydrous Magnesium Chloride by Dehydration of Ammonium Magnesium Chloride Hexahydrate 247
Hyoung-Choon Eom, Hyungkyu Park, Chul-Joo Kim, Ho-Sung Yoon

Supported Gold Nano-Catalysts by Spray Synthesis and Solution Route. A Comparative Study 248
Sachit Chopra, Jagdeep Singh, Jeroen van Bokhoven, Gregory Beaucage

Characterization of Turbulent Fluidization Regime with Geldart Group B Particles by Wavelet Analysis of Pressure Fluctuation Signals 249
Lii-ping Leu, Yung-Sz Yu

Ex-Situ Surface Modification of BaTiO3 Nanoparticles Prepared by Hydrothermal Synthesis 250
Mehrnoosh Atashfaraz, Mojtaba Shariaty-Niassar, Tadafumi Adschiri

Simulation of Dry Coating Process Using a DEM Method 251
Peng Weng, Costas G. Gogos, Linjie Zhu, Ming-Wan Young

A Novel Process for Simultaneous Milling and Coating Particulates 252
Peng Weng, Costas G. Gogos, Linjie Zhu, Ming-Wan Young, Peter Bonnett

Molecular Layer Deposition of Ultra-Thin Polymer Films on Large Quantities of Cohesive Nanoparticles 253
Xinhua Liang, David M. King, Steven M. George, Alan W. Weimer

Design of High Temperature, In-Situ Coating of Aerosols: Efficiency & Texture 254
Beat Buesser, Sotiris E. Pratsinis

Mechano-ChEmical Stability of Gold Nanoparticles Coated with Alkanethiolate SAMs 255
Brian J. Henz, Takumi Hawa, Michael Zachariah

Quantum Confinement In Core-Shell Particles Via Atomic Layer Deposited Zinc Oxide Films 266
David M. King, Xinhua Liang, Samantha Johnson, Alan Weimer

In-Situ Coating of Flame-Made TiO2 Nanoparticles by SiO2 267
Alexandra Teleki, Martin C. Heine, Frank Krumeich, Kamal M. Akhtar, Sotiris E. Pratsinis

Vesicle Templating of Hollow Oxide Nanoparticles 268
Spyros Monastiriotis, Alexander Couzis

Application of Coarse Grained Drag Law In Computational Fluid Dynamics Simulations of Fluidized Beds 269
Alvin U. ChEn, E. Nicholas Jones

Cfd Modeling of Cold-Flow Fluidized Beds and Validation with X-Ray Computed Tomography 270
Jian Min, Joshua Drake, R. O. Fox, Theodore Heindel

Industrial Application of Cpfd to Biomass Gasification 275
Charles Churchman

Large Scale Multiphase Modeling: Continuum Vs. Discrete Approach 285
Chris Guenther

Three-Dimensional Gas-Solid Fluidized Bed Simulation Applying the Cfd Technique 286
Milton Mori, Fabio Marini, Maria das Graças E. Silva, Maximilian J Hodapp

George Klinzing - Educator, Author, Researcher, Friend 294
Paul Solt

Unloading Pressure Discharge Trucks 297
Francisco J. Cabrejos, Tom G. Troxel

Acoustic Measurement and Monitoring in Dry Particulate Systems: Inspiration, Education, Application, Appreciation 306
Clive E. Davies, Steve J. Tallon, Emile S. Webster

Dilute Phase Pneumatic Conveying of Blended Plastic Pellets 315
Nestor A. Vasquez, Shrikant Dhodapkar, Karl Jacob, Joerg Theuerkauf

New Perspectives on Fluid-Phase Turbulence Modulation by Particles 316
Jennifer S. Curtis, Kunn Hadinoto-Ong

Local Friction Forces Between Plug and Pipe Wall in Dense Phase Pneumatic Conveying 317
Jae Bum Pahk, George E. Klinzing

Transformative Research in Pneumatic Conveying 326
George E. Klinzing

Development of Ion-Selective Thermo-Reversible Polymers 340
Sjoerd F. G. M. van Nispen, Leo J.P. van den Broeke, Johan P. A. Custers, J.T.F. Keurentjes

Co-Formulation of API with Polymer Using Pca Process 342
Ranjit Thakur, Maike Lohrmann

Engineering Porous Drug Particles by Emulsification-Diffusion for Propellant-Driven Oral Inhalation Formulations 343
Libo Wu, Sandro R. P. da Rocha

Effect of Poly-(N-vinyl pyrrolidone) Coating on the Synthesis of Magnetic Particles for the Formulation of Nanoscale Magnetorheological Fluids 344
Kaustav Sinha, Olivia A. Graeve

Optimal SiO2 Doping for Enhanced SnO2 Sensitivity and Thermal Stability 345
Antonio Tricoli, Markus Graf, Sotiris E. Pratsinis

Polymer Microsphere Synthesis by Membrane Emulsification 346
Jo M. M. Simons, Jan Meuldijk, Maartje F. Kemmere, Jos T.F. Keurentjes

Preparation and Characterization of NOVEL Polymer-MAGNETIC COLLOIDS 348
Marco Furlan, Marco Lattuada, Massimo Morbidelli

Functionalization of Monodisperse Magnetite Nanoparticles with Fluorescent Amino-Silane-Carboxymethyl Dextran for Cancer Treatment 349
Adriana P. Herrera, Carola Barrera, Vanessa Ayala, Magda Latorre, Madeline Torres-Lugo, Carlos Rinaldi

Silica-Coated Quantum Dot-Iron Phosphide Nano-Composites 350
Shuang Deng, Jessica O. Winter

Mimicking Selectin-Mediated Cell Motion with a Synthetic Receptor System 351
Maria M. Santore, Surachate Kalasin, Jeffrey,M. Davis, Ranojoy D. Duffadar

Functional Nanoparticles for Enantioselective Separations 352
Nisha Shukla, Andrew J. Gellman, Melissa Bartel

Preparation, Assembly and Application of Surface Anisotropic (Janus) Particles 353
Jingqin Cui, Ilona Kretzschmar

Cfd Modeling of Polymerization Fluidization Reactor 354
Ram G. Rokkam, R. O. Fox, Michael E. Muhle

Cfd Simulation of the Turbulent Flow of Pulp Fibre Suspensions 355
Maria da Graca Rasteiro, Carla A.F Ventura, Fernando A. P. Garcia, Paulo J. Ferreira

Heterogeneous Gas ChEmistry in the Cpfd - Eulerian-Lagrangian Numerical Scheme (Ozone Decomposition) 366
Dale Snider

Numerical Simulation of Pneumatic Drying Using Computational Fluid Dynamics 378
Tarek Jamaleddine, Madhumita Ray

Calibration of DEM Material Models 379
John Favier

Evaluation of Contact Models for Discrete Element Method Simulations 380
Joerg Theuerkauf, D.V.N Prasad

The Effect of Particle Shape on Bulk Powder Friction In a Schulze Ring Shear Tester Using the Discrete Element Method 381
William R. Ketterhagen, Jennifer S. Curtis, Carl R. Wassgren, Bruno C. Hancock

Force Model Considerations for Glued Sphere Dem Simulations 382
Madhusudhan Kodam, Rahul Bharadwaj, Jennifer Sinclair Curtis, Bruno C. Hancock, Carl R. Wassgren

Effect of Dem Parameters on the Collision Distribution of a Small Shearing Region beneath the Surface of a Rotating Drum.......383
Ben Freireich, James D. Litster, Carl R. Wassgren

Controlled Patterning of Nanostructured Surfaces Using Self-Assembly.......384
ChEtana Singh, Sharon C. Glotzer

Molecularly Engineered Organic Particulate Films: Preparation, Structure and Performance.......385
Ozge Guney-Altay, Dmitry Pestov, Gary Tepper

Nanostructured Coatings by Combined Flame Spray Synthesis of Nanoparticles and Cluster Beam Deposition.......387
Simone Vinati, Alessandro Antonini, Paolo Piseri, Paolo Milani, Luca Belforte, Nello Li Pira, Karsten Wegner

Factors Affecting Strength of Porous Films Made by Gas to Particle Conversion/ Thermophoretic Deposition of Nanoparticle Agglomerates.......388
Oluwatosin Ogunsola, Ratna Oetama, Tridib Bhowmick, Seonmin Kim, Sheryl Ehrman

Psri Lectureship in Fluidization Award: Fluidization of Nanopowders.......389
Robert Pfeffer

Thomas Baron Award: Rheology and Processing of Particle-Loaded Complex Fluids at Particle Concentrations That Approach the Maximum Packing Fraction.......390
Dilhan M. Kalyon

VOLUME 2

Syngas ChEmical Looping Process.......391
Fanxing Li, Deepak Sridhar, Hyung Rae Kim, Liang Zeng, Fei Wang, L. -S. Fan

Heterogeneous Flow Structure in Gas-Solid Risers: Continuous MODELING.......401
Dawei Wang, Jun You, Chao Zhu

Design, Development and Operation of a Novel Recirculating Liquid/solids Reactor Based on a Type 2 Draft Tube Spout Fluid Bed.......409
John Paccione, Howard Littman, Lealon L. Martin, Joel L. Plawsky, David M. Follansbee

Mass Transfer Resistance in Downer Reactors.......411
Subbarao Duvvuri

Modeling and Multi-Objective Optimization of Continuous Protein Recovery in Liquid-Solid Circulating Fluidized Bed Ion Exchange System.......419
Jahirul I. Mazumder, Jesse Zhu, Ajay K. Ray

Local Holdups and Phase Propagation Velocity Measurement in Glscfb Riser Using Electrical Resistance Tomography and Optical Fibre Probe.......430
Shaikh Razzak, Shahzad Barghi, Jesse Zhu

Use of Co-Milling to Improve Physical Stability of Amorphous Salbutamol Sulphate.......438
Wai Kiong Ng, Vivian S.Y. Wong, Prashant N. Balani, Sui Yung Chan, Reginald B. H. Tan

Modeling of Changes In Particle Size-Shape Distribution Due to Breakage.......442
Priscilla J. Hill, Devkant S. Gandhi

Investigation of Sub-Micron Milling Process of Poorly Water Soluble Products Using Factorial Design.......443
Stephen Hennart, G.M.H. Meesters, W.J. Wildeboer, Maria Cristina Domingues

Milling of Roller Compacted API with Excipients: Model Identification and Verification 448
Pavan Kumar Akkisetty, NandKishor Nere, Ryan McCann, Kenneth Morris, Doraiswami Ramkrishna, Gintaras V Reklaitis, Venkat Venkatasubramanian

Solving Inverse Problems In Fragmentation / Breakage 449
R. Bertrum Diemer Jr., Jon H. Olson, Yulong Ding, Mojtaba Ghadiri, Dolapo Olusanmi

Delamination of Graphite Sheets in Stirred Media Mills 450
Catharina Knieke, Wolfgang Peukert

Theoretical Study of Granule Friability during Pneumatic Conveying 453
Peter Spicka, Pavol Rajniak

Particle Breakage in the Nanometer Range 454
Catharina Knieke, Wolfgang Peukert

Nanoparticle Milling and Encapsulation in Supercritical Carbon Dioxide Mixtures 462
Scott M. Paap, Michael S. Wolfe, Jefferson W. Tester

Stability of Silica Nanoparticle Agglomerates 463
Martin Seipenbusch, Sonja Rothenbacher, Martin J. Kirchhof, Hans-Joachim Schmid, Alfred P. Weber, Gerhard Kasper

High Shear De-Agglomeration of Nanoparticle Agglomerates in Stirred Tanks 465
Micaela Caramellino, Piero M. Armenante

Development of a Visco-Plastic Constitutive Law for Dense Flow of Granular Matter 466
Mehrdad Kheiripour Langroudi, Gabriel I. Tardos

Predicting Discharge Dynamics from a Rectangular Hopper Using the Discrete Element Method (DEM) 467
Anshu Anand, Jennifer S. Curtis, Carl R. Wassgren, Bruno C. Hancock, William R. Ketterhagen

Use of the Attainable Region (AR) Analysis to Optimize Particle Agglomeration and Breakage Operations 468
Matthew J. Metzger, Benjamin J. Glasser, David Glasser, Brendon Hausberger, Diane Hildebrandt, Ngangezwe Khumalo

Classification of Fine Particles Using a Taylor-Couette Device 469
Navina D. Tungapindi, Kimberly H. Henthorn

Power Consumption and Granular Flow In a Vibro-Fluidized Stirred Granular Bed 470
Kenneth J. Ford, James F. Gilchrist, Hugo S. Caram

Density Measurements In a Vibro-Fluidized Deep Granular Bed 471
Kenneth J. Ford, James F. Gilchrist, Hugo S. Caram

Spray Drying of Large Particles of Film Forming Materials 472
P. Threlfall-Holmes, R. Ocone

A Kinetic-Based Model for a Non-Isothermal Granular Gas with Mono- and Bi-Dispersed Particles 473
Alberto Passalacqua, Prakash Vedula, Christine M. Hrenya, R. O. Fox

Comparison of Euler-Euler and Euler-Lagrange Simulations of Finite-Stokes-Numbers Gas-Particle Flows in a Lid-Driven Cavity 474
Rahul Garg, Alberto Passalacqua, Shankar Subramaniam, R. O. Fox

Combination of Monte Carlo Simulation of Population Balances and Cfd for Modeling of Nanoparticle Formation 475
Einar Kruis, J. Wei

Coarse-Grid Simulation of Fluidized Gas-Particle Flows 476
Yesim Igci, Sankaran Sundaresan

Representation of a Continuous Particle Size Distribution Using a New Polydisperse Kinetic Theory 478
Harish Viswanathan, Christine M. Hrenya, R. O. Fox

Solution of the Boltzmann Equation by the Fcmom 479
Matteo Strumendo, Hamid Arastoopour

Direct Numerical Simulation of Nanoparticle Evolution In Turbulent Flames 481
Yonduck "James" Sung, Venkatramanan Raman, R. O. Fox

Determining the Density and Moisture Content of Powders Using Microwave Dielectric Spectroscopy 482
Talesha T. Hall, Michael T. Harris

Correlating Retention and Drainage in Papermaking with Flocculation: Effect of Polyelectrolyte Branching 483
Elisabete Antunes, Fernando A.P. Garcia, Paulo Ferreira, Maria da Graca Rasteiro

Experimental Investigation into Shielding Effect of Magnetic Powder on Drilling Fluid 492
S. Ding, Biplab Data, a. Saasen, P.a. Amundsen

Solids Drying Process Simulation by Solidsim --- a New Tool for Flow Sheeting of Solids Processes 498
Ting(Tim) Han, Joerg Theuerkauf

Granulation Time in Fluidized Bed Granulators 499
Papiya Roy, Rajesh Khanna, Subbarao Duvvuri

Size Resolved Kinetics of Nickel Nanoparticle Oxidation 511
Lei Zhou, A. Rai, N. Piekiel, X. Ma, Michael Zachariah

Preparation and Characterization of Metal-Bound Hybrid Reactive Particles 516
Alexandre Ermoline, Mirko Schoenitz, Mikhaylo A. Trunov, Edward L. Dreizin

Reactive Nanocomposite Materials Prepared by Cryogenic Milling 517
Carlo Badiola, Edward L. Dreizin, Mirko Schoenitz, Xiaoying Zhu

Effect of Mechanical Activation on Reaction Kinetics In Gasless Heterogeneous Systems 527
Jeremiah D.E. White, Alexander S. Mukasyan

Mechanical Alloying and Reactive Milling In a High Energy Planetary Mill 528
Xianjin Jiang, Mikhaylo A. Trunov, Mirko Schoenitz, Rajesh Dave, Edward L. Dreizin

Pyrophoric Films Based on Nano-Sized Iron 529
Zac Doorenbos, Alok Vats, Jan Puszynski, Rajesh Shende, Deepak Kapoor, Darold Martin, Chris Haines

Directed Release of ChEmical Energy In Single Walled Carbon Nanotube Films 537
Wonjoon Choi, Michael S. Strano

A Flame-Pyrolysis Reactor for Carbon Black Production: Yield from Thermal Decomposition of Heavy Hydrocarbon Fuels 538
X. Ma, Anshuman Lall, Dale Huang, Michael Zachariah

Large-Scale Gas Phase Synthesis of Zinc Sulfide Nanoparticles 539
Evagelos K. Athanassiou, Robert N. Grass, Wendelin J. Stark

Synthesis and Characterization of Zinc Oxide Nanoparticles for Dye-Sensitized Solar Cells 541
Renata Marczak, Fabian Werner, Jan-Frederik Gnichwitz, Andreas Hirsch, Dirk M. Guldi, Wolfgang Peukert

Simultaneous Molecular and Microemulsion Templating of Inorganic Oxide Particles with Hierarchal Porosity 555
Nick Carroll, Dimiter N. Petsev

Use of Calorimetry Model and Batch Control Technique for Scale-up of Unseeded Batch Cooling Crystallization of Polyhydroxybenzophenon 556
Kwang Soon Lee, Huiyong Kim, Youkyung Bang

Atomistic Simulation Studies of Stability Enhancement of Griseofulvin Nanocrystal Aqueous System by Adding Surfactant, Polymer, and Surfactant-Polymer Mixtures 557
M. Silvina Tomassone, Wusheng Zhu, Frank Romanski

Characterization of Elastomer Particle Aggregation and Implications for Industrial Processing 558
Cornelius Gauer, Miroslav Soos, Hua Wu, Massimo Morbidelli

Reactor Analysis of Coal Pyrolysis to Acetylene In Hydrogen Plasma Reactor 559
Jiaqi ChEn, Yi ChEng

Multi-Scale Modeling and Control of Particulate Cvd for Production of Solar Grade Silicon In Fluidized Beds 560
B. Erik Ydstie, Juan Du

A Study of the Effect of Feeding Device Choice on Conversion In An Aerosol Flow Reactor 561
Todd M. Francis, Peter Kreider, Paul Lichty, Alan W. Weimer

Fluidized Bed Mixing of Nanosized Particles 562
James V. Scicolone, Rajesh Dave

Experimental Investigation of a Rotating Fluidized Bed in a Static Geometry 563
Rahul P. Ekatpure, Geraldine J. Heynderickx, Axel De Broqueville, Guy B. Marin

Effect of Baffles on Gas Bypassing In a 0.9-M-Diameter Unit 566
Allan Issangya, Reddy S. B. Karri, Ted M. Knowlton

Advanced Materials Development at Rdecom-Ardec 567
Paul Redner, Deepak Kapoor, Rajen Patel, Chris Haines, Darold Martin, Moon Chung

Development of High-Efficient Nanoenergetic Gas-Generators 568
Karen S. Martirosyan, Leizheng Wang, Arol Vicent, Dan Luss

Aluminum Burn Rate Modifiers Based on Reactive Nanocomposite Powders 569
Demitrios Stamatis, Xianjin Jiang, Ervin Beloni, Edward L. Dreizin

Tunability of Nanoenergetic Materials 579
Chris J. Bulian, Jan A. Puszynski, Jacek J. Swiatkiewicz

Fabrication and Characterization of Bimetallic Al-Ni Reactive Nanostructures as Nanoheaters 586
Qingzhou Cui, Julie ChEn, Zhiyong Gu

Metastable Intermolecular Composites (MIC) for Small Caliber Cartridges and Cartridge Actuated Devices 587
Peter P. Ostrowski, Magdy Bichay, John M. Hirlinger, Jan Puszynski, Todd Michael Allen

Filled Nanocomposite Materials for Biocidal Applications 594
Mirko Schoenitz, Demitrios Stamatis, Edward L. Dreizin, Sergey A. Grinshpun

Design for Demilitarization - Be Creative Now or Pay Me Later 595
Jack L. Johnston

Characterizing and Modeling of Particle Size Distributions from Propellant Open Burning or Open Detonation (OB/OD) 601
Mitchell Lindsay, Jonathon Koenig, Chris Merrill, Glenn Palmer

Analysis of Transport Phenomena & Modeling the Behavior of Single Particle during Drying Process 610
Srikanth Bellur Ramachandra, Basavaraju Agasanapur Nagendra, Charles J. Coronella

Gas-Solid Heat Transfer In Rotating Fluidized Beds In a Static Geometry 611
Axel de Broqueville, Juray De Wilde

Particle Size Optimization for Syngas ChEmical Looping Process 615
Fei Wang, Fanxing Li, Andrew Tong, Deepak Sridhar, L. -S. Fan

Computational Fluid Dynamics of Gas Bypassing In Fluidized Beds with Imposed Solids Flux 622
Ray A. Cocco, Allan Issangya, S.B. Reddy Karri, Ted Knowlton

Equilibrium Moisture Content and Drying Analysis of Biosolids 623
Srikanth Bellur Ramachandra, Charles J. Coronella, Victor R. Vasquez

Precipitation of Green Tea Polyphenols Using a Supercritical Anti-Solvent Process 624
Darja Meterc, Maria Jose Cocero, Ángel Martín Martínez, Eckhard Weidner

Nanoencapusulation Using Water-in-Supercritical CO2 Microemulsion Technology 625
Yuichiro Otsuji, Masahiro Ikeda, Chihiro Fushimi, Atsushi Tsutsumi

Precipitation of Ultra-Fine Particles of Fenofibrate with Controlled Size Distribution by Resolv: Rapid Expansion Vs Stabilizer Concentrations 626
Sameer Dalvi, Dhananjay Singh, Rajesh Davé

A Study of Production of Micronized Powders Using Supercritical Fluid Atomization Techniques 627
Sriram Kiran Annapragada, M. P. Sharma, William Lindberg

Effect of Co-Precipitation on Crystallinity of Pharmaceutical Compounds Using Supercritical Anti-Solvent Process 628
Tau Yee Lim, Wai Kiong Ng, Reginald B. H. Tan

Re-Engineering Particle Sizes by Precipitation with Compressed Antisolvent (PCA) Process 629
Candy S. Lin, Ka M. Ng, Christianto Wibowo

Hydrodynamics Modeling and Analysis of Rapid Expansion Systems of Supercritical Solutions (RESS) 630
Yudi Samyudia, Freddie Panau, Yuh Inn Liong

Energetic Materials, What Are They? 643
Bruce Cranford Jr., PE

History of Energetic Materials 664
Boyd Hill

Incorporation of Pressure Dependent Wall Slip to the Simulation of the Continuous Processing of Energetic Suspensions 665
Dilhan Kalyon, Moinuddin Malik

Prediction of Conditions Under Which Flow Instabilities Are Onset to Give Rise to Extrudate Distortions during the Extrusion of Highly Filled Energetic Suspensions 666
Suzanne Prickett, Hansong Tang, Halil Gevgilili, Dilhan Kalyon

Twin Screw Extrusion Processing of a Nanoalumina Based Formulation in Conjunction with a Molding Powder Preparation Methodology 667
Rajen Patel, Paul Redner, Halil Gevgilili, Dilhan Kalyon

Inverse Problem Solution of Squeeze, Rectangular Slit, Steady Torsional and Capillary Flows to Enable the Determination of Parameters of Wall Slip and Shear Viscosity for Energetic Suspensions 687
Hansong Tang, Halil Gevgilili, Dilhan Kalyon

Mesoscale Modeling of Energy Release Processes: Dissipative Particle Dynamics Simulations of Shear Initiated Behavior 688
John K. Brennan, Martin Lisal, Josep Bonet Avalos

Molecular Dynamics Simulation of the Kinetic Reaction of Ni and Al Nanoparticles 689
Brian J. Henz, Takumi Hawa, Michael Zachariah

Experimental Study of Ignition of Magnesium Powder by Electro-Static Discharge 699
Ervin Beloni, Edward L. Dreizin

T-Jump Mass Spectrometry to Probe Condensed State Reactive Processes 725
Lei Zhou, Nicholas Piekiel, Snehaunshu Chowdhury, Michael Zachariah

Thermophysical Property Prediction of Energetic Materials with Atomistic Computer Simulations 735
Nandhini Sokkalingam, Marybeth H. Ketko, Jeffrey J. Potoff

Monte Carlo Simulations of 1,3,5-Triamino-2,4,6-Trinitrobenzene (TATB): Pvt Data, Phase Equilibria, and Solubility 736
J. Ilja Siepmann, Neeraj Rai, John L. Lewin, Divesh Bhatt, Laurence E. Fried

Prediction of Physicochemical Properties of Energetic Materials 737
Siri Chakka, Veera M. Boddu, Stephen W. Maloney, Reddy Damavarapu

Mechanism and Kinetics of Thermal Decomposition of Cl-20: 747
Olexandr Isayev, Leonid Gorb, Mo Qasim, Jerzy Leszczynski

Identifying Emulsion Drop Breakage Mechanisms for Population Balance Equation Models of High Pressure Homogenization 752
Neha B. Raikar, Surita R. Bhatia, Michael F. Malone, Cristhian P. Almeida-Rivera, Peter Bongers, Michael A. Henson

Population Balance Equation Modeling of Emulsion Drop Coalescence in High Pressure Homogenization 753
Jason D. Rosenberg, Neha B. Raikar, Michael F. Malone, Michael A. Henson

A Generalized Geometric Population Balance Model for Asphaltene Aggregation 754
Tabish Maqbool, Sasanka Raha, H. Scott Fogler

A Combined Mechanistic Model for Nucleation, Aggregation and Breakage In a Population Balance Model of Granulation 755
Rohit Ramachandran, Charles D. Immanuel, James D. Litster, Francis J. Doyle III, Frantisek Stepanek

The Elasticity and Breakage of Colloidal Aggregates In Shear and Turbulent Flows 758
Alessio Zaccone, Hua Wu, Marco Lattuada, Massimo Morbidelli

Study on Aggregation of Gold Nanoparticles by Tuning the Surface Properties 759
Varun Kumar, Chungyin ChEng, R. O. Fox, Robert K. Prud'homme

Effect of Particle Surface Charge Groups on the Aggregation of Elastomer Particles Under Static and Flow Conditions 760
Miroslav Soos, Cornelius Gauer, Hua Wu, Massimo Morbidelli

Distinguishing Between TiO2 Aggregates and Agglomerates by High Pressure Dispersion 761
Alexandra Teleki, Robert Wengeler, Hermann Nirschl, Sotiris E. Pratsinis

Use of Asymmetric Flow-Field Flow Fractionation to Characterize Aggregating Colloidal Dispersions 762
Marco Lattuada, Carlos Olivo, Cornelius Gauer, Giuseppe Storti, Massimo Morbidelli

A Structural Model for Aggregation and Its Application to Small-Angle Scattering 763
Gregory Beaucage

Size and Shape Engineering of ZnO Nanoparticles 764
Renata Marczak, Robin Klupp Taylor, Michael Voigt, Doris Segets, Wolfgang Peukert

Multifunctional Particles: Magnetic Nanocrystals and Gold Nanorods Coated with Fluorescent Dye-Doped Silica Shells 774
Danielle K. Smith, Andrew T. Heitsch, Brian Korgel

Phospholipid-Assisted Formation and Dispersion of Aqueous Nano-C60 775
Yanjing ChEn, Geoffrey D. Bothun

Synthesis of Doped Zinc Selenide Quantum Dots Using Microemulsions as Templates 780
Tracy Heckler, Qi (Grace) Qiu, Jun Wang, T. J. Mountziaris

Zeolite Crystal Growth: A High-Resolution Cryo-TEM Study 781
Sandeep Kumar, R. Lee Penn, Michael Tsapatsis

Synthesis of Size Distribution Controllable Zeolite Nanocrystals Via a Novel Confined Space Strategy 782
Yi Huang, Anita J. Hill, Huanting Wang

Author Index

27a Characterization of Crystal Size and Shape

Michel Kempkes, Jan Eggers, Thomas Vetter, and Marco Mazzotti. Institute of Process Engineering, ETH Zurich, Sonneggstrasse 3, Zurich, Switzerland

Crystal shape is a ubiquitous property, affecting not only many downstream processing steps in industrial crystallization such as filtration and drying, but also the bioavailability of pharmaceuticals. It is however also an elusive property: unlike the characterization of particle populations with constant shape, which can be described with a single characteristic size L, few techniques have been developed for the characterization of particle populations with non-constant shape.

The characterization of a population of particles with different shapes requires the measurement of a multi-dimensional particle size distribution (PSD). In the simplest approach, particles are described by two independent sizes L1 and L2. Such a bidimensional particle model allows for the description of a wide variety of particles, ranging from parallelepipeds to e.g. cylindrical particles or more complex geometries where the other dimensions are related to L1 and L2. When considering a population of these particles, it needs to be characterized using a two dimensional PSD n(L1, L2). For certain particle geometries it is possible to directly measure L1 and L2 using for example optical microscopy. In the works of Puel et al. [1, 2], bidimensional PSDs were successfully measured in this way. Such an approach is however requires extensive sample preparation to ensure a proper distribution of the particles on the microscope stage and is rather time consuming, leading to the measurement of a very limited number of particles.

In this work, we use a specially constructed view cell through which we pump a suspension of crystals. The view cell is positioned between a bright xenon flash lamp and an optical system with a digital camera. This setup enables us to acquire high quality images of the randomly oriented crystals in suspension at a rate of 10 images per second. These images are then analyzed by a fully automated image analysis routine that identifies the particles in the suspension and measures, amongst other properties, the major and minor axis length of an ellipse with an equal area moment of inertia. This yields a 2D measurement called Axis Length Distribution (ALD). Clearly, the ALD is related to the 2D PSD, but this relation is not straightforward. A measurement model has been developed [3], allowing for the calculation of an ALD for a given 2D PSD. The inverse operation, i.e. the restoration of a 2D PSD from a measured ALD, is obviously of great interest, but is highly ill-posed. To solve this problem, a genetic algorithm was developed [4] that enables us to find the discretized 2D PSD from a measured ALD.

We applied this procedure in real crystallization processes to monitor the bidimensional PSD. In these processes, samples were drawn during the crystallization, which were immediately quenched with a large volume of saturated solution and were analyzed using the view cell to obtain the ALD, with a Coulter counter to obtain the particle volume distribution and by SEM to study particle geometry. From the bidimensional PSDs restored from the measured ALDs, particle volume distributions were calculated. These were found to agree very well with the Coulter counter measurements.

The restored 2D PSDs revealed a great deal of information on interesting aspects of crystallization such as fines generation, bidimensional particle growth and the effects of operating conditions such as supersaturation and stirring rate on particle size and shape. Currently, our research efforts are aimed at the study of agglomeration and breakage using image analysis.

References

1. Puel F., Févotte G., Klein J. P., "Simulation and analysis of industrial crystallization processes through multidimensional population balance equations. Part 2: a study of semi-batch crystallization", Chem. Eng. Sci. 58 (2003) 3729-3740.

2. Oullion M., Puel F., Févotte G., Righini S., Carvin P., "Industrial batch crystallization of a plate-like organic product. In situ monitoring and 2D-CSD modelling: Part 1: Experimental study", Chem. Eng. Sci. 62 (2007) 820-832.

3. Kempkes M., Eggers J., Mazzotti M., "Measurement of particle size and shape by FBRM and in-situ microscopy", (2007), Chem. Eng. Sci. doi: 10.1016/j.ces.2007.10.030

4. Eggers J., Kempkes M., Mazzotti M., "Measurement of size and shape distributions through image analysis", (2008) submitted to Chem. Eng. Sci..

* Author to whom correspondence should be addressed.

Phone: +41-44-632 24 56. Fax: +41-44-632 11 41. Email: marco.mazzotti@ipe.mavt.ethz.ch

27b Modeling the Manipulation of Crystal Morphology Distributions

Christian Borchert[1], Nandkishor Nere[2], Andreas Voigt[1], Kai Sundmacher[1], and Doraiswami Ramkrishna[2]. (1) Max Planck Institute, Magdeburg, Germany, (2) School of Chemical Engineering, Purdue University, 480 Stadium Mall Drive, West Lafayette, IN 47907-2100

In the chemical industries in general and in the specialty and fine chemical industries, in particular, the crystal shape and its distribution are of considerable interest because it can remarkably influence the product downstream processing and usability. Also bioavailability and dissolution behavior of active pharmaceutical ingredient is governed by the nature of prevailing crystal faces. Consequently it is important to predict, optimize and control shape distributions for industrial applications.

The shape of a convex crystal can be completely defined by a set of surfaces (planes) along with their normal distances from the crystal center (Zhang et al., 2006). One can then obtain boundaries in the shape space, which stand for the appearance/ disappearance of distinct morphological properties of a crystal (Borchert et al., 2008). An evolutionary crystal shape reaching such a boundary looses a face (or faces of similar kind) on the crystal surface in the sense that it grows out of the crystal. Therefore its dynamics is dimensionally smaller by one than it was before because only the real faces on the crystal surface determine the evolution in shape space. This imparts the system a hybrid character with varying degrees of freedom. This makes computations of single crystal shape evolution trajectories more complex than for purely continuous processes.

Based on the systematic procedure of Borchert et al. (2008), we identify morphological subspaces and define corresponding crystal number densities. In order to describe the dynamics of crystal population in these subspaces the identification of multidimensional population balance equations for each of the subspaces is required. Clearly, since a crystal can evolve from one shape to another, meaning transitions of number densities from one sub-space to another and vice versa, the dynamics of these number densities is coupled. This aspect requires the specification of clear boundary conditions and exchange terms, which will be the significant part of discussion. There are two different cases of how the crystals can be exchanged between morphological domains one by continuous influx into a lower or equally dimensional domain and second by discontinuous (instantaneous) jump from a lower to a higher dimensional domain. Due to this peculiarity one obtains a varying set of partial differential equations that describes the evolution of crystal populations.

Clearly the dynamics is governed by the supersaturation which is the result of the joint crystal growth. The dependency of boundary conditions and exchange terms on supersaturation is an interesting aspect. Since supersaturation determines the growth behavior of the individual faces one can manipulate the crystal shape towards a desired direction in shape space by changing supersaturation. Hence a thorough analysis of the joint state space, which consists of the shape space and the supersaturation coordinate, is necessary to keep track of possible discrete morphological changes, which will be a key part of presentation. We also present an interesting model case study that illustrates the proposed methodology.

References

Zhang, Y., Sizemore J. P. & Doherty M. F. (2006). Shape Evolution of 3-Dimensional Faceted Crystals. AIChE Journal, 52, 1906-1915.

Borchert, C. B., Nere, N. K., Ramkrishna, D., Voigt, A., and Sundmacher, K. On the Prediction of Crystal Shape Distributions in a Steady State Continuous Crystallizer. Chemical Engineering Science, under revision, 2008.

Borchert ,C. B., Nere, N. K., Ramkrishna, D., Voigt, A., and Sundmacher, K. Evolution of Crystal Shape Distributions and Morphology Classification. ISIC17, Maastricht, 2008.

27c Parameter Estimation of Preferential Crystallization Using a Detailed Mathematical Model

Yoshiaki Kawajiri[1], Ivan Angelov[2], Grzegorz Ziomek[2], Martin Peter Elsner[2], and Andreas Seidel-Morgenstern[2]. (1) School of Chemical & Biomolecular Engineering, Georgia Institute of Technology, 311 Ferst Drive, Atlanta, GA 30332-0100, (2) Max Planck Institute for Dynamics of Complex Technical Systems, Sandtorstr. 1, Magdeburg, 39106, Germany

Separation of chiral compounds is of large interest because most of the organic molecules are chiral and usually only one of the enantiomers exhibits the desired properties with regard to therapeutic activities or metabolism, whereas the other enantiomer may be inactive or may even cause some undesired effect. For this reason, enantiomeric separations have become increasingly important and their application ranges from the pharmaceutical to the agricultural industry. An attractive separation technique for chiral separation is the so-called preferential crystallization[1], where homochiral crystals are seeded to a supersaturated solution of the racemic mixture, and only one of the enantiomers is preferentially produced. It is important to note that this process must be stopped before the equilibrium is reached; the preferred enantiomer of a high purity can be obtained only before the other (counter) enantiomer starts to crystallize. Therefore, an accurate prediction method is in crucial need for efficient design and operation of this process.

Recently Elsner et al.[2] have proposed a detailed physical model which takes into account the size-dependent crystal growth as well as the primary and secondary nucleation, which closely reflects the underlying thermodynamics of the crystallization process. Due to the complexity of this mathematical model, however, it is necessary to resort to a numerical parameter estimation technique to identify some of the model parameters. Furthermore, we require that both enantiomers have identical parameter values, adhering closely to the underlying physics; this poses a further challenge in obtaining a good fit to the experimental data.

In this study, we carry out a simple batch experiment changing the seed mass amounts, where the mass fractions of the both enantiomers in the liquid phase are measured. For the multiple experimental data sets obtained, an efficient parameter estimation technique is applied based on a Newton-based optimization approach. Here, the population balance equation is turned into a set of ordinary differential equation using the quadrature method of moments (QMOM)[3]. The QMOM model is then discretized into a large number of nonlinear algebraic constraints. Finally, the parameter estimation problem for the multiple experimental data sets is formulated as an optimization problem constrained by the nonlinear algebraic constraints. We show that this problem can be solved efficiently for the multiple data sets, and the mathematical model with the obtained parameters can be a powerful tool for process optimization.

[1] Elsner, M.P., Ziomek, G., Seidel-Morgenstern, A.: Simultaneous preferential crystallization in a coupled, batch operation mode - Part I: Theoretical analysis and optimization, Chemical Engineering Science, 62, 4760-4769 (2007)

[2] Elsner, M.P., Seidel-Morgenstern, A : One suitable kinetic model for enantioselective crystallization of theonine in water, Crystal Growth & Design (in preparation)

[3] McGraw, R: Description of Aerosol Dynamics by the Quadrature Method of Moments, Aerosol Science and Technology, 27, 255-265(1997)

27d Dynamic Modelling of Particle-Droplet Collisions

M.J. van der Hoeven[1], Tony Howes[1], **James D. Litster**[2], W.J. Wildeboer[3], G.M.H. Meesters[3], and I.T. Cameron[1]. (1) School of Engineering, University of Queensland, Brisbane, 4072, Australia, (2) Chemical Engineering and Industrial Pharmacy, Purdue University, Forney Hall of Chemical Engineering, Room 2154, 480 Stadium Mall Drive, West Lafayette, IN 47907-2100, (3) Research and Development, DSM Food Specialties, P.O. Box 1, Delft, 2600 MA, Netherlands

Spray drying is widely used for producing particulate products directly from a liquid feed. Commonly fines are recycled into the spray zone, with the aim to collide them with just formed droplets to increase their size. These droplets will dry as they move away from the atomiser. The outcome of the collisions depends on the moisture content of the droplets. Close to the nozzle full penetration is obtained, and with increasing distance and subsequent increasing viscosity of the droplets the particle will penetrate less. Once the droplets have become particles, collisions will result in bouncing.

A model has been developed that computes the penetration depth of the particle into the droplet. The characteristic parameters that describe the collision outcome are the Weber and Ohnesorge numbers. Collision outcome maps based on the numbers have been plotted for different contact angles. The different outcomes that have been observed are full penetration, partial penetration, penetration to a force equilibrium, oscillating behaviour and ejection.

27e Shear-Rate Dependence of the Initial Cluster Aggregation Kinetics of Polystyrene Latices Under Turbulent Conditions in Stirred Tank

Lyonel Ehrl[1], **Miroslav Soos**[2], and Massimo Morbidelli[1]. (1) Institute for Chemical and Bioengineering, Department of Chemistry and Applied Biosciences, ETH Zurich, Wolfgang-Pauli-Str. 10, HCI F 133, Zurich, 8093, Switzerland, (2) Institute for Chemical and Bioengineering, Department of Chemistry and Applied Biosciences, ETH Zurich, Wolfgang-Pauli-Str. 10, Zurich, 8093, Switzerland

When producing or processing sub-micron particles, e.g., in the case of waste-water treatment, emulsion polymerization, or nano-particle precipitation, knowledge of the dependence of the aggregation kinetics on shear rate is required for process design and control. In addition, the correct description of the aggregation kinetics is a key to properly model and simulate such processes, as well as the basis for the development and validation of any breakage model. Therefore, the aggregation kinetics of model colloidal systems consisting of polymer latexes with different primary particle diameter (120nm, 420nm, and 810nm) were investigated under turbulent conditions in stirred tank. It was found that the integral quantities of the cluster population measured by small-angle static light scattering, i.e., the root-mean-square radius of gyration, the zero-angle intensity of scattered light, and obscuration (extinction) all scale with a dimensionless time of the form $\alpha \times \langle G \rangle \times \phi \times t$. The prefactor represents an experimentally obtained aggregation efficiency which according to experimental data follows a power law of the form $\alpha \propto \langle G \rangle^n$, with the value of the exponent, n, being negative and dependent on the primary particle size. The broad range of primary particles sizes allowed investigating the transition from diffusion-affected aggregation to purely shear-induced aggregation. The results of this work together with literature data provide a relation between the experimentally obtained aggregation efficiency and a characteristic dimensionless group, N_f, which is proportional to the ratio of the Hamaker constant normalized by thermal energy over the Péclet number. These findings can help in the practical design of coagulation processes as well as to benchmark models on the aggregation and breakage of colloidal systems under turbulent conditions.

27f Shear Induced Aggregation Rate of Colloidal Nanoparticles and Clusters In the Presence of Repulsive Interactions

Marco Lattuada and Massimo Morbidelli. Institute for Chemical and Bioengineering, Department of Chemistry and Applied Biosciences, ETH Zurich, Wolfgang-Pauli-Str. 10, HCI F 133, Zurich, 8093, Switzerland

Coagulation of colloidal nanoparticles in the presence of shear is a key step in processing of many particulate materials, polymeric nanoparticles, food products etc. In the case of charged nanoparticles, coagulation is usually induced out by adding sufficient amount of electrolytes to completely screen all electrostatic interactions. In addition, the suspension is usually sheared to accelerate the coagulation process. However, for certain applications, the addition of large amounts of electrolytes is not beneficial for subsequent processing of the material and removal of electrolytes might be necessary. Therefore, operating with lower amounts of electrolytes, and in particular lower than the critical coagulation concentration, might be desirable. The major obstacle is given by the high sensitivity of the particle stability to the electrolyte concentration, and by the poor understanding of the mechanism of shear induced aggregation in the presence of a repulsive barrier.

In this work, we have performed detailed simulations of the aggregation rate of colloidal nanoparticles and fractal clusters in the presence of both shear flow and repulsive interactions, by numerical solution of the convection-diffusion equation for the pair probability function, i.e. the probability of finding two particles (respectively two clusters) at a given relative position. The equation has been solved for all the most important linear flow fields (e.g. simple shear, elongational flow, etc.).

In order to gain better physical insight of the interplay between the various mechanisms affecting the aggregation, a simple but effective model is presented and used to interpolate the results of the rigorous calculations. This approach provides a simple expression for the aggregation rate that can be used in population balance equation calculations, and is alos provides a simple criterion to estimate the relative contributions to the aggregation rate due to shear and to repulsive interactions.

References

Melis S, Verduyn M, Storti G, Morbidelli M, Baldyga J, AIChE J. 45, 1383-1393 (1999)

Babler MU, Sefcik J, Morbidelli M, Baldyga J, Physics of Fluids 18, 013302 (2006)

28a Insights in Scaling and Optimization of Ball Mill Operations - a Discrete Element Method Based Approach

Joerg Theuerkauf, Solids Processing, The Dow Chemical Company, Freeport, TX 77566 and Kaustubh S. Mujumdar, Dow Chemical International Private Limited,, Pune, India.

This paper aims to provide an understanding on the dynamics of grinding media in the ball mills using a Discrete Element Method (DEM) approach. A three dimensional DEM slice element model was developed to analyze the grinding media motion in ball mills.

The developed framework was used to predict the center of mass of the charge in the mill to subsequently predict the power draw of the mill. These predictions compare favorably with theoretical and in plant observations for ball mills over a range of fill levels and rotational mill speeds. The simulations show that the inclusion of the center of mass term in the scale-up equations is important for the correct prediction of the power draw when scaling ball mills. Furthermore, the time averaged collision intensity and frequency of the grinding media was analyzed. The performance of the mill was gauged by spatial contact force distributions in regard to possible wear at the ball mill cylinder when the fill level, rotational mill speed, media diameter and lifter shapes were varied. These results were used to identify optimal operating conditions for the mill.

28b Characterization of Granular Flow In a Bladed Mixer Using the Discrete Element Method

Brenda Remy[1], Johannes G. Khinast[2], and Benjamin J. Glasser[1]. (1) Dept. of Chemical and Biochemical Engineering, Rutgers University, 98 Brett Road, Piscataway, NJ 08854, (2) Institute for Process Engineering, Graz University of Technology, Inffeldgasse 21/A, Graz, A-8010, Austria

Bladed mixers are commonly used in a variety of areas, ranging from the bulk chemical to the food and pharmaceutical industries. Despite their extensive use, the flow of granular materials in agitated devices and the mechanisms by which particle motion is generated are still not fully understood. Experimental examination of granular flows in these mixers has been limited to off-line analysis of location-specific samples, and to on-line measurement of a limited set of process parameters. Numerical simulation techniques have the potential of bridging the knowledge gap since they allow for the study of experimentally difficult-to-measure parameters.

Utilizing the discrete element method the 3-dimensional flow of cohesionless particles in a cylindrical vessel agitated by a 4-blade impeller was studied. Instantaneous, averaged and fluctuating velocity fields show that the particle movement is governed by the angular motion of the blades. The granular bed deforms by forming heaps where the blades are present and valleys between blades passes. The presence of these heaps leads to the formation of a 3-dimensional recirculation zone in front of the blade, where particles flow upwards by the wall and downwards by the impeller shaft. This 3-D recirculation leads to enhanced mixing by promoting radial and vertical mixing. The particle's frictional characteristics are shown to strongly influence the flow structure and mixing kinetics within the mixer. At low friction coefficients the 3-D recirculation in front of the blade is not present reducing the convective mixing inside the mixer. Higher friction coefficients lead to an increase in granular temperature which is associated with an increase in diffusive mixing. Collisional stresses within the mixer showed periodic behavior indicating that the granular material compresses as the blade passes and dilates in between the blades. The long-time averages of the normal and shear stresses were found to vary with mixer height with maximum values near the bottom plate. Additionally, a strong dependence between the magnitude of the shear stresses and the friction coefficient of the particles was found. These stress tensor characteristics indicate that the granular flow in this bladed mixer occurs in the quasi-static regime.

Mixing analysis of binary system of particles in an agitated bed by Discrete Element Method

Ajit Mujumdar[a], Masayuki Horio[b], Toshio Maetani[c], and Naomichi Nakamura[c]

[a] New Jersey Center for Engineered Particulates, New Jersey Institute of Technology, Newark, NJ 07102, USA
[b] Tokyo University of Agriculture & Technology, Nakamachi, Tokyo, 184-8588, Japan
[c] JFE Steel Corporation, Steel Research laboratory, Chuo-ku, Chiba, 260-0835, Japan

Abstract

A three dimensional numerical code based on the discrete element method (DEM) is applied to simulate the mixing characteristics of binary system of particles in an industrial agitator. The device consists of blades rotating in the range of 60 -140rpm enabling the rigorous mixing of particles. In order to overcome the computational limitations on using the number of particles, simulation scale is magnified in the range of 10 to 40. Effect on the quality of mixing of particles for various magnification factors is compared. Concentration of particles is analyzed by taking a sectional cut in the agitated bed in 3D planes. Quantitatively, it is found that the magnification factor has a little effect on the estimated concentration of particles. Effect of withdrawal of the blades on the mixing of particles is also analyzed. Experimental comparison with the numerically simulated results is found to be in good agreement.

Introduction and simulation methodology

Application of discrete element method for modeling of powders in industries has been a focus of research for the past several decades. Apart from analyzing the applicability of various contact models and computational algorithms, the discrete element modeling faces stiff challenges in simulating the real life systems, which typically consists of millions of particles. In order to quantify the mixing of binary or ternary system of particles, simulations are performed for the smaller systems and appropriate scalable techniques are used for the interpretation of such results for the bigger systems. However, even a small size system consists of millions of particles therefore a realistic simulation can be performed only by making certain assumptions by which a small size system can be modeled. In this work, simulations of an industrial agitator are performed using discrete element model for the smaller size system consisting of binary system of particles agitated by the blades mounted on a rotating shaft. A magnification factor is estimated, which represents the real size system and the results are compared with the experimentally observed characteristics of the particles in the mixed bed.

Mixing vessel is a conical frustum of 4.5 cm diameter and a vertical height of 7.0 cm. Numerical simulations are performed for the magnification factors in the range of 10 to 40. Agitation is provided to the vessel by rotating the blades at speeds of 60, 120 and 140 rpm. Densities of particles are 7.8 g/cc and 8.9 g/cc, respectively. A well-known Cundall Stack model is applied for the calculation of normal and tangential forces. Initially, lighter particles are allowed to settle by gravity and the heavier particles are places on the top of lighter particles. Rotation is applied and the blades are withdrawn from the mixture at the end of simulation period. Figure 1 illustrates the initial settled position of particles in the vessel.

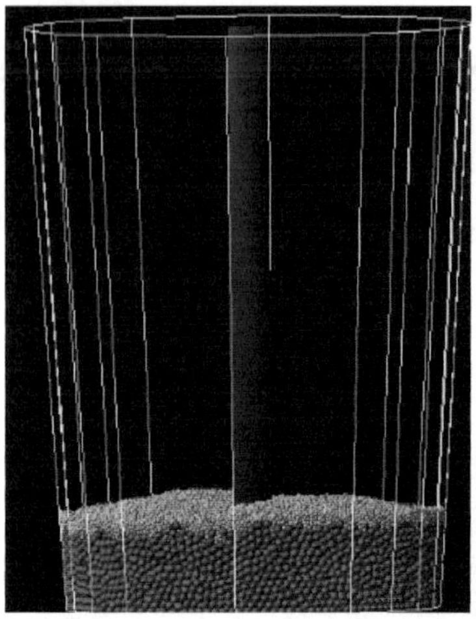

Figure 1. Initial position of settled particles in the mixing vessel

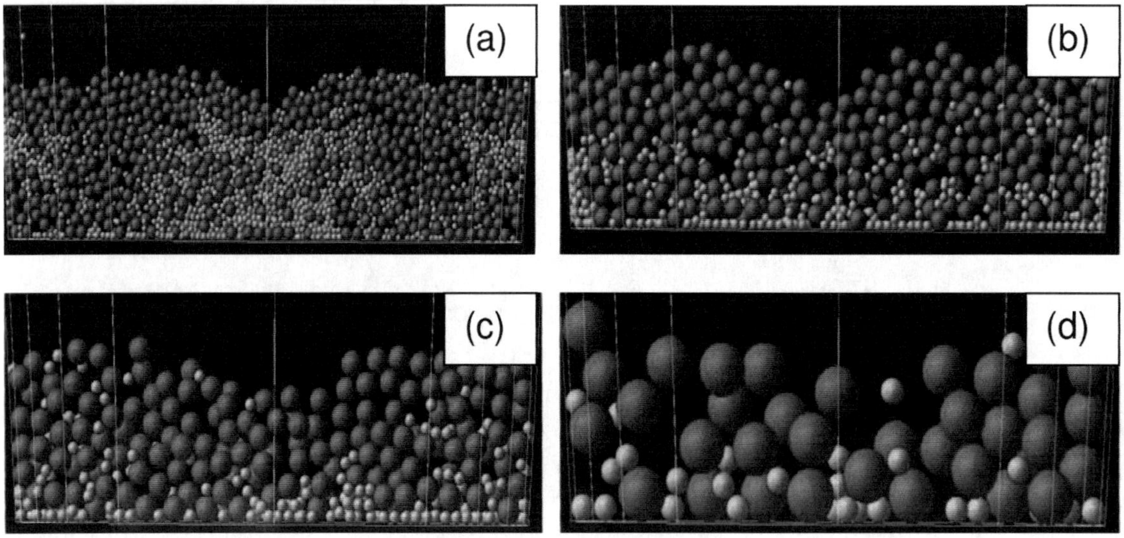

Figure 2. XY-Cut, 2.5 revolutions, 60 rpm, (a) M10, (b) M15, (c)M20 and (d)M40,

Snapshots from the simulations for 2.5 revolutions of the blades rotating at 60 rpm for the various magnification factors are shown in Figure 2. As seen from sectional cut in X-Y plane of the snapshots, heavier particles appeared to be reaching the bottom of the vessel. With an increase in the magnification factor, the capability of discrete element model results clearly shows the inability to qualitatively analyze the flow pattern of the particulate systems; however, quantitative analysis of such results does not seem to differ significantly. Average concentration of bigger size (lighter) particles is found to be in close agreement for all the magnification factors.

Comparison of the simulation case of magnification factor 10 found to be in good qualitative agreement with the experimental results (performed at JFE facilities) as shown in Figure 3. Also, for the ideal concentration of 0.3623 in a binary mixture, the estimated concentration values are found to be different for the different sectional cuts taken for the analysis (c.f. Figure 4).

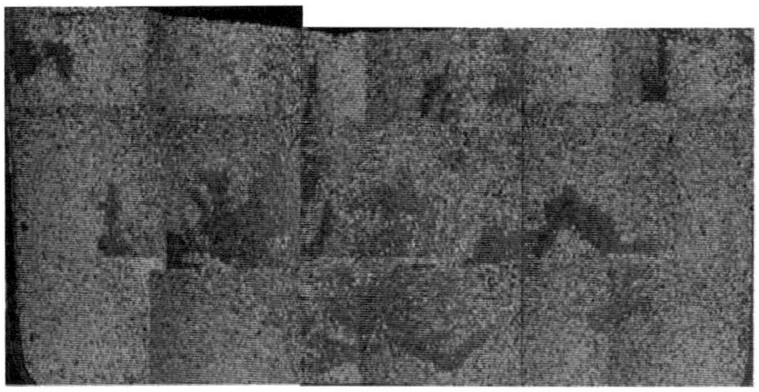

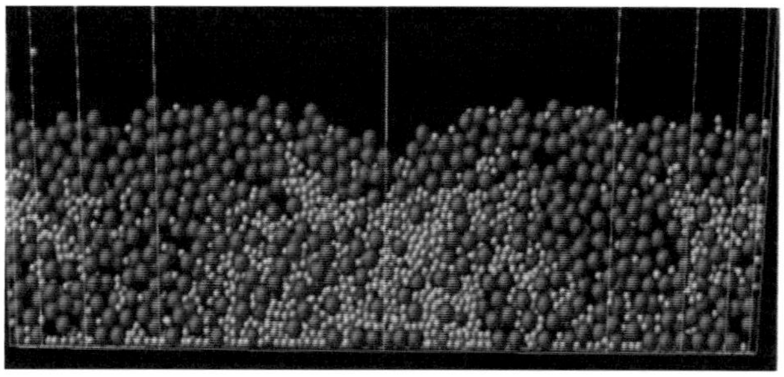

Figure 3. Experimental and simulation comparison of binary system of particles after 2.5 revolutions of the blades.

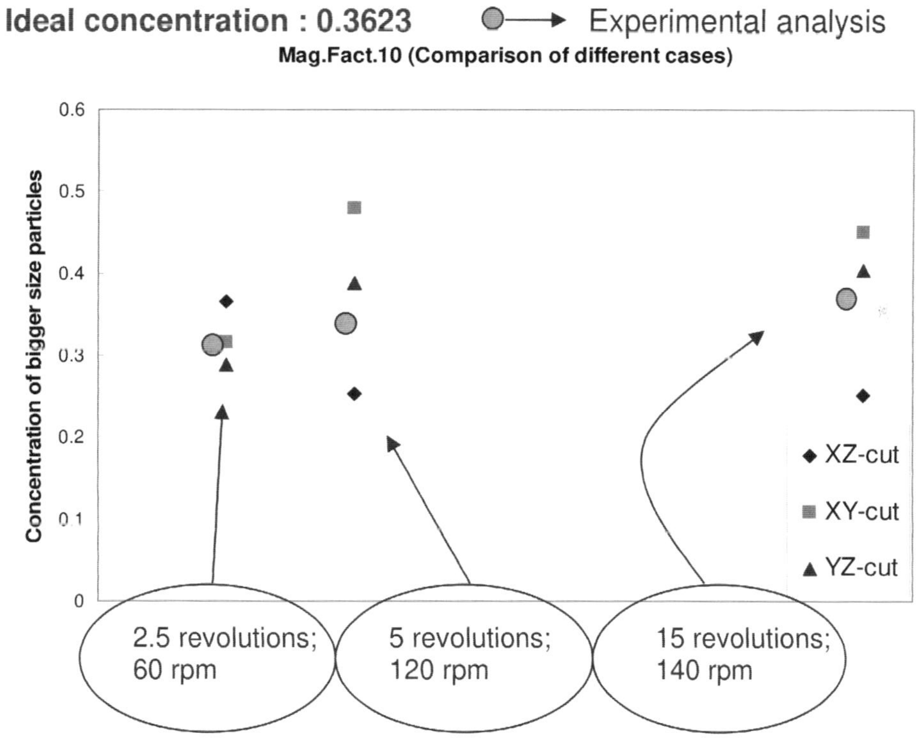

Figure 4. Concentration of bigger size particles for different rotational speeds (Magnification factor: 10)

Acknowledgements

Authors acknowledge Japanese Society for Promotion of Science (JSPS) for the partial support of this work. Also, valuable comments and support from Prof. Rajesh Davé for the work are greatly appreciated.

28d Rheological Behavior of Quasi-Static Granular Flows Under Unsteady Simple Shear

Jin Sun, Lee R. Aarons, and Sankaran Sundaresan. Chemical Engineering, Princeton University, Princeton, NJ 08544

In our previous work, we have studied steady simple shear flow of nearly homogeneous assemblies of uniformly sized, spherical particles in periodic domains under constant volume and constant normal stress conditions using the discrete element method [1, 2]. In addition to confirming the flow regime map for non-cohesive granular materials as reported in [3, 4], we found a new scaling which permits collapse of all steady shear data at various strengths of cohesion and shear rates into a single master curve for each particle volume fraction and coefficient of friction.

As an extension to more complex flow conditions, we now investigate the stress responses of quasi-static granular flows subjected to unsteady shear under similar computational settings in the present work. In these simulations, the shear rate varies with time according to either a square wave or a sinusoidal wave function. We find that the stress response is rate-independent as shown by the collapsed stress-strain curves at different shear rates and patterns. More interestingly, strong history dependence of the stress response is found. It is evidenced by two phenomena. Firstly, the normal and shear stresses undergo a transition after shear reversal that requires a shear strain of order unity (between one and two) to evolve. In contrast, they reach the previous steady state stress level rapidly when shear is resumed in the same direction. This characteristic strain of order unity is a robust scale independent of shear rate, volume fraction, strength of cohesion and fiction coefficient. Secondly, for unsteady shear at strain amplitudes smaller than the characteristic strain, the stress after reversal is affected by shear history in the preceding half-cycle. We also study the microscopic origin of the history dependence. We characterize the microstructure by a fabric tensor, whose shear component is used as a measure of the microstructure anisotropy. The development of shear-induced anisotropy and the correlation of the stress transition after shear reversal with the microstructure rearrangement are observed. The small strain amplitude effect can also be explained by the different microstructure evolution pattern. Comparison of the simulation data with the hypoplastic constitutive model [5, 6] predictions shows that the model without incorporating the microstructure evolution cannot correctly capture the characteristic strain after shear reversal. Thus this study also directly provides a justification and a database for developing fabric-incorporated anisotropic constitutive models for quasi-static granular flows.

References:

[1] L. Aarons and S. Sundaresan. Shear flow of assemblies of cohesive and non-cohesive granular materials. Powder Technology, 169(1):10–21, 2006.

[2] L. Aarons and S. Sundaresan. Shear flow of assemblies of cohesive granular materials under constant applied normal stress. Powder Technology, 183(3):340–355, 2008.

[3] C. S. Campbell. Granular shear flows at the elastic limit. Journal of Fluid Mechanics, 465:261–291, 2002.

[4] C. S. Campbell. Stress-controlled elastic granular shear flows. Journal of Fluid Mechanics, 539:273–297, 2005.

[5] W. Wu, E. Bauer, and D. Kolymbas. Hypoplastic constitutive model with critical state for granular materials. Mechanics of Materials, 23(1):45–69, 1996.

[6] D. Kolymbas. Introduction to Hypoplasticity. A. A. Balkema, 2000.

Parametric Study on the Reactive Extraction of Rapeseed Oil for Biodiesel Production in a Batch Reactor.

Adam Harvey, Rabitah Zakaria

School of Chemical Engineering and Advanced Materials, University of Newcastle upon Tyne, United Kingdom.

1. Introduction

In the conventional process to make biodiesel, the raw material is oil that has been pre-extracted from the seeds, and usually degummed and refined. However, it has been shown that transesterification of the oil can actually be carried out directly from the plant seeds without prior extraction [1]. In this scheme, macerated seeds are contacted directly with methanol and catalyst to produce biodiesel without the need of other solvents. This route will simplify the overall biodiesel process scheme and has the potential to improve the efficiency of the process as well as to reduce the production cost. Using this technique, the use of hexane to extract the oil can be eliminated as the extraction is being carried out by the reactant it self. This provides an added environmental benefit as hexane is classified as a hazardous air pollutant and contributes to the production of smog and global warming.

The overall objective of this research is to develop an integrated technology combining the extraction of vegetable oil from oilseeds with reaction of that oil to biodiesel in a one-step extraction/reaction process. The steps and parameters involved in the integrated extraction/reaction scheme differ from that of the conventional process in many ways. The fact that the reaction system is heterogeneous could alter the kinetics of the transesterification reaction. The mechanism on how the reaction and extraction occurs in the seeds is also not well understood. The parameters that affect extraction such as seeds pretreatment, particle size, moisture level, and solvent amount will also affect the yield of ester. The presence of other compounds in the seeds might also interfere with the transesterification reaction and the final ester and glycerol product might contain compound from the seeds that is soluble in methanol. Hence, the optimum values for methanol/oil molar ratio, catalyst concentration and temperature may differ between conventional transesterification and the combined reaction/extraction scheme.

In this study, investigations were carried out to understand the process parameters that affect the efficiency of the *in situ* process. Specifically, the effect of operating parameters such as molar ratio of alcohol to oil, particulate size, catalyst concentration, and moisture content were investigated.

2. Methods and Analysis

Materials

The rapeseeds were cultivated in the university farmland for the Newcastle University Agricultural Department. The seeds contain 45.9 wt% oil. The free fatty acid of the oil, was 3.35 wt % and the moisture content of the seeds was 6.7 wt %. Methanol used was of analytical grade with an assays of 99.95%.

Experimental Setup and Procedures

The experimental apparatus consisted of a two-neck round bottom flask containing the seed and the solvent, which was immersed in a temperature controlled water bath. An overhead stirrer with variable speed provided stirring for the system. An overhead condenser was placed in one of the flask neck to condense the evaporated solvent back into the system.

50 grams of rapeseed was grounded and sieved to a specific particle size. The seeds were then placed in the round bottom flask with a specific amount of methanol in which NaOH has been dissolved. The temperature used for the reaction is 60°C and the methanol has been previously heated to this temperature before the start of the reaction. The stirring rate was kept at 150 rpm through out all the studies. After the reaction time was reached, glacial acetic acid was added to the reaction mixture to stop the reaction. The liquid was separated from the seeds using vacuum filtration. The seeds were washed with an additional 60 ml of alcohol to recover any products that stick to the seeds during the filtration step. The extract was then transferred to a rotary evaporator where the solvent was removed under vacuum at a temperature ranging from 60 – 90 °C . After the entire methanol was evaporated, the product consists of an upper layer of ester phase and a lower layer of glycerol phase. In order to separate between the two layers, the products were transferred to a separatory funnel where the lower layer was withdrawn from the bottom of the funnel. The mass of the total extract and the ester phase was recorded
The composition of the ester phase for each reaction time was analyzed using a gas chromatograph according to BS EN 14105:2003.

3. Results and Discussion

3.1 Progress of Reactive Extraction

Figure 1 shows the progress of the reactive extraction process. The extraction was very fast in the first 5 minutes whereby 50% of the oil is extracted out during this period. This could be due to very fast reaction of the oil at the particle surface with the excessive amount of methanol and the washing of the ester to the bulk phase. Analysis of the ester phase during the first 5 minutes shows that it contains some amount of mono and diglycerides but very little triglycerides. The absence of significant amount of triglycerides at the early reaction times suggests that the reaction of oil to ester occurs either in the seeds particles or at the seeds surface which is then followed by the extraction of that ester or mono and diglycerides into the bulk methanol phase. After the first 5 minutes the extraction rate slows down. This could correspond to diffusional leaching of the ester from inside the solids particles which is a slower mass transfer step. The ester yield become almost constant after about 1 hour indicating that the equilibrium is reached between the concentration of ester in the particles and in the bulk methanol. The composition of the ester shows that it contains less than 2% mono and diglycerides which indicates that the conversion of oil to ester is very high in the reactive extraction process.

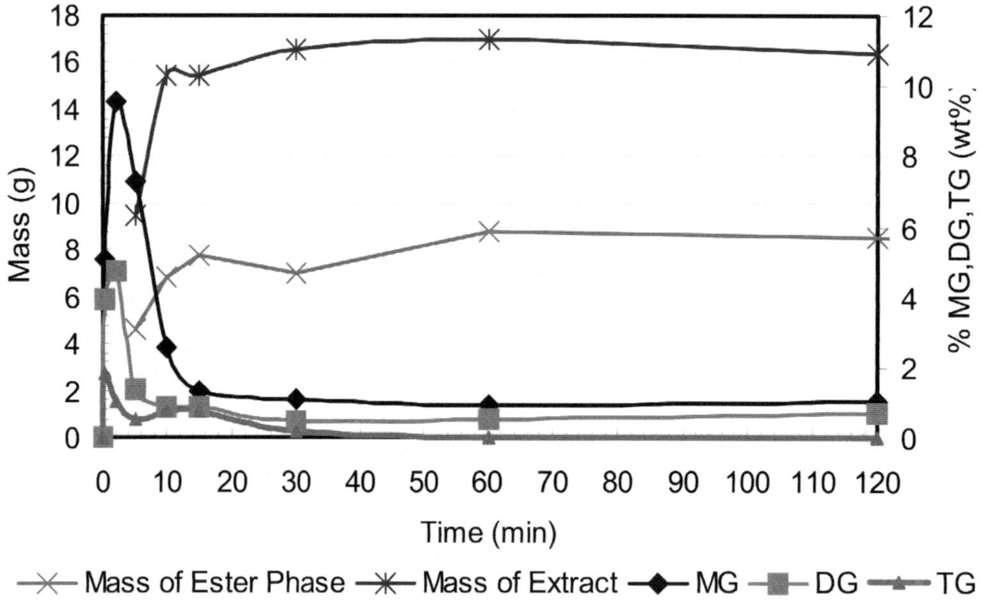

Figure 1- Progress of Reactive Extraction

3.2 Effect of Solvent to Oil Ratio

From the time study above it can be seen that under the particular process conditions, the equilibrium concentration of ester in the methanol phase should be reached by 1 hour. The effect of solvent to oil ratio is therefore studied after 1 hour reaction as shown in Figure 2. It can be seen that the amount of ester extracted increases with the amount of methanol used. As the diffusion process depends on the concentration gradient between the bulk phase and inside the particles, the more solvents that are used, the higher the concentration gradient and the more ester is extracted out. Using a lower ratio than 95 does not result in any significant biodiesel being produced. The ester yield reaches a maximum at 650:1 molar ratio. Further than this, emulsion exists between the ester and the glycerol phase and the amount of ester phase that can be physically separated from the glycerol phase reduces. This could be due to the high amount of catalyst when high amount of solvent is used. However, if the emulsion is extracted using hexane, the yield of ester from the higher solvent to oil molar ratio is similar to the maximum achievable yield.

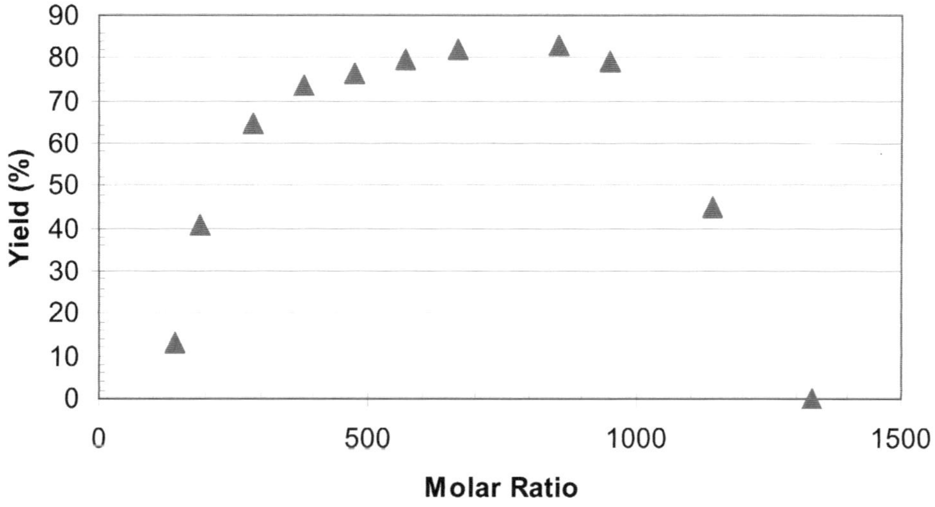

Figure 2 - Effect of Methanol to Oil Molar Ratio

3.3 Effect of Catalyst Concentration

The effect of catalyst concentration after a 1 hour reaction is shown in Figure 3. Using 0.05 m NaOH concentration, the maximum ester yield is achieved at about 1250 molar ratio whereas the maximum yield can be achieved at 650 molar ratio using 0.1 m NaOH. Therefore less solvent is required to achieve similar yield within the one hour reaction time for 0.1 m compared to 0.05 m NaOH concentration. Increasing the catalyst concentration to 0.15 m however causes a reduction in the yield. As shown in Figure 3, the amount of total extract for 0.1 and 0.15 m is similar but the ester yield is lower for 0.15 m. This could be caused by either the saponification reaction associated with high NaOH concentration or partial emulsification of the ester with the glycerol phase. Also it can be seen that at 0.15 m, the extract is completely emulsified at a much lower solvent to oil ratio than at 0.1m. Using 0.05 catalyst concentration on the hand does not result in any emulsification in the range of solvent to oil ratio studied.

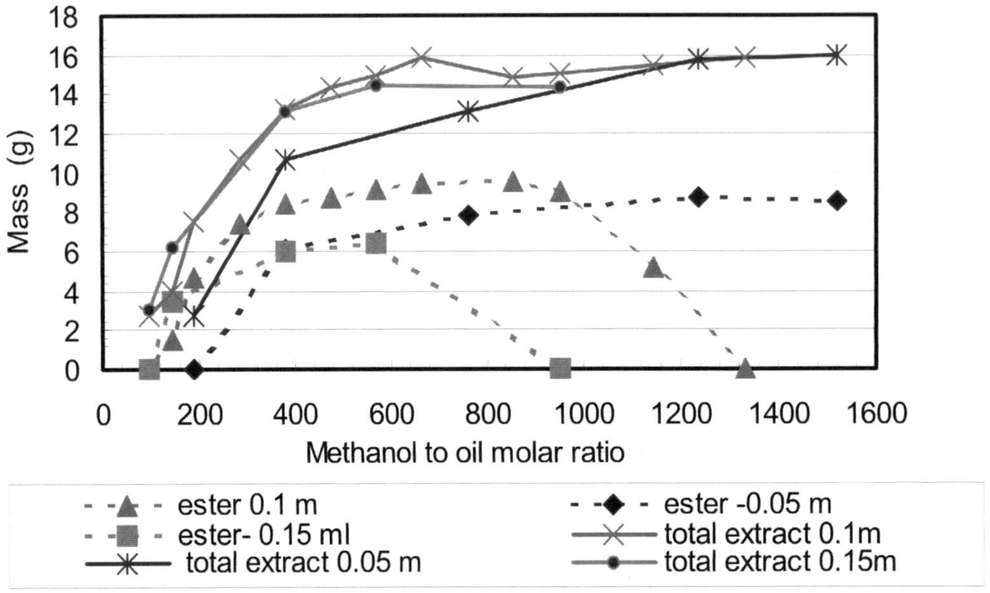

Figure 3- Effect of Catalyst Concentration

3.4 Effect of Particle size

The effect of particle size is seen in Figure 4 and 5. Reducing the particle size reduces the distance for the oil to diffuse out from the seeds particles and hence increases the rate of extraction. Therefore within the one-hour reaction, a lower particle size will give a higher yield as seen in Figure 4. However, at a longer time (5 hour), the yield for the larger particle size is similar to that of the smaller particles size as seen in Figure 5. This shows that particle size is important in determining the rate of diffusion but does not affect the final equilibrium yield for the rapeseeds particles

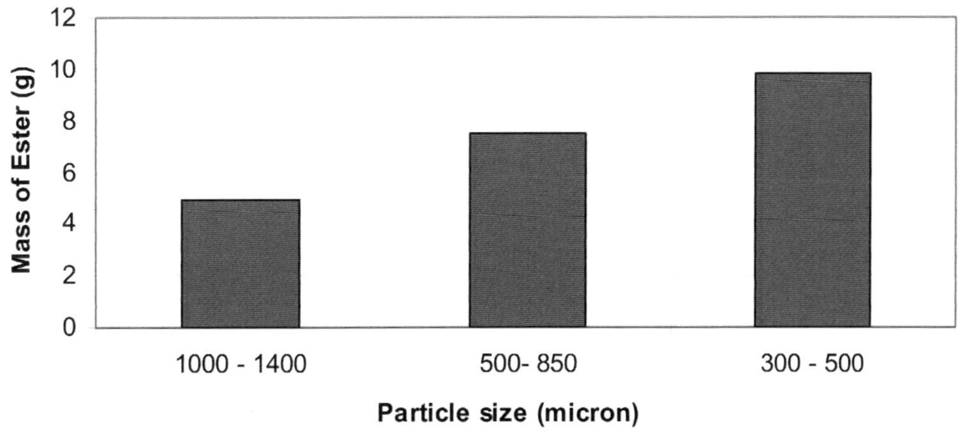

Figure 4- Effect of Particle Size

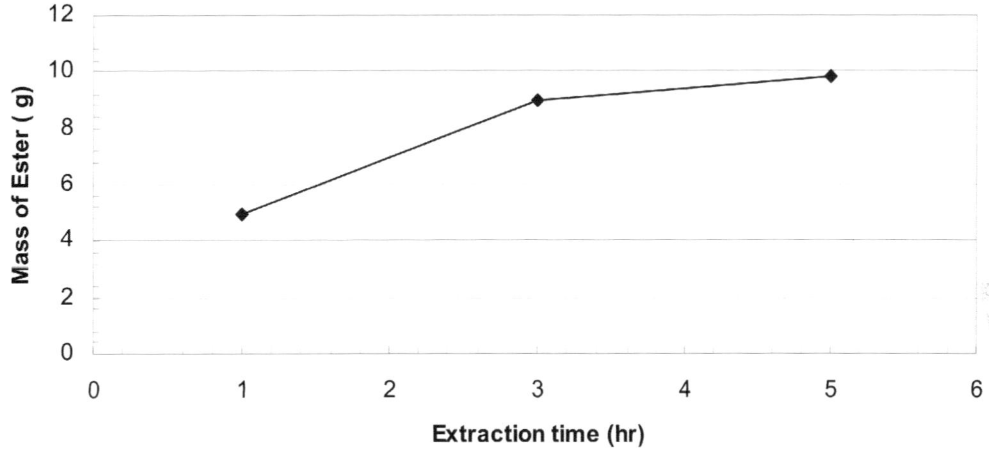

Figure 5- Yield of Ester versus Time for particle size 1000-1400 micron.

3.5 Effect of Moisture Content in Seeds

Previous literatures have shown that water adversely affect the transesterification reaction when NaOH is used [2]. The presence of water causes the saponification of both ester and triglycerides to soap and therefore reduces the yield of the reaction. Water was also reported to reduce the solvent capacity of some solvent such as ethanol [3]. It is therefore speculated that by drying the seeds to 0% moisture, the ester yield could be increased. The seeds initially have 6.7% moisture and were dried to 0% moisture in an oven at 104°C until constant weight. The result of drying is shown in Figure 6. Contrary to the prediction, drying the seeds to 0% moisture does not result in an increase in ester yield. The amount of solvent needed to achieve maximum yield also does not reduce when using dry seeds. This indicates that the amount of water in the seeds (6.7%) does not affect the transesterification and extraction process to any significant extent.

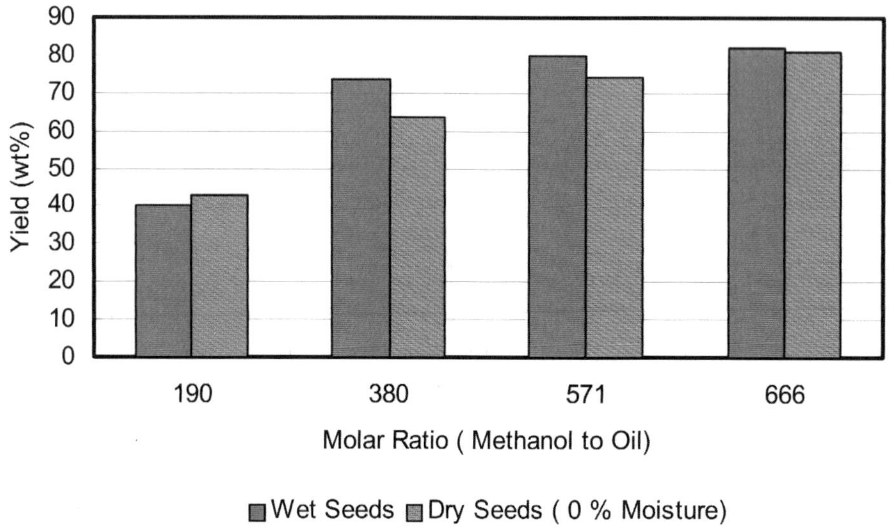

Figure 6 – Effect of Drying

4.0 Conclusions

The studies show that methanol to oil molar ratio and catalyst concentration affects the ester yield significantly. The equilibrium yield of ester is mainly determined by the methanol to oil molar ratio. The maximum ester yield is achieved at 650:1 molar ration and 0.1 m NaOH concentration. Smaller particle size affect the kinetics of the extraction but does not seems to alter the final equilibrium yield of ester. Reducing the water content of the seeds from 6.7% to 0% does not seem to present any significant effect to the ester yield.

References

[1] Harrington, K.J and D'Arcy-Evans, C., A comparison of Conventional and In Situ Methods of Transesterification of Seed Oil from a Series of Sunflower Cultivars, Journal of American Oil Chemist Society , v 81, n 2, p 83-89

[2] Haas, M.J., Scott, K., Moisture removal substantially improves the efficiency of in situ biodiesel production from soybeans, Journal of American Oil Chemist Society , v 84, 2007, p 197b-204

[3] Hron, R.J., M.S. Kuk, G.Abraham, and P.J.Wan, Ethanol extraction of oil, gossypol and aflatoxin from cottonseed, Journal of the American Oil Chemists' Society, v 71, n 4, 1994, p 417-421

28f Granular Mixing and Segregation In Zig-Zag Chute Flow

Suman K. Hajra, Deliang Shi, Watson L. Vargas, and J. J. McCarthy. Department of Chemical and Petroleum Engineering, University of Pittsburgh, 1249 Benedum Hall, Pittsburgh, PA 15261

By introducing periodic flow inversions, we have shown both experimentally and computationally that forcing with a value above a critical frequency can effectively eliminate both density and size segregation. The critical frequency is the key to applying this technique and is directly related to the inverse of the characteristic time of segregation. In this work, we study size segregation and propose a generalized size segregation model to determine the critical forcing frequency in these systems. The influences on mixing of the length of each leg of the zig-zag chute and the size ratio of the particles are examined. Mixing is observed instead of segregation when $L<U_{avg}t_S$, where L, U_{avg}, and t_S denote the length of each leg of the zig-zag chute, the average stream wise flow velocity of the particle, and the characteristic time of segregation, respectively.

28g Discrete Element Method Simulations of Dry Impregnation Processes in Catalysts

M. Silvlna Tomassone, Charles Radeke, and Arthur W. Chester. Chemical and Biochemical Engineering, Rutgers University, 98 Brett Road, Piscataway, NJ 08854

Catalyst impregnation is one of the most crucial steps for preparing industrial catalysts. In this process, metal complexes are dissolved in an aqueous solution and contacted with a porous oxide catalyst support such as alumina (Al2O3) or silica (SiO2). Generally, the process of impregnation is performed in rotating vessels with sprinklers that pour the aqueous solution with the metal precursor into the catalyst support. As the liquid penetrates into the dry powder, the particles start adsorbing the fluid, increasing the density of the particles, until they become saturated, after which the excess liquid forms liquid bridges between particles, which causes wet cohesive forces that disrupt flow and affect granular mixing processes. This can adversely affect the homogeneous dispersion of the metal precursor in the solid. Some of the open questions are (i) how mixing and flow are affected when the particles have a certain degree of moisture or are saturated with liquid, (ii) whether the liquid is homogeneously distributed into the powder bed and (iii) the extent of dead zones for a given impregnator configuration.

We performed DEM (Discrete Element Method) simulations of the dry impregnation process with different nozzle configurations that distribute an aqueous solution into the catalyst support. A novel water transfer algorithm was implemented in the simulations to account for the actual distribution of water in the particle bed. We computed several parameters, which include the residence time distribution (time that the particles spend under the spraying zone), the extent of the dead zones (zones that do not receive any water during the impregnation process) and the spatial distribution of the sprayed water throughout the powder bed in different cut planes. Our results show that the residence time distribution of the particles in the spraying zone is not significantly affected when we vary the speed of the blender. For four, eight and ten revolutions the particles remain almost a similar amount of time in the spraying zone. The distribution of water is inhomogeneous across different planes parallel to rotation axis reflecting poor radial mixing typical of double cone blenders. When we increase the number of nozzles the extent of dead zones significantly reduces and 70 to 80 percent of particles show a level of saturation that exceeds 50 percent after ten revolutions. For all impregnator configurations the liquid is never distributed completely homogeneously throughout the powder bed. However, if we correctly manipulate the variables and the configuration of the spraying guns we can obtain a nearly homogeneous distribution.

A DE numerical study upon particulate solids flow from model silos

Guoxian Xiao [1, 2], Delong Xu [2] and S Ding [3]
1 College of Chemistry, Chemical Eng. and Material Sci., Soochow Univ., Suzhou, China
2 Institute for Powder Eng., Univ. of Archi. & Tech., Xi'an, China
3 Postec dept. av Tel-Tek, Kjølnesring, Porsgrunn, Norway

Introduction

The importance of avoiding flow disruptions and quality variations associated with funnel flow requires a mass flow rather than a funnel flow in silo design (Jenike, 1964; Drescher, 1992). Quite often, a silo designed to perform a mass flow turns into a funnel flow silo after a certain period of service, or later on when being used to store materials with changed flow properties. Some efforts have been expended in order to obtain mass flow in silos with a reasonably flat hopper; among which are to install a flow aid device such as inserts in a certain position in a silo (Johansen, 1982). A prediction to whether such inserts function or are suitable for a given application is therefore desirable.

However, one must admit that no perfect analytical methods have been developed to carry out such predictions because of complexity caused by a presence of inserts in silos (Nedderman, 1992; Drescher, 1991). Instead numerical approaches are usually adapted (mostly the Finite element (FE) method and the Decreased element(DE) mthod) to seek solutions or clues (Rotter at al., 1998; Karlsson et al., 1998; Ding et al., 2003; Eibl & Rombach, 1987; Haussler & Eibl, 1984; Eibl & Rombach, 1987; Cundall & Strack 1979 etc.). In the paper the Discrete Element (DE) models were created in an effort to investigate the flow mode of particulate solids discharged for a plane silo. Effects of a cone-in-cone insert on such flow modes were also addressed. The influence of the fluid filled voids of particulate solids on the flow mode was also attempted with modification to the DE model after considering the forces exerted on the particulate solids due to fluid bridge.

Object of the DE model and conditions

A two-dimensional model silo was considered in the present study. It consisted of a vertical bin section and an inclined hopper section. A cone-in-cone insert was fitted into the silo. The key dimensions for both silo and insert, and the configuration of the insert and silo were as shown as in Figure 1. A 2000 particulate solids as spheres of a 5 mm in diameter were filled into the silos. The solids have a density of a 2700 kg/m^3, and are pretty soft with an elastic Young's module of a 40000 Pa and a Poisson ratio of 0.3. The interaction between the particulate solids and solids against the surface of silo walls (/insert) was modelled as friction with a coefficient of 0.3.

Interaction of particulate solids considered in the model

In the model the particulate solids stored in a silo are treated as a set of discrete elements (Cundall and Strack, 1973), where interaction between the contacting elements could be determined through contacting forces. Such forces depend on the deformations of solids, and are usually resolved into the normal and tangential components (Johnson, 1994; Mindlin and Deresiewicz, 1953). The Newton's equation of motion of a particle is given by:

$$\frac{d\vec{V}}{dt} = (\sum_{j=1}^{k_j} \vec{F}_{cj} + m\vec{g}) / m \tag{1}$$

The rotational motion of a particle caused by the contact force is governed by:

$$I\frac{d\vec{\omega}}{dt} = \sum_{j=1}^{k_j} \vec{r} \times \vec{F}_{cj} \tag{2}$$

Equations as given above can be solved numerically by a finite difference method. The trajectories and the transient forces of the individual solids can then be determined. As a result, particulate solids will undergo translational and rotational motion, and to be discharged from silos.

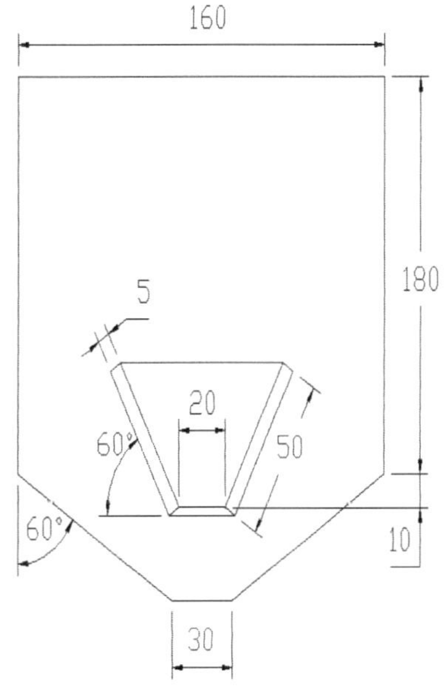

Figure 1. The key dimensions and a configuration of a silo, a cone-in-cone insert considered in the models

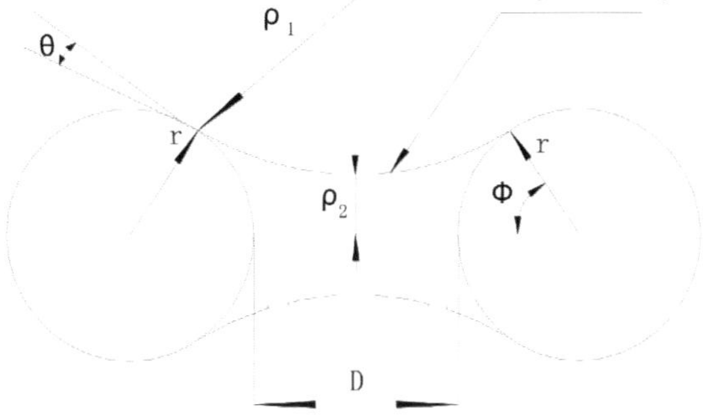

Figure 2 A pair of central moving spheres with interstitial Newtonian fluid.

When there exists interstitial fluid in the voids among the solids, liquid bridges would form in the gap between particles as illustrated in Figure 2, exerting adhesive force and viscous force up the particulate solids.

Fisher (1926) gave the adhesive force as follows:

$$F_{cap} = \pi\rho_2^2\Delta p + 2\pi\rho_2\sigma_s = 2\pi\rho_2\sigma_s(1+\frac{\Delta p}{2\sigma_s}\rho_2) = 2\pi\rho_2\sigma_s(1+\frac{\rho_2-\rho_1}{2\rho_1\rho_2}\rho_2) \tag{3}$$

$$= \pi\sigma_s\rho_2(1+\rho_2/\rho_1)$$

Where $\rho_1 = [r(1-\cos\phi)+D/2]/\cos(\phi+\theta)$ (4)

$$\rho_2 = r\sin\phi - [1-\sin(\phi+\theta)]\rho_1 \tag{5}$$

While the viscous forces were given by Adans and Perchard (1985) as:

$$F_{vn} = 6\pi\eta r^* V_n \frac{r^*}{D} \tag{6}$$

for the normal component, and

$$F_{vt} = (\frac{8}{15}\ln\frac{r^*}{D} + 0.9588)6\pi\eta r^* V_t \tag{7}$$

for the tangential component, where $r^* = r/2$ for spheres of the same diameter.

Lian et al. (1994) showed that the distance of liquid bridge stability is proportional to the cube root of the volume of liquid bridge. This representation was used in the current study.

Incorporating a liquid bridge force into DE Model simulations, the program was further modified to investigate the effect of liquid bridge on the particulate solid flow mode during silo discharge.

Effect of a cone-in-cone insert on flow mode during discharge

The flow modes of particulate solids discharged from silo were simulated. Thin layers of particles with colour were set at different levels as tracers to visualize the development of the flow patterns. The results are given in Figure 3.

As seen in Figure 3 (a) via the coloured particles, the particulate solids in the central zone started to drop, while those close to the walls sections remained intact at the beginning of discharge when no insert was inserted. With the discharging going on, those solids slid into the central moving zone and to be discharged. According to the shape of the coloured particles layer, it is obvious that the solids were discharged in a typical funnel flow mode.

With a presence of a cone-in-cone insert in silos (see Figure 1), one can see in Figure 3 (b) that all solids in silos were in move; and were in fact discharged rather evenly at all crossed sections at the commencement of discharge. With the discharging going on, the solids above the insert were slowed down, and the evenness of discharge lost, but solids in the silo were still all in motion until totally discharged.

In brief, the funnel flow phenomenon is apparent from the motion of the coloured layers of particles when no inset was installed, whereas such funnel flow was turned into a mass flow in a presence of a cone-in-cone insert for the configuration as presented in Figure 1.

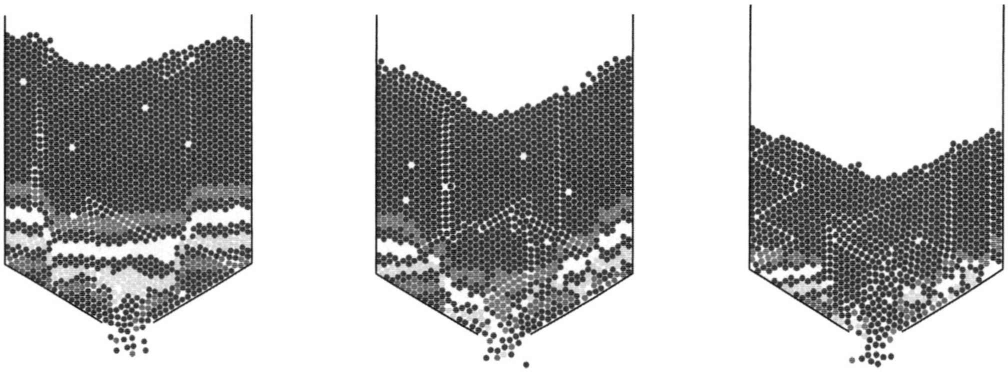

t=0.4 s t=1.3 s t=2.4 s

a.) without an insert

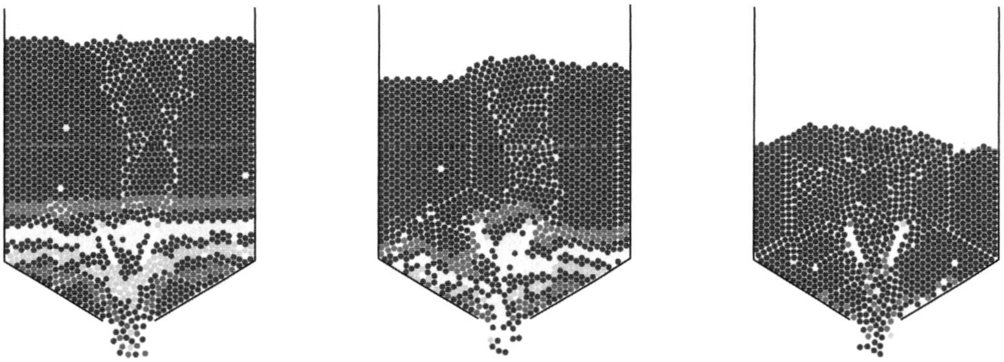

t=0.4 s t=1.3 s t=2.4 s

b.) with a cone-in-cone insert

Figure 3 Flow mode of particulate solids discharged from a silo

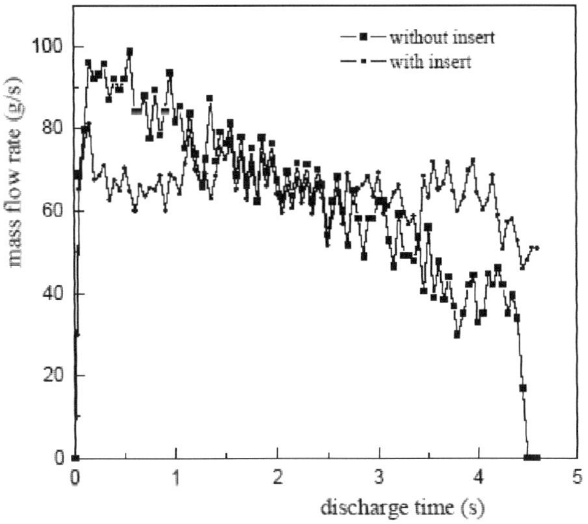

Figure 4 Development of discharging mass rate

The mass rate of discharging is respectively given in Figure 4 for the cases in a presence or an absence of the cone-in-cone insert. It is evident that the discharging rate fluctuated in both simulations. It could also be seen that an installation of a cone-in-cone insert decreases the fluctuations of the discharging rate, and become more stable.

Influence of water content to particles flow

Modification to the DE model was made in order to focus on investigation of the effect of interstitial fluid content on the discharge mode. In the modified model, the silo had an half angle of a 60°, with a width of a18 mm for its bin section and an outlet of a 6 mm; A 1000 particles with a 1 mm in diameter were considered; the liquid surface tension was assumed to be 0.0725 N/m and viscosity of 0.001Pa·s. The numerical results are shown in Figure 5. As seen in Figure 5, the water content has a distinctive affect on the flow of particulate solids in the silo. The discharge was uniform and steady from the silo when the water content is 0 (see Figure 5 (a)); the discharge took place, and the particles was discharged in agglomerate when the water content increased to 1%, Figure 5(b); however, an arch was developed at the outlet when the water content increased to 2%, Figure 5 (c), i.e., no discharge occurred.

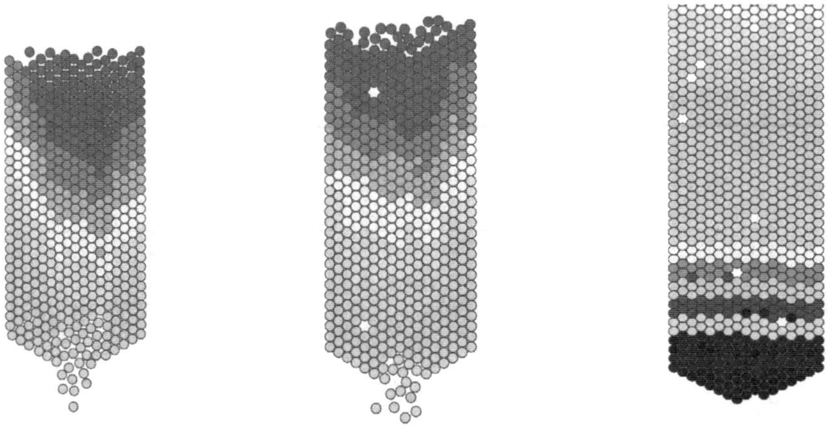

(a) Water content=0%　　(b) Water content=1%　　(c) Water content=2%

Figure 5 Influence of water contents on the mode of silo discharge

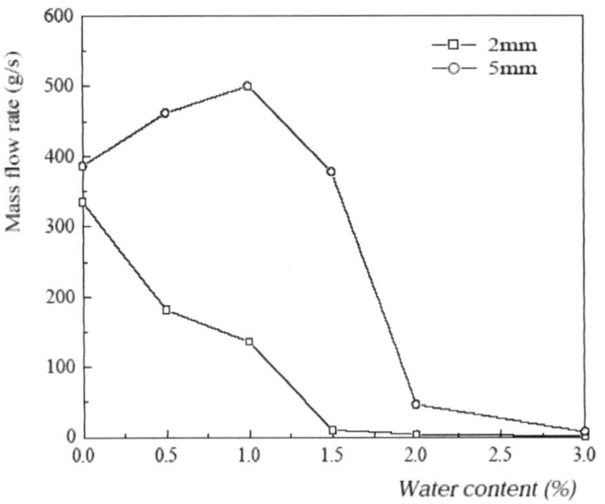

Figure 6 Effect of water contents on mass flow-rate

Efforts on the affect of the fluid content on the mass flow rate were also made. It was found that the mass flow rate decreased with the increase of the water content if particulate solids were small (2 mm in diameter); the affect of water content on the mass flow rate varied for particles of a 5mm in diameter. Typical results with such regards are given in Figure 6.

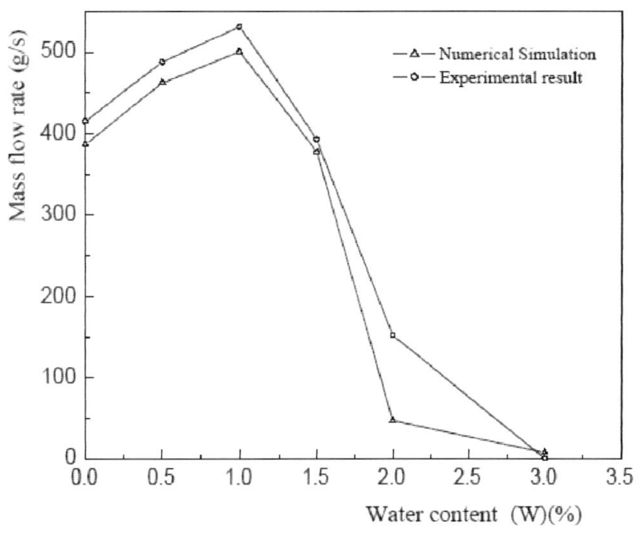

Figure 7 Comparison of calculated results and experimental data for mass flow-rate

Verification was conducted to demonstrate the influence of fluids content to the mass flow rate. It is shown in Figure 7 with the data obtained in experiments carried out with a silo in dimensions as used in the models; the particulate solids used were 5mm in diameter. One can see the good agreement between the predicted results and the experimental data.

Conclusions

The program developed with the discrete element method was used to predict the particulate solid discharge mode from silos. The prediction showed that the cone-in-cone insert can effectively change the particles flow mode from funnel flow to mass flow. It also showed that the interstitial fluid has a distinctive affect on the flow of particles; higher fluid content (with the value considered in the present study) might cause stoppage of silo discharge. The mass flow rate decreases with the increase of the fluid content for small particles. The existence of certain amount of fluid facilitate the discharge when the particle is relative large, but the mass flow rate of discharge would also decrease when the fluid exceeds a limitation.

References

Adans M. J. & Perchard V., 1985, The cohesive forces between particles with interstitial liquid, Institution of chemical Engineering Symposium, No.91:pp. 147-160.

Cundall P. A. & Strack O. D. L., 1979, A discrete numerical model for granular assemblies, Geotechnique, No. 29: pp. 47-65.

Ding S., de Silva S. R. and Enstad G. G., 2003, Effect of Passive Inserts on the Granular flow From Silos Using Numerical Solutions, *Particulate Science and Technology*, 21, 211-226.

Drescher A., 1991, Analytical Methods in Bin-Load Analysis, Elsevier Science Publishers B.V. 1991.

Drescher A., 1992, On the criteria for mass flow in hoppers, *Powder Technology*, 73, p. 251-260.

Eibl J. and Rombach G., 1987, Stress and Velocity Fields at Discharging of Silos, Proc. of NUMETA Conference, Swansea, p. D1-D12

Fisher R. A., 1926, On the capillary forces in an ideal soil correction of formulae given by W. B. Haines, Journal of Agricultural Science, No.16:pp. 492-505.

Haussler U. and Eibl J., 1984, Numerical Investigation of Discharging Silos, *Journal of Engineering Mechanics*, 110(6), P. 957-971.
Jenike W., 1964, Storage and Flow of Solids, Bulletin,123, Utah Engineering Experiment Station, University of Utah.
Johansen J. R., 1982, Controlling flow patterns in bins by use of inserts, *Bulk solid handling* 2 (3), 495-498.
Johnson K. L., 1994, Contact mechanics, Cambridge University Press, Cambridge,
Karlsson T., Kilsinski M. and Runesson K., 1998, Finite Element Simulation of Granular Material Flow in Plane Silos with Complicated Geometry, Powder Technology, 1998.
Lian G., Thornton C. & Adans M. J., 1993, A theoretical study of the liquid bridge forces between two rigid spherical bodies, J Colloid Interface Sci., No.161: pp. 138-147.
Mindlin R. D. & Deresiewicz H., 1953, Elastic spheres in contact under varying oblique forces, Journal of Applied Mechanics, No.20, pp. 327-344.
Nedderman R. M., 1992, Statics and Kinematics of Granular Materials, Cambridge University Press.
Rotter J.M., Holst J. M. F. G., Ooi J. Y. and Sanad A M., 1998, "Silo Pressure Predictions using Discrete-Element and Finite-Element Analyses," Philosophical Transactions of the Royal Society of London: Series A-Mathemat, 356(1747): P. 2685-2712.

Nomenclatures

D	–	spherical surface distance, (m)
F_{cap}	–	the adhesive force, (N)
$\vec{F}_{cj}$	–	contact force acting on the particle, (N)
F_{vn}	–	normal component of the viscous force, (N)
F_{vt}	–	tangential component of the viscous force, (N)
$\vec{g}$	–	acceleration of gravity, (m/s^2)
I	–	moment of inertia, (kg m^2)
m	–	mass of particle, (kg)
Δp	–	capillary suction pressure, (Pa)
r	–	radius of particle, (m)
t	–	time, (s)
$\vec{V}$	–	translational velocity of particle, (m/s)
V_n	–	normal component of relative velocity, (m/s)
V_t	–	tangential component of relative velocity, (m/s)
$\vec{\omega}$	–	angular velocity of particle, (rad/s)
ρ_1	–	radii of curvature of liquid bridge surface, (m)
ρ_2	–	radii of neck of liquid bridge, (m)
φ	–	half-filling angle, (rad)
θ	–	contact angle, (rad)
σ_s	–	liquid surface tension, (N/m)
η	–	dynamic viscosity, (pa s)

ACKNOWLEDGEMENTS

The authors would like to thank the Institute of Powder Engineering, Xi'an University of Architecture & Technology for providing the support for this study.

78a Measuring Particle Attrition In a Jet Cup

Ray A. Cocco, S.B. Reddy Karri, Yeook Arrington, and Ted Knowlton. Particulate Solid Research, Inc., 4201 West 36th Street, Suite 200, Chicago, IL 60632

Particle attrition is usually a detriment to product quality and process cost. It is an issue in virtually every solids processing unit operation. Thus, it is important to know how particle attrit under relevant operating conditions. Jet cup attrition test devices (such as the Davison Jet Cup) are typically used to "infer" relative particle attrition for fluidized bed and riser operations. Jet cup attrition studies typically provide a relative comparison between two or more materials that can be used to generate a relative ranking of the material attrition indices. Ideally, the attrition rates measured in these laboratory units will provide a relative indication of how the materials will behave in the commercial unit. However, Particulate Solid Research, Inc. (PSRI) has found that a cylindrical jet cup attrition measurement may not be effective in providing accurate attrition rankings. Attrition index rankings from a cylindrical jet cup and a 12-inch (0.3-meter) diameter, pilot-plant fluidized bed unit did not correlate with each other.

Barracuda™ computational fluid dynamics (CFD) simulations of a traditional cylindrical jet cup suggest that the results from a cylindrical jet cup design may not be completely accurate. CFD studies show that 30 to 50% of the particle sample will not be exposed to the solid stresses needed for accurate particle attrition measurements. This is because many of the solids are nearly stagnant, even at high jet velocities. This was subsequently confirmed with cold flow studies at PSRI in Plexiglas jet cup models. As a results, it is unlikely that relevant attrition rankings can be determined from traditional jet cup studies as a significant portion of the particle sample is not exposed to sufficient solid stresses to cause attrition. Only by insuring that the entire sample is under similar amount of stresses can attrition be accurately linked to inlet jet velocity and directly compared with different materials.

This paper discusses the development of a jet cup device that allows all of the sample particles to experience similar amounts of solids stresses. The rankings of the attrition indices from the new jet cup were found to correspond to the same rankings observed from pilot-plant attrition tests. The agreement in rankings obtained with the new cup was not observed with the traditional cylindrical jet cup.

78b Shear and Composition Effects on the Flow and Electrostatic Properties of Pharmaceutical Powders

Alisa Vasilenko, Kalyana Pingali, and Fernando J. Muzzio. Chemical and Biochemical Engineering, Rutgers University, 98 Brett Rd, Piscataway, NJ 08854

Most drug products sold on the market today are solid dosage forms, making it largely a powder technology industry. The complex behavior of pharmaceutical powders is not well understood and characterized; specified processing conditions and formulations are determined by extensive trial-and-error procedures, and even the small changes in the said parameters can cause great variations in powder flow and electrostatic properties. There is a growing interest in developing the correlations between the input conditions, processing parameters and the resulting product properties.

In this regard, a parametric study was undertaken to characterize the effects of formulation concentration and processing conditions on the flow and electrostatic properties of pharmaceutically relevant powders. Twelve blends with various concentrations of active ingredient, magnesium stearate as a lubricant and silica as a glidant were subjected to four different shear rates and levels of total strain in a v-blender with an intensifier bar. The resulted formulations were characterized by measuring flow index, dilation volume, yield strength, and impedance. The results showed substantial effects of total strain and, to a lesser extent, the shear rate on the flowability and electrical properties of the studied blends; the optimal levels of shear and additives to achieve the desired flowability were determined.

78c Experimental and Numerical Analysis of Stress Generation In Non-Circular Pharmaceutical Tablets - a Manufacturing Case Study

Stephen L. Conway[1], Adam Procopio[2], Marty J. Smith[3], and Michael Gentzler[1]. (1) Center for Material Science and Engineering, Merck & Co., Inc., WP75B-210, PO Box 4, West Point, PA 19486, (2) Materials Characterization & Technology Assessment, Merck & Co., Inc., WP75B-210, PO Box 4, West Point, PA 19486, (3) Pharmaceutical Operations, Merck & Co., Inc., WP69-2, PO Box 4, West Point, PA 19486

A detailed understanding of the evolution of mechanical stresses in pharmaceutical tablets is critical in achieving product performance requirements, development of appropriate process control strategies, and in reducing manufacturing variability. Extensive literature exists to describe the rigorous solution of stress fields in simplified tablet geometries (most commonly, circular, flat-faced tablets), but considerably less addresses the complex tablet shapes of marketed pharmaceutical products.

Experimental and numerical results are presented for the case of a pharmaceutical product manufactured by compressing lubricated granulation in rotary tablet presses. We investigate potential causes of increased compression force requirements in full-scale manufacture, leading to premature punch tool wear. After verification that lubricant maldistribution is an unlikely cause of tensile strength depression (via NIR methods) we explore differences in tablet hardness control between two product images of differing shape. Combining instrumented hydraulic compaction simulation and hardness test data from the laboratory with two-dimensional finite element method (FEM) numerical simulations, a coherent representation of tensile stress generation and failure modes for the two tablet images is developed. Qualitative agreement between laboratory observations and FEM simulations of diametrical compression simulations is achieved. By scaling intrinsic tensile strength data developed for circular tablets, the effectiveness of varying compression force to control tablet hardness in the two manufacturing processes is determined. The analysis illustrates why compression of one image may struggle to meet existing hardness specifications readily achieved for another tablet shape. We find that the established hardness test methods for the two tablet images are not equivalent and cause significant diversions in principle stress fields and their temporal development.

Extending the FEM simulations enables the virtual testing of a modified tablet hardness test apparatus. Maximum principle stress development is examined for several options and designs evolve to avoid tablet pivoting, excessive bending moments, or localized tensile stresses at the tablet edge. A simple modification to the tablet testing apparatus is fabricated and confirmed to provide greater equivalency between the two tablet shapes. Contributors to test variability are considered. Revalidation of the manufacturing process is expected to confirm that tablet hardness specifications can be met with reduced compression forces, with no impact on tablet appearance, friability and dissolution. Greater consistency between tablet images is predicted, along with more effective exploitation of press-force adjustments as a process control variable. While extensions to the FEM simulation approach are proposed to enhance quantitative accuracy, the utility of fit-for-purpose FEM techniques in a pharmaceutical manufacturing setting is demonstrated.

78d Scale up and Optimization of Fluid Bed Coating and Layering Processes Using In-Situ Particle Characterization

Andrew Birkmire[1], Kim T. Walter[1], **Eric Dycus**[2], Benjamin Smith[2], and Frederic Prulliere[2]. (1) Niro Pharma Systems, Columbia, MD 21045, (2) Particle Systems Characterization, METTLER TOLEDO, Redmond, WA 98052

Fluidized pellet coating processes are commonly used when developing a novel solid formulation. Yet there are often process inefficiencies and inconsistencies during scale-up. A successful fluid bed coating process establishes a repeatable endpoint particle distribution to achieve repeatable content uniformity and dissolution profiles. However, inconsistencies occur during scale-up with varying raw materials or changing process dynamics such as segregation, agglomeration, or attrition. The application of in-situ particle monitoring techniques, Lasentec FBRM and PVM is demonstrated to track a series of spray coating batches while varying the batch volume, column diameter, spray rate, humidity, temperature, atomization pressure and excipients. FBRM and PVM allow a user to link process control parameters to the particle size distribution and product quality. Target particle distributions were obtained by controlling critical control parameters during scale up.

Web Page: www.mt.com/lasentec

78e Quantitative Large-Scale Validation of Image-Based Sensors for Online Particle Size Characterization during Crystallization

Ying Zhou, Process Science and Modelling, Institute of Chemical and Engineering Sciences (ICES), 1 Pesek Road, Jurong Island, Singapore, 627833, Singapore, Rajagopalan Srinivasan, Chemical and Biomolecular Engineering, National University of Singapore, 10 Kent Ridge Crescent, Singapore 119260, Singapore, Singapore, and L Samavedham, Chemical & Biomolecular Engineering, National University of Singapore, 10 Kent Ridge Crescent, Singapore 119260, Singapore, Singapore.

Crystallization is a critical operation in pharmaceutical, fine chemical, petrochemical, food and semiconductor industries. In crystallization, particle size and shape are important specifications of product quality that need to be well controlled. In pharmaceutical crystallization process, the dissolution rate, bioavailability and therapeutic effects of the drugs crystallizing out depends significantly on the particle size and shape. A narrow particle size distribution with specific particle shape is indicative of good product quality.

Real-time monitoring and control of the crystallization process is important to ensure that the desired final product quality is achieved. Traditionally, the control of crystallization processes has relied extensively on empirical experience. Complex chemistries, non-availability of detailed models, and the lack of in-situ sensors to directly measure product quality have been the main reasons for this state of affairs. Although technologies for offline particle size and shape measurements such as microscopy have been available and widely used, it is but recently that in-line measurements are becoming possible. Technologies such as Focused beam reflectance measurement (FBRM) and Particle vision and measurement, PVM (Wilkinson et al., 2000) from Lasentec / Mettler-Toledo, are widely used in manufacturing units to monitor particle size distribution and shape variation.

The development of image-based sensors has evoked recent interest in the use of image-analysis based approaches to estimate crystal size and shape in real-time and in situ (Zhou et al., 2006; Larsen et al., 2006 & 2007; Wang et al., 2008). In spite of the increased research activity in this area, there is little or no work that demonstrates the success of the image analysis (IA) techniques to any reasonable degree. While image analysis techniques are well developed, the quality of images from inline sensors is variable and often poor, leading to incorrect estimation of the process state. The lack of alternate approaches to compare the results has prevented large-scale validation. The primary objective of the present paper is to fill this void by addressing a key step in IA viz. segmentation. Also, segmentation involves several steps such as image selection, image enhancement, edge detection and morphology operation. Any IA algorithm involves the use of a series of user-specified parameters in each of the above steps. Our second objective is to estimate the sensitivity of the results to these parameters.

Our solution strategy uses manual segmentation of particles as an independent measure of the CSD. To quantify the similarities between the manual and automated segmentations, we have devised metrics which compare the particle sizes from the two segmentation techniques. These metrics provide a quantitative approach to estimate the quality of results as well as form the basis to suggest "optimal" parameter settings for the IA technique. Mono-sodium Glutamate (MSG) seeded cooling crystallization process is used to illustrate that, with proper settings, IA can be used to accurately track the CSD in real-time. As additional evidence, we also compare the results from IA to those from alternate sensors such as FBRM. Our results show that, with proper settings, the CSD estimates from IA can be obtained within ~10% error.

References

Larsen, P.A., Rawlings, J.B. & Ferrier, N.J., (2006), An algorithm for analyzing noisy, in situ images of high-aspect-ratio crystals to monitor particle size distribution. Chemical Engineering Science 61, 5236-5248.

Larsen, P.A., Rawlings, J.B. & Ferrier, N.J., (2007), Model-based object recognition to measure crystal size and shape distributions from in situ video images. Chemical Engineering Science 62, 1430-1441.

Ma, C. Y., Wang, X. Z. & Roberts, K. J., (2007), Multi-dimensional population balance modeling of the growth of rod-like L-glutamic acid crystals using growth rates estimated from in-process imaging. Advanced Powder Technol., Vol. 18, No. 6, pp. 702-723.

Wang, X. Z., Roberts, K. J. & Ma, C., (2008), Crystal growth measurement using 2D and 3D imaging and the perspectives for shape control. Chemical Engineering Science 63 (2008) 1173-1184.

Wilkinson, M. J., Jennings, K. H. & Hardy, M., (2000), Non-invasive video imaging for interrogating pharmaceutical crystallization. Microscopy and Microanalysis, Vol (6) (Supplement), 996-997.

Zhou, Y., Doan, X. T. & Srinivasan, R., (2006), Real-time imaging and product quality characterization for control of particulate processes. Joint 16th ESCAPE and 9th PSE: Computer-Aided Chemical Engineering 21A, W. Marquardt and C. Pantelides (Eds), 775-780. Elsevier, Garmish-Partenkirchen.

78f Characterization of Density of Cohesive Granular Materials Using An X-Ray CT Scanner

M. Silvina Tomassone, Rutgers University, Department of Chemical & Biochemical Engineering, 98 Brett Road, Piscataway, NJ 08854, **Charles Radeke**, Chemical and Biochemical Engineering, Rutgers University, 98 Brett Road, Piscataway, NJ 08854, and Divya Jonnavittula, Department of Chemical and Biochemical Engineering, Rutgers University, 98 Brett Road, Piscataway, NJ 08854.

The tendency of powders to flow has a significant effect on a large number of industrial applications. Different experimental techniques to measure density of powders show dramatic differences in the results. We propose a reproducible procedure to measure density using a CT scanner. We performed tap density measurements using an acrylic cylinder containing different cohesive powders in an electromechanized shaker. After tapping, the cylinder was placed inside a new X-ray CT scanner Skyscan -1172. The image reconstruction and the analysis were performed with the Skyscan's NRecon and CTAn softwares. We compared measurements of the tap density using the microCT scanner to other more traditional methods to measure tap density. Results show a significant overestimation of the tap density using conventional methods: The bulk density shows a linear behavior as a function of time, and frequency of tapping. The percent of change in the slope between conventional methods to measure tap density and microCT scanner measurements varies from 33.33 percent at 3 HZ to 54.54 percent differences at 10 Hz.

Tap density as a function of frequency of tapping, and level of cohesion of the powder were computed. Our results show that after a transient the tap density is independent of the frequency of tapping. Density variations as a function of the height of the cylinder were also analyzed using the micro CT scanner. Our results show that, in general, as the height increases, the density of the powders decreases. For the least cohesive powders there is a density difference of 22.40% for a height difference of 1.45cm for a tapping frequency of 3 Hz. For the highly cohesive powders the density difference is not as pronounced; (it is 5.2% for the same conditions.) We plotted the high (tap density) to low density value for different cohesive powders as a function of the flow index. We observed that the higher the ratio between the "high" and the "low" , the more cohesive the powder is and the poorer is their flowability.

Application of Helmholtz Resonance to Determination of Bulk Density and Fill Fraction of dry particulate solids

Emile S. Webster and Clive E. Davies

School of Engineering and Advanced Technology
Massey University, New Zealand

INTRODUCTION

Helmholz resonance is the term for resonance occurring in a cavity linked to the surrounding atmosphere via a constricted neck or necks. The resonant frequency is principally determined by the volume of the cavity and the dimensions of the neck (s). An object placed inside a resonator reduces its effective volume, and it follows that for a well defined system, the volume of the object can be estimated.

Such a system is thus potentially useful for characterizing bulk volumes of particulate solids, and hence the degree of fill in a container or bin, though we anticipate constraints and limitations with large bins, dictated by the low resonant frequencies of large volumes.

Experiments have been carried out with a 3 liter resonator using a variety of particulate materials including glass ballotini and sand. Degree of fill is calculated and compared with values estimated from direct measurements of material mass and particle density measured off-line.

THEORY

The resonant frequency of a Helmholtz resonator is given by Equation 1:

$$Freq_{res} = \frac{c}{2\pi}\sqrt{\frac{s_p}{V_c l_p}} \qquad (1)$$

where $Freq_{res}$ is resonant frequency, c is velocity of sound; s_p is cross sectional area of the port; V_c is volume of the empty chamber; and l_p is corrected length of the port.

Rearrangement of Equation 1 gives Equation 2, which relates the volume, w, of an object placed in the resonant chamber to the empty chamber volume:

$$w = V_c - \frac{s_p}{l_p\left(\frac{2\pi Freq_{res}}{c}\right)^2} \qquad (2)$$

EXPERIMENTAL

All experiments were conducted with a 3 L chamber made of 140 mm inside diameter Perspex tube. The chamber had a single port with an internal diameter of 22 mm and was 170 mm long. The acoustic signals were sensed by two PCB103A sound pressure microphones; one was located in the centre of the base, and the other was located at the top of the 170 mm port. The sound source was an 8 inch infinite baffle loud speaker with a flat linear response over the frequency range 80 Hz to 500 Hz expected in this investigation. Figure 1 is a SolidWorks drawing of the resonator chamber, and Figure 2 is a photograph of the loudspeaker and resonator.

Signal and data acquisition were carried out using a National Instruments PCI 6024 M series DAQ. Audio generation and acquisition were carried out simultaneously at 100 kHz. A pink noise source was first applied to the resonator so that the approximate value of the resonant frequency could be established. A further sweep was then made to refine the estimate to within $\pm$ 2 kHz. This was followed by a very narrow sweep, enabling the detection of resonant frequency to within $\pm$ 0.005 Hz. Air temperature in the chamber was also recorded.

The test materials used were water, sand with a Saunter mean diameter of 166 μm, glass ballotini with a Sauter mean diameter of 261 μm, glass ballotini with a Sauter mean diameter of 713 μm, marbles with a mean diameter of 15 mm and marbles with a mean diameter of 24 mm. Experiments were carried out by placing a known mass of material in the chamber, calculating associated volume and then and finding the resonant frequency of the chamber, and then calculating the volume of the object.

A fuller description of the experimental apparatus and method is given by Webster and Davies (2008).

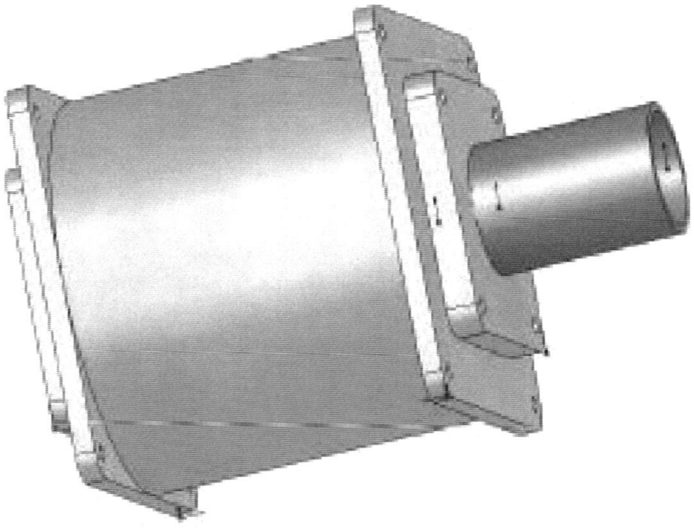

Figure 1 SolidWorks drawing of 3 L resonator.

Figure 2 Photograph of loud speaker and 3 L resonator.

RESULTS AND DISCUSSION

Figure 3 shows the results of a calibration test run with water and Figure 4 shows a comparison of actual volume and the predicted volume of the material placed in the chamber. It can be seen that the measured and predicted resonant frequencies when water is used to alter chamber volume are in excellent agreement. Likewise there is very good agreement between experiment and theory for the large marbles, small marbles and water. However for the smaller materials, *i.e.* the glass ballotini and sand, there is a considerable difference between measured and predicted values, and the relationship between these is non-linear, particularly at higher volumes.

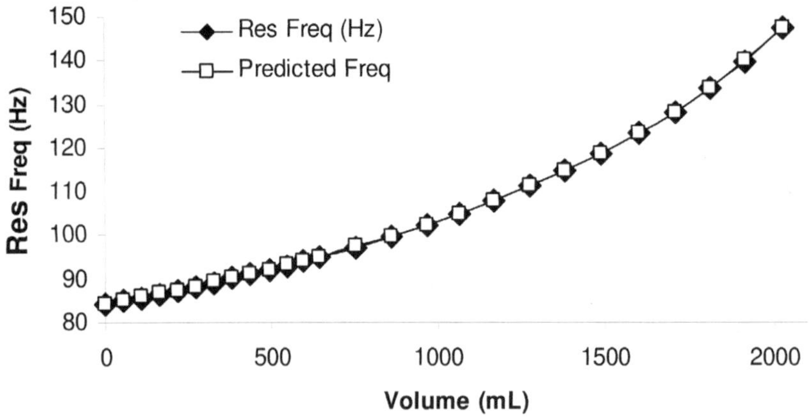

Figure 3 Actual and predicted resonant frequency shown as a function of volume of water in chamber, for the 3 L resonator.

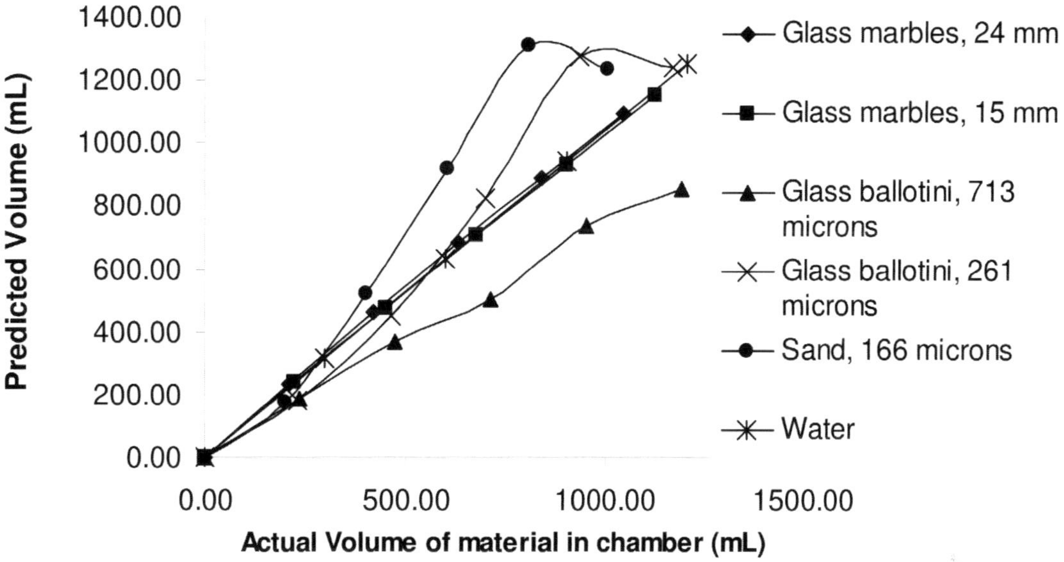

Figure 4 Actual and predicted volumes of material in 3 L resonator.

ACKNOWLEDGEMENT

ESW acknowledges financial support from a Massey University postgraduate scholarship.

REFERENCE

Webster, E. S. and Davies, C. E., 2008. "Determination of Bulk Density and Fill Fraction of Dry Particulate solids Using Helmholtz Resonance", SCENZ Conference, Hamilton, New Zealand, 20 October.

82a Newton's Cradle Undone: Experiments and Collision Models for the Normal Collision of Three Solid Spheres

Carly M. Donahue, Christine M. Hrenya, Kenshiro Nakagawa, and Alexandra Zelinskaya. Department of Chemical and Biological Engineering, University of Colorado, UCB 424, Boulder, CO 80309-0424

Molecular dynamics is an essential tool in the field of solid particulate flows. However, the variety of simplified force laws built in to the simulations has not been tested experimentally beyond two-particle collisions. Using an apparatus inspired by Newton's cradle, high-speed photography is utilized to examine the simultaneous, normal collision between three solid spheres. Namely, an initially touching, motionless pair of particles (doublet) is impacted on one end by a third particle. Surprisingly, the impacting particle reverses its direction and separates from the middle particle after collision. This reversal is not observed if the "doublet" particles are separated by a small distance (not in contact) initially, though a separation still occurs between the impacting and middle particle, with both particles traveling in the same direction after collision. In other words, the expected Newton's cradle outcome of a motionless, touching particle pair at the bottom of the pendulum arc is not observed in either case. A subsequent implementation of a variety of hard-sphere and soft-sphere collision models indicate that a three-body (soft-sphere) treatment is essential for predicting the velocity reversal, consistent with the experimental findings. Finally, a direct comparison between model predictions and measurements of post-collisional velocities reveal the relative accuracy of existing collision models for three-body interactions.

82b Stokesian Dynamics for Rigid Body Motion of Particle Aggregates

Yogesh M. Harshe[1], Marco Lattuada[1], and Massimo Morbidelli[2]. (1) Chemistry and applied biosciences, Institute for chemical and bioengineering, ETH Zurich, Wolfgang-Paulistr. 10, HCI-F133, Zurich, Switzerland, (2) Institute for Chemical and Bioengineering, Department of Chemistry and Applied Biosciences, ETH Zurich, Wolfgang-Pauli-Str. 10, HCI F 133, Zurich, 8093, Switzerland

Abstract:

The motion of colloidal clusters made of spherical particles in the presence of complex flow fields has not been thoroughly investigated thus far. Nevertheless, processing of colloidal dispersions under the conditions where coagulation is induced, leading to formation of complex fractal clusters, and their aggregation kinetics is strongly influenced by their interactions with the flow field. In order to better understand the nature of these interactions, we have used Stokesian Dynamics simulations to derive rigid body equations for clusters of spherical particles under both sheared and non-sheared flow conditions. To estimate the rigid body resistance matrix the grand resistance matrix obtained from the Stokesian dynamics approximation, including many-body interactions and lubrication forces, is used. The dependence of the aggregate motion related coefficients on cluster structure, defined by the fractal geometry and number of primary particles are established and scaling laws are derived.

The derived equations are first used to study the motion of aggregate under gravity. It was found that aggregates always tend to align themselves in the orientation with the least drag. The developed model is further utilized to simulate the motion of aggregates under sheared conditions.

Keywords:

Aggregates, rigid body, Stokesian dynamics

82c Silica Precipitation during Analcime Dissolution

Elizabeth A. Gorrepati and H. Scott Fogler. Chemical Engineering, University of Michigan - Ann Arbor, 2300 Hayward Ave,, Ann Arbor, MI 48109

It is estimated that 32% of matrix acidization treatments fail to increase permeability, and in fact, the treatments often-times decrease permeability of formations. One of the main reasons for these failures is cited as poor treatment fluid selection. An example of poor fluid selection occurred during acidization of Gulf of Mexico reservoirs that contain analcime, a common, naturally-occuring zeolite [1]. Analcime reacts with hydrochloric acid to form silicon reaction products, which subsequently polymerize, precipitate, and damage formations. Our studies focus on describing the fundamental reactions and phenomenon that govern analcime dissolution and the subsequent silicon precipitation.

We show that dissolution rate of analcime in hydrochloric, hydrobromic, or nitric acid is the same in the proton concentration range [H+] = 0.1 to 6M. That is, the anions Cl-, Br-, and NO3- do not affect analcime dissolution rate, and dissolution rate depends only on proton concentration. Thus, using HBr or HNO3 will not dissolve analcime any faster than using the standard hydrochloric acid.

We also show that silicon dissolved from analcime behaves macroscopically the same as a solution of sodium-metasilicate-nonahydrate. Therefore, we used the s-m-n to study silicon precipitation, rather than using analcime (the s-m-n is less costly, and the solutions are more well-defined). We show that the s-m-n forms a silicon "monomer", which disappears extremely rapidly in HCl via a second-order reaction. The polymerization products are silicon oligomers, which then coalesce to form spherical silica particles. These particles grow exponentially in size, as measured by dynamic light scattering. The particle growth mechanism is conjectured to be flocculation or Ostwald ripening. We studied systems of 2, 4, and 8M HCl, and found higher acid concentration results in much higher monomer disapperance rates as well as particle growth rates. The monomer disapperance and particle growth rate constants were measured, and are hypothesized to be the same for silicon dissolved from analcime.

[1] Underdown, D.R., Hickey, J.J., and Kalra, S.J. "Acidiation of Analcime-Cemented Sandstone, Gulf of Mexico", 1990, SPE 20624.

COMPARISON OF THE USE OF DIFFERENT SOLUBILITY MODELS FOR THE COOLING CRYSTALLIZATION OF ACETAMINOPHEN

David Widenski[a], *Ali Abbas*,[b] *and Jose Romagnoli*[a]
[a]*Louisiana State University, USA*
[b]*University of Sydney, Australia*

Abstract

In this paper we evaluate the sensitivity of using different solubility models on the cooling crystallization of acetaminophen. Empirical, correlative thermodynamic (namely NRTL) and predictive thermodynamic (namely MOSCED) models are considered. Equilibrium solubility model prediction determines the predicted supersaturation profile. The different solubility equations are used within a population balance model for prediction of crystal size properties. Incorrect prediction of the supersaturation profile results in incorrect prediction of crystal size distribution. The NRTL model was found to be more accurate at predicting equilibrium solubility and consequently crystal size. After the solubility sensitivity is evaluated, two methods are proposed to make the crystallization model more robust against solubility model errors.

Introduction

Crystallization is a traditional and widely used industrial process for the production of particulates. These particulates may include agrochemicals, biological proteins, fine chemicals, and pharmaceuticals. A key advantage of utilizing crystallization as a separation process is its role in the production of high purity products, important for specialized industries. Making the crystallization process more relevant is the FDA's strict regulations on the required purity for biological proteins and pharmaceuticals.

The thermodynamic driving force for crystallization is the difference in the chemical potential between the solute and the solution. Since chemical potential is difficult to quantify, solution supersaturation, a more readily measurable quantity, is used conveniently as an approximation. The trajectory the supersaturation follows affects the final crystal's size, crystal size distribution (CSD), habit (shape), and purity. Not only can the crystal's size and habit affect the particle product's performance, as in the case of the biological availability/activity of pharmaceuticals, but the production of inadequately sized particles can cake the filter downstream causing operational problems.

Traditionally, industrial crystallizers were run either on a trial and error basis, or by utilizing 'thumb rules". These methods commonly produced suboptimal recipes which resulted in failed batches and economic losses. In addition, these methods often proved to be less reliable when used during scale up. There has recently been increased interest in the development of crystallization models to predict operating conditions that produce crystals with desirable characteristics (Nowee et al., 2007). The use of a crystallization models allows the development of optimal recipes without the use of excessive laboratory time or resources. In addition, the crystallizer model in combination with the recent availability of robust *in situ* crystallization equipment can be implemented into a model based control scheme in order to keep the crystallizer operating on the correct trajectory.

In this work, the cooling crystallization of acetaminophen in ethanol is investigated. Acetaminophen also known as paracetamol, is the active ingredient in a commonly used painkiller, Tylenol®. Acetaminophen is advantageous to use because not only is it inexpensive, it also has been studied heavily in the literature. The temperature dependent solubility of acetaminophen in ethanol was investigated by Romero et al. (1996), Fernandez (1999), Granberg and Rasmuson (1999), and Worlitschek and Mazzotti (2004). Its thermodynamic properties were studied by Hojjati and Rohani (2006). Although the cooling crystallization of acetaminophen was previously studied by several groups including Hendrickson and Grant (1998) and Fujiwara et al. (2002), very few works such as that by Worlitschek and Mazzotti (2004) looked into simulating the crystal size distribution (CSD), or used a thermodynamically-based solubility model. However, the latter workers did not evaluate the effect of different solubility models on the CSD prediction, leaving the open question as to which of these available solubility models is most appropriate This paper compares the use of empirical, thermodynamical, and generalized thermodynamic solubility models to evaluate how these affect the resultant supersaturation profile and CSD, and then details how to make the crystallization model more robust against solubility model error.

Supersaturation and Solubility

As stated previously, the driving force for crystallization is the difference in chemical potential between the solid and liquid phases. Supersaturation is usually used instead of chemical potential because it is easier to measure. Supersaturation can be defined as either a relative or an absolute quantity, where the former is represented by the ratio of the solvent concentration to the equilibrium concentration, and the latter represented by the difference between the solution concentration and the equilibrium concentration. Supersaturation can be generated by one of three primary methods, namely evaporation, cooling, and antisolvent addition. In evaporative crystallization, the solution is heated which causes the solvent to evaporate. This loss of solvent from the solution makes the solution more concentrated which simultaneously causes the generation of supersaturation. Cooling crystallization is reliant on the fact that most solutes experience a decrease in solubility as temperature drops. Finally, in antisolvent crystallization, supersaturaion is generated by the addition of a carefully chosen antisolvent that reduces the solubility of the solute in the original solvent. This antisolvent may either be a liquid, gas, or a supercritical fluid. Two or more of these mentioned techniques can be combined in the same operation enabling enhanced results.For instance, adding antisolvent to a cooling crystallization operation provides that operation with an extra degree of freedom, where a calculated antisolvent addition can work as a seeding mechanism. .

A good solubility model accurately predicts how the equilibrium concentration of the solute changes over the course of the crystallization batch. This accurate solubility prediction is required for a crystallization model to in turn become accurate in predicting crystal product properties like size. Solubility models can be based on either empirical or thermodynamic foundations. An empirical solubility model is an equation fitted to experimental solubility data, and typically has no underlying physical meaning, while on the other hand, a thermodynamic solubility model both fits the data and has physical meaning. Common types of thermodynamic models include those based on excess Gibbs energy such as Wilson, NRTL, or UNIQUAC. In addition, generalized thermodynamic models such as the MOSCED (Modified Separation of Cohesive Energy Density) model can also be used. The MOSCED model is a thermodynamic model used to calculate infinite dilution activity coefficients. The advantage of the MOSCED model is that no experimental data is needed to calculate the infinite dilution activity coefficients. The MOSCED model further calculates temperature dependent infinite dilution activity coefficients, such that a temperature-dependent activity coefficient model is not required. The MOSCED model was originally developed for binary liquid solutions, but was later extended to liquid-solid systems by Drauker et al. (2007). The MOSCED model carries 5 fitting parameters: λ, α, β, q, and τ (Lazzaroni et al., 2005). In this work, values for individual MOSCED parameters for acetaminophen and ethanol are adopted from

Lazzaroni et al. (2005). Once the two infinite dilution activity coefficients are known, they can be substituted into an excess Gibbs energy model to find the binary interaction parameters for that system. With the MOSCED model, no experimental data is needed to calculate these binary interaction parameters. In order to calculate an equilibrium solute concentration, the activity coefficient equation is solved simultaneously with a solid solubility equation to calculate both the mole fraction and the activity coefficient of the solute. We present full details of these solubility models and their implementation in the crystallization model in a separate paper (Widenski et al., 2008).

Comparison of Solubility Models

Several different solubility models were used for the purpose of evaluating the effect of these models on the predicted final crystal properties. The first on the list is an empirical correlation solubility model developed by Fernandez (1999) for acetaminophen in ethanol. We next evaluate excess Gibbs energy models, in activity coefficient form. These activity coefficient models considered are the Van Laar, Wilson, and NRTL. All three have binary interaction parameters while the NRTL model has a third parameter, α, the nonrandomness parameter. The last solubility model considered is the generalized MOSCED model combined with the Wilson activity coefficient model. The computer package gPROMS (Process Systems Enterprise Ltd, London) with its parameter estimation facility gEST was used to estimate the optimal values of the binary interaction parameters for each activity coefficient model.

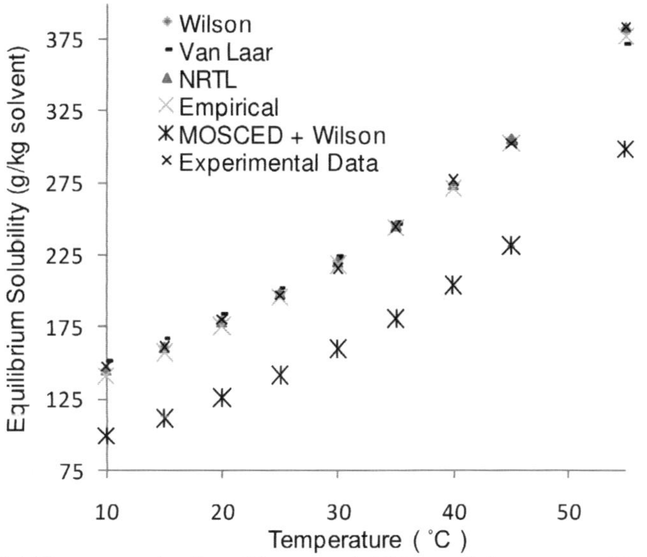

Figure 1. Equilibrium solubility curves for the different solubility models compared to experimental data from Worlitschek and Mazzotti (2004).

With the newly calculated binary interaction parameters, the prediction of the different models can now be compared. This is done graphically in Figure 1 against literature experimental data. The empirical model and each of the activity coefficient models all appear to provide good fits to the experimental data while the MOSCED model systematically underestimates the solubility over the entire temperature range of interest. The reason for the poor fit of the MOSCED model is because it is a generalized model. The tabulated MOSCED parameters are averaged over a wide variety of solvents in order to make the model as applicable to as many systems as possible. In the case where experimental solubility data is not available, the MOSCED model gives a first estimate to the solubility. However, if experimental data is available, it is preferable to fit binary interaction parameters of an activity coefficient model to get a more accurate solubility prediction.

Since crystallization is a particulate process, a population balance is used to account for the number of crystals during the batch. The population balance for a constant volume batch crystallizer with no agglomeration or attrition, and where crystal growth follows McCabe's Law is written as:

$$\frac{dn(L,t)}{dt} + G\frac{dn}{dL} - B = 0 \qquad (1)$$

Where *G* is the growth rate, *B* is the nucleation rate, *L* is the length, and *n(L,t)* is the crystal distribution. The population balance is solved numerically in gPROMS using a backward finite difference discretization method.

Secondary nucleation is the dominant nucleation mechanism when supersaturation levels are less than the primary nucleation metastable limit., Secondary nucleation is caused by the presence of suspended particles in solution through several different mechanisms. It can occur due to the solvent washing away weakly held surface crystals (fluid shear), from severe crystal collisions (attrition), or due to weak collisions with crystallizer equipment or other crystals (contact nucleation). Crystal growth occurs when the solution's relative supersaturation is greater than one. Countering growth, dissolution occurs when the solution's relative supersaturation is less than one. Depending on the nature of the system, growth can either be diffusion or surface reaction controlled, while dissolution is normally diffusion controlled. Worlitschek and Mazzotti (2004) modeled acetaminophen crystal growth as being surface reaction limited with an Arrhenius function, and dissolution was modeled as a diffusion limited process. In this work, we use the constitutive equations for nucleation and growth presented in Worlitschek & Mazzotti (2004).

Simulation Results

The crystallization kinetics developed by Worlitschek and Mazzotti (2004) only consider secondary nucleation, and not primary nucleation. This implies that each crystallization batch must be seeded with crystals for either crystal growth or nucleation to occur. From their CSD data, the seed was approximated to being normally distributed with a mean size of 190 microns and standard deviation of 30. The amount of seed they use is not reported. It is important to have the correct amount of seed to use in the simulation because both the initial nucleation rate and the solute mass balance depend on the second moment of the seed. It was found that if insufficient seed was used, the supersaturation profile peaked too high, and if an excessive amount of seed was used the supersaturation rose too slowly. An optimum seed amount was found to be 1.15 x 10^{11} μm^2, providing the supersaturation data fit of Figure 2.

Of the solubility models tested, the NRTL model was chosen because it was the most accurate activity coefficient model, and should give the best representation of experimental data. The empirical and MOSCED models were also chosen to depict how the magnitude of error in the solubility model affects the final CSD. These three models are compared under the linear cooling profile used by Worlitschek and Mazzotti (2004). This linear cooling profile decreases linearly from 30 °C to 10 °C over 10 hours and is then held constant at 10 °C for 4 hours.

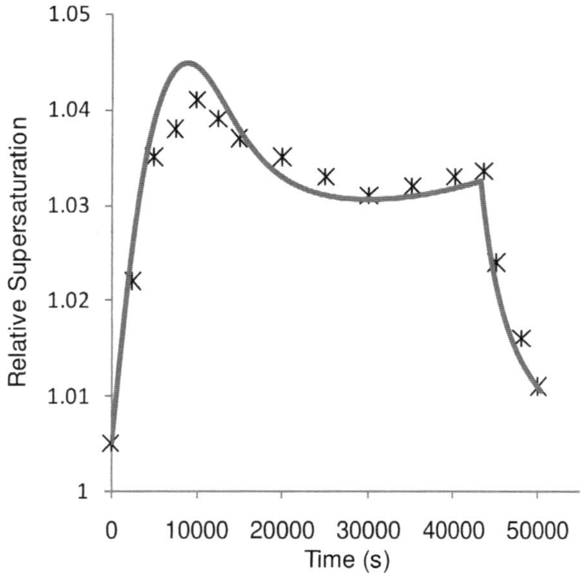

Figure 2. Relative supersaturation for linear cooling where the simulated profile using the NRTL model is plotted with experimental data from Worlitschek & Mazzotti (2004).

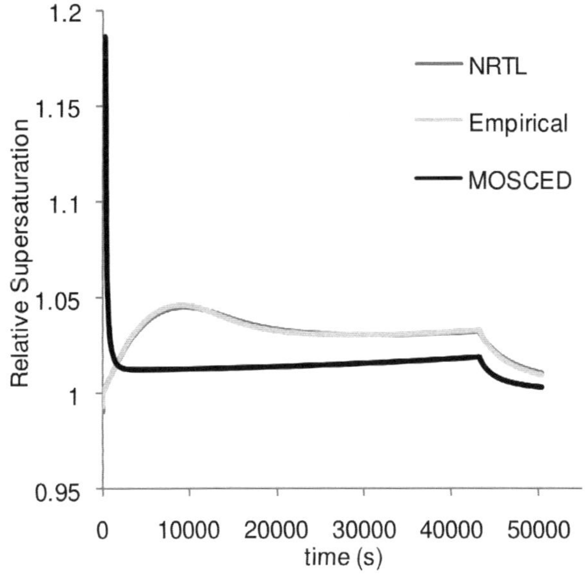

Figure 3. Relative supersaturation.

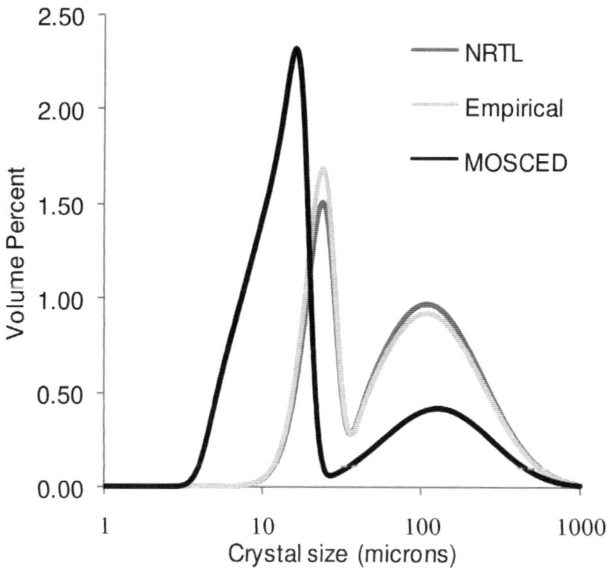

Figure 4. Volume percent CSD.

For the linear cooling temperature profile there is a large difference between the MOSCED and the other solubility models as seen in Figure 3. The MOSCED model greatly overestimates the initial relative supersaturation due to the model's bias for under-predicting the equilibrium solubility. After 5000 seconds the NRTL and empirical models are almost indistinguishable from each other while the MOSCED model is a poor prediction throughout the run.

By inspecting the final volume averaged CSD for each model (Figure 4), it can be seen that the NRTL and empirical models give approximately the same results, while again the MOSCED model being less accurate than the others. The reason for the slight differences in the first peak between the NRTL and empirical models is due to the relative supersaturation curve peaking at a higher value than the others causing increased secondary nucleation. The MOSCED model predicts a large peak of small crystals due to excessive secondary nucleation in the beginning of the batch. This is due to the high initial supersaturation which causes the large amount of secondary nucleation. Since the MOSCED model predicts secondary nucleation earlier than the other three models, and peaks at a higher relative supersaturation value than the others, it also should predict that the first peak's maximum should have a larger crystal size than the others. However, this is not the case because the growth kinetics was modeled as being a function of absolute supersaturation. For a fixed relative supersaturation, the absolute supersaturation can have varying values depending on the value of the equilibrium concentration. Since the MOSCED model under-predicts the solubility of acetaminophen it will have a corresponding absolute supersaturation that is lower than the other models. This lower absolute supersaturation causes less growth of the crystals. The errors between the different solubility models do not result in large differences in the final CSD except for the MOSCED model which only is qualitatively accurate.

Robustness against Solubility Model Errors

Since crystallization models are often used to predict or to obtain optimal cooling profiles it would be advantageous to make the model more robust against solubility model errors or biases. One way to make the model more robust is to specify the initial supersaturation condition instead of an initial solution concentration. This will remove the solubility's model effect on the initial supersaturation. As seen earlier, this can be detrimental to the model. It is reasonable to be able to specify the initial supersaturation because that will be known when doing crystallization experiments. Also the relative supersaturation is usually kept small to stay within the metastable region. If the relative supersaturation is too high and the metastable region is exceeded, uncontrollable primary nucleation will occur. When the initial supersaturation is specified, each solubility model does a much better job at predicting the relative supersaturation profiles seen below in Figure 5.

Another way to make the model more robust is to make the growth kinetics a function of relative supersaturation instead of absolute supersaturation. The absolute supersaturation can thus be rewritten as:

$$\Delta c = c_{eq}^{ref}(S-1) \quad (2)$$

In order to make this change of variables a reference equilibrium concentration needs to be chosen. Since the crystallization cooling batch varies from 30°C to 10°C the reference equilibrium concentration was taken to be 176 g Acetaminophen/kg solvent at 20°C. Substituting this into the growth kinetics resulted in the CSD profile shown in Figure 6. Now the MOSCED model does a much better job at predicting the final CSD. The grown secondary nucleates now have approximately the same size distribution for each solubility model. By making the growth kinetics a function of relative supersaturation instead of absolute supersaturation, the crystallization model is much more robust against solubility model errors.

Conclusions

Different models for solubility were presented and evaluated within a population balance crystallization model. The effect of the error of solubility models on the crystallization final CSD predictions was evaluated using four different solubility models. It was shown that small solubility model errors do not greatly affect the final CSD. However, large errors in the solubility model can be detrimental to the prediction as in the use of the MOSCED model. However, there are two ways to modify the model to make it more robust against solubility errors. By specifying the initial relative supersaturation and by making the crystallization kinetics a function of relative supersaturation the crystallization model is much more robust against solubility model errors. With these changes to the model, the generalized MOSCED model now can predict the final CSD more accurately. The next step in the solubility sensitivity testing of crystallization models will be to evaluate the effect of these different solubility models on the predictions of antisolvent crystallization.

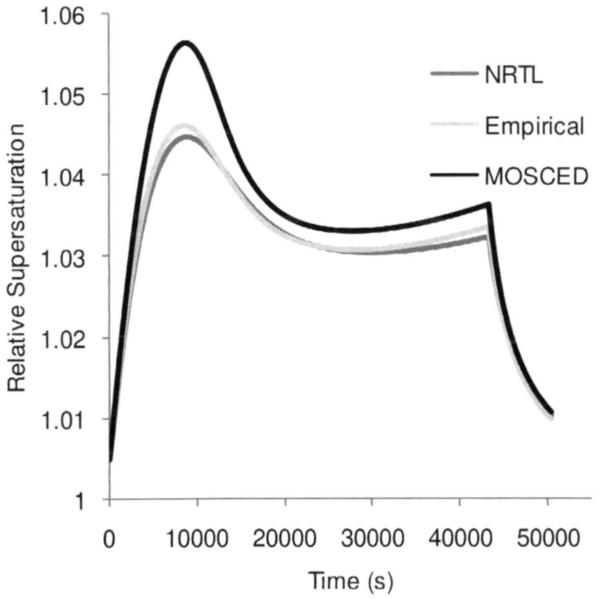

Figure 5. Relative supersaturation with a fixed initial relative supersaturation.

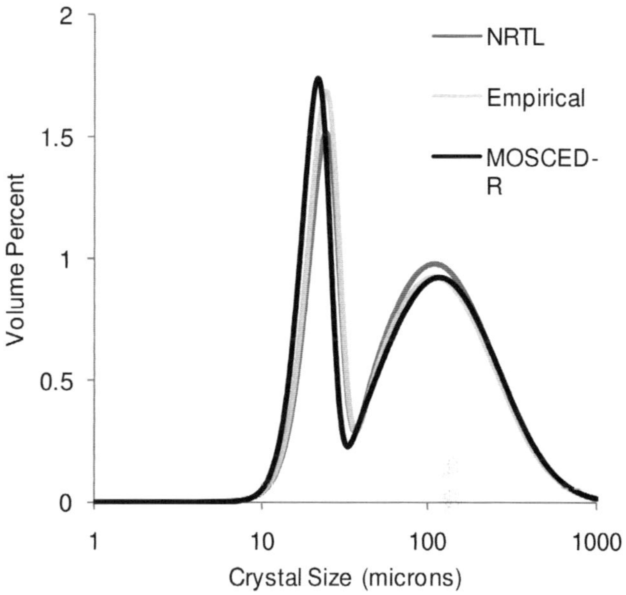

Figure 6. Volume percent CSD for the robust MOSCED crystallization model.

References

Abbas, A., *A model-based framework for advanced operation of crystallisation processes : a pilot-scale study.* Ph.D. Thesis, Graduate School of Engineering, University of Sydney, 2003.

Draucker, L.; et al., Experimental Determination and Model Prediction of Solid Solubility of Multifunctional Compounds in Pure and Mixed Nonelectrolyte Solvents, *Industrial & Engineering Chemistry Research* **2007**, 46, 2198-2204

Fernandez, C., Effect of the nature of the solvent on the crystallization of paracetamol. In *Proceedings of the 14th International Symposium on Industrial Crystallization, Cambridge, UK*; IChemE: Rugby, UK, 1999.

Fujiwara, M.; Chow, P. S.; Ma, D. L.; Braatz, R. D., Paracetamol Crystallization Using Laser Backscattering and ATR-FTIR Spectroscopy: Metastability, Agglomeration, and Control, *Crystal Growth and Design,* **2002**, *2*, 363-370.

Granberg, R. A.; Rasmuson, A. C., Solubility of Paracetamol in Pure Solvents, *J. Chem. Eng. Data* **1999**, *44*, 1391-1395.

Hojjati, H.; Rohani, S., Measurement and Prediction of Solubility of Paracetamol in Water-Isopropanol Solution. Part2. Prediction, *Organic Process Research & Development* **2006**, 10, 1110-1118.

Hendriksen B.; et al., Crystallisation of Paracetamol (acetaminophen) in the Presence of Structurally Related Species, *Journal of Crystal Growth* **1998**, 183, 629-640.

Lazzaroni, M.; Bush, D.; Eckert, C., Revision of MOSCED Parameters and Extension to Solid Solubility Calculations, *Industrial & Engineering Chemistry Research* **2005**, 44, 4075-4083

Nagy, A.; Fujiwara, M.; Braatz, R.D., Recent Advances in the Modelling and Control of Cooling and Antisolvent Crystallization of Pharmaceuticals, In *Proceedings of the 8th International IFAC Symposium on Dynamics and Control of Process Systems, Cancun, Mexico;* 2007

Nowee, S.M.; Abbas, A.; Romagnoli, J.A.; Yeo, P., Optimal Strategies for Controlling Particle Size in Antisolvent Crystallization Operations. In *Proceedings of the 8th International IFAC Symposium on Dynamics and Control of Process Systems, Cancun, Mexico*; 2007

Nowee, S.M.; Abbas, A.; Romagnoli, J.A.; Yeo, P., Optimization in seeded cooling crystallization: A parameter estimation and dynamic optimization study, *Chemical Engineering and Processing* **2007**, 46, 1096-1106

Perry, R.; Green, D., Perry's Chemical Engineer's Handbook, 7th ed.; McGraw Hill: New York City, NY, 1997.

Prausnitz, J.; Lichtenthaler, R.; de Azevedo, E., *Molecular Thermodynamics of Fluid-Phase Equilibria*; Prentice Hall PTR: Upper Saddle River, NJ, 1999.

Randolph, A.; Larson, M., *Theory of Particulate Processes*, 2nd ed.; Academic Press: San Diego, CA, 1988.

Romero S. et al., The Behavior of Paracetamol in Mixtures of Amphiprotic and Amphiprotic-Aprotic Solvents. Relationship of Solubility Curves to Specific and Nonspecific Interactions. *Chem. Pharm. Bull.* **1996**, *44* (5), 1061-1064

Sandler, S., *Chemical and Engineering Thermodynamics*; John Wiley & Sons: New York City, NY, 1999.

Widenski, D., Abbas, A., Romagnoli J.A., Comparison of different solubility equations for modeling Acetaminophen cooling crystallization, In preparation.

Worlitschek, J.; Mazzotti, M., Model-Based Optimization of Particle Size Distribution in Batch-Cooling Crystallization of Paracetamol. *Crystal Growth and Design* **2004**, 4, 891-903

82e Development and Validation of History-Dependent Inter-Particle Bonding Terms In Compacted Powder Beds

Athanas A. Koynov, Rutgers University, 98 Brett Rd., Piscataway, NJ 08854 and Alberto Cuitino, Mechanical Engineering, Rutgers University, 98 Brett Rd., Piscataway, NJ 08854.

The compaction of granular materials is a process encountered in a wide variety of systems – both naturally occurring and man-made. A clear understanding of the dependence of the mechanical properties of the solid product on both the characteristics of the starting materials and the system dynamics during the process itself is crucial for the effective design, control and scale-up of powder compaction.

In order to enable existing numerical tools to better predict the mechanical properties of finished composite products, a model has been developed, simulating the inter-particle bonding during compaction. The outcome of the model is dependent on pressure, history of the interaction and properties of the constituent powders.

The results of the model are validated using a Digital Image Correlation (DIC) technique, coupled with the mechanical compression of the finished compact. These allow for the local resolution of the composite's mechanical properties and their comparison with those predicted by simulations.

82f Discrete Element Method Simulation of a Dense Phase Conveying System

Bing Du[1], Ting(Tim) Han[2], Nestor A. Vasquez[3], and Joerg Theuerkauf[2]. (1) The Dow Chemical Company, 2301 N. Brazosport Blvd., Freeport, TX 77541, (2) Solids Processing, Engineering & Process Science, The Dow Chemical Company, Freeport, TX 77566, (3) R&D, The Dow Chemical Company, 1319 Bldg., Midland, MI 48667

In the chemical industry many conveying systems are used to move granular material in between different process steps. The most common types of pneumatic conveying modes are dilute and dense phase. This paper focuses on the modeling of a dense phase conveying system. The simulation is done by coupling a flow solver with a Discrete Element Method solver.

The presented results show the correlations for plug stability, length and porosity as a function of the conveying velocity. The numerical results are compared with experimental results.

82g Discrete Element Method Simulation of a Dilute Conveying System

Bing Du[1], Ting(Tim) Han[2], Nestor A. Vasquez[3], and Joerg Theuerkauf[2]. (1) The Dow Chemical Company, 2301 N. Brazosport Blvd., Freeport, TX 77541, (2) Solids Processing, Engineering & Process Science, The Dow Chemical Company, Freeport, TX 77566, (3) R&D, The Dow Chemical Company, 1319 Bldg., Midland, MI 48667

In the chemical industry many conveying systems are used to move granular material in between different process steps. The most common types of pneumatic conveying modes are dilute and dense phase. This paper focuses on modeling of dilute phase conveying systems. For the simulations commercial software is used. A fluid flow solver is coupled with a Discrete Element Method solver enabling a 2 way coupling.

The presented results show the influence of particle properties e.g. the coefficient of restitution on the motion of the particles. The simulations focus on the motion of granular material in a straight pipe as well as in a pipe with a 90 degree bend. Besides the particle motion the pressure drop is investigated as a function of gas flow rate and the particle loading. Moreover, particle wall impact forces and frequencies are studied.

The numerical findings are validated with experimental results.

Computation of turbulence and low Sherwood numbers in fluidized beds

Benjapon Chalermsinsuwan[1], Pornpote Piumsomboon[1], Dimitri Gidaspow[2], Mayank Kashyap[2] and Ronald W. Breault[3],

1. Chemical Technology, Chulalongkorn University, Phayathai Road, Patumwan, Bangkok, 10330, Thailand, (2) Chemical and Biological Engineering, Illinois Institute of Technology, 10W 33rd St., Perlstein Hall, Chicago, IL 60616, (3) Gasification and Combustion Projects Division, National Energy Technology Laboratory, 3610 Collins Ferry Rd, Morgantown, WV 26507

Abstract

We have already shown (Gidaspow, 1994) that the kinetic theory based CFD code is capable of predicting the turbulent behavior of fluidized bed risers (Jiradilok et al., 2006). For FCC particles, it predicted the Kolmogorov -5/3 law at high frequencies. In dimensionless form the computed spectrum was that of single phase flow at a Reynolds number of 21,500. To get agreement with experiments we had used the modified EMMS drag law. We used a similarly modified drag law for modeling the flow in the PSRI riser "challenge problem one" summarized in this study.

Dispersion coefficients

The objective of this study is to develop predictive theories for the dispersion and mass transfer coefficients and to measure them in the turbulent fluidization regime, using existing facilities.

The dispersion coefficient is a measure of the quality of mixing. We have identified two types of solids dispersion coefficients: those due to random particle oscillations, "laminar" type, and those due to cluster or bubble motion, "turbulent" type. A literature review (Breault, 2006) shows that dispersion coefficients in fluidized beds differ by more than five orders of magnitude. To understand the phenomena, two types of hydrodynamics models that compute turbulent and bubbling behavior were used to estimate radial and axial gas and solids dispersion coefficients. The autocorrelation technique was used to compute the dispersion coefficients from the respective computed turbulent gas and particle velocities.

The computations show that the gas and the solids dispersion coefficients are close to each other in agreement with measurements. The simulations show that the radial dispersion coefficients in the riser are two to three orders of magnitude lower that the axial dispersion coefficients, but less than an order of magnitude lower for the bubbling bed at atmospheric pressure. The dispersion coefficients for the bubbling bed at 25 atmospheres are much higher than at atmospheric pressure due to the high bed expansion with smaller bubbles. The computed dispersion coefficients are in reasonable agreement with the experimental measurements reported over the last half century and those measured at IIT and in the NETL riser in Morgantown (Jiradilok et al., 2007, 2008).

Some typical measurements for the IIT two dimensional bed shown in Figure 1 are shown in Figure 2 and Table 1. Computations for this configuration are in progress.

Figure 3 shows a summary of granular temperatures. For the two dimensional bed, the measured granular temperature is low due to small particle velocity. For the IIT riser, the granular temperature is high. The computed and experimental values agree.

Figure 4 shows the axial and radial solids dispersion coefficients. Consistent with the granular temperature, the dispersion coefficients are low in the two dimensional bed due to low solid velocity. For the IIT riser, the computed and measured dispersion coefficients agree with literature values.

Mass transfer coefficients

It was known for half a century that the Sherwood and Nusselt numbers in fluidized beds are often three orders of magnitude lower than the classical diffusion controlled limit of two. We have shown (Chalermsinsuwan et al., 2008a, 2008b) that our kinetic theory based computer codes correctly compute low Sherwood numbers in agreement with published experimental data. For tall fluidized bed risers the computed behavior is similar to that for convective diffusion in a channel, but with a greatly reduced mass transfer.

Figure 3 shows our comparison of the computed Sherwood number to Kato et al. (1970) results. Clearly, our CFD code predicts these low Sherwood numbers. However, the low Sherwood number has nothing to do with the diffusion to the particles as is implied in its definition. It simply reflects the difference in the radial concentration distributions in the fluidized bed of fine particles.

References:

1 Breault, R.W., (2006). A review of gas-solid dispersion and mass transfer coefficient correlations in circulating fluidized beds. Powder Technology, 163, 9-17
2. Chalermsinsuwan, B., P. Piumsomboon, and D. Gidaspow, "Kinetic theory based computation of PSRI riser- Part I: Estimate of mass transfer coefficient", Submitted for publication in Chemical Engineering Science, 2008a
3. Chalermsinsuwan, B., P. Piumsomboon, and D. Gidaspow, "Kinetic theory based computation of PSRI riser- Part II: Computation of mass transfer coefficient with chemical reaction", Submitted for publication in Chemical Engineering Science, 2008b
4. Gidaspow, D., (1994). Multiphase Flow and Fluidization: Continuum and Kinetic Theory Description. Academic Press, Boston
5. Jiradilok, V., Gidaspow, D., Damronglerd, S., Koves, W.J., Mostofi, R., (2006). Kinetic theory based CFD simulation of turbulent fluidization of FCC particles in a riser. Chemical Engineering Science, 61, 5544-5559
6. Jiradilok, V., Gidaspow, D., Breault, R.W., (2007). Computation of gas and solid dispersion coefficients in turbulent risers and bubbling beds. Chemical Engineering Science, 62, 3397-3409.
7. Jiradilok, V., D. Gidaspow, R.W. Breault, L.J. Shadle, C. Gunther, and S. Shi, "Computation of turbulence and dispersion of cork in the NETL riser", Chemical Engineering Science 63 (2008) 2135-2148
8. Kato, K., Kubota, H., Wen, C.Y., (1970). Mass transfer in fixed and fluidized beds. Chemical Engineering Progress Symposium Series, 105(66), 87-99.

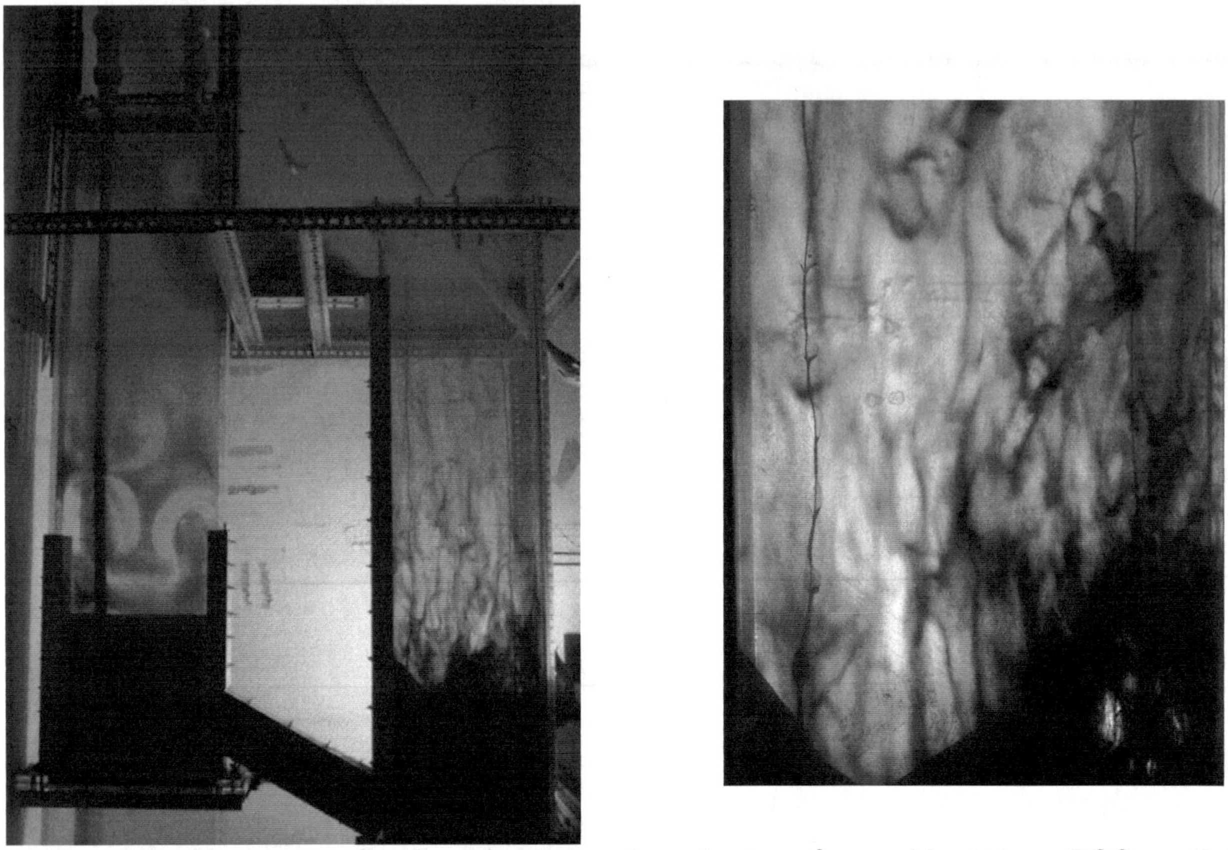

Figure 1: Circulating fluidized bed showing clusters formed by 75 µm FCC particles

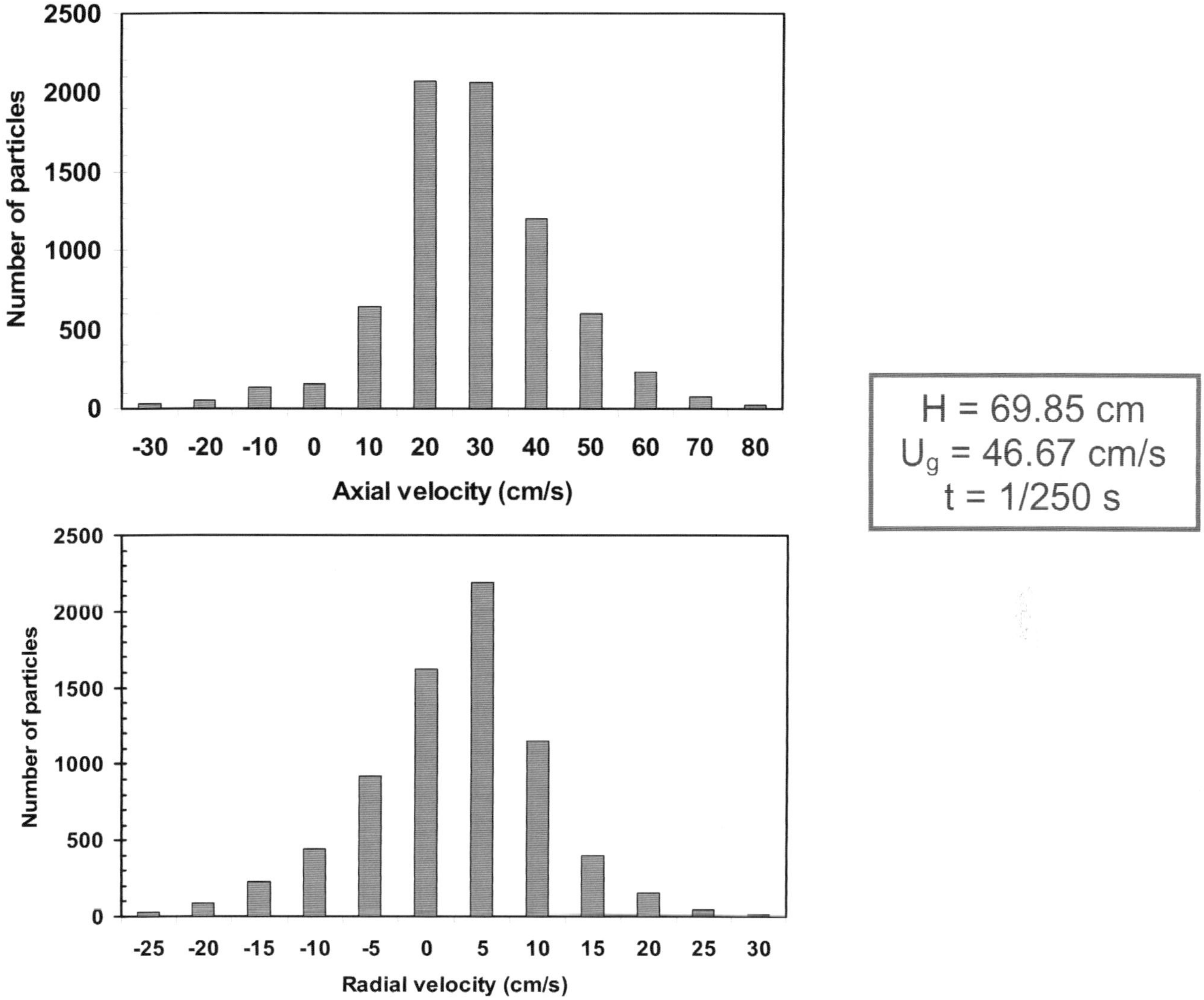

Figure 2: Overall instantaneous velocity distribution for 75 μm FCC particles in the system shown in Figure 1

Table 1: Comparison of laminar and turbulent granular temperatures in IIT 2- D circulating fluidized bed and IIT riser

Granular Temperature, m^2/s^2			
System	Radial Position	Laminar due to individual particle oscillations	Turbulent due to cluster oscillations
2-D CFB, 75 μm FCC particles	Center	1.27×10^{-2}	6.73×10^{-3}
2-D CFB, 75 μm FCC particles	Right Wall	6.67×10^{-3}	2.54×10^{-3}
IIT Riser, 1093 μm	Wall	9.48×10^{-2}	2.61×10^{-2}

Mixing is on the level of particles

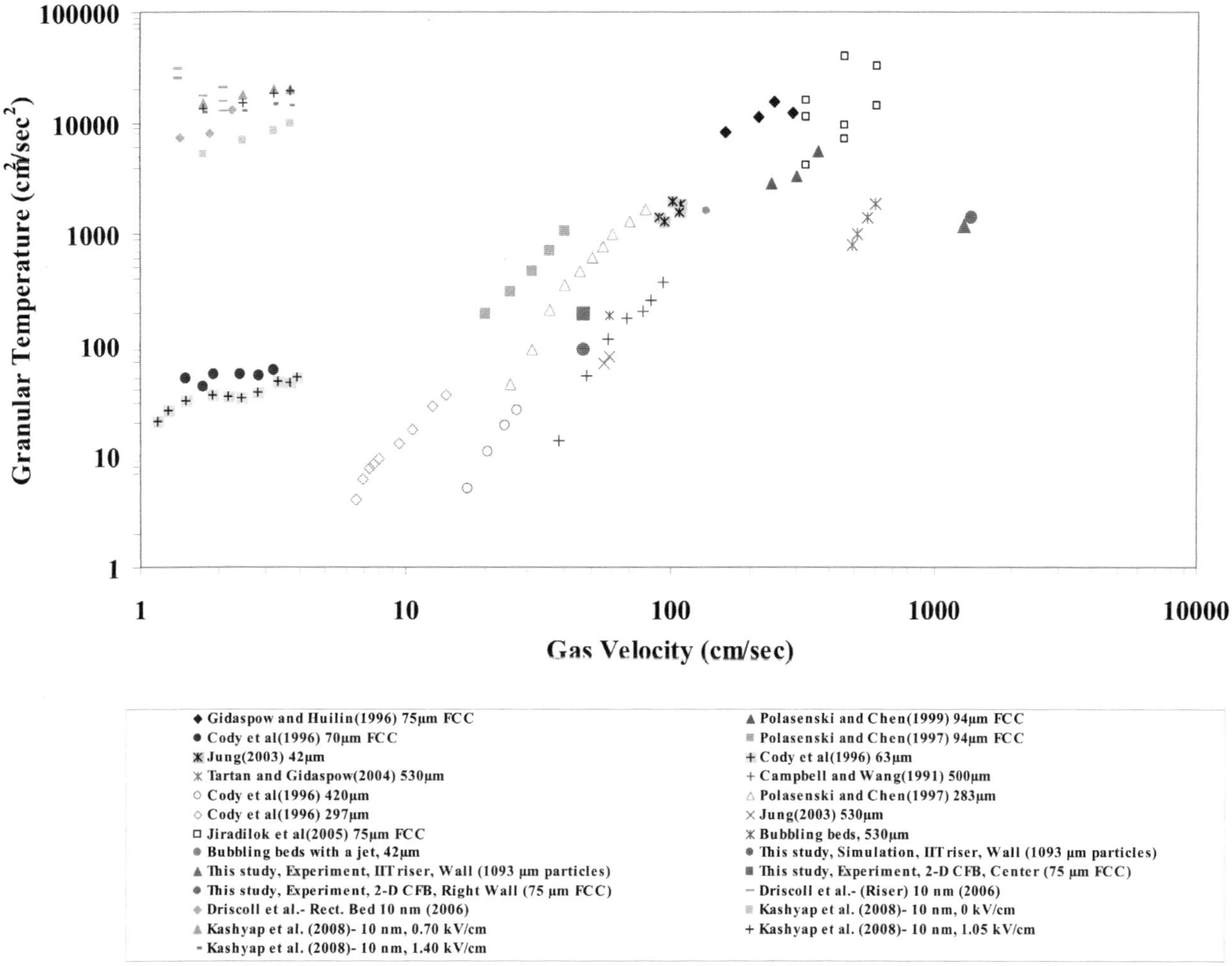

Figure 3: Comparison of granular temperatures versus gas velocities for various particles

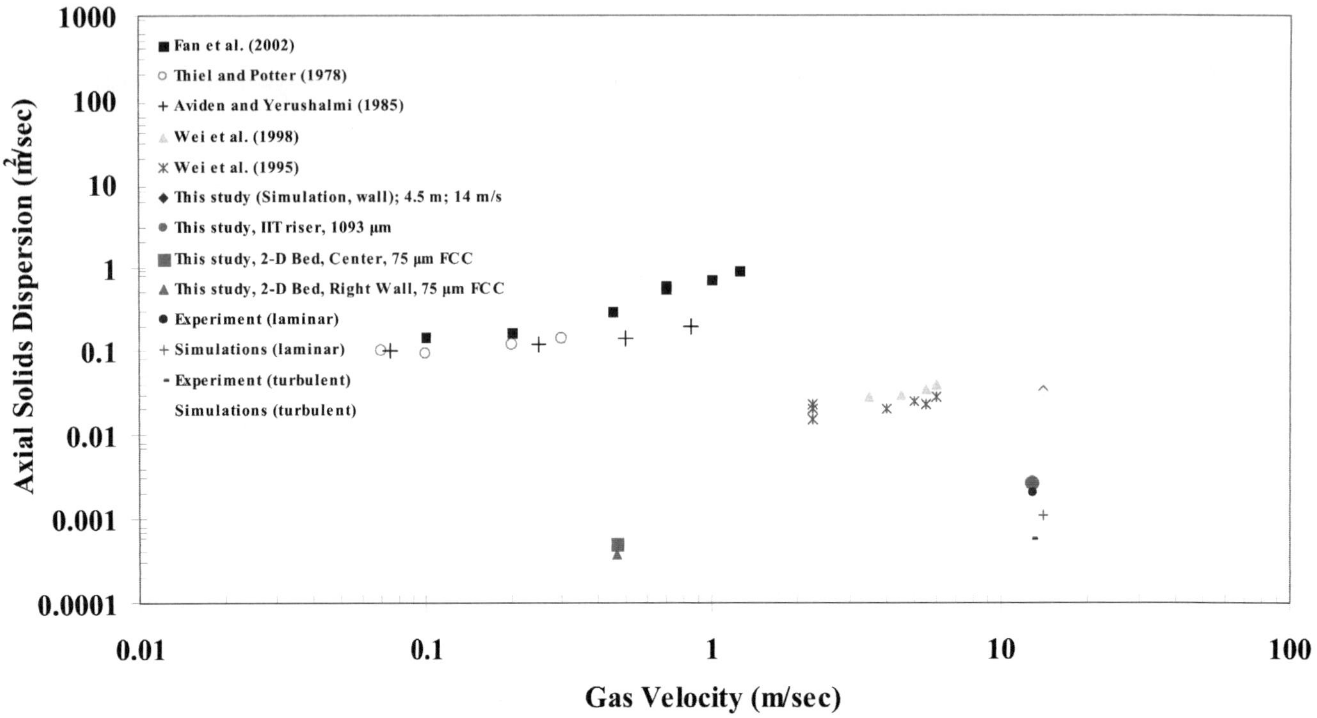

(A)

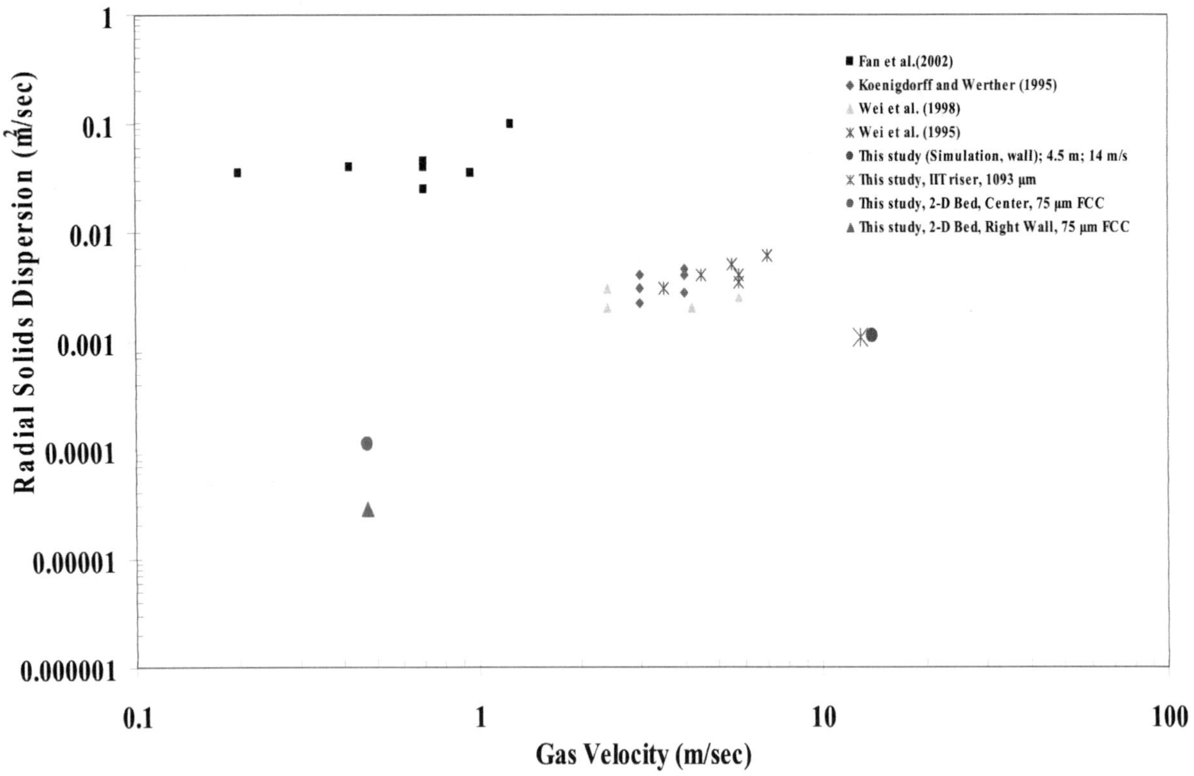

(B)

Figure 4: (A) Axial and (B) radial solids dispersion coefficients for various particles

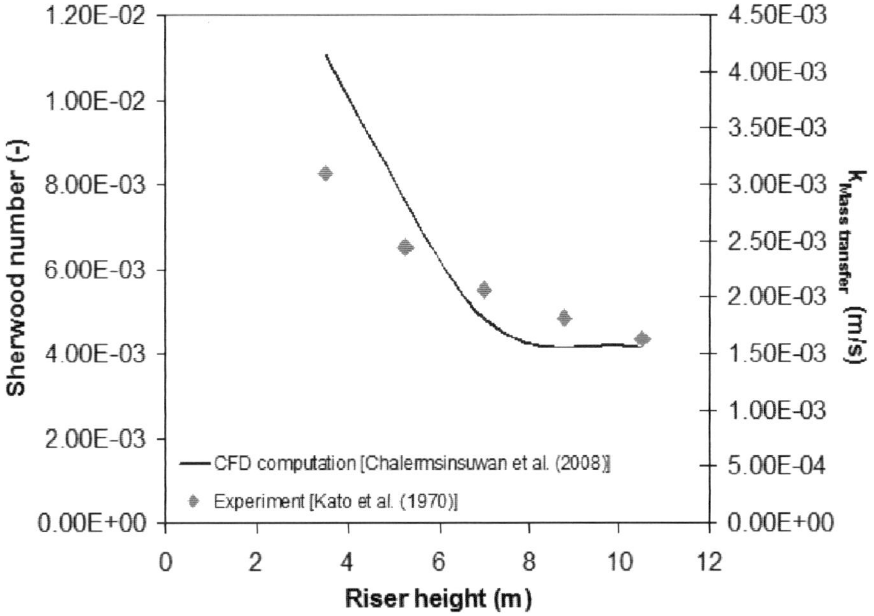

Low mass transfer coefficients in PSRI riser with $k_{reaction}$ = 39.6 s^{-1}

Figure 5: Variation of Sherwood number with height in the PSRI riser

Influence from Particle Size Distributions on the CFD Simulations and Experiments of Bubbling Fluidized Beds

Sanoja A. Jayarathna[a], Chameera K. Jayarathna[a], Wen Jun Wu[a] and Britt M. Halvorsen[a,b]
a Telemark University College, Porsgrunn, Norway
b TelTek, Porsgrunn, Norway

Abstract

Industrial fluidized beds in common normally use powders with size distributions. The size and size distributions of particles used in the bed may lead to different bubble behaviours. Because of that it is important to study the influence on bubble behaviour from particle size distribution in fluidized beds.

In addition to that it is important to study the influence on the simulations of fluidized beds from particle size distributions. That is because, in modelling of fluidized beds a mean particle diameter is often used, and important information about flow behaviour can therefore be lost.

A series of experiments are performed in order to check the effect from particle size distribution on bubble behaviour. Several simulations also carried out in order to analyse the influence from the particle size distribution on the simulated results. The particle size distribution is accounted for by including multiple particle phases.

The computational and experimental results are analyzed separately and compared with each other with respect to the volume fraction changes along the bed with time, particle segregation and bubble frequency.

The simulations and experiments show that the particle size distributions significantly influence on the bubble behaviour and particle segregation.

Introduction

Fluidized beds are widely used in industrial operations. Good mixing and large contact area between phases, enhances chemical reactions, heat transfer and mass transfer. Industrial fluidized beds normally use powders with particle size distributions. The efficiency of fluidized beds depends on bubble behavior and bubble characteristics in the particle phase. The bubble characteristics are very important in the design of fluidized beds because they govern hydrodynamics and efficiency of the operation for which the bed is used [1]. The size and size distributions of particles used in the bed may lead to different bubble behaviors. Because of that it is important to study the influence of particle size distribution on bubble behavior in fluidized beds. Both simulations and experiments have a major role when it comes to the noted aspect. In simulations of fluidized beds a mean particle diameter is often used, and due to this important information about flow behavior can be lost.

Computational fluid dynamics is the key role of simulations and has been improved significantly during the last decades. Two different approaches are used in modelling of fluidized beds, the Euler-Lagrange and the Euler-Euler approaches. The Euler-Euler approach is most commonly used and is also used in this work. In the Euler-Euler model, all the phases are considered as a continuum. The Euler-Euler approach is useful and computationaly cost effective when the volume fractions of the phases are comparable, or when the body forces such as gravity act to separate phases, or when the interaction within and between the phases plays a significant role in determining the hydrodynamics of the system [2]. Halvorsen, B. [3] has used the Euler-Euler approach in the simulations of bubbling fluidized beds. Patil et al [4] and Patil et al [5] have used Euler-Euler approach with two different closure models. Those are the constant viscosity model and a model based on the kinetic theory of granular flow. Enwald et at [6] have presented a model using Euler-Euler approach as well as the application of the model in the simulations of bubbling and circulating fluidized beds. Huilin et al [7] has used both the Langrangian and the Eulerian approaches separately and the results have been compared to experimental data. Boemer et al [8] have developed a computer code to simulate the fluid dynamics of fluidized beds using Eularian approach. Arastoopour, H. [9] has used Eularian approach for the simulations he used to compare the predicted flow parameters with large scale experimental data of fluidized beds. Few studies have being carried out including particle size distributions in simulations. Among the few, Huilin et al [10] has used multi fluid Eulerian CFD model with closure relationships according to the kinetic theory of granular flow to study the motion of particles in a gas bubbling fluidized bed with the binary mixtures. Mathiesen et al. [11],[12],[13] presented both computational and experimental data from their work with circulating fluidized beds.

A lot of effort has been devoted to experimental studies of bubbles in a fluidized bed. For practical reasons the experimental studies have been performed on cold beds. Different measurement systems have been developed and used for studying bubble behavior in fluidized beds. Werther and Molerus [14] used a miniaturized capacitance probe for measuring parameters characterizing local state of fluidization in cylindrical beds. Gidaspow et al. [15],[16], Kuipers et al. [17], Halvorsen and Mathiesen [18],[19], Bokkers et al [20], Patil et al. [5] studied bubbles in two dimensional beds with a jet. Patil et al. [4] also studied bubbles in two dimensional beds with uniform air distribution. In these studies photographic techniques as movie camera or digital video camera were used to measure bubble sizes and bubble velocities and the experimental data have been used to validate CFD models.

In this work glass particles with different size distributions are used to study how particle size distributions influence bubble behavior in a 2-D fluidized bed with uniform air distribution.

Experiments and Simulations

A series of experiments is performed to study the influence of particle size distribution on bubble behavior and particle segregation in fluidized beds. A 2-D lab scale fluidized bed with a uniform air distribution is constructed, and a video camera is used to record the bubble behavior in the bed. The width, height and depth of the bed is 0.20 m, 0.80 m and 0.025 m respectively. The experimental data are used to validate computational CFD model. Several simulations are carried out in order to analyze the influence of the particle size distribution on the computational results. The commercial CFD code Fluent 6.3 is used for the simulations. A 2-D wire frame mesh with the same dimensions as the experimental set-up is used. A model combination finalized by Ariyarathna [21] is used in this study. The particle size distribution is accounted for by including multiple particle phases in the simulations.

Spherical glass particles with density 2485 kg/ m^3 and mean particle diameter of 488 μm are used in all the experiments and the corresponding simulations. The superficial gas velocity is 0.134 m/ s and the initial bed height is 28 cm. In the experiments glass powders with mean particle diameter of 153 μm , 488 μm and 960 μm are used, and different compositions of the three powders are used to give mixtures with mean diameter of 488 μm . The compositions of the mixtures are given in Table 1.The simulations are performed with one, two, three and four particle phases. The composition of the computational particle phases are presented in Table 2. Simulations 1 and 2 correspond to experiments 1 and 2 respectively. Simulations 3, 4 and 5 are performed with two, three and four particle phases, and the simulations correspond to experiment 3.

Table 1: Mean particle diameters and the compositions used in the experiments

Experiment	Particle phase			Particle phase		
No:	1	2	3	1	2	3
	Mean diameter, [μm]			Composition, [%]		
1	488			100		
2	153	960		58.5	41.5	
3	153	488	960	29	50	21

Table 2: Mean particle diameters and the compositions used in the simulations

Simulation	Particle phase				Particle phase			
No:	1	2	3	4	1	2	3	4
	Mean diameter, [μm]				Composition, [%]			
1	488				100			
2	153	960			58.5	41.5		
3	153	625			29	71		
4	153	488	960		29	50	21	
5	153	425	578	960	29	30.5	10.5	21

Analysis of the Results

The results from the simulations and the experiments are analyzed with respect to the bubble behavior and the particle segregations. Picture frames from the movies are used to analyze the experimental results and the contours along with the data recorded by the monitors are used to analyze the simulated results.

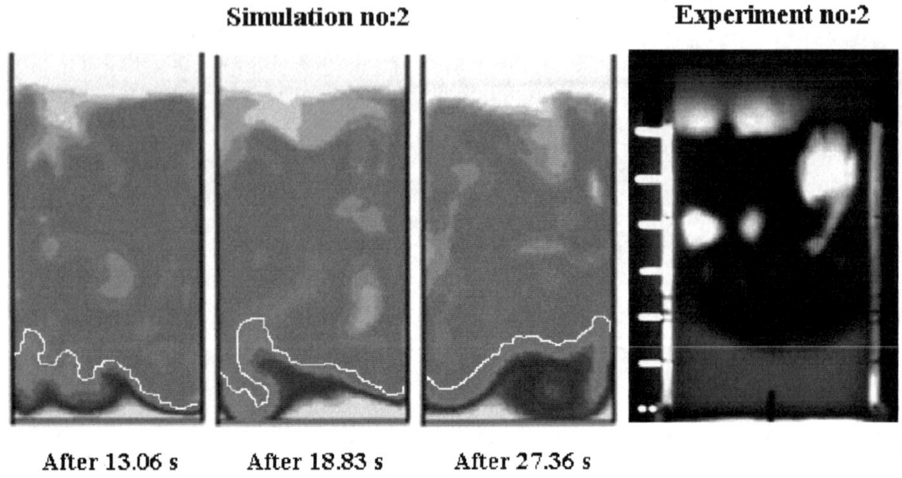

Figure 1: Representation of the particle segregation in simulation 2 and experiment 2

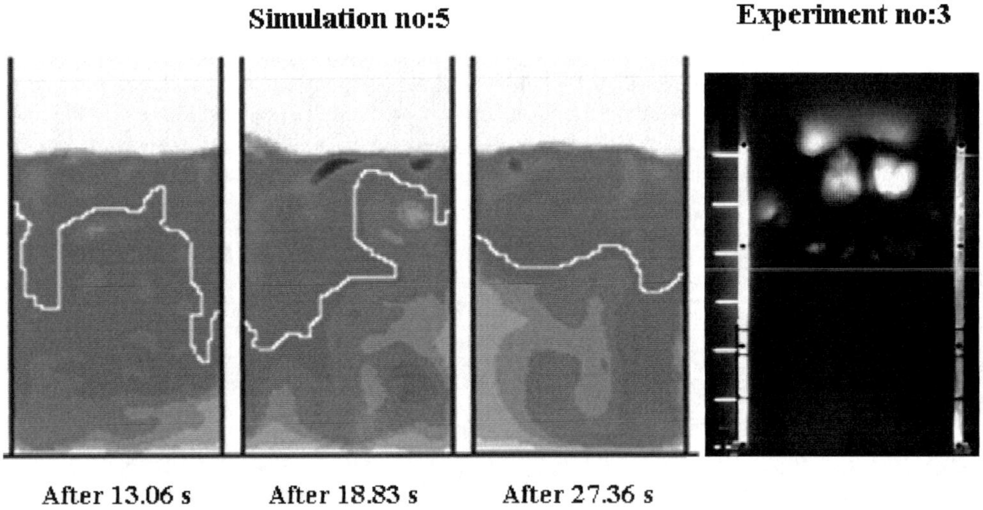

Figure 2: Representation of the particle segregation in simulation 5 and experiment 3

Particle Segregation

Experiment 1 is performed with a powder with a particle diameter range of 400-600 μm. The corresponding simulation is carried out with one particle phase. No bubbles are observed in the experiment and the simulation. The superficial gas velocity, 0.134 m/ s, is well below the theoretical minimum fluidization velocity for particles with diameter 488 μm, and bubble formation is therefore not expected.

The simulations 1, 3, 4 and 5 were analyzed with respect to particle segregation and bubble appearance in an earlier study by Jayarathna et al. [22]. The computational results were compared to experimental data performed by Wu et al. [23]. The comparison of the results showed that the deviation between simulations and experiments decrease significantly with increasing number of particle phases in the simulations.

Experiment 2 is performed with a mixture of the powders with mean diameter of 153 μm and 960 μm, and the mixture includes a wide range of particle sizes. The experiment corresponds to simulation 2 which is carried out with two particle phases. The comparison between experimental and computational results is shown in Figure 1. Both the experiment and simulation predict a very clear segregation of particles. The segregation influence on flow behavior and this effect can only be predicted by using multiple particle phases in the simulations.

In Figure 2, the particle segregation in experiment 3 and simulation 5 is compared. The thickness of the layer of small particles on the top of the bed is about the same in the simulation as in the experiments. In the

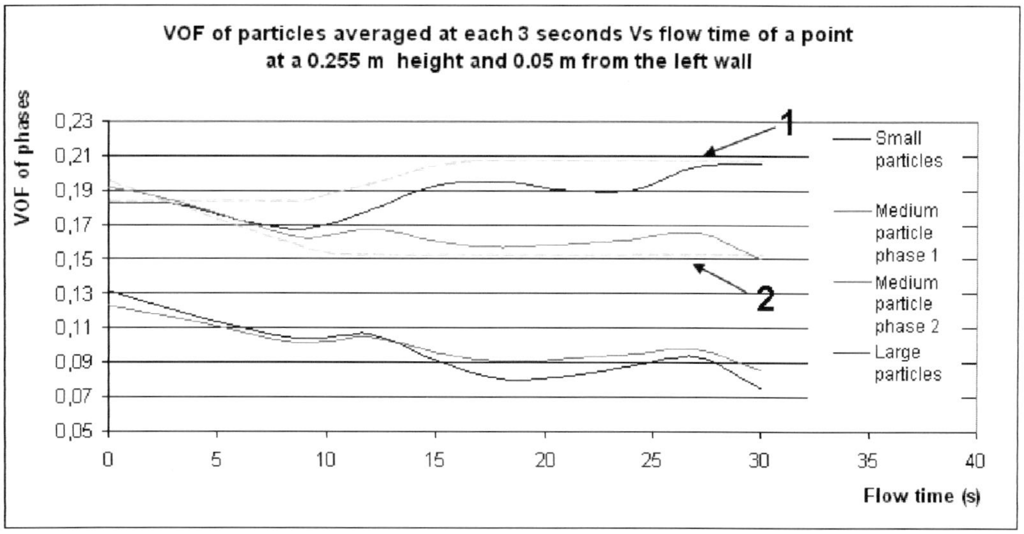

Figure 3: Particle segregation as a function of time at a selected point in the particle bed of the simulation 5.

experiment the boundary between small and large particle layer can be assumed to be at the same level as the bubbles appear. The simulation 5 is performed with 3 particle phases.

Segregation as a function of time

Figure 3 shows the segregation of particles as a function of time observed in simulation 5. The analysis is performed at position 0.05 m from the wall and height 0.255 m in the bed. The simulation is performed with 4 particle phases, small particles, medium particles 1, medium particles 2 and large particles. The particle diameters for the phases are given in Table 2. In the figure the volume fraction of the different phases are averaged over 3 seconds, and the values are given from time 0 to time 30 s. It can be seen that the tendency is that the fraction of small particles increases with time, the medium 1 particles decreases slightly, and the medium 2 and the large particles decreases significantly with time at the current position. The dashed lines marked with 1 and 2 are presenting the segregation tendencies of the small and medium particles 1.

Bubble velocity

Bubble velocity (rise velocity) of the simulations and the experiments are compared using an averaged velocity values. Rise velocities is useful to study the dynamics in the particle bed and also to compare the prediction of the simulations with the reference experiments to evaluate how close the simulations are to the experiments.

Rise velocities of the bubbles in the simulations are calculated using contours of VOF of the gas phase that shows some of the bubbles raised in the particle bed at each simulation. For the experiments, movie frames were used. To calculate the rise velocity, one or more bubbles are selected and the change of the position of the bubble with time is measured. Bubble position variation of two selected bubbles with time from the simulation 5 is presented in Figure 4. Those two bubbles have 0.25 $\mathrm{m\,s^{-1}}$ and 0.33 $\mathrm{m\,s^{-1}}$ respectively as their average rise velocities. The calculated averaged velocities are presented in Table 3.

Table 3: Averaged velocities.

Simulation	Averaged rise velocity, m/s			Experiment	Averaged rise velocity, m/s		
No:	Bubble1	Bubble2	Bubble3	No:	Bubble1	Bubble2	Bubble3
1	No bubbles			1	No bubbles		
2	0.35	0.39	0.41	2	0.21	0.42	0.59
3	0.29	0.22		3	0.31	0.47	0.41
4	0.39	0.46	0.26				
5	0.25	0.33					

Since the simulation 3, 4 and 5 represents the particle distribution in the experiment 3 but with different number of particle phase, velocity values found in those are compared with the respective experiment. Results from the simulation 2 is compared with the experiment 2. The comparison showed that all the bubbles that

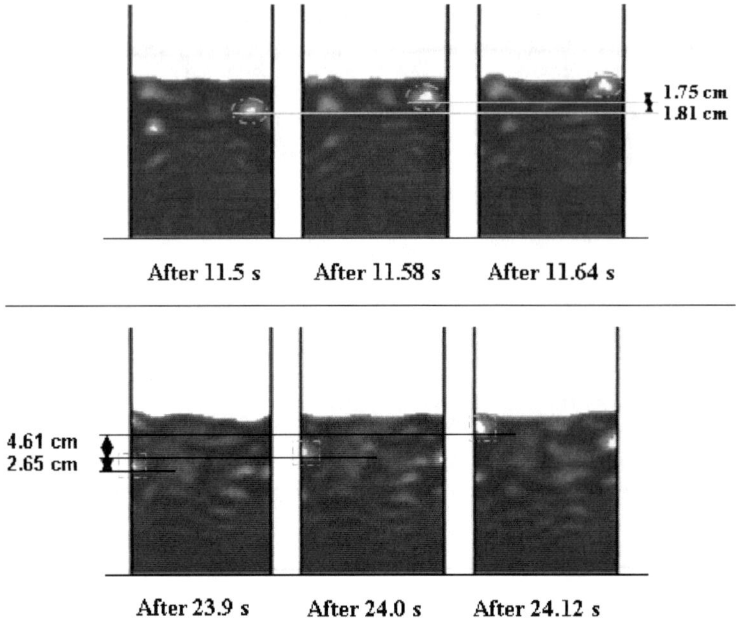

Figure 4: Change of the position of the bubbles with time in the particle bed of the simulation 5.

are analyzed have velocities values which are very much different from each other. In addition, it was clearly noticeable that the bubbles are growing larger with time and speeds up as the bubbles grow. Since the bubble dimensions are not analyzed during this study, it is not possible to give a conclusion regarding the bubble velocities. But, when the rise velocities are compared with the emulsion gas velocity, it is clear that all of the analyzed bubbles are fast moving bubbles.

Bubble Appearance

Bubble appearance in the bed is also analyzed both in the simulations and the experiments. The lowest position of bubble appearance should depend on particle segregation present in the bed. Figure 5 shows the lowest position for bubble appearance in the particle bed in the simulation 2 and 5 and in the experiment 2 and 3. Comparison of the details in the picture shows that the simulation 2 have similar predictions to the experiment 2 while the simulation 5 have predicted the results similar to the experiment 3.

By including multiple particle phases in the simulations, the particle size distribution present in the experiments are accounted for in the simulation, and the flow behavior can be predicted without loosing important information.

Bubble frequency

The assumption of solid free bubbles is an oversimplification of what actually happens in the particle bed [24]. To accept some particles inside the bubbles, a volume fraction of gas of 0.7 is selected as the criterion for a bubble. Bubble frequencies are calculated for all the five simulations, and it is found that the bubble frequency is lowest close to the walls and along the central axis of the bed. It is observed that the bubble frequency increases with increasing height in the bed. Figure 6 shows the bubble frequency as a function of radial position at two heights in the bed. The figure represents the results from simulation 4 which is performed with three particle phases.

Similar observations are done with the experiments 2 and 3. Figure 7 provides evidence for the above statement.

Conclusions

Experiments and simulations are carried out on a 2-D bubbling fluidized bed with uniform air distribution. The aim of the work is to study the influence of particle size distribution on flow behavior in a fluidized bed. The

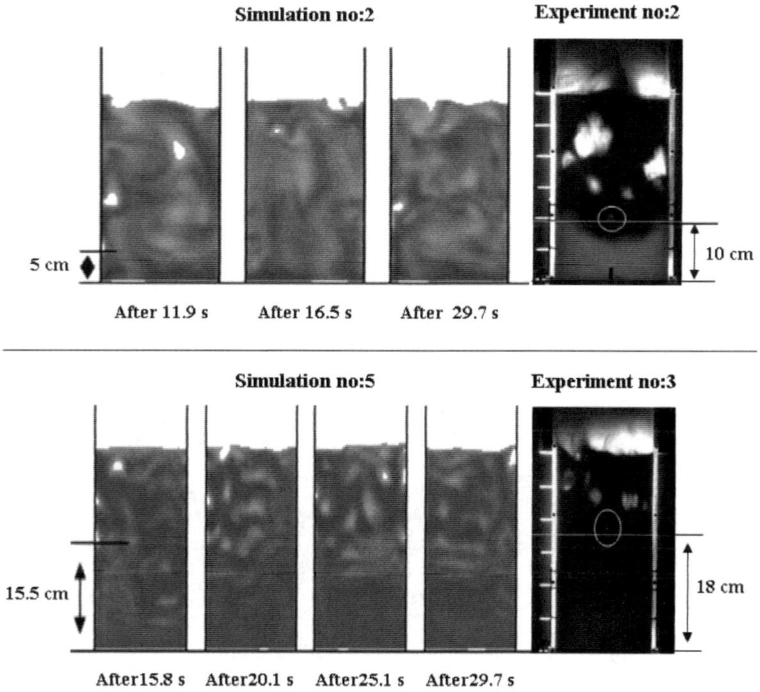

Figure 5: Lowest position for bubble appearance in the particle bed of the simulations 2, 5 and experiments 2, 3.

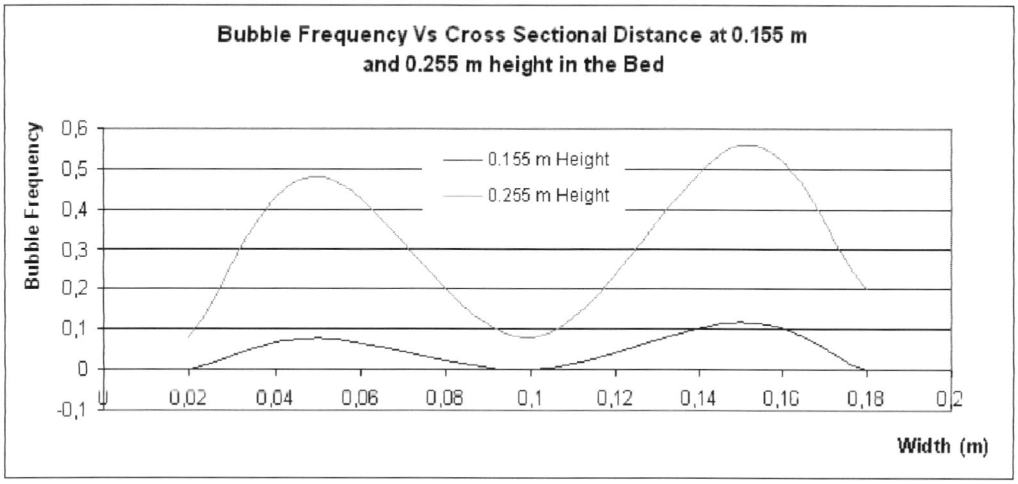

Figure 6: Bubble frequency in two selected levels in the particle bed of the simulation 4.

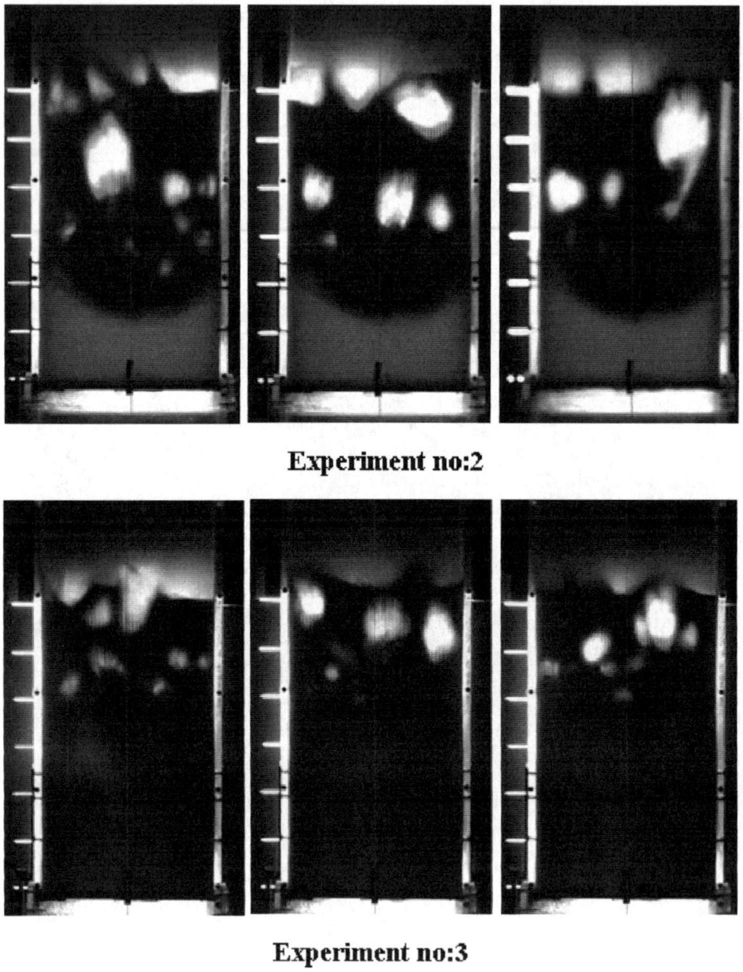

Figure 7: Bubble appeatence in the particle bed of the experiment 2 and 3.

comercial CFD software Fluent 6.3 is used in the computational study.

Three experiments are performed with glass particles with the same mean particle diameter but different particle size distribution. The mean diameter is 488 μm and the superficial gas velocity is 0.134 m/s. The superficial velocity is well below the theoretical minimum fluidization velocity for particles with diameter 488 μm. The results from the experiments show that when particle mixtures with a wide range of particle sizes are used, the particles will segregate, and bubbles will appear although the superficial velocity is below the calculated minimum fluidization velocity. Segregation influences significantly on the flow behavior and the position of bubble appearance in the bed. It is therefore important to account for the particle size distribution in the simulations in order to get an acceptable prediction of flow behavior in fluidized beds.

Simulations are performed with the same conditions as used in the experiments, and in order to study the effect of particle size distribution, multiple particle phases are included. The simulations are performed with one, two, three and four particle phases, and the results are compared to each other and to the experimental data. It was found that the simulations with the highest number of particle phases give a good agreement with experiments. The results of the simulations show the same tendency of particle segregation as is found in the experiments. Segregation of particles as a function of time is analyzed in one selected position in the bed. The analysis shows that the larger particles tend to settle close to the bottom of the bed whereas the smaller particles tend to accumulate in the upper part of the particle bed with time.

The results from the simulations and experiments support to conclude that the particle size distribution has a significant effect on the flow behavior in fluidized beds, and that this has to be accounted for in the simulations. The analysis proved that the validity of the simulated results depends on the accuracy of representing the particle size distribution.

References

[1] Busciglio, A., Vella, G., Micale, G. & Rizzuti, L. (2008). Analysis of the bubbling behavior of 2D gas solid fluidized beds, Part I. Digital image analysis technique. *Chemical Engineering Journal.* [Online]. Available from: http://www.elsevier.com [15th March 2008].

[2] Massah, H. & Oshinowo, L. (2000). Advanced Gas-Solid Multiphase Flow Models Offer Significant Process Improvements: JA112. *JOURNAL ARTICLES BY FLUENT SOFTWARE USERS,* p. 2. Fluent .Inc

[3] Halvorsen. B. M. (2005). An Experimental and Computaional Study of Flow Behavior in Bubbling Fluidized Beds. **In** *Thesis for the degree of Dr. Ing,* pp. 32-45.

[4] Patil, D.J., Annaland, M.V.S. & Kuipers, J.A.M. (2005 (a)). Ctritical comparison of hydrodynamic models for gas-solid fluidized beds–Part II: freely bubbling gas-solid fluidized beds. *Chemical engineering science 60.* [Online]. Available from: http://www.elsevier.com [10th February 2008].

[5] Patil, D.J., Annaland, M.V.S. & Kuipers, J.A.M. (2005 (b)). Ctritical comparison of hydrodynamic models for gas-solid fluidized beds–Part I: bubbling gas-solid fluidized beds operated with a jet. *Chemical engineering science 60.* [Online]. Available from: http://www.elsevier.com [10th February 2008].

[6] Enwald, H., Peirano, E. & Almstedt, A.E. (1996). EULERIAN TWO-PHASE FLOW THEORY APPLIED TO FLUIDIZATION. *Multiphase Flow,* 22 pp. 21-66

[7] Huilin, L., Yunhua, Z., Ding, J., Gidaspow, D. & Wei, L. (2007). Investigation of mixing / segregation of mixture particles in gas-solid fluidized beds. *Chemical engineering science 62.* [Online]. Available from: http://www.elsevier.com [15th February 2008].

[8] Boemer, A., Qi, H. & Renz, U. (1998). Verification of Eulerian simulation of spontaneous bubble formation in a fluidized bed. *Chemical Engineering Science,* 53 (10) pp. 1835-1846

[9] Arastoopour, H. (2001). Numerical simulation and experimental analysis of gas / solid flow systems: 1999 Flour Daniel Plenary lecture. *Powder Technology,* 119 pp. 59-67

[10] Huilin, L., Yurong, H. & Gidaspow, D. (2003). Hydrodynamic modelling of binary mixture in a bas bubbling fluidized bed using the kinetic theory of granular flow. *Chemical Engineering Science 58.* [Online]. Available from: http://www.elsevier.com [5th March 2008].

[11] Mathiesen, V, Arastoopour, H., Hjertager, B.H. (1999). Experimental and Computational Study of Multi-phase Gas/Particle Flow in a CFB Riser. *AIChE Journal,* 45(12) pp. 2503-2518.

[12] Mathiesen, V, Solberg, T, Hjertager, B.H. (2000a). Prediction of gas/particle flow including a realistic particle size distribution. *Powder Technology,* 112 pp. 34-45.

[13] Mathiesen,V., Solberg,T., Hjertager, B.H. (2000b). An experimental and computational study of multiphase flow behavior in a circulating fluidized bed. *Int. Journal of Multiphase Flow,* 26(3) pp. 387-419.

[14] Werther, J., Molerus, O. (1973b). The local structure of gas fluidized beds -II. The spatial distribution of bubbles. *Int. Journal of Multiphase flow,* 1 pp. 123-138.

[15] Gidaspow, D. (1994). The Fluidized State. **In** *Multiphase Flow and Fluidization, Continuum and Kinetic Theory Descriptions,* Boston: Academic press,Inc.

[16] Gidaspow, D. Lin, C., Seo, Y.C. (1983). Fluidization in Two-Dimensional Beds with a Jet, Part I. Experimental porosity distribution. *Ind. Chem. Fundam.*, 22(2) pp. 187-193.

[17] Kuipers, J. A. M. ,van Duin, K. J., van Beckum, F. P. H.,van Swaaij, W. P. M, (1993). Computer simulation of the hydrodynamics of a two-dimensional gas-fluidized bed. *Computers chemical Engineering,* 17(8) pp. 839-858.

[18] Halvorsen, B., Mathiesen, V. (2002(a)). CFD Modelling and simulation of a lab-scale Fluidized Bed. *Modeling, Identification and Control,* 23(2) pp. 117-133.

[19] Halvorsen, B., Mathiesen, V. (2002(b)). A comparison of first and second order spatial scheme for simulation of a bubbling fluidised bed. *Advances in Fluid Mechanics IV,* Wit Press, pp. 569-578.

[20] Bokkers, G.A., van Sint Annaland, M., Kuipers, J.A.M. (2004). Mixing and segregation in a bidiperse gas-solid fluidised bed: a numerical and experimental study, *Powder Technology,* 140 pp. 176-186.

[21] Ariyarathna, D.G.A.S.U (2008). Recommendation of a Model for Simulating & Analysis of the Influence of Particle Size Distribution on the Simulations of Bubbling Fluidized Beds. **In** *Thesis for the degree of MSc. Ing,* pp. 22-46.

[22] Jayarathna, S.A (2008). VERIFICATION OF THE IMPORTANCE OF INTRODUCING PARTICLE SIZE DISTRIBUTIONS TO BUBBLING FLUIDIZED BED SIMULATIONS. Accepted for: 49th SIMS Conference on Simulation and Modelling, SIMS 2008, October 7-8 2008, Oslo, Norway.

[23] Wu, W (2008). EXPERIMENTAL STUDY OF EFFECTS OF PARTICLE SIZE DIZTRIBUTION ON BUBBLE BEHAVIOR FOR VALIDATION OF CFD MODELLING OF BUBBLING FLUIDIZED BEDS. Accepted for: 49th SIMS Conference on Simulation and Modelling, SIMS 2008, October 7-8 2008, Oslo, Norway.

[24] Cui, H., Mostoufi, N & Chaoki, J. (2000). Characterization of dynamic gas-solid distribution in fluidized beds.*Chemical Engineering Journal 79.* [Online]. Available from: http://www.elsevier.com [8th March 2008].

87c Jetting Observations In a Cold-Flow Fluidized Bed

Nathan P. Franka, Iowa State University, 2025 Black Engineering Building, Ames, IA 50011 and **Theodore J. Heindel**, Department of Mechanical Engineering, Iowa State University, 2025 Black Engineering Building, Ames, IA 50011.

Jetting in a 10.2 cm diameter cold-flow fluidized bed with a perforated plate gas distributor is observed using X-ray computed tomography (CT). The bed is filled with one of three 500-600 micron materials (glass beads, ground walnut shell, or ground corncob), and operated over a range of superficial gas velocities ($1.25U_{mf} - 3U_{mf}$, where U_{mf} is the minimum fluidization velocity). Local time-averaged gas holdup is determined in the entire bed from the CT data. Regions of high time-averaged gas holdup are observed directly above the perforated plate showing jet penetration into the fluidized bed.

The perforated plate is composed of 62 1-mm diameter holes arranged in a circular pattern. The CT images clearly show that not all aeration holes are active at low superficial gas velocities and the number of active aeration holes increases with increasing U_g. The more dense glass bead bed has less active holes at the same U_g when compared to the less dense beds of ground walnut shell or ground corncob. However, all aeration holes are active, for all bed materials, at the highest superficial gas velocities of this study. The CT images also show that the jets in the center of the bed dissipate before those near the bed walls. Eventually, all jets lose their identify.

Imaging Horizontal Gas/Solid Jet Penetration in A Gas-solid Fluidized Bed Using Electrical Capacitance Volume Tomography

Fei Wang, Qussai Marashdeh, W. Wasito, and Liang-Shih Fan*

Department of Chemical and Biomolecular Engineering
The Ohio State University
140 W 19th Ave, Columbus, OH 43210, USA

Abstract

The horizontal gas-solid jet penetration of a gas-solid fluidized bed is studied using electrical capacitance volume tomography (ECVT). ECVT is a non-invasive imaging technique developed by the authors and it involves providing 3D concentration maps from capacitance sensors.. The fluidized bed under investigation is of 0.3m (12 inch) ID with a 0.5m ID disengagement section and has a total height of 2.4m. A gas-solid jet side injection, with 1 inch ID, penetrates the wall of the fluidized bed in the middle of the ECVT test region. Solids in the side injection are supplied through and blown into the fluidized bed using high pressure gas. Fluid catalytic cracking catalysts (Group A) are used in this study. The superficial gas velocity from the distributor of the gas-solid fluidized bed employed ranges from 0.03 to 0.24 m/s and the gas velocity from the side jet penetration ranges from 0 to 16 m/s. In this study, we observe and analyze gas and gas-solid side injection in the bubbling regime of the fluidized bed, variation in bubble shape, formation, and coalescence.

Introduction

Gas-solid fluidized beds have been employed extensively in chemical, petrochemical, metallurgical, food, and pharmaceutical industries. Fluidized beds provide good mass and heat transfer characteristics and high mixing rates between solid particles and gas. In industrial fluidized bed reactors, reactants of gas and solids are often introduced into reactors by horizontal jets at a certain height above the bed distributor to enhance the lateral mixing of solids. Although considerable research efforts have been placed on the phenomenon of the horizontal gas-solids jet penetration in gas-solid fluidized beds, further understanding on the hydrodynamics of horizontal gas-solids jets penetration are still needed for improving reactor design. Among the past several decades, little work has been done on horizontal gas-solids jet penetration phenomenon. Kozin and Baskakov (1967) studied horizontal gas jets using a specific design of a cap-air distributor and provided a penetration correlation. Merry (1971) measured penetration depths of horizontal air jets into three fluidized beds of different diameters using sand kale seed and steel shot to propose a correlation of the horizontal jet penetration depth. Hong et al. (1997) used a semi-theoretical correlation to predict horizontal-inclined jets

* To whom correspondence should be addressed.

penetration by visual observation of a 314 mm by 25 mm two-dimensional gas-solid fluidized bed with an angular changeable inclined jet at the side of the bed. However, accurate and simultaneous 3D measurements of the instantaneous shape and motion of volume solid fractions from jets and the hydrodynamics of coalescence of horizontal jet and bubble from the bottom of a distributor are needed to better understand side injection effect. Such measurements need to be carried out with three-dimensional non-intrusive techniques. In this study, Electrical Capacitance Volume Tomography (ECVT) is applied for the measurements of the instantaneous shape and motion of solids in jets.

Experimental Studies

Experimental setup

Figure 1 is the schematic diagram of the gas-solid fluidized bed with a horizontal gas/solid jet penetration system used in this study. The fluidized bed column made from Plexiglas is of 0.3 m ID with a 0.5 m ID disengagement section. A two-stage cyclone is installed at the freeboard of the fluidized bed to separates solids from gas. The distributor of the fluidized bed is made of a porous plate with a pore size of 20 μm and a fractional free area of 60%. The total bed height of the fluidized bed is 2.4 m. A horizontal tube with a 0.0254 m ID is mounted on the wall of the testing section of the fluidized bed, 0.3 m above the distributor to inject the mixture of solids and gas from the side. Solids are fed into the horizontal tube by a solids-feeder and a flow controller. A gas jet at a high speed velocity is mixed with the solids in the horizontal tube and the mixture of gas and solids is injected into the fluidized bed. FCC particles (Geldart group A) with a mean diameter of 60 μm and a particle density of 1400 kg/m^3 are used both as the fluidized solids in the fluidized bed and the injected solids from the side horizontal tube. The testing region of the fluidized bed with the horizontal gas-solid injection is mounted with the ECVT sensor, which will be discussed in the next section.

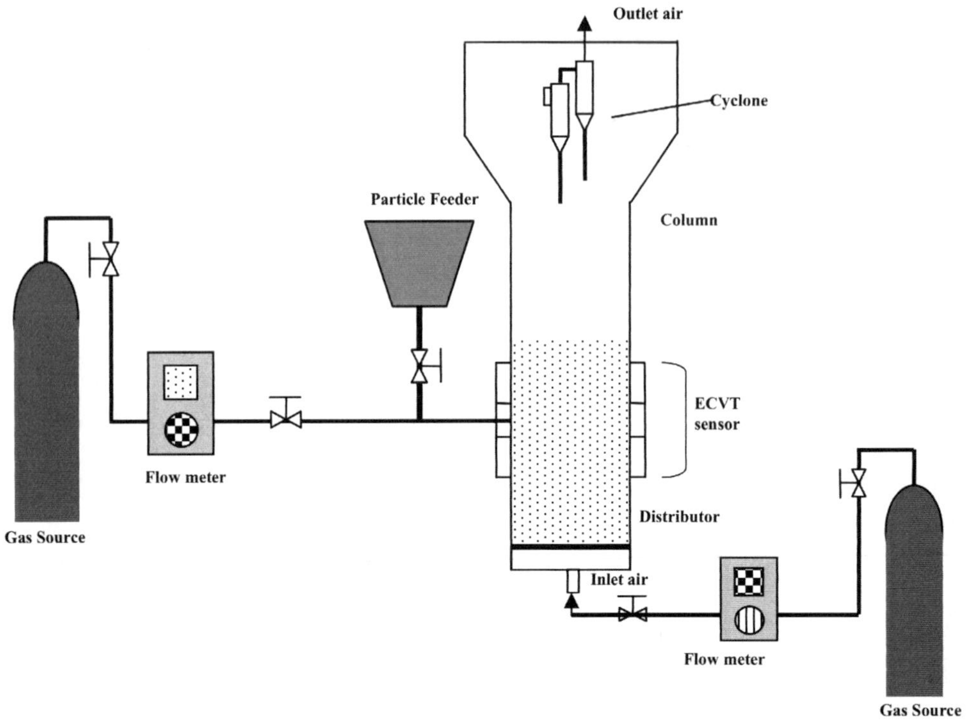

Figure 1. Schematic diagram of the gas-solid fluidized bed with a horizontal gas/solid jet penetration system

Electrical capacitance volume tomography

ECVT system consists of three basic parts: (1) a three-dimensional capacitance sensor, (2) a data acquisition system to obtain capacitance signals, and (3) a computer system to reconstruct and view the images. Figure 2 is a schematic diagram of the ECVT system used for the measurement of horizontal gas-solid penetration in the gas-solid fluidized bed. The volume tomography technology acquires 3D images directly from the measured capacitance data. (Warsito and Fan, 2005; Warsito, et al., 2007; Marashdeh et al. 2008). The image reconstruction is based on optimizing a set of objective functions related to both the measured capacitance and reconstructed image itself. The configuration of the three-dimensional capacitance sensor is illustrated in Figure 3. It has three layers with four electrodes in each layer and a total of twelve channels, each electrode has a rectangular plate shape. The edges of two side-by-side electrodes in the middle layer are modified to accommodate the penetration tube. The acquisition frequency is 80 Hz and the reconstruction resolution is of $20 \times 20 \times 20$ pixels for the three-dimensional images.

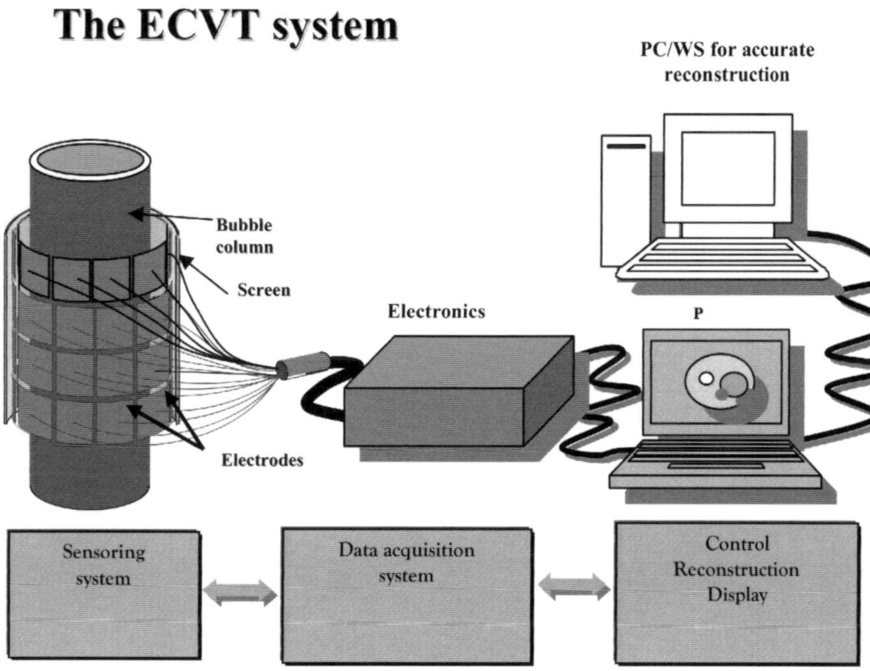

Figure 2. ECVT system

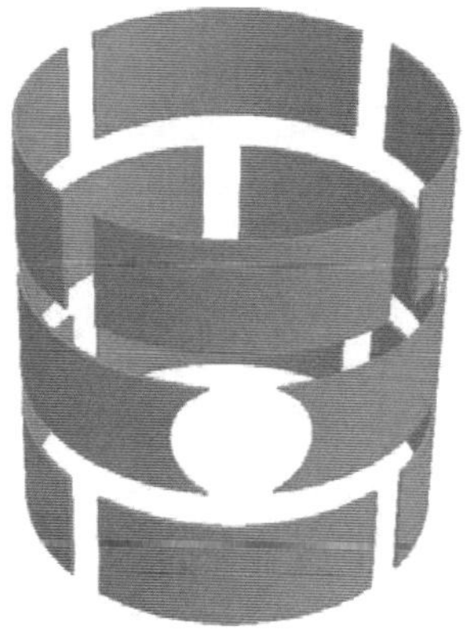
Figure 3. ECVT sensor configuration

Results and Discussion

Comparisons between the ECVT, ECT and the optical fiber probe measurements

Figure 4. shows the comparison between the time-averaged cross-sectional solids concentrations obtained by ECT, optical fiber probe (Bing Du et al., 2005), and volume solids concentration by ECVT for the 12-in gas-solid fluidized bed with FCC fluidized particles. The results obtained by ECVT, ECT and the optical fiber probe are comparable. It is observed that the solids concentrations obtained by the ECVT technique are slightly higher than those measured by ECT which is attributed for the high spatial resolution of ECT(32×32) compared to that of ECVT ($20 \times 20 \times 20$).

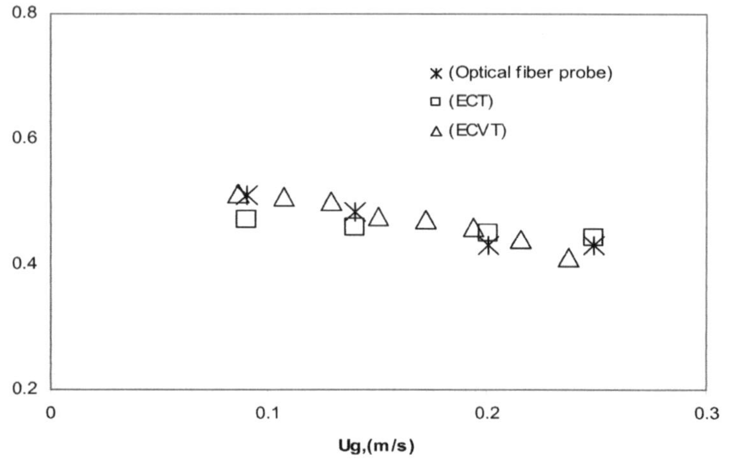

Figure 4. Comparison between the time-averaged cross-sectional solids concentrations obtained by ECT, optical fiber probe and volume solids concentration by ECVT for a 12-in gas-solid fluidized bed

Horizontal Gas/Solid Jet Penetration

Figure 5 shows the three-dimensional and two-dimensional views (from left to right, 3D Concentration map, axial cross-sectional maps at top, middle, and bottom of the monitoring region, X-Z concentration map, axial cross-sectional map in the middle, respectively) of jet penetration from the side horizontal tube of the fluidized bed. The fluidized bed is operated at two different superficial gas velocities: 0.032 m/s (Figure 5a) and 0.108 m/s (Figure 5b). The gas jet velocity is 15.5 m/s for both cases. The gas jet penetration length at high superficial gas flow rate is greater than that at low superficial gas flow rate. Figure 6 shows the images of penetration of gas-solids mixture jet from the side horizontal tube with Ug 0.032 m/s, gas jet velocity 16 m/s, and solid jet velocity 16 m/s. The horizontal penetration length of the gas-solids mixture is high and the jet penetrates the center of the column. Figure 7 shows the coalescence of bubbles from fluidized bed distributor and gas jet penetration with Ug 0.108 m/s and gas jet velocity 15.5 m/s. The jet penetrates the gas-solid fluidized bed and meets the bubble from the bottom of the distributor to form a "large" rising bubble toward the bed surface.

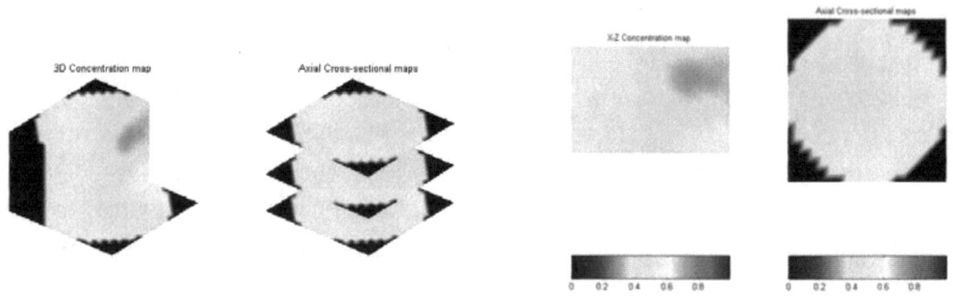

(a)

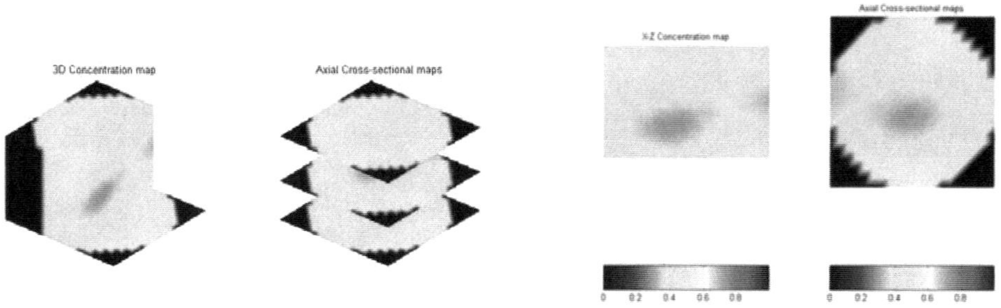

(b)

Figure 5. views of fluidized bed with gas jet penetration from the side horizontal tube: (a) Ug=0.032 m/s, gas jet velocity=15.5 m/s; (b) Ug=0.108 G/s, gas jet velocity=15.5 m/s.

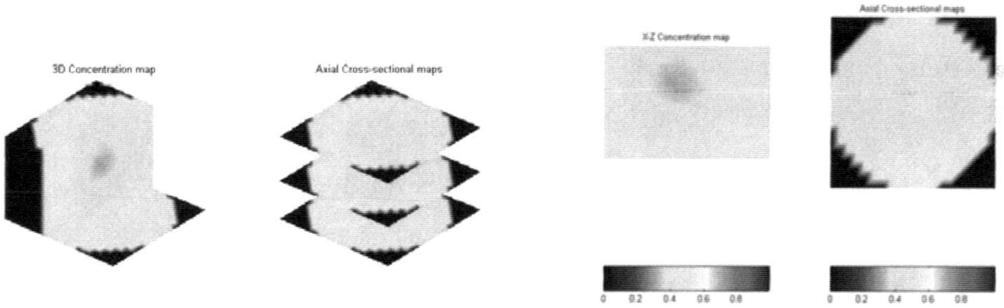

Figure 6. Images of fluidized bed with gas-solid jet penetration mixture jet from the side horizontal tube: Ug=0.032 m/s, gas jet velocity=16 m/s, solid jet velocity=16 m/s.

t=0 t=25 ms t=50 ms t=75 ms t=100 ms

Figure 7. Coalescence of bubbles from fluidized bed distributor and gas jet penetration: Ug=0.108 m/s, gas jet velocity=15.5 m/s

Conclusion

ECVT is successfully applied for the visualization of horizontal gas-solid jet penetration of a gas-solid fluidized bed. Three-dimensional images of the horizontal gas jet and gas/solids mixture flow structures are obtained by ECVT. The accurate, instantaneous, and quantitative volume solids fraction of the gas-solid fluidized bed is provided. The images show the process of coalescence of horizontal gas jet and bubbles from the distributor in the gas-solid fluidized bed. ECVT shows its robustness and powerful capability for the three-dimensional visualization and quantitative measurements of gas-solid fluidization system and horizontal gas-solid penetration system.

Acknowledgements

The support of US. Department of Energy under Grant # DE-AC26-04NT41817.670.01.07 is gratefully acknowledged.

References

Du, B. Warsito, W. and Fan, L.-S., "ECT Studies of Gas-Solid Fluidized Beds of Different Diameters," *Ind. Eng. Chem. Res.*, **44**, 5020 (2005).

Hong, R. Y. Li, H. Z. Li, H. B. and Wang, Y., "Studies on the Inclined Jet Penetration Length in A Gas-Solid Fluidized Bed," *Powder Technol.*, **92**, 205 (1997).

Kozin, V. E. and Baskakov, A. P., *Khim i Tekhnol. Topliv i Masel.*, **3**, 4 (1967).

Marashdeh, Q. Fan, L.-S. Du, B. and Warsito, W., "Electrical Capacitance Tomography – A Perspective," *Ind. Eng. Chem. Res.*, **47**, 3708 (2008).

Merry, J. M. D. "Penetration of A Horizontal Gas Jet into A Fluidized Bed," *Trans. Instn Chem. Engrs*, **49**, 189 (1971).

Warsito, W. Marashdeh, Q. and Fan, L.-S., "Electrical Capacitance Volume Tomography (ECVT): Sensor Design and Image Reconstruction," *Proceedings of the 4th World Congress on Industrial Process Tomography, Aizu, Japan*, 827 (2005).

Warsito, W. Marashdeh, Q. Fan, L.-S., "Electrical Capacitance Volume Tomography (ECVT)," *IEEE Sens. J.*, **7**, 525 (2007).

87e A Rotating Chimney to Increase the Flexibility of Rotating Fluidized Beds

Juray De Wilde, Materials and Process Engineering Dept. (IMAP), Universite catholique de Louvain (UCL), Reaumur, Place Sainte Barbe, 2, Louvain-la-Neuve, B - 1348, Belgium

Abstract:

Introduction and Concept

To increase the flexibility of rotating fluidized beds, the new concept of a rotating chimney (Figure 1(a)) is introduced and experimentally investigated using 1G-Geldart D- and B-type particles. The rotating chimney consists of multiple blades separated by openings to the chimney and rotates in the same sense as the rotating particle bed. The rotational speed of the chimney and of the particle bed may, however, differ. In particular, chimney rotational speeds higher than the average particle bed rotational speed are possible. The rotating chimney then allows to increase the centrifugal force locally, i.e. in the vicinity of the chimney, making it more difficult for the particles to leave the fluidization chamber with the fluidization gas to the chimney.

Experimental set-up and Results

The action of the rotating chimney was experimentally investigated using a dual step response technique (Figure 1(b)). The responses of the rate of solids losses via the chimney and of the pressure drop over the fluidization chamber to step changes in the solids feeding rate were measured. With a well chosen step, this dynamic method allows obtaining quasi steady state data over the entire solids loading range from one single experiment. The influence of the chimney rotational speed and the fluidization gas flow rate were measured.

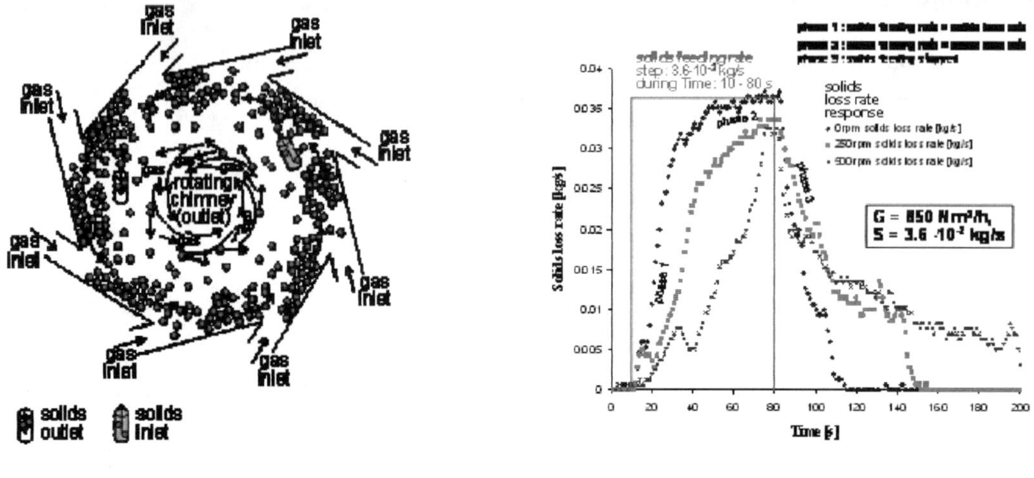

(a) (b)

Figure 1: (a) Schematic representation of a rotating chimney in a rotating fluidized bed; (b) Dual step response technique. Steps in the solids feeding rate given (red curve) and responses in the rate of solids losses via the chimney detected. Illustration with the solids loss rate curves from a 24-cm diameter, 11.5 cm long fluidization chamber. Fluidization gas flow rate = 850 Nm^3/h, 1G-Geldart D-type solids feeding rate = 3.6×10^{-2} kg/s, chimney rotational speeds: 0, 250, and 500 rpm.

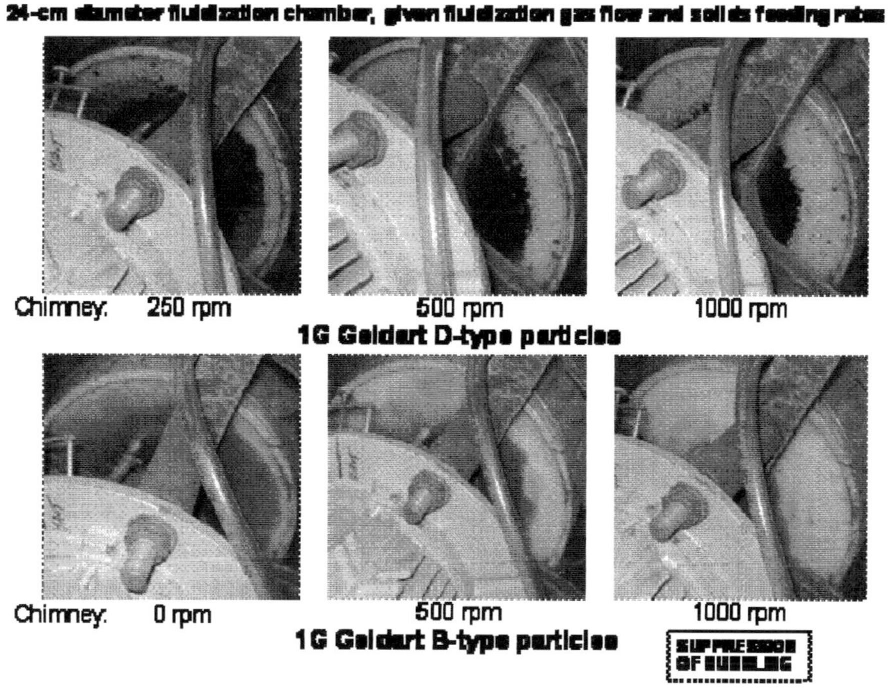

Figure 2: Quasi-steady state solids loading in a 24-cm diameter, 11.5 cm long fluidization chamber at given fluidization gas flow rate of 850 Nm^3/h and solids feeding rate of 3.6×10^{-2} kg/s as a function of the chimney rotational speed. Top: 1G-Geldart D-type particles; Bottom: 1G-Geldart B-type particles (suppression of bubbling at higher solids loadings).

At solids loadings below the theoretical maximum solids loading in the fluidization chamber, the rate of solids losses via the chimney was found to be proportional to the solids loading in the fluidization chamber and inversely proportional to the square of the chimney rotational speed. Hence, at high chimney rotational speeds, the solids loss rate is drastically reduced and becomes almost independent of the solids loading in the fluidization chamber over a broad solids loading range. As the solids loading in the fluidization chamber approaches the theoretical maximum, the rate of solids losses via the chimney drastically increases to compensate the solids feeding rate. The fluidization gas flow rate was shown to have only a minor effect on the solids losses via the chimney. Nevertheless, at solids loadings below the theoretical maximum solids loading, the rate of solids losses via the chimney was found to decrease with increasing fluidization gas flow rate, whereas at solids loadings approaching the theoretical maximum, the rate of solids losses via the chimney was found to increase with increasing fluidization gas flow rate.

As illustrated in Figure 2, increasing chimney rotational speeds allow to increase the quasi steady state solids loading in the fluidization chamber at given fluidization gas flow and solids feeding rates, up to the theoretical maximum. The latter is easily reached at chimney rotational speeds higher than the average particle bed rotational speed. Alternatively, at given fluidization gas flow rate and solids loading in the fluidization chamber, the chimney rotational speed can be used to control the particle residence time in the fluidization chamber. Finally, the pressure drop over the rotating chimney was found to be relatively small, that is, at least an order of magnitude smaller than the total pressure drop over the fluidization chamber under the conditions investigated.

87f The Effect of Bed Depth on the Hydrodynamics of a Fluidized Bed of Fcc Catalyst

Todd S. Pugsley and **Shayan Karimipour**. Chemical Engineering, University of Sasktachewan, 57 Campus Drive, Saskatoon, SK S7N5A9, Canada

Fluidized beds have broad applications in many chemical, pharmaceutical, and mineral processing industries. The bed depth, which is directly related to the material inventory of the fluidized bed, is an important operating parameter in these various applications. Achieving specific efficiencies or throughput can lead to the necessity of employing deep fluidized beds. In these cases, keeping sufficiently long gas residence times, low particle entrainment, and a good fluidization quality possess special importance.

In this work, pressure fluctuations as a measure of the fluidized bed hydrodynamics has been used to perform a comparative analysis of the influence of different bed depths. For this purpose, the absolute and differential pressure fluctuations have been measured in several locations along the fluidized bed column for various combinations of gas velocities and bed depths and also two distributor designs. Both classical and advanced signal analysis methods have been employed to extract and evaluate useful information regarded to the fluidized bed behavior in different bed depths.

89a New Nanopowders and Nanostructured Materials Derived from Liquid-Feed Flame Spray Pyrolysis (LF-FSP)

Richard M. Laine[1], Min Kim[2], Jose Azurdia[2], and Bingyuan Huang[1]. (1) Materials Science and Engineering, University of Michigan, 2300 Hayward St, Ann Arbor, MI 48109-2136, (2) Mse, University of Michigan, 2300 Hayward St, Ann Arbor, MI 48109-2136

LF-FSP of metal alkoxide and carboxylate complexes provide access to a variety of novel mixed-metal oxide nanopowders. We discuss here recent studies on the production of core-shell nanopowders and the study of their properties. We find very novel sintering behavior for these materials that will be discussed here.

89b Generating Nanoparticle Test Aerosol Using Liquid Flame Spray

Mikko Aromaa[1], Jyrki M. Mäkelä[1], Antti Rostedt[1], Thomas J. Krinke[2], Kauko Janka[3], Marko Marjamäki[1], and Jorma Keskinen[1]. (1) Aerosol Physics Laboratory, Tampere University of Technology, PO Box 692, FIN-33101, Tampere, Finland, (2) TSI GmbH, Neuköllner Str. 4, D-52068, Aachen, Germany, (3) Dekati Ltd., Osuusmyllynkatu 13, FIN-33700, Tampere, Finland

Liquid Flame Spray (LFS) is a specific gas phase synthesis method of nanoparticles. It has recently been used in producing e.g. antimicrobial nanoparticle coatings. However, nanoparticles include a potential health risk for humans when inhaled. One can use respirator masks to protect oneself from inhaling the nanoparticles. There are also devices that measure the number and active surface area concentration of particles in room air.

LFS is a flame based method for generating nanoparticles in gas phase. Several different flame methods are widely used in material synthesis. In this study our point is, however, to show that LFS can be used to generate a well defined test aerosol for devices measuring nanoparticles. The aim is to simply make a stable concentration of a test aerosol in room air or in an aerosol chamber. The test aerosol can then be used in instrument testing or even in toxicological experiments. LFS consists of a high-temperature H_2-O_2 flame. One of the combustion gases atomises the liquid precursor into the flame where it evaporates and eventually forms nanoparticles. Using LFS metal and metal oxide nanoparticles can be processed with the production rate in the order of g/min. In this study we produced ferric oxide, titanium dioxide and metallic silver nanoparticles.

A fume chamber and a small room were used as test aerosol generation chambers. These two spaces were filled with nanoparticles. We measured the nanoparticles using several aerosol measurement devices: Electrical Low Pressure Impactor (ELPI), Ultrafine Condensation Particle Counters (working fluids butanol and water, UF-CPC and UF Water-CPC, respectively), Scanning Mobility Particle Sizer (SMPS), Nanoparticle Surface Area Monitor (NSAM), Engine Exhaust Particle Sizer (EEPS) and a prototype of an industrial hygiene monitor based on the escaping current technology (ECT-sensor). All the instruments measured the same room aerosol. Using the measurement data we show that LFS can be widely used in nanoparticle test aerosol generation. In this study we generated a steady distribution of nanoparticles ranging from 5 to 60 nm with surface area concentrations from 1 to 10000 $\mu m^2/cm^3$ in the room air. We also show that a bimodal distribution can be generated using the same or two different materials.

89c Flame Synthesis of Nano-Scale Nio/ysz and NiO/CeO2/YSZ Powders for Solid Oxide Fuel Cell Anode Catalyst Applications

Ranjan Pati[1], Osifo Akhuemonkhan[1], Michael B. Pomfret[2], Hillary Sadoff[1], Robert A. Walker[2], and Sheryl Ehrman[3]. (1) Chemical and Biomolecular Engineering, University of Maryland, 2113 Building 90, College Park, MD 20742, (2) Department of Chemistry and Biochemistry, University of Maryland, 0107 Chemistry Building, Building 091, College Park, MD 20742, (3) Department of Chemical and Biomolecular Engineering, University of Maryland, Bldg. # 090, College Park, MD 20742

Nickel oxide/yttria stabilized zirconia (NiO/YSZ) cermets are the most commonly used anode materials for high temperature solid oxide fuel cells (SOFCs). The electrochemical performance of the NiO/YSZ cermets strongly depends on the microstructure and distribution of the Ni and YSZ phases in the cermets. Addition of CeO2 in the NiO/YSZ cermets not only improves the electrochemical performance but also protects the anode from carbon deposition during fuel reformation. With this motivation, homogeneous mixtures of nanocrystalline NiO/YSZ, NiO/CeO2/YSZ cermet powders were prepared by flame synthesis method from a mixture of nickel acetate, cerium nitrate yttrium nitrate and zirconyl oxynitrate precursors. In this process, aqueous solutions of precursors were atomized using a nebulizer. The atomized solution was then passed through the flame reactor. In the reactor, the precursors react in the solid phase, forming the final product. The particles were collected by thermophoresis onto a water-cooled surface. The as prepared materials were characterized by using x-ray diffraction (XRD), high resolution transmission electron microscopy (HRTEM) and BET surface area measurement. X-ray diffraction study shows the formation of cubic and tetragonal phases of the as prepared powders along with the presence of NiO and CeO2. The particles are crystalline with size in the range of 20-30 nm.

To test the anode performance, a SOFC electrolyte was prepared by pressing 8 mole% YSZ at room temperature. The resulting disk was heated at 1500°C for 3 hrs. Anodes were prepared by tape casting the as prepared nanoparticles over an electrolyte disk. The assembly was then sintered at 1000°C for 12 hrs to ensure adherence to the electrolyte. Electrochemical performance showed that the maximum power density increases as the amount of CeO2 in the nanoparticle sample decreases. This result is expected as Ni is a better fuel decomposition catalyst and therefore more active in SOFCs. The maximum power density obtained was 8.3 mW/cm2 for NiO/YSZ anode material with NiO:YSZ mole ratio of 50:50. However, CeO2 is an excellent oxygen storage material, and a maximum power density of 40.2 mW/cm2 was obtained for NiO/CeO2/YSZ having a NiO:CeO2:YSZ mole ratio of 20:30:50, emphasizing the role of CeO2 in the electrochemical performance of the cermet materials.

89d Synthesis of ZnS Nanoparticles by Spray Pyrolysis

Sha Liu, Hongwang Zhang, and Mark T. Swihart. Chemical and Biological Engineering, The University at Buffalo (SUNY), 303 Furnas Hall, Buffalo, NY 14260-4200

Zinc sulfide (ZnS) is a II-VI semiconductor with a large direct band gap in the near-UV region. It has potential applications in areas such as solar cells, lasers, sensors and displays. ZnS nanoparticles (or quantum dots) exhibit size-dependent optical and electronic properties. We have prepared ZnS nanoparticles through spray pyrolysis of two different sets of precursors, and both approaches will be described in this presentation. The single source precursor zinc diethyldithiocarbamate has been used alone, with no need to match flow rates of separate zinc and sulfur precursors to produce stoichiometric particles. In this approach, the precursor evaporation and decomposition and the nucleation of particles occur sequentially in the furnace. X-ray diffraction shows that cubic ZnS crystals are produced in this approach. Transmission electron microscopy shows that the average particle size is below 20 nm. They exhibit blue photoluminescence under UV excitation. We have also prepared ZnS nanocrystals using zinc acetate and thiourea as separate Zn and S sources. With a precursor molar ratio of 1:2, ZnS can be produced using the same reactor system that was used for producing ZnS particles from a single source precursor, by premixing these two precursors in one gas assisted atomizer. This produces larger, spherical particles characteristic of incomplete precursor evaporation. In addition, two ultrasonic atomizers have been used to separately atomize the two precursors and separately evaporate them in the furnace, then finally mix them to nucleate particles from the vapor phase.

89e Catalysis by Flame Synthesized Supported Gold Nano-Particles

Sachit Chopra[1], Jeroen van Bokhoven[2], Jagdeep Singh[2], and Gregory Beaucage[1]. (1) Department of Chemical and Materials Engineering, University of Cincinnati, Cincinnati, OH 45221-0012, (2) Institute for Chemical and Bioengineering, ETH Zurich, HCI E1127, Zurich, 8093, Switzerland

We investigated the catalytic properties of supported gold nanoparticles synthesized using Spray Flame Pyrolysis (SFP) for the CO oxidation reaction. The ability of supported Au nanoparticles to catalyze low-temperature oxidation of CO is of interest from both theoretical and practical view point. We evaluate the flame synthesized supported gold nanoparticles on TiO_2, Al_2O_3, SiO_2 and Fe_2O_3 supports, and compare their performance against the deposition-precipitation synthesized catalysts. We present the effect of oxidizing and reducing pretreatments on the catalytic activity. We used in-situ XAS to understand the reaction pathway of CO and O_2 over gold in order to determine the oxidation state of the most active gold species. The supports produced in the flame appear to be markedly different from those used in the solution methods. Comparison is also made of the nano-structure using SAXS and TEM as well as the molecular structure using XAFS. Anomalous SAXS is used to distinguish between the ceramic supports nano-structure and the nanostructure of the supported gold particles.

Supported by the US NSF through International Research and Education in Engineering (IREE)and CBET 0626063 grants as well as by the Swiss NSF.

89f Flame Aerosol Synthesis of Functional Composite Nanoparticles

Bing Guo and Hoon Yim. Department of Mechanical Engineering, Texas A&M University, College Station, TX 77843

Magnetic and fluorescent composite nanoparticles are useful for biomedical imaging applications. Flame aerosol synthesis may be used for the manufacturing of such nanoparticles, with advantages such as simplicity and high energy efficiency. However, producing the desired particle properties in flame aerosol synthesis may prove challenging. In flame aerosol synthesis, both thermodynamics and kinetics play an important role. Better understanding of these factors would greatly improve our ability to design efficient routes for the synthesis of smart nanoparticles.

In this study, flame aerosol processes were used to synthesize M-SiO_2 composite nanoparticles (M=Fe, Eu, Tb, Dy, Tm). These particles are characterized by TEM, XRD, and fluorescence spectroscopy. The Fe/SiO_2 nanoparticles contain separate phases of amorphous SiO_2 and crystalline gamma-Fe_2O_3. The lanthanides exist in the silica matrix as a separate amorphous phase. All lanthanide/silica particles are fluorescent, with photoluminescence peak wavelength at 544, 574, 616, and 657 nm (Tb, Dy, Eu and Tm respectively). The particles generally do not have a core-shell structure. The iron- or lanthanide-rich phases randomly distribute in the silica matrix. When on the edge of the nanoparticles, these phases may be dissolved in 4% sulfuric acid at room temperature.

In conclusion, flame aerosol synthesis can be used to generate silica-based composite nanoparticles for magnetic or fluorescent properties. These particles can be potentially used in biomedical imaging. However, with this method making core-shell structures with full encapsulation by silica remains a challenge. The toxicity of these particles needs to be examined before they can be used in biomedical applications.

Web Page: www1.mengr.tamu.edu/bguo/index.html

139a Investigation of Ligand Solvation Effects for Alkyl Thiol Stabilized Metal Nanoparticles In CO2 Expanded Liquids Using Small Angle Neutron Scattering

Gregory Von White II and Christopher L. Kitchens. Chemical and Biomolecular Engineering, Clemson University, 127 Earle Hall, Clemson, SC 29634

Recent advances in metallic nanoparticle synthesis and processing have opened the door for new opportunities taking advantage of their distinct properties varying from bulk metals. For many of these nanoparticle applications, monodispersity is critical because of their size-dependent functionality. Utilizing CO_2 as an anti-solvent for organic solvent dispersed metal nanoparticles, nanoparticles 10nm in diameter or less can be controllably size-fractionated into monodisperse populations. Here, the stabilizing ligands play a major role in the synthesis and application of metallic nanoparticles and allows for dispersion in various solvents. Manipulation of the nanoparticle surface chemistry through synthesis or ligand exchange, affords significant opportunity for solution based "green" processing and impacts their potential applications; the solvent – ligand interactions are deterministic of the nanoparticle behavior in solution. We have recently used an interaction energy model to predict the dispersibility of nanoparticles in gas-expanded liquids (GXLs) as a function of the solvent – ligand interactions; however, questions remain unanswered regarding the stabilizing ligand structure. We have examined the structural properties of alkane thiol stabilizing ligands using Small Angle Neutron Scattering (SANS) as a function of CO_2 pressure and thus the nanoparticle dispersibility. SANS was used to determine the local solvation effects of nanoparticles ligands as well as the effective ligand solvation length as a function of the solvent properties.

139b A New Ion-Mobility Method for Probing Reaction Kinetics at the Nanoscale

Lei Zhou[1], X. Ma[1], Anshuman Lall[1], G. M. Mulholland[1], and Michael Zachariah[2]. (1) University of Maryland-College Park, College Park, MD 20742, (2) University of Maryland and NIST, 2125 Glenn L Martin Hall, College Park, MD 20742

A novel tandem ion-mobility method for probing reaction kinetics at nanoscale is developed. The experimental system consists of two different ion-mobility schemes in series. The first mobility characterization is to size select particles with a differential mobility analyzer (DMA). The second mobility characterization employs an aerosol particle mass analyzer (APM) and measures changes in mass resulting from a controlled reaction of nanoparticles. The DMA, used in this experiment for size selection, consists of an annular region between two concentric cylinders, with the center cylinder held at high voltage and the outer one at ground. Charged particles of the right polarity feel an attractive force toward the center electrode and move radically inward at an electrophoretic velocity determined by the particle charge, and drag force which is a function of particle size. When charged particles flow between the cylinders the electric force on the particle is balanced by the drag force, and at a fixed voltage all particle exiting the instrument have (to the resolution of the instrument) equivalent mobility sizes. The APM can determine the particle mass distribution based on particle mass to charge ratio, and is used in our experiment to monitor changes to the particle mass resulting from evaporation. The APM consists of two concentric cylindrical electrodes that rotate together at a controlled speed. An electrical field is created by applying high voltage on the inner electrode while the outer one is held at ground. Charged particles flowing within the concentric cylinders experience opposing centrifugal and electrostatic forces and as a result particles exiting the instrument at fixed voltage and rotation speed all have the same nominal mass. By scanning either the voltage or the rotation speed, the particle mass distribution (independent of particle shape) can be determined. Based on operating conditions for the DMA and APM and the root-sum-square (RSS) method, we estimate the uncertainty of density calculation ~5%.

We demonstrate the use of the DMA-APM technique to measure the inherent density of nanoparticles, measuring the surface energy of unsupported nanocrystals, and the kinetics of aluminum and nickel nanoparticle oxidation.

139c Functionalization and Characterization of Gold Nanoparticles

Sirnegeda D. Techane, Department of Chemical Engineering, University of Washington, P.O. Box 351750, Seattle, WA 98195-1750, Lara J. Gamble, Department of Bioengineering, University of Washington, P.O. Box 351750, Seattle, WA 98195-1750, and David G. Castner, Departments of Bioengineering and Chemical Engineering, University of Washington, P.O. Box 351750, Seattle, WA 98195-1750.

Gold nanoparticles (AuNPs) have a high percentage of surface atoms that gives them special electronic properties and reactivities. Since these properties are dependent on AuNP size, shape and surface chemistry one can tune the AuNPs for optimal performance in nanomedicine, microarray and biosensor applications. In this research, AuNPs 14, 25 and 40 nm in diameter were synthesized via the citrate reduction method and then functionalized with various chain length (6, 8, 11, and 16 CH_2 units) carboxylic acid terminated alkyl thiols by direct displacement of the citrate ions. The size, shape and size distribution of the AuNPs were characterized with transmission electron microscopy (TEM) and ImageJ. The size distribution increased with increasing particle size, going from a 3 σ value of 2.5 nm for the 14 nm AuNPs to a 3 σ value of 20nm for the 40 nm AuNPs. Surface chemistries of the alkyl thiol self assembled monolayer (SAMs) were characterized with X-ray photoelectron spectroscopy (XPS) and time-of-flight secondary ion mass spectrometry (ToF-SIMS). The XPS and ToF-SIMS results were used to optimize the synthesis and functionalization procedures so contaminants (e.g., Na) and unbound thiols were not detected. In particular, it was found dialysis was the best method for removing unbound thiol molecules. Comparison for SAMs of 16-mercaptohexadecanoic acid on the three different AuNPs sizes and flat Au surface were done to investigate effect of particle size and surface curvature on the SAM surface chemistry. In addition, 14nm AuNPs and flat Au surfaces were functionalized with the four different chain length COOH-SAMs to investigate effect of SAMs chain length on the measured SAMs chemistry. In the first case, changing the substrate while holding the SAM constant, the XPS C/Au atomic ratio decreased as the AuNPs size increased and was the smallest on the flat Au. The changing C/Au atomic ratio could be due several effects, including the curvature of the AuNP surfaces and the packing of the thiol molecules on those surfaces. Work is currently in progress to develop XPS data analysis methods to account for AuNP surface curvature effects. In the second case, changing SAMs chain length while holding substrate constant, the XPS C/Au atomic ratio decreased as the chain length decreased for both AuNPs and flat Au surfaces, as expected since the Au signal will be more attenuated as the thickness of the SAM layer increases. The ToF-SIMS data were analyzed using principal component analysis (PCA) and the results supported the XPS results. Intensity ratios of hydrocarbons ions to gold containing ions exhibited similar trends as the XPS C/Au atomic ratios for all samples. We are also investigating the SAM packing with FTIR and the AuNP stability with UV/VIS.

139d The Quality of SiO2 Coatings on Flame-Made TiO2-Based Nanoparticles

Alexandra Teleki, Department of Mechanical and Process Engineering, Swiss Federal Institute of Technology Zurich, Particle Technology Laboratory, Sonneggstrasse 3, Zurich, CH-8092, Switzerland, Kamal M. Akhtar, Millennium Inorganic Chemicals (a Cristal Company), 6752 Baymeadow Drive, Glen Burnie, MD 21060, and Sotiris E. Pratsinis, Particle Technology Laboratory, Institute of Process Engineering, Department of Mechanical and Process Engineering, ETH Zurich, Sonneggstrasse 3, Zurich, CH-8092, Switzerland.

Silica-coated titania nanoparticles are important in sunscreens, UV filters and optical nanocomposites. The surface characteristics (extent of coverage, texture and thickness) of in-situ SiO_2-coated, mostly rutile titania nanoparticles made in one-step by flame spray pyrolysis (FSP) were compared to FSP-made mixed (co-oxidized) SiO_2/TiO_2 ones by transmission electron microscopy (TEM), Raman, FT-IR spectroscopy, electrophoretic mobility and isopropanol chemisorption followed by mass spectroscopy [1]. Increasing the silica content shifted the isoelectric point (IEP) toward that of pure silica for externally-mixed SiO_2 and TiO_2, co-oxidized SiO_2/TiO_2 and low SiO_2 content (< 10 wt%) silica-coated TiO_2 nanoparticles. At higher SiO_2 contents, SiO_2-coated titania exhibited negative zeta potentials at all pH (and thus no IEP) indicating hermetic or complete coverage of the TiO_2 surface by SiO_2 as was confirmed by isopropanol chemisorption. Co-oxidized particles containing segregated TiO_2 and SiO_2 domains exhibited Si-O-Ti and Si-O-Si bonds of higher IR intensity than hermetically-coated TiO_2. The latter exhibited a peak at 1225 cm^{-1} attributed to Si-O-Si asymmetric vibrations in contrast to the former. So electrophoretic behavior and isopropanol chemisorption of these silica-coated TiO_2 could distinguish between partial (patchy) and complete (hermetic) coatings of TiO_2 and when combined with FT-IR it might be possible to identify separate SiO_2 domains (rough coatings) or particles.

[1] Teleki, A., M.K. Akhtar, and S.E. Pratsinis, "The quality of SiO_2 coatings on flame-made TiO_2-based nanoparticles," J. Mater.Chem. in press (2008).

139e Properties and Characterization of Pharmaceutical Powders Modified by Dry Coating

Lauren Beach[1], Yuhua Chen[2], Laila Jallo[1], Ajit Mujumdar[1], and Rajesh Dave[3]. (1) Chemical Engineering, New Jersey Institute of Technology, 138 Warren Street, YCEES 208, Newark, NJ 07102, (2) Mechanical Engineering, New Jersey Institute of Technology, 138 Warren Street, YCEES 208, Newark, NJ 07102, (3) New Jersey Center for Engineered Particulates, New Jersey Institute of Technology, 138 Warren Street, YCEES 208, Newark, NJ 07102

Dry coating is an innovative technique in which nano particles are mechanically coated onto the surface of larger (from a micron to few microns) host particles to impart useful properties to the final product, which are engineered particles. Dry coated engineered particles can have improved flow and handling properties, and hence are very useful for a number of industrial applications. In this paper, we present preparation, characterization and applications of such engineered particles to pharmaceutical applications.

Positron Emission Particle Tracking (PEPT) was used as a tool to determine the differences in mobility of powders with no additive compared to powders that had been coated with a specific level of additive using the Magnetically Assisted Impaction Coater . PEPT is a tool derived from the positron emission tomography (PET) nuclear technology used in medical imaging. Instead of imaging as with PET, PEPT uses the same triangulation principle to track a positron emitting tracer particle. The triangulation principle works using the pairs of gamma rays emitted from the particle to track the exact location of the particle at that time. Previous results indicate that increasing the level of additive caused an increase in the flowability and mobility of the powder. PEPT was used as a tool to quantify the increase in mobility of the powder.

139f Prediction of Inter-Particle Adhesion Force from Surface Energy and Roughness

Laila Jai Jallo[1], Yuhua Chen[2], Lauren Beach[3], Ajit Mujumdar[3], and Dave Rajesh[3]. (1) New Jersey Center for Engineered Particulates, New Jersey Institute of Technology, 138 Warren Street, YCEES 208, Newark, NJ 07102, (2) Mechanical Engineering, New Jersey Institute of Technology, 138 Warren Street, YCEES 208, Newark, NJ 07102, (3) Chemical Engineering, New Jersey Institute of Technology, 138 Warren Street, YCEES 208, Newark, NJ 07102

Abstract

It is well known that fine particles do not flow well due to the dominant interparticle adhesion force over the gravitational force. In order to reduce the adhesion force two techniques, dry particle coating and surface silanization, were used to improve the flow behavior of fine cohesive powders (5 ƒÝm aluminum). To quantify the inter-particle adhesion force of the pure and surface modified aluminum powders, surface energies of these powders were determined using inverse gas chromatography (IGC). Surface roughness of the particles was obtained with atomic force microscopy (AFM). The Inter-particle adhesion force is calculated using the Derjaguin, Muller, and Toporov (DMT) model. The results indicated that surface modification reduced the surface energy and altered the surface roughness,leading to a dramatic reduction in the inter-particle adhesion force. Meanwhile, flow characterization results of these powders also demonstrated the drastic decrease of interparticle adhesion force. The polar interfacial contributions of the surface free energy showed that the surface modification process increased basicity of the powders in the same trend that the flow properties were improved.

149a Axial Particle Concentration Distribution In Two-Phase and Three-Phase Fluidized Beds with Binary Solid Mixtures

Dong Hyun Lee[1], Byung Soon Chun[1], Yoong Lee[1], Ah-Hyung Alissa Park[2], Sang Done Kim[3], John R. Grace[4], and Norman Epstein[4]. (1) Department of Chemical Engineering, Sungkyunkwan university, 300 chunchun, jangan, Suwon, South Korea, (2) Earth and Environmental Engineering and Chemical Engineering, Columbia University, 918 S.W. Mudd Hall, MC 4711, 500 W. 120th Street, New York, NY 10027, (3) Department of Chemical and Biomolecular Engineering & Energy and Environment Research Center, Korea Advanced Institute of Science and Technology, 373-1 Guseung, Yuseong, Daejeon, South Korea, (4) Chemical and Biological Engineering, University of British Columbia, 2360 East Mall, Room 218, Vancouver, BC V6T 1Z3, Canada

The axial particle concentrations before and after layer inversion were measured in two-phase (liquid-solid) and three-phase (gas-liquid-solid) fluidized beds. The experiments were carried out in a using a half-tube acrylic column (210-mm diameter) having a height of 1.8 m. 3.2-mm polymer beads (1,280 kg/m3) and 0.385-mm glass beads (2,500 kg/m3) were selected as a binary solid mixture of lighter/larger particles and denser/smaller particles, respectively. Water was the continuous phase in both cases, whereas air is the dispersed phase for the three-phase system. The axial particle concentration distribution was determined by measuring the axial pressure gradient and sampling the denser/smaller particles along the bed height. In three-phase fluidized beds of the particle mixture, the total solid phase holdup decreased with increasing liquid velocity, but the gas phase holdup remained the same at a given superficial gas velocity. After air injection into the two-phase fluidized bed of the binary mixtures, there were initial contractions of the upper layer of the bed above the liquid velocity corresponding to the inversion point.

PHASE TRANSFER IN A DROPLET-SOLID SPHERE COLLISION

Dawei Wang[1], Zheng Shen[1], Chao Zhu[1]*

1 New Jersey Institute of Technology, Newark, NJ 07102, USA

Abstract - Heat and mass transfer of evaporation by collisions between droplets and solid particles can be significantly different under different conditions and could be affected by many factors, such as, the diameter ratio of solid and droplet, the hydrodynamic characteristics of the droplet, temperature difference between solids and droplet, the geometry curvature, the bridge effect of two or three balls and off-center condition. However, the reported studies on the droplet impacting onto the solid sphere are very limited. Thus a systemic research is very important and meaningful.

This study presents the effect of impact of a droplet upon the solid sphere under different conditions and compare the experimental results with modeling to see whether it can be validated. Both experimental methods and analytical modeling approaches have been included in this study.

1. INTRODUCTION

Heat and mass transfer of evaporation by collisions between droplets and solid particles can be significantly different under different conditions and could be affected by many factors, such as, the diameter ratio of solid and droplet, the hydrodynamic characteristics of the droplet, temperature difference between solids and droplet, the geometry curvature, the bridge effect of two or three balls and off-center condition. However, the reported studies on the droplet impacting onto the solid sphere are very limited. Thus a systemic research is very important and meaningful.

The hydrodynamics of a liquid droplet impinging on a hot surface have been extensively investigated using theoretical, numerical and experimental method. In the previous study, most experimental studies have been performed using high-speed camera, which showed the deformation process of liquid droplets impacting on a hot flat surface.

The boil heat transfer regimes could be divided into four basic regimes, i.e. free convection regime, nucleate boiling regime, transition boiling regime and film boiling regime. The characteristics of these regimes are detailed and established in Nukiyama's famous experiment in 1934.

Numerical efforts in the collision phenomena could be well addressed the fluid dynamics of the droplet impinging process. Numerical calculations with new simulation codes for two incompressible and non-viscous fluids including surface tension effects are compared with experimental results. The vapor cushion between the droplet and hot wall was simulated with special boundary conditions. Many researchers contributed on the isothermal droplet collision processes (Harlow and Shannon, 1967; Tsurutani et al. 1990; Hatta et al. 1993; Fukai et al. 1993; Hatta et al. 1995).

Recently, very little systemic research has been reported on the liquid attachment upon a collision of droplet and solids under different operational conditions such as center-to-center collision and off-center collisions, the bridge effect of multiple particles, the curvature effect of particles and heat and mass transfer of collision at temperature above Leidenfrost point.

Experimental approaches are conducted in this study to investigate the liquid attachment of a droplet colliding on a solid sphere and heat and mass transfer when there is a temperature difference during the collision. The effects of the droplet size, droplet velocity and the off-center condition on the outcome of the collision are analyzed. Also, the geometry curvature and bridge effect of two and three balls are illustrated. Some analytical explanations are given.

2. EXPERIMENTAL APPROACHES

To investigate the effect of different key parameters, such as droplet velocity, droplet-solids size ratio, center-to-center distance and temperature difference between droplet and solid, on the liquid attachment, heat and mass transfer during droplet-solid collision, two sets of experimental system are established and a series of experiment conditions are investigated, which include center-to-center collision without heat transfer, center-to-center collision with heat transfer, small droplet on large solid (a liquid droplet impinging onto a hot copper wall surface), comparable size collision (a liquid droplet impacting onto a hot solid ball) , off-center collision without heat transfer, droplet impacting with two solid balls and droplet impacting with three solid balls.

2.1 Liquid attachment upon collision without heat transfer

The liquid attachment during a process of collision between droplet and solid without temperature difference is acquired by measuring the mass of droplet before and after collision as shown in Figure 1(a). A pipette with

* corresponding author: zhu@adm.njit.edu

adjustable droplet size is adopted to create tiny droplet. The pipette and solid particles whose material is aluminum and with a diameter of 0.6mm are located vertically center-to-center with 3-D stages respectively. After dripping from pipette tip, the droplet is accelerated by forces of gravity and drag force from ambient air, and thus obtained a certain velocity before colliding with solid particle. The mass of droplet is calculated by the pipette-controlled droplet size and the mass after collision is measured by a Ohaus model E-400 scale under a plastic cup which is located under solid particle to collect droplet. The velocity of droplet upon collision was controlled by the adjusting the vertical distance between pipette tip and solid particle.

The pipette and solids then could be moved horizontally by adjusting the stages to make a off-center collision with a certain off-center distance to investigate the effect of off-center distance on the liquid attachment. The schematic diagram of off-center collision is shown in Figure 1(b).

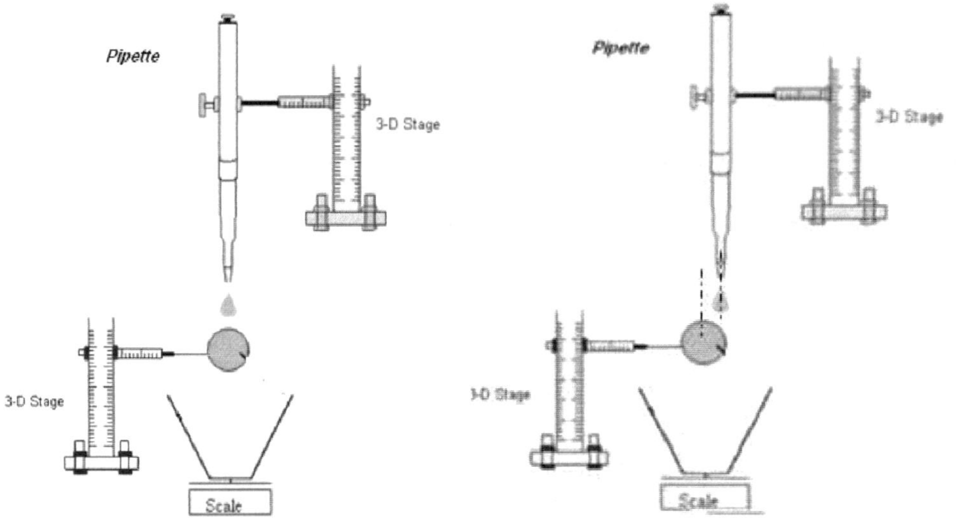

(a) center-to-center collision (b) off-center collision

Figure 1 schematic diagrams of collision without heat transfer.

2.2 Bridge effect of droplet colliding with Multiple Solid Balls

The impact of the droplet on a solid sphere is quite different from that impinging on several solid spheres. To study the bridge effect of the droplet impinging on the solid surface, the collision of the water droplet is investigated in two or three balls. Schematic diagrams of the set-up are given in Figure 2(a) and 2(b) respectively. The system is very similar with collision with single solid except that more solids are located closely with each other by stages. The distances among solids could be adjusted to investigate bridge effect on liquid attachment under different conditions. Three different distances between two solid particles were investigated at different droplet velocities on two particles. Two different distances between three solid balls were studied at different droplet velocities on three particles. Similarly, the velocity was controlled by distance between pipette tip and particles.

2.3 Heat and mass transfer upon Center-to-center Collision with temperature difference

The heat and mass transfer during a collision between droplet and hot solid particle with large temperature difference is also investigated. Two sets of experiments are used according to the size ratio of droplet and solid. During a collision, when the particle size is much larger than droplet size, it could be treated as a droplet colliding on a plate surface. Thus, a system of hot droplet colliding on smooth inner surface of a big hot copper cylinder is designed (Figure 3 (a)). The copper cylinder could be heated up to a temperature of 500oc which is above Leidenfrost point of droplet with a propane flame. The tiny water droplet flow is injected with a pipette onto inner surface of cylinder with a velocity of 6 cm/s and is collected at bottom of cylinder with collection bins which is located at top of a scale to measure hot droplets after collision. The temperature of hot droplets and copper cylinder are measured with type K thermocouples.

The experiment of comparable size collision with heat transfer is measured by heating up the solid particles of experiment in section 2.1. The type K thermocouple is adopted to measure solid temperature before and after collision to calculate total heat transfer during collision. The schematic diagram is given in Figure 3(b).

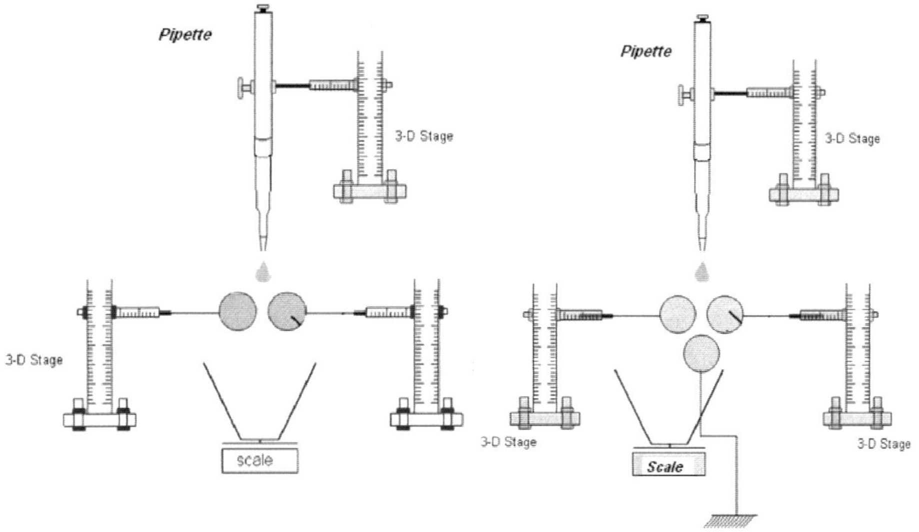

(a) two particle collision (b) three particle collision

Figure 2 schematic diagram of bridge effect by multiple particles

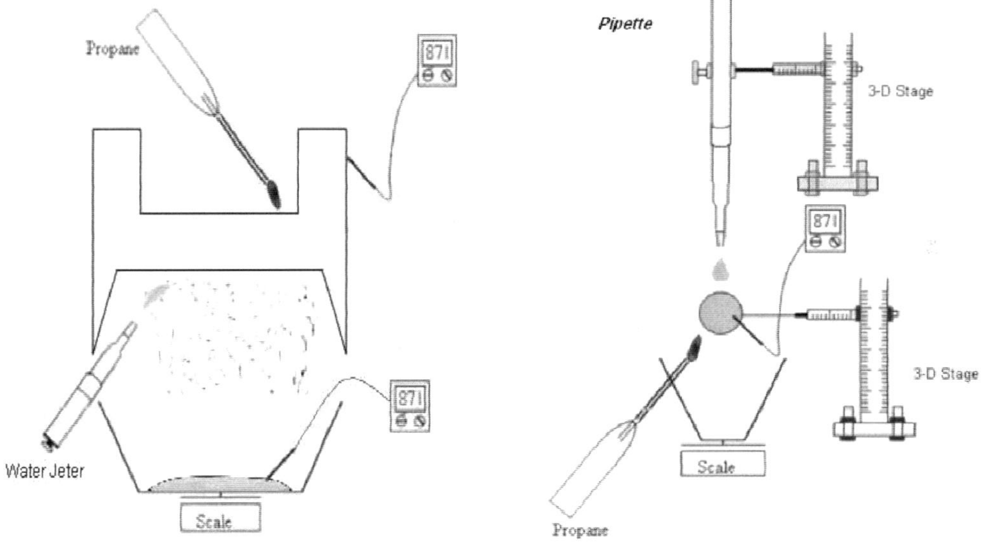

(a) collision between small droplet and large particle (b) comparable size collision

Figure 3 schematic diagram of center-to-center collision with heat transfer.

3. RESULTS AND DISCUSSIONS

The experiments are conducted for the liquid attachment of droplet colliding with single particle center-to-center and off-center for different droplet size and velocity. The bridge effect of multiple particles on the liquid attachment is also studies with different particle-particle distance. The heat and mass transfer when a droplet collides with large particle and comparable size particle are also studied for different temperature difference between droplet and solid.

3.1 Liquid attachment of droplet colliding with solids without temperature difference

Firstly, the liquid attachment of center-to-center collision of droplet and particle is measured and analyzed. The off-center collision is then be investigated and compared with that of center-to-center collision. The liquid attachment of multiple particles is also compared with that of collision with single particle to investigate bridge effect.

3.1.1 Liquid attachment upon a center-to-center collision

When a droplet falls onto a solids particle center-to center, in general, it could be divided into three different conditions based on their sizes, i.e., small droplet on large solid, large droplet on small solid and comparable size collision. For a solid with certain size in a collision with droplet, there is a maximum attach-ability which should be about half volume of itself, if it is not trapped by the droplet. When the droplet size is over the maximum attach-ability of the solid, the amount of attachment would keep constant instead of increasing with the droplet size. While, when the droplet size is much less than the solid size, the droplet may completely attached to the solid until the size reach to a critical droplet size which is related to the Webber number of the droplet upon collision and could be solved by the mass and energy conservation of droplet deformation process during the collision. When droplet size is between critical size and maximum attach-ability of solids, the liquid attachment shall be proportional to the droplet size. The trend of liquid attachment on the droplet size is given in Figure 4.

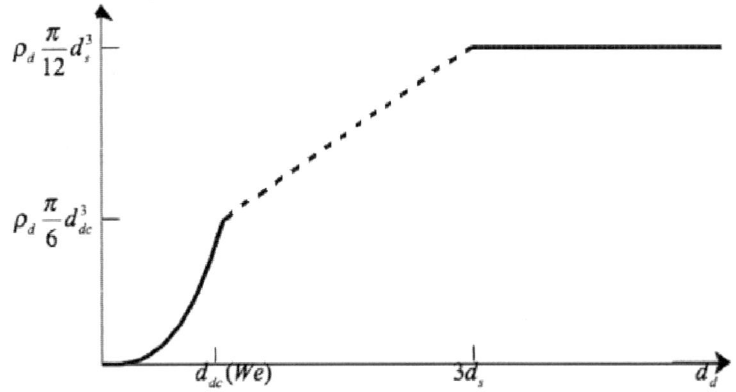

Figure 4 Liquid attachment v.s. droplet sizes

The experiment investigated the effect of Webber number of droplet upon collision on percentage of attached droplet mass with adjusting the impact velocity of droplet which is controlled by the height difference between pipette tip and solid particle. The droplet velocity is calculated by the momentum balance of droplet with effect of gravity and drag force of ambient air. Table 1 listed the velocities of droplets with two different sizes upon collision in the experiments.

The experimental results are showed in Figure 5. The attachment percentage at different impact velocities for two droplet sizes at the room temperature was studied. It is indicated that the variation trend of attachment percentage is similar for different droplet sizes when the impact velocity changes. As the impact velocity increases, the percentage of the attachment decreases. The theoretical results are shown in the figure with lines which are well matched with experimental data.

Table 1 Weber Number at Different Impacts Velocity and Droplet Sizes

Impact velocity (m/s) H (cm)	0.280 (0.4)	0.313 (0.5)	0.443 (1.0)	0.542 (1.5)	0.626 (2.0)	0.700 (2.5)
d_1=0.284 (cm)	**3.1**	**3.9**	7.7	11.6	15.5	19.3
d_2=0.337 (cm)	**3.7**	4.6	9.2	13.8	18.3	22.9

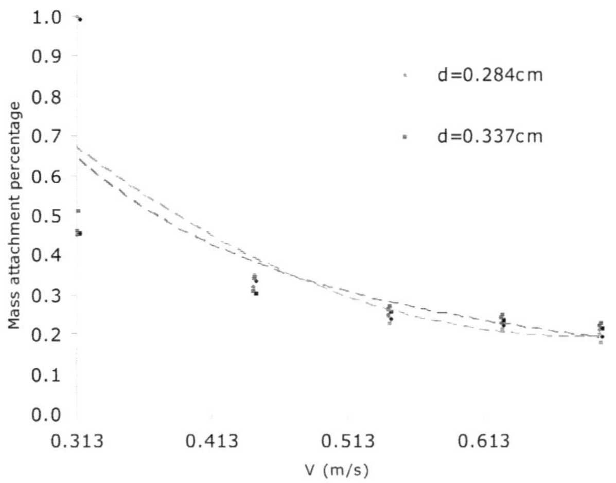

Figure 5 Liquid attachment percentage of center-to-center collision with a solid

4.1.2 Liquid attachment of off-Center Collision without Heat Transfer

When the droplet collide off-center with a solid, the attachment would be less than that under the center-to-center condition and its amount shall be dependent on the off-center distance (Δd). In the experiment, the attachment percentage at two different off-center distances (Δd_1=2mm and Δd_2=3mm) are measured and compared with that of center-to-center condition. Figure 6 shows a schematic diagram of relationship between liquid attachment and off-center distance.

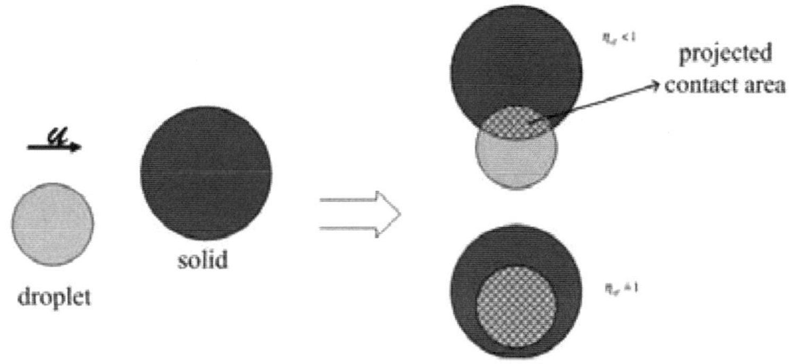

Figure 6 schematic diagram of off-center collision between droplet and solid

An off-center-collision efficiency could be defined to express the effect of off-center collision on liquid attachment, which is given by

$$\eta_{off} \equiv \frac{Liquid - attachment\,@\,off - center - collision}{Liquid - attachment\,@\,center - to - center - collision}$$

We could easily conclude that the off-center-collision is determined by offer-center distance of collision which could be estimated by the projected contacted area between droplet and solid during a collision, which is shown in Figure 7. Thus, the off-center-collision efficiency is obtained as:

$$\begin{cases} \text{if} \quad \frac{|d_1 - d_2|}{2} < Z < \frac{|d_1 + d_2|}{2}, & \eta_{off} < 1 \\ \text{if} \quad Z \le \frac{|d_1 - d_2|}{2}, & \eta_{off} = 1 \\ \text{if} \quad \frac{|d_1 + d_2|}{2} \le Z, & \eta_{off} = 0 \end{cases}$$

Figure 8 listed the experimental results of liquid attachment percentage of different off-center distances. It can be seen that, the effect of the off-center to the balls for the droplet impacting process is significant. The attachment percentage drops much when the collision changes from center-to-center to off-center and the offer-center distance increases. It is also indicated when efficiency (η_{off}) is less than 1, the droplet couldn't be all attached in our experimental data range and the attachment of which is less than the center-to-center condition. Note that the lines in the figure are the theoretical results calculated by above analysis under same operational conditions with experiment, which are in good agreement with experimental results.

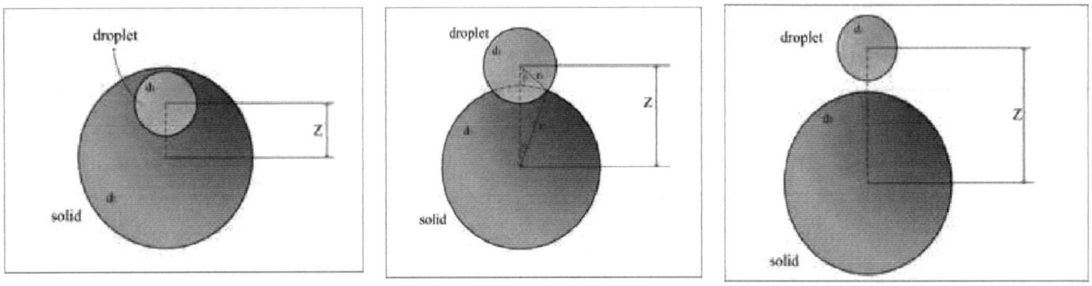

η_{off} =1 η_{off} <1 η_{off} =0

Figure 7 off-center efficiency under different offer-center

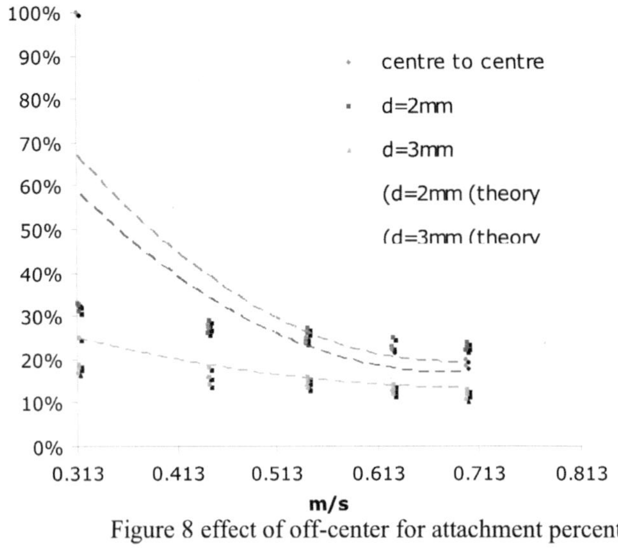

Figure 8 effect of off-center for attachment percentage

4.1.3 Bridge Effect of multiple solids during collision

When there are more than one solids located closely and the droplet collide with them simultaneously, more liquid tends to be held and stay on and between solids due to bridge effect of multiple solids. The amount would vary according to the amount of solids and the distances among the solids. In this study, the attachment percentage at three different distances under the configurations of two solid balls and three balls are measured. The results are compared with liquid attachment of single solid which is shown in Figure 9. It can be seen that, the bridge effect has a significant effect on the attachment percentage. When the solid number changes from one to two, the liquid attachment is nearly doubled and increases further when there are three solids.

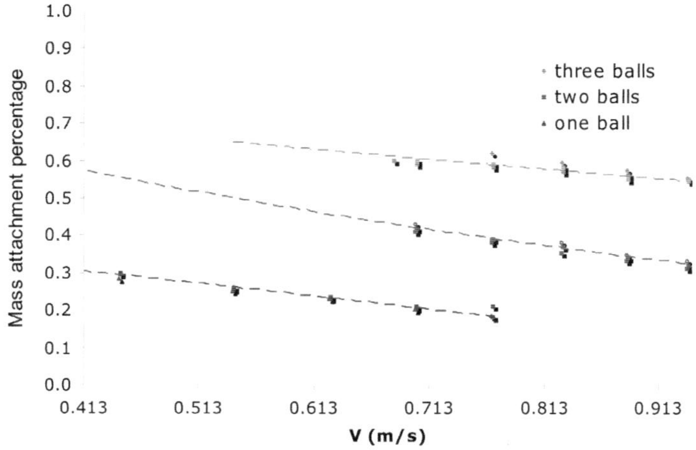

Figure 9 Bridge effect on liquid attachment

The liquid attachments under different distances among solids with two balls and three balls are shown in Figure 10 and 11 respectively. We could easily draw a conclusion that under both conditions, when the distance between solid balls increases, the percentage of the attachment decreases.

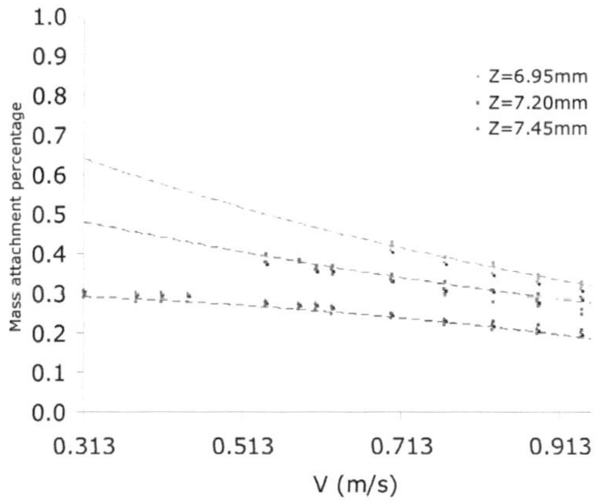

Figure 10 liquid attachment with different distance between two balls

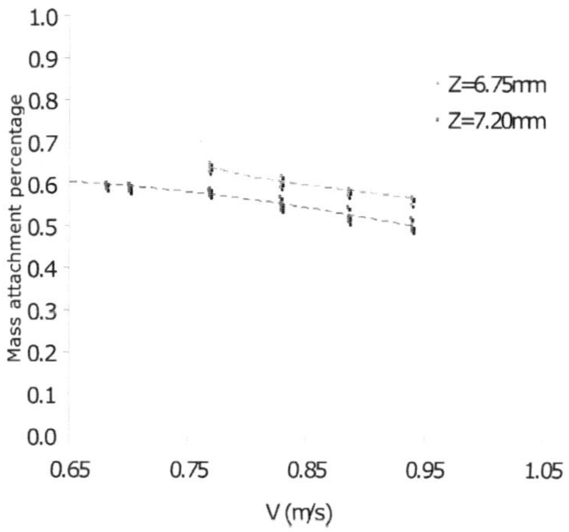

Figure 11 liquid attachment with different distance among three balls

4.2 Heat and mass transfer upon collision with temperature difference

When droplet collides with a hot solid, there is intensive heat transfer between droplet and solid. Part of heat transferred to the droplet is then conducted to the inner side of droplet and used to heat up the droplet. The other part of heat is rapidly absorbed by the surface of the droplet and is used to evaporate a portion of liquid on the droplet surface. Here, we define the percentage of heat used for evaporation as $f=Qe/Q$ and percentage of evaporated droplet mass as $fm=me/m$. Two different experiments are conducted to measure heat and mass transfer upon collision with temperature difference when small droplet colliding on large solid and comparable size collision.

4.2.1 Small Droplet on Large Solid

In this section, the evaporation process of a liquid droplet impinging onto inner surface of a hot copper cylinder has been investigated and the evaporation amount at various surface temperatures was measured. Each value has been obtained as an average of 5 experimental results, under the same operative conditions.

Figure 12 gives the mass and energy evaporation percentage at different temperatures. From the figure, it could be easily found that at the lower temperature, the total heat transfer during collision is very high and gets to a peak value when temperature reaches to about 175oc. As the temperature increases further until the temperature reaches to Leidenfrost point, the rate of heat transfer decreases because of the transition boiling regime. In this regime, the direct contact between the liquid and the surface is decreasing due to the vapor, which has a lower thermal conductivity than of liquid, accumulated on the droplet surface.

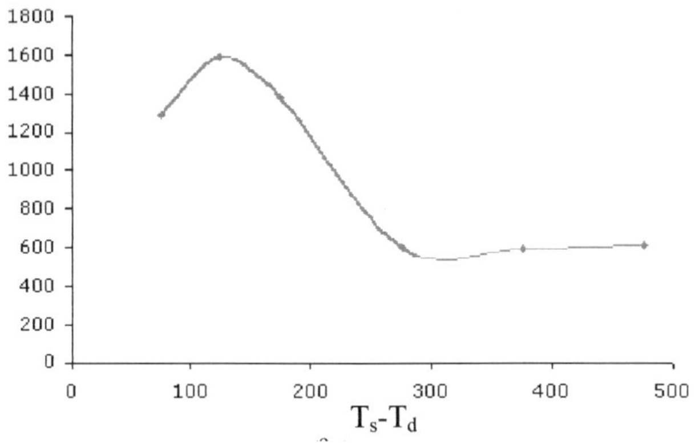

Figure 12 Total heat transfer during collision

Figure 13 shows the change of heat percentage for evaporation and evaporated mass percentage with temperature difference change. It seems that there have similar tendency with that of total heat transfer. When total heat transfer is higher, more heat percentage is used for evaporation. When heat transfer is lower, more heat tends to be conducted into inside of droplet for heating up.

4.2.2 Comparable Size Collision

The experimental results of heat and mass transfer upon comparable size collision is shown in Figure 14. From the figure, it could be easily found that at the lower temperature, the percentage of the evaporation mass is increasing and gets to a peak value when solid partice temperature increases to a value about 175oc. This could be explained as the heat transfer is in the nucleate boiling regime, the heat transfer rate tends to increase with temperature increases. As the temperature increases further, the heat transfer changes into transition boiling regime and the rate tends to decrease until the temperature reaches to Leidenfrost point.

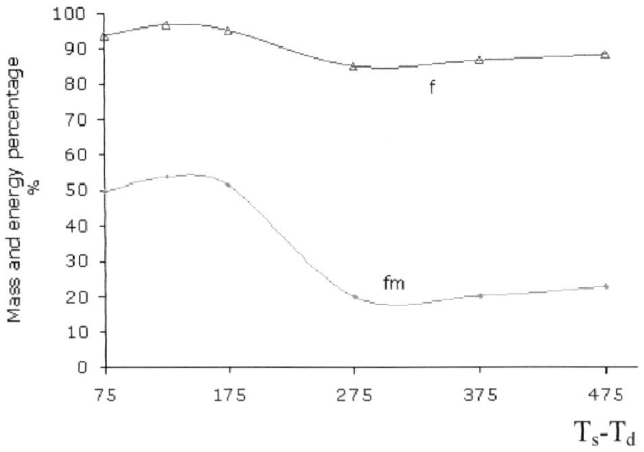

Figure 13 Evaporated mass and heat percentage for evaporation at different temperature

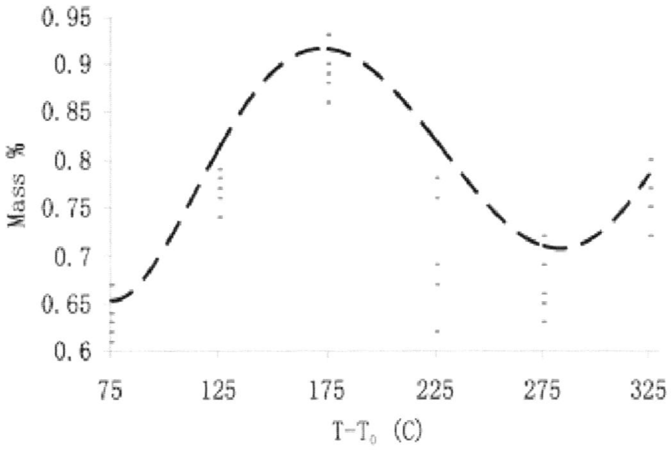

Figure 14 The percentage of evaporation mass at different temperatures.

4. Conclusion

A series of experimental systems are set up and a lots of experiments are conducted to investigate the heat and mass transfer process of droplet-solid collision under different conditions. The droplet velocity, size ratio between the droplet and solid, temperature difference on the heat and mass transfer during droplet-solid center-to-center collision, distance between multiple particles are changed to make parametric studies on the effects of these factors.

When the droplet collides velocity increases, the percentage of liquid attachment decreases. The off-center collision always has less liquid attachment compared with that of center-to-center collision and is proportional to projected contact area of collision. The bridge effect of multiple particles tends to increase liquid attachment during collision and decreases with increasing particle-particle distance.

The heat transfer and evaporated mass percentage changes when temperature difference between droplet and solid particle changes. Two different regimes were investigated by varying the tested surface temperature, a bubble boiling regime, for solid temperature larger than saturation temperature and lower than Leidenfrost temperature, a film boiling regime, for solid temperature larger than Leidenfrost temperature. A highest heat transfer rate is obtained at about 175oc and at the Leidenfrost temperature which is about 275oc, a lowest heat transfer rate is reached.

149c Enhanced Fluidization of Agglomerates of Nanoparticles

Jose A. Quevedo and Robert Pfeffer. Department of Chemical Engineering, New Jersey Institute of Technology, Newark, NJ 07102

Nanoparticles produced by a pyrogenic process cannot be fluidized in a gas suspension individually but in the form of large porous agglomerates. Some nanopowders show agglomerate particulate fluidization (APF) behavior characterized by the absence of bubbles, low minimum fluidization velocity, and a large bed expansion. When conventionally fluidized, the bed expansion is several times (about 5) the initial bed height before reaching the minimum bubbling velocity. Nanopowders that show this behavior are Aerosil® 200, R974, 300, R972 among others. However some nanopowders such as Aeroxide® TiO2 P25, Aeroxide Alu C and Aerosil® 90 show agglomerate bubbling fluidization (ABF) behavior because of the presence of bubbles at gas velocities higher than the minimum fluidization velocity. These nanopowders show a limited bed expansion (less than double initial bed height) and a high minimum fluidization velocity when conventionally fluidized, and an abundance of bubbles that makes the fluidized bed interface unclear.

In the present work, a secondary flow using a jet is introduced which greatly improves the fluidization behavior of agglomerates of nanoparticles. Agglomerates that show an APF behavior show a bed expansion more than double that observed by conventional fluidization. The larger bed expansion signifies a much better dispersion of the agglomerates in the dilute phase. In some cases, a 10-fold increase in bed expansion is obtained while maintaining the particulate fluidization behavior, showing a well defined interface and the absence of bubbles. Furthermore, when agglomerates of ABF behavior are exposed to the jet, the fluidization behavior is changed to APF type showing a bed expansion much larger the initial bed height and the absence of bubbles.

In addition, average agglomerate sizes and agglomerate size distributions of Aerosil R974 (APF behavior) and Aerosil 90 (ABF behavior) nanopowders were measured in-situ in a gas fluidized bed by using a FBRM® probe, and images were obtained with a PVM® probe both from Lasentec. The measured average agglomerate sizes are relatively smaller than the values reported in the literature for Aerosil R974.

149d Emulsion Phase Density In Deep Fluidized Beds of Geldart Group a Materials

Allan Issangya, Reddy S. B. Karri, and Ted Knowlton. Particulate Solid Research, Inc., 4201 West 36th Street, Suite 200, Chicago, IL 60632

Fluidization behavior in bubbling fluidized beds has been widely studied. Solids mixing and bubble characteristics studies have shown a fluidized bed to be well mixed with bubble voidage fraction profiles that are more or less symmetrical about the central axis of the column. Recent studies have shown, however, that deep fluidized beds of Group A particles can have a severe bypassing of gas. Knowlton (2001) presented videos of fluidization tests conducted with 4 and 11% fines (<44 micron) FCC catalyst particles in a transparent 0.3-m-diameter unit. The 4% fines bed was observed to have a significant gas bypassing in a form of a high velocity air jet that preferentially flowed up near the column wall while the rest of the bed stayed defluidized or poorly fluidized. The air jet moved around the wall but it occasionally anchored or stayed longer at the cyclone dipleg return location. When the fines content was increased to 11% the bed fluidized uniformly with no gas bypassing.

Wells (2001) using FCC catalyst particles in a semi-circular Plexiglas unit, for bed heights of 2.4 to 4.9 m, observed that a "snake" of streaming gas formed a short distance above the grid and passed through the bed, bypassing the mostly stagnant bed. Later studies (Karri et al., 2004, Issangya et al., 2006, Cocco et al., 2006 and Issangya et al., 2008) identified various parameters that can influence gas bypassing. These included fines content, bed height, superficial gas velocity, baffles, system pressure and imposed solids flux. Issangya et al. (2007) suggested that the standard deviation of differential pressure fluctuations measured across short intervals around the column can diagnose gas bypassing and can also give the approximate location(s) of where gas bypassing was occurring. If gas bypassing occurs in industrial beds, it can result in poor gas/solids contacting and poorly fluidized entrances to standpipes or discharge regions of cyclone diplegs. Gas bypassing appears to be a result of gas compression by the pressure head generated in the deep beds that is significant enough to cause defluidization of solids, similar to what occurs in standpipes. Being unable to rise as well distributed bubbles through the nearly defluidized solids mass the air then flows up in a stream (or streams) of bubbles via any lesser resistive path. The objective of this study was to determine if the emulsion phase density is affected by the added pressure head in deep beds.

Measurement of fluidised bed dryer by different frequency and different normalisation methods with electrical capacitance tomography

H G Wang and W Q Yang

School of Electrical and Electronic Engineering, The University of Manchester, Manchester, UK, haigang.wang@manchester.ac.uk

Abstract

An electrical capacitance tomography (ECT) sensor has been mounted near the bottom of a fluidised bed dryer. An ECT system based on an HP4128 impedance analyser has been used to measure capacitance and loss conductance between the electrode pairs in the sensor. It has been found that the capacitance depends on not only the particle moisture but also the frequency. With low moisture content, the relationship between capacitance and frequency is simple and linear. With high moisture content, however, the relationship becomes more complex and non-linear. For image reconstruction, different normalisation models have been used: series, parallel and Maxwell. The results show that with low moisture content, the three models give nearly the same image error. With the increase in moisture content, the difference between the three models becomes more and more obvious. With different gas-solids flow patterns, the three models also give slightly different images.

Keywords: Fluidised bed dryer, Particle moisture, Spectroscopy, Normalisation model, Electrical capacitance tomography

Introduction

Fluidised bed dryers have been widely used in the pharmaceutical, food, chemical process and other industries for many years (Vojtěch *et al.* 1966). To investigate the solids distribution during a drying process, electrical capacitance tomography (ECT) has been applied to a lab-scale fluidised bed dryer (Chaplin *et al.* 2005a, 2005b, Wang *et al.* 2008). While some advantages of using ECT to investigate fluidisation processes over the conventional methods (Makkawi and Wright 2004, Liu *et al.* 2002), the application of ECT in fluidised bed dryers has encountered some difficulties because of the moisture content changes during the drying process (McKeen and Pugsley 2002, Wang *et al.* 2008). Typically, during a drying process, the moisture content of granule changes from say 30% to 1%, which results in the change in the dielectric properties of material (Robinson *et al.* 1999, Sacilik *et al.* 2006), including permittivity, dielectric loss and conductivity. The changes in permittivity and conductivity results in the change in measured capacitance and conductance. In the past, the frequency is normally kept constant and also in the low frequency range, typically between 100 kHz and 200 kHz (Chaplin *et al.* 2005a, 2005b, Wang *et al.* 2008). With a fixed low frequency, good results can be obtained when the moisture content is lower than a certain level, say 18% (Wang *et al.* 2008). With a high moisture content, however, images are reconstructed less accurately and sometimes a wrong solids distribution may be given.

As well known, the dielectric properties of many materials are frequency-dependent and a function of material moisture and density (Mamishev *et al.* 2002, Kraszewski and Nelson 1989). To improve the measurement accuracy, it is necessary to use an ECT system with a suitable frequency.

The ultimate objectives of this research are to provide accurate measurement of fluidised bed drying and to control a fluidised bed dryer effectively for improved operation efficiency.

Dielectric properties and normalisation methods

The dielectric properties to be considered include permittivity and loss factor. The permittivity ε' refereed to a free space is the capacitance of a unit volume of matter divided by the permittivity of free space. The loss factor ε'' can be defined as the conductance of a unit volume of matter divided by $\omega\varepsilon_0$. The complex permittivity and the loss tangent of the material are

$$\varepsilon = \varepsilon' - j\varepsilon'' \tag{1}$$

$$\tan\delta = \varepsilon''/\varepsilon' \tag{2}$$

Where δ is the loss angle of the dielectric material.

The dielectric properties of wet bulk granule depend upon the density and moisture content of the material as well as the frequency. The effects of moisture content and frequency on capacitance and loss conductance are considered. In general, the dielectric properties of granule are known to be dependent on moisture content, bulk density, temperature and frequency (Sacilik *et al.* 2006, Mamishev *et al.* 2002). In this study the dielectric properties of static samples are measured at different frequency in the range of 50 kHz-13 MHz with moisture content in the range of 1.5-38%. The effect of bulk density of solids on the dielectric properties is neglected in this research. Some details about the effect of bulk density can be read in (Kraszewski and Nelson 1989, Brosseau and Beroual 2001, Mamishev *et al.* 2002).

In this research, three normalisation models are used: series, parallel and Maxwell, which are given by

$$C_{NS} = \frac{C_M - C_L}{C_H - C_L} \tag{3}$$

$$C_{NP} = \frac{\frac{1}{C_M} - \frac{1}{C_L}}{\frac{1}{C_M} - \frac{1}{C_L}} \tag{4}$$

$$C_{NM} = \frac{C_{NS} \cdot (2.0 + k)}{(3.0 + C_{NS}) \cdot (k - 1.0)} \tag{5}$$

In equation (5), k is the ratio of high and low permittivity values, C_M is the measured capacitance, C_L is the capacitance when the ECT sensor is empty and C_M is the capacitance when the sensor is filled with wet granule.

The high permittivity is for wet granule and the low permittivity is for hot air. The relationship between the measured capacitance and the normalised capacitance for different distribution was presented in a previous paper (Yang *et al.* 1999). They are used to evaluate the effects of frequency on image reconstruction for different solids distributions with different moisture content.

The Landweber iteration has been used to reconstruct solids distribution (Yang *et al.* 1999, 2002). Because of the non-linear problems, a relaxation factor is often employed to ensure convergence to seek an iterative solution to strongly nonlinear equations (Patankar 1980, Yang and Peng 2003). A relaxation factor is used in the Landweber iteration algorithm to avoid divergence and the algorithm is written as

$$G^{n+1} = G^{n} \cdot (1.0 + \omega) + \alpha \cdot S^{T} \cdot (C_{N} - S \cdot G^{n}) \cdot (1 - \omega) \tag{6}$$

Where α is the step length or gain factor and ω is the relaxation factor.

An optimum method can be used to choose the step length (Liu *et al.* 1999). However, there are no general rules for choosing the best value for the relaxation factor. The optimum value depends upon a number of factors, e.g. the nature of the problem, the number of grid points, the grid spacing, and the iteration procedure used (Patankar 1980). When ω is set to be zero, the iteration becomes the conventional Landweber iteration. If it is negative, the iteration is called under-relaxation. If it is positive it is called over-relaxation (Patankar 1980). In this research, the value is decided by tried-and-error.

Experiment method

It is well known that the gas-solids behaviour in fluidised beds is more complicated than other gas-solids process, e.g. pneumatic conveying and spouted beds (Davidson *et al.* 1985, Kunii and Levenspiel 1991). To verify the above assumption, an ECT sensor has been mounted near the bottom of a lab-scale fluidised bed dryer from Sherwood (Tornado M501). An ECT system based on an HP4128 impedance analyser (Yang 2007) has been used to measure capacitance and loss conductance between electrode pairs in the ECT sensor. During experiment, the frequency was changed from 50 kHz to 13 MHz and the moisture content changed from 1.5-38%.

The choice of normalisation model is important for image reconstruction of different distributions (Yang *et al.* 2004). Series, parallel and Maxwell models are used to normalise the measured capacitance. Four typical gas solids distributions, i.e., annular, core, stratified and mixing distributions were tested with the normalisation models and the effects of frequency on imaging reconstruction. To evaluate the image quality, two parameters, namely image error and capacitance residual, were used (Yang and Peng 2003).

A type of wet granule, semolina, is used. The properties of this type of granule can be found in a paper (Erbas *et al.* 2005). For preparation of the wet granule with different moisture content, water was sprayed onto dry material and mixed in a low-speed food mixer (KENWOOD CHEF) for 10-20 minutes. The granulation product was then sieved through a 1.5 mm screen to remove large particles due to agglomeration of wet fine particles and make sure each batch to start from the same bulk density. The moisture content was measured by an HB43 moisture meter from Mettler-Toledo as reference. The impedance analyser was programmed to measure the dielectric properties of granule at different frequencies in the range from 50 kHz to 13 MHz. Details of the system hardware and operation were given in a PhD thesis (Yang 2007). The ECT sensor has 8 electrodes, with the wall of the fluidised beds dryer as the frame of the sensor, and is enclosed by an earthed screen to eliminate external interface. An opposite electrode pair, 1 and 5, is used to determine the dielectric properties. The permittivity, conductivity and dielectric loss can be deduced from the measured capacitance and conductance.

The experiments were carried out in a laboratory with a room temperature of 20±2ºC. To reduce the effect of bulk particle density, for each batch measurement, the samples were poured into the glass tube and then fluidised for half minutes by cold air at the minimum fluidisation condition. After switching off the air supply, the sample settles down freely and a natural packing condition can be achieved with almost the same bulk particle voidage fraction of 65%.

Effects of frequency and moisture content on dielectric property measurements

The variations in capacitance, conductance, dielectric loss and loss tangent with the frequency in the range of 100 kHz-10 MHz for different moisture content are shown in Fig. 1. In the frequency range of 50 kHz-4 MHz, the measured capacitance decreases with an increase in frequency when the moisture content is lower than 20%. With high moisture content, the measured capacitance increases with the increase in frequency. In general, the capacitance with high moisture content is smaller than that with low moisture content when the frequency is lower than 1 MHz. When the frequency is higher than 4 MHz, the capacitance increases with the increase in frequency for all the cases with different moisture content. From Fig. 1 (b), it can be seen that the conductivity increases with the increase in frequency. The difference is that the change becomes faster and faster with the increase in moisture content. The capacitance as shown in Fig. 1 (a) is less regular than the results measured with a parallel plat capacitor (Sacilik *et al.* 2006), because of different geometry.

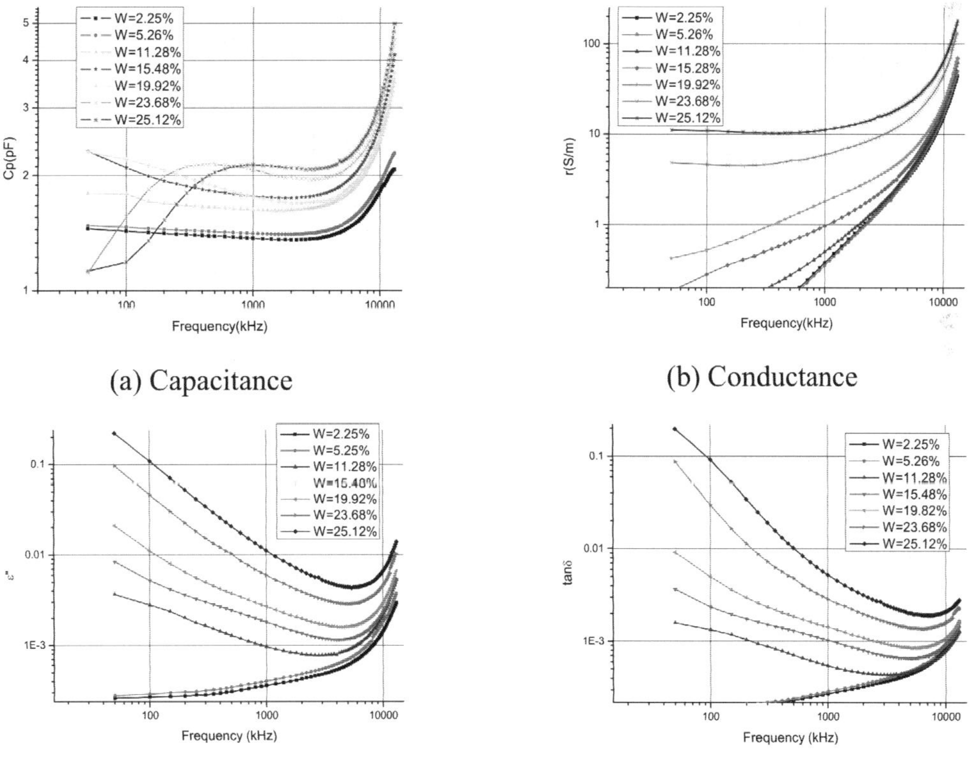

(a) Capacitance

(b) Conductance

(c) Dielectric loss

(d) Loss tangent

Fig. 1 Variation in dielectric properties with frequency at different moistures

The dielectric loss is shown in Fig. 1 (c). With low moisture content, the loss increases with the increase in frequency. With high moisture content, however, the value first decreases with the increase in frequency and then increases with the increase of frequency. The loss tangent has the same trend as dielectric loss. At a low frequency, e.g. 100 kHz, the dielectric loss for 25.1% water is more than 100 times bigger than that of 5.1% water. At a high frequency, say 10 MHz, this value is close to each other, with only a few times difference. These characteristics can be clearly seen in Fig. 1 (c).

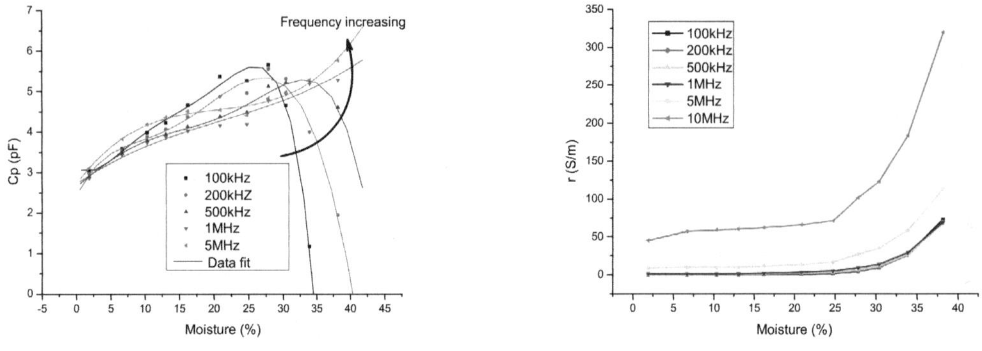

(a) Capacitance

(b) Conductivity

Fig. 2 Variation in capacitance and conductivity with moisture content and frequency

200 kHz

10 MHz

Core

Annular

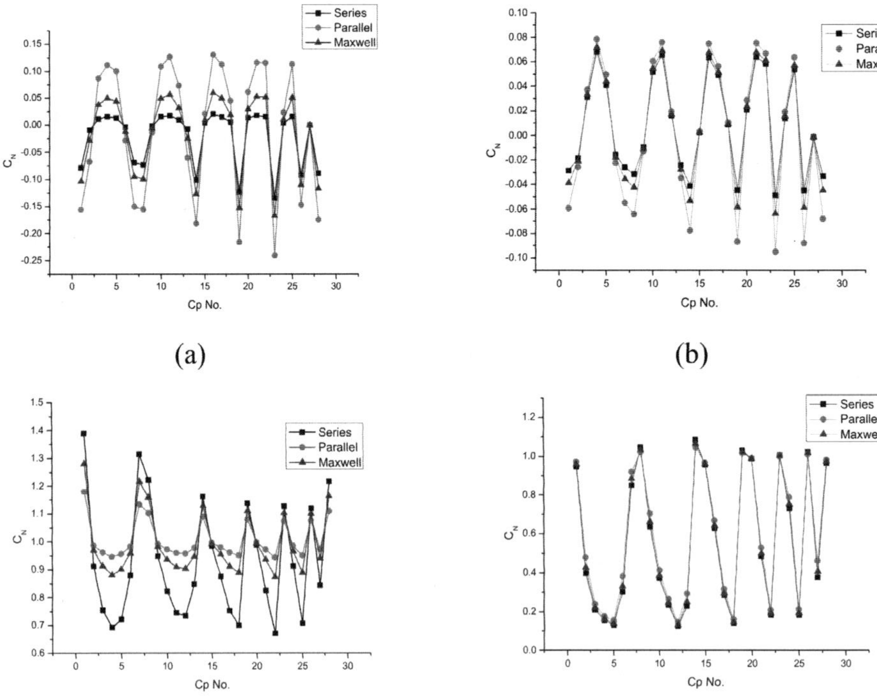

(a)

(b)

(c)

(d)

Stratified

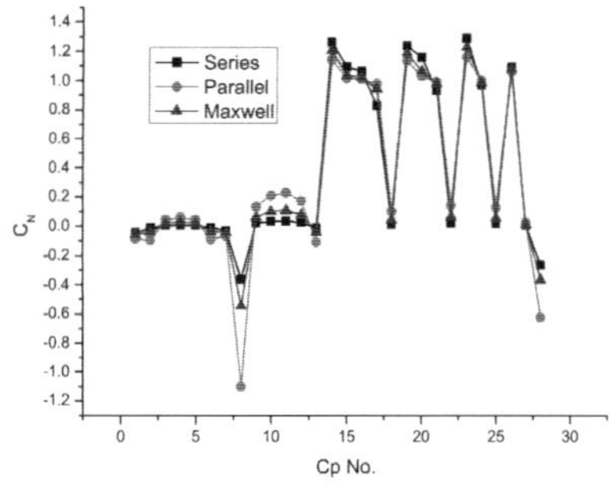

(e)

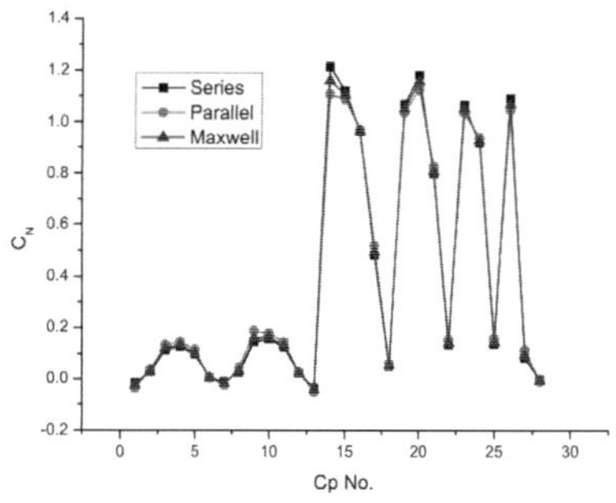

(f)

Fig. 3 Normalised capacitance

Fig. 2 shows the change in capacitance and conductance with moisture content at different frequency. It is clear that the trend for capacitance is more complex than that for conductance. With the increase in frequency, the relationship between capacitance and moisture changes from non-linear to linear as shown in Fig. 2 (a). With low moisture content, the relationship between capacitance and moisture content is non-linear. With high moisture content, the relationship becomes simple and linear. The change in conductance has the same trend for all the moisture content levels as shown in Fig. 2 (b). All these complex characteristics of dielectric properties are mostly due to the effect of water in such porous material.

From the above results, it can be seen that the moisture content and frequency strongly affect capacitance. In a fluidised bed dryer, in the first drying stage, especially in the beginning of drying process, the moisture content normally is more than 20.0%. In such a condition, the conventional measurement methods of ECT with a fixed frequency may not be suitable. Therefore, it may be necessary to use an ECT system with programmable frequency and with different normalisation models.

Image reconstruction with different normalisation model

Fig. 3 shows the normalised capacitance with different models for different distribution at two different frequencies. At low frequency, the three models give different normalised values. At high frequency, however, they nearly give the same results. It is clear to see which model is better from the reconstructed images.

Fig. 4 gives the image reconstruction results for three distributions with different normalisation models. For stratified and core distributions, the parallel model gives the best results. For an annular distribution, the series model gives better images than other two models. It seems that the reconstructed images are much better if the span between the minimum and maximum capacitance is bigger. For an annular distribution, this trend is clear with the series model. For core and stratified distributions, the parallel and Maxwell models are better than the series model.

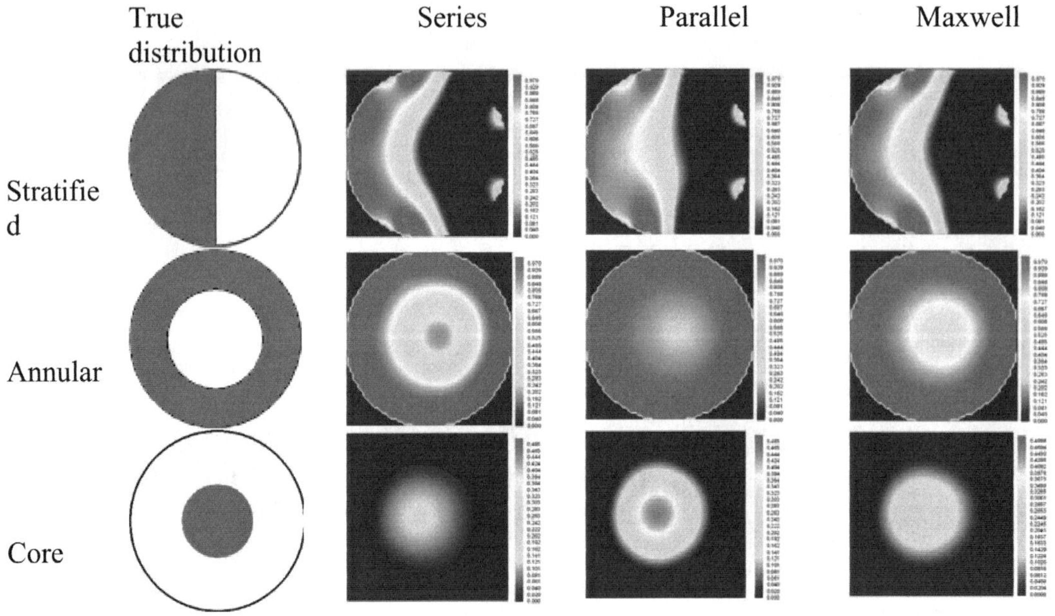

Fig. 4 Reconstructed images for 5.0% water at 200 kHz

To evaluate image quality, two parameters are used: capacitance residual and image error. The definition of these parameters can be found in (Yang and Peng 2003). Fig. 5 gives the capacitance residual for core distributions. At low frequency, the parallel model gives the minimum value and the series model gives the maximum values. At high frequency, in the end of iteration, these three models give similar results.

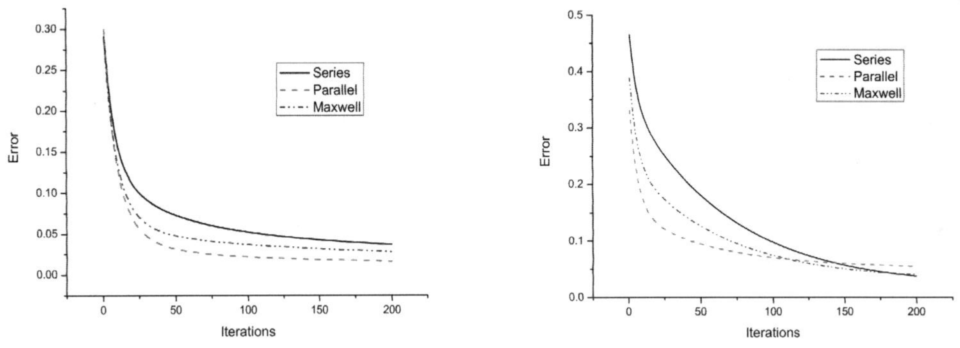

(a) 200 kHz (b) 4 MHz

Fig. 5 Capacitance residual with different normalised models and frequency for core distribution

Figs. 6 shows the effect of frequency and moisture content on image reconstruction with different normalisation models. Fig. 6 (a) gives the result for annular-core distributions and Fig. 6 (b) shows three robs distribution. For these two types of distributions, the best results are obtained for all the moisture content with 4 MHz. At low frequency, only with low moisture content, image reconstruction is good. Fig. 7 gives the capacitance residual for three rob distribution at different frequency. Fig. 8 gives the image error for this type of distribution. From the capacitance residual results, it can be seen that high frequency gives a low capacitance residual. From Fig. 8, it can be seen that with a relaxation factor, the iteration process becomes more stable and trends to converge.

Also it can be seen that for the three rob distribution, the parallel and Maxwell models gives more accurate results than the series model. All these results are summarised in Tables 1 to 3.

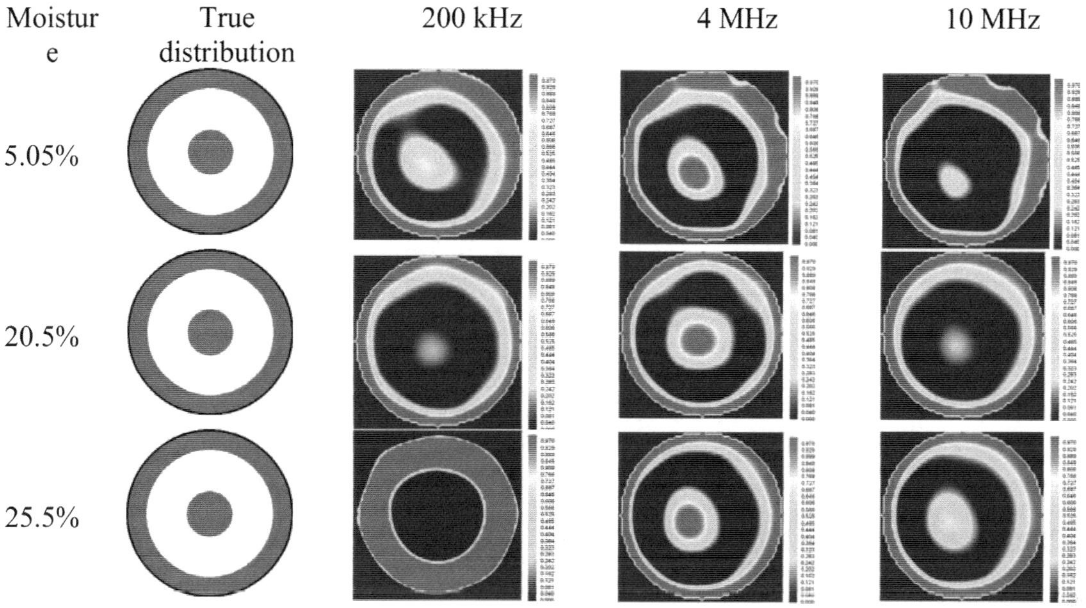

(a) Annul-core distribution

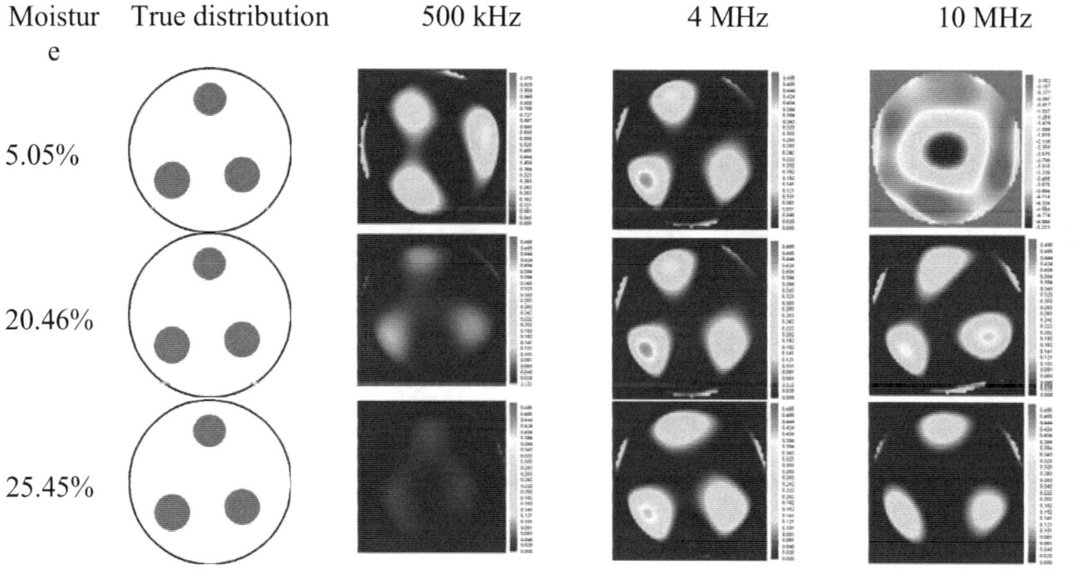

(b) Three robs distribution

Fig. 6 Image reconstruction of annul-core and three robs distributions

From Tables 1 and 2, it can be seen that for different moisture contents, 500 kHz frequency always gives slightly bigger image error than 1 or 4 MHz. With the increase in frequency, the image error becomes smaller and smaller. However, with the further increase in frequency, the image error increases again, because the stray capacitance of ECT sensor increases with the increase in frequency (Yang 2007). It can also be seen from Tables 1 and 2 that the Maxwell model gives the best results compared with the series and parallel models for these two complex distributions.

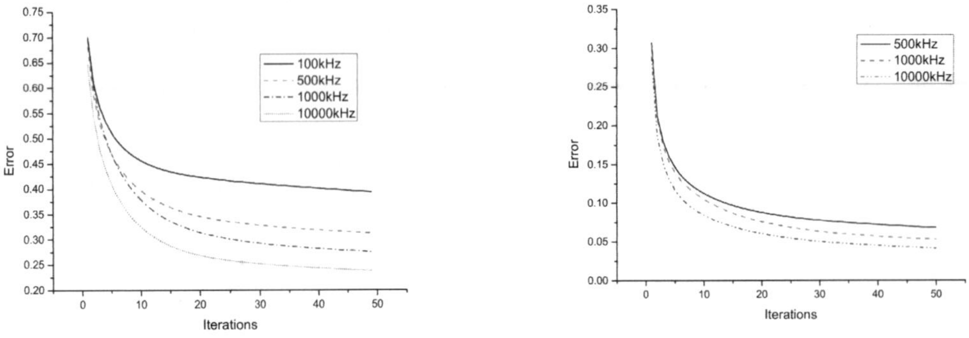

(a) W = 5.67%

(b) W = 29.7%

Fig. 7 Error with different frequency vs. iteration for core distribution

Table 1 Image error for annular-core distribution

	W = 5.05%			W = 24.7%		
	Series	Parallel	Maxwell	Series	Parallel	Maxwell
500 kHz	79.1	68.7	65.2	98.3	59.6	51.1
1 MHz	56.6	61.0	62.5	45.2	51.1	45.2
10 MHz	70.2	72.0	72.6	56.9	60.5	61.7

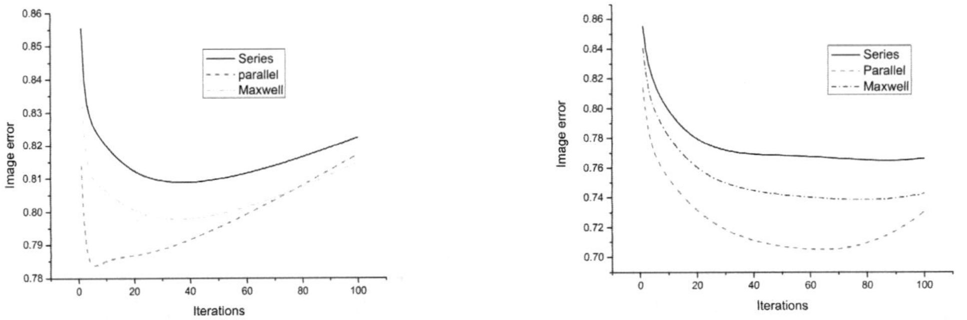

(a) Conventional Landweber

(B) Landweber with relaxation factor

Fig. 8 Image error with different normalisation models

Table 2 Image error for three robs distribution

	W = 5.05%			W = 24.7%		
	Series	Parallel	Maxwell	Series	Parallel	Maxwell
500 kHz	85.0	79.4	78.6	94.5	96.0	94.8
4 MHz	72.0	71.2	70.9	75.9	72.6	71.3
10 MHz	97.4	97.4	97.4	78.8	76.9	76.1

Table 3 shows the results of capacitance residual for annular-core distributions. For different moisture contents, low frequency always gives the biggest capacitance error. With the increase in frequency, the error tends close to zero. Note that small capacitance error does not necessarily mean good image (Yang and Peng 2003). This can be seen clearly from the image error as listed in Table 1.

Fig. 9 gives the variations in image error with different frequency for different moisture content and normalisation models. This result shows that with low moisture content, the image error nearly keeps the same level in the low frequency range. For high moisture content, the image error is lower in the middle frequency range and high in both low and high frequency ranges. With the same moisture content, in the low frequency range, the result with the series model is worse than the parallel and Maxwell models. At high frequency, these three models give nearly the same image errors. From these results, it can be seen that the choice of a normalisation model play an importance role in imaging reconstruction in the low frequency rang.

Table 3 Capacitance residual for annular-core distribution

	W = 5.15%	W = 20.7%	W = 24.7%
200 kHz	7.15	11.7	21.0
500 kHz	2.46	2.13	18.7
1 MHz	2.75	2.85	12
4 MHz	2.45	2.18	3.02
6 MHz	2.16	2.12	2.32
10 MHz	2.60	2.22	2.18

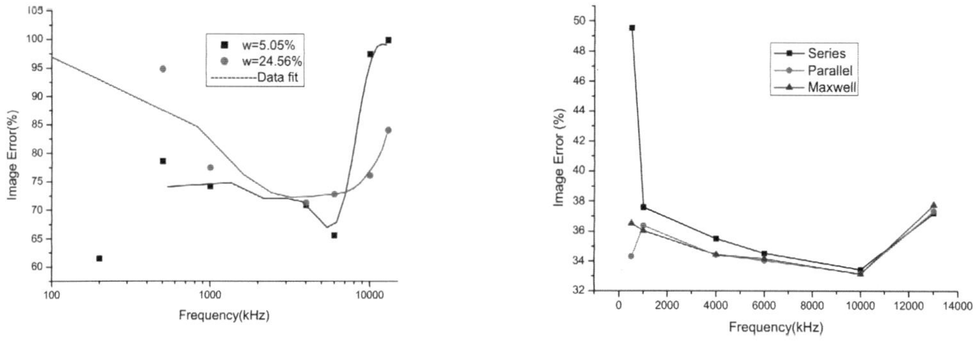

(a) Three robs distribution

(b) Core distribution

Fig. 9 Image error vs. frequency

Conclusions

An ECT system based on an HP 4128 impedance analyser has been used to measure capacitance from a multi-electrode sensor mounted near the bottom of a fluidised bed dryer, showing that capacitance, loss factor, loss tangent, and conductance are functions of the moisture content and frequency. The change in capacitance is dominated by moisture content. The change in capacitance with frequency is higher with larger moisture content. The loss factor and loss tangent depend on frequency more regularly than capacitance. The conductance increases more rapidly with the increase in frequency than the increase in moisture content.

For stratified and core distributions, the parallel model gives better results than the series and Maxwell models. For an annular distribution, the series model gives the best result. For a complex distribution, e.g. annular-core and multi objects distributions, the Maxwell model gives the best images than other two normalisation models. However, the difference between these three models is only obvious when the frequency is lower than 1 MHz. With the increase in frequency, they nearly give the same results.

In fluidised bed dryers, the solids distribution is complex and dynamic. It is useful to combine these three normalisation models together to image the solids distribution. Also, the frequency should be considered when ECT is used to measure a drying process, especially in the warming-up and constant drying periods due to high moisture content.

Acknowledges

The authors would like to thank EPSRC (GR/T29383/01, GR/T29376/01, EP/G005702/1) and the University of Manchester Intellectual Property (UMIP) Ltd for financially supporting this research. Peter Senior and Rambali Sundara Raghavan from School of Chemical Engineering and analytical Science are thanked for working on a joint project. Oxford University, GEA Process Engineering Ltd, Sherwood Scientific Ltd, DuPont, AstraZeneca and Sensatech Research Ltd are also thanked for their collaboration.

References

1. Brosseau C and Beroual A (2001), Effective permittivity of composites with stratified particles, *Journal of Physics D: Applied Physics*, 34, pp 704-710
2. Chaplin G, Pugsley T, van der Lee L, Kantzas A and Winters C (2005), The dynamic calibration of an electrical capacitance tomography sensor applied to the fluidized bed drying of pharmaceutical granule, *Meas. Sci. Technol.*, 16, pp 1281-1290
3. Chaplin G and Pugsley T (2005), application of electrical capacitance tomography to the fluidized bed drying of pharmaceutical granule, *Chemical Engineering Science*, 60, pp 7022-7033
4. Davidson J F, Clift R and Harrison D (1985), Fluidization, Academic Press, London, UK, pp 331
5. Erbas M, Ertugay M F and Certel M (2005), Moisture adsorption behavior of semolina and farina, *Journal of Food Engineering*, 69, pp 191-198
6. Kraszewski A and Nelson S O (1989), Composite Model of the complex permittivity of cereal grain, *J. Agric. Engng Res.*, 43, pp 211-219
7. Kunii D and Levenspiel O (1991), Fluidization Engineering, Butterworth-Heinemann, Boston
8. Liu S, Wang H G, Jiang F, Yang W Q (2002), A new image reconstruction method for tomographic investigation of fluidized beds, *AIChE J.*, 48, pp 1631-1638
9. Makkawi Y T and Wright P C (2004), Electrical capacitance tomography for conventional fluidized bed measurements—remarks on the measuring technique, *Powder Technology*, 148, pp 142-157
10. Mamishev A V, Takahashi A R, Du Y, Lesieutre B C and Zahn M (2002), Parameter estimation in dielectrometry measurements, *Journal of electrostatics*, 56, pp 465-492
11. McKeen T R and Pugsley T S (2002), The influence of permittivity models on phantom images obtained from electrical capacitance tomography, *Meas. Sci. Technol.*, 13, pp 1822-1830
12. Robinson D A, Gardner C M K and Cooper j D (1999), Measurement of relative permittivity in sandy solids using TDR, capacitance and theta probes: comparison, including the effects of bulk soil electrical conductivity, *Journal of Hydrology*, 223, pp 198-211
13. Sacilik K Tarimci C and Colak A (2006), Dielectric properties of flaxseeds as affected by moisture content and bulk density in the ratio frequency range, *Biosystems Engineering*, 93, pp 153-160
14. Suhas V. Patankar (1980), Numerical heat transfer and fluid flow, Hemisphere Publishing Corp, Washington, DC, USA, pp 67-68
15. Vojtěch V C Sc, Markvart M and Drbohlav R (1966), Fluidized bed drying, Leonard Hill, London

16. Wang H G, Yang W Q, Senior P, Raghavan R S and Duncan S R (2008), Investigation of batch fluidized bed drying by mathematical modelling, CFD simulation and ECT measurement, *AIChE J.*, 54, pp 427-444
17. Warsito W and Fan L S (2003), ECT imaging of three-phase fluidized bed based on three-phase capacitance model, *Chemical Engineering Science*, 58, pp 823-832
18. Yang M (2007), Development of electrical tomography systems and application for milk flow metering, PhD thesis, The University of Manchester, Manchester, UK
19. Yang W Q, Chondronasios A, Nattrass S, Nguyen V T, Betting M, Ismail I and McCann H (2004), Adaptive calibration of a capacitance tomography system for imaging water droplet distribution, *Flow Measurement and Instrumentation,* 15, pp 249-258
20. Yang W Q, Nguyen T, Betting M, Chondronasios A, Okimoto F and McCann H (2002), Imaging wet gas separation process by capacitance tomography, *IS&T/SPIE 14 Symposium; Electronic Imaging*, San Jose, USA, pp 347-358
21. Yang W Q and Peng L H (2003), Image reconstruction algorithms for electrical capacitance tomography, *Meas. Sci. Technol.*, 14, pp R1-R13
22. Yang W Q, Spink D M, York T A and McCann H (1999), An image-reconstruction algorithm based on Landweber's iteration method for electrical-capacitance tomography, *Meas. Sci. Technol.*, 10, pp 1065-1069

149f Tomographic Imaging during Semi-Batch Reactive Precipitation of Barium Sulphate In a Stirred Vessel

Thomas L. Rodgers[1], David Stephenson[1], Michael Cooke[1], Trevor A. York[2], and Reginald Mann[1]. (1) School of Chemical Engineering and Analytical Science, University of Manchester, Po Box 88, Sackville Street, Manchester, M60 1QD, United Kingdom, (2) School of Electrical and Electronic Engineering, University of Manchester, Po Box 88, Sackville Street, Manchester, M60 1QD, United Kingdom

The work reported in this paper has illustrated, for the first time, the capability of three-dimensional EIT to monitor reactive systems under realistic process conditions and scale. The semi-batch precipitation of barium sulphate from barium chloride and sodium sulphate has been investigated in a 200 litre stirred tank reactor. Experiments have been performed using both Rushton and pitched blade turbines under a range of agitation rates with a surface feed of barium chloride. The repeatability of EIT in monitoring the reaction progress has been illustrated. The visualisation of plume structure and evolution during the semi-batch reaction has been examined in 3D. The quantification of the effect of changes in reactor level on reconstructed images has also been examined suggesting this is a limitation to the commonly used finite element approach, as only a 2 % change can cause significant effects.

Time-resolved X-ray tomography of a fluidized bed

R.F. Mudde*
Kramers Laboratorium voor Fysische Technologie
Delft University of Technology
Pr.Bernhardlaan 6, 2628 BW Delft, The Netherlands.

September 15, 2008

Abstract

This paper discusses first results of bubbles moving through a fluidized bed imaged with an X-ray Tomographic Scanner. The scanner is made of 3 medical X-ray sources equipped with 30 $CdWO_4$ detectors each. The fluidized bed has a diameter of 23cm and is filled with Geldart B powder. The scanner measures the attenuation in a thin slice perpendicular to the column axis at a sampling frequency of 2500Hz.
The data collected during 2 seconds are reconstructed using the SART algorithm with a one-step-late correction. The reconstructions show the distribution of the bubbles in the 2-dimensional cross-section. By stacking these images, a 3-dimensional view of the bubbles in the column is represented.

Key words: X-ray tomography, time resolved, spatial resolution, SART, bubbling fluidized bed

1 Introduction

The simplest form of a fluidized bed is a cylinder containing a large quantity of small particles through which gas is flowing upwards. The gas velocity is such that the particles 'float' on the gas stream, creating a gas-particle mixture that has fluid like properties. Our understanding of fluidization is still far from complete, although during the last century computational methods have increased our knowledge significantly. However, due to their opaqueness, fluidized beds create serious problems for experimentalists. The well-known laser based techniques developed for single phase flows are inadequate for fluidized systems. Consequently, experiments on local phenomena are rare. However, solids are to some extend transparent to radiation such as X or γ rays and a variety of nuclear techniques have been developed during the last decade. In (Duduković 2000) the use of nuclear techniques in opaque multiphase systems is reviewed.
Radiation can be used to measure the volume fraction of gas (or equivalent of the solids) in a fluidized bed. If tomographic techniques are used, the volume fraction distribution in a cross section of the fluidized bed can be found. This is referred to as 'nuclear densitometry'. Several papers have appeared in literature, not only dealing with densitometry in fluidized beds, see e.g. (Kumar et al. 1995), (Kumar et al. 1997), (Kumar and Duduković 1997). For instance, (Shollenberger et al. 1997) measured the volume fraction in a 48cm-diameter bubble column, using a single beam-detector pair to perform horizontal, unidirectional scans over a cross-sectional plane assuming cylinder symmetry. (Mudde et al. 1999) followed a similar procedure to measure the solids

*e-mail: r.f.mudde@tudelft.nl

volume fraction in a turbulent fluidized bed. (Kumar et al. 1995) reported the use of a fan beam from a single γ source with multiple detectors. By rotating the source-detector combination around the column about 4000 projections could be taken, resulting in a spatial resolution of 5mm. As the total measuring time is about one hour, a time averaged volume fraction is obtained. This type of tomographic reconstruction is referred to as Computed Tomography or CT. In medical imaging, the use of X-ray tomography is standard practice. The source-detector system again rotates around the patient. This way high resolution, static images can be produced.

An alternative to the nuclear radiation is offered by electro-magnetic fields, like employed in Electrical Impedance Tomography. These EM-methods are fas, giving a high temporal resolution. images can be produced at rates of several hundreds per second(see e.g. (Reinecke and Mewes 1996), (Kühn et al. 1996), (Beck et al. 1998)). However, the EM-techniques work with so-called soft fields: a change in the electro-magnetic properties in one location changes the field everywhere in the domain. This makes reconstruction difficult and the resolution the center of the object is usually relatively poor. Nuclear densitometry relies on hard fields. Hence, they do not suffer from this problem. But, the nuclear techniques are rather slow. Due to inherent noise, the frame-rate (temporal resolution) that can be obtained is usually low.

A very fast X-ray tomographic system has been developed by Hampel and co-workers, see e.g. (Bieberle and Hampel 2006), (Bieberle et al. 2007). They generate the X-rays by moving an electron beam over a Tungsten element. They can create tomographic images up to a rate of 10,000 frames/s. This is evidently showing that nuclear techniques can also be fast.

In the present paper, we focus on the use of X-rays for tomographic imaging of bubbles in a 23cm fluidized bed. Our system consists of 3 X-ray sources that simultaneously radiate each a fan beam through the bubbling bed. The radiation is detected by 3*30 detectors. From the data collected, the solids distribution is reconstructed in the imaging plane. Consecutive images are stacked, giving an impression of the bubbles, their shape and spatial distribution. As far as we know, these are the first data of this kind.

2 Experimental Setup

The fluidized bed is formed by a 23cm inner-diameter tube (perspex, wall thickness 5mm) filled with 0.56mm sized polystyrene particles (st.dev. of particle size distribution is 0.16mm). The powder is of Geldart type B. The polystyrene has a density of $1.102 \cdot 10^3$kg/m^3. The mean density of loosely packed polystyrene particles is 625 kg/m^3. The minimum fluidization and minimum bubbling velocity are virtually identical for this powder: 0.12m/s. In the bubbling regime of fluidization, voids travel upwards through the powder mass. The ungassed height of the bed is 61cm. The air (room temperature) is via a wind box supplied to the bed through a porous plate (sintered bronze). A schematic of a fluidized bed with two bubbles in the measuring plane of the densitometer is given in fig.1.

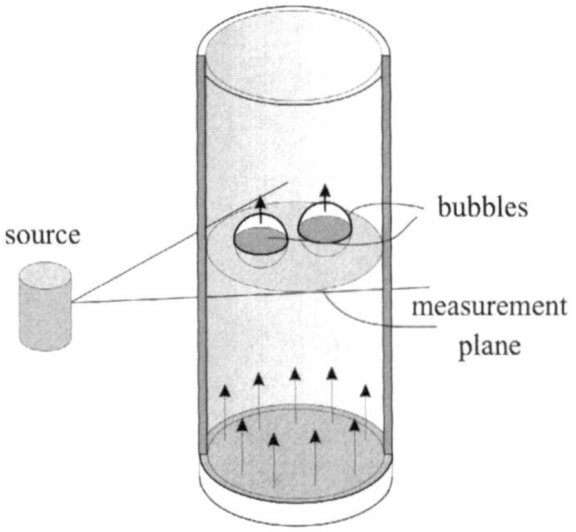

Figure 1: Fluidized bed with two bubbles in the measuring plane of the X-ray densitometer.

The X-ray sources are placed at 120° around the fluidized bed. Three arrays of sensors are placed opposite of the source. A photo of the scanner is shown in fig.(2). The distance from the X-ray-target to the center of the fluidized bed is 68.5cm, from X-ray target to the detectors is 138.6cm. The measuring plane is 50cm above the distributor plate.

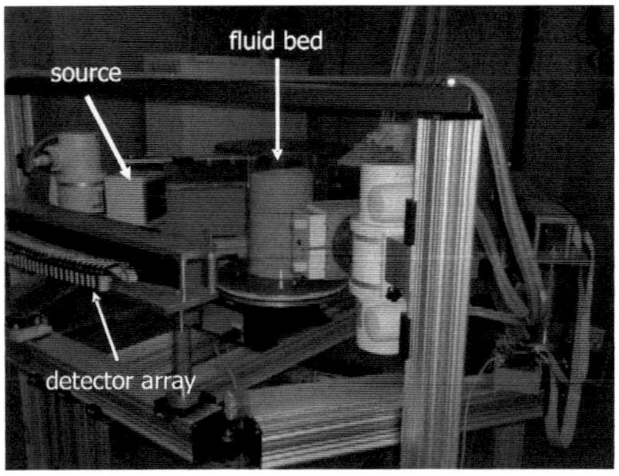

Figure 2: Photo of the tomographic X-ray scanner.

The X-ray source used is of standard medical type with a maximum X-ray energy of 150keV for source 1 and 133keV for the other two. We operate them at the lowest possible energy flux, which is in our case a tube current of 10mA. The detectors all consist of a $CdWO_4$ scintillation crystal optically coupled to a PIN photo diode. They are manufactured by Hamamatsu (type: S 1337 - 1010BR). Their crystal size is 10mm*10mm*10mm.
A curved plastic casing holds a horizontal array of 32 detectors. The curvature of the array is such that the distance to the focal point of the source is equal for all detectors. Of these detectors 30 are used to produce data for tomographic reconstruction, the two outermost detectors are used to monitor the alignment of the object vessel. The data are collected simultaneously at a sampling frequency of 2500Hz. The measured data are read out using a 12 bit ADC-card. Using thin copper plates (thickness varying from 1 to 3mm) in front of the detectors, the radiation incident on the detectors is reduce to a level that maximizes the dynamic range. Without this additional attenuation, detectors measuring rays traveling through the wall region would clip. Moreover, the same could happen to the other detectors if a very large bubble passes the measuring plane. The entire process is controlled via a workstation that send out the trigger signals to the sources and reads out the detectors.

3 Tomographic Reconstruction

3.1 Measuring Principle & Calibration

Each detector measures the attenuation of the X-rays in a thin cone. In the reconstruction this is treated as a thin line. If such a line if mono-energetic γ- or X-rays is transmitted through a closed system containing a particle-gas two-phase mixture, the number of photons registered per second, R, follows from the Lambert-Beer law:

$$R = R_0 exp\left[-\left((1-\alpha)\mu_p + \alpha\mu_g\right) d\right] \quad (1)$$

where R_0 is the number of photons registered per second when the system is in vacuum; μ_p and μ_g denote the linear absorption coefficient of the particle and gas phase; α is the volume fraction of the gas phase; d is the inner diameter of the system. It should be noted that the attenuation characteristics of the fluidized bed wall is incorporated in R_0. Both μ_p and μ_g are functions of the photon energy E. An X-ray source generates a wide spectrum of photon energies. In order to deal with this, we have calibrated all detector individually by placing various amounts of packed powder in between a source and its detectors. This way a calibration curve is made, of the form $A_{cal} + B_{cal} \cdot exp(-x/C_{cal})$, with x the distance traveled by the beam through the powder phase. Using this calibration, the data measured in an experiment can be converted to a measured distance, x_{meas}, the beam travels through the powder mass.

Tomographic Reconstruction
Various reconstruction techniques exist. For instance, the Filtered Back Projection algorithm can be used. A second class is formed by the algebraic techniques, which are iterative and reconstruct the object function on a discretized domain. Although significantly slower, algebraic methods offer more flexibility in terms of limited data sets and are more appropriate for the CT configuration system under consideration here. Detailed accounts of reconstruction techniques can be found in (Brooks and DiChiro 1976), (Herman 1980) or (Kak and Slaney 1988).

We would like to reconstruct the gas fraction, or equivalently the solids fraction as they add up to one. We use the calibration curve to convert the measured line-averaged attenuation into a line averaged solids fraction. We will reconstruct the solids fraction $\alpha(x, y)$ in a pixel representation of the cross-section of the fluidized bed. The number of pixels is an independent parameter that we can select. A good value for our case is a square pixel array of 65*65 pixels. The circle representing the cross-section of the fluidized bed exactly fits in this square. All pixels outside the circle have a solids fraction of zero. For a given ray, traveling through the object, the total solids fraction on the line, p_i, referred to as *ray sum*, can be estimated as

$$\tilde{p}_i = \sum_{k=1}^{N} W_{ik}\alpha_k \tag{2}$$

with α_k the pixel-based value of the solids fraction distribution and W_{ik} the weighing factor for pixel k for the ith ray through the object. We use a linear weighing matrix W. Hence, the weighing factor W_{ik} is the length of ray i through pixel k (see fig.(3)).

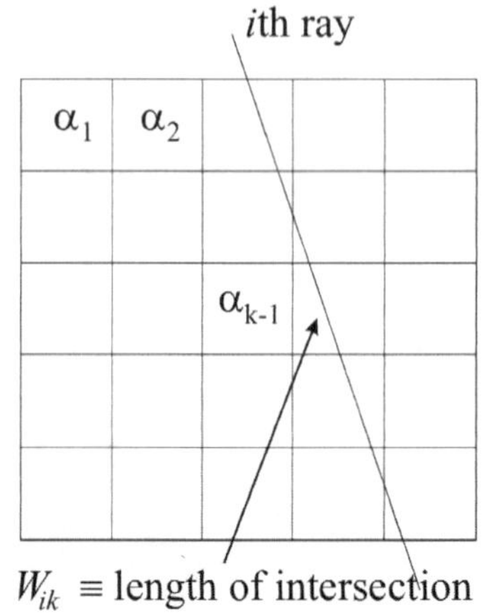

Figure 3: coordinate system

To reconstruct the image we need to solve the unknown pixel-averaged solids fraction α_k from eq.(2) for M different rays on N pixels. As the number of independent measurements is only 3*30=90 and the number of unknown pixels is easily 1000 or more, the problem is ill-posed. Moreover, there will be measuring noise in the data. We write formally:

$$\vec{p} = \mathbf{W} \cdot \vec{\alpha} + \vec{\epsilon} \tag{3}$$

where $\vec{\epsilon}$ contains all errors. The Algebraic Reconstruction Techniques (ART, see e.g. (Brooks and DiChiro 1976)) are designed to minimize the mismatch between the data $\vec{p}$ and $\mathbf{W} \cdot \vec{\alpha}$. They are iterative methods that solves $\vec{p} = \mathbf{W} \cdot \vec{\alpha}$. We use the Simultaneous Algebraic Reconstruction Technique (SART) (Andersen and Kak 1984). Instead of sequentially updating the pixels on a ray-by-ray basis, SART simultaneously applies to a pixel the average of the corrections generated by all rays. This offers a reduction in the amplitude of the salt and pepper noise that is usually present in ART. However, it goes at the expense the computation time. The SART algorithm is given by:

$$\alpha_k^{n+1} = \alpha_k^n + \frac{w}{W_{+,k}} \frac{p_i - \tilde{p}_i}{W_{i,+}} W_{ik} \tag{4}$$

with $W_{+,k} = \sum_{i=1}^{N} W_{ik}$, $W_{i,+} = \sum_{k=1}^{N} W_{ik}$ and w a relaxation parameter that should be between {0,2}; we have set it to 1. Originally, SART is part of the ART family. However, recently (Jiang and Wang 2003) have shown that SART is a maximum likelihood estimator for the mapping of α in case the error in eq.(3) is purely stochastic with a Gaussian distribution.

Still, pepper and salt noise will be present in the images. This can be reduced by using a so-called one-step-late algorithm (see (Green 1990)). We invoked an algorithm based on the median root function (suggested first by (Alenius and Ruotsalainen 1997)). It effectively removes pepper and salt noise, but keeps the edges of larger objects sharp enough. In the reconstruction algorithm an extra step is added to eq.(4):

$$\alpha_{OSL,k}^{n+1} = \frac{1}{1 + \beta \frac{\alpha_{OSL,k}^{n} - Med(\alpha_{OSL,k}^{n})}{Med(\alpha_{OSL,k}^{n})}} \alpha_k^{n+1} \tag{5}$$

with α_k^n the value of pixel k after the nth SART step and $\alpha_{OSL,k}^n$ the same after the nth one-step-late correction.

The median filter $Med(\alpha_k)$ sets the value of the kth pixel equal to the median of the pixel values contained in the kth pixel neighborhood. The definition of this pixel neighborhood is the only parameter of the median filter. We use a 3×3 neighborhood; a 5×5 neighborhood causes too much smearing. The parameter β in eq.(5) again controls the weight of the correction.

4 Experimental Results

4.1 Calibration

All detectors are individually calibrated. For each calibration point a known amount of packed solid particles is placed inside the fluidized bed. This includes a completely empty and a completely filled bed. Seven different points are measured this way. The X-ray sources are switched on for one second and the data collected are averaged. An exponential curve is fitted through the averaged data points. An example of a calibration curve and the corresponding data is given in fig.(4).

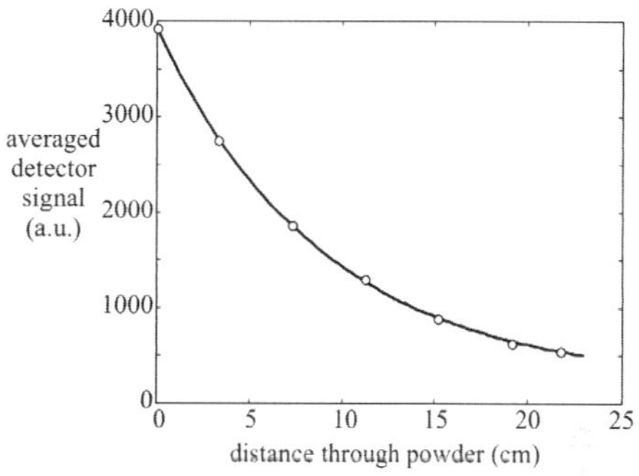

Figure 4: Calibration curve of one of the detectors with its corresponding beam going almost through the bed center

4.2 Pre-processing of tomographic data

The X-ray sources all three suffer from a ripple on their high-voltage supply. This shows up in the measured data as a saw tooth behavior of the data in time. The frequency of this saw tooth depends on the source: for source one it is about 400Hz, for the other two around 90Hz. In principle, the data can be corrected for this fluctuation in source strength, e.g. by using the two outer detectors of each detector array. They measure outside the fluidized bed and thus directly see the fluctuations of their source. In the present work, we have averaged the data first over one period of the 28 data points and moved in steps of 10 time steps through the data series for each new reconstruction. This obviously reduces the time resolution of the images from 2500 frames per second to about 100 frames per second. This is, however, not a principle reduction of the time resolution of the tomographic scanner, but for the moment a practical solution. The averaged data are than, using the calibration curves, in a few steps converted to a line-averaged solids fraction as measured by a detector. Note that the thickness of the perspex walls of the vessel is via the calibration removed from the data.

4.3 Results

A fluidization experiment is run in the bubbling regime. Experiments are run at a superficial gas velocity of 1.3 and 1.6 times U_{mf}. data are taken for a period of 2 seconds. This generates potentially 5000 images. However, due to the pre-processing, only 500 images are reconstructed that are not fully independent. The number of independent images is 180. In fig.(5) the response of one of the detectors is shown. The corresponding path for the X-rays is almost through the center of the bed. Clearly, the passage of bubbles is recorded. The base line has the same value as the minimum fluidized bed case.

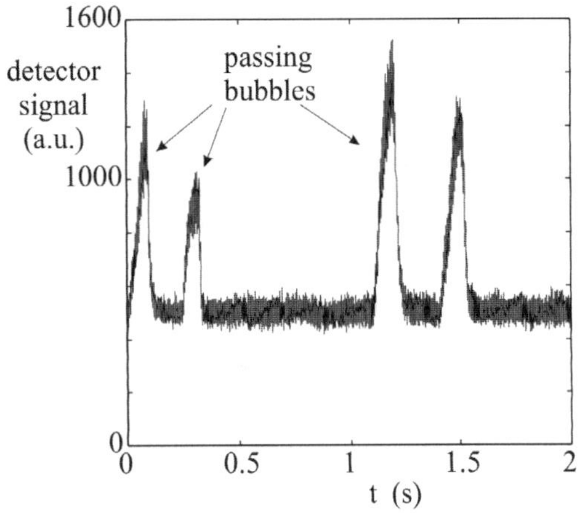

Figure 5: Raw signal of one of the detectors ($U_{sup} = 1.3U_{mf}$
. The corresponding X-rays travel almost through the center of the bed.

The raw signals are pre-processed as described above and fed to the reconstruction algorithm. Each reconstruction starts from an initial guess: a uniform fluidized bed with no bubbles. On a reconstruction grid of 65*65 pixels, a single image reconstruction takes about 7 seconds of computer time (programs are written in FORTRAN and run on a single PC with processor speed of 1.2GHz). Figure (6) shows an image cutting through a bubble with a diameter of about 8cm. This image is a filtered version of the corresponding raw image. A Gaussian filter (the filter kernel is based on a 3*3 pixel representation) is used for this.

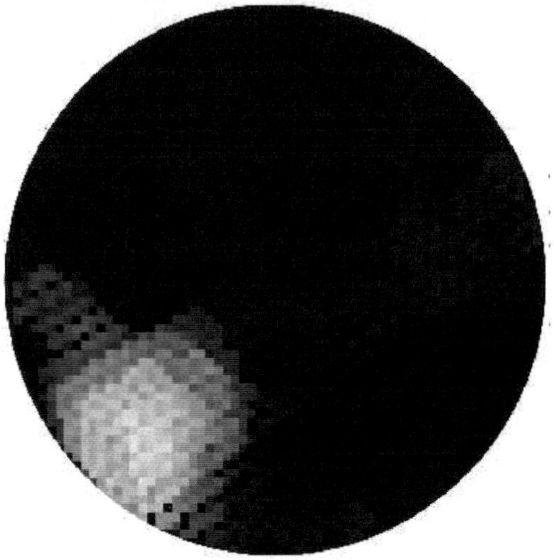

Figure 6: Image showing the passage of a bubble.

In the figure, a 'tail' can be observed going from the bubble towards the upper right corner of the figure. This is an artifact of the reconstructions and most likely caused by the use of only three sources. From simulations (Mudde et al. 2005) we found, that a 5 source system suffers less from this effect. This artificial tail is taken out by thresholding the images: all pixel values below 125 are set to zero. The images have gray values ranging from 0 (black) to 255 (white).

The present set-up does not allow for velocity measurements. That can be relatively easy made by stacking a second ring of detectors above the present ones. This way a second measuring plane is formed and velocity measurements are possible. The two planes will not be exactly parallel, but slightly diverse making the interpretation less straight forward. This is a step for future work. Here, we have estimated the velocity of the bubbles to be about 50cm/s (a single bubble with an equivalent diameter of 6cm has a rise velocity of this magnitude according to the Davidson relation $v_b = 0.71\sqrt{gd_{eq}}$ with g the acceleration of gravity and d_{eq} the equivalent diameter of the spherical-cap shaped bubble). The time interval between two consecutive images we have reconstructed is 1/250s. The bubble shown in fig.(6) is visible in 20 images, corresponding to a time interval of 80ms. Combined with the estimated velocity, we conclude that the bubble is 4cm in height. We have stacked the images giving a quasi 3-dimensional picture of the flow in the column. The scaling is chosen such that the bubble of fig.(6) are about 4cm high. This is shown in fig.(7)

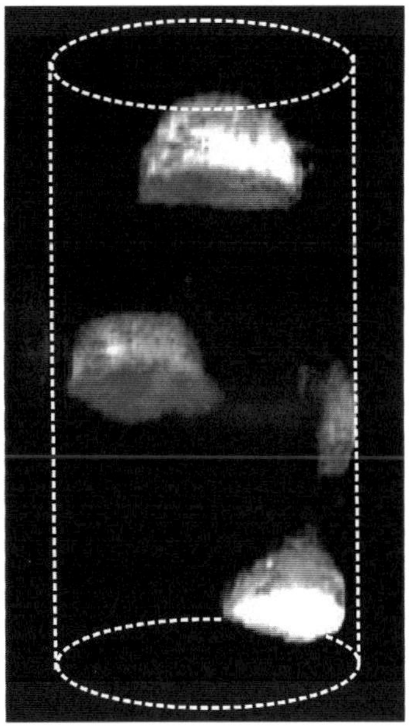

Figure 7: Stacked image showing four bubble. The dotted lines represent the column of the bed. Superficial velocity $\approx 1.3U_{mf}$.

The bubbles are clearly visible: they are of the spherical cap type with flat bottoms. One smaller bubble, most to the right) is moving along the column wall. The entire figure is 168 images, which is equivalent to 0.67s.

A similar experiment is run at $1.6U_{mf}$. Again 2 seconds of data are recorded and reconstructed. The bubbles captured are shown in fig.(8). As expected, the bubbles are larger. The vertical trace of gray pixels is an artifact of the reconstruction and probably caused by noise.

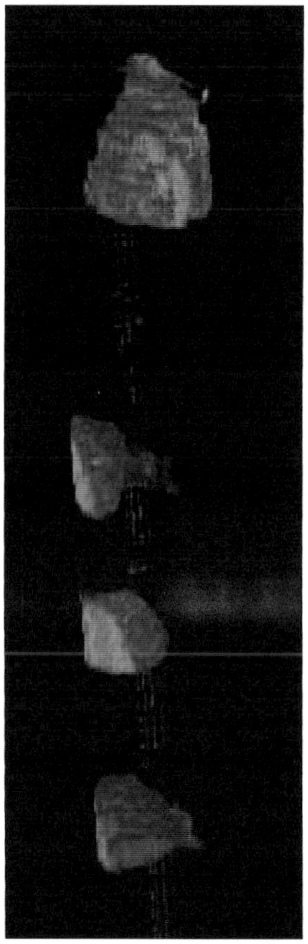

Figure 8: Stacked image showing four bubble. The three lower bubble move along the column wall. Superficial velocity $\approx 1.6U_{mf}$.

Concluding Remarks

In this paper, we present first results of the bubbles in a fluidized bed made with an X-ray based tomographic scanner. The scanner consists of three standard medical X-ray tubes with 30 detectors containing $CdWO_4$ scintillator crystals. The data are collected at a frequency of 2500Hz. Due to a voltage ripple in the sources the full speed can not be used. Therefor the data are averaged over 28 images and every 10 images a new average is calculated. Thus, the effective frequency of the results presented is 250Hz. This, however, not a principle limitation.

We have reconstructed the images on a 65*65 grid with a pixel size of 3.5mm. the bed diameter is 23cm and clearly spherical cap like bubbles are found. We conclude that an X-ray based tomographic scanner has lots of potential to investigate the motion of bubbles ion a fluidized bed in a tomographic way.

References

Alenius, S. and U. Ruotsalainen (1997). Bayesian image reconstruction for emmission tomography based on median root prior. *Eur. J. of Nucl. Med. 22*.

Andersen, A. H. and A. C. Kak (1984). Simultaneous algebraic reconstruction technique (sart): a superior implementation of the art algorithm. *Ultrasonic Imaging 6*, 81–94.

Beck, M. S., T. Dyakowski, and R. A. Williams (1998). Process tomography - the state of the art. *Trans. Inst. Meas. and Control 20(4)*, 163–177.

Bieberle, M., F. Fischer, E. Schleicher, U. Hampel, D. Koch, K. Aktay, H.-J. Menz, and H.-G. Mayer (2007). Ultrafast limited-angle-type x-ray tomography. *Appl. Phys. Lett. 91*, 123516/1–123516/3.

Bieberle, M. and U. Hampel (2006). Evaluation of a limited angle scanned electron beam x-ray ct approach for two-phase pipe flows. *Meas. Sci. Technol. 17*, 2057–2065.

Brooks, R. A. and G. DiChiro (1976). Principles of computer assisted tomography (cat) in radiographic and radioscopic imaging. *Phys. Med. Biol. 21*, 689.

Duduković, M. P. (2000). Opaque multiphase reactor: Experimentation, modeling and troubleshooting. *Oil & Gas Sci. and Techn. - Rev. IFP 55*, 135.

Green, P. J. (1990). Bayesian reconstruction from emmission tomography data using a modified em algorithm. *IEEE Trans. on Med. Imag. 9*, 84–93.

Herman, G. T. (1980). *Image reconstruction from projections - the fundamentals of computerized tomography*. Academic Press.

Jiang, M. and G. Wang (2003). Convergence of the simultaneous algebraic reconstruction technique (sart). *IEEE Trans. on Image Proc. 12*, 957–961.

Kak, M. and M. Slaney (1988). *Principles of computerized tomographic imaging*. New York: IEEE Press.

Kühn, F. T., J. C. Schouten, R. F. Mudde, C. M. Van den Bleek, and B. Scarlett (1996). Analysis of chaos in fluidization using electrical capacitance tomography. *Meas. Sci. & Techn. 7*, 361.

Kumar, S., D. Moslemian, and M. Duduković (1995). A gamma ray tomographic scanner for imaging void fraction distribution in bubble columns. *Flow. Meas. Instr. 6*, 61.

Kumar, S., D. Moslemian, and M. Duduković (1997). Gas holdup measurments in bubble columns using computed tomography. *AIChE J. 43*, 1414.

Kumar, S. B. and M. P. Duduković (1997). *Computer assisted gamma and X-ray tomography: applications to multiphase flow systems*, In: *Non-invasive Monitoring of Multiphase Flows, Chaouki, J. and Larachi, F. and Duduković, M.P. (eds.), Ch.2.*, pp. 43–103. Elsevier.

Mudde, R. F., P. R. P. Bruneau, and T. H. J. J. Van der Hagen (2005). Time-resolved gamma-densitometry imaging within fluidized beds. *Ind. & Eng. Chem. Res. 44*, 6181–6187.

Mudde, R. F., W. K. Harteveld, H. E. A. Van den Akker, T. H. J. J. Van der Hagen, and H. Van Dam (1999). Gamma radiation densitometry for studying the dynamics of fluidized beds. *Chem. Eng. Sci. 54*, 2047.

Reinecke, N. and D. Mewes (1996). Recent developments and industrial research applications of capacitance tomography. *Meas. Sci. & Techn. 7*, 233.

Shollenberger, K. A., J. R. Torczynski, D. R. Adkins, T. J. O'Hern, and N. B. Jackson (1997). Gamma-densitometry tomography of gas holdup spatial distribution in industrial-scale bubble columns. *Chem. Eng. Sci. 52*, 2037.

157a Improvement of Granular Mixing In Rotating Cylinders

Suman K. Hajra, Tathagata Bhattacharya, and J. J. McCarthy. Department of Chemical and Petroleum Engineering, University of Pittsburgh, 1249 Benedum Hall, Pittsburgh, PA 15261

We have found that periodic flow inversions of free-flowing materials may be used to effectively eliminate both size and density segregation in free-surface flows. In a rotating cylinder, these inversions may be induced by placement of an axially located baffle within the flow. In this work, we continue our experimental study of radial mixing and segregation for binary mixtures of granular materials in a rotating cylinder both with and without an axially-located baffle. Qualitative and quantitative measures of the segregation/mixing are shown via digitized images and the intensity of segregation, respectively. We define a mixing improvement factor (Ψ) to quantify the relative impact of the axial baffle for a variety of baffle placements and geometries. We find that the maximum extent of mixing is achieved when the baffle size is equal to the radius of the cylinder and fixed either at the free surface or within the shear layer. In contrast, we show that segmenting the baffle, baffle placement outside the shear layer (in fixed bed), or placement at the periphery of the cylinder resulted in an insignificant increase in mixing. Mixing was found to be independent of the shape of a baffle.

157b Using Dem to Identify Mixing Mechanisms In Bladed Mixers

Stefan Radl[1], Benjamin J. Glasser[2], and Johannes G. Khinast[1]. (1) Institute for Process Engineering, Graz University of Technology, Inffeldgasse 21a, Graz, A-8010, Austria, (2) Rutgers University, Department of Chemical and Biochemical Engineering, 98 Brett Road, Piscataway, NJ 08854

Flow and aggregation of powders is of central interest for applications like powder mixing, granulation or drying in the pharmaceutical industry. Despite intensive research has been carried out in this field, the description of mixing in that type of equipment is still a great challenge. This is because experimental techniques to measure powder flow and mixing quality have still significant limitations. Especially the complex flow in equipment with rotating internals, e.g. bladed mixers, is poorly understood from a theoretical point of view.

By using specialized simulation techniques it is possible to get a deep insight into granular flow and mixing. In our work we have used the discrete element method (DEM) to simulate the flow and mixing in a bladed mixer. The main goal was to investigate the effects of powder moisture and stirrer speed on mixing mechanisms in a wet powder bed. The method is used to explore phenomena that have been observed in the experimental work of Lekhal et al.[1]

By using sophisticated particle interaction models we were able to study both dry and wet bulk solid flow behavior. Because DEM is currently limited to a relatively small number of particles, we have used dimensionless quantities to effectively simulate the motion of the powder bed. The cohesive forces have been characterized by the granular Bond number Bo_g and the Collision number Co recently introduced by Li and McCarthy.[2] The results show that mixing in radial direction is strongly influenced by the powder moisture. However, the integral mixing quality in the mixer seems to be quite unaffected by the water content of the powder. Using the detailed information from our simulations we were able to investigate the mechanism behind this phenomena. Hence, we present the differences in granular diffusion coefficients between dry and wet powders and their effect on mixing pattern.

(1) Lekhal A, Conway SL, Glasser BJ, Khinast JG. Characterization of granular flow of wet solids in a bladed mixer. AIChE Journal. 2006;52:2757-2766.

(2) Li HM, McCarthy JJ. Cohesive particle mixing and segregation under shear. Powder Technology. 2006;164:58-64.

157c Understanding Mixing to Enhance Coating Performance In a Pancoater

Ekneet Sahni, Department of Pharmaceutical Sciences, University of Connecticut, 69 North Eagleville Road, Storrs, CT 06269, Robert Yau, Department of Chemical Engineering, University of Connecticut, 69 North Eagleville Road, Storrs, CT 06269, and **Bodhisattwa Chaudhuri**, Pharmaceutical Sciences, University of Connecticut, 69 North Eagleville Road, Storrs, CT 06269.

Coating of particles (tablets) is done to mask the unpleasant taste or odor of the drug, to control the bioavailability of the API, and to increase shelf-life. The coating solution is sprayed in specific locations of the granular bed and uniformity of coating is achieved by interparticle collisions and overall mixing behavior. Thus, the knowledge of particle flow and mixing in a pan coater is critical to optimize the design and operation of coating equipment. Our study focuses on the fundamental mechanisms of flow, mixing and their relationship to coating performance.

In a lab scale oblate spheroid coater, mixing studies have been performed with different colors of lactose nonpareils. Discrete pocket samplers are used to collect samples as a function of time and space. Discrete Element Method (DEM) based model is also developed to simulate the flow, mixing and coating of particles. Digitally recorded mixing states from experiments are used to fine tune the model. A parametric study is subsequently performed to quantify the effect of vessel speed, tilt of the vessel, fill level on mixing and coating behavior. Similar to the experiments, our model predicts that the axial mixing enhances with the increase in tilt of the coater and that there is slowest axial mixing at the vertical orientation of the coater. In general, the radial convection is estimated to be faster than axial dispersion transport. The speed of the rotating vessel has no prominent effect on the rate of mixing in both the experiments and simulations. Vessel fill level has no effect on the rate of mixing.

At the optimal mixing conditions, a black coating dye is sprayed at different levels of concentrations and flow rates to a granular bed of white nonpareils to study the coating behavior. Coating on the particles is measured by the change in diameter of the particles using vernier calipers. The coating performance is found to be directly proportional to the concentration of the dye and the spray rate. The higher tilt of the vessel, lower vessel speed and higher spray rate enhances coating performance. Moreover, numerical simulation provides the residence time distribution of all the particles passing through the spray zone.

157d Using Dem Models to Understand Segregation In Mixtures with Continuous Grain Size Distribution In Long Cylinders

Atul Dubey[1], Fernando J. Muzzio[1], and Maria. S Tomassone[2]. (1) Chemical and Biochemical Engineering, Rutgers University, 98 Brett Road, Piscataway, NJ 08854, (2) Department of Chemical and Biochemical Engineering, Rutgers University, 98 Brett Road, Piscataway, NJ 08854

Understanding granular flow dynamics has been and continues to remain an interesting problem for engineers and physicists. Granular materials attract special attention due to their ability to behave either as solids, as liquids, or neither - depending on the surrounding conditions. Segregation of granular particles has been of particular interest mostly due to its undesirable role in mixing processes although segregation can be desirable in certain applications. The quality of end products in many industrial applications such as pharmaceutical drug manufacturing depend on efficient mixing of ingredients of various shapes, sizes, densities, charge distributions etc. However, mixtures of materials like pharmaceutical excipients and active pharmaceutical ingredients (API) segregate in a variety of ways such as based on size, density, charge, shape etc. The most common modes of segregation encountered are size and charge based segregation. A number of studies have been performed with an attempt to enhance the understanding of particle movement in size-differentiated mixtures. It has been observed that segregation depends on the geometry and environment. Hence it has been studied in relation to specific geometries, for example, in a mixture flowing from a chute, or when being agitated by a moving bar. Study of mixing in long cylindrical geometries is of great interest. In some industrial applications, it is desirable to mix the ingredients slowly without much agitation from baffles etc. Size segregation is seen in the form of banding in such cases. There is both radial as well as axial segregation.

Computational methods using discrete element modeling and simulation (DEM) have been shown to be very useful in complementing experimental analyses in this field. However due to limitations in computational power and effective parallel algorithms, either the particle to vessel diameter ratios had to be kept high or extremely low fill levels had to be used, or the size differential among the particles had to be kept unrealistically high. These constraints, including the limitation on the number of particles that can be simulated (usually kept around 10,000) have limited the scope of computational study in this field. However with increasing computational power and more efficient algorithms becoming accessible some of these limitations have largely been reduced. In this study we apply DEM techniques using a fast, parallel algorithm to simulate the mixing and segregation in a 100,000 particles in a long cylinder with length to diameter ratio of 2.5 and cylinder to particle diameter ratio of 220. The size differential among the particles was kept according to a more realistic normal size distribution with a 20% standard deviation instead of a binary size distribution which was found to segregate very easily. Further studies with mixtures of particles of non-spherical shapes were also performed. The effect of cohesion, rotational speed, wall effects and their absence due to periodic boundary conditions, and fill level were also studied. A summary of the results will be presented with some more insights into the modes of segregation and the effect of the above mentioned variables on it along with a discussion on ways to prevent undesirable particulate segregation.

Mixing of Nanosized Particles by Magnetically Assisted Impaction Mixing in Dry and Fluid Suspension

James V. Scicolone, **Ajit Mujumdar** and **Rajesh Dave**, New Jersey Center for Engineered Particulates, New Jersey Institute of Technology, Newark, NJ 07102

Abstract

The mixing of nano-scale particles using a novel dry mechanical mixing technique called magnetically assisted impaction mixing (MAIM) has been studied experimentally in this work. Intensity of Segregation was evaluated at the micron length scale based on field emission scanning electron microscopy (FESEM) images coupled with the energy dispersive x-ray spectroscopy (EDS). In order to achieve the homogeneous mixture of nano-particles, MAIM process was optimized by studying the effects of magnet-to-sample ratio (1:2, 1:1, 2:1, 5:1, 10:1), processing time (5 to 120 mins), magnet size (size ranges of 2360 to 1700 μm, 1400 to 850 μm, and 1000 to 600 μm), and constituents of the mixture (binary systems of SiO_2+TiO_2 and SiO_2+Al_2O_3).

Introduction

Powder mixing is an important and extensively researched area in many industrial sectors such as chemicals, pharmaceuticals, food, cosmetic, ceramic, and electronics during the past several decades [1-2]. However, there are serious challenges faced in handling and homogeneous mixing due to high cohesion and tendency to form large, difficult to break agglomerates that form due to strong inter-particle forces. In fact, conventional methods for powder mixing cannot mix at scales smaller than about a few microns or even hundreds of microns because they fail to break the primary aggregates [3-8].

The focus of this paper is on the Magnetically Assisted Impaction Mixing, which is an environmentally benign mixing technique. In MAIM, a magnetic field is created from the surrounding electromagnetic coil and the magnetic particles undergo agitation. The magnetic particles undergo rotational and translational motion, inside the container, creating a fluidized state for the nanoparticles. Magnetic particles collide with the agglomerates of nanoparticles, and other magnetic particles or the walls of the container, transferring the energy from the generated momentum. It is believed that the collisions between magnetic particles and the agglomerates under appropriate operating conditions should contain enough energy to deagglomerate the nanoparticle agglomerates and promote mixing.

A LEO 1530 VP Field Emission Scanning Electron Microscope equipped with an Oxford UTW X-ray detector was used to obtain quantitative analysis of the Intensity of

Segregation. Two representative areas of 55 x 40 μm at 5000 times magnification were randomly chosen on both the tablet surfaces. The EDS was used to obtain normalized compound composition from 100 points, arranged in a 10 x 10 grid, for each of the areas. Four hundred total points obtained for each mixture were used to calculate average concentrations (μ_a and μ_b) and variance (σ^2). As a basis to compare the homogeneity of a mixture, the Intensity of Segregation, a dimensionless number, is employed in this work. Intensity of Segregation, which was originally developed by Danckwerts [9], is calculated by dividing the variance by the two mean values of each component compound percent. The Intensity of Segregation for a perfect random mixture would be 0, while for completely unmixed mixtures the Intensity of Segregation it would be 1.

Results

Mixing can be controlled primarily through the processing time and the amount of magnets. Figure 1 is a graph of intensity of segregation versus time for multiple magnet-to-sample weight ratios. When more magnets are introduced into the system, and a longer processing time, the intensity of segregation decreased, indicating a more dispersed mixture. When graphing the intensity of segregation versus the combination of time and the number of magnets used in each of the conditions, the data points fall together, forming a single line, Figure 2. The same trend occurs when comparing different sizes of magnets, in Figure 3.

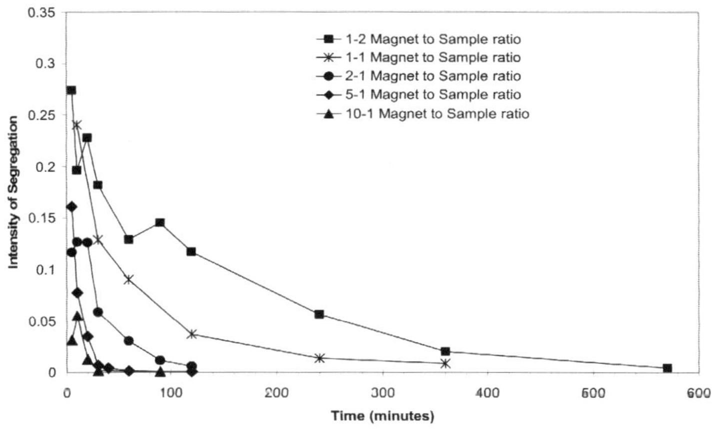

Figure 1. Intensity of Segregation versus time for various mixing times for SiO_2/TiO_2 mixtures and a magnet size range of 1400-850 microns.

Powders that are more cohesive compact on the bottom of the container forming a cake, while the magnets remain on top. In this situation, only limited mixing can take place before there is no longer interaction between the constituents and the magnets. In such cases, we looked into mixing the constituents in a liquid medium, water. Studying the same magnet-to-sample weight ratios, similar trends can be seen with the wet mixing as with the dry mixing. When comparing Intensity of Segregation with time and number of magnets, the different magnet-to-sample weight ratios condensed into a single line. The wet mixing data is presented in Figure 5. By controlling the mixing time and the number of magnets used, a specific intensity of segregation can be obtained referring to approximate agglomerate sizes in the mixture.

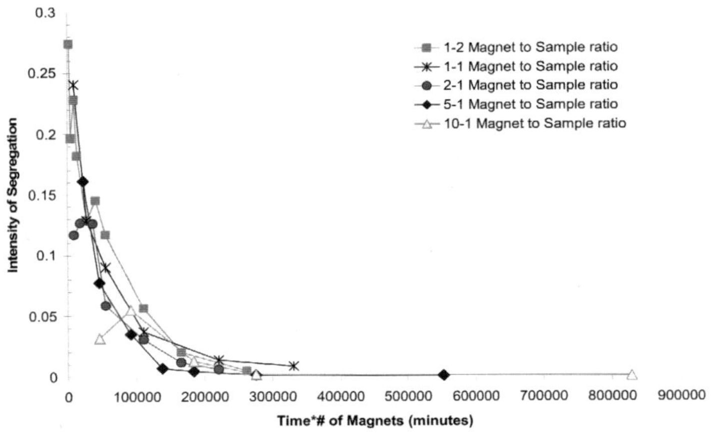

Figure 2. Intensity of Segregation versus time*mass of magnets/mass of powder sample for the magnet to sample ratios of 1:2, 2:1, 5:1, and 10:1 with a magnet size range of 1400-850 microns for SiO_2/TiO_2 mixture.

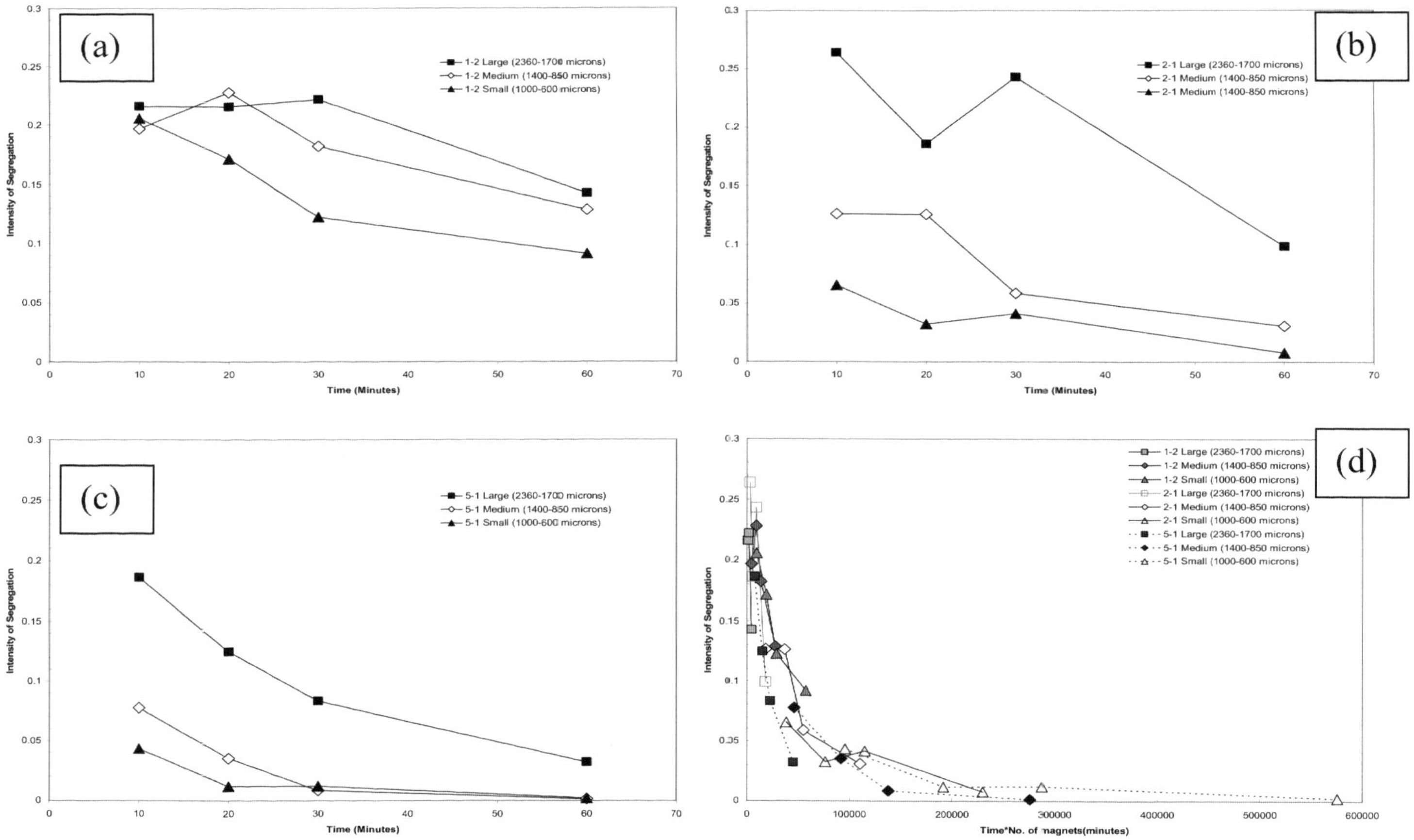

Figure 3. Intensity of Segregation versus time for SiO_2/TiO_2 mixtures with various magnet sizes and a magnet-to-sample ratio of (a) 1 to 2, (b) 2 to 1, (c) 5 to 1. Intensity of Segregation versus time*number of magnets for magnet-to-sample ratios of 1:2, 2:1, and 5:1 (d).

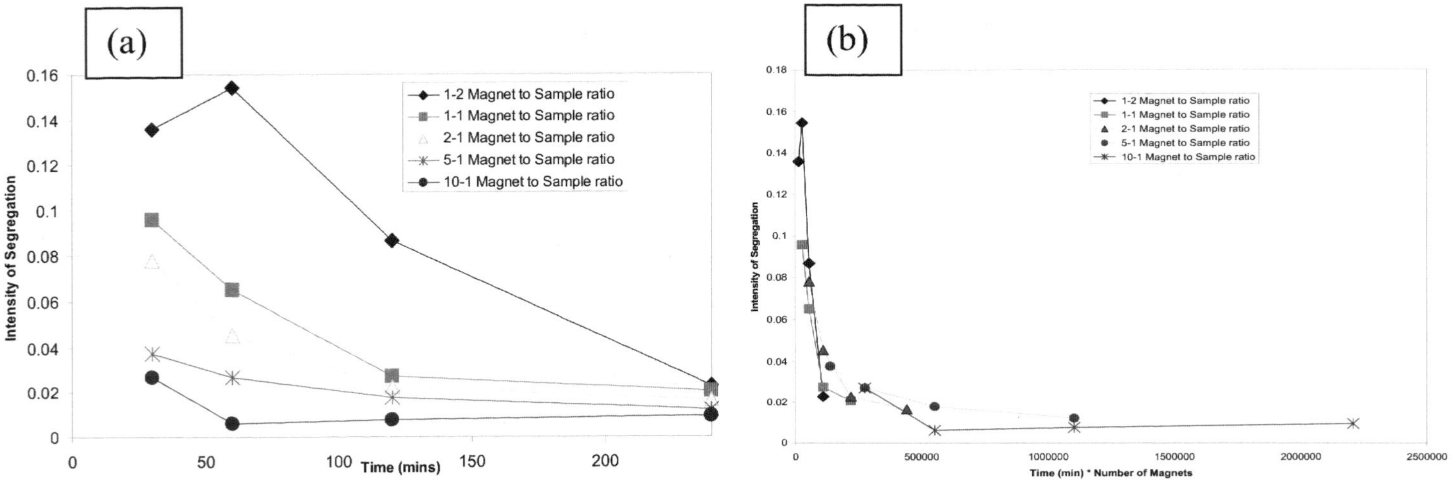

Figure 5. Intensity of Segregation versus time for magnet-to-sample ratios of 1:2, 2:1, and 5:1 with a magnet size range of 1400-850 microns for wet mixtures of SiO_2/TiO_2 (a). Intensity of Segregation versus time*number of magnets for magnet-to-sample ratios of 1:2, 2:1, and 5:1 with a magnet size range of 1400-850 microns for wet mixtures of SiO_2/TiO_2 (b).

Reference:

[1] Wei D, Dave RN, Pfeffer R. "Mixing and characterization of nanosized powders: An assessment of different techniques", Journal of Nanoparticle Research, 4 1-2, pp. 21-41, 2002.

[2] Yang J, Wang Y, Dave RN, Pfeffer R. "Mixing of nano-particles by rapid expansion of high-pressure suspensions", Advanced Powder Technology, 14 (4), pp. 471-493, 2003.

[3] Venables H and Wells JI. "Powder mixing", Drug Development and Industrial Pharmacy, 27(7), pp. 599-612, 2001.

[4] Van der Wel, Peter G. J., "Powder Mixing", Powder Handling Process., 11(1), pp. 83-86, 1999.

[5] Peciar, Marian et al., "Mixing of finely powdered materials", Chem. Prum., 42(5-6), pp. 124-6, 1992.

[6] Naganuma, Noriyuki et al., "Mixing powder with liquid", Netsu Shori., 31(2), pp. 95-9, 1991.

[7] Vaizoglu, Orhan et al., "Assessment of the degree of mix of powder mixtures", Turk. J. Phys., 23(1), pp. 97-104, 1999.

[8] Kaye BH. "Powder Mixing", Powder Technology Series, Chapman & Hall, pp. 19-35, 77–131, 1997.

[9] Danckwerts PV, "The definition and measurement of some characteristics of mixtures", Appl. Sci. Res. A3, pp. 279-296, 1952.

157f Simulation of a Continuous Granular Mixer: Effect of Operating Conditions on Flow and Mixing

Avik Sarkar and Carl R. Wassgren. School of Mechanical Engineering, Purdue University, 500 Central Drive, Potter 209-A, West Lafayette, IN 47906

Continuous particulate blending has received much attention in the recent decade as it provides a more economical and efficient alternative over batch blending for large scale manufacturing. The influence of operating conditions in continuous blenders is not well understood quantitatively, even though qualitative parallels can be drawn with batch blending for similar geometries. This work investigates the influence of fill level and impeller rotation rate (Froude number) in a horizontal bladed continuous blender using the discrete element method (DEM). Particle flow within the blender was found to be strongly dependent on the Froude number and fill level, with distinct fluidized and quasi-static regimes at smaller fills. The axial flow rates showed significant variation with Froude number and fill, and also showed considerable variation over the course of a shaft revolution. Favorable mixing was obtained at smaller impeller rotation rates for larger fills, but at larger impeller rotation rates for smaller fills.

157g Nano-Mixing of Dipyridamole Drug and Excipient Nanoparticles by Sonication in Liquid CO2

Ganesh Sanganwar and Ram B. Gupta. Chemical Engineering, Auburn University, Auburn, AL 36849-5127

Nanoparticles (about 200 nm thick and 600-12000 nm long flakes) of dipyridamole, a poorly-water-soluble anti-thrombosis drug, are produced by supercritical antisolvent solvent with enhanced mass transfer (SAS-EM) method. Mixing of drug nanoparticles (in order to prevent reagglomeration during storage and to improve the handling properties) is carried out with three different excipients: silica nanoparticles, lactose microparticles, and polyvinylpyrrolidone (PVP) nanoparticles. To intimately mix at nanoscale, macro mixtures of dipyridamole and excipient particles are sonicated in liquid carbon dioxide. Effect of ultrasonic energy, amplitude, and component weight ratio were studied for binary mixtures. Qualitative analysis of deagglomeration and mixing of drug/silica, drug/PVP, and lactose/silica mixtures were done using scanning electron microscopy (SEM) while drug/lactose mixture was analyzed using both SEM and blend homogeneity (i.e., relative standard deviation in the drug concentration). Upon mixing, the handling properties are significantly improved as measured by compressibility index and Hausner ratio. Liquid CO2 offers an environmentally benign media for mixing. In addition, the mixture obtained does not contain any residual solvent as compared to the sonication in organic liquids. Upon depressurization, CO2 easily evaporates off the mixture providing a facile recovery of the product.

Web Page: www.eng.auburn.edu/users/gupta/

159a Forces Between Functionalized Silica Nanoparticles

Ahmed E. Ismail[1], J. Matthew D. Lane[2], Michael E. Chandross[3], Kelly Anderson[4], Chris Lorenz[5], and Gary S. Grest[2]. (1) Performance Assessment and Decisions Analysis Department, Sandia National Laboratories, 4100 National Parks Hwy, Carlsbad, NM 88220, (2) Surface and Interface Sciences, Sandia National Laboratories, PO Box 5800, MS 1395, Albuquerque, NM 87185-1395, (3) Computational Materials Science and Engineering, Sandia National Laboratories, PO Box 5800, MS 1395, Albuquerque, NM 87185-1395, (4) Modeling and Simulation, Proctor & Gamble, 11810 East Miami River Road, Cincinnati, OH 45252, (5) Division of Engineering, King's College, London, Strand, London, WC2R 2LS, United Kingdom

Polymer-coated nanoparticles have a wide variety of applications including drug delivery, adhesives, coatings, and magnetics. Although their complexity precludes atomistic simulations of large numbers of nanoparticles in solution, using large-scale molecular dynamics simulations, it is possible to study the interaction between pairs of nanoparticles in an explicit solvent and express these interactions in terms of empirical forces such as solvation, depletion, and lubrication forces. From these simulations, we can compute the forces exerted on nanoparticles treated with explicit-atom models, which can be used in coarse-grained simulations at larger length and time scales. In particular, we present results between two bare or polymer-grafted silica nanoparticles as a function of chain length, core size, and approach velocity. We show the work required to bring together two bare silica nanoparticles in water, alkylsilane-coated nanoparticles in decane, or poly(ethylene oxide)-coated nanoparticles in water, and compare them to the expected behaviors of nanoparticles in solution. We also examine the "equilibrium" forces on a pair of stationary nanoparticles, as well as the work required to retract the particles following an approach.

159b Molecular Modeling of Particle Nucleation In Emulsion Droplets: Towards An Improved Understanding of Nanoparticle Production Processes

Andreas Voigt, Process Systems Engineering, University of Magdeburg, Universitaetsplatz 2, Magdeburg, 39106, Germany, Michael Fricke, PCP, MPI Magdeburg, Sandtorstr. 1, Magdeburg, 39106, Germany, and K. Sundmacher, Process Systems Engineering, Otto-von-Guericke-Univ. Magdeburg, Universitaetsplatz 2, Magdeburg, 39016, Germany.

The production of nanoparticles with a well-defined size, narrow size distribution and particular shapes is of great importance in recent and future industrial applications and materials of commercial interest. This nanoparticle production process represents a major technological challenge due to the complexity of basic physical and chemical phenomena involved. Corresponding fundamental research, both experimentally and theoretically, is necessary and we present our current steps towards a improved understanding of the whole process. In recent experiments it has been shown that nanoparticle of different size and shape can be produced using an emulsion-based particle precipitation process [1,2]. The microemulsion droplets are volume-limited nanoreactors where due to the small-scale limitations the nanoparticle building process can be tailorized for size, distribution and even shape. The fundamental understanding of the all phenomena in such a process is the important prerequisite to analyze and later on control the particle production process according to the needs of the application and product. Recent dynamical experiments of Barium Sulfate nanoparticle precipitation in a microemulsion in our group [3] showed interesting shape and size variations under different process conditions. We believe that a fundamental step towards improved understanding of these effects lays in the understanding of the particle nucleation in such small confined volumes. Molecular modeling approaches can indeed be applied to this problem and we will focus in our presentation on important aspects which govern the dynamical evolution of particle populations in such systems. [1] D. Adityawarman, A. Voigt, P. Veit and K. Sundmacher Precipitation of BaSO4 Nanoparticles in a Non-ionic Microemulsion - Identification of Suitable Control Parameters, Chemical Engineering Science 60, p.3371-3383 (2005). [2] F. Rauscher, P. Veit, K. Sundmacher Analysis of a technical-grade w/o-microemulsion and its application for the precipitation of calcium carbonate nanoparticles, COLLOIDS AND SURFACES A-PHYSICOCHEMICAL AND ENGINEERING ASPECTS 254, p.183 (2005). [3] Niemann B., Veit P. and Sundmacher K. Nanoparticle precipitation in reverse microemulsions: Particle formation dynamics and tailoring of particle size distributions, LANGMUIR 8, p.4320 (2008).

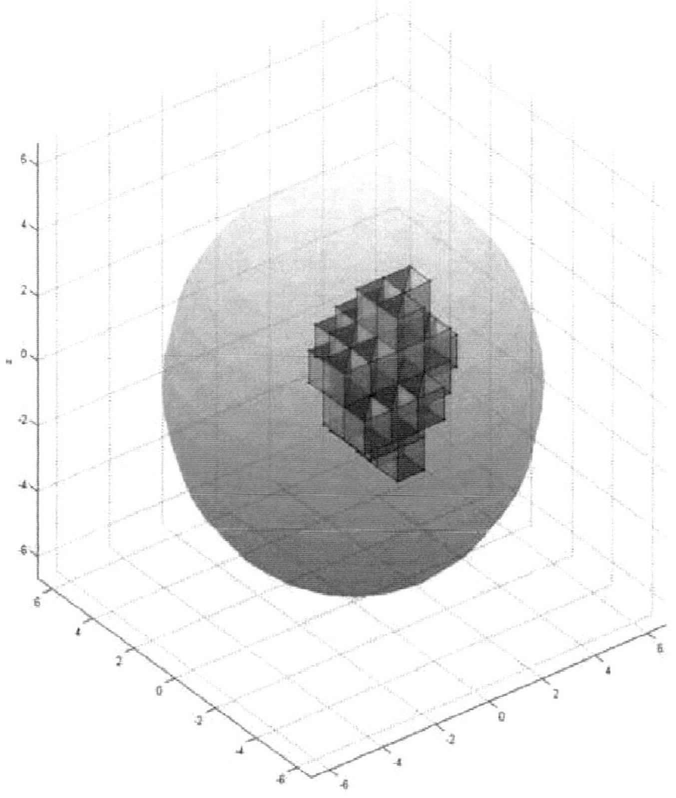

Web Page: www.uni-magdeburg.de/ivt/svt/person/av

159c Computational Modeling of Silicon Nanoparticle Synthesis In a Laser-Driven Reactor

Hongyi Dang, University at Buffalo (SUNY), 303 Furnas Hall, Buffalo, NY 14260-4200 and **Mark T. Swihart**, Chemical and Biological Engineering, The University at Buffalo (SUNY), 303 Furnas Hall, Buffalo, NY 14260-4200.

Synthesis of silicon nanoparticles is of great interest because of their unique optical and electronic properties. Fundamental understanding of the various interconnected mechanisms involved in the particle formation, such as gas phase and gas-surface phase chemical kinetics and particle size/ morphology evolution through nucleation, growth, coagulation and coalescence, would be of great value in designing and optimizing processes for producing silicon nanoparticles. In this work, we modeled silicon nanoparticle synthesis by silane decomposition in a six-way cross laser driven aerosol reactor. This corresponds to a reactor system used in our laboratory. Because of the complexity of the problem at hand, the simulation was carried out via several sub-models. First, the chemically reacting flow inside the reactor was simulated in three dimensions in full geometric detail, but with no aerosol dynamics and with highly simplified chemistry. This was done using MPSalsa, a finite element program developed by Sandia National Laboratories. Second, the reaction zone was simulated using an axisymmetric 2-D MPSalsa model, whose boundary conditions were obtained from the first step. Last, a quasi-2-D aerosol dynamics model was used to study the silicon nanoparticle formation using more complete silane decomposition chemistry, together with the temperature and velocities extracted from the reaction zone simulation. Both moment-based and sectional models were used to describe the aerosol dynamics, and results will be compared. In addition, a bivariate moment model that models finite rate coalescence was applied, allowing realistic predictions of primary particle size at high particle number concentration.

159d Scale-up of a Nanoparticle Size-Selective Fractionation Process Using CO_2-Expanded Liquids

Steven R. Saunders and Christopher Roberts. Department of Chemical Engineering, Auburn University, 212 Ross Hall, Auburn, AL 36849

We have recently scaled up a process from the benchtop test tube level to application scale that allows for the size-selective fractionation of ligand-stabilized metal and semiconductor nanoparticles that utilizes the pressure tunable physico-chemical properties of CO_2 gas expanded liquids (GXL's). The nanoparticle size separation technique is based on the controlled reduction of the solvent strength of organic phase nanoparticle dispersions through increases in concentration of the anti-solvent CO_2 via pressurization. These changes in solvent strength affect the subtle balance between the osmotic repulsive forces (due to the solvation of the stabilizing nanoparticle ligand tails) and the van der Waals forces of attraction between different sized particles necessary to maintain a stable dispersion. Through modest changes in CO_2 pressure, increasingly smaller sized nanoparticles can be controllably precipitated from the dispersion, resulting in the separation of particle dispersions typically ranging from 2nm to 12nm into ±1nm fractions.

Previously, we have demonstrated an effective separation of metal (Au and Ag) and semiconductor (CdSe/ZnS) nanoparticles at the bench scale using a novel glass spiral tube apparatus within a high pressure vessel. However, this initial apparatus was only capable of separating mg quantities of nanoparticle dispersions. It was desirable to design a new apparatus capable of processing larger quantities of monodisperse nanoparticles for applications in catalysis, sensors, semiconductors, and optics. This new scaled up apparatus consists of three vertically mounted high pressure vessels connected in series with high pressure needle valves that allow for sequential isolation and separation of the fractionated nanoparticle dispersions. This process, operated at room temperature and CO_2 pressures between 28 bar and 49 bar, can result in a batch or semi-continuous size selective separation of larger volumes of the nanoparticle dispersions, resulting in a scale up factor of more than 200 over the previous process. Performing separations at this scale illustrates that it is possible to separate large quantities of nanoparticle dispersions using this technique. This process can be easily modified by increasing the number of fractionation vessels, by increasing the size of the vessels, by adjusting the ΔP at each stage, or through changes in ligand type or solvent choice. The efficiency of separation of several nanoparticle systems such as metals and semiconductors will be discussed including the effects of different nanoparticle ligands and other design parameters.

159e Kinetic Study and Modeling of Metal Particle Formation

Roberto Irizarry[1], Madeline Leon[2], and Miguel Castro[2]. (1) Electronic Technologies, DuPont, 14 TW Alexander Drive, Research Triangle Park, Raleigh, NC 27709, (2) Chemical Imaging Center, Department of Chemistry, University of Puerto Rico, Car. 108 km 2.0, Mayaguez, PR 00680

Determination of kinetic parameters of metal precipitation is very complex because it is a multiple scale phenomena with a very fast kinetics. The kinetic data is obtained using a stopped flow reactor to eliminate the effect of micro-mixing. The change of size with time is determined using surface plasmon theory. The characteristics of the surface plasmon, is also utilized to determine changes in morphology during growth. The inverse problem is solved to infer nucleation parameters from experimental kinetic data. The dynamic of the metal formation is simulated using MC simulation. The inverse problem is solved using the coarse-grained methods (PEMC and t-PEMC) recently introduced in the literature ([1], [2]) and the artificial chemical plant paradigm for global optimization [3].

[1] R. Irizarry, Fast Monte Carlo Methodology for Multivariate Particulate Systems-I: Point Ensemble Monte Carlo (2008) Chemical Engineering Science 63, 95-110.

[2] R. Irizarry, Fast Monte Carlo Methodology for Multivariate Particulate Systems-II: t-PEMC (2008) Chemical Engineering Science 63, 111-121.

[3] R. Irizarry, LARES: An Artificial Chemical Process Approach for Optimization (2004) Evolutionary Computation Journal, 12 (4), 435-460.

159f Efficient Optimization of Particulate Systems Described by Monte Carlo Method: Determination of Nucleation Parameters from Kinetic Data

Roberto Irizarry, Electronic Technologies, DuPont, 14 TW Alexander Drive, Research Triangle Park, Raleigh, NC 27709

In this paper the inverse problem is solved to determine silver and gold nucleation parameters from kinetic data. The kinetic data is obtained using a stopped flow reactor and the change of size with time is determined using surface plasmon theory. The dynamic of the metal formation is simulated using MC simulation. Solving the inverse problem of this system is very difficult because the model is given in a form of a MC simulation, which accurate solutions are time consuming. To make optimization feasible two approaches are presented here: (a) Use of coarse-grained methods (PEMC and t-PEMC) recently introduced in the literature ([1], [2]) and (b) Modify the stochastic algorithm to optimize systems with noise (make rough MC simulations). The algorithm used to solve the optimization problems is based on artificial chemical plant paradigm [3]. Comparison of the results, in terms of CPU time and parameters obtained are presented using both approaches.

[1] R. Irizarry, Fast Monte Carlo Methodology for Multivariate Particulate Systems-I: Point Ensemble Monte Carlo (2008) Chemical Engineering Science 63, 95-110.

[2] R. Irizarry, Fast Monte Carlo Methodology for Multivariate Particulate Systems-II: t-PEMC (2008) Chemical Engineering Science 63, 111-121.

[3] R. Irizarry, LARES: An Artificial Chemical Process Approach for Optimization (2004) Evolutionary Computation Journal, 12 (4), 435-460.

159g Large Scale Synthesis of Luminescent Rare-Earth Doped Oxide Nanoparticles

Yuanbing Mao and **Jane P. Chang**. Department of Chemical and Biomolecular Engineering, University of California at Los Angeles, Los Angeles, CA 90095

Intensive research effort devoted to the preparation of nanomaterials has resulted in impressive progress in their synthesis, in order to access their unique properties. One of the remaining challenges is to develop large scale synthetic methods. The molten salt synthesis (MSS) is one of these methodologies in which a molten salt is used as the reaction medium for reactant dissolution and product precipitation. The characteristics of the products by MSS are generally affected by the synthesis temperatures, the type of salts used, the precursor compositions and the solubilities of the reactive constituents in the molten salts.

As energy becomes a scarce commodity, common low efficiency lighting device must be replaced with more efficient alternatives, such as light emitting diodes, where metal oxides are well suited. Since the crystallite size, composition and presence of dopants greatly affect the optical performance of these materials, in this work, we report the synthesis and characterization of luminescent rare-earth doped oxide nanoparticles (NPs) with a focus on controlling the spatial distribution of the dopants in metal oxide hosts and assessing its effect on the device performance.

In this talk, we present our recent work on the large scale synthesis of RE doped oxide NPs, including Er: Y_2O_3 and Er:$La_2(Zr_xHf_{1-x})_2O_7$, by MSS. The formation of these NPs was probed in-situ by time-resolved synchrotron x-ray diffraction and absorption spectroscopy to delineate the process-structure-property relations. The as-synthesized NPs were further characterized by electron microscopy and various spectroscopy to be single crystalline, with well controlled size distributions. The erbium coordination number and local bonding environment were shown to dictate the measured photoluminescent characteristics, including photoluminescence (PL) and cathodoluminescence (CL). Specifically, these erbium-doped oxide NPs have sharp and well-resolved PL behavior in the near-infrared region, outstanding green and red upconversion emissions, and excellent CL properties. These properties make these NPs promising for applications in display, bioanalysis and telecommunications.

189a Solids Drying Process Simulation by Solidsim - New Tool for Flow Sheeting of Solids Processes

Ting(Tim) Han and Joerg Theuerkauf. Solids Processing, Engineering & Process Science, The Dow Chemical Company, Freeport, TX 77566

Solidsim is a new Microsoft WindowsTM based flow sheet simulation software for complex solids processes which can be used to predict the performance of a processing plant and to simulate plant extensions, modifications or designs.

In this paper, Solidsim was used to simulate the processes of two types of convective dryers: the co-current flow dryer and the cross-current flow dryer, respectively. The material and energy balances for each process was calculated by Solidsim. The results were evaluated. In addition, by changing input and output data of the dryers, Solidsim has been further investigated on predicting and optimizing the performance of the convective dryers.

189b Characterization of Nano-Sized Ni/ce(x)Zr(1-x)O2 Catalyst for the Auto Thermal Reforming

Text Not Available.

189c Characterization of Ni/Ce(x)Zr(1-x)O2 Catalyst for Autothermal Reforming

Jin-Hwa Kong[1], Dong-Ju Moon[2], Jong-Ho Kim[3], Nam-Cook Park[3], Jae-Soon Shin[3], and Young-Chul Kim[3]. (1) Department of advanced chemical engineering and the research institute for catalysis, Chonnam National University, 300 Yongbong-dong, Gwang-ju, 500-757, South Korea, (2) Clean energy research center, Korea Institute of Science and Technology, Hawalgock-dong Seongbuk-gu, Seoul, 136-791, South Korea, (3) School of applied chemical engineering and the research institute for catalysis, Chonnam National University, 300 Yongbong-dong, Gwang-ju, 500-757, South Korea

The catalytic performance and characterization of Ni/CexZr1-xO2 were investigated using an autothermal reforming(ATR) process for hydrogen production. The Ni/CexZr1-xO2 catalysts were prepared using the following methods: the water method (W), urea water method (UW), urea ethylacetate method (UE) and urea water - alcohol method (UWA). The effect of the support composition as well as the metal loading on the ATR reaction was examined in a fixed-bed flow reactor, over a temperature range of 300 to 700 „aC. The catalysts were characterized using various techniques. The characterization results prevented carbon deposition on the surface of Ni/CexZr1-xO2. SEM indicated that the surface carbon was denser on the Ni/CZ - UW used catalyst than on the Ni/CZ - UWA used catalyst.

TGA showed that the amount of carbon deposition on the Ni/CeZr - UWA was approximately 19% lower than on the Ni/CZ - UW catalyst. SEM indicated that the surface carbon was denser on the Ni/CZ - UW used catalyst compared with the Ni/CZ - UWA used catalyst. In addition, there was a higher level of hydrogen production on the Ni/CZ - UWA catalyst than on the other catalyst. However, further studies will be needed to confirm the stability of the Ni/CZ - UWA catalyst.

189d Generation and Geometrical Analysis of Dense Clusters with Variable Fractal Dimension Providing a Basis for the Interpretation of Data from Optical Particle Characterication Techniques

Lyonel Ehrl[1], Marco Lattuada[2], Miroslav Soos[3], and Massimo Morbidelli[1]. (1) Institute for Chemical and Bioengineering, Department of Chemistry and Applied Biosciences, ETH Zurich, Wolfgang-Pauli-Str. 10, HCI F 133, Zurich, 8093, Switzerland, (2) Chemistry and applied biosciences, Institute for chemical and bioengineering, ETH Zurich, Wolfgang-Paulistr. 10, HCI-F133, Zurich, Switzerland, (3) Institute for Chemical and Bioengineering, Department of Chemistry and Applied Biosciences, ETH Zurich, Wolfgang-Pauli-Str. 10, Zurich, 8093, Switzerland

In the solid-liquid separation of colloidal matter, e.g., in the case of waste-water treatment, in the downstreaming of emulsion polymers or wet born nano-particles, usually smaller particles are aggregated to from larger clusters to enhance and facilitate the separation step. By doing so, initially formed open clusters are restructured and reversibly broken and aggregated by the shearing of the fluid to produce dense clusters. A methodology to artificially generate dense clusters with variable fractal dimension is proposed. It is based on the algorithm of Thouy and Jullien [1], which provides clusters with a fractal dimension up to 2.5, where an additional densification routine was implemented that utilizes voronoi tessellation in order to provide clusters of densest random packing (for the given framework). In this way, cluster populations with fractal dimensions up to the Euclidian limit of 3 can be obtained. Hence, cluster populations with fractal dimensions ranging from 2.2 to 3 were generated and, consequently, analyzed with respect to common geometric measures that are typical for optical particle characterization techniques, such as microscopy, light scattering, and focused beam reflectance. To assure the representativeness of the provided averages sufficiently large cluster populations were analyzed, which was particularly important for the smaller clusters. Relations between fractal dimension, cluster mass, and certain geometric measures are discussed and compared to data available in literature. Special attention is given to the effect of blur that is unavoidable when the objects fine characteristics become comparable with the wave length of the light source used. Such results can serve as a basis for the interpretation of data from optical particle characterization techniques for systems containing aggregated matter.

Reference:

[1] R. Thouy, R. Jullien, A Cluster-Cluster Aggregation Model with Tunable Fractal Dimension. Journal of Physics a-Mathematical and General. 1994, 27, 2953-2963.

189e The Breakage Matrix Approach In Evaluating Inadvertent Breakage of Dense Particulate Mixtures

A. Abu-Nahar, Faculty of Engineering&Physical Sciences, University of Surrey, Guildford, GU2 7XH, United Kingdom and U Tüzün, Faculty of Engineering&Physical Sciences, Faculty of Engineering&Physical Sciences, University of Surrey, Guildford, GU2 7XH, United Kingdom.

In general, breakage matrix is an analysis technique, which is potentially useful in describing any degradation process. The application is widespread and unconstrained to model an intentional degradation process such as the study of the material breakage efficiency in a milling operation or capable of identifying which parts in a transport system contribute most to attrition problems. Breakage matrix approach is however, often limited to linear breakage assumption neglecting the nature of the many body of inter-particle interaction of individual species and the surrounding particles within mixture assembly.

Linear breakage assumption has received to open criticism for more than a century and as matter of fact, it is not necessarily held true in many degradation scenarios. Broadbent and Calcott, 1956 a&b observed linear or independent breakage for their case of coal transporting system even tough they remarked earlier, "..Breakage in an assembly is partly due to interaction between particles in the assembly". Therefore, it can be reasoned at small extent, inter-particle interaction may result in abnormal breakage effects of the output size distribution. The successfulness of breakage matrix as a prediction tool is no doubt for self-governing condition, in fact, Campbell and Webb, 2001 in their work essentially suppose the breakage matrix approach is worthless for significant interacting mixture. Non-interacting conditions are sensible for some degradation scenarios for example during roller milling process in which each particle passing through the two rotating rolls and break independently, or in dilute phase pneumatic conveyor where the particles impact on pipe bends one at a time. Consider an example of dense phase condition where dense assemblies are subjected under compression loads/ shear, the breakage of individual particles within the mixture obviously far from negligible and are supposedly influenced by the presence of particles in other size classes.

This paper attempts to demonstrate the usefulness of breakage matrix as a more versatile tool in predicting degradation than it has been appreciated to date with additional feature of coefficient of interaction as a function of micro-structural limit fines composition and size ratio to account for the role of inter-particle interaction in influencing the binary mixture output size distribution after degradation. The prediction capability of this proposed more flexible breakage matrix approach is verified with some degradation data set obtained from a compression test bed of dense particulate mixtures.

References:

1-) Broadbent, S.R., Callcott, T.G.,(1956-a) J. Inst Fuel., pp: 524-528

2-) Broadbent, S.R., Callcott, T.G.,(1956-b) J. Inst Fuel., pp: 528-539

3-) Campbell G.M., Webb C., (2001), J. Powder Technol., 115, pp: 243-255

4-) Tuzun U., Artega P., (1992), Chem. Eng. Sci., 47, pp:1619-1630

Modeling of an Industrial Vibrating Double-Deck Screen of a Urea Granulation Circuit

Paper #123584

Ivana M. Cotabarren, José Rossit, Juliana Piña, Verónica Bucalá
Chem. Eng. Department, Universidad Nacional del Sur, Bahía Blanca, Argentina

Introduction

Urea granulation is a complex operation that cannot be carried out in a single device; it is rather achieved by a combination of process units with specific functions constituting a granulation circuit (Figure 1).[1] The main unit is the granulator where small urea particles, known as seeds (generally out of specification product), are continuously introduced and sprayed with a concentrated solution of the fertilizer (melt). The seeds grow through deposition of the melt droplets onto the solids surface, followed by urea solidification and water evaporation.[2] The granules that leave this size enlargement unit are cooled down and subsequently classified by double-deck screens into product, oversize and undersize streams. The product is transported to storage facilities, while the oversize fraction is fed to double-roll crushers for size reduction. The crushed oversize particles are then combined with the undersize granules and returned to the granulator as seeds.[3]

Many authors found that the operation of the screening and crushing units has a decisive influence on the recycle stream and hence on the circuit stability.[4-8] In view of this and the recognized important role of plant simulation to predict and optimize the granulation circuit operation,[9,7,10] reliable models for all the process units should be available. Recently, a validated mathematical model for the industrial double-roll crusher of a urea granulation circuit has been reported.[3]

The fertilizer industry requires, as aforementioned, a classification step that is usually performed by double-deck vibrating screens. These devices have been extensively studied by numerous authors in the context of the mineral processing industry.[11,-16] Models in the literature can be classified as phenomenological, empirical and numerical; being based on theory of the screening process, empirical data and computer solutions of Newtonian mechanics, respectively.[17]

Within the phenomenological models two different approaches, the kinetic and probabilistic, have been used to represent these screening operations. The probabilistic approach[12,14] is based on the probability of a particle passing through the aperture of the screen. On the other hand, the kinetic approach defines the screening performance as a rate process that varies with the distance along the screen and depends on the amount and particle size distribution of the material being processed. The empirical models aim to predict the quantity of undersize that can pass the screen. Among these, some are based on a theoretical capacity which is affected by a set of correction factors accounting for the effect of oversize, half-size, and near-size material; whether the screen is a top or lower deck on a multi-deck unit; type of aperture; material density, etc.[17] Due to their simplicity, the empirical models are preferred for the simulation of industrial screens. In fact, many commercial simulators such as Aspen Plus,[18] Modsim,[19] and Moly-Cop Tools[20] have empirical models implemented in their routines.

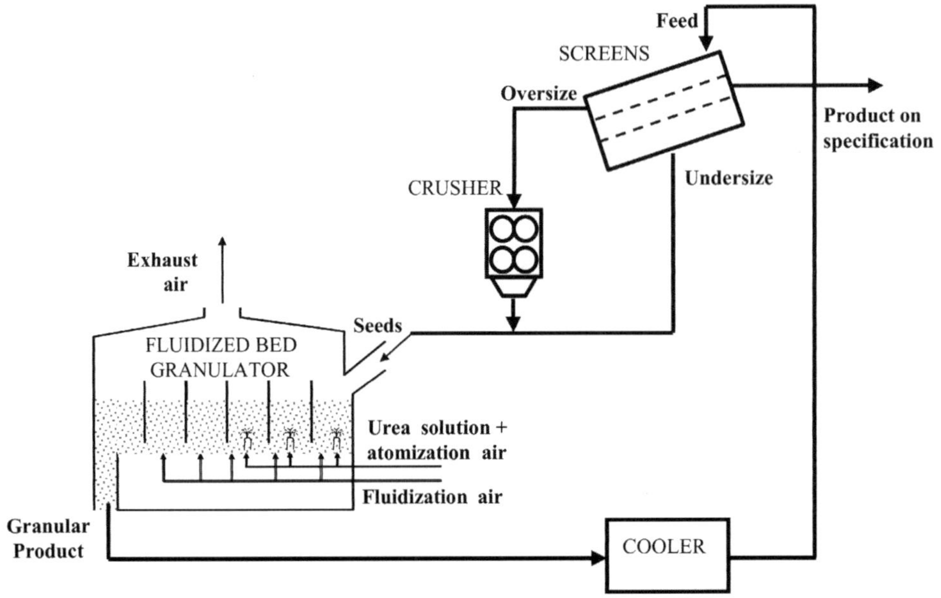

Figure 1. Typical urea granulation circuit.

In this contribution, a fitted model for a large-scale double-deck vibrating screen of an industrial urea granulation plant is presented. After thorough examination of several empirical, probabilistic and kinetic models, it was concluded that the model proposed by Karra[11] is the most suitable to describe the screening operation in terms of the available industrial data. This empirical approach results of great simplicity and high efficiency since it manages to predict all the experimental data (approximately 940 points) by just fitting three parameters. Besides, it is usually recommended for predicting the performance of industrial vibrating screens[21] and has demonstrated to be overwhelming respect to other empiric models to predict the urea granules classification.

In order to estimate the screen performance, all the flowrates around the equipment and their corresponding particle size distributions must be known. It is of common knowledge that significant errors may occur when attempting to measure solid streams flowrates. In fact, the only reliable and available experimental mass flow for this work was that corresponding to the product stream. The remaining and unmeasured individual mass flows were determined through a data reconciliation procedure, which simultaneously allows modifying the most unreliable measured variables to satisfy the material mass balances for every size interval.[4,22] Indeed, the experimental results confirm the need of using solid streams data reconciliation to have industrial data consistency.

Data reconciliation

In the present work, the screen classification parameters were fitted using industrial data from a plant of high capacity. The experimental data were collected from two large-scale double-deck vibrating screens of identical characteristics. They operate with a small angle of inclination and with fixed vibration. Samples by duplicate, of the feed (*F*) and of the oversize (*O*), product (*P*) and undersize (*U*) streams of each of the two screens (A and B) were collected and granulometrically analyzed (Figure 2).

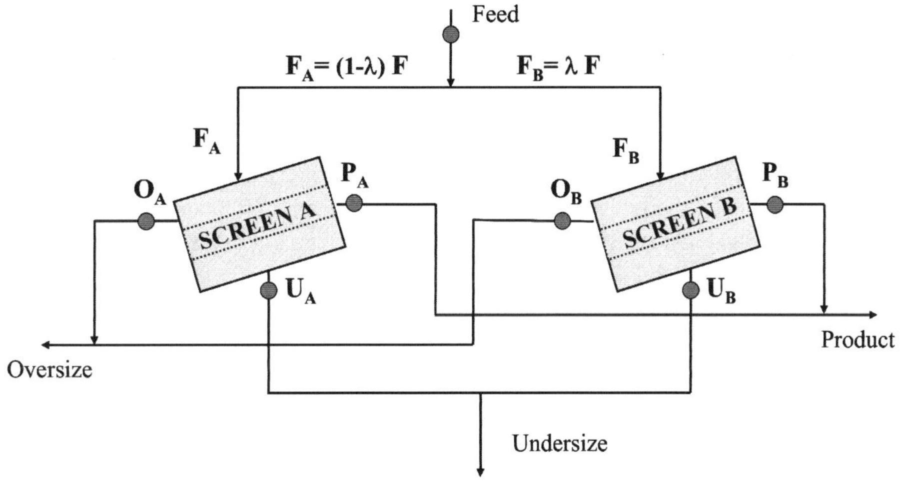

Figure 2. Schematic diagram of the experimental screening sector. (●) Sampling points.

Data reconciliation was formulated as a constrain optimization problem and solved by means of the Athena Visual Studio Software.[23] The goal of this optimization problem was to minimize the total difference between measured and estimated mass fractions for each size interval, weighted by the variance of the measurements using the least square method.[29] Defining F, O, P and U as the feed, oversize, product and undersize mass flows and $X_{F_{Ai}}$, $X_{F_{Bi}}$, $X_{O_{Ai}}$, $X_{O_{Bi}}$, $X_{P_{Ai}}$, $X_{P_{Bi}}$, $X_{U_{Ai}}$ and $X_{U_{Bi}}$ as the mass fractions for the size class i of the feed, oversize, product and undersize streams related to screens A and B; the objective function was expressed as:

$$f_o = \sum_{j=1}^{M} \sum_{i=1}^{N} \left(X_{ji} - \overline{X_{ji}}\right)^2 / W_{ji} \tag{1}$$

where N was the number of size intervals (i.e. 13) and M was the number of streams around the screens with different particle size distributions (i.e. 7: O_A, P_A, U_A, O_B, P_B, U_B and F, since F_A and F_B have the same mass fractions); $\overline{X_{ji}}$ represented the average value of duplicate measurements; X_{ji} corresponded to the estimated value; and W_{ji} was the variance associated to each $\overline{X_{ji}}$.

To completely formulate the data reconciliation, the objective function (f_0) was subjected to the following restrictions:

$$F_A X_{F_{Ai}} = O_A X_{O_{Ai}} + P_A X_{P_{Ai}} + U_A X_{U_{Ai}} \tag{2}$$

$$F_B X_{F_{Bi}} = O_B X_{O_{Bi}} + P_B X_{P_{Bi}} + U_B X_{U_{Bi}} \tag{3}$$

$$\sum_{i=1}^{13} X_{ji} = 1 \qquad j = 1..7 \tag{4}$$

$$F_A/(1-\lambda) = F_B/\lambda \tag{5}$$

$$P_A + P_B = P \tag{6}$$

where λ was the fraction of the total feed derived by a diverter to screen B (Figure 2) and *P* corresponded to the measured product flowrate, which was not a variable to be reconciled. The parameter λ was fitted during data reconciliation to estimate the actual flowrate being fed to screens A and B during the experimental tests.

Figure 3 condenses the correspondence between the reconciled and experimental mass fractions for each particle size and solid stream. The observed differences are as expected, considering the well known uncertainties when sampling and analyzing solid streams. From the 1638 total experimental points (Test 1: 13 size classes x 7 streams x 10 samples; Test 2: 13 size classes x 7 streams x 8 samples), the 92% is included within the 20% deviation line, and 81% within the 10% one.

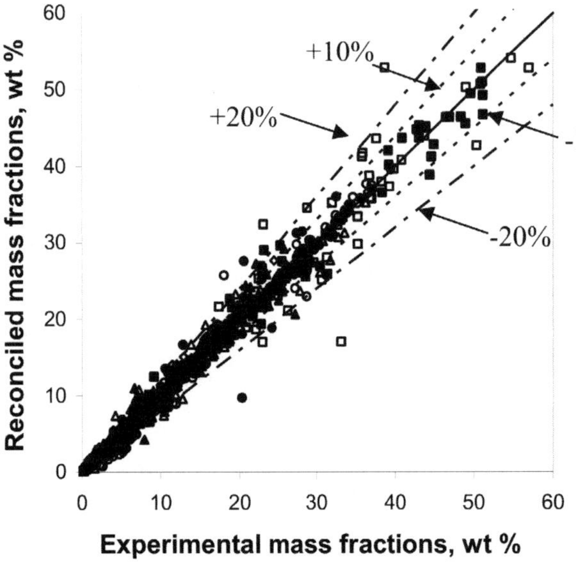

Test 1:

(♦) F, (■) O_A and O_B, (▲) P_A and P_B,
(●) U_A and U_B.

Test 2:

(◇) F, (□) O_A and O_B, (△) P_A and P_B,
(○) U_A and U_B

Figure 3. Reconciled vs. experimental mass fractions of all the size classes.

The feed flow split parameter (*λ*) varied from 0.43 to 0.56, being its average value 0.48. These results indicated that the diverter was almost capable to split the feed evenly, in good agreement with its initial position selected to perform the experiments. The feed flowrates (*F*) estimated through the reconciliation procedure were in good agreement with the values calculated from the energy balance of the cooler (see Figure 1) located downstream the granulation unit. The reconciliation step was essential to have consistent industrial experimental data to fit the screen models.

Screen Model

An ideal classification step would separate perfectly the solid stream, i.e. the particles bigger than the mesh aperture would be retained while the smaller ones should pass through the screen. However, in practice, separation processes are not perfect and the probabilities of reaching the screen and passing through it are not the same for different particle sizes.

The most realistic measurement of screen performance is provided by the partition curves. To represent a non-ideal classification operation, like the actual screening process, Karra[11] used oversize partition coefficients for each size interval. For each class size i, the oversize partition coefficient (T_i) is defined as the amount of oversize within interval i divided by the amount of material of that size in the feed.

$$T_i = (O\,X_{Oi})/(F\,X_{Fi}) \tag{7}$$

Once all T_i are known, the particle size distributions of the oversize and undersize streams can be calculated through simple mass balances for the solids belonging to each size class i.

$$O\,X_{Oi} = T_i\,F\,X_{Fi} \tag{8}$$

$$U\,X_{Ui} = (1 - T_i)\,U\,X_{Ui} \tag{9}$$

The oversize partition values (T_i) are usually plotted against the corresponding geometric mean particle size:

$$d_i = \sqrt{(d_1 d_2)_i} \tag{10}$$

where d_1 and d_2 are the lower and upper limits for size interval i, respectively.

In order to develop an empirical model capable of describing the screening performance, a mathematical formulation of the partition curve is needed. Such formulation requires determining the functionality with respect to the d_{50} parameter. This cut size (median) is defined as the size corresponding to the 50% partition curve, and thus represents the size at which particles have equal chance of staying on the screen or passing through. The cut size is a measure of the screening efficiency; d_{50} values close to the mesh size indicate high efficiencies. Nevertheless, the cut size is always less than the aperture.[17] It has been established that process parameters such as particle to aperture ratio, composite nature of the feed, deck location, feed rate and bulk density, affect the d_{50} significantly.[11] This dependence was also observed analyzing the available experimental data.

Karra[11] proposed the following d_{50} correlation to represent industrial screening data:

$$d_{50} = h\left(U^T / SK\right)^a \tag{11}$$

where h is the screen aperture (mm), U^T are the theoretical undersize tons per hour fed to the screen (fed material whose size is smaller than the aperture size); S is the screen surface (m^2); K is a product of factors that correct the screen basic capacity to represent the deviation from the standard conditions and adjustment to the specific conditions and a is a fitting parameter.

The variable K is expressed as follows:

$$K = A\,B\,C\,D\,E\,F_D\,G \tag{12}$$

where A is the basic capacity, defined as ton/h of undersize that a particular screen can process per unit of screen area. B is the amount of oversize in the feed (percentage of material with $d_i > h$). Screens that handle great amount of oversize operate more efficiently because that material is directly recovered over the screen. C is the amount of half-size under in the feed (percentage of material with $d_i < 0.5\ h$). Feeds containing a large proportion of material considerably smaller than the aperture size, manage to manipulate it more easily. D is the location of the deck. Lower decks receive material harder to handle than those decks that take fresh feeds; therefore the capacity decreases with position. E is for wet screening. If dry screening is performed, as in the urea process, its value is set to one. F_D is the bulk density factor; denser materials are separated more easily than lighter materials. G is the amount of near-size material in the feed (percentage with material of $0.75\ h < d_i < 1.25\ h$). Feeds with large quantities of particles close to the aperture size present more difficult separation because the passage of undersize material is inhibited.[24]

Karra[11] used the standard expressions given in the Nordbeg Process Machinery Reference Manual[25] for parameters A, B, C, D, E and F_D and developed his own correlation for the near-size material factor G. The corresponding equations are presented in Table 1.

Table 1. Equations for screen correction factors.[11]

Factor	Equation		Reference
A	$A = 12.1286\,h^{0.3162} - 10.2991$	$h < 50.8\,mm$	h =opening aperture (mm)
	$A = 0.3388\,h + 14.4122$	$h \geq 50.8\,mm$	
B	$B = -0.012\,Q + 1.6$	$Q \leq 87$	Q = % of oversize in feed to deck
	$B = 0.0425\,Q + 4.275$	$Q > 87$	
C	$C = 0.012\,R + 0.7$	$R \leq 30$	R = % of half-size in feed to deck
	$C = 0.1528\,R^{0.564}$	$30 < R < 55$	
	$C = 0.0061\,R^{1.37}$	$55 \leq R < 80$	
	$C = 0.05\,R - 1.5$	$R \geq 80$	
D	$D = 1.1 - 0.1L$		L = deck location top = 1, bottom = 2
E	$E = 1.0$	$T < 1$	$T = 1.26h$
F_D	$F_D = \rho_B / 1602$		ρ_B = bulk density (kg/m^3)
G	$G = 0.975\left(1 - X^{0.511}\right)$		X = % of near-mesh in the feed

Once d_{50} is determined, its relationship with the T_i function has to be defined. Several standard functional forms are available to describe it, Karra[11] proposed:

$$T_i = 1 - \exp\left[-0.693\left(d_i / d_{50}\right)^m\right] \tag{13}$$

This correlation, recognized as the Plitt equation, is expressed in terms of the well known Plitt's adjustable parameter m, which defines the classification sharpness.[26]

Parameter estimation

In this work, the Karra[11] model was adapted to describe the performance of two large-scale double-deck screens from an industrial urea granulation plant, once all the experimental data were reconciled.

Each deck was considered as a separate screen; therefore a set of parameters was adjusted for the top deck and another for bottom deck. Following the procedure described in the previous section, the correction factors A to G were determined for both decks using the expressions given in Table 1. The screen factor E, corresponding to wet screening, was equaled to 1.

For the special case under consideration, the bulk density corresponding to each industrial deck was not a measurable property because it varied along the screen length with the quality of the material being classified. For this reason, it was set as a fitting parameter for both decks.

Once the theoretical undersize U^T and all the correction parameters that defined K were evaluated for each test hour and deck, the d_{50} were calculated by means of Equation (11) and the corresponding partition coefficients curves were established.

The adjustment of the bulk density (ρ_B), the Plitt parameter (m) and the exponent in d_{50} (a) for each deck, was performed by using the Athena Visual Studio Software[23] with the aim of reproducing the respective cumulative undersize and oversize streams. The undersize from the top deck was considered as the feed to the bottom deck.

Results

Table 2 reports the estimated model parameters. The fitted values are in agreement with the usual operation of both decks. In fact and due to the greater amount of oversize material (i.e. higher bed porosity), the bulk density for the top deck results lower than the one corresponding to the bottom deck. The solid urea density is 1330 kg/m^3, so the bulk densities found for the top and bottom decks indicate bed porosities of about 58% and 47% respectively, which are expected values. The separation sharpness as a function of the feed quality is clearly represented through the adjusted Plitt parameter. The top deck handles material with a greater proportion of oversize than the bottom deck, being more effective in the classification process. Thus, the higher efficiency of the first deck, given by a significant higher m value, is confirmed.

The exponent in the d_{50} correlation (a) can be considered as an indicator of the cut size evolution during the operation. According to the fitting results, the d_{50} for the top deck remains almost equal to the screen aperture (for very high m values T_i tends to 1, see Equation 13). This result is in agreement with the higher efficiencies found for this first deck. Contrary, the d_{50} for the bottom deck is a strong function of the operating conditions and differs considerably from the corresponding h value. Indeed, the highest d_{50}/h ratio over time obtained for the top deck was 0.4% against 31.9% for the bottom one.

Table 2. Estimated parameters for the top and bottom deck screens.

Parameter	Top deck	Bottom deck
Bulk density (ρ_B) (kg/m^3)	560	705
Power of d_{50} equation (a)	-0.0023	-0.2862
Plitt's parameter (m)	25.097	3.758

Figure 6 shows the experimental and predicted oversize (O) and undersize (U+P) particle size distributions (*PSD*s) for the top deck. Two samples from screen A (Figures 6 a and c) and two from screen B (Figures 6 b and d) among the 36 acquired, have been included in order to illustrate the goodness of the model. The samples presented in Figure 6 c and d represent the worst *PSD*s prediction obtained for screens A and B, respectively. Considering the relatively high number of experimental points (i.e. 468, 18 samples x 13 classes x 2 screens) to be predicted for each stream (i.e., undersize and oversize) by estimating only 3 parameters, it can be concluded that the fitted model satisfactorily reproduces the available experimental data. In fact, for the oversize stream, 75% of the predicted points have less than 5% of error, 81% less than 10% and 92% less than 20%. For the undersize stream the results are even better, 97.7% of the data are within 5% of error, 98.8% within 10% and 99.6% within 20%.

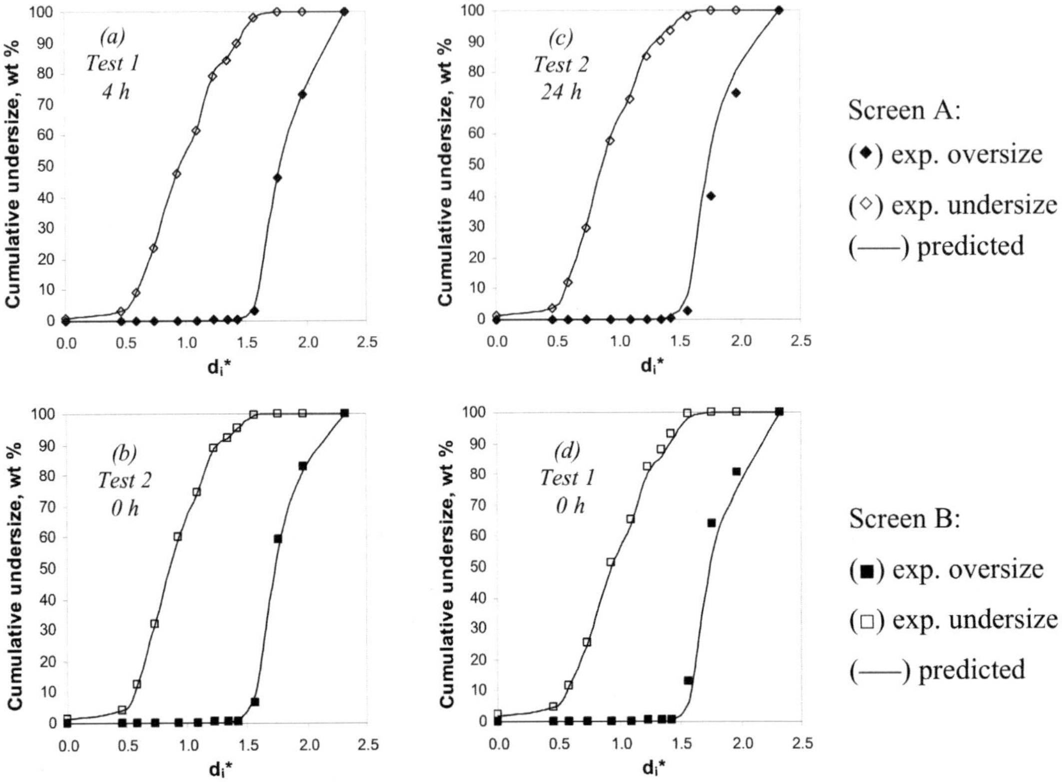

Figure 6. Some selected experimental and predicted size distributions for the top deck.

Figure 7 presents some selected experimental and predicted *PSD*s of the oversize (*P*) and undersize streams (*U*) for the bottom decks of screens A and B. The oversize corresponds to the product stream and the undersize to the fraction of fines recycled directly from the double-deck screens to the granulator. Again, two distributions from each screen are shown, being Figures 7 c and d representative of the worst predictions. Regarding these bottom decks, 84% of the predicted points for the oversize stream have less than 30% error and for the undersize stream 81% have less than 30% error. In general, the prediction is not as good as that obtained for the top deck. Nevertheless, and taking into account the errors inherent to sample collection, it can be considered to be in reasonable agreement with the experimental data.

The model previously described, together with the parameters presented in Table 2, can be considered a suitable mathematical representation of the industrial double-deck screens. It is worth to mention that this model is capable of satisfactorily predict changes in the *PSD* and mass flowrates of the feed stream. Besides, it gives valuable information about the influence of the screening operation on the particle size distributions of the fines recycled to the granulation unit, the oversize fed to the double-roll crushers and the product derived to storage facilities.

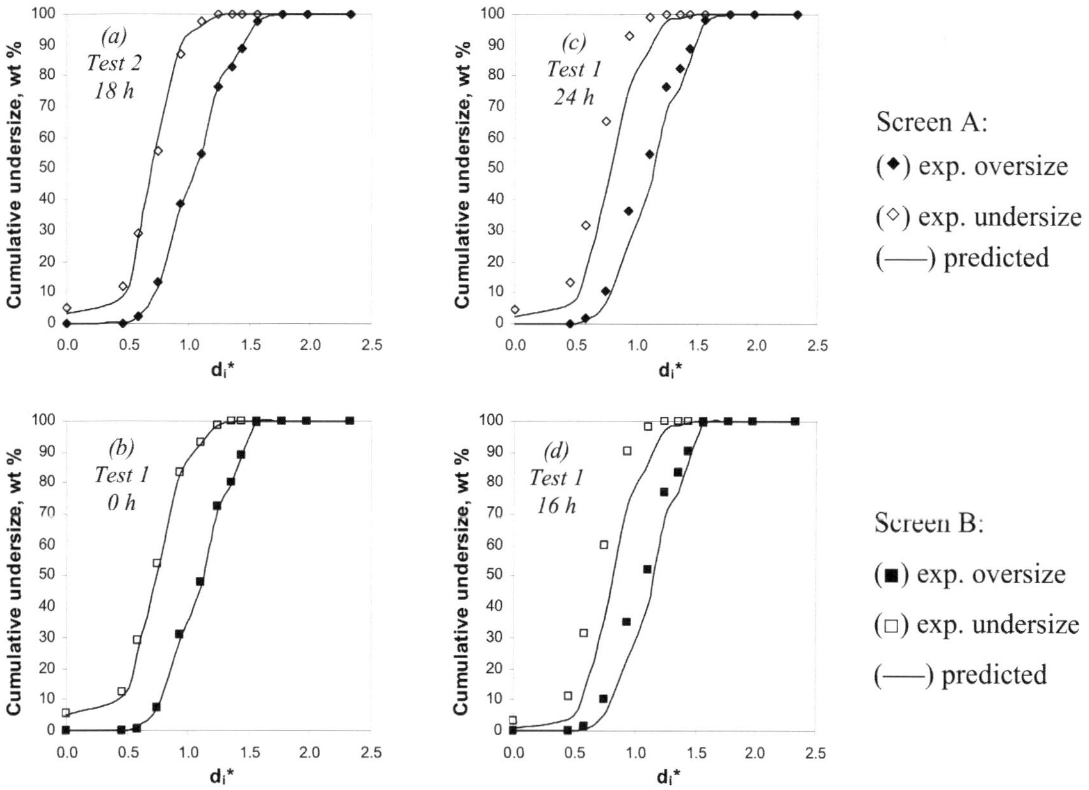

Figure 7. Selected experimental and predicted size distributions for the bottom deck.

Conclusions

The simple empirical model proposed by Karra,[11] for the mineral processing industry, was satisfactorily adapted for predicting the classification performance of large-scale double-deck screens from an industrial urea granulation plant. The data reconciliation procedure was found as an unavoidable step to estimate the unknown mass flow rates and correct the experimental measured data to satisfy the material balances for every particle size interval. By applying this new technique to the industrial granulation plant under study, it was possible to obtain a set of reliable and consistent experimental data for subsequent modeling, supporting its usefulness for handling data belonging to solid processes.

The presented model allowed predicting successfully the experimental d_{50}, a parameter found to be a key model variable. The goodness of the fitted model was widely confirmed over a large number of samples collected from the industrial units, operating with different feed *PSDs* and mass flowrates.

Market conditions for agricultural commodities and consequently for the urea are exceptionally favorable since the first half of 2007. The demand of urea is forecast to grow about 40% in less than three years. In this context, knowledge improvements to operate more efficiently urea granulation plants will be extremely worthy. In line with the projected urea market conditions, the presented mathematical model is a valuable contribution for the understanding of the urea classification step and the improvement of the existing knowledge in granulation circuits. It is useful not only to predict the particle size distribution of the product, crusher feed and undersize recycle for specified conditions, but also to determine the effect of different operating variables (such as deck apertures) on the screen performance. In addition, it represents an accurate module for being used in plant simulators of the complete urea granulation circuit.

Literature Cited

(1) Korotkiy I.; Brazgovka A.; Kremers G. *Tenth Stamicarbon Urea Symposium*, **2004**.

(2) Bertin, D.E.; Mazza, G.D.; Piña, J.; Bucalá, V. *Ind. & Eng. Chem. Res.*, **2007**, *46*, 7667-7676.

(3) Cotabarren, I.; Schulz, P.G.; Bucalá, V.; Piña, J. *Powder Technology*, **2008**, *183* (2), 224-230.

(4) Balliu, N.E. PhD. Thesis, University of Queensland, Australia, **2005**.

(5) Heinrich, S.; Peglow, M. ; Mörl, L. *Powder Technology*, **2003**, *130*, 154-161.

(6) Zhang, J.; Litster, J.D.; Wang, F.Y.; Cameron, I.T. *Powder Technology*, **2000**, *108*, 122-129.

(7) Drechsler, J.; Peglow, M.; Heinrich, S.; Ihlow, M.; Mörl, L. *Chem. Eng. Sci.*, **2005**, *60*, 3817-3833.

(8) Radichkov, R.; Müller, T.; Kienle, A.; Heinrich, S.; Peglow, M.; Mörl, L. A Numerical Bifurcation Analysis of Continuous Fluidized Bed Spray Granulation with External Product Classification. *Chem. Eng. & Proc.*, **2006**, *45*, 823-837.

(9) Adetayo, A.A.; Litster, J.D.; Cameron I.T. *Computers Chem. Eng.*, **1995**, *19*, 383-393.

(10) Wildeboer, W.J. Master Thesis, Delft University of Technology, The Netherlands, **1998**.

(11) Karra, V. K. *CIM Bull*, **1979**, *72* (804), 167-171.

(12) Whiten, W. J., PhD Thesis, University of Queensland, Australia, **1972**.

(13) Ferrara, G.; Preti, U.; Schena, G.D. *Int. Journal of Mineral Processing*, **1988**, *22*, 193-222.

(14) Subasinghe, G. K. N. S.; Schaap, W.; Kelly, E. G. *Powder Technology*, **1989**, *59*, 37-44.

(15) Subasinghe, G. K. N. S.; Schaap, W.; Kelly, E. G. *Int. Journal of Mineral Processing*, **1990**, *28*, 289-300.

(16) Soldinger, M. *Minerals Engineering*, **1999**, *12* (5), 497-516.

(17) Wills, B.; Napier-Munn, T. *Will's Mineral Processing Technology, Seventh Edition: An Introduction to the Practical Aspects of Ore Treatment and Mineral Recovery*; Butterworth-Heinemann: **2006**, UK.

(18) Aspen Plus 12.1. Unit Operation Models. Aspen Technology Inc. Cambridge, MA, June **2005**.

(19) Modsim Simulator. Mineral Technologies Inc. Salt Lake City, UT, USA. http://www.mineraltech.com/MODSIM.

(20) Moly-Cop Tools. Moly-Cop Chile S.A., Santiago, Chile. http://www.molycop.cl.

(21) Standish, N.; Bharadwaj, A.K.; Hariri-Akbari, G. *Powder Technology*, **1986**, *48*, 161-172.

(22) Reimers, C.; Werther, J.; Gruhn, G. *Chem. Eng. & Proc.*, **2008**, *47*, 138-158.

(23) Athena Visual Studio, Software for Modeling and Parameter Estimation, Athena Visual Software, Inc., Naperville, IL ,USA. http://www.AthenaVisual.com.

(24) King, R.P. *Modeling and Simulation of Mineral Processing Systems*; Butterworth-Heinemann: **2001**, Oxford.

(25) *Nordberg Process Machinery: Reference Manual*, Renox, 1st Edition, **1976**.

(26) Plitt, L.R. The Analysis of Solid-Solid Separations in Classifier. *CIM Bull*, **1971**, *64*, 42-47.

189g Simulation of Continuous Granular Mixing

Avik Sarkar and Carl R. Wassgren. School of Mechanical Engineering, Purdue University, 500 Central Drive, Potter 209-A, West Lafayette, IN 47906

Continuous particulate blending has received much attention in the recent decade as it provides a more economical and efficient alternative over batch blending for large scale manufacturing. The influence of operating conditions in continuous blenders is not well understood quantitatively, even though qualitative parallels can be drawn with batch blending for similar geometries. This work investigates the influence of fill level and impeller rotation rate (Froude number) in a horizontal bladed continuous blender using the discrete element method (DEM). Particle flow within the blender was found to be strongly dependent on the Froude number and fill level, with distinct fluidized and quasi-static regimes at smaller fills. The axial flow rates showed significant variation with Froude number and fill, and also showed considerable variation over the course of a shaft revolution. Favorable mixing was obtained at smaller impeller rotation rates for larger fills, but at larger impeller rotation rates for smaller fills.

189h Coagulation of Highly Concentrated Aerosols

Beat Buesser and Sotiris E. Pratsinis. Particle Technology Laboratory, Institute of Process Engineering, Department of Mechanical and Process Engineering, ETH Zurich, Sonneggstrasse 3, Zurich, CH-8092, Switzerland

Coagulation in aerosol and other particulate or colloidal processes is described typically by the Smoluchowski theory. Corrections to that have been developed for van der Waals forces, charged or non-spherical particles and other particle interactions without altering, however, its basic theoretical framework. By definition, the Smoluchowski equation cannot describe coagulation at high particle concentrations (effective volume fraction above 1%). Such concentrations can be encountered in large scale aerosol manufacture of materials as fumed silica (1) and titania (2) as well as alumina and carbon black or soot.

The present work extends the coagulation theory for concentrated aerosols to the free molecule and transition regimes (3). Subsequently, Langevin dynamics (LD) simulations tracing individual particle trajectories were carried out for polydisperse, spherical aerosols at high concentrations. The influence of the particle concentration on coagulation rate and size distribution is investigated with aerosols growing continuously from the free molecule through transition into the continuum regime where coagulation rates are 2 - 30 times higher than that of classic Smoluchowski theory. A relationship is developed for the aerosol coagulation rate of highly concentrated aerosols from the free molecule to continuum regime.

Applications of this theory are in materials synthesis at high temperatures and/or low pressure. The evolution of the enhancement of coagulation rate and the particle size distribution under the influence of particle concentration and size is discussed. Furthermore deviations from ideal behavior in the free molecular regime are presented and discussed.

1. M.C. Heine, S.E. Pratsinis, "High Concentration Agglomerate Dynamics at High Temperatures", *Langmuir*, **22**, 10238-10245 (2006).

2. M.C. Heine, S.E. Pratsinis, "Agglomerate TiO_2 Aerosol Dynamics at High Concentration", *Part. Part. Syst. Charact.*, **24**, 56-65 (2007).

3. M.C. Heine, S.E. Pratsinis, "Brownian coagulation at high concentrations", *Langmuir*, **23**, 9882-9890 (2007).

189i Preparation of Dextran Particles by Using Sas Process

Dong-Yuk Kang[1], Dong-Lyun Cho[1], and **Choon-Hyoung Kang**[2]. (1) Department of Advanced Chemicals, Chonnam National University, 300 Yongbong-Dong, Buk-ku, Gwangju, 500-757, South Korea, (2) Chemical Engineering Department, Chonnam National University, 300 Yongbong-dong, Buk-ku, Kwangju, South Korea

In this work, micro-sized particles of dextran, which has recently been focused as one of the candidate materials for the Drug Delivery System(DDS), were prepared by means of the Supercritical Antisolvent (SAS) process with CO2. With dimethyl sulfonate (DMSO) as the solvent, effects of the operating variables such as temperature (308.15K~323.15K), pressure (90bar~130bar), solute concentration (10 mg/ml~20 mg/ml), and the molecular weight of the solute(Mw=37,500, 450,000) on the size and morphology of the resulting particles were thoroughly examined. While the higher solute concentration resulted in the larger particles, the injection velocity of the solution and pressure did not show significant effects on the size of the resulting particles. For dextran of the lower molecular weight (Mw=37500), the smallest particle size was obtained at 313.15K. On the other, the size of the particles resulted from the high molecular weight dextran ranged between 0.1~0.5μm. Further, the higher temperature and pressure the larger the particles. For the solute concentration of 5 mg/ml, the lower molecular weight dextran did not form discrete particles while aggregation of the particles appeared when the solute concentration exceeded 15mg/ml for the higher molecular weight dextran. It is believed that if the solute concentration is too low, the degree of the supersaturation in the recrystallization chamber would not be sufficient for initiation of the nucleation and growth mechanism. Instead, the spinodal decomposition mechanism leads to formation of the island-like phase separation which appears similar to aggregation of the discrete particles. This effect would be more pronounced for the smaller molecular weight polymer system due to the narrower phase-splitting region.

189j Effects for TiO2 Structures In Dye Sensitized Solar Cell

Byoung-Jun Min[1], Bok-Min Kim[1], Seon-Gyun Rho[2], Sung-Yong Cho[1], Seung-Jai Kim[1], and **Choon-Hyoung Kang**[3]. (1) Department of Environmental Engineering, Chonnam National University, 300 Yongbong-Dong, Buk-ku, Gwangju, 500-757, South Korea, (2) BK21 Development of Envrionmentally Firendly Core Materials and Processes, Chonnam National University, 300 Yongbong-Dong, Buk-ku, Gwangju, 500-757, South Korea, (3) Department of Applied Chemical Engineering, Chonnam National University, 300 Yongbong-Dong, Buk-ku, Gwangju, 500-757, South Korea

Nanocrystalline TiO2 is now widely applied to dye-sensitized solar cells(DSSC) as a nanocrystalline structure which effectively delivers electrons injected from optically excited dye to the back contact. Further, more efficient photoelectrochemical performance of DSSC was demonstrated with the anatase TiO2 compared with the rutile-based DSSC. In this study, we investigated the effects of TiO2 morphology on the photovoltaic conversion efficiency of DSSC solar cell in a quantitative way.

For this purpose, TiO2 nanoparticles with varying content of rutile and anatase were prepared by using the sol-gel method from Titanium(IV) isopropoxide(TTiP) solution. DSSC with purely anatase DSSC showed by 2% higher conversion efficiency than the pure rutile DSSC. This is attributed to the less adsorbed amount of the dye due to the smaller surface area and the lower degree of interparticle connectivity of rutile TiO2. DSSC fabricated with TiO2 paste prepared by mixing anatase and rutile in the ratio of 9:1 showed a more efficient photoelectrochemical performance than DSSC with the TiO2 prepared in the ratio of 7:3.In General, the higher the anatase content, the higher the photovoltaic conversion efficiency. However, it is noteworthy that conversion efficiency levels off as the oxide morphology approaches the pure anatase form. In other words, the maximum photovoltaic conversion efficiency for varying contents of rutile and anatase was attained when the oxide morphology is somewhat between pure forms. This result can be explained in the way mentioned above.

189k The Role of Gas-Aerosol Mixing during In-Situ Coating of Flame-Made Titania Particles

Alexandra Teleki, Department of Mechanical and Process Engineering, Swiss Federal Institute of Technology Zurich, Particle Technology Laboratory, Sonneggstrasse 3, Zurich, CH-8092, Switzerland, Martin C. Heine, Particle Technology Laboratory, Swiss Federal Institution of Technology Zurich, Sonneggstrasse 3, Zurich, 8092, Switzerland, Frank Krumeich, Laboratory for Inorganic Chemistry, ETH Zurich, Wolfgang-Pauli-Str. 10, Zürich, CH-8093, Switzerland, Kamal M. Akhtar, Millennium Inorganic Chemicals (a Cristal Company), 6752 Baymeadow Drive, Glen Burnie, MD 21060, and Sotiris E. Pratsinis, Particle Technology Laboratory, Institute of Process Engineering, Department of Mechanical and Process Engineering, ETH Zurich, Sonneggstrasse 3, Zurich, CH-8092, Switzerland.

Rutile TiO_2 particles of 40 nm average diameter were made by flame spray pyrolysis (FSP) and in-situ coated with ultrathin (2 – 4 nm) SiO_2 layers. The spray flame was enclosed by a quartz glass tube while hexamethyldisiloxane (HMDSO) vapor-laden N_2 was injected in crossflow to and downstream of the flame through a metal torus pipe ring. The as-prepared powders were characterized by transmission electron microscopy (TEM), X-ray diffraction and nitrogen adsorption. The coating quality of titanium dioxide was evaluated further by photocatalysis of isopropanol to acetone under UV light using suspensions of the above particles. The effect of mixing intensity on product particle morphology was investigated by varying the exit velocity or the number of jet outlets for HMDSO-laden N_2 gas. At high mixing intensity between the TiO_2 aerosol and HMDSO-laden jet streams, the particles were encapsulated by smooth and homogeneous coatings. Poor mixing between HMDSO vapor and TiO_2 aerosol resulted in separate SiO_2 and poorly-coated TiO_2 particles. The effect of mixing TiO_2 aerosol with the HMDSO vapor stream jets was elucidated also by computational fluid dynamics: low flowrates or few jet outlets of the HMDSO vapor stream resulted in incomplete mixing across the radius of the reactor, consistent with the experimental data. Well-coated TiO_2 particles produced with multiple coating vapor jets or high mixing intensity exhibited limited photooxidation indicating the effectiveness of their SiO_2 coatings.

189l New Flame Pyrolysis Reactor for Carbon Black Synthesis: Soot Surface Growth and Oxidation

X. Ma[1], **Anshuman Lall**[2], Dale Huang[2], and Michael Zachariah[3]. (1) University of Maryland-College Park, College Park, MD 20742, (2) Department of Mechanical Engineering, University of Maryland-College Park, College Park, MD 20742, (3) University of Maryland and NIST, 2125 Glenn L Martin Hall, College Park, MD 20742

A new relatively large-scale laboratory reactor (up to ~100 g/hr) for carbon black synthesis via improved soot surface growth is developed. The reactor consists of a diffusion flame on which the heated fuel/liquid precursor is sprayed in gas/droplet form with nitrogen as carrier gas. The sprayed fuel gets partially pyrolyzed to carbonaceous soot particles and other gaseous byproducts. The resulting soot aerosol is maintained at a high temperature for a certain time to allow soot surface growth onto previously formed seed soot particles. However, the overall soot yield is a complex interplay of competing soot surface growth [Yozgatligil, A. and Zachariah, M. R. (2008) Measurement of soot surface growth kinetics. Combust. Sci. and Tech., 180: 941–949] and post-oxidation kinetics. Our goal is to optimize soot yield by improving the soot surface growth. In our preliminary measurements, it is found that a higher reactor temperature leads to lower yield due to higher soot post-oxidation rate. The results are supported by the increase in total CO and CO2 concentrations.

The soot yield measurements are made by using a novel online aerosol method of measuring soot yield in the gas phase. The online method enabled us to measure soot yield without knowing the total product flow rate for a number of process parameters and operating conditions such as the overall and burner equivalence ratio, and the time-temperature history. Our preliminary measurements indicate that the soot yield increases with the overall equivalence ratio at a given burner equivalence ratio. The CO and CO2 emissions from the reactor are monitored and suggestions are made to reduce emissions of the greenhouse gases.

189m Preparation of CeO_2/SiO_2 Nano-Abrasive Particles

Sang-Ho Chung[1], In-Seon Park[1], Young-Moo Park[1], Dae-Won Lee[2], and Kwan-Young Lee[1]. (1) Chemical and Biological Engineering, Korea University, Anam-dong Seongbuk-Gu, Seoul, 136-701 Korea, Seoul, South Korea, (2) Research Institute of Clean Chemical Engineering, Korea University, Anam-dong Seongbuk-Gu, Seoul, 136-701 Korea, Seoul, South Korea

Chemical Mechanical Polishing (CMP) is a surface planarization method in which a wafer rotated against a polishing pad in the presence of abrasive slurries while applying pressure. The common abrasive particles used in the CMP are usually irregularly shaped and of broad size distributions. It is, therefore, of general interest to carry out polishing processes with slurries, consisting uniform particles of morphologies, sizes, and other properties.

It is well known that the ceria particles are used in CMP as abrasive particles and represent high removal rates of oxide films than silica particles due to the ability to interact with the surface of the SiO_2 layer. The ceria abrasive nanoparticles are also desirable for the CMP of shallow trench isolation (STI), due to their high selectively of silicon dioxide over silicon nitride. However, the ceria particles have irregular shape and are easy to aggregate each other due to their high specific gravity. Therefore, the ceria abrasives used in CMP have negative effects on the polished films such as excessive scratches and defects.

To make up for the weak points of the ceria abrasives, there is a method to coat CeO_2 on the SiO_2 nanoparticles that have spherical shape and low specific gravity. In this study, the microemulsion method was introduced to make CeO_2 thin film deposition on the SiO_2 nanoparticles. Microemulsions are isotropic, clear or translucent, thermodynamically stable dispersions of oil, water, emulsifier, and often a small amphiphilic molecule called a cosurfactant. These surfactant-covered entities (referred to as Water-in-Oil (W/O) microemulsion droplets) offer a unique microenvironment for inorganic precipitation reactions. In this experiment, the silica nanoparticles were synthesized first, and ceria was thin coated on the silica by adding ceria precursor to microemulsion which contain silica nanoparticles inside the micelles. In this work, we conducted a systematic investigation of the effect of the surfactants species, the amount of cosurfactants and the pH of the microemulsion system.

Web Page: catalysis.korea.ac.kr

189n Shear-Induced and Brownian Coagulation

Max L. Eggersdorfer, Beat Buesser, and Sotiris E. Pratsinis. Particle Technology Laboratory, Institute of Process Engineering, Department of Mechanical and Process Engineering, ETH Zurich, Sonneggstrasse 3, Zurich, CH-8092, Switzerland

Smoluchowski (1) developed collision frequencies for these two cases of coagulation. One case describes the collision of particles by Brownian motion (perikinetic coagulation) and the other neglects diffusion as particle collisions are induced by the surrounding fluid flow (orthokinetic coagulation). In both cases inter-particle forces are neglected. When both mechanisms are present, the combination affects the collision rate and the resulting particle size distribution.

Swift & Friedlander (2) concluded that the collision frequencies for Brownian motion and laminar shear are independent and additive for a wide range of conditions. Contradictory Van de Ven & Mason (3) showed by solving the convective diffusion equation for conditions with a low influence of shear ($Pe \ll 1$) that the two mechanisms are coupled and are not simply additive. Shear enhances the coagulation rate proportional to the square root of the Peclet number, which represents the ratio between the convective (inertial) and diffusive terms. Feke & Schowalter (4) studied the regime for high Peclet numbers ($Pe \gg 1$) where the dominant mechanism is shear. Their analysis of binary collisions also revealed a dependency between the contribution of Brownian motion and shear. They concluded that the combination of both can either increase or decrease the collision frequency. All of these models treat collisions between two particles by calculating the flux of a particle to a reference sphere. The flux is typically computed from the pair probability function, which gives the probability to find two particles at a certain position.

Ernst and Pratsinis (5) showed by population balance simulations that shear-induced coagulation leads to pseudo self-preserving distributions that end to gelation. Heine and Pratsinis (6) simulated Brownian coagulation at high concentrations with Langevin dynamics (LD) and showed that classic Smoluchowski equation can not describe highly concentrated systems (particle volume fraction above 1%). They found out that for complete particle coalescence upon collision, the self-preserving size distribution is still obtained for high concentrations. In the limiting cases, where either peri- or orthokinetic aggregation dominates, the process of coagulation is understood well. The transition regime on the other hand is not well understood.

Here Brownian and shear-induced coagulation are investigated by LD simulations for a multi-particle system of coalescing spheres. Initially all particles are monodisperse. Simulations are done for monodisperse and polydisperse systems. In monodisperse simulations particles are redistributed after a collision. If particles collide in polydisperse simulations, full coalescence takes place. Collision frequencies and particle size distributions are obtained for a wide range of *Pe*, including the two limiting cases of pure shear-induced coagulation and pure Brownian coagulation.

1. Smoluchowski, M. V., (1917). *Z. Phys. Chem.*, **92**, 129-.

2. Swift, D. L., & Friedlander, S. K. (1964). *J. Colloid Sci.*, **19**, 621-647.

3. van de Ven, T. G. M. & Mason, S. G. (1977). *J. Colloid Polymer Sci.*, **255**, 794-804.

4. Feke, D. L., & Schowalter, W. R. (1985). *J. Colloid Interface Sci.*, **36**, 203-214.

5. Ernst, F. O., & Pratsinis, S. E. (2006). *J. Aerosol Sci.*, **37**, 123-142.

6. Heine, M. C., & Pratsinis, S. E. (2007). *Langmuir*, **19**, 9882-9890.

THREE-DIMENSIONAL GAS-SOLID FLUIDIZED BED SIMULATION USING A LATTICE BOLTZMANN BASED DRAG CORRELATION.

Maximilian J. Hodapp, School of Chemical Engineering, University of Campinas, Campinas-SP, Brazil.
Milton Mori, School of Chemical Engineering, University of Campinas, Campinas-SP, Brazil.
Maria G. E. Silva, School of Chemical Engineering, University of Campinas, Campinas-SP, Brazil.

ABSTRACT

From the pool of industrial gas-solid fluidization processes those using circulating fluidized bed (CFB) technology are implemented in a wide variety of fields, including fluid catalytic cracking (FCC), combustors, powder generation and mineral processing. As this process requires a huge economical investment for building a new or upgrading an existing unit, there is an especial interest in understanding the fluid dynamic behavior of CFBs. Unfortunately there is no sufficiently accurate analytical model that describes all the complex phenomena that occur in such units. In this context the computational fluid dynamic (CFD) is a very promising tool, allowing the simultaneous modeling and numerical solution of mass, energy and momentum transfer as well as the hidrodynamics and collisional phase interactions even on very complex geometries. This work has as aim to study the implementation of a drag force correlation in three-dimensional gas-solid CFD simulations, performed with a commercial CFD code (Ansys CFX 10.0), in a laboratory scale. This correlation was developed by [4], which was obtained not by the usual empirical data, but was instead formulated on lattice Boltzmann simulations data and later extended by [2]. A Fortran subroutine was developed in order to include in the commercial CFD code a set of drag correlations, based on data from lattice-Boltzmann simulations. The numerical results were compared with the laboratory scale CFB experimental work of [10]. The apparatus consist of a 1 m vertical column with an internal diameter of 0.032 m. For the numerical mathematical modeling, the Eulerian-Eulerian approach was used to describe the gas-solid flow in a CBF. For the gas phase the k-epsilon model was applied to represent the turbulence, and a laminar model was used for the particles phase. Also, the results of this work were compared with simulations carried out by [8] using the hybrid drag correlation suggested by [3].

Keywords: gas-solid fluidization, CFD, lattice Boltzmann drag correlation

1. INTRODUCTION

The main hydrodynamic phenomena to be accounted for in a gas-solid fluidization process in order to correctly model it , is the thorough contact between the solid particles and the fluid phase. A large number of experimental and numerical researches have been conducted and reported in the literature [1,7,9,10,11,12]. The reason for such an interest in this process is due to its large application in several branches of industries, including chemical, petrochemical, metallurgical and food industries, taking advantage of the favorable mass and heat transfers properties of such units. Still, their design and scale-up remains very difficult not only because of the complex hydrodynamic, but also of the other phenomena that may be involved. Even so, only the modeling of the multiphase flow is by itself a challenging accomplishment.

Different approaches have been developed considering specific spatial scales for these problems. As described by [5], three approaches are commonly used for the mathematical modeling, these being the lattice Boltzmann method (LBM), the Eulerian-Lagrangean or Discrete Particle Model (DPM) and the Eulerian-Eulerian or Two-Fluid Model (TFM) approach. The lattice Boltzmann is the most elementary methodology, describing the fluid-particle interaction at a microscopic scale, where both the fluid and the solids are represented as finite particles. According to [6], the LBM is based on a discrete particle kinetic, which involves collision and propagation of particles in a lattice. As an advantage, the simplicity of the algorithm and the accuracy of the mass and momentum conservations are cited. Recent articles [4,5] made use of this method to derive a drag law for gas-solid flow, since by doing so there is no need for empirical data to be used in order to formulate a drag correlation. The DPM calculates the fluid phase through the Navier-Stokes equations and the particles through Newton's law of motion. This method can produce very precise results for the particulate phase although it requires great computational processing and data storage capability, which depending on the amount of particles to be simulated, becomes prohibitive. The third approach, TFM, considers that the solid phase can be modeled as a fluid, resolving thus both phases through the Navier-Stokes equations. This is a very popular CFD methodology, requiring less computational effort, even though generating quite accurate results for a variety of multiphase flows.

In the present work the drag correlation developed by [4] and later extended by [2] was used in a turbulent gas-solid riser CFD simulation. The results were compared to experimental data from the work of [11] and to the CFD simulation of [8].

2. MODEL DESCRIPTION

All the numerical simulations were carried out using the commercial ANSYS CFX 10.0 CFD code. This code is based on the Finite Volume Method for the discretization of groups of partial differential equations. The well known k-epsilon turbulence model was applied for the gas phase. The particulate phase was modeled using a laminar approach. The multiphase flow was modeled with the two-fluid model, for which formulation and mathematical details can be found elsewhere [1,10]. The main closure equations, which defined this work, were briefly described.

Closure Equations:

The drag law applied was presented by [2]. The Friction Coefficient β is given by

$$\beta = \frac{3}{4}\frac{Cd(1-\phi)\rho_g\phi\left|\overrightarrow{v_g}-\overrightarrow{v_s}\right|}{d_s} \qquad (1)$$

where ρ_g is the gas density, $\emptyset$ is the solid volume fraction, $\vec{v_g}$ is the gas velocity vector, $\vec{v_p}$ is the solid velocity vector, d_s is the particle diameter and the adimensional drag coefficient Cd can be expressed as

$$Cd = 12\frac{(1-\phi)^2}{Re}F \tag{2}$$

where *Re* is the Reynolds number (Equation 3) and *F* is the drag force coefficient (Equation 4)

$$Re = \frac{\rho_g(1-\phi)\left|\vec{v_g} - \vec{v_s}\right| d_s}{2\mu_g} \tag{3}$$

here μ_g is the gas phase viscosity.

$$F = \begin{cases} 1 + \frac{3}{8}Re \ \ for\ \phi \leq 0.01,\ \ Re \leq \frac{(F_2 - 1)}{(3/8 - F_3)} \\ F_0 + F_1 Re^2\ for\ \phi > 0.01,\ \ Re \leq \frac{F_3 + \sqrt{F_3^2 - 4F_1(F_0 - F_2)}}{2F_1} \\ F_2 + F_3 Re\ \ for \begin{cases} \phi \leq 0.01\ ,\ \ Re > \frac{(F_2 - 1)}{(3/8 - F_3)} \\ \phi > 0.01\ ,\ \ Re > \frac{F_3 + \sqrt{F_3^2 - 4F_1(F_0 - F_2)}}{2F_1} \end{cases} \end{cases} \tag{4}$$

The Coefficients F_0, F_1, F_2, F_3 in the equations above are given by

$$F_0 = \begin{cases} (1-w)\left[\frac{1 + 3\sqrt{\frac{\phi}{2}} + \left(\frac{135}{64}\right)\phi ln(\phi) + 17.14\phi}{1 + 0.681\phi - 8.48\phi^2 + 8.16\phi^3}\right] + w\left[10\frac{\phi}{(1-\phi)^3}\right], 0.01 < \phi < 0.4 \\ 10\frac{\phi}{(1-\phi)^3}, \phi \geq 0.4 \end{cases} \tag{5}$$

$$F_1 = \begin{cases} \frac{\sqrt{\frac{2}{\phi}}}{40}, 0.01 < \phi \leq 0.1 \\ 0.11 + 0.00051e^{11.6\phi}, \phi \geq 0.4 \end{cases} \tag{6}$$

$$F_2 = \begin{cases} (1-w)\left[\dfrac{1+3\sqrt{\frac{\phi}{2}}+\left(\frac{135}{64}\right)\phi ln(\phi)+17.89\phi}{1+0.681\phi-11.03\phi^2+15.41\phi^3}\right]+w\left[10\dfrac{\phi}{(1-\phi)^3}\right], \phi < 0.4 \\ 10\dfrac{\phi}{(1-\phi)^3}, \phi \geq 0.4 \end{cases} \tag{7}$$

$$F_3 = \begin{cases} 0.9351\phi + 0.03667, \phi < 0.0953 \\ 0.0673 + 0.212\phi + \dfrac{0.0232}{(1-\phi)^5}, \phi \geq 0.0953 \end{cases} \tag{8}$$

In Equations (5) and (7) w is a weighting factor

$$w = e^{\left(\frac{\frac{-10}{0.4-\phi}}{\phi}\right)} \tag{9}$$

The drag relation used in the work of [8], for dense (Equation 10) and dilute (Equation 11) as functions of solid volume fractions

$$\beta = 150\frac{\phi^2}{(1-\phi)}\frac{\mu_g}{d_s^2} + 1.75\phi\frac{\rho_g|\vec{v_s}-\vec{v_g}|}{d_s}, \phi > 0.2 \tag{10}$$

$$\beta = \frac{3}{4}\frac{Cd(1-\phi)\rho_g\phi|\vec{v_g}-\vec{v_s}|}{d_s}(1-\phi)^{-2.65}, \phi < 0.2 \tag{11}$$

$$Cd = (1-\phi)^{-1.65} max\left[\frac{24}{Re(1-\phi)}\left(1+0.15(Re(1-\phi))^{0.687}\right), 0.44\right] \tag{12}$$

Physical Description

The experimental setup reported by [11] consisted of a 1.0 m long riser with an internal diameter of 0.032 m. At 0.05 m of its heights there is a secondary entrance with a diameter of 0.008 m. The density of the particles was 1600 kg/m^3 with a Sauter mean diameter of 0.06 mm. Three different superficial gas velocities were used in the experiment, 0.36 m/s, 0.71 m/s and 1.42 m/s, all at 25 ºC. A constant velocity of 0.05 m/s was held at the secondary entrance. The initial bed height of the solids was 0.05 m. Results for radial profiles of axial velocity were reported for three different heights of the column, namely 0.16 m, 0.32 m, 0.48 m.

An overview of the geometry design and numerical mesh can be seen in Figure 1.

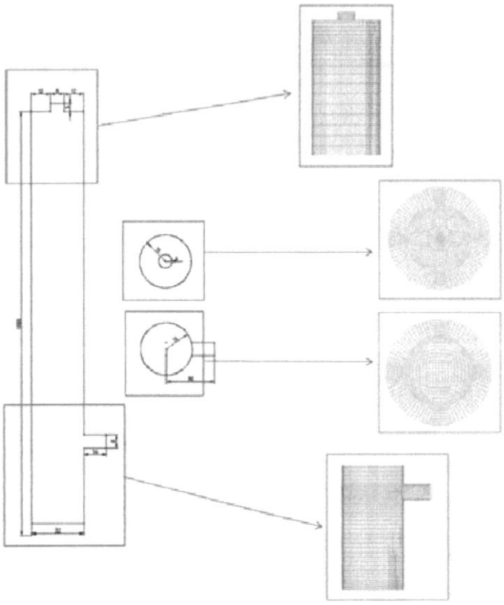

Figure 1: Geometry and details of the *Riser*'s mesh.

Initial and Boundary Conditions

No velocity profile was applied to the primary entrance due to the presence of the distributor in the experimental column. The constant velocity for each case, 0.36 m/s, 0.71 m/s and 1.42 m/s, was set at this boundary. For the secondary entrance a function that represented the recirculation of the solids was necessary. It was set as equal to the mass flux of the solid at the exit. Also, a constant air entrance velocity of 0.05 m/s was fixed there. An opening condition for the exit was preferred over an outlet due to numerical stability. The conditions at the walls were no slip for the gas phase and free slip for the solid phase. As an initial condition, the bottom of the column was filled with a bed height of 0.05 m. The time step of the numerical solver ranged from $1x10^{-5}$ s to $1x10^{-3}$ s. The RMS convergence criterion for the mass and momentum equations was set to $1x10^{-4}$. Higher order discretization schemes were selected.

3. RESULTS AND DISCUSSION

A mesh test was carried out in order to obtain an acceptable numerical accuracy with the smallest computational cost.The results showed that a mesh with about 300000 hexahedral volumes of control was suitable for this geometry.

The computational results obtained in the present work correspond to a time average of the last 7 s of simulation, hence assuming the initial bed expansion would have no influence on the results and that a pseudo-stationary state was achieved. The velocity profiles were presented in Figures 2, 3 and 4.

All the numerical results using the drag law presented by [3] were obtained by [8]. Both simulations showed the typical core-annulus flow, where a low solid volume fraction region moves upward in the center of the flow and a ring region with a higher particle volume fraction falls downward near the walls.

For the 0.36 m/s inlet velocity case, the current simulations showed that a strong internal particle re-circulating pattern was present. Due to this pattern the predicted radial particle velocity profiles were higher for the 0.16 m and 0.32 m height and lower for the 0.48 m measurement, as seen in Figure 2. Just as reported by [11] the profiles were not symmetric, especially near the secondary entrance.

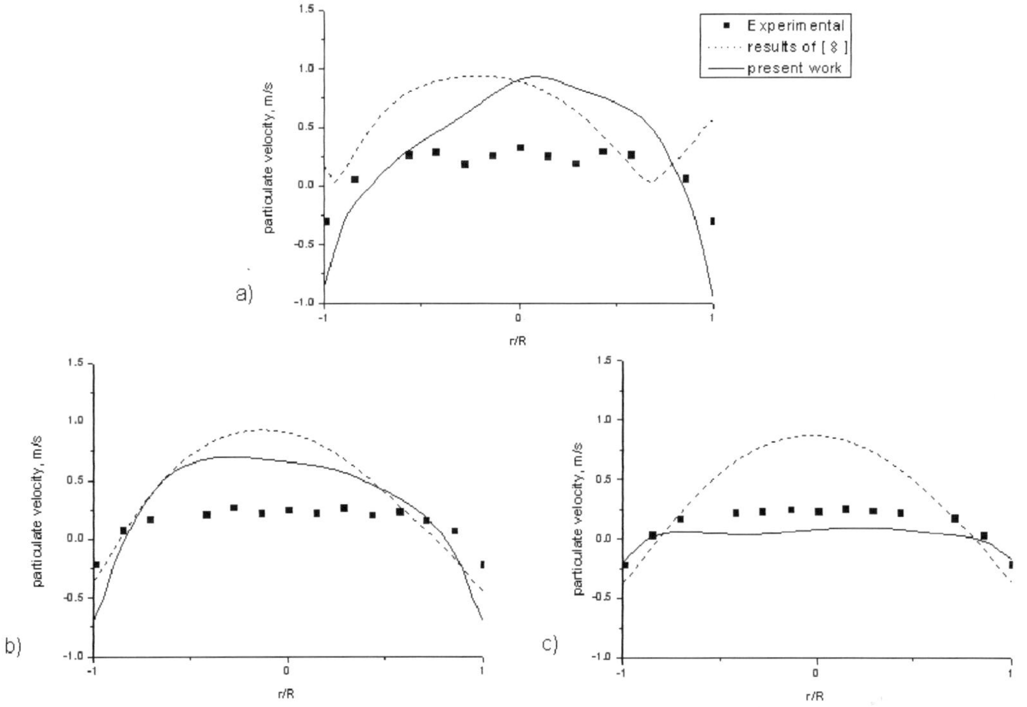

Figure 2: Experimental and simulated radial particle velocity for $v_{sup} = 0.36$ m/s, at heights of a) 0.16 m; b) 0.32 m; c) 0.48 m.

Figure 3 shows the 0.71 m/s inlet velocity case. These simulations were more consistent with the experimental data. However there was an overprediction of the core velocity, which was compensated by higher downflow velocities near the walls. The comparison of the two drag correlations showed only small differences between them, having both generated higher velocity profiles than those obtained experimentally.

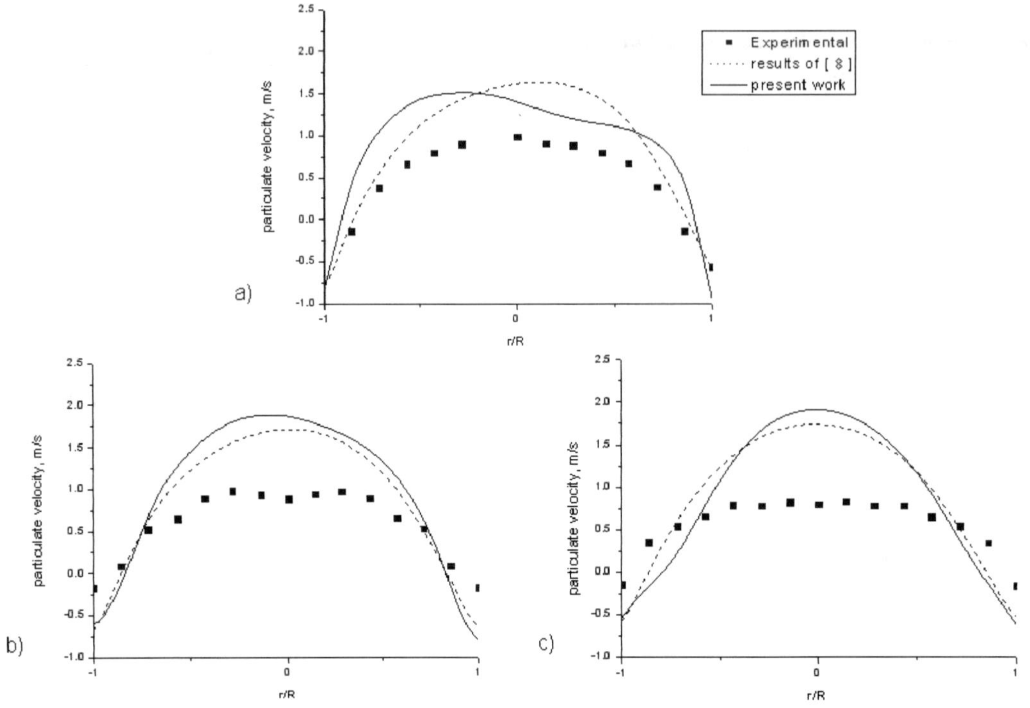

Figure 3: Experimental and simulated radial particle velocity for $v_{sup} = 0.71$ m/s, at heights of a) 0.16 m; b) 0.32 m; c) 0.48 m.

The 1.42 m/s inlet velocity case, Figure 4, also shows a similar solid velocity profile, when comparing the two drag correlations. Both have the same characteristics, producing a steeper profile at the center, especially at 0.32 m and 0.48 m, but a closer approximation with experimental data at the annulus region. At a hight of 0.16 m the predicted radial solids velocities showed good agrement with the measurements in both the core and annulus regions for the present work simulation.

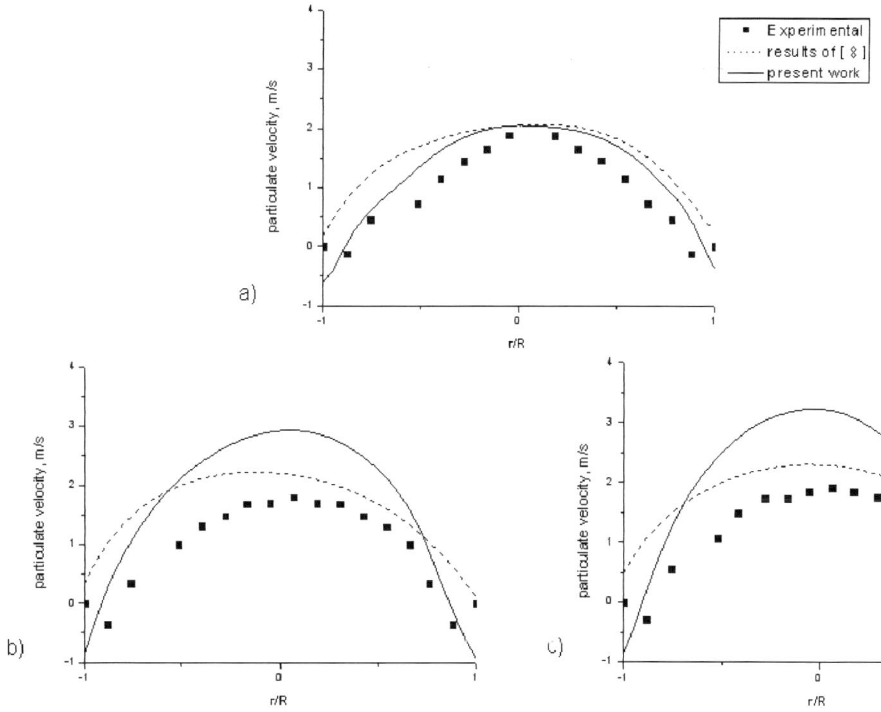

Figure 4: Experimental and simulated particle velocity for v_{sup} = 1.42 m/s, at heights of a) 0.16 m; b) 0.32 m; c) 0.48 m.

4. CONCLUSION

The study of a lattice Boltzmann based drag correlation was accomplished trough the transient 3D CFD simulations of a gas-solid CFB Riser. The results of these simulations were compared to the experimental work of [11] and the CFD simulation reported by [8]. The predicted radial solids velocities at the hights of 0.16 m and 0.48 m and the superficial velocities of 1.42 m/s and 0.36 m/s, respectively show good agreement with the measuments in both the core and annulus regions. For other cases studied, the results overpredicted the core flow region, but a set of drag correlations, based on data from lattice-Boltzmann simulations showed, in the regions near the wall, similar profiles to those found experimentally.

NOMENCLATURE

Cd	drag coefficient
d_s	particle diameter, m
F	drag force coefficient
μ_g	viscosity of gas phase, kg/m.s
$\rho_{g,p}$	density of gas or particulate phase, kg/m^3
Re	Reynolds number
$\vec{v}_{g,p}$	velocity vector gas or particulate phase, m/s
$\emptyset$	volume fraction of particulate phase

ACKNOWLEDGEMENTS

The authors of this research are thankful to PETROBRAS and to the government agency CAPES for the financial support.

REFERENCES

1. BASTOS, J. C. S. C.; ROSA, L. M.; MORI, M.; MARINI. F; MARTIGNONI, W. P. (2008), Modeling and simulation of a gas-solids dispersion flow in a high-flux circulating fluidized bed (HFCFB) riser. *Catalysis Today*, 130 (2/4), p. 462-470.
2. BENYAHIA, S.; SYAMLAL, M.; O'BRIEN, T. J. (2006), Extension of Hill-Koch-Ladd drag correlation over all ranges of Reynolds number and solids volume fraction. *Powder Technology*, v. 162, p. 166-174.
3. GIDASPOW, D. (1994), Multiphase Flow and Fluidization – *Continuum and Kinetic Theory Descriptions*, Academic Press, Inc., San Diego, California.
4. HILL, R. J.; KOCH, D. L.; LADD, A. J. C. (2001) The first effects of fluid inertia on flows on ordered and random arrays of spheres. *Journal of Fluid Mechanics,* v. 448, p. 213-241.
5. HOEF, M. A.; ANNALAND, M. S.; KUIPERS, J. A. M. (2005), Computational fluid dynamics for dense gas-solid fluidized beds: a multi-scale modeling strategy. *China Particuology*, v. 3, p. 69-77.
6. INAMURO, T. (2006), Lattice Boltzmann methods for viscous fluid flows and for two-phase fluid flows. *Fluid Dynamics Research. v.*38, p. 641-659.
7. KNOWLTON, T. M. (2005), Experimental modeling of gas-solid flow systems. In: *Workshop Latino Americano de CFD Aplicado À Indústria de Petróleo (CFD OIL)*. Rio de Janeiro.
8. MARINI, F. (2008), Simulation of a fluidized bed applying the CFD technique based in the kinetic theory of granular flow, *M.Sc Thesis*, University of Campinas, Campinas, 2008.
9. MEIER, H. F.; ALVES, J. J. N.; MARTIGNONI, W.; MORI, M. (2000), Prediction of the fluid dynamics of circulating fluidized bed reactors by numerical simulation. *Rio Oil & Gas Expo and Conference.*
10. ROPELATO, K.; MEIER, H.; CREMASCO, M. A. (2005), CFD study of gas-solid behavior in downer reactors: An Eulerian-Eulerian approach. *Powder Technolog.* v. 54, p. 179-184.
11. SAMUELSBERG, A.; HJERTAGER, B. H. (1996), An experimental and numerical study of flow patterns in a circulating fluidized bed reactor. *International Journal of Multiphase Flow*, v. 22, p. 575–591.
12. SILVA, M. G. E.; MARINI, F.; MORI, M. (2008), Three dimensional gas-solid fluidized bed simulation based on the kinetic theory of granular flow. CHEMPOR, Portugal.

189p Optimization of Particle Agglomeration and Breakage Operations with the Attainable Region (AR) Theory

Matthew J. Metzger[1], Benjamin J. Glasser[2], Ngangezwe Khumalo[3], Brendon Hausberger[4], Diane Hildebrandt[5], and David Glasser[5]. (1) Chemical and Biochemical Engineering, Rutgers University, 98 Brett Rd, Piscataway, NJ 08854, (2) Dept. of Chemical and Biochemical Engineering, Rutgers University, 98 Brett Road, Piscataway, NJ 08854, (3) University of the Witwatersrand, School of Process and Materials Engineering, Wits University, P. Bag 3, Wits, Johannesburg, 2050, South Africa, (4) Centre of Material and Process Synthesis, School of Chemical and Metallurgical Engineering, University of the Witwatersrand, Johannesburg, 1 Jan Smut Avenue, Johannesburg, South Africa, (5) COMPS, School of Chemical and Metallurgical Engineering, Wits University, Private Bag x3, PO Wits, Johannesburg, 2050, Johannesburg, 2050, South Africa

Tight particle size distribution control is vital to the success of any process involving particles. However, even controlling crystal size during crystallization - a process with a relatively well-known mechanism - is a daunting chore. In other operations central to the pharmaceutical and chemical industries, such as agitated drying and granulation, the task of size control is complicated due to the competition between particle agglomeration and breakage. A tool that with the ability to encompass all aspects of particle agglomeration and breakage within one framework is needed to advance the development of solid processes.

Presented here is the application of the Attainable Region (AR) approach to optimizing the simplest case of an industrial sizing process: milling. Previously, the AR approach has been used to optimize complex reactor networks and separation systems. Similarities between species in chemical reaction networks and particle sizes in agglomeration and breakage processes, suggest the AR is a capable tool for determining the optimal control strategy for a milling project. Though the optimization done in this work is for a batch ball mill and a specific material, the concepts can be applied to any system consisting of breakage, agglomeration, and classification.

We have experimentally determined the optimal control strategy for achieving a certain size distribution with the shortest amount of time and most efficient energy usage. The control strategy recommends a variation of speed throughout the process - fast speed at first followed by a lower speed - to achieve the maximum amount of material of a desired size. Benefits of this approach can be realized in any size alteration operation, as a size profile over time can be predicted. Therefore, a design methodology can be constructed without complicated mathematical tools.

189q Force Model Concerns for Glued Sphere Dem Simulations

Madhusudhan Kodam[1], Rahul Bharadwaj[2], Jennifer Sinclair Curtis[3], Bruno C. Hancock[2], and Carl R. Wassgren[1]. (1) Mechanical Engineering, Purdue University, 500 Central Drive, Room # 209, Potter Building, West Lafayette, IN 47907, (2) Pfizer Inc., MS 8156-007, Eastern Point Road, Groton, CT 06340, (3) University of Florida, Chemical Engineering Department, P.O. Box 116005, Gainesville, FL 32611-6005

Despite knowing that particle shape plays a significant role in the dynamics of powder flow, most discrete element method (DEM) simulations utilize spherical particles. The reasons for using spherical particles are that (a) the contact detection scheme for spherical particles is simple, and (b) the contact force models for contacting spheres are well known (e.g. a Hertzian contact).

Several schemes for modeling non-spherical particles have been proposed including those that concern polyhedra, ellipsoids, sphero-cylinders, and superquadrics. Perhaps the most common approach for modeling non-spherical particles, however, is using "glued spheres," in which irregular particle shapes are produced by rigidly connecting individual, and possibly overlapping, spheres. The advantage of the glued spheres approach is that even for complex particle shapes the simple spherical contact detection algorithm may be retained.

Recent publications have focused on how approximating a given particle shape using a glued sphere geometry affects the rebound of colliding particles (e.g. Price et al., 2007 and Kruggel-Emden et al., 2008). These investigations have focused on the errors introduced by approximating the geometry of the true particle shape. What has not been investigated, however, is how the spherical particle derived force models used in glued sphere particle geometries influence the response of particle collisions. This paper demonstrates that in instances where more than a single component sphere in a glued sphere model is involved in a contact, a modified force model must be used to produce an accurate force-deflection response.

189r Effect of Dem Parameters on the Collision Distribution of a Small Shearing Region below the Surface of a Rotating Drum

Ben Freireich, School of Mechanical Engineering, Purdue University, 500 Central Drive, Rm. 209a Potter Building, West Lafayette, IN 47907, James D. Litster, Chemical Engineering and Industrial Pharmacy, Purdue University, Forney Hall of Chemical Engineering, Room 2154, 480 Stadium Mall Drive, West Lafayette, IN 47907-2100, and Carl R. Wassgren, Purdue University, 585 Purdue Mall, West Lafayette, IN 47907.

Traditionally, DEM simulations are validated via particle trajectories and velocity fields using experimental techniques such as PIV or PEPT. However, some researchers use DEM simulations to extract particle-scale measurements such as collision frequency. It is unclear whether a validation of a macroscopic velocity field also implies a validation of particle-scale measurements. In this study, measurements of solid fraction, shear rate, and collision frequency and impact velocity are made in a small region just below the free surface in a rotating drum. The effects of periodic drum length, particle stiffness, coefficient of restitution, and particle size are investigated. It is found that the solid fraction and shear rate within the studied region are unaffected by variations in particle stiffness and coefficient of restitution over the range of values studied. However, the rate of collisions increases for both increasing particle stiffness and coefficient of restitution. Also, the mean collision speed decreases with increasing stiffness, and decreasing coefficient of restitution. The shear rate within the region decreases as particle size increases, while the mean collision velocity remains unaffected. These findings suggest that care must be taken in choosing DEM force models and model parameters in order to accurately model collision distributions, while particle size need not be matched exactly.

189s Behavior of the Elastomer Particles with Different Surface Charge Groups during Aggregation Under Static and Flow Conditions

Miroslav Soos[1], Cornelius Gauer[1], Hua Wu[1], and Massimo Morbidelli[2]. (1) Institute for Chemical and Bioengineering, Department of Chemistry and Applied Biosciences, ETH Zurich, Wolfgang-Pauli-Str. 10, Zurich, 8093, Switzerland, (2) Institute for Chemical and Bioengineering, Department of Chemistry and Applied Biosciences, ETH Zurich, Wolfgang-Pauli-Str. 10, HCI F 133, Zurich, 8093, Switzerland

Aggregation (coagulation or floculation) is a commonly used process to separate colloidal particles from liquid, where the particles after addition of appropriate amount of coagulant are aggregated to form large clusters. Besides Brownian motion, which is the controlling mechanism for sub-micron particles or clusters, effect of shear, induced by stirring of the suspension, contributes additionally to the cluster growth as well as it controls final size of formed aggregates. The final size of produced aggregates is strongly connected to the material and surface properties of the primary particles, internal structure of aggregates and inter-particles forces between primary particles composing an aggregate. When primary particles undergo coalescence the situation becomes even more complicated.

In this work we investigate aggregation behavior of elastomer particles with sub-micron size and different surface charge groups under both, static and flow conditions. In all experiments the amount of added coagulant (acid and various types of salts) was well above the critical coagulation concentration. Depending on the size of produced aggregates either light scattering (dynamic as well as static light scattering) or image analysis were used for characterization. Moreover, for selected conditions cryogenic scanning electron microscopy and confocal scanning microscopy were used to analyze the cluster structure. Simulation of the aggregation kinetics with comprehensive, physical models (population balances) provides useful information about the physics and chemistry governing the aggregation process.

It was found that depending on the nature of the particle surface charge groups as well as elastomer bulk properties, coalescence between primary particles can occur, resulting in significantly different steady state aggregate sizes. Moreover, depending on the experimental conditions one may obtain fractal aggregates under static conditions while under flow conditions particles exhibit coalescence. This will be demonstrated by the series of experiments carried out at different pH as well as at different temperatures.

SIMULATION OF A HIGH-FLUX RISER-REACTOR FLOW USING CFD TECHNIQUES

Leonardo Machado da Rosa, University of Campinas, Campinas, SP, Brazil
Jaci Carlo Schramm Camara Bastos, University of Campinas, Campinas, SP, Brazil
Milton Mori, University of Campinas, Campinas, SP, Brazil
Waldir Pedro Martignoni, Cenpes/Petrobras, Rio de Janeiro, RJ, Brazil

Abstract

The application of Computational Fluid Dynamics (CFD) in industrial operations has been increased not only in the project of new equipments but also in the optimization of the existing ones. The present work shows a set of three-dimensional two-phase and adiabatic conditions, using the two-fluids model for the calculation of gas-solids flow in a high-flux risers of two different geometries, one with 10m of height and 76mm of internal diameter, and the second with 18m of height and 50mm of internal diameter. The k-epsilon model was applied to consider the influence of turbulence over the gas phase. The four-lump and ten-lump schemes kinetic models were used to predict the catalytic cracking reactions in the two different reactors. The first riser had two entrances, one for particles and one for the organic materials. The mesh size for this case was 584,000 control volumes. The second studied riser had one entrance for vapor, one for organic materials, and a third for particles. The mesh generated over this geometry had 650,000 control volumes. The numerical results of gas-solids dispersion flow in a high-flux were compared with the experimental data from the work of Pärssinen and Zhu [8], the four lump scheme for the mass concentration of chemical species was compared with the work of Ali *et al.* [2], and the ten lump scheme was compared with the work of Nayak *et al.* [7]. The catalytic cracking reactions models show an important role in the prediction of all phenomena present in the reactor. The simulated results show that the kinetics models are important not only because of the variation of the gas phase and the temperature profiles inside the reactor but also because these parameters interfere directly in the dynamic behavior of the gas-solid flows.

KEYWORDS: CFD, gas-solid, catalytic cracking

Introduction

A fluidized catalytic cracking unit (FCCU) consists of a riser reactor, where the catalytic reactions occur, and a regenerator, where the catalyst is reactivated. The main purpose of the riser is to put into contact hydrocarbons and catalyst particles, causing the catalytic cracking reactions. The cracking process converts the hydrocarbons fed into products with a higher commercial value. Although this equipment is widely used, phenomena in risers have not yet been entirely explained. They are operated with a high concentration of particles in a fluidized bed regime (fast transport regime). The adoption of fluidized beds is related to the catalytic cracking of gasoil in gasoline; previously, cracking had been done in fixed bed reactors.

Multiphase flows are more difficult to predict and analyze because their phases can assume complicated configurations [3, 5]. The phases do not mix uniformly and interactions between the phases (on a small scale) can have a strong effect on the macroscopic properties of the flow. The formation of heterogeneous structures in gas-particle systems (e.g., clusters) strongly affects the gas-solid contact and transport processes in dense fluidized beds. These structures depend on the overall forces acting on the particles. The drag force plays a fundamental role in the coupling of gas and particles. Zhang and Reese [9] propose a new equation for the mean drag force that considers not only the solids volumetric fraction, but also random fluctuations in particle movement.

In the modeling of chemical reactions, the use of groups of chemical species (or *lumps*) is common. A lump is the representation of a group of chemical species that have similar chemical properties. Thus, the catalytic cracking reactions that occur concurrently between numerous chemical species can be modeled as reactions between a fewer number of lumps [4]. The advantage of this procedure is the small number of chemical species and of reaction rates that have to be calculated separately, requiring a data bank for their calculation, and proportional computational resources. In the literature simulations that apply lumps to model the catalytic cracking of hydrocarbons, in gas-solid reaction systems [1, 2], or three-phase systems [7] are mainly one-dimensional. However, one-dimensional simulations cannot provide realistic solutions for radial distributions and are not able to properly handle some phenomena, such as particle reflux and swirls in a turbulent flow, and one-dimensional balances cannot be applied at the entrance region due to the high nonhomogeneity and turbulence in this region.

This article shows the influence of cracking models on the fluid flow through simulations with and without calculating reactions and with different cracking models. The finite volume method was utilized.

Mathematical Modeling

For simulation of the flow inside the reactor, the flow of two phases was considered: the particulate phase, consisting of catalyst particles, and the gaseous phase, a mixture of water vapor and organic compounds. The formation of new products and consumption of the existing ones alter the physicochemical properties of the gaseous phase. Both phases are fed at different temperatures; thus, at the bottom of the reactor, high temperature gradients exist.

Gas-solid systems can be modeled by several techniques. One consists in modeling each phase separately with the established equations for one-phase flows with a moving surface between them. However, when the phases are well mixed, as in many industrial operations, these cases require an average procedure to make the solution of the equations possible.

Ishii [6] divides the models for solution of fluid flow problems into two categories: diffusion models and two-fluid models. A diffusion model is formulated considering the mixture to have one transport equation for the continuity, momentum and energy, and a diffusion equation. A two-fluid model consists in two equations for the continuity, momentum and energy, one for each phase. In this work, the Eulerian-Eulerian approach was also applied. The *k-epsilon* model was applied to take account of the turbulent behavior of the flow. The equations of the model are given below.

Momentum and Continuity Conservation Equations

The continuity conservation transient equations for each phase (gas and particulate solids) are as follow:

$$\frac{\partial}{\partial t}\left(\varepsilon_g \rho_g\right)+\nabla\cdot\left(\varepsilon_g \rho_g u_g\right)=0 \tag{1}$$

$$\frac{\partial}{\partial t}\left(\varepsilon_p \rho_p\right)+\nabla\cdot\left(\varepsilon_p \rho_p u_p\right)=0 \tag{2}$$

where ε_g and ε_s are the volume fractions, ρ_g and ρ_s are the density of both phases, gas and particulate solids respectively; *u* is the velocity vector, and S_i^ρ represents the source term of continuity for each phase. The right hand side of the Equations 1 and 2 are null, due to the assumption of no mass transfer between phases. The momentum conservation equations are as follow:

$$\frac{\partial}{\partial t}(\varepsilon_g \rho_g u_g)+\nabla\cdot(\varepsilon_g \rho_g u_g u_g)=-\varepsilon_g \nabla p_g+\nabla\left(\varepsilon_g \mu_g\left(\nabla u_g+(\nabla u_g)^T\right)\right)+\varepsilon_g \rho_g g+\beta_{gp}^m(u_p-u_g) \quad (3)$$

$$\frac{\partial}{\partial t}(\varepsilon_p \rho_p u_p)+\nabla\cdot(\varepsilon_p \rho_p u_p u_p)=-G\nabla \varepsilon_p+\nabla\left(\varepsilon_p \mu_p\left(\nabla u_p+(\nabla u_p)^T\right)\right)+\varepsilon_p \rho_p g+\beta_{pg}^m(u_g-u_p) \quad (4)$$

where μ is the viscosity. *G* represents the elasticity modulus, β^m the momentum transfer coefficient, and *g* the gravity vector.

Chemical Species Transport

The concentration of the chemical species are calculated with another transport equation given in Equation 5:

$$\frac{\partial}{\partial t}(\varepsilon_g \rho_g c_i)+\nabla\cdot(\varepsilon_g \rho_g u_g c_i)=\nabla(\varepsilon_g \Gamma_i \nabla c_i)+S_g^i \quad (5)$$

where S_g^i represents the formation or consumption of each specie.

Energy Transport

$$\frac{\partial}{\partial t}(\varepsilon_g \rho_g H_g)+\nabla\cdot(\varepsilon_g \rho_g u_g H_g)=\nabla(\varepsilon_g \lambda_g \nabla T_g)+\beta_{gp}^h(T_p-T_g)+\sum_r \Delta H\frac{\partial c_r}{\partial t} \quad (6)$$

$$\frac{\partial}{\partial t}(\varepsilon_p \rho_p H_p)+\nabla\cdot(\varepsilon_p \rho_p u_p H_p)=\nabla(\varepsilon_p \lambda_p \nabla T_p)+\beta_{pg}^h(T_g-T_p) \quad (7)$$

Closure Equations

The term $G\nabla\varepsilon_p$ in Equation 4 is a simple model for the adjustment of the solids pressure (fluidized bed case), which is related to all other spatial directions. The convenient form to express *G* is given by Equation 8:

$$G=\exp(600(\varepsilon-0.62)) \quad (8)$$

The coefficients of friction or drag between fluid and particles are obtained from standard correlations with the negligence of acceleration. The momentum transfer coefficient between gas phase and particles is in accordance with Equation 9:

$$\beta_{gp}^m=\frac{3}{4}\mathrm{Cd}\frac{|u_p-u_g|\varepsilon_p \rho_p}{d_p}f(\varepsilon) \quad (9)$$

where the f is the volume fraction correction, $f = \varepsilon^{-1.65}$. The drag coefficient, *Cd*, applied to Equation 9, is a function of the Reynolds number and behaves (for $Re_p < 1000$) according to

$$\mathrm{Cd} = \frac{24}{\mathrm{Re}}\left(1 + 0.15\mathrm{Re}^{0.687}\right) \tag{10}$$

or (for nonlaminar flows, with $Re_p > 1000$)

$$\mathrm{Cd} = 0.44 \tag{11}$$

For dense flows (i.e., when $\varepsilon_p > 0.2$), the momentum transfer coefficient between gas and particles is

$$\beta_{gp}^{m} = 150\frac{\varepsilon_p^2 \mu_g}{\varepsilon_g d_p^2} + 1.75\frac{|u_p - u_g|\varepsilon_p \rho_p}{d_p} \tag{12}$$

For the calculation of the heat transferred between phases, the Nusselt number obtained from the Ranz-Marshall model is used:

$$\mathrm{Nu} = 2 + 0.6\mathrm{Re}_p^{1/2}\mathrm{Pr}^{1/3} \tag{14}$$

In Equation 6, the formation and consumption of chemical species are considered at the last term of the equation. The reactions can behave as a 2nd order kinetic model (Equation 15) or as a 1st order kinetic model:

$$\frac{\partial c_r}{\partial t} = -k_r \varepsilon_s c_i^2 \tag{15}$$

$$\frac{\partial c_r}{\partial t} = -k_r \varepsilon_s c_i \tag{16}$$

The k term in Equations 15 and 16 is adjusted with an Arrhenius equation,

$$k_r = k_0 \exp\left(\frac{-E_r}{RT}\right) \tag{17}$$

Chemical Kinetic Models

Several models are being proposed for the calculation of catalytic cracking reactions. The advances in both the analytical techniques and mathematical routines for determining the parameters lead to the progression from simple models with three and four lumps, to more complex models with ten, twelve, and more lumps.

The four-lump kinetic model classifies the hydrocarbon fractions into four groups, as illustrated in Figure 1. Oil cracking follows a second-order kinetics, while gasoline cracking follows a first-order kinetics. Parameters for the determination of reaction rates and heat were found in the work of Ali *et al.* [2].

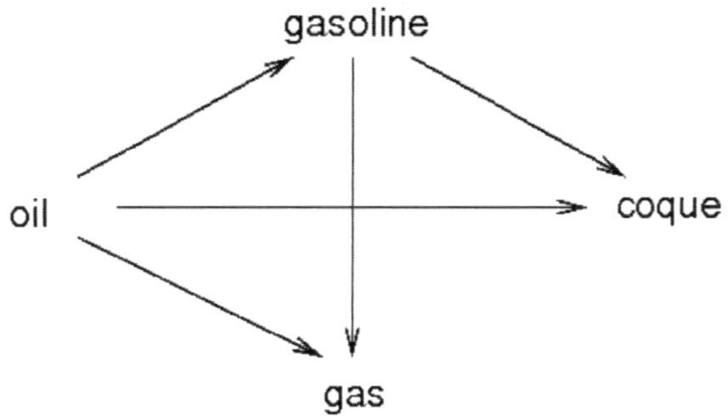

Figure 1. Four lumps kinetic model

Analogously, the ten-lump model describes the relationship of ten groups of hydrocarbons – four light hydrocarbon groups (paraffins, naphthenics, aromatics and aromatic substituents groups), four heavy hydrocarbon groups, coque and gasoline (Figure 2). Parameters for the determination of reaction rates and heat were found in the work of Nayak *et al.* [7].

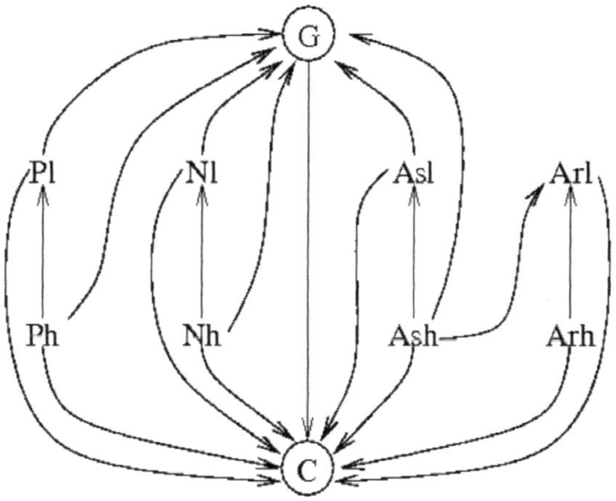

Figure 2. Ten lumps kinetic model

Meshes

Several three-dimensional mesh tests using hexahedral control volumes were elaborated for two different geometries. For all test cases a first control volume (at the walls) thickness of about 1mm, which provided the required value of y+ for the turbulence model. The first geometry, shown in Figure 3a, had a height of 10m and an internal diameter of 76mm, and was based on the equipment presented by Pärssinen and Zhu [8]. It is a combination of a cylindrical duct with a lateral junction that has the purpose of injecting the catalyst, and ends in a curve. A generation of mesh known as "O-Grid" was applied for the generation of a higher quality mesh. Meshes of different sizes were tested, and the mesh composed of 584,000 control volumes was chosen.

The second geometry, shown in Figure 3b, has a height of 18m and an internal diameter of 50mm. It is also a combination of a cylindrical duct with a lateral junction, and terminates with another junction that forms a 90 degrees angle with the main duct. It has two more entrances inside the reactor for vapor and feed inlet. Because of the complexity of this geometry an "L-Grid" generation of mesh was applied in the curved inlet. The final mesh chosen for the second geometry was composed of 650,000 control volumes.

The generated meshes for each of the geometries are detailed in Figures 4 and 5, where it can be seen the distribution of the control volumes at (a) the inlet region, (b) the outlet region, and (c) a transversal section. A more refined grid was applied near the walls, and at the locations where strong gradients of both momentum and temperature exist.

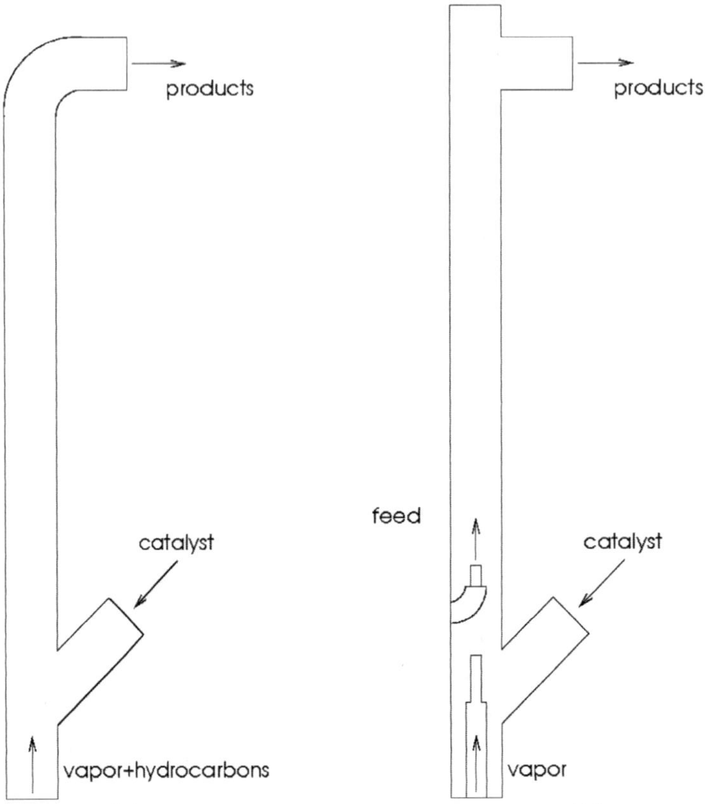

(a) (b)

Figure 3. Geometries utilized

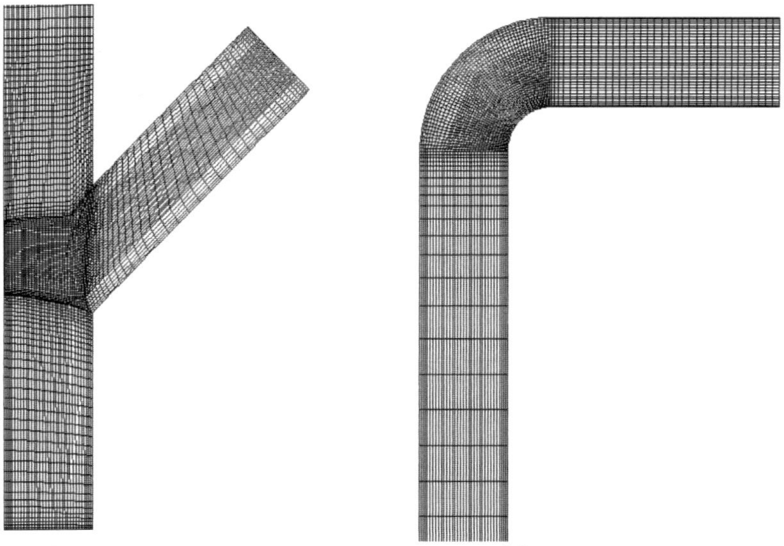

(a) (b) (c)

Figure 4. Mesh details for Geometry A

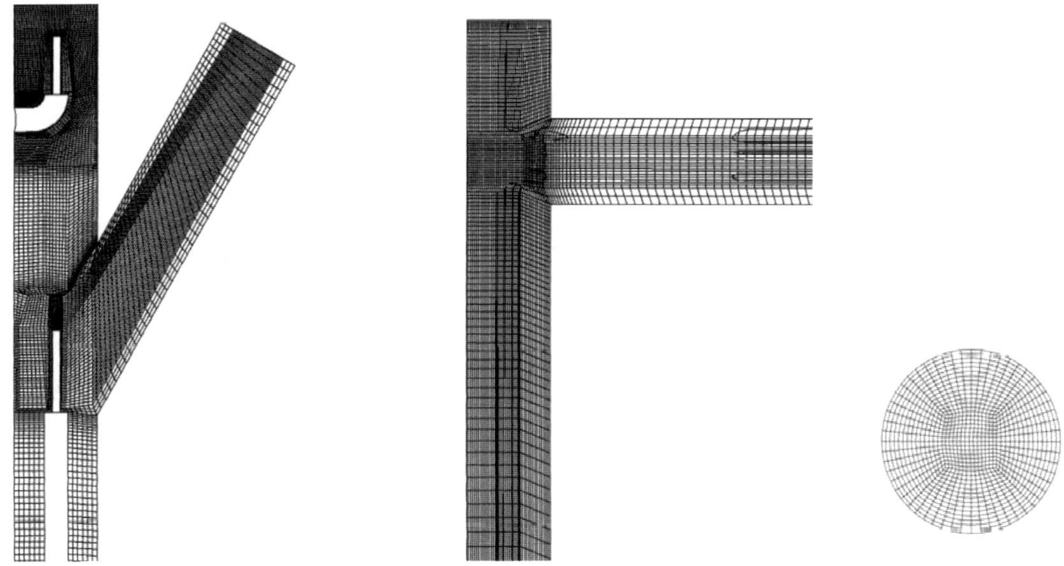

(a) (b) (c)

Figure 5. Mesh details for Geometry B

Results and Discussion

The following results were obtained with the simulation of a gas-solid system flowing at 5s. An initial time step of 0.0001s was used. The fluid considered in the simulations was a mixture of water vapor, air and hydrocarbons, with its physical properties varying with its composition. The particulate phase was a 1500 kg/m^3 catalyst with a diameter of 67μm.

Figure 6 shows the gas-phase temperature maps calculated for the two geometries. In both cases it could be seen that these values are influenced by the temperature of the particle phase. In Figure 6a, more heat is transferred to the gas phase, whereas the temperature after the contact with hot catalysts is higher than the temperature in Figure 6b.

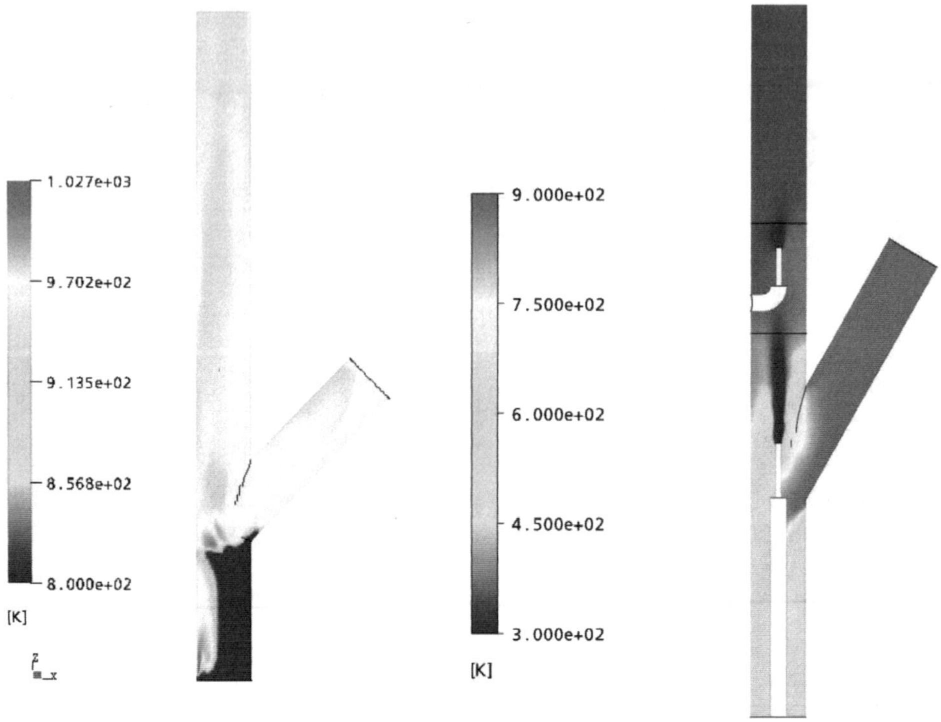

(a) (b)

Figure 6. Temperature maps along axial plane

Figure 7 shows the concentration profile of hydrocarbons calculated with the four-lump model for geometry A. The values obtained follow the same trends as the experimental ones (given by Ali *et al.* [2]), but the conversion was a little lower than expected. One possible explanation for this behavior could be the heat transfer model in which the chemical reactions are greatly influenced by the temperature.

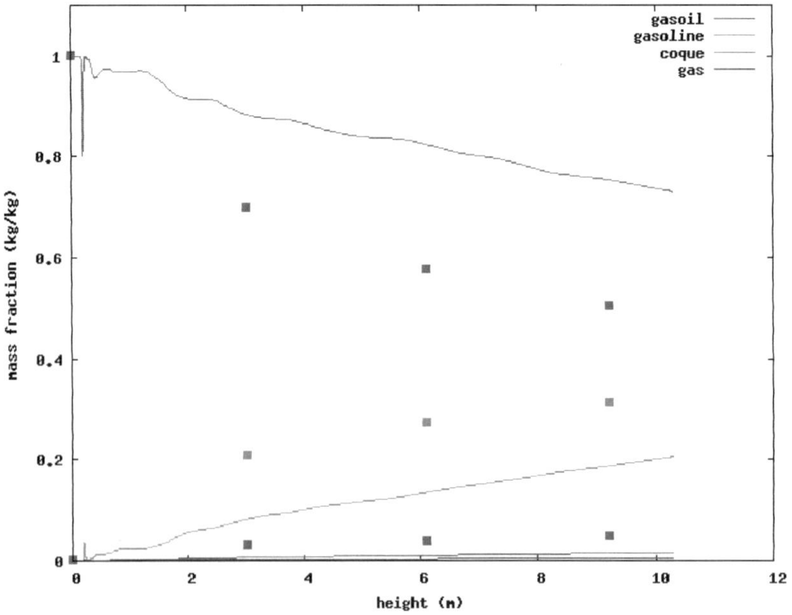

Figure 7. Hydrocarbons conversion along the axis

The behavior of the flow with reaction compared with the flow without reaction are shown in Figures 8 and 9. When the composition of the gas phase varies, its properties are not the same because they depend on the properties of each chemical species. Figure 8 shows the flow in the lower section of the reactor at a height of 1.53m. There are few differences between the profiles, but these are greater at higher heights (Figure 9) because there are more products formed in this region.

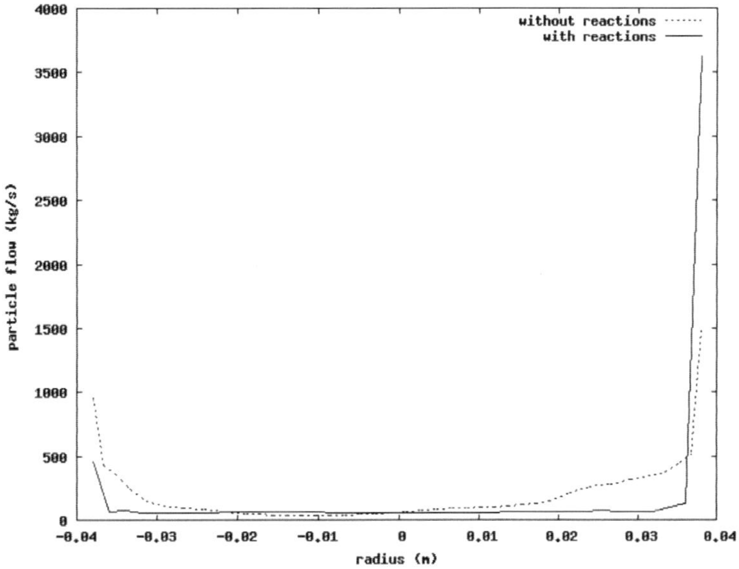

Figure 8. Comparison of flows at a height of 1.53m

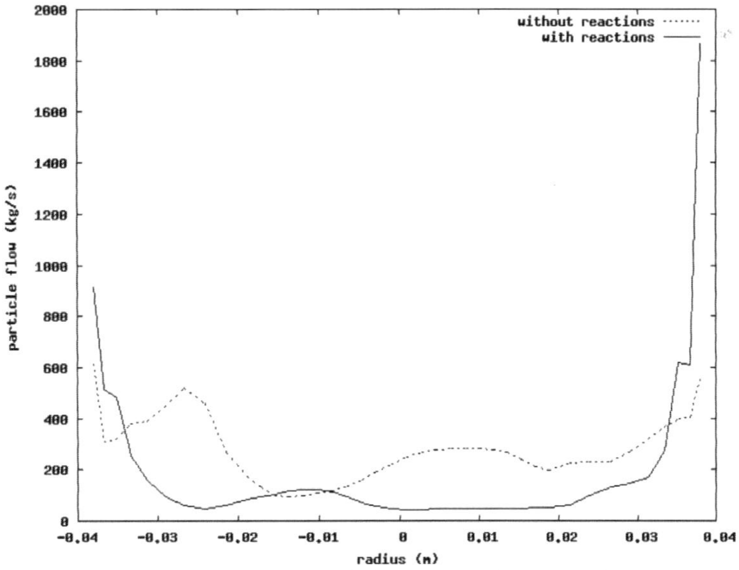

Figure 9. Comparison of flows at a height of 9.42m

Conclusions

The simulation of a reacting gas-solid fluid flow, coupling the four-lumps and ten-lumps kinetic models with the two-fluids model, was calculated for two different riser configurations in the CFX package. Simulation of the reacting flows resulted in conversions lower than what was expected, but with the same trends. Further researches are in progress to extend this work to other heat transfer coefficients and more accurate analysis for the fluid dynamics of the flow at high temperatures. Results with and without chemical reactions showed differences, indicating the need of coupling these phenomena, even when comparing just the simulated fluid-dynamic with experimental data.

Acknowledgments

The authors would like to thank FAPESP and PETROBRAS for their financial support.

Nomenclature

c	species concentration	[kgmol/m^3]
Cd	drag coefficient	[–]
d_p	particle diameter	[m]
E	activation energy	[J/mol]
g	gravity	[m/s^2]
G	elasticity modulus	[Pa]
H	enthalpy	[J/kg]
k	reaction rate coefficient	[s^{-1}]
Nu	Nusselt number	[–]
P	pressure	[Pa]
R	gas universal constant	[J/(mol.K)]
Re	Reynolds number	[–]
Pr	Prandtl number	[–]
S	species source	
t	time	[s]
T	temperature	[k]
u	velocity	[m/s]
β	interphase transfer coefficient	[kg/(m^{2}s)]
ε	volumetric fraction	[–]
λ	thermal conductivity	[W/(m.K)]
Γ	diffusivity	
μ	viscosity	[kg/(m.s)]
ρ	density	[kg/m^3]

References

1. Ali, H., Rohani, S. (1997), "Dynamic modeling and simulation of a riser-type fluid catalytic cracking unit," *Chem. Eng. Technol.*, n. 20, pp. 118-130.
2. Ali, H., Rohani, S., Corriou, J.P. (1997), "Modelling and Control of a riser type fluid catalytic cracking (FCC) unit," *Trans. IchemE.*, 75A, pp. 401-412.
3. Bastos, J.C.S.C., Rosa, L.M., Mori, M., Marini, F., Martignoni, W.P. (2008), "Modelling and Simulation of a gas-solids dispersion flow in a high-flux circulating fluidized bed (HFCFB) riser," *Catalysis Today*, n. 130, pp. 462-470.

4. Farkas, G. (1999), "Kinetic lumping schemes," *Chemical Engineering Science,* 54, pp. 3909-3915.
5. Hanratty, T.J., Theofanous, T., Delhaye, J.M., Eaton, J., McLaughlin, Prosperetti, A., Sundaresan, S., Tryggvason, G. (2003), "Workshop findings," *International Journal of Multiphase Flow,* n. 29, pp. 1047-1059.
6. Ishii, M. (1975), *Thermo-Fluid Dynamic Theory of Two-Phase Flow*. Eyrolles.
7. Nayak, S.V., Joshi, S.L., Ranade, V.V. (2005), "Modeling of vaporisation and cracking of liquid oil injected in a gas-solid riser," *Chemical Engineering Science,* n. 69, pp. 6049-6066.
8. Pärssinen, J.H. and Zhu, J.X. (2001), "Particle velocity and flow development in a long and high-flux circulating fluidized bed riser," *Chemical Engineering Science,* 56, pp. 5295-5303.
9. Zhang, Y., Reese, J.M. (2003), "The drag force in two-fluid models of gas-solid flows," *Chemical Engineering Science*, v. 58, pp. 1641-1644.

Synthesis of Doped TiO_2 Metal Oxide Nanoparticles in Supercritical CO_2

Gabrielė Aksomaitytė, **Esther Alonso**, and Maria José Cocero.
Dpto. Ingeniería Química, Universidad de Valladolid,
Prado de la Magdalena s/n, Valladolid, Spain

TiO_2 nanoparticles are interesting for many applications due to their photocatalytic properties. However, TiO_2 activity is reduced to the region of UV light (wavelength around 360 nm). In order to harness the abundant and safe potential of solar energy, the development of processes to enhance and shift the absorption of TiO_2 to the visible region is of the great importance. One of the ways to extend the absorbance of TiO_2 is transition metal doping in the lattice of TiO_2 [1]. In this work, supercritical CO_2 is used as reaction medium for the synthesis of doped TiO2 nanoparticles with Fe.

Keywords: doped TiO_2 nanoparticles, synthesis, supercritical CO_2.

Introduction

Titanium dioxide is a semiconductor which exhibit catalytic properties. It is widely used in environmental application as a photocatalyst for oxidation reactions of organic compounds, in alcohol thermal decomposition reactions, in photocatalytic decomposition of water for hydrogen production, in photovoltaic cells and etc [1].

Generally, the photocatalysis process over a semiconductor involves the absorption, band gap excitation, separation of the photoexcited electron/hole pairs, and redox reactions on the semiconductor surface [2].

As a photocatalyst TiO_2 can act as sensitizers for light – reduced redox processes due to their electronic structure, which is characterized by a filled valence bad and an empty conduction band. When the photon with an energy of hv exceeds the bandgap, E_g, of the semiconductor, an electron e^-, is promoted from the valence band, VB, into the conduction band, CB, leaving a hole h_{vb} behind. Graphically it can be expressed in Fig.1 [3].

TiO_2 (anatase) has wide band gap, $E_g \approx 3{,}2$eV, thus only light below 400 nm is absorbed and capable of forming the e^-/h^+ pair [4]. The dyes, impurities and dopants can sensitize TiO_2 to react to a much larger visible light region [1]. The different dopants may not have the same effect on trapping electrons or holes on the surface during the interface charge transfer because of the different positions of the dopant in the host lattice. In general photocatalytic activity of TiO_2 depends of many factors, such as, specific surface area, the absorption affinity and capacity for organic contaminants, electron-hole recombination processes in the bulk and on the surface of the catalyst, intensity and spectral distribution of the illuminating light, crystal morphology, intrinsic solid stated defects, stoichiometry of the catalyst oxide, pH, the presence of electron acceptors, the concentration of the pollutant [4].

The main aim of the research was focused on the efforts to synthesise doped TiO_2 nanoparticles with Fe atoms in order to change catalytic properties of TiO_2 nanoparticles which were tested in the methyl orange oxidation reactions.

Experimental

In the experimental procedure, diisopropoxititanium bis(acetylacetonate), (DIPBAT), is used as titania precursor and Iron (III) acetylacetonate and $Pd(hfa)_2$ for the doping metal ion. Under the supercritical conditions metal oxide particles are formed as a result of the metal salt hydrolysis and followed dehydration reaction under the supercritical conditions. The great advantage of this technology is that the nanoparticle doping takes place during the synthesis process. The synthesis of doped TiO_2 nanoparticles in batch and continuous flow reactors are presented. The wide range reactions at temperature (200 – 400 °C) and pressure (200 -250 bar) have been tested. Photocatalytic activities for the degradation of methyl orange under visible light irradiation are tested for the doped TiO_2 and they are compared to the non-doped samples.

Results

The experimental results showed that TiO_2 particles with high (5%) or low Fe content (1%) can be synthesised from DIPBAT precursor using two different alcohols (isopropanol and ethanol) as the co-solvents and reactants. The addition of Fe in to the titanium precursor allows TiO_2 –Fe particles to be synthesized with the improved photocatalytic activity using visible light compared to TiO_2. The particles with lower content of Fe (1 %) exhibit more efficient photocatalytic effect that those obtained with higher content of Fe (5 %).

References

[1] M. Kitano, M. Matsuoka, M. Ueshima and M. Anpo, *Applied Catalysis, A. General,* 1 (2007), 325.
[2] D. Wang, J. Ye, T. Kako, T. Kimura, *J. Phys. Chem. B,* 32 (2006) 110.
[3] M.R. Hoffmann, S. T. Martin, W. Choi, D. W. Bahnemannt, *Chem. Rev.* 95, (1995), 69-96.
[4]. S. Al-Qaradawi, S. R. Salman, *Journal. of Photochemistry and Photobiology A: Chemistry,* 148, (2001), 161-168.

SIMULATION OF A CONFINED TURBULENT TWO PHASE JET FLOW

Jaci C. S. C. Bastos[1], Rodrigo K. Decker[1], Udo Fritsching[2], Milton Mori[1]

(1) School of Chemical Engineering, University of Campinas, Brazil
(2) IWT, University of Bremen, Bremen, Germany

Abstract

This paper deals with 3D mathematical modelling of a confined turbulent two-phase jet flow. The model treats the gas and the solid phases from an Eulerian approach. The closure of the averaged time transport equations have been accomplished by using a k-ε turbulence model. The accuracy of the model predictions in a particle-laden confined jet with time averaged characteristics as well as turbulence correlation coefficients have been improved. Radial mean velocity profiles for the particulate phase were computed on eight axial levels, subdivided in two cases of four profiles. The results were compared with experimental data obtained with Phase Doppler Anemometry on a plexiglass chamber with three nozzle entrances in the top, two of them for the gas flow and the other one for a mixture between gas and particles. This last one was located in the central position. The mean particle diameter used for the two phase mixture was 75 μm. The inlet velocities used in the central nozzle were 4 and 4.5m/s. The gas superficial velocities in the two adjacent nozzles were different for each case. The interparticle collisions occur frequently in the local regions with higher particle concentration of the flow field. Under the influence of the local accumulation and the turbulent transport effects, the variation of the average interparticle collision number with the Stokes number allows a complex non-linear relationship. The particle distribution is more uniform as a result of interparticle collisions, and the lateral and the spanwise dispersion of the particles considering interparticle collision also increase. The model predicts a flow development similar to that found experimentally.

Keywords: Gas-Solid Flows, Jets Simulation, Computacional Fluid Dynamics (CFD)

1. Introduction

Two phase jet flows are extensively used in various emgineering applications, more exactly fluid flows containing solid particles are encountered in, and are fundamental importance to, manufacturing in the chemical, pharmaceutical, healthcare, biomedical, fuel, personal product, mineral and new materials sectors, and will increasingly grow in importance as new techniques and applications, such as functional nanomaterials, are developed. In all these applications, a fundamental understanding of how particles interact with the fluid flows is necessary to allow the use of computacional fluid dynamic (CFD) models in the optimization and performance improvement of existing equipament and processes, the identification and solution of operating problems, the evaluation of retrofit options, and the design of new equipament, systems and plant, incluiding process scale-up. Of specific interest in the present research are flows containing particle-laden confined jet.

Multiphase flow equations have been developed and analyzed by many researchers, such as He and Rudolph [8], Theologos and Markatos [19] and Ali and Rohani [1]. However, Soo [18] is credited with the mathematical approach to this type of flow. Rietema and vand der Akker [17] presented a detailed derivation of the momentum equations for disperse two phase systems, Michaelides [14] developed a model based on the phenomenological methods for the behavior of the variable density in pneumatic conveying lines and Crowe [5] reviewed the numerical models for dilute gas particle flow.

Currently, two of the most well-known methods for the mathematical modelling of the particulate solids phase in numerical simulations of gas-solids flows are the discrete particle simulation and the two-fluid approach. In both approaches the gas phase is described as a local average of the Navier-Stokes equation and both phases are usually connected by a drag force. At first, the dense gas-solids problem could be resolved using just the Newton motion equations for each suspended particle and the Navier-Stokes equation for the gas phase, Fairweather and Hurn [6] presented this approach. However, due to the large number of particles, the resultant number of equations would be very large to allow a direct solution, at least with the computer capacity currently available. Thus, the particulate solids phase is also treated as a continuous phase, exposed to the analogous conservation equations for the fluid phase. The Eulerian-Eulerian approach (two fluids approach) proved to be capable of predicting the gas-solids flow, as seen in the research realized by Meier and Mori [11], Alves et al. [3] and Bastos et al. [4].

In general, the performance of the current models critically depends on the accuracy of the formulation of the drag force, which has been extensively analyzed in many studies on multiphase flows. However, as the drag force follows empirical models, it is of extreme importance that this model be evaluated in accordance with the equipment employed. Gidaspow [7] presented a model for the gas-solids drag correlation, which proved to be sufficiently satisfactory for cyclone flows ([13] and [11]) and for riser-type reactors ([2], [3], [4]).

Intending to analyze the physical properties of the gas-solids flow inside equipament only along height, several authors, such as Ali and Rohani [1] and Martignoni and de Lasa [10], have presented unidimensional models. These models allowed the simulation up to 500 minutes and evaluation of the dynamic response of the system. Nieuwland et al. [16] proposed the use of a cylindrical bidimensional geometry, where only one sector of the equipament was calculated. In the research of Neri and Gidaspow [15], both Cartesian and cylindrical bidimensional geometries (with and without axial symmetry surfaces) were compared. They observed that the results obtained with Cartesian bidimensional geometry with axial symmetry were similar to those obtained with cylindrical bidimensional geometry. The effect of turbulence on particle motion in gas-solids suspension was analyzed by Yoshida and Masuda [20], Crowe [5], Alves et al. [3] and Zhang and Reese [21].

2. Mathematical model

The equations used for the gas and particulate solids phases, which are included in the description of the mathematical model, were developed with an Eulerian–Eulerian phenomenological approach, assuming a continuous and interpenetrating representation of the phases that ignores the material discontinuity and its composition; in other words, molecular interactions are rejected. Differential equations (Navier- Stokes) in the balance are formulated for mass and momentum treating the gas phase as incompressible. In this particular research, the mass source terms are null due to the assumption of no mass transfer between phases. According to this approach, different phases can divide the same control volume at the same instant. The k–ε model was applied to determine the influence of turbulence on the gas phase. Appropriate constitutive equations are specified for the description of the rheological physical properties of each phase and for the closure of the conservation equations.

2.1. Turbulence

Turbulence is the elemental key in a rapid particles–fluid mixture flow. It is characterized, according to Reynolds, as the appearance of instabilities in an originally stable flow, which increase as a nonlinear process and finally develop in a turbulent regime. The k–ε turbulence model, in which the velocity and length scales are determined with separate transport equations, is used extensively for the determination of gas-phase turbulence because it offers good agreement between numeric effort and computational accuracy. Turbulent kinetic energy k is defined as the variation in velocity fluctuations. The edge turbulent dissipation ε is defined as the rate at which the velocity fluctuations dissipate and it has dimensions of k per time unit.

The exact determination of the effective viscosity of the particulate solids phase is fundamental in attaining the radial distribution of the particles, and consequently, all the fluid dynamics variables [2]. The effective viscosity is the sum of dynamic viscosity and turbulent viscosity:

$$\mu_{g,ef} = \mu_g + \mu_{g,t} \tag{1}$$

where $\mu_{g,t}$ is the turbulence viscosity.

The model assumes that turbulent viscosity depends on the turbulent kinetic energy and the turbulent kinetic energy dissipation, through the relation proposed by Prandtl and Kolmogorov:

$$\mu_{g,t} = C_\mu \rho_g \frac{k^2}{\varepsilon} \tag{2}$$

where C_μ is a constant of the turbulent model with a value of 0.09.

The values of k and ε are given directly by solution of the differential transport equations for the kinetic energy and turbulent dissipation rate (Eqs. (3) and (4), respectively):

$$\frac{\partial}{\partial t}\left(\alpha_g \rho_g k\right) + \nabla \cdot \left(\alpha_g \rho_g v_g k\right) = \nabla \cdot \left[\left(\mu_g + \frac{\mu_{g,t}}{\sigma_k}\right)\nabla k\right] + P_k - \alpha_g \rho_g \varepsilon \tag{3}$$

$$\frac{\partial}{\partial t}\left(\alpha_g \rho_g \varepsilon\right) + \nabla \cdot \left(\alpha_g \rho_g v_g \varepsilon\right) = \nabla \cdot \left[\left(\mu_g + \frac{\mu_{g,t}}{\sigma_\varepsilon}\right)\nabla \varepsilon\right] + \frac{\varepsilon}{k}\left(C_{\varepsilon 1} P_k - C_{\varepsilon 2} \alpha_g \rho_g \varepsilon\right) \tag{4}$$

where $C_{\varepsilon 1}$, $C_{\varepsilon 2}$, σ_k and σ_ε are constants of the k–ε turbulence model with values of 1.44, 1.92, 1.0 and 1.3, respectively.

P_k is the production of turbulence due to the viscous forces and buoyancy forces, expressed by

$$P_k = \mu_{g,t} \nabla u_g \cdot \left(\nabla v_g + \nabla v_g^{\,T}\right) - \frac{2}{3} \nabla \cdot v_g \left(3\mu_{g,t} \nabla \cdot v_g + \alpha_g \rho_g k\right) + P_{kb} \tag{5}$$

2.2. Constitutive equations

2.2.1. Continuity between the phases

$$\sum_i \alpha_i = \alpha_g + \alpha_s = 1 \qquad (6)$$

2.2.2. Interphase momentum exchange

The coefficients of friction or drag between fluid and particles are obtained from standard correlations with the negligence of acceleration.Without acceleration, friction on the wall or gravity, the unidimensional momentum balance for the gas phase is

$$\alpha_g \frac{\partial p_g}{\partial x} - \beta_{gs}^m \left(v_g - v_s\right) = 0 \qquad (7)$$

The momentum transfer coefficient β_{gs}^m is obtained by comparison with Eq. (7), which results in the Ergun equation [9]:

$$\frac{\Delta p}{\Delta x} = 150 \frac{\alpha_s^2 \mu_g U}{\alpha_g^3 \left(\theta_s d_p^2\right)} + 1.75 \frac{U^2 \alpha_s \rho_g}{\alpha_g^3 \left(\theta_s d_p\right)} \qquad (8)$$

where U is the superficial velocity, $U = \alpha_g (v_g - v_s)$, and θ_s is the particle sphericity; in this particular case, $\theta_s = 1$ was used.

The comparison between Eqs. (7) and (8) shows that for dense regimes (with $\alpha_g < 0.8$), the momentum transfer coefficient between gas phase and particles is in accordance with the following equation:

$$\beta_{gs}^m = 150 \frac{\alpha_s^2 \mu_g}{\alpha_g d_p^2} + 1.75 \frac{\left|v_s - v_g\right| \alpha_s \rho_g}{d_p} \qquad (9)$$

where d_p is the particle diameter of the catalyst.

For porosities higher than 0.8, the expression for pressure drop results in the following equation for the interphase momentum transfer coefficient (Crowe [6]); this was also used in the simulation of Meier and Mori [12]:

$$\beta_{gs}^m = \frac{3}{4} Cd \frac{\left|v_s - v_g\right| \alpha_s \rho_g}{d_p} \alpha_g^{-2.65} \qquad (10)$$

The drag coefficient, Cd, applied to Eq. (10), is a function of the Reynolds number and behaves according to (for $\mathrm{Re}_p < 1000$)

$$Cd = \alpha_g^{-1.65} \max\left[\frac{24}{\mathrm{Re}_p \alpha_g}\left(1 + 0.15(\mathrm{Re}_p \alpha_g)^{0.687}\right)\right] \quad (11)$$

or (for nonlaminar flows, with $\mathrm{Re}_p > 1000$)

$$Cd = 0.44 \quad (12)$$

3. Simulation

The simulated model basically consists of a turbulent flow, where two phases enter into contact; a gas phase formed by air at 25°C with a controlled humidity of 70% and a particulate solids phase composed of catalysts. The particles are considered smooth, spherical and inelastic, with an average diameter of 75μm, a density of 2500kg/m^3 and 1.85 x 10^{-5}kg/m.s for solids viscosity. Table 1 shows three initial velocities conditions for each jet and case.

Due to the force of the fluid phase, which is responsable for the effective particles distribuition, these accelerate and move in the flow direction, characterizing the particle laden confined jet regime in a short space of time along the chamber. In a vertical plexiglass chamber, is shown in Fig. 1, the phases are fed at the top, forming angles of 0° in relation to this.

Table 1. Initial velocity condition and total time in the simulations.

	Jet 1 (gas) *U0* (m/s)	**Jet 2 (gas + particles)** *U0* (m/s)	**Jet 3 (gas)** *U0* (m/s)	Time (s)
Case A	8.0	**4.5**	8.0	7s
Case B	6.0	**4.0**	10.0	7s

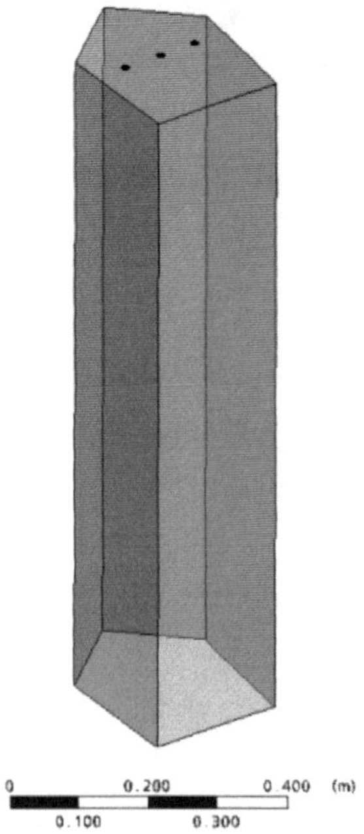

Fig. 1. Analyzed chamber.

The main characteristics of the chamber are three inlets at the top each with a diameter of 12mm (left, central and right), one outlet at the base (entire base) and a height of 1m. Both lateral nozzles were used to supply air at different velocities in order to investigate the flow instabilities due to the presence of different local velocities in the chamber, and consequently local flow structures.

Inside this chamber, different radial profiles were obtained at four different axial distances from the jet nozzles [*z* (mm) = 150, 180, 210 and 240], providing data of interest.

3.1. Boundary conditions

At the entrance, all velocities and concentrations of both phases are specified. The gas-phase pressure, treated as incompressible, was defined at the exit, assuming atmospheric pressure. As initial conditions, the superficial gas velocity is specified as 4 and 4.5m/s and the solids flow as 15 x 10^{-3}kg/s. At the walls, the gas-phase velocity is zero; the particulate solids phase velocity has a free-slip condition.

3.2. Mesh and computacional code

The nonstructured mesh is composed of 162,000 control volumes. The details of its refinement are presented in Fig. 2. The time step is on the order of 10^{-3}s. Adaptation of the mathematical model for the numerical model generation was achieved with the use of the ANSYS CFX 11.0 commercial simulator, which is based on the finite volume method, incorporating the higher upwind interpolation scheme.

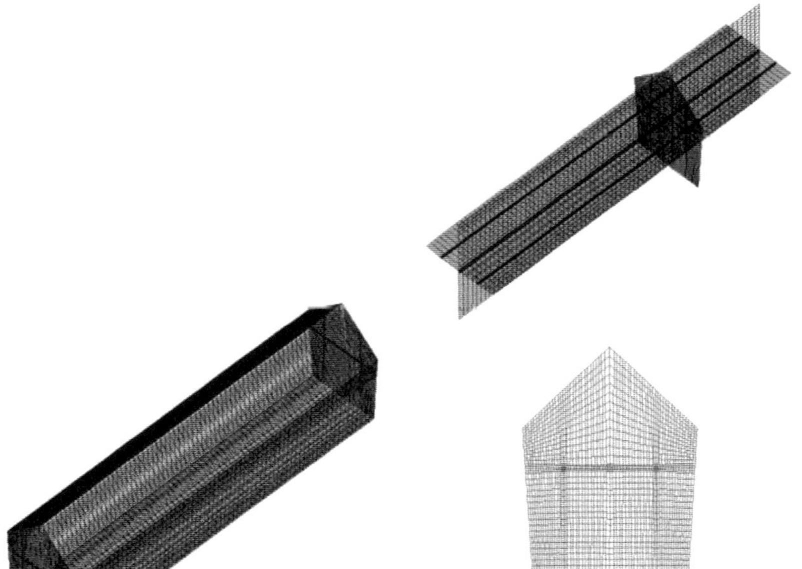

Fig. 2. Numerical mesh details at the entrances and the exit of the chamber.

4. Results and discussion

4.1. Mesh tests

Initially the numerical meshes were tested and the one that had the best behavior in the standard flow would be used to begin the simulations for the real process. The particle velocity radial profiles for the two phase confined jet flow were analyzed, using the same conditions as those proposed by Bastos et al. [22]. In the dependency evaluation of the flow with the numerical mesh, it is known that an adequate number of control volumes are of extreme importance to avoid numerical errors, which is not possible with less refined meshes.

Thus, seven meshes, with differents number of control volumes were tested. The mathematical model used for numerical mesh simulation was monophase, in accordance with the practice normally adopted, i.e., the flow inside the chamber is purely gaseous, subjected to conservation equations of continuity and momentum for each phase. Then the mesh composed of 162,000 control volumes was chosen; this was shown to be in accordance with the established standards (i.e., there is not a significant difference between the calculated flow of this mesh and those of more refined ones).

4.2. Dispersion model verification

The simulations carried out throughout this research showed that in each one of the transversal sections, there are two distinct zones: a uniform velocity zone (i.e., a zone where the particle velocity is high and quasi-constant), that extends from the radial center ($r = 0$) to approximately $r = 7$mm and a stage zone (i.e., a zone where the particle velocity is moderate and decreases), it is located at $r = 7$mm until $r = 20$mm (end of measurement line). In the axial direction, an unique region was observed. The Figs. 3-6 show the profiles of the solid phase velocity between the two cases comparing in the same section [z (mm) = 150, 180, 210, 240].

The radial mean velocity profiles for the above mentioned axial positions are presented for the two cases under analysis. It is interesting to observe that with the change in the adjacent velocities in jets 1 and 3, the radial profiles with particles loaded in jet 2 tend to the same behaviour. This happens with the change in the jet 2 too. The model shows great agreement with the experimental data.

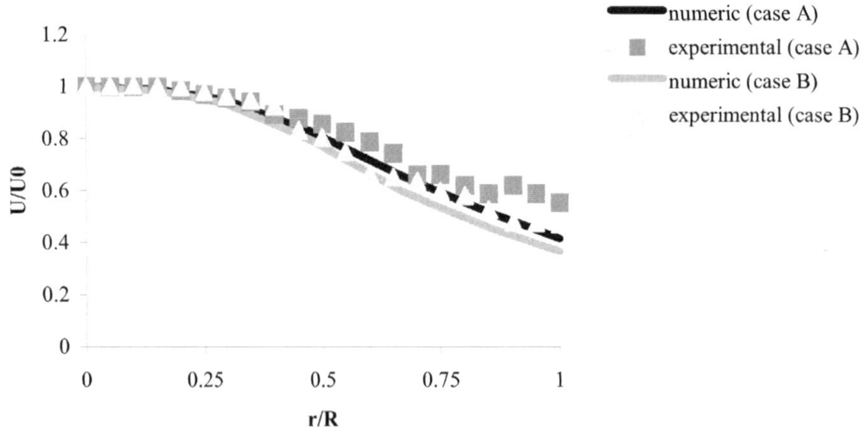

Fig. 3. Solids velocity profile case A versus case B comparison at 150mm.

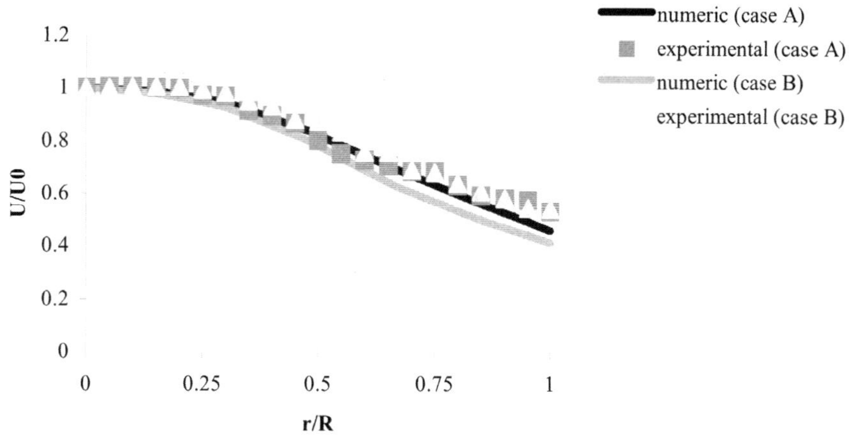

Fig. 4. Solids velocity profile case A versus case B comparison at 180mm.

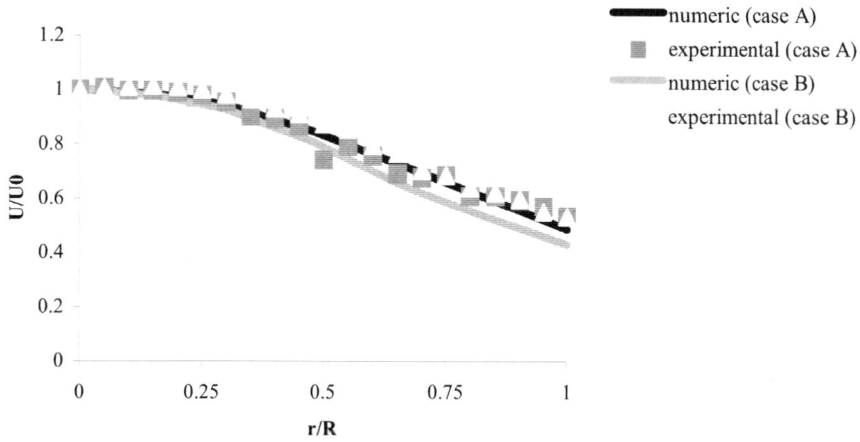

Fig 5. Solids velocity profile case A versus case B comparison at 210mm.

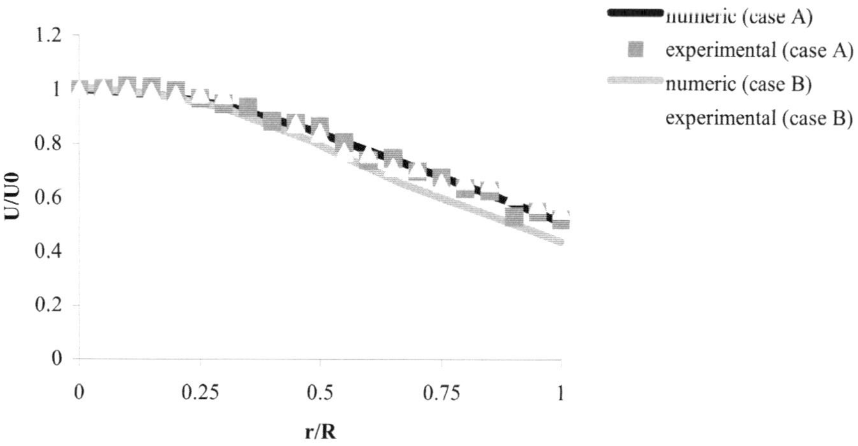

Fig. 6. Solids velocity profile case A versus case B comparison at 240mm.

4.3. Radial profiles of the solids phase velocity

These profiles were computed numerically at 7s of real time, (sufficient time to the developed flow) using a constant value for the solids phase viscosity (1% higher than the gas phase viscosity) and the averaging procedure at each iteration. This procedure showed good agreement between the solid phase velocity and the experimental data found in [22].

Figs. 7-10 show the experimental data versus numeric data, respectively, to the four transversal sections [*z* (mm) = 150, 180, 210 and 240] of the case A.

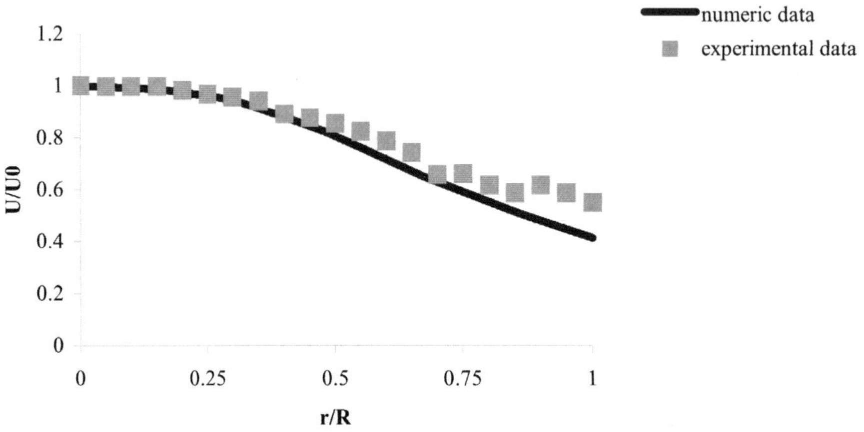

Fig. 7. Solids velocity profile to the case A at 150mm.

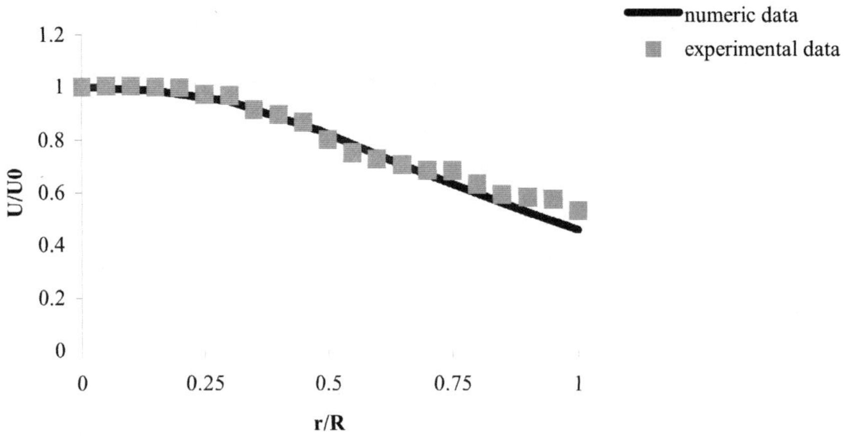

Fig. 8. Solids velocity profile to the case A at 180mm.

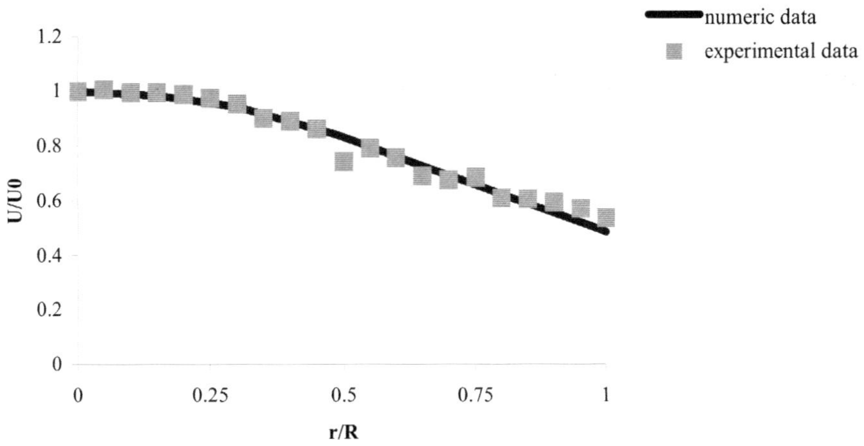

Fig. 9. Solids velocity profile to the case A at 210mm.

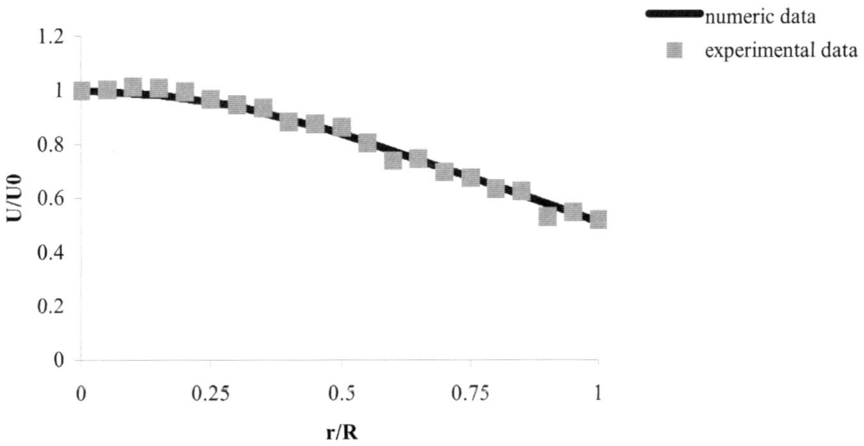

Fig. 10. Solids velocity profile to the case A at 240mm.

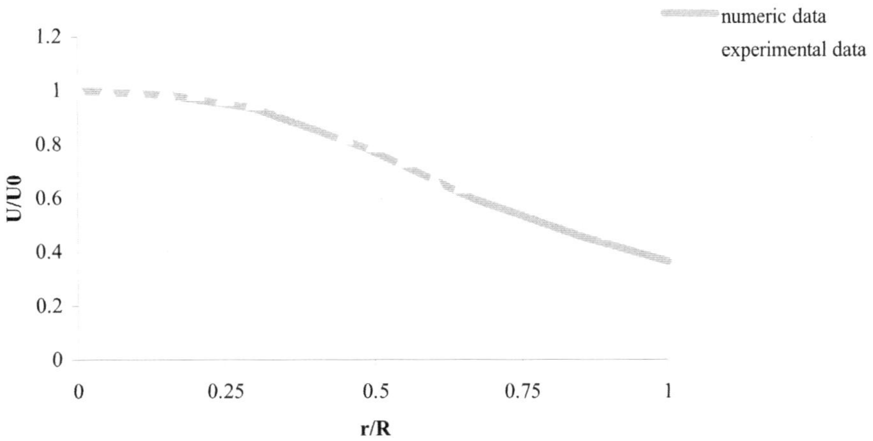

Fig. 11. Solids velocity profile to the case B at 150mm.

It is possible to observe the same behaviors, as quantitative as qualitative, at the different radial and axial positions. Figs. 11-14 show the experimental data versus numeric data, respectively, to the four transversal sections [z (mm) = 150, 180, 210 and 240] of the case B. Both cases presented to the uniform zone an excellent concordance between the experimental and numeric data. An insignificant deviation, at the transversal sections (z = 150mm) in the stage zone was observed.

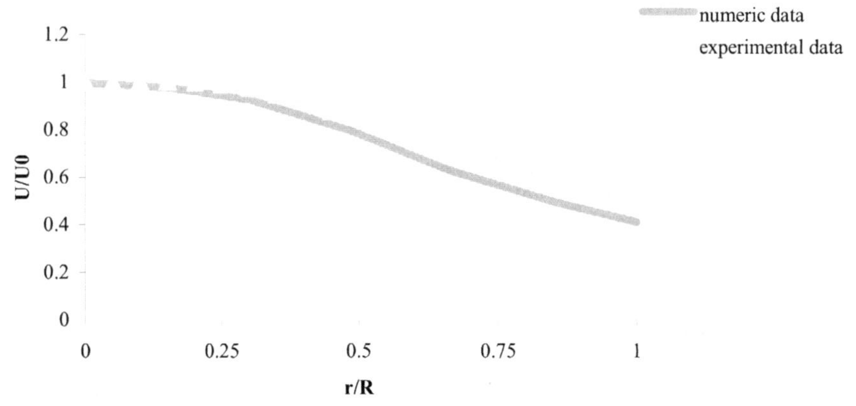

Fig. 12. Solids velocity profile to the case B at 180mm.

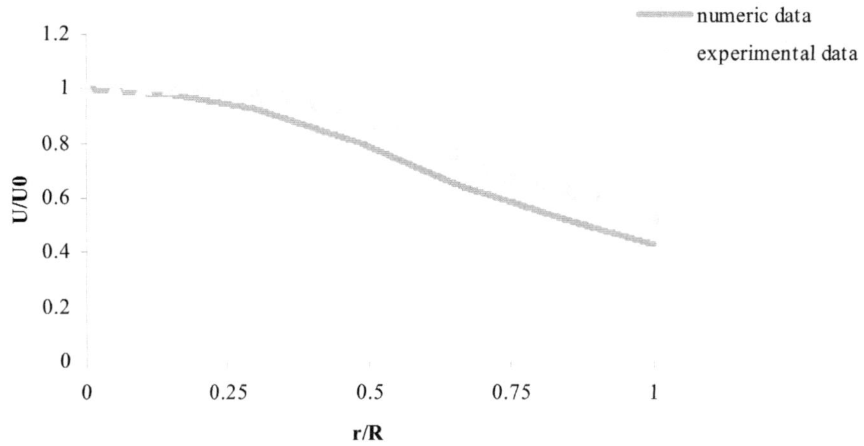

Fig. 13. Solids velocity profile to the case B at 210mm.

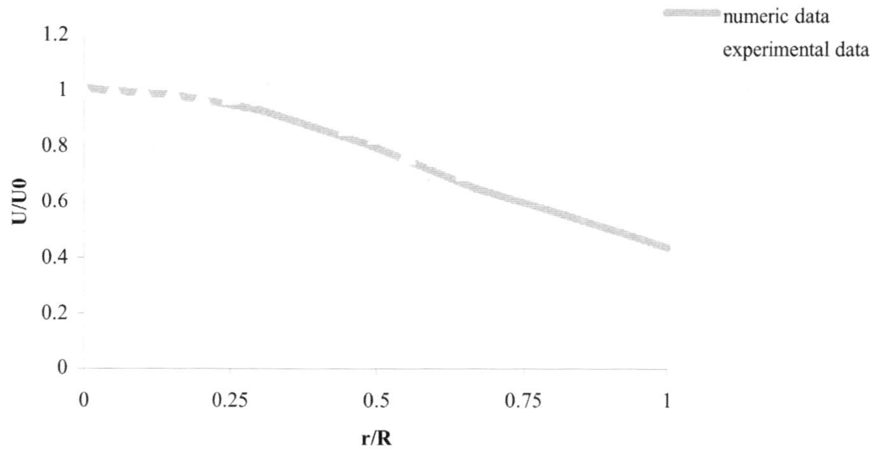

Fig. 14. Solids velocity profile to the case B at 240mm.

5. Conclusions

Research dealing with a confined turbulent two phase jet flow is required to improve and optimize the existing industrial equipaments. The *k-ε* gas-particle turbulence model is compared against experimental radial solids velocities on four axial levels in a height of 1m in a plexiglass pentagonal chamber of 15 x 10^{-3}kg/m².s and superficial velocities of 4 and 4.5m/s. The flow is divided into two zones with a zone where the particle velocity is high and quasi-constant, that extends from the radial center to approximately r = 7mm and a zone where the particle velocity is moderate and decreases, it is located at r = 7mm until r = 20mm. In the axial direction, an unique region was observed.

The predicted radial solids velocities show excellent agreement with the measurements in both flow zones to the two study cases. The model predicts a two phase (gas-solid) confined jet flow similar to that found experimentally. The discrepancies between experimental and predicted results in the stage zone, at the 150mm are insignificants.

Nomenclature

List of symbols

Cd drag coefficient;

C_μ, $C_{\varepsilon 1}$, $C_{\varepsilon 2}$, σ_k, σ_ε constants of the turbulence model;

d_p particle diameter (m);

k turbulent kinetic energy (m^2/s^2);

ε edge turbulent dissipation (m^2/s^3);

P_k turbulence production (m^2/s^3);

U superficial velocity (m/s);

$U0$ initial velocity;

v velocity vector (m/s);

Re Reynolds number;

Greek Letters

α volume fraction;

β_{gs}^{m} momentum transfer coefficient (kg/m^2s);

μ laminar viscosity (Pa.s);

μ_t turbulent viscosity (Pa.s);

ρ density (kg/m^3);

θ_s particle sphericity;

Subscripts

g related to the gas phase;

s related to the particulate phase;

References

[1] ALI, H., ROHANI, S., Dynamic Modelling and Simulation of a Riser-Type Fluid Catalytic Cracking Unit, *Chemical Engineering Technology*, n° 20, pp. 118-130, 1997.

[2] ALVES, J. J. N., MORI, M., Fluid Dynamic Modelling and Simulation of Circulating Fluidized Bed Reactors: Analyses of Particle Phase Stress Models, *Computers and Chemical Engineering*, vol. 22, pp. S763-S766, 1998.

[3] ALVES, J. J. N., MEIER, H. F., MARTIGNONI, W. P., MORI, M., The Effect of Turbulence on the Flow Pattern of Circulating Fluidized Bed Reactors, *Proceedings of ENPROMER*, vol. 3, pp. 1351-1356, 2001.

[4] BASTOS, J. C. S. C., ROSA, L. M., MORI, M., MARINI, F., MARTIGNONI, W. P., Modelling and Simulation of a Gas-Solids Dispersion Flow in a High-Flux Circulating Fluidized Bed (HFCFB) Riser, *Catalysis Today*, vol. 130, pp. 462-470, 2008.

[5] CROWE, C. T., On Models for Turbulence Modulation in Fluid-Particle Flows, *International Journal of Multiphase Flow*, n° 26, pp. 719-727, 2000.

[6] FAIRWEATHER, M., HURN, J.-P., Validation of an Anisotropic Model of Turbulent Flows Containing Dispersed Solid Particles Applied to Gas-Solid Jets, *Computers & Chemical Engineering*, vol. 32, pp. 590-599, 2008.

[7] GIDASPOW, D., Multiphase Flow and Fluidization – Continuum and Kinetic Theory Descriptions, *Academic Press*, Inc., San Diego, California, 1994.

[8] HE, Y., RUDOLPH, V., Gas-Solids Flow in the Riser of a Circulating Fluidized Bed, *Chemical Engineering Science*, vol. 50, n° 21, pp. 3443-3453, 1995.

[9] KUNII, D., LEVENSPIEL, O., Bubbling Bed Model for Kinetic Processes in Fluidized Beds, *Industrial and Engineering Chemistry Process Design and Development*, vol. 7, n° 4, pp. 481-492, 1968.

[10] MARTIGNONI, W. P., de LASA, H. I., Heterogeneous Reaction Model for FCC Riser Units, *Chemical Engineering Science*, vol. 56, pp. 605-612, 2001.

[11] MEIER, H. F., MORI, M., The Eulerian-Eulerian-Lagragian Model for Cyclone Simulation, *Proceedings of the Fourth European Computational Fluid Dynamics Conference*, vol. 1 (part 2), pp. 1206-1211, 1998.

[12] MEIER, H. F., MORI, M., Anisotropic Behavior of the Reynolds Stress in Gas-Solid Flows in Cyclones, *Powder Technology*, vol. 101, pp. 108-119, 1999.

[13] MEIER, H. F., ALVES, J. J. N., MORI, M., Comparison Between Staggered and Collocated Grids in the Finite Volume Method Performance for Single and Multiphase Flows, *Computers and Chemical Engineering*, vol. 23, pp. 247-262, 1999.

[14] MICHAELIDES, E. E., A Model for the Flow of Solid Particles in Gases, *International Journal Multiphase Flow*, vol. 10, n° 1, pp. 61-77, 1984.

[15] NERI, A., GIDASPOW, D., Riser Hydrodynamics: Simulation Using Kinetic Theory, *AIChE Journal*, vol. 46, n° 1, 2000.

[16] NIEUWLAND, J. J., ANNALAND, M. S., KUIPERS, J. A. M, SWAAIJ, W. P. M., Hydrodynamic Modelling of Gas/Particle Flows in Riser Reactors, *AIChE Journal*, vol. 42, n° 6, pp. 1569-1582, 1996.

[17] RIETEMA, K., van der AKKER, H. E. A., On the Momentum Equations in Dispersed Two-Phase System, *International Journal Multiphase Flow*, vol. 9, n° 1, pp. 21-36, 1983.

[18] SOO, S. L., Fluid Dynamics of Multiphase Systems, Waltham MA: Blaisdell Publishing Co., 1967.

[19] THEOLOGOS, K. N., MARKATOS, N. C., Advanced Modeling of Fluid Catalytic Cracking Riser-Type Reactors, *AIChE Journal*, vol. 39, n° 6, pp. 1007-1016, 1993.

[20] YOSHIDA, H., MASUDA, H., Model Simulation of Particle Motion in Turbulent Gas-Solid Pipe Flow, *Powder Technology*, vol. 26, pp. 217-220, 1980.

[21] ZHANG, Y., REESE, J. M., Gas Turbulence Modulation in a Two-Fluid Model for Gas-Solid Flows, *AIChE Journal*, vol. 49, n° 12, 2003.

[22] BASTOS, J. C. S. C., DECKER, R. K., FRITSCHING, U., MORI, M., Experimental Analysis of Particle Interaction in a Confined Two-Phase Jet, *submitted.*

STUDY OF DIFFERENT APPROACHES FOR MODELING CYCLONES USING CFD

Daniel de Brito Dias, School of Chemical Engineering, University of Campinas, Campinas, SP, Brazil
Milton Mori, School of Chemical Engineering, University of Campinas, Campinas, SP, Brazil
Waldir Pedro Martignoni, Cenpes/Petrobras, Rio de Janeiro, RJ, Brazil

Abstract

Due to their relative robustness and low cost, cyclones are widely used in industries as gas-solid and liquid-solid separators. Their performance is usually measured by two parameters: pressure drop and collection efficiency. While this equipment is simple in concept, studies have shown that the confined vorticial flow inherent to cyclones is in fact quite complex. Because of this, computational fluid dynamics (CFD) has been employed to model flows and solve them numerically, and has shown relative success in dealing with phenomena such as vortex breakdown, reversal flow, and high turbulence intensity.

Cyclone simulations are usually expensive in terms of computer processing, mainly due to the fact that an anisotropic turbulence model must be used. Several studies have addressed that issue, and Reynolds Stress Model (RSM) has been considered by many researchers as being the most suitable model. In order to reduce computational times while still acquiring reliable results, other aspects of the simulation must be studied, such as the modeling approach to the particulate phase, which is the subject of this work.

While the Eulerian approach is generally applied for the gas, different schemes can be considered for the particulate phase. In this work, two different methods are explored: the classic Eulerian-Eulerian (E-E), where a mean diameter is considered to calculate the entirety of the flow field of the particulate phase; and Eulerian-(Eulerian)$_n$ (E-E$_n$), where particles with different diameters are included in the same simulation, each as a distinct phase. The fractional collection efficiency curve is obtained for both methods. An analysis is made regarding accuracy and computational costs.

The simulations make use of ANSYS CFX, a commercial software package, running in parallel mode in a high performance computing cluster. The fractional and overall collection efficiencies comparison with the experimental data showed that E-E$_n$ achieved more accurate results.

KEYWORDS: CFD, cyclone, gas-solid

Introduction

Cyclones are the most commonly used gas-solid separation equipments in industry, mainly due to their relatively low manufacturing and operational costs. Although the operation principles involved are simple, the fluid dynamics are not: the dust-laden gas enters a cylindrical chamber tangentially and spirals downwards. The particles, being denser, concentrate on the wall due to their inertia and are collected at the bottom. The clean gas reverses its flow and exits through a central opening known as a vortex finder.

The result is a complex confined double vortex which is part of the reason why cyclone design remained, for many years, restricted to empirical or semi-empirical methods which are still in use. Aiming to maximize collection efficiency while minimizing pressure drop, design models in which the geometrical parameters of the equipment are defined as a function of its diameter were created. Also, several algebraic models for estimating velocities, pressure drop and collection efficiency are available and have been compiled from literature in reviews [1, 2].

Recent advances in computer hardware, numerical codes and parallel processing have made possible the use of computational fluid dynamics (CFD), a technique that proposes the solution of the equations of conservation of mass, momentum and energy via numerical methods such as finite volumes. In the field of cyclones it has proven the complexity of the phenomena involved with the need for unsteady simulations, higher order discretization schemes and anisotropic turbulence models such as the Reynolds Stress Model (RSM) [3,4]. Even more difficulties arise when taking multiphase flow into account. This can be done initially considering two different approaches: Eulerian-Eulerian, in which the phases are considered continuous and interpenetrating and Eulerian-Lagrangian, where the solid phase is considered discrete and each individual particle is tracked along the continuous phase. Each method has its particular pros and cons, and both have been applied to cyclone modeling with relative success, which can be exemplified by references [5] and [6].

One particular disadvantage of the traditional Eulerian-Eulerian approach is that the solid phase is given an average diameter, so that any information about particle size distribution is lost. Such information is very important in cyclone analysis because it provides design engineers with more complete information on collection efficiency in the form of the fractional collection efficiency curve. This work analyzes and compares two different methods for obtaining the mentioned curve while still retaining the E-E scheme. The first method (E-E) consists on performing a fixed number of simulations, with one particulate phase per simulation, varying particle diameter. Each time, it is considered that the single particulate phase represents the entire solid load fed to the cyclone, obtaining one point in the fractional collection efficiency curve per simulation. In the second method (referred to as E-E_n) one single simulation considering multiple solid phases of different diameters weighted according to the particle distribution is performed, thus directly obtaining an approximation of the fractional collection efficiency curve. A comparison on the accuracy and applicability of each method is made, and the results are validated by experimental data by Zhao [7].

The authors hope that this discussion adds to the cyclone CFD literature, and helps establish it as a tool not only for analysis but proposal of new designs, as it has been in recent years [8, 9]. The simulations make use of ANSYS CFX 11, a commercial software package, running in parallel mode in a high performance computing cluster.

Mathematical Modeling

The conservation equations of mass and momentum solved in this work for phase α (out of n total phases) are described by Equations 1 and 2, respectively.

$$\frac{\partial \rho_\alpha}{\partial t} + \nabla(\xi_\alpha \rho_\alpha v_\alpha) = 0 \qquad (1)$$

$$\frac{\partial(\xi_\alpha \rho_\alpha v_\alpha)}{\partial t} + \nabla(\xi_\alpha \rho_\alpha v_\alpha v_\alpha) = \nabla\{\xi_\alpha \mu_\alpha [\nabla v_\alpha + (\nabla v_\alpha)^T]\} - \xi_\alpha \nabla p + \xi_\alpha \rho_\alpha g + \sum_{i \neq \alpha}^{n} \beta(v_i - v_\alpha) \qquad (2)$$

Where ξ is the volumetric fraction, ρ is the density, v the velocity, p the pressure, g the acceleration due to gravity, μ the viscosity and β is the inter-phase momentum transfer coefficient.

As for closure equations, the continuity of volumetric fractions was guaranteed by Equation 3, and the inter-phase drag was calculated by Equations 4 and 5. Due to dilute flow conditions, solid pressure force and particle-particle drag were neglected. Solid phases were considered to be inviscid.

$$\sum_i^n \xi_i = 1 \tag{3}$$

$$\beta = \frac{3}{4} Cd \frac{|v_g - v_p| \xi_p \rho_\alpha}{d_p} \tag{4}$$

$$Cd = max\left[\frac{24}{Re}(1 + 0.15 Re^{0.687}), 0.44\right] \tag{5}$$

Turbulence Model

For the complex turbulent flow inherent to cyclones, a robust model such as RSM must be considered. First, the Reynolds average, which splits a variable into a time-average and a fluctuating part ($i.e.\ v = \overline{v} + v'$) is applied to the conservation equations. Transport equations for the individual Reynolds stress ($\rho\overline{v'v'}$) components obtained (Equations 6 and 7) and for the turbulence dissipation rate ε (Equation 8) are then solved, rather than adhering to the eddy viscosity hypothesis.

$$\frac{\partial\left(\rho\overline{v_i'v_i'}\right)}{\partial t} + \frac{\partial\left(U_k\rho\overline{v_i'v_i'}\right)}{\partial x_k} = P_{ij} + \phi_{ij} + \frac{\partial}{\partial x_k}\left[\left(\mu + \frac{2}{3}c_s\rho\frac{k^2}{\varepsilon}\right)\frac{\partial\left(\overline{v_i'v_i'}\right)}{\partial x_k}\right] - \frac{2}{3}\delta_{ij}\rho\varepsilon \tag{6}$$

$$P = -\rho\left[\overline{v'v'}(\nabla v)^T + (\nabla v)\overline{v'v'}\right] \tag{7}$$

Where P is the exact production term and ϕ is the pressure-strain correlation (Equations 8 through 12).

$$\frac{\partial(\rho\varepsilon)}{\partial t} + \frac{\partial(\rho U_k\varepsilon)}{\partial x_k} = \frac{\varepsilon}{k}(c_{\varepsilon 1}P - c_{\varepsilon 2}\rho\varepsilon) + \frac{\partial}{\partial x_k}\left[\left(\mu + \frac{\mu_t}{\sigma_\varepsilon}\right)\frac{\partial\varepsilon}{\partial x_k}\right] \tag{8}$$

$$\phi_{ij} + \phi_{ij1} + \phi_{ij2} \tag{9}$$

$$\phi_{ij1} = -\rho\varepsilon\left[C_{s1}a + C_{s2}\left(aa - \frac{1}{3}a{\cdot}a\delta\right)\right] \tag{10}$$

$$\phi_{ij2} = -C_{r1}Pa + C_{r2}\rho kS - C_{r3}\rho kS\sqrt{a{\cdot}a} + C_{r4}\rho k\left(aS^T + Sa^T - \frac{2}{3}a{\cdot}S\delta\right) + C_{r5}\rho k\left(aW^T + Wa^T\right) \tag{11}$$

$$\begin{cases} a = \frac{\overline{v\prime v\prime}}{k} - \frac{2}{3}\delta \\ S = \frac{1}{2}[\nabla v + (\nabla v^T)] \\ W = \frac{1}{2}[\nabla v - (\nabla v^T)] \end{cases} \tag{12}$$

Where a is the anisotropy tensor, S is the strain rate and W is the vorticity. From the many varieties of RSM available, Speziale-Sarkar-Gatski (SSG) was chosen, as its quadratic relation for the pressure-strain correlation is recommended for swirling flows [10]. The model constants are as follows:

$$c_s = 0.22; c_{\varepsilon 1} = 1.45; c_{\varepsilon 2} = 1.83 \qquad (13)$$
$$C_{s1} = 1.7; C_{s2} = -1.05; C_{r1} = 0.9; C_{r2} = 0.8; C_{r3} = 0.65; C_{r4} = 0.625; C_{r5} = 0.2$$

The RSM approach makes the overall model far more computationally expensive but also theoretically more suited to predict complex anisotropic flows.

Numerical Methods

The aforementioned equations were solved using commercial CFD code ANSYS CFX 11, by means of the finite volume method. A numerical algorithm based on Rhie-Chow was applied to the pressure-velocity coupling and the higher upwind high order interpolation scheme was considered. A very detailed description of these methods can be found in [11]. A time step of 0.0001 seconds was used to simulate, for each case, five seconds of real time. The error criterion was 0.0001 for the RMS residue for all equations.

Case Description

The simulation attempted to numerically reproduce the experimental work of Zhao [7] (specifically the tangential inlet cyclone data), which has been previously studied [12]. Equipment dimensions are summarized in Table 1. The 3-D geometry and its corresponding numerical grid (consisting of approximately 350,000 hexahedrical elements) were built using meshing code ICEM CFD, and are displayed in Figure 1.

The fluid considered in the simulations was air at 25°C. The particulate phase was a talcum powder of density 2700 kg/m^3 obeying a log-normal size distribution with a mass-mean diameter of 5.97μm and geometric deviation of 2.08. The concentration of the solids was of 5.0 g/m^3. Four different inlet velocities were simulated: 12, 16, 20 and 24 m/s. The air exit was set to an atmospheric pressure opening. Wall conditions were set to no slip for the gas and free slip for the talcum, and a dust hopper was added at the bottom, with the same wall boundary conditions applied to it.

Three diameters were taken into account, so that three repetitions for the E-E simulation and one E-E$_3$ simulation were performed. Considered particle sizes were the mean diameter (5.97 μm), the mean diameter divided by the geometric deviation (corresponding to 2.87 μm) and the mean diameter multiplied by the geometric deviation (corresponding to 12.42 μm). In the E-E$_3$ simulation they correspond to 68, 16 and 16 percent in volume of the solid load respectively.

Table 1. Geometrical Parameters of the Studied Cyclone

	D	De	h	H	B	S	a	B
Size [mm]	300	150	450	1200	112.5	150	150	60

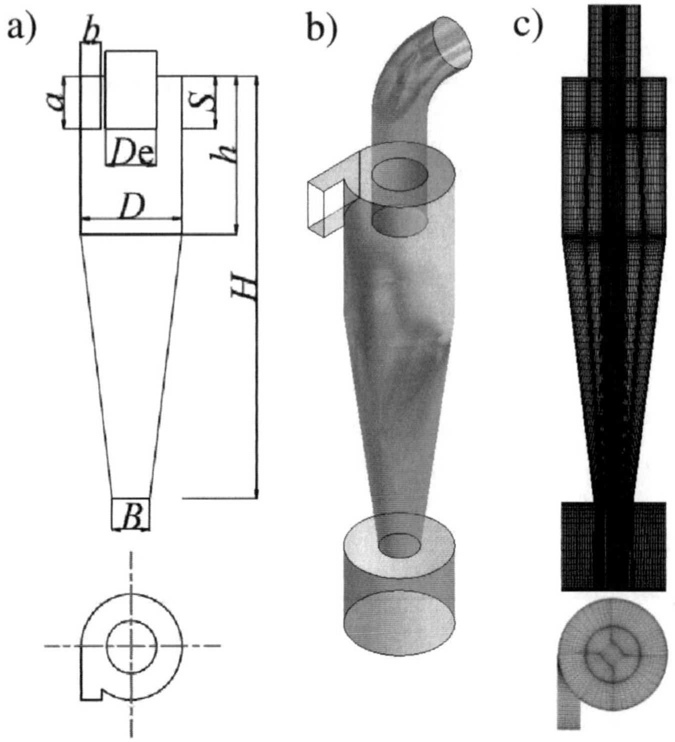

Figure 1. Studied cyclone and its a) Geometrical parameters b) CAD Geometry c) Hexa mesh

Results and Discussion

The pressure and velocity profiles for all simulations, exemplified in axial cut planes in Figure 2, came to good agreement with what is expected of cyclone flow.

Along the axis, in the central zone of the equipment, the formation of an ascending low pressure zone can be observed, contrasting with the downward high pressure zone closer to the walls. The negative and positive tangential velocities, peaking at almost two times the inlet velocity, form a quasi-symmetrical plane along the central axis and indicate the swirling motion of the double vortex: the left (negative) section represents fluid *entering* the plane and the right (positive) section *leaving* the plane.

This adequate description of the fluid dynamics is validated by the good agreement with experimental data, showed in Table 2 and Figures 3 and 4. Overall and fractional efficiencies (η) are calculated as functions of the mass flows in the two boundaries of the system and analyzed.

$$\eta = \frac{\dot{m}_{in} - \dot{m}_{out}}{\dot{m}_{in}} \tag{14}$$

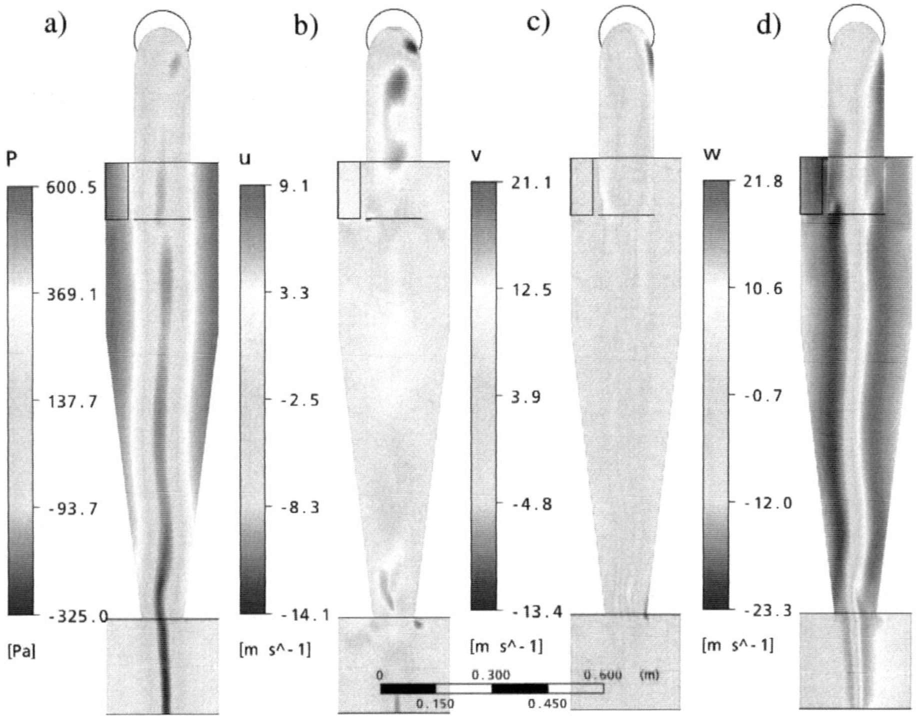

Figure 2. Profiles for a) Pressure b) Radial Velocity c) Axial Velocity and d) Tangential Velocity for the E-E_3 case with an inlet velocity of 12 m/s.

Table 2. Comparison of fractional and overall collection efficiencies

v_{in} = 12 m/s				v_{in} = 16 m/s			
	Zhao [7] [%]	**E-E [%]**	**E-E_3 [%]**		**Zhao [7] [%]**	**E-E [%]**	**E-E_3 [%]**
2.87 μm	70.2	57.6	63.4	**2.87 μm**	82.8	81.9	80.4
5.97 μm	98.0	99.3	98.8	**5.97 μm**	98.8	99.5	99.2
12.42 μm	99.7	99.8	99.1	**12.42 μm**	99.8	99.9	99.3
Overall	91.0	92.8	93.3	**Overall**	94.7	96.8	96.2
v_{in} = 20 m/s				v_{in} = 24 m/s			
	Zhao [7] [%]	**E-E [%]**	**E-E_3 [%]**		**Zhao [7] [%]**	**E-E [%]**	**E-E_3 [%]**
2.87 μm	88.1	92.4	87.1	**2.87 μm**	92.7	96.3	90.1
5.97 μm	99.0	99.6	99.4	**5.97 μm**	99.1	99.7	99.5
12.42 μm	99.9	99.9	99.5	**12.42 μm**	99.9	99.9	99.6
Overall	95.8	98.5	97.4	**Overall**	95.7	99.2	98.1

At a first glance, it can be seen that both methods predict the separation reasonably well. Experimentally the increase of velocity also sees an increase in collection efficiency, and this holds true for the CFD runs. Nonetheless, the E-E_3 approach proved to have a more acceptable accuracy in the numerical results. This is most likely due to the fact that the gas-solid momentum transfer in this case is closer to experimental data. The particles with different diameters are included in the inter-phase drag equation and consequently three number ranges are taken into account.

The overall collection efficiency, showed in Figure 4, agrees as well as the numerical results of fractional collection efficiency, and gives a better visualization on how the E-E_3 method was superior in representing the separation phenomena.

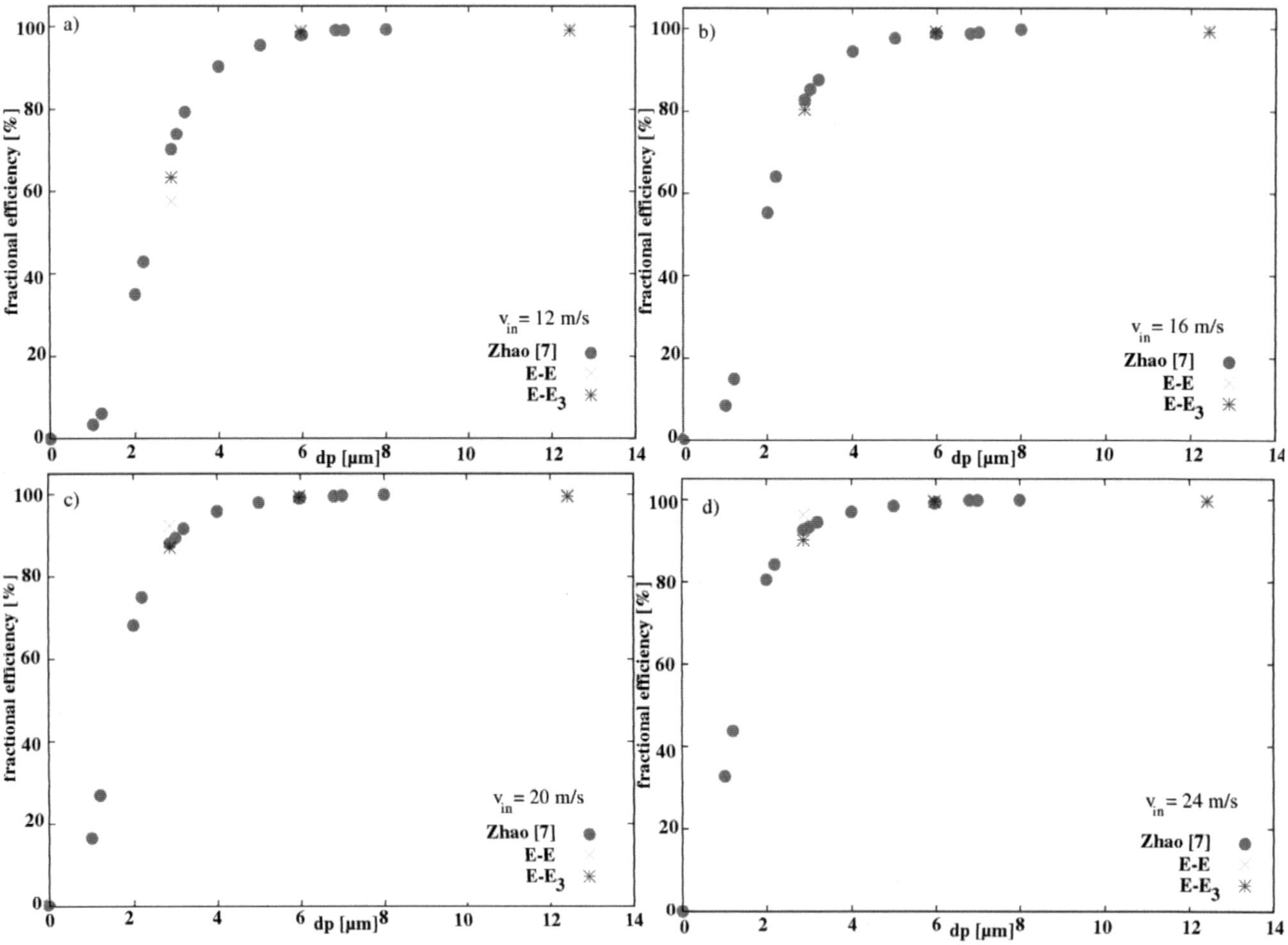

Figure 3. Fractional collection efficiency curves compared for both methods with inlet velocities of a) 12 m/s; b) 16 m/s; c) 20 m/s; d) 24 m/s.

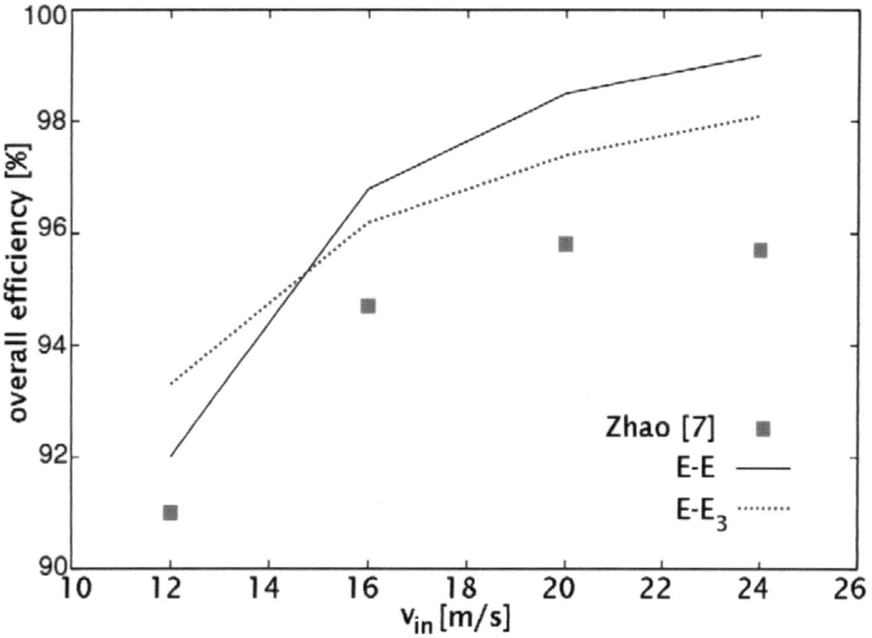

Figure 4. Overall collection efficiency comparison.

As a final analysis, the processing time and scalability of each method was tested. The average time taken per iteration was measured varying the number of parallel processes used for the run. Since three E-E simulations gave us the equivalent amount of data of one E-E_3 simulation, its processing time was multiplied by three. The simultaneous multiple solid phase approach performed far better in this analysis, with not only smaller simulation times but also better scalability. This happens because of two facts. First, it is only solving the gas phase once as opposed to three times and, second, more sets of differential equations are being solved simultaneously and consequently the solver can make better use of the additional parallel processes.

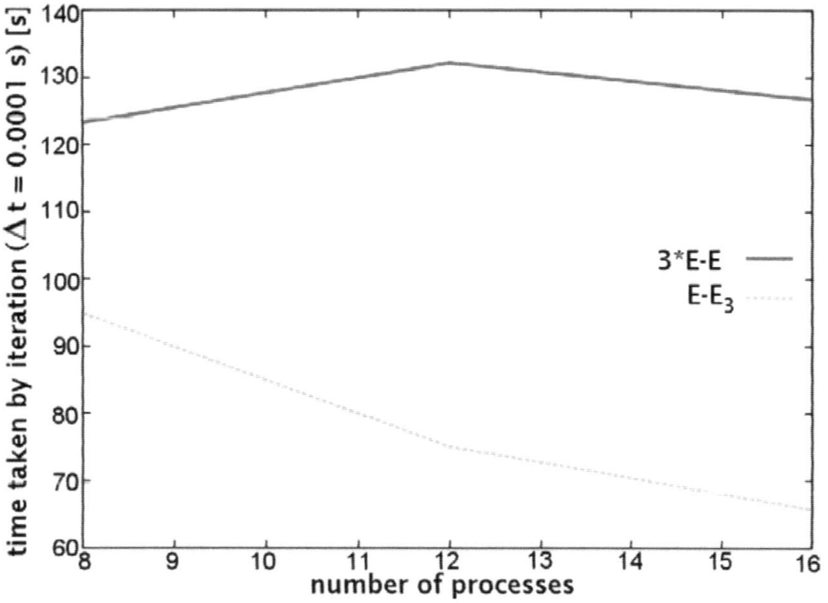

Figure 5. Processing time per iteration taken by each method *versus* number of processes used

Conclusions

The collection efficiency and flow characteristics of a gas-solid cyclone were obtained by numerical analysis. Two different approaches of modeling the solids have been studied in this work, aiming to obtain the fractional collection efficiency curve. These methods were referred to as E-E and E-E_n. The first consists in simulating several times a regular Eulerian-Eulerian flow with a different particle diameter each time. The second is achieved by performing one single simulation simultaneously involving multiple solid phases, with different diameters.

While both schemes correctly reproduced the complex fluid dynamics inside cyclones (which was confirmed by validation with experimental data [7]), E-E_n proved itself not only more accurate, but also faster and more scalable, acting as a better alternative for obtaining the fractional collection efficiency curve for the case studied in this work.

A suggestion for future studies is the application of these methods to a case with a higher concentration of solids.

Acknowledgments

The authors would like to thank PETROBRAS for their financial support.

Nomenclature

Cd	drag coefficient	-
$c_s, c_{\varepsilon1}, c_{\varepsilon2}, C_{s1}, C_{s2}, C_{r1}, C_{r2}, C_{r3}, C_{r4}, C_{r5}$	RSM constants	-
dp	particle diameter	m
g	gravitational acceleration	$m \cdot s^{-2}$
k	turbulence kinetic energy	$m^{-2} s^{-2}$
$\dot{m}$	mass flow	$kg \cdot s^{-1}$
P	exact production	$kg \cdot m^{-1} \cdot s^{-3}$
p	pressure	Pa
Re	Reynolds number	-
t	time	s
v	velocity	$m \cdot s^{-1}$
x	dimension	m

Greek Letters

β	inter-phase momentum transfer coefficient	$kg \cdot m^{-2} \cdot s^{-1}$
δ	Dirac delta	-
ε	turbulent dissipation rate	$m^{-2} \cdot s^{-3}$
η	collection efficiency	-
μ	viscosity	$kg \cdot m^{-1} \cdot s^{-1}$
ξ	volumetric fraction	-
ρ	density	$kg \cdot m^{-3}$
ϕ	pressure-strain	$kg \cdot m^{-1} \cdot s^{-3}$

Subscripts

i,j,k	indexes
in	inlet
n	total number of phases
out	outlet
t	turbulent
α	phase indicator

References

1. Hoffmann, A. C. and Stein, L. E. (2002), "Gas Cyclones and Swirl Tubes: Principles, Design and Operation", ed. Berlin, Springer.
2. Cortés, C. and Gil, A. (2007), "Modeling the Gas and Particle Flow Inside Cyclone Separators", *Progress in Energy and Combustion Science,* 33, pp. 409-452.
3. Boysan, F.; Ayers, W. H. and Swithenbank, J. A. (1982), "A Fundamental Mathematical Modeling Approach to Cyclone Design", *Institution of Chemical Engineers*, 60, pp. 222-230.
4. Meier, H. F. and Mori, M. (1999), "Anisotropic behavior of the Reynolds Stress in Gas and Gas-Solid Flows in Cyclones", *Powder Technology*, 101, pp. 108-119.
5. Meier, H. F. and Mori, M. (1998), "Gas-Solid Flow in Cyclones: the Eulerian-Eulerian Approach", *Computers and Chemical Engineering*, 22, pp. 641-644.
6. Derksen, J. J.; van den Akker, H. E. A. and Sundaresan, S. (2008) "Two-Way Coupled Large-Eddy Simulations of the Gas-Solid Flow in Cyclone Separators", *AIChE Journal*, 54, pp. 872-885.
7. Zhao, B.; Shen, H.; Kang, Y. (2004), "Development of a Symmetrical Spiral Inlet to Improve Cyclone Separator Performance", *Powder Technology*, 145, pp. 47-50.
8. Bernardo, S.; Mori, M.; Peres, A. P.; Dionísio, R. P., "3-D Computational Fluid Dynamics for Gas and Gas-Particle Flows in a Cyclone with Different Inlet Section Angles", *Powder Technology*, 162, pp. 190-200.
9. Noriler, D.; Vegini, A. A.; Soares, C.; Barros, A. A. C.; Meier, H. F.; Mori, M. (2004), "A New Role for Reduction in Pressure Drop in Cyclones Using Computational Fluid Dynamics Techniques", *Brazilian Journal of Chemical Engineering*, 21, pp. 93-101.
10. "ANSYS CFX Reference Guide", *ANSYS CFX Release 11.0*, December, 2006.
11. Patankar, S. V. (1980), "Numerical Heat Transfer and Fluid Flow", ed. New York, Hemisphere Pub Co.
12. Vegini, A. A. (2007), "Eulerian-Eulerian Approach for Numerical Simulation of Cyclones", *PhD Thesis*, University of Campinas, Campinas, 2007.

189x Investigating Size-Resolved Nanoparticle Reaction Kinetics: A Novel Tandem Ion-Mobility Method

Lei Zhou[1], X. Ma[1], Anshuman Lall[1], G. M. Mulholland[1], and Michael Zachariah[2]. (1) University of Maryland-College Park, College Park, MD 20742, (2) University of Maryland and NIST, 2125 Glenn L Martin Hall, College Park, MD 20742

A novel tandem ion-mobility method for probing reaction kinetics at nanoscale is developed. The experimental system consists of two different ion-mobility schemes in series. The first mobility characterization is to size select particles with a differential mobility analyzer (DMA). The second mobility characterization employs an aerosol particle mass analyzer (APM) and measures changes in mass resulting from a controlled reaction of nanoparticles. The DMA, used in this experiment for size selection, consists of an annular region between two concentric cylinders, with the center cylinder held at high voltage and the outer one at ground. Charged particles of the right polarity feel an attractive force toward the center electrode and move radically inward at an electrophoretic velocity determined by the particle charge, and drag force which is a function of particle size. When charged particles flow between the cylinders the electric force on the particle is balanced by the drag force, and at a fixed voltage all particle exiting the instrument have (to the resolution of the instrument) equivalent mobility sizes. The APM can determine the particle mass distribution based on particle mass to charge ratio, and is used in our experiment to monitor changes to the particle mass resulting from evaporation. The APM consists of two concentric cylindrical electrodes that rotate together at a controlled speed. An electrical field is created by applying high voltage on the inner electrode while the outer one is held at ground. Charged particles flowing within the concentric cylinders experience opposing centrifugal and electrostatic forces and as a result particles exiting the instrument at fixed voltage and rotation speed all have the same nominal mass. By scanning either the voltage or the rotation speed, the particle mass distribution (independent of particle shape) can be determined. Based on operating conditions for the DMA and APM and the root-sum-square (RSS) method, we estimate the uncertainty of density calculation ~5%.

We demonstrate the use of the DMA-APM technique to measure the inherent density of nanoparticles, measuring the surface energy of unsupported nanocrystals, and the kinetics of aluminum and nickel nanoparticle oxidation.

189y Nanofluidization as Affected by Vibration and Electrostatic Fields

Daniel Lepek[1], Robert Pfeffer[1], Rajesh Dave[2], Miguel Angel Sanchez Quintanilla[3], Jose Manuel Valverde[3], and Antonio Castellanos[3]. (1) Department of Chemical Engineering, New Jersey Institute of Technology, Newark, NJ 07102, (2) New Jersey Center for Engineered Particulates, New Jersey Institute of Technology, 138 Warren Street, YCEES 208, Newark, NJ 07102, (3) University of Seville, Seville, Spain

The behavior of a fluidized bed of silica nanoparticles under the influence of externally applied vibrations and electrostatic field will be presented. It has been observed that the application of these fields separately has opposite effects on bed expansion. On one hand, vertical vibrations enhances bed expansion as the vibration intensity is increased up to a critical value. On the other hand, an electrostatic field applied in the horizontal direction, hinders bed expansion. In previous research papers, it has been suggested that the size of nanoparticle agglomerates could be affected either by vibration or by the action of the electric field. Using the modified Richardson-Zaki method to analyze our experimental data, we find that vertical vibration tends to decrease the average agglomerate size in agreement with previous research. However, in this work we look further into the physical mechanisms which affect the response of the fluidized bed. Our results suggest that both vibration and the electric field produce a significant perturbation to the flow of agglomerates on the fluidized bed. Vibration transmits a vertical motion to the agglomerates that enhances bed expansion until vibration velocity becomes of the order of the expected rising velocity of macroscopic bubbles. At this critical point, bubbles growth is stimulated by vibration. A horizontal electrostatic field produces a drift of the charged agglomerates towards the walls that gives rise to fluidization heterogeneity and bed collapse. When both fields are jointly applied these opposed effects can be practically compensated.

Alexander Samokhvalov, Sachin Nair, Zenda Davis, and Bruce Tatarchuk. Center for Microfibrous Materials Manufacturing, Department of Chemical Engineering, Auburn University, 230 Ross Hall, Auburn, AL 36849

Preparation And Characterization Of Nanodispersed Ag-Based Sorbent For Removal Of Sulfur From Logistic Fuels

The Ag/TiO2-based adsorbent for selective removal of sulfur from logistic fuels was prepared by incipient wetness impregnation of titania with AgNO3, with the subsequent calcination, and characterized. The survey XPS spectra obtained with the "calcined" sorbents with typical nominal loadings from 0.5% to 8% wt. Ag show the similar features. The following elements were identified in the specimens: Ag, Ti, O and spurious carbon as expected. No residual nitrogen was detected in the calcined samples, thus indicating that complete decomposition of AgNO3 deposited onto the sample during the incipient wet impregnation step does occur during the calcination step. To determine the valence state of Ag in the calcined samples of the sorbent, XPS line scans for Ag were performed and standard curve fitting procedures executed. Binding energy of the XPS peaks of Ag, including the spectroscopically most important $3d_{3/2}$ and $3d_{5/2}$ lines, is known to be insensitive to the valence state and coordination environment of Ag. Therefore, the Auger parameter was used to determine the oxidation state of Ag: Ag metal vs Ag+. The Auger parameter α is known to be especially useful for processing XPS spectra of the insulating samples e.g. titania-supported sorbents, since its value is independent on the electrostatic charging of the specimens. The Ag Auger lines $M_4N_{45}N_{45}$ and $M_5N_{45}N_{45}$ lines and XPS lines $3d_{3/2}$ and $3d_{5/2}$ were used to calculate the Auger parameter for Ag. Moreover, model sorbents Ag/TiO2 were prepared, containing only Ag+ (freshly impregnated and dried specimen) and only Ag metal (calcined specimen reduced in hydrogen). Auger parameter α was calculated to find the oxidation state of Ag in the dried, calcined and reduced samples. The dried samples have the smallest value of α, the reduced samples – the highest, and the calcined samples have the intermediate value. Literature indicates that metal Ag shows the highest value of α, while Ag+ shows the smallest value. Thus, the obtained data indicate that in the calcined samples that are commonly used for desulfurization, significant fraction of Ag is present as metal silver. Another indication of the presence of metal silver in the "calcined" samples is their low aptitude towards electrostatic charging under XPS conditions. "Dried" samples showed significant charging (up to 3 eV), while both the "reduced" and "calcined" samples showed rather small and similar charging (less than 1 eV). Dispersion of silver on titania was studied by XRD and SEM, and occurs at nanoscopic dimensions. Studies of adsorption and desorption pathways of several sulfur-containing polyaromatic hydrocarbon (S-PAH) molecules typically present in logistic fuels – Thiophene (T), Benzothiophene (BT) and Dibenzothiophene (DBT) - on the surface of the sorbent was performed by Temperature-Programmed Desorption (TPD) and Temperature-Programmed Reaction Spectroscopy (TPRS). The adsorption of selected S-PAHs onto the sorbent was carried out from their solutions in octane that represent chemical composition of sulfur-containing liquid fuels. Next, the sorbent was excessively rinsed with octane, to remove the excess of S-PAHs, either entrapped in the pores of the sorbent or weakly bound to the surface sites. The sorbent was then dried, and studied by the TPD and TPRS in presence of oxygen under High Vacuum (HV) conditions. The data show that desorption pathways proceed differently for S-PAHs molecules with different molecular structure. For thiophene (T), significant fraction of adsorbed molecules binds reversibly to the surface sites, and desorbs without decomposition ("molecular desorption") at about 350 K. The rest of adsorbed thiophene reacts chemically with the surface, forming products of C-S bond scission (butadiene) and surface-supported silver sulfide. The presence of the latter is found due to high-temperature TPRS band of sulfur dioxide SO2 that is formed from silver sulfide in presence of molecular oxygen. In contrast, products of molecular desorption were not identified for DBT, yet C-S bond scission products (biphenyl) were found. Upon TPRS in presence of oxygen, strong SO2 desorption band at ~725 K suggests that in case of DBT, chemisorption pathway with formation of titania-supported silver sulfide takes place.

189aa Synthesis of Nanoscale Cobalt-Iron Spinel Oxides Via Atomic Layer Deposition and Their Applications for Solar Thermochemical Water Splitting

Jonathan R. Scheffe[1], Nathan P. Siegel[2], Mark D. Allendorf[2], and Alan W. Weimer[1]. (1) Department of Chemical and Biological Engineering, University of Colorado, 1111 Engineering Dr., ECCH 111, Boulder, CO 80309-0424, (2) Solar Technologies, Sandia National Laboratories, PO Box 5800, Albuquerque, NM 87185

Nano-thick cobalt-iron spinel oxides (cobalt ferrites) have been synthesized via atomic layer deposition (ALD) on yttria-stabilized zirconia (YSZ) and Al_2O_3 nanoparticles. Multilayers of cobalt(II) oxide and iron(II,III) oxide were deposited and subsequently heat treated at 900 °C. X-ray diffractometry (XRD) and x-ray photoelectron spectroscopy (XPS) were utilized to determine film crystallinity and oxidation state, and induced coupled plasma - atomic emission spectroscopy (ICP-AES) was verified the relative amounts of cobalt and iron.

ALD of ferrites has the advantage of being able to precisely control the thickness and Co/Fe ratio on the atomic level, which is significant because it has been suggested in the literature that this two-step cycle water splitting cycle is a surface dominated reaction. The kinetics of the water splitting step has been investigated and results indicate that the temperature required is decreased due to a reduction of diffusion limitations. Additionally, results were directly compared to samples synthesized via traditional methods (coprecipitation, etc), and the amount of hydrogen generated per mole of ferrite was shown to be greater. Potential factors affecting the water splitting step that were studied included film thickness, Co/Fe ratio, and the substrate material (YSZ and Al_2O_3), and each was shown to be statistically significant. Hydrogen evolution was measured in situ using a residual gas analyzer, and changes in crystallinity were measured using a powder x-ray diffractometer.

189ab Effect of Interactions on Nanoparticle Dispersion In a Polymer Solution

Deepika R. Gollamandala and Ileana C. Carpen. Department of Chemical Engineering, Tennessee Tech University, Box 5013, Cookeville, TN 38505

Addition of nanosized particles to a polymer solution/matrix can lead to new and improved properties over conventional composites using larger filler particles. Various factors, including size of nanoparticles, nanoparticle-polymer interaction, polymer chain length, and relative volume fractions, affect the dispersion of nanoparticles in the polymer matrix. The objective of this study is to investigate the influence of nanoparticle-polymer interactions on the overall dispersion behavior using Brownian dynamic simulations. The shape of the nanoparticles was considered to be spherical and the polymer architecture was modeled as bead spring chains. The degree of dispersion and agglomeration in the system were characterized by mean square displacement.

189ac Numerical Simulation of Granular Flow In a Bladed Mixer

Brenda Remy[1], Johannes G. Khinast[2], and Benjamin J. Glasser[1]. (1) Dept. of Chemical and Biochemical Engineering, Rutgers University, 98 Brett Road, Piscataway, NJ 08854, (2) Institute for Process Engineering, Graz University of Technology, Inffeldgasse 21/A, Graz, A-8010, Austria

Bladed mixers are commonly used in a variety of areas, ranging from the bulk chemical to the food and pharmaceutical industries. Despite their extensive use, the flow of granular materials in agitated devices and the mechanisms by which particle motion is generated are still not fully understood. Experimental examination of granular flows in these mixers has been limited to off-line analysis of location-specific samples, and to on-line measurement of a limited set of process parameters. Numerical simulation techniques have the potential of bridging the knowledge gap since they allow for the study of experimentally difficult-to-measure parameters.

Utilizing the discrete element method the 3-dimensional flow of cohesionless particles in a cylindrical vessel agitated by a 4-blade impeller was studied. Instantaneous, averaged and fluctuating velocity fields show that the particle movement is governed by the angular motion of the blades. The granular bed deforms by forming heaps where the blades are present and valleys between blades passes. The presence of these heaps leads to the formation of a 3-dimensional recirculation zone in front of the blade, where particles flow upwards by the wall and downwards by the impeller shaft. This 3-D recirculation leads to enhanced mixing by promoting radial and vertical mixing. The particle's frictional characteristics are shown to strongly influence the flow structure and mixing kinetics within the mixer. At low friction coefficients the 3-D recirculation in front of the blade is not present reducing the convective mixing inside the mixer. Higher friction coefficients lead to an increase in granular temperature which is associated with an increase in diffusive mixing. Collisional stresses within the mixer showed periodic behavior indicating that the granular material compresses as the blade passes and dilates in between the blades. The long-time averages of the normal and shear stresses were found to vary with mixer height with maximum values near the bottom plate. Additionally, a strong dependence between the magnitude of the shear stresses and the friction coefficient of the particles was found. These stress tensor characteristics indicate that the granular flow in this bladed mixer occurs in the quasi-static regime.

Verification and validation study of continuum granular frictional flow theories using a discrete technique and experimental data*

Sofiane Benyahia

National Energy Technology Laboratory, Morgantown, WV 26505

Introduction

Two frictional flow theories are evaluated for dense granular flows. We assume, similar to Johnson and Jackson,[2] that the total stresses acting on a granular assembly are the sum of stresses due to instantaneous binary collisions derived from kinetic theory and frictional stresses due to enduring contact between layers of particles. We describe a frictional model,[3,4] which has been traditionally used in the MFIX computer code (www.mfix.org). We also use another frictional model developed by Srivastava and Sundaresan,[5] who implemented it in the MFIX code and conducted validation studies of bin discharge. A simplified version of this model is described here. By implicitly expressing the divergence of solids velocity in this model, it is no longer necessary to relax the stresses by solving an additional transport equation for the ratio of frictional to critical pressure. The frictional stresses in this model affect the granular assembly at solids concentration lower than packing as proposed by Johnson and Jackson.[2] Validation studies of dense frictional flow theories are available in the literature.[8,9] Such studies include other complications in the flow physics, such as interstitial fluid, complex geometries, and cohesive particles. In contrast, this study focuses only on dense frictional flows in a vacuum, so no effect of the fluid is introduced in the model equations. Furthermore, we only study relatively simple systems such as bin discharge, similar to Srivastava and Sundaresan,[5] and compare results of the discharge rate to the Beverloo et al.[10] correlation. For a detailed quantitative comparison of the dense frictional models, we use further simplifications by studying a gravity-free granular shear flow between two parallel plates. Computer data generated using a discrete particle method with a soft-sphere contact model (DEM) are used for comparison with continuum models. Discrete techniques using both soft-sphere and hard-sphere contact models are powerful tools for validating continuum models, as shown by many studies in the literature.[11-13] Although most of these studies are aimed at validating granular kinetic theories, discrete techniques are also useful to study dense frictional flows and are used here to validate two frictional flow theories commonly used in continuum modeling.

Granular flow model

The granular kinetic theory model used in this study is essentially the same as that derived by Lun et al.[1] This includes conservation equations for flow of solids in a vacuum and the constitutive relations for the solids stresses based on granular kinetic theory. In this study, the total stress of the granular assembly is assumed to be the sum of collisional (derived from granular kinetic theory) and frictional stresses.[2]
Conservation of mass for constant solids density:

---

* An extended version of this study was submitted for publication to I&ECR.

$$\rho_s\left[\frac{\partial \varepsilon_s}{\partial t}+\nabla\cdot\left(\varepsilon_s \mathbf{v_s}\right)\right]=0 \tag{1}$$

Conservation of linear momentum:

$$\rho_s\left[\frac{\partial \varepsilon_s \mathbf{v_s}}{\partial t}+\nabla\cdot\left(\varepsilon_s \mathbf{v_s}\mathbf{v_s}\right)\right]=\nabla\cdot\left(\boldsymbol{\tau}_\mathbf{k}+\boldsymbol{\tau}_\mathbf{f}\right)+\varepsilon_s\rho_s\,\mathbf{g} \tag{2}$$

Translational granular energy conservation equation:

$$\frac{3}{2}\rho_s\left[\frac{\partial \varepsilon_s\Theta_s}{\partial t}+\nabla\cdot\left(\varepsilon_s\Theta_s \mathbf{v_s}\right)\right]=-\nabla\cdot\mathbf{q}+\boldsymbol{\tau}_\mathbf{k}:\nabla\mathbf{v_s}-\rho_s J_s \tag{3}$$

Solids kinetic-collisional and frictional stress terms:

$$\boldsymbol{\tau}_\mathbf{k}=\left[-P_s+\eta\mu_b\nabla\cdot\mathbf{v_s}\right]\mathbf{I}+2\mu_s\mathbf{S_s} \tag{4}$$

$$\boldsymbol{\tau}_\mathbf{f}=-P_f\,\mathbf{I}+2\mu_f\mathbf{S_s} \tag{5}$$

$$\mathbf{S_s}=\frac{1}{2}\left(\nabla\mathbf{v_s}+\left(\nabla\mathbf{v_s}\right)^T\right)-\frac{1}{3}\nabla\cdot\mathbf{v_s}\,\mathbf{I} \tag{6}$$

Solids pressure:

$$P_s=\varepsilon_s\rho_s\Theta_s\left[1+4\eta\varepsilon_s g_0\right] \tag{7}$$

The compressibility factor (Z) derived by Carnahan and Starling[15] can be used to express the radial distribution function at contact (g_0):

$$g_0=\frac{1-0.5\varepsilon_s}{\left(1-\varepsilon_s\right)^3} \tag{8}$$

Solids viscosity model:

$$\mu_s=\left(\frac{2+\alpha}{3}\right)\left[\frac{\mu}{g_0\eta(2-\eta)}\left(1+\frac{8}{5}\eta\varepsilon_s g_0\right)\left(1+\frac{8}{5}\eta(3\eta-2)\varepsilon_s g_0\right)+\frac{3}{5}\eta\mu_b\right] \tag{9}$$

$$\mu=\frac{5}{96}\rho_s d_p\sqrt{\pi\Theta_s}\,,\quad \mu_b=\frac{256}{5\pi}\mu\varepsilon_s^2 g_0 \tag{10}$$

Granular energy flux and conductivity:

$$\mathbf{q}=-\kappa_s\,\nabla\Theta_s \tag{11}$$

$$\kappa_s=\left(\frac{\kappa}{g_0}\right)\left[\left(1+\frac{12}{5}\eta\varepsilon_s g_0\right)\left(1+\frac{12}{5}\eta^2(4\eta-3)\varepsilon_s g_0\right)+\frac{64}{25\pi}(41-33\eta)\eta^2\left(\varepsilon_s g_0\right)^2\right] \tag{12}$$

$$\kappa=\frac{75\rho_s d_p\sqrt{\pi\Theta_s}}{48\eta(41-33\eta)} \tag{13}$$

Collisional dissipation of granular energy:

$$J_s=\frac{48}{\sqrt{\pi}}\eta(1-\eta)\frac{\varepsilon_s^2 g_0}{d_p}\Theta_s^{3/2} \tag{14}$$

Frictional-collisional wall boundary condition:[2]

$$\frac{\mathbf{v_{sl}}}{\left|\mathbf{v_{sl}}\right|}\cdot\left(\boldsymbol{\tau}_\mathbf{k}+\boldsymbol{\tau}_\mathbf{f}\right)\cdot\mathbf{n}+\frac{\phi\pi\,\rho_s\varepsilon_s g_0\sqrt{\Theta_s}}{2\sqrt{3}\varepsilon_s^{\max}}\mathbf{v_{sl}}+\left(\mathbf{n}\cdot\boldsymbol{\tau}_\mathbf{f}\cdot\mathbf{n}\right)\tan\delta=0 \tag{15}$$

$$\mathbf{n}\cdot\mathbf{q} = \frac{\phi\pi\left|\mathbf{v_{sl}}\right|^2 \rho_s \varepsilon_s g_0 \sqrt{\Theta_s}}{2\sqrt{3}\varepsilon_s^{\max}} - \frac{\sqrt{3}\pi \rho_s \varepsilon_s g_0 \left(1-e_w^2\right)\sqrt{\Theta_s}}{4\varepsilon_s^{\max}}\Theta_s \tag{16}$$

Syamlal et al. (S-R-O) frictional model

This model has been traditionally used in the MFIX code and was described in detail by Syamlal et al.[4] The model equations as presented in this study were first written by Schaeffer,[3] who described the plastic flow of a granular material and related the shear stress to the normal stress. The Syamlal et al.[4] (S-R-O) model expresses the frictional stresses by the following equations:

$$P_f = P_c = \begin{cases} 10^{25}\left(\varepsilon_s - \varepsilon_s^{\max}\right)^{10} & \varepsilon_s > \varepsilon_s^{\max} \\ 0 & \varepsilon_s \le \varepsilon_s^{\max} \end{cases} \tag{17}$$

$$\mu_f = \begin{cases} \dfrac{P_c \sin(\delta)}{2\sqrt{I_{2D}}} & \varepsilon_s > \varepsilon_s^{\max} \\ 0 & \varepsilon_s \le \varepsilon_s^{\max} \end{cases} \tag{18}$$

In this case, the critical solids pressure is a power law function[4] of the solids volume fraction that allows for some compressibility near the packing limit similar to other plastic flow theories.[7,17,18]

Srivastava-Sundaresan (S-S) frictional model

The frictional model used in this study was proposed by Srivastava and Sundaresan[5] who gave expressions of the frictional stresses for a compressible granular assembly. This model is expressed by the following equations:

$$P_c = \begin{cases} 10^{25}\left(\varepsilon_s - \varepsilon_s^{\max}\right)^{10} & \varepsilon_s > \varepsilon_s^{\max} \\ Fr\dfrac{\left(\varepsilon_s - \varepsilon_s^{\min}\right)^r}{\left(\varepsilon_s^{\max} - \varepsilon_s\right)^s} & \varepsilon_s^{\max} \ge \varepsilon_s > \varepsilon_s^{\min} \\ 0 & \varepsilon_s \le \varepsilon_s^{\min} \end{cases} \tag{19}$$

$$\frac{P_f}{P_c} = \left(1 - \frac{\nabla\cdot\mathbf{v_s}}{n\sqrt{2}\sin(\delta)\sqrt{\mathbf{S_s}:\mathbf{S_s} + \Theta_s/d_p^2}}\right)^{n-1} \tag{20}$$

$$\mu_f = \frac{\sin(\delta)}{\sqrt{2}}\frac{P_f}{\sqrt{\mathbf{S_s}:\mathbf{S_s} + \Theta_s/d_p^2}}\left\{n - (n-1)\left(\frac{P_f}{P_c}\right)^{\frac{1}{n-1}}\right\} \tag{21}$$

Application of the frictional models to a discharge of 1 mm particles from a 2D bin
The first application of the frictional models presented in this study is to a 2D bin discharge. This is the same example that was studied by Srivastava and Sundaresan:[5] a 2D rectangular bin 8 cm wide and 100 cm high with an open top and an orifice centered at the bottom. Two issues were reported by Srivastava and Sundaresan[5] that are addressed in this section: First, the cause of the grid-scale flutter in the solids volume fraction observed initially in the simulation was determined and eliminated in this study, and second, lower discharge rates were computed in this study, which resulted in better agreement with the Beverloo correlation.

Temporal profiles of discharge rate
Figure 1 shows the temporal variation of the discharge rates for the four orifice diameters of 1.4, 1.6, 1.8 and 2 cm.
A comparison of the discharge rate for the case of a bin with an orifice diameter of 1.4 cm with the simulation results of Srivastava and Sundaresan[5] (see their figure 4) shows that our computed discharge rate (shown in figure 2) was lower. The reason for the disagreement could be a limitation in the frictional viscosity. The current model does not set any limit for computed values of the frictional viscosity for either S-S or S-R-O models. In the previous publication, Srivastava and Sundaresan[5] may have used an upper limit to the frictional viscosity, which value is not known. This is suspected because in one simulation using the S-S model, we limited the frictional viscosity to100 poise, which resulted in almost double the discharge mass flow rate.
Figure 1 shows that lower values of discharge rate are computed using the Syamlal et al.[4] (S-R-O) frictional model as compared to the S-S predictions. This is due to a fundamental difference between these two frictional models: the S-S model friction starts at a lower solids volume fraction ($\varepsilon_s^{\min}$) so that the computed normal and shear stresses are lower than those computed using the S-R-O model where friction starts at maximum packing.

Verification of Beverloo correlation
Numerical data obtained using the S-S and S-R-O frictional models were compared to the Beverloo et al.[10] correlation for estimating the discharge rate from hoppers and bins, which was written by Srivastava and Sundaresan[5] for a 2D bin discharge as:

$$W = C\,\rho_B\,g^{0.5} D_o^{1.5} H \tag{23}$$

Figure 2 shows that the discharge rate is a function of $D_o^{1.4}$ and $D_o^{1.9}$ for the S-S and S-R-O models, respectively. This is in agreement with the results of Srivastava and Sundaresan,[5] who found that the discharge rate scales as $D_o^{1.4}$ using their frictional model. The S-R-O frictional model over-predicts the value for the exponent of the orifice width.
According to Figure 2, the computed value of C (the empirical constant in the Beverloo correlation) was calculated as 1.08 and 0.61 for the S-S and S-R-O models, respectively. Srivastava and Sundaresan[5] computed a larger value $C = 1.6$, due to their larger computed discharge rate. This current study, however, found the value of C to be closer to the experimental measurements using the S-S model. The S-R-O frictional model predicted

lower discharge rates for all widths of the orifice and, thus, a value of C that is in better agreement with the Beverloo correlation.

The next section is proposed to help us highlight more accurately the distinctions between these two frictional continuum models.

Validation of the frictional models for 1D granular Couette flow

We applied the frictional flow theories discussed previously to model a gravity-free granular flow (in a vacuum) in a Couette shear cell. To conduct the validation study, we generated computer data of the granular flow using the soft-sphere discrete element method[20] (DEM) available in MFIX. In this study, the mass, m, of a particle is equal one. The normal and tangential spring constants were equal, with a value of $k/m = 8\times10^{10}$ s^{-2}. The same value for particle-particle and particle-wall friction coefficients was used in this study: $\mu_p = \mu_w = 0.5$. Other physical parameters such as restitution coefficients ($e = e_w = 0.8$), were the same as those used in the study of Karion and Hunt.[21]

Figure 3 shows a snapshot (after 20 sec) of particle position and velocity in the periodic shear cell. Particles with high velocity are observed at the top and bottom walls due to friction with the moving walls. Near the walls, a relatively dilute flow is observed due to the high energy particles that tend to push nearby particles toward the center of the channel where the flow is clearly denser. Layering of particles is observed at the center of the channel, although some pockets of void can also be seen—perhaps due to large shear (and energy) applied at the frictional walls.

Figure 4 shows a comparison of time-averaged DEM data and the results obtained using the continuum frictional flow theories discussed in this study. Although we used the transient continuum models, the numerical simulations quickly reached a steady-state after only few seconds of simulation. It was expected that transient continuum gas-solids simulations reach steady-state for flows with no gravity as demonstrated by Benyahia et al.,[22] who verified that the frequency of oscillations scales with $\sqrt{g}$. Furthermore, we found identical results in both 1D and 2D periodic geometries, and therefore, only 1D continuum simulation results are reported in this study. Figure 4 shows that both the continuum collisional-frictional models used in this study are able to reproduce the qualitative trends computed using DEM. Due to the high granular temperature gradient between the walls and center of the channel, solids moves to the center in order to balance the collisional pressure. This migration of solids to the center can be limited with frictional pressure, which, in the case of S-R-O model, does not occur until maximum packing is reached. This is the reason why this model predicts packed regions ($\nu_s^{max} = 0.9$) at the center of the channel. This is not the case with the S-S model where friction starts at $\nu_s^{min} = 0.6$ so that a better agreement is obtained at the center. The continuum models predict a more dilute flow near the walls due to the large values of granular temperature computed in that region because of wall boundary conditions. We should mention that one parameter, the specularity coefficient ϕ in the granular boundary condition proposed by Johnson and Jackson,[2] was adjusted in order to fit the near-wall velocity data obtained using the S-S model. A value of $\phi = 0.052$ is used to obtain the continuum data shown in figure 4. It is possible that the cause of the disagreement observed for the granular temperature profiles is mainly due to the fact that we are only solving for the translational granular energy, and, thus, a good agreement is not expected in the dilute regions of this highly frictional flow. Nevertheless, it is clear from figure 4 that a better quantitative agreement with DEM data is obtained using Srivastava-Sundaresan (S-S) frictional model.

Conclusion

We present in this study a granular kinetic theory model derived by Lun et al.,[1] and conduct a comparative study of two frictional flow theories, one proposed by Syamlal et al.,[4] and one by Srivastava and Sundaresan.[5] These two continuum frictional flow theories are validated for a granular bin discharge with an empirical correlation developed by Beverloo et al.[10] and show reasonable agreement for both the Srivastava-Sundaresan (S-S) and Syamlal et al.[4] models. A more detailed comparison of the two continuum frictional models is conducted in the second part of this study. We validate these two theories with computer simulation data obtained using a discrete element method (DEM) for a granular Couette flow. Both frictional continuum theories show similar trends as DEM data. The solids area fraction was lower at the walls due to the higher granular temperature generated from friction with the moving walls. Overall, the S-S frictional model shows better agreement with DEM data.

Literature Cited

(1) Lun, C.K.K.; Savage, S.B.; Jeffrey, D.J.; Chepurniy, N. *J Fluid Mech.* 1984;140;223-256.

(2) Johnson PC, Jackson R. *J Fluid Mech*, 1987;176:67-93.

(3) Schaeffer, D.G. *J. Diff. Eq.*, 1987;66:19-50.

(4) Syamlal, M., Rogers, W.A., O'Brien, T.J., 1993. MFIX documentation: theory guide, DOE/METC-94/1004 (DE94000087). Also available electronically from: http://mfix.org/documentation/Theory.pdf

(5) Srivastava, A; Sundaresan, S. *Powder Tech.* 2003;129:72-85.

(6) Ng, B.H.; Ding, Y.L.; Ghadiri, M. *Chem. Eng. Science* 2008;63;1733-1739.

(7) Bouillard, J.X.; Lyczkowski R.W.; Gidaspow, D. *AIChE J.* 1989;35:908-922.

(8) Makkawi, Y.T.; Wright, P.C.; Ocone, R. *Powder Tech.* 2006;163:69-79.

(9) Huilin, L.; Yurong, H.; Wentie, L.; Ding, J.; Gidaspow D.; Bouillard, J. *Chem. Eng. Science* 2004;59;865-878.

(10) Beverloo, W.A.; Leniger, H.A., van de Velde, J. *Chem. Eng. Science* 1961;15;260-269.

(11) Lun, C.K.K.; Bent, A.A. *J Fluid Mech.* 1994;258;335-353.

(12) Galvin, J.E.; Dahl, S.R.; Hrenya, C. M. *J Fluid Mech.* 2005;528;207-232.

(13) Ye, M., van der Hoef, M.A., Kuipers J.A.M. *Chem. Eng. Research and Design* 2005;83;833-843.

(14) Jenkins JT, Zhang C. *Phys. Fluids*, 2002;14-3:1228-1235.

(15) Carnahan, N.F.; Starling, K.E. *J. Chemical Physics* 1969;51-2;635-636.

(16) Bravo Yuste, S.; Lopez de Haro, M.; Santos, A. *Physical Review E* 1996;53-5;4820-4826.

(17) Jenike, A.W. *Powder Tech.* 1987;50:229–236.

(18) Tardos, G.I. *Powder Tech.* 1997;92:61–74.

(19) Savage, S.B. 1998 *J Fluid Mech.* 1998;377:1-26.

(20) Boyalakuntla, D.S., Ph.D. Thesis 2003, Carnegie Mellon University. Available electronically at: http://mfix.org/open_citations/uploads/Jay_PhD_Thesis.pdf

(21) Karion, A.; Hunt, M.L. *Powder Tech.* 2000;109:145–163.

(22) Benyahia, S.; Syamlal, M.; O'Brien T.J. *AIChE J.* 2007; 53:10;2549-2568.

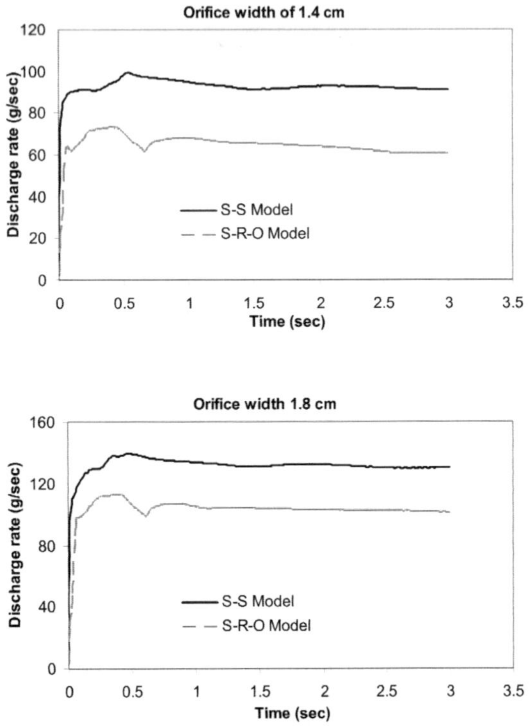

Figure 1: Temporal variation of the solids discharge rate for 4 different orifice widths using the Srivastava and Sundaresan[5] (S-S), and Syamlal et al.[4] (S-R-O) frictional models.

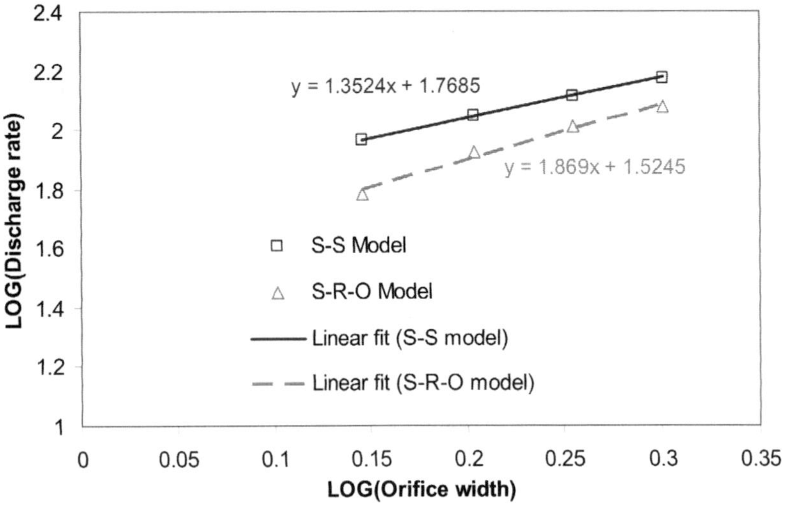

Figure 2: Verification of the Beverloo correlation using both S-S and S-R-O frictional models. The linear best fits of the computed data and the corresponding equations are also plotted.

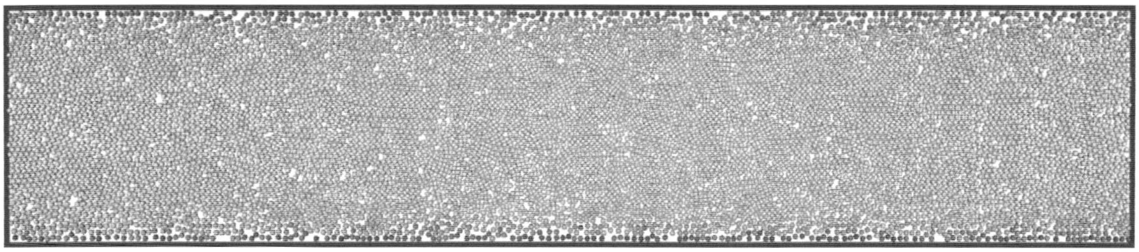

Figure 3: Instantaneous particle position and velocity in the periodic shear cell. Red color indicates higher velocity with maximum of about 9 m/s obtained at the top and bottom moving walls. Periodic boundaries apply to the right and left sides. Particles are monodisperse with an average solids area fraction of $\nu_s = 0.75$ and $h/d_p = 40$.

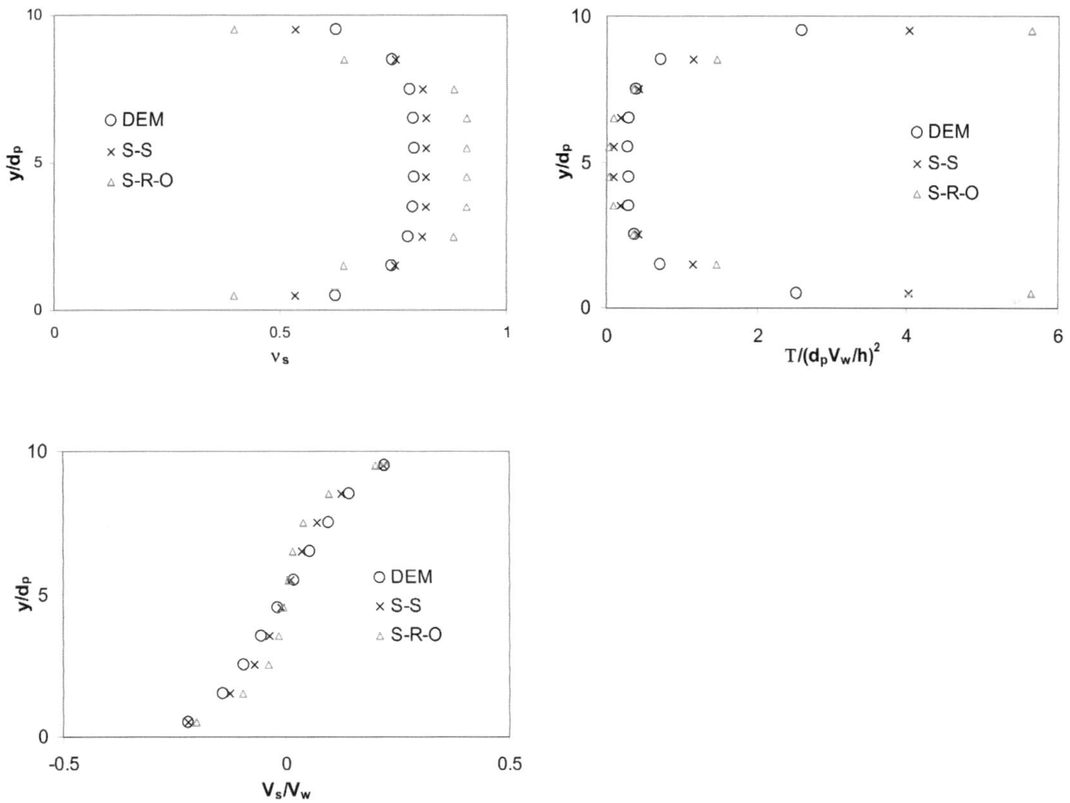

Figure 4: Comparison between the time-averaged DEM results and the steady continuum collisional-frictional models (using S-S and S-R-O frictional models) for the solids area fraction, granular temperature, and velocity.

189ae Development of An Automatic Fluidization Rig

S. Ding[1], Are Dyrøy[2], and Morten Karlsen[2]. (1) Postec, Tel-Tek, Kjølnesring, Porsgrunn, Norway, (2) Hydro Aluminium Metal, Porsgrunn

A fluidization rig was built in order to provide basic data required by the Hydro Aluminum metal. The rig is made of a stainless steel column, with an internal diameter of 250 mm and 1000 mm in height. Four pressure gauges were mounted to the rig; they were installed from the bottom of the column, with a 200 mm in increment along a generator. In addition, two loads cells were installed diametrically opposite to the pressure gauges, one on the level of lowest gauge, another on the level of highest gauge. An auto control system was developed to introduce a pressurized to fluidize Aluminum powders filled in the rig; a data logging unit were also implemented in the system register in line the signals measured by the pressure gauges and loads cells with NI-Labview platform. The parameters for minimum velocity, the expansion of bed and the de-aeration time etc. were given with a series runs of measurement. Comparisons were also made between results measured with gauges and the loads cells. The properties changes of aluminium powders from a packing stage to a fluidised stage were assessed.

189af Effect of Process Variables in the Preparation of Bohemite Sol on the Microstructure of Gamma-Alumina Particles

Chul-Joo Kim, Ho-Sung Yoon, Sung-Don Kim, Hyungkyu Park, and Jung-Soo Sohn. Minerals and Materials Processing Division, Korea Institute of Geoscience and Mineral Resources, 92 Gwahangno, Yuseong-gu, Daejeon, South Korea

Nanocrystalline boehmite[gamma-AlO(OH)] is a cost-effective material for the production of gamma-Al2O3 finding many industrial applications as catalyst or catalyst support, membranes and adsorbents. Therefore the effects of experimental variables in the preparation process of bohemite[gamma-AlO(OH)] sol were investigated in order to control the microstructure of gamma-Al2O3 particles in this study. Using aluminum isopropoxide as a precursor, gamma-AlO(OH) sol solution was prepared through the process of hydrolysis/condensation, aging, peptization and then gamma-Al2O3 particles were manufactured at high temperature through thermal treatment. As a result of this study, aging process among the sol preparation steps had an importance effect upon the crystallinity and microstructure of gamma-AlO(OH). And the microstructure of gamma-AlO(OH) particles controlled by aging had an influence upon the microstructure of gamma-Al2O3 particles such as specific surface area and pore size distribution. Hence, it could be known that the aging process was the most important step to control the microstructure of gamma-Al2O3 particles.

189ag Preparation and Growth of Silica Sol from Sodium Silicate Solution

Ho-Sung Yoon, Chul-Joo Kim, Sung-Don Kim, and Hee Dong Jang. Minerals and Materials Processing Division, Korea Institute of Geoscience and Mineral Resources, 92 Gwahangno, Yuseong-gu, Daejeon, South Korea

Colloidal silica which has high surface area and excellent surface properties are chemically stable inorganic materials and used for various applications in industry. Therefore the effects of the experimental conditions such as the content of silica in aqueous solution, pH of solution and reaction temperature on the formation of silica sol were investigated for the preparation process of colloidal silica sol from sodium silicate solution in this study. The properties of sodium silicate solution and silicate solution removed sodium ion by ion exchange was characterized by yellow silicomolydate method. The SiO2 content of 2% in sodium silicate solution was proper to oxidize sodium silicate with sulfuric acid. After the removal of sodium ions in sodium silicate solution, the pH of silicate solution had to be controlled above 9 for a stable silicate solution. The silica sol, which size is about 10 nm, could be prepared by heating the mixed solution of sodium silicate and silicate solution removed sodium ions at pH 10 and 80°C. And the silica sol grew into about 50 nm as silicate solution was added to silica sol solution.

189ah Preparation of Anhydrous Magnesium Chloride by Dehydration of Ammonium Magnesium Chloride Hexahydrate

Hyoung-Choon Eom[1], Hyungkyu Park[2], Chul-Joo Kim[2], and Ho-Sung Yoon[2]. (1) Resources Recycling, Korea University of Science and Technology, 92 Gwahang-no, Yuseong-gu, Daejeon, 305-350, South Korea, (2) Minerals and Materials Processing Division, Korea Institute of Geoscience and Mineral Resources, 92 Gwahangno, Yuseong-gu, Daejeon, South Korea

Anhydrous magnesium chloride, the dehydration product from magnesium chloride hydrate ($MgCl_2 \cdot nH_2O$), is a conventional raw material to prepare electrolytic magnesium. Because the dehydration of magnesium chloride hydrate can be accompanied by hydrolysis leading to the production of undesirable hydroxy chloride and oxide compounds of magnesium, it should be carried out in hydrogen chloride gas atmosphere. However, it is difficult to use HCl gas in operation, and the presence of large quantities of corrosive HCl is an obvious disadvantage. Consequently, alternative approaches have been tried for the production of anhydrous magnesium chloride. In this work, the dehydration characteristics of ammonium magnesium chloride hexahydrate ($NH_4MgCl_3 \cdot 6H_2O$) were studied, which was considered as an intermediate material for the dehydration of magnesium chloride hydrate by the alternative method for using HCl gas to produce anhydrous magnesium chloride. The results indicated that $NH_4MgCl_3 \cdot 6H_2O$ decomposed through three steps into $MgCl_2$ below about 290℃, and higher temperatures can cause the magnesium chloride to oxidiz into magnesium oxide. From this work, it was confirmed that dehydration of ammonium magnesium chloride hexahydrate is an effective way for producing anhydrous magnesium chloride even in air atmosphere. It was expected that this process could reduce the production cost and simplify the process of dehydration.

189ai Supported Gold Nano-Catalysts by Spray Synthesis and Solution Route. A Comparative Study

Sachit Chopra[1], Jagdeep Singh[2], Jeroen van Bokhoven[2], and Gregory Beaucage[1]. (1) Department of Chemical and Materials Engineering, University of Cincinnati, Cincinnati, OH 45221-0012, (2) Institute for Chemical and Bioengineering, ETH Zurich, HCI E1127, Zurich, 8093, Switzerland

Supported gold nanoparticles have been shown to catalyze oxidation of CO to CO_2 at room temperature as well as at higher temperatures. For the most part these supported nanoparticles have been produced using solution routes. In this study we compare solution and spray flame synthesis of support gold catalysts using titania, alumina, zirconia and iron oxide supports. Comparison will be made of the nano-structure using SAXS and TEM as well as the molecular structure using XAFS. Anomalous SAXS is used to distinguish between the ceramic supports nano-structure and the nanostructure of the supported gold particles. The temperature dependence of catalyst efficiency and specificity will be presented.

Supported by the US NSF through International Research and Education in Engineering (IREE) and CBET 0626063 grants as well as by the Swiss NSF.

189aj Characterization of Turbulent Fluidization Regime with Geldart Group B Particles by Wavelet Analysis of Pressure Fluctuation Signals

Lii-ping Leu and Yung-Sz Yu. Chemical Engineering, National Taiwan University, no.1, Roosevelt Rd., Sec. 4, Taipei, 10617, Taiwan

The pressure fluctuation signals were measured in a 0.108 m i.d. × 7m high circulating fluidized bed for using Geldart group B powders. By means of the multi-resolution analysis of wavelet transform, the wavelet energy distribution profile of absolute and differential pressure fluctuations in three fluidization regimes were calculated. In bubbling and turbulent fluidization, the energy distribution was dominated by detail signals D6 and D7. Moreover, when the superficial gas velocity reached the transition velocity Uc, the wavelet energy distribution at D6 + D7 reached the maximum value. In the dense lower section of fast fluidization regime, the energy distribution was dominated by detail signals D8 and S8; in the dilute upper section of fast fluidization regime, the energy distribution was dominated by detail signals D4 and D5. When the superficial gas velocity reached the transition velocity Uk, the value of the wavelet energy distribution at D6 + D7 decreased quickly. When the superficial gas velocity exceeded the transition velocity Uk, the energy distribution was dominated by detail signals D6 + D7 or D8 + S8. In this way the regime transition could be found by wavelet analysis.

189ak Ex-Situ Surface Modification of $BaTiO_3$ Nanoparticles Prepared by Hydrothermal Synthesis

Mehrnoosh Atashfaraz, Faculty of Chemical Engineering, Tehran University, Enghelab Ave., 16 Azar street, Tehran, Iran, Mojtaba Shariaty-Niassar, Chemical Eng., Tehran University, Enghelab Ave., 16 Azar street, Tehran, Iran, and Tadafumi Adschiri, Institute of Multidisciplinary Research for Advanced Materials, Tohoku University, Katahira 2-1-1, Aoba-ku, Sendai, 980-8577, Japan.

Controlling the surface of nanoparticles is increasingly becoming important to develop next generation of nanoparticles. In this work, surface modification of $BaTiO_3$ nanoparticles after hydrothermal synthesis was examined by adding modifier reagent(phenol)to the hydrothermally synthesized $BaTiO_3$ nanoparticles. The results showed that reagents successfully modified and controlled the surfaces of the $BaTiO_3$ nanoparticles.

Simulation of Dry Coating Process Using a DEM Method

Peng Weng[1], Costas G. Gogos[1], Linjie Zhu[2], and Ming-Wan Young[2]. (1) Otto H. York Department of Chemical, Biological and Pharmaceutical Engineering, NJIT, Newark, NJ 07102, Albania, (2) Polymer Processing Institute, GITC Building 3901, NJIT, Newark, NJ 07102

Our laboratory recently developed a novel process for simultaneous milling and coating particulates using fluid energy mills. The novel process dramatically improves the production efficiency and eliminates usage of organic solvents. In this study, the particle surface coating process is simulated using software PFC 3D 3.1 (Minneapolis, Itasca, MN). The dependence of coating efficiency on the bond strength, particle size and particle number is investigated. The simulation results indicate that bond strength plays a vital role in the coating process

A Novel Process for Simultaneous Milling and Coating Particulates

Peng Weng[1], Costas G. Gogos[1], Linjie Zhu[2], Ming-Wan Young[2], and Peter Bonnett[3]. (1) Otto H. York Department of Chemical, Biological and Pharmaceutical Engineering, NJIT, Newark, NJ 07102, (2) Polymer Processing Institute, GITC Building 3901, NJIT, Newark, NJ 07102, (3) ARDEC, Picatinny Arsenal, Piscataway, NJ 08854

Ascorbic acid particulates are simultaneously milled and coated with wax in this study using a novel process based on fluid energy milling. A premix of ascorbic acid particulates and wax powders are fed into a fluid energy mill, inside which ascorbic acid particulates frequently collide with each other and with wax powders. Consequently ascorbic acid particulates are broken down to smaller sizes and coated with wax simultaneously. This simultaneous milling and coating process has several advantages compared to traditional separated milling and coating processes such as dramatically improved production efficiency, elimination of solvent usage and prevention of particle agglomeration. The wax coating slows the dissolution rate of ascorbic acid in water, especially when the wax content passes certain threshold value.

220a Molecular Layer Deposition of Ultra-Thin Polymer Films on Large Quantities of Cohesive Nanoparticles

Xinhua Liang, Chemical and Biological Engineering, University of Colorado, 1111 Engineering Drive, Campus Box 424, Boulder, CO 80309-0424, David M. King, Department of Chemical Engineering, University of Colorado, 1111 Engineering Dr., Campus Box 424, Boulder, CO 80309, Steven M. George, Chemistry and Biochemistry, University of Colorado, Boulder, CO 80309, and Alan W. Weimer, Department of Chemical and Biological Engineering, University of Colorado, 1111 Engineering Drive, Boulder, CO 80309-0424.

This work describes the conformal coating of ultra-thin aluminum alkoxide (alucone) polymer films on primary silica and titania nanoparticles using molecular layer deposition (MLD) in a fluidized bed reactor from 100 °C to 160 °C. In situ mass spectroscopy revealed that the growth of alucone MLD films was self-limiting as a function of the individual trimethylaluminum (TMA) and ethylene glycol (EG) exposures. The composition and highly conformal alucone films throughout the surface of both silica and titania nanoparticles were confirmed. The highest growth rate was observed at the lowest sample temperature. Primary nanoparticles were coated individually despite their strong tendency to aggregate during fluidization. This work presents herein the first attempt at placing ultra-thin conformal polymer films on nanoparticles at large scale. MLD has the advantage of low deposition temperature, precision thickness control, excellent conformality, and high film growth rate.

220b Design of High Temperature, In-Situ Coating of Aerosols: Efficiency & Texture

Beat Buesser and Sotiris E. Pratsinis. Particle Technology Laboratory, Institute of Process Engineering, Department of Mechanical and Process Engineering, ETH Zurich, Sonneggstrasse 3, Zurich, CH-8092, Switzerland

Aerosol coating is a promising one-step process to modify nanoparticle surfaces. Here gas-phase coating of TiO_2 particles with SiO_2 films is investigated theoretically, accounting for SiO_2 monomer generation, coagulation, sintering and surface growth by HMDSO (hexamethyldisiloxane) precursor. The evolution of the SiO_2 coating particle population is simulated distinguishing the fine and coarse modes of SiO_2 coating particles. The fine mode deposits fast on TiO_2 particles forming a smooth coating while the coarse deposits further downstream and may contribute to rough coatings depending on SiO_2 sintering kinetics (1).

Coating starts by injecting the HMDSO after the TiO_2 particles have reached their optimal primary particle size that no longer changes by sintering (2). That way TiO_2 primary particles and aggregates have attained a self-preserving distribution that can be represented by a single size facilitating the simulations. Sintering may take place on the surface of the TiO_2 particles and transform rough into smooth coatings depending on process conditions.

The influence of TiO_2 particle size and concentration, process temperature, cooling rate, mixing time and HMDSO precursor concentration on SiO_2 coating thickness and texture are investigated. Especially the evolution of SiO_2 coating thickness and surface area of the rough SiO_2 coating on the TiO_2 core particles surface are discussed.

1. A. Teleki, S.E. Pratsinis, K. Wegner, R. Jossen, F. Krumeich "Flame-coating of titania particles with silica", *J. Mater. Res.*, **20**, 1336-47 (2005).

2. A. Teleki, M.C. Heine, F. Krumeich, M.K. Akhtar, S.E. Pratsinis, "The Role of Gas-Aerosol Mixing during In-Situ Coating of Flame-Made Titania Particles", *Ind. Eng. Chem. Res.*, accepted for publication (2008).

Mechano-Chemical Stability of Gold Nanoparticles Coated with Alkanethiolate SAMs

Brian J. Henz[1], Takumi Hawa[2,3], and Michael R. Zachariah[2,3]

[1]U.S. Army Research Laboratory
Aberdeen Proving Ground, MD

[2]Departments of Mechanical Engineering and of Chemistry and Biochemistry
University of Maryland
College Park, MD

[3]National Institute of Standards and Technology
Gaithersburg, MD

Abstract

Molecular dynamics simulations are used to analyze the structure and stability of alkanethiolate self-assembled monolayers (SAMs) on gold nanoparticles. We have observed that the surface of gold nanoparticles become highly corrugated by the adsorption of the SAMs. Furthermore, as the temperature is increased, the SAMs dissolve into the gold nanoparticle, creating a liquid mixture at temperatures much lower than the melting temperature of the gold nanoparticle. By analyzing the mechanical and chemical properties of gold nanoparticles at temperatures below the melting point of gold, with different SAM chain lengths and surface coverage properties, we have determined that the system is metastable. The model and computational results that provide support for this hypothesis are presented.

1. Introduction and Background

Nanoparticles are often used in applications where a high surface area to volume ratio is desired. Some of the typical applications include catalysis, biosensors, and use as drug delivery vehicles.[1,2] A major limitation of the expanded use of these nanoparticles is that the fabrication of nanoparticles with the desired morphology and structure is challenging.[3,4,5] Gold is a commonly used material because of its resistance to oxidation and its interesting electrical, magnetic, optical, and physical properties[6]. Furthermore there is a considerable knowledge base on functionalizing gold surfaces that are being ported to analogous particles.[7] For this reason gold nanoparticles are commonly chosen for basic research because of the availability of experimental[8,9] and atomistic simulation data.[10,11] Using molecular dynamics (MD) simulations Lewis et al.[12] were able to determine the melting process for gold nanoparticles. This last result is important because the authors found that the surface atoms melt first, and at a temperature below that of the atoms in the core of the nanoparticle. The implications of this for coated nanoparticles is that the surface atoms, where binding of the monolayer takes place, is affected by a lower temperature than is expected from the melting temperature of the whole nanoparticle. The results of these past efforts are built upon in this work in order to analyze, and in larger gold nanoparticle and in particular the nature and role of surface functionalization.

Coating of gold surfaces and nanoparticles with a self-assembled monolayer (SAM) creates many additional applications including lithography, lubrication, catalysis, biocompatible materials, and biosensors.[6,13] There have been numerous experimental[14,15,16] and numerical[13,17,18] studies performed that have attempted to characterize the properties of alkanethiolate SAM coated gold surfaces. Simulations investigating alkanethiolate SAM coated gold surfaces accurately predict the c(4x2) superlattice structure of alkanethiols on an Au(111) surface. Additional work has predicted the segregation of SAM chains adsorbed on a gold surface by length, surface frictional forces[19], and phase behavior.[20] Efforts to characterize the alkanethiolate SAM coated gold nanoparticle system have been limited to atomistic simulations considering a single alkanethiol chain, and a small (less than 100 atom) gold cluster[17,21] or MD simulations of small nanoparticles that ignore gold-gold interactions.[6,18]

During the fabrication of nanoparticles the propensity to agglomerate and sinter must be inhibited by the passivation of the nanoparticle surface. The use of coatings for surface passivation has been found to significantly reduce the degree to which agglomeration and sintering will occur.[22] For instance, Hawa and Zachariah found that a hydrogen surface passivation coating used with a silicon nanoparticle will prevent reaction with other particles when collided at thermal velocities or will retard the process.[23] Other affects related to the passivation layer investigated computationally, include changes in the internal pressure and surface tension.[24]

2. Computational Modeling and Numerical Procedure

There are numerous documented MD studies of alkanethiolate SAMs on gold surfaces.[13,17,19] However, these studies have been limited to flat gold surfaces, or to fixed gold atoms in a nanocrystallite.[18,6] In addition to, and in support of, these previous MD simulations, *ab-initio* quantum chemical calculations have been used to simulate gold clusters with short alkanethiolate SAMs, primarily to determine potential parameters for MD or MC computations. The *ab-initio* calculations show that the gold crystal lattice is perturbed by the adsorbed ligand.[25] The consequence of the perturbation is an increase in Au-Au bond length of up to 20%. The increase in bond length is observed around the adsorption site with relaxation propagating to the second layer. These results suggest that the MD simulations should include dynamic gold atoms in order to accurately consider the nanoparticle/SAM system.

Alkanethiol chains adsorbed on a flat surface made of immobile gold atoms are limited to diffusion in only two dimensions, whereas, chains adsorbed onto a nanoparticle surface with mobile gold atoms may diffuse in three dimensions. This disparity in model detail is expected to provide more insight into the SAM coated nanoparticle system, than previously available. The inclusion in the simulation of a surface constructed of dynamic gold atoms should be able to more accurately predict alkanethiol chain movement than the limited movement possible on a flat surface of fixed gold atoms.

The size of the model system investigated here is determined by approximating experimentally realistic systems. The manufacture of consistently sized small diameter nanoparticles becomes more difficult at diameters of less than 50Å, therefore, in this work 50Å diameter gold nanoparticles (4093 gold atoms) with and without SAMs are considered. The SAM considered consists of alkanethiolate chains that contain a sulfur atom head group that binds to the gold surface, and a carbon backbone with 3 to 18 carbon atoms comprising the alkyl chain.

The MD simulations in this work are carried out using the LAMMPS (Large-scale Atomic/Molecular Massively Parallel Simulator)[26] software.

2.1. Simulating the Gold Substrate

In this effort the gold substrate atoms are explicitly considered as dynamic atoms, requiring that an accurate potential be used to simulate these atomic interactions. The embedded atom method[11,27,28] (EAM) has received much attention for accurately simulating atomic gold interactions and the interactions of other fcc lattice metals. This method uses empirical embedding functions that originate from density functional theory (DFT) and have been successfully used to model many of the physical properties of gold. We chose the EAM for modeling the gold substrate because of its reported accuracy, and the potentially important extensions developed for the EAM for modeling surfaces, such as the extended EAM[29] (XEAM), and the surface EAM (SEAM).[30] The actual EAM potential used in this work is taken from LAMMPS.

2.2. Modeling the Alkanethiolate Chain

The alkanethiolate polymer chain may be simulated with a range of computational complexity depending upon the resources available, and the accuracy of results required. For MD simulations, the three methods available for modeling the alkanethiol chain are the high-accuracy/high computational requirement all-atom model, the less accurate/lower computational requirement united atom method, and the coarse grained bond model. These methods can be compared by considering the CH_2 molecule that makes up the backbone of the alkanethiolate chain. When using the all-atom model, the numerical simulation must consider three particles for each molecule and the interactions between each. Using the united atom model only one particle is simulated for each molecule[31], which eliminates the intermolecular interactions, and limits the number of intermolecular interaction potentials required. Finally, with the coarse grained model, multiple molecules are clustered together into one particle with perhaps five or more CH_2 molecules grouped into one simulated particle.[32,33,34] For the simulations in this work we have chosen the united atom method, which is computationally more efficient than the all atom method, yet provides sufficient accuracy for the current analysis.

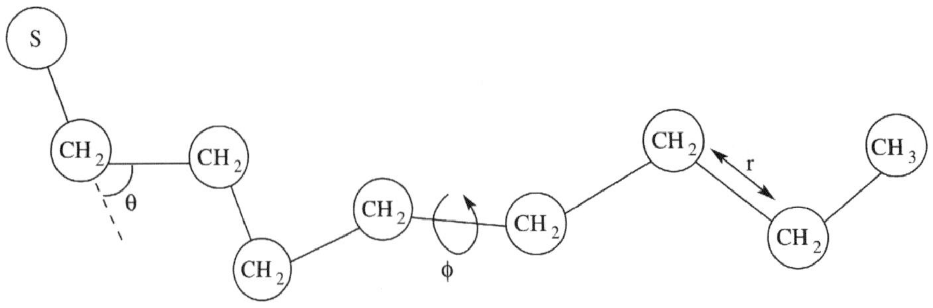

Figure 1. Sample alkanethiolate chain with 8 carbon atoms along the backbone.

The potentials and parameter values for the bending angle and dihedral angle from Figure 1 and used in this work are given in Shevade et al., the bond stretch parameters are given in Rai et al.[35]

2.3. Binding of Sulfur to Gold

The binding potential of the head sulfur atom of an alkanethiol chain to the gold substrate is of great interest as it affects the location, orientation, movement, and desorption of chains from the gold surface.[6,36,37] There has been much effort devoted to finding accurate potentials for simulating the sulfur to gold binding energy from ab-initio methods.[6,36] The most commonly published potentials resulting from these calculations are the 12-3 potential,[38] used most commonly in Monte Carlo (MC) simulations of alkanethiolates on gold[13,19] and the Morse potential, used in many of the MD simulation studies.[19,17,36,39] In addition to the issue of which potential best describes the binding between SAM headgroup and substrate, the possibility of SAM mobility must also be considered. Since the Morse potential, Eq. 1, can mimic a partially covalent bond, it is used in MD simulations along with the parameters given by Zhang et al to model the Au-S bond. The parameters for Eq. 1 are $D_e = 13.3$ kJ/mol, $\alpha = 1.38$, and $r_0 = 2.903$ Å.

$$U_{Morse}(r) = D_e\left[e^{-2\alpha(r-r_0)} - 2e^{-\alpha(r-r_0)}\right] \tag{1}$$

The binding energy of the alkanethiol to the gold surface has been determined to be around 184.1 kJ/mol when all of the sulfur-gold interactions are considered. Although this reported value is often referenced, their are several studies that report the sulfur-gold bond energy to be closer to 126.0 kJ/mol.[40,41] This discrepancy in the binding energy is large, and so in this work we will investigate using both binding energies in order to understand the affect that binding energy has on the computed results. For the sulfur gold binding energy of 126.0 kJ/mol, the Morse potential parameter $D_e = 9.108$ kJ/mol.

3. Simulation Results

In this section, the results from the MD simulations are presented and analyzed. In order to study the affects of the alkanethiolate SAM on the gold nanoparticle we have computed many system properties. These properties include diffusion coefficients, radial pressure and density distributions, melting temperature, and the pair correlation function. A corrugation factor that describes the depth to which the gold surface is modified by the adsorbed alkanethiol chains is also defined in this section. Each of these results is a data point of the changes the nanoparticle surface experiences with the adsorption of the alkanethiolate SAM. For instance, the diffusion coefficient is used to compute the mobility of the SAM chains that are adsorbed on the surface of the nanoparticle, and the mobility of the gold atoms near the surface of the nanoparticle. The combination of these two coefficients will partially define the phase of the materials at and on the nanoparticle surface and also help in predicting if nucleation of the SAM chains will occur. Nucleation may occur under low surface coverage conditions when the chains form dense groups and uneven surface coverage. The density distribution is used to find the depth to which the sulfur atoms penetrate the surface of the gold nanoparticle and to compute the corrugation factor.

The simulation results will also be compared for the two sulfur-gold binding energies considered in this work. The computed results will be compared and discussed in order to better understand how the head group binding energy affects the properties of the nanoparticle and anticipated trends for these properties. It is expected that the lower binding energy will have less of an affect on the properties of the gold nanoparticle than the high binding energy. On the other hand some processes such as desorbtion will possibly be observed only at the lower binding energy because the temperature required for desorbtion will be lower. A detailed discussion of the model initialization can be found in Henz et al.[42]

3.1. Pair Correlation Function

The pair correlation function, $g(r)$, is defined as the number of atoms a distance r from a given atom compared with the number of atoms at the same distance in an ideal gas at the same density.[43] The pair correlation function for a bare 50Å diameter gold nanoparticle is shown in Figure 2 for a range of temperatures.

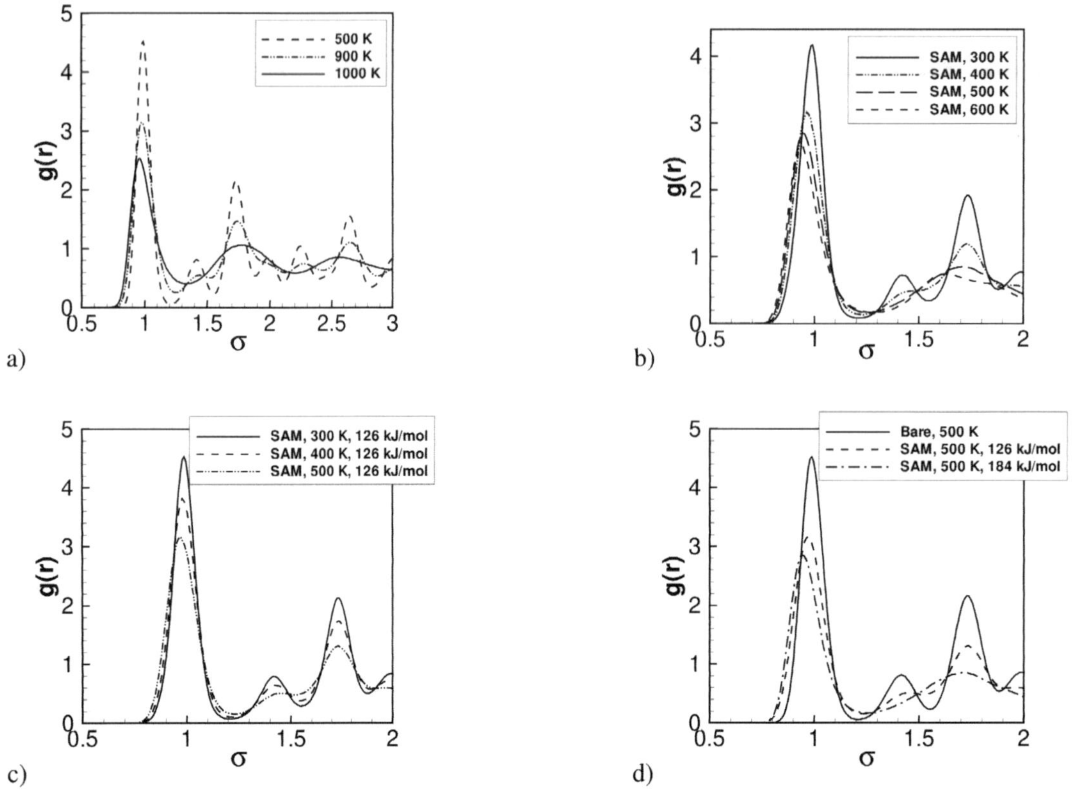

Figure 2. a) Pair correlation function for bare 50Å gold nanoparticle. Note the change in shape and the location of peaks between 900 K and 1000 K indicating a phase change from solid to liquid. **b)** Pair correlation function results for the gold atoms in a 50Å gold nanoparticle coated with an alkanethiolate SAM at various temperatures with a sulfur binding energy of 184 kJ/mol. **c)** The pair correlation function results for the SAM coated gold nanoparticle with the lower sulfur binding energy of 126 kJ/mol. **d)** a comparison of the pair correlation results at 500 K for the bare nanoparticle and high (184 kJ/mol) and low (126 kJ/mol) alkanethiol binding energies.

The results in Figure 2a indicate that a phase transition occurs in the nanoparticle from a crystalline solid at 900 K to a liquid at 1000 K, and as expected, this temperature range is below that of the bulk melting temperature of 1340 K.[44]

For a spherical nanoparticle the melting temperature can be crudely predicted from experimental data using Eq. 2, where ρ_S and ρ_L are the specific mass of the solid and liquid phases, respectively. γ_S and γ_L are the surface energies of the solid and liquid phases, respectively, D_p is the particle diameter, L is the heat of fusion, and T_∞ and $T(D_p)$ are the bulk and size dependent melting temperatures, respectively. Using published data for gold the predicted melting temperature of a 50Å diameter gold nanoparticle is about 995 K. This result is within the range of our MD simulation prediction for the melting temperature.

$$\frac{T(D_p)-T_\infty}{T_\infty} = \frac{4}{\rho_S L D_p}\left[\gamma_S - \gamma_L(\rho_S/\rho_L)^{2/3}\right] \tag{2}$$

Next we compare the pair correlation function results for the uncoated gold nanoparticle to the results for an alkanethiolate coated gold nanoparticle. The coating considered is a full coating, meaning one chain per 15.4Å^2 of surface area on the gold nanoparticle[45], and is about 30% denser than the packing density on a flat surface (21.4Å^2). The pair correlation function results for the alkenthiolate SAM-coated gold nanoparticle show a nanoparticle with a crystalline structure for temperatures below 400 K, and an amorphous structure above 600 K, see Figure 2b. The pair correlation function is computed for the gold atoms only in Figure 2b. By only considering the gold atoms in computing the pair correlation results, the structure of the nanoparticle is analyzed separately from the SAM. At 500 K, the structure of the gold nanoparticle has changed to a less ordered structure, similar to that observed during the melting of the uncoated gold nanoparticle. This change in the pair correlation function indicates that a phase change occurs in the gold nanoparticle between 400 K and 500 K. Hence the melting temperature of the gold nanoparticle is lowered by about 500 K by mixing with the alkanethiolate chains.

In the next set of simulations we considered the lower gold-sulfur binding energy of 126 kJ/mol. In these simulations the chain-chain interactions use the same interaction potentials discussed earlier. In Figure 2c the pair correlation results for the fully coated gold nanoparticle are plotted for temperatures of 300K, 400K, and 500K. Notice that at 500K the nanoparticle is still highly structured. This indicates that the nanoparticle does not undergo a phase change at 500K as is observed in the high binding energy results. In Figure 2d the pair correlation results for the gold nanoparticle at 500K is compared for the uncoated, and coated nanoparticles with high and low binding energy. In this figure the affect of the high binding energy is clearly evident. The lower degree of crystallinity with the higher binding energy indicates a more liquid-like structure at 500K.

3.2. Diffusion

From the results of the previous section it is apparent that the alkanethiolate SAM coating has an appreciable affect on the properties of the gold nanoparticle. In order to gain a better understanding of the SAM, and how it interacts with the gold nanoparticle we have computed the diffusion coefficient for the SAM and the gold atoms. The alkanethiolate chain is primarily bound to the gold nanoparticle through the sulfur-gold interactions at the head of the chain. The diffusion coefficient of the sulfur atoms is of particular interest because it will provide insight into the mobility of the alkanethiol chains in relation to the surface gold atoms. The diffusion coefficient of the alkanethiolate chain is calculated in the MD simulations by computing the mean squared displacement of the sulfur head group. The diffusion coefficient, D, is then evaluated using Eq. 3.

$$\frac{\partial\langle r^2(t)\rangle}{\partial t} = 2dD \tag{3}$$

In Eq. 3, the number of dimensions available for atomic diffusion, d, is assumed to be 3, t is time, and $\langle r^2(t)\rangle$ is the mean squared displacement (MSD) of the atoms being tracked. For diffusion on a surface, d would typically be 2. However, for a nanoparticle, as opposed to a flat surface, the gold atoms near the nanoparticle surface are less strongly bound to the core atoms, and thus more mobile, allowing for the sulfur atoms to move in a third dimension normal to the nanoparticle surface. This movement normal to the nanoparticle surface has not previously been considered in the simulation of SAM coated gold nanoparticles, but is expected for the reasons discussed previously.

The diffusion coefficient of the sulfur head group has been computed for various system temperatures and presented in an Arrhenius plot in Figure 3. There are two linear regions with slopes of -285.9 and -3477.6, indicating two distinct phases of the alkanethiolate SAM in the temperature ranges of 100 K – 500 K and 500 K – 800 K, respectively.

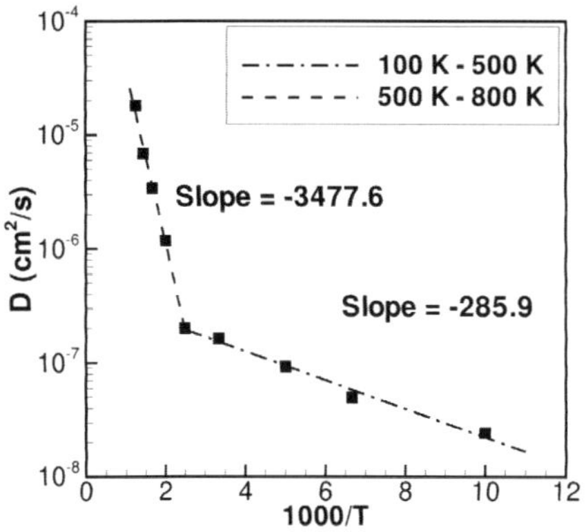

a)

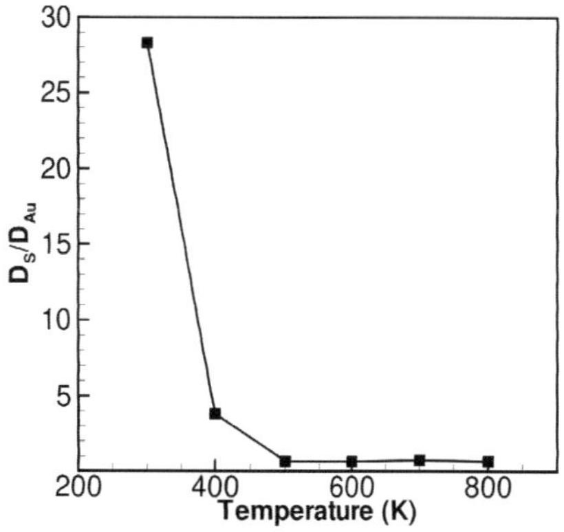

b)

Figure 3. a) Arrhenius plot of diffusivity for sulfur atoms. The activation energy for SAM mobility is estimated by the slope of $\ln(D)$ vs. $1/T$ between 100 K and 800 K. **b)** Ratio of sulfur diffusivity to gold diffusivity in the alkanethiol coated gold nanoparticle material system.

The activation energy for the low and high temperature regions is 2.4 kJ/mol and 29.0 kJ/mol respectively. This is the expected behavior for heterodiffusion, where one type of atom diffuses on another.[46,47] In the low temperature regime the observed activation energy corresponds to the alkanethiol chains hopping between adsorption sites, which is observed as diffusion. For flat gold surfaces, particularly the Au(111) surface, the adsorption binding energy surface has been computed using atomistic simulations.[17,48] These atomistic computations have determined that the difference between the minimum binding energy (face-centered cubic) site, and the maximum binding energy (atop) site is between 25.1 kJ/mol and 16.9 kJ/mol. The diffusion barrier is the energy required for an adsorbate atom to move to the next adsorption site. In Zhang et al. the authors have calculated the binding energy at a bridge site that would provide a possible diffusion path between adsorption sites with a diffusion barrier of 12.2 kJ/mol. Each of these published results are for flat surfaces, and report a somewhat higher activation energy than the computed activation energy from the MD simulations for the low temperature regime. One possible explanation for the discrepancy in the calculated value is that for a curved nanoparticle surface their may be lower energy and irregular diffusion paths available for the alkanethiolate chains to diffuse along. More likely however is that all the observed activation energies are small and in part at least differences may by associated with the accuracy of the interatomic potential. Nevertheless, is does indicate a significant difference between a low temperature diffusion process, and that occurring at higher temperature.

Typically, as the temperature rises the dominant mechanism for diffusion shifts from adatoms to vacancies. However, what is occurring here is that the diffusion of the alkanethiolate chain is dominated by dissolution into the gold nanoparticle, resulting in a more highly activated process. In both instances the activation energy required for diffusion is increased, but in the former process the change in activation energy is observed to be a factor of two or more. In this case the change is about one order of magnitude. The driving force for dissolution will be discussed in the next section. We notice however, in Figure 3b, that the diffusivity of the gold atoms is more than an order of magnitude lower than the sulfur head group atoms at low temperatures, below 500 K. This difference suggests that the alkanethiol chains are moving freely over the gold nanoparticle surface, but at around 500K this is no longer the case, and the diffusivity of the sulfur atoms closely tracks that of the gold nanoparticle atoms. Above 500K the diffusivity of the gold and sulfur atoms is similar, supporting the observation that mixing of the two materials has occurred.

3.3. Radial Pressure Distribution

The radial pressure distribution is used in this work to quantify the affect of the adsorbed alkanethiolate SAM on the gold nanoparticle. In very small droplets the internal pressure can be much larger than the surrounding environment. One method of evaluating the radial pressure distribution is to use the normal component of the Irving-Kirkwood (IK) pressure tensor.[49] The IK pressure tensor comprises two terms, corresponding to a kinetic $P_K(r,T)$ and a configurational $P_U(r,u)$ contribution as shown in Eq. 4.

$$P_N(r) = P_K(r,T) + P_U(r,u) \tag{4}$$

The kinetic pressure term is a function of temperature and the computed radial density distribution. The configurational term is computed from the interactions, u, between pairs of particles. For each shell of radius r, the forces between particles whose line of interaction intersects the shell is considered when computing the normal pressure component. The normal pressure at the surface of a nanoparticle or droplet is typically positive, indicating a compressive surface tension. Near the melting temperature of the gold nanoparticle the surface tension is computed to be nearly the published data of 742 *mN/m*.

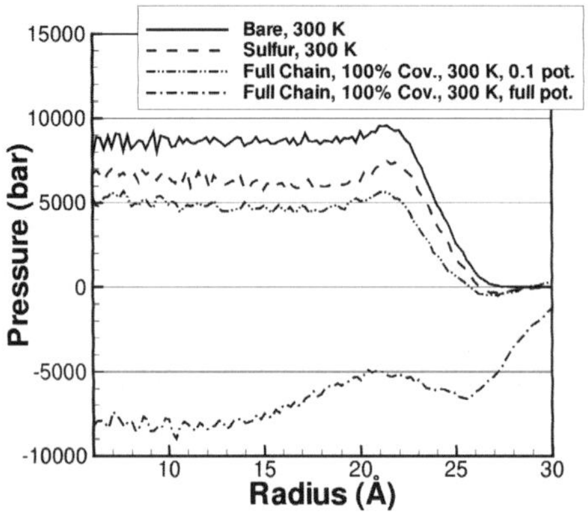

Figure 4. Radial pressure distribution for 50Å gold nanoparticle with alkanethiolate SAMs with differing interaction coefficients.

From past efforts it is expected that a surface coating will modify the nanoparticle internal pressure, through the surface tension, of the gold nanoparticle. This possibility is considered here by comparing the radial pressure distribution for gold nanoparticles with varying degrees of surface coatings. The systems considered include a bare gold nanoparticle, a gold nanoparticle with only sulfur atoms adsorbed onto the surface, and finally a gold nanoparticle with alkanethiolate chains adsorbed on the surface, but with varying interaction potentials. For bare and coated gold nanoparticles with weak SAM chain-chain interactions the internal pressure is positive, meaning a positive surface tension. This result is expected, and the trend is also reasonably expected from past research involving hydrogen passivated silicon nanoparticles. As observed in Figure 4, the pressure trends lower, and becomes negative on the interior of the fully coated nanoparticle as the chain-chain and chain-gold interaction potential parameter values are increased to the published values, indicated by "Full Chain, 100% Cov., 300 K, full pot.".

As a first interpretation of this negative pressure result, one could consider the volume integral of the internal pressure distribution. Since the volume integral of the pressure distribution is negative, it indicates a mechanically unstable system, or a nanoparticle under tension. As seen in Figure 4, the negative pressure is observed in the interior of the nanoparticle, and only when full chain-chain and chain-gold interactions are considered. This negative pressure result indicates that the gold nanoparticle is experiencing a tensile stress and will change its configuration over time or with increases in temperature in order to accommodate the positive stress or negative pressure. The chain-chain interactions apparently have a strong influence on the surface tension, and therefore the internal pressure of the nanoparticle. The strength of the interaction potential that will flip the internal pressure from positive and mechanically stable, to negative and mechanically unstable is still under investigation but is between 10% and 100% of the published potentials used for the alkanethiolate SAM interaction potential. Other factors that may affect the surface tension are the surface area per adsorbed alkanethiol chain, and temperature.

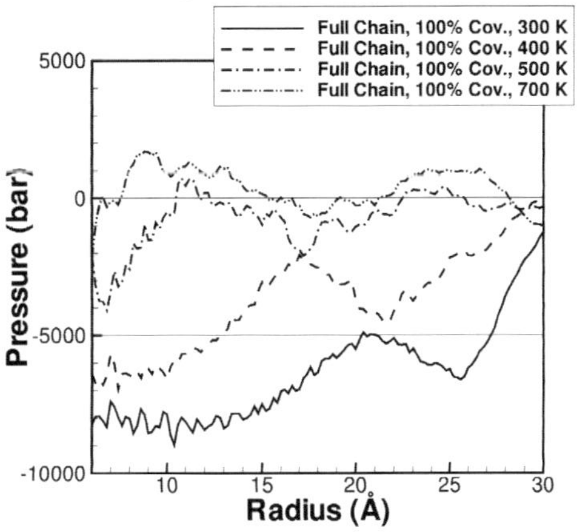

Figure 5. Radial pressure distribution for fully coated gold nanoparticle for range of temperatures with high binding energy (184.1 kJ/mol).

In Figure 5 the radial pressure profile is plotted for a fully coated gold nanoparticle with a sulfur gold binding energy of 184.1 kJ/mol. Notice that as the temperature increases the pressure that is primarily negative at 300 K becomes positive at 700 K. As the system is held at a temperature above the melting temperature of the solution the pressure continues to increase and will most likely equilibrate to an average pressure above 0.

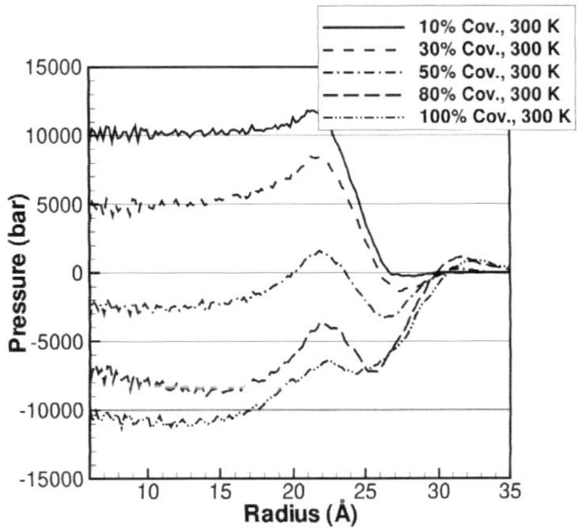

Figure 6. Radial pressure distribution for alkanethiol coated 50Å gold nanoparticle with 126 kJ/mol sulfur-gold binding energy.

A negative surface tension resulting in a tensile stress inside the nanoparticle is not stable for a liquid nanoparticle, and will result in deformation. In a solid nanoparticle the tensile stress will exist for a time before creep[50] or an increase in temperature will lower the yield strength allowing the nanoparticle to yield and flow. For these reasons the solid gold nanoparticle with an alkanethiolate coating is meta-stable at low temperatures, but will stabilize at higher temperatures by changing shape. In this particular case, we observed as discussed in the section on diffusion that the alkylthiol chains dissolve into the particle. This result has not been previously observed in computer simulations because the gold nanoparticle was assumed to maintain its shape, with the interaction of the alkanethiolate SAM constrained to the surface. Using the lower, 126 kJ/mol, binding energy we initially observe desorbtion of alkanethiolate chains from the nanoparticle surface as the temperature is increased, until the temperature is above 600K at which time the remaining alkanethiolate chains dissolve into the nanoparticle.

Notice in Figure 6 that the internal pressure of the fully coated nanoparticle with the 126 kJ/mol sulfur binding energy is very similar to the internal pressure of the fully coated gold nanoparticle in Figure 4, where the binding energy of the sulfur-gold bond is 184 kJ/mol. Recall that only the sulfur-gold binding energy has been changed in Figure 6, whereas in Figure 4 the "0.1 pot." curve represents data collected from simulations where the sulfur-gold and the chain-chain interactions have both been modified. By comparing these results we can conclude that the chain-chain interactions have a greater affect on internal pressure than the sulfur-gold binding energy. This is an important result because although the sulfur-gold binding energy has some affect on the structure of the gold nanoparticle and diffusivity, it is not the primary factor in determining stability of the nanoparticle. Additionally, the affect of chain length on the radial pressure distribution is also considered. After computing the radial pressure distribution for alkanethiol chain lengths of 3 and 18 carbon atoms, we have found that the shorter chains result in an internal pressure about 5000 bar higher than the previously considered 9 carbon atom chains. Alternatively, the longer chains don't appear to have a greater affect on the internal pressure of the nanoparticle than the 9 carbon atom chains studied here. This leads to the conclusion that for the number of monomers considered, a longer chain has little affect on the stability of the nanoparticle, whereas a shorter chain will lower the total chain-chain interaction sufficiently to affect the nanoparticle stability.

4. Phase Behavior and SAM Solubility

The computed potential energy (PE) of the alkanethiolate SAM coated gold nanoparticle system is used to determine the critical temperature at which mixing and phase change will occur. The computed slope of the PE versus temperature curve from the MD simulation results are used to determine the heat capacity of the system.

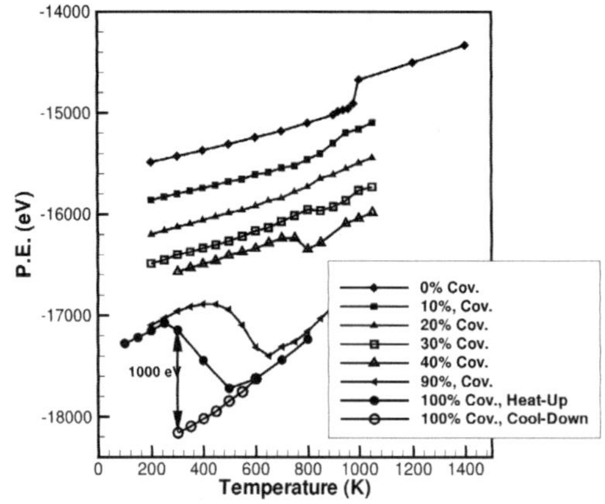

Figure 7. Potential energy versus temperature for the alkanethiolate SAM coated gold nanoparticle with various amount of surface coverage and the 184 kJ/mol sulfur binding energy.

In Figure 7, the system PE is plotted versus temperature for a 50Å gold nanoparticle coated with increasing densities of the alkanethiolate SAM. For the bare nanoparticle there is a discontinuity around 980 K, which corresponds to a phase change. The SAM coated nanoparticle demonstrates a very different behavior. Initially, for small numbers of adsorbed alkanethiolate chains, the melting temperature of the gold nanoparticle decreases, and does not occur at a unique temperature but rather over a range of temperatures. This behavior is expected for a mixture of two materials. The appearance of the SAM on the particle surface decreases the surface tension and internal pressure. As the coverage percentage is increased further, another behavior is observed. At higher surface coverages, above 70%, the PE of the system begins to decrease as the temperature is increased, and then after a small temperature range the PE resumes its linear increase. This behavior is explained by the mixing that occurs between the alkanethiolate chains and the gold atoms. The mixing of the SAM chains with the gold atoms is possible because of the lower binding energy of surface Au atoms. The lower binding energy of the surface atoms allows these atoms to become mobile at a lower temperature than the interior gold atoms. The surface atoms begin to mix with the alkanethiol chains, and as the chains penetrate deeper into the nanoparticle they are able to interact with an ever increasing number of gold atoms. Once complete mixing has occurred, the PE resumes increasing monotonically.

As we found with the internal pressure in Figure 4, the structure of the SAM coated gold nanoparticle is metastable. The increase of both the surface coverage, and the temperature aids this particle in transforming into a mixed structure, Figure 8. This result has not been previously predicted, as it was previously assumed that the SAM chains would desorb at these temperatures, as occurs on flat surfaces. If the SAM was to desorb, it is expected that the structure of the underlying gold nanoparticle would not be affected. This is reflected in the pair correlation data discussed earlier that shows that the gold nanoparticle with SAM coating is crystalline at a higher temperature with the lower 126 kJ/mol sulfur binding energy than with the 184 kJ/mol binding energy.

a) b)

Figure 8. Cross section of 100% coated gold nanoparticle at 300K (a) and 600K (b), showing the dissolution of the alkanethiol chains at high temperatures. In these images the blue spheres represent gold atoms, light blue are sulfur, yellow are CH_2 and red spheres represent CH_3.

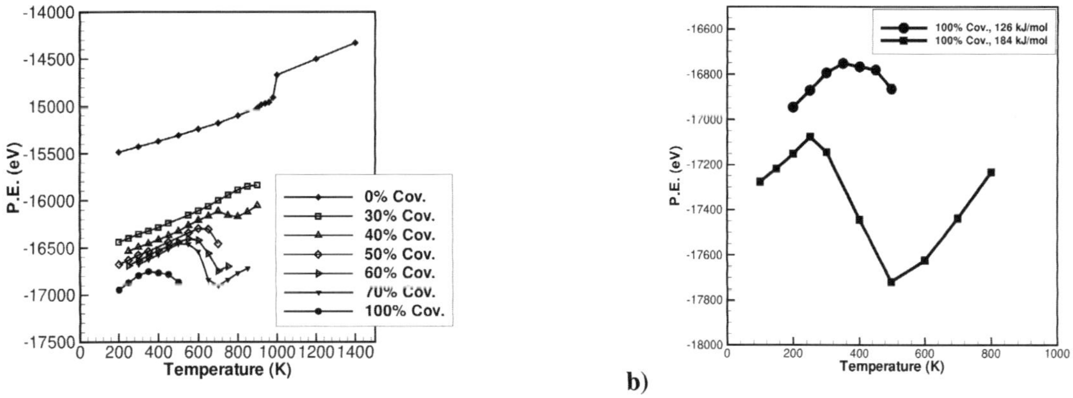

a) b)

Figure 9. a) Potential energy versus temperature for the alkanethiolate SAM coated gold nanoparticle with various amount of surface coverage and 126 kJ/mol binding energy. b) Comparison of potential energy versus temperature for fully coated nanoparticle with 184 kJ/mol and 126 kJ/mol binding energy.

In Figure 9a, the PE for the 126 kJ/mol sulfur binding energy simulation is plotted. During these simulations, as opposed to the 184 kJ/mol binding energy simulations, we observed some desorbtion of alkanethiol chains from the gold surface. This desorbtion process is irreversible because of the low density of desorbed chains in the simulation volume, a realistic assumption for coated nanoparticles in a vacuum or near vacuum. In Figure 9b it is interesting to note that the temperature at which the decrease in PE occurs for the 126 kJ/mol sulfur binding energy is higher than the 184 kJ/mol binding energy simulations. This result confirms the observation that when considered with a lower sulfur binding energy that the alkanethiol chains have a lower propensity to dissolve into the gold nanoparticle and a higher probability of desorbing from the gold surface.

The enthalpy of solution, H_{sol}, can be computed from the PE data. The enthalpy of solution is computed as the difference in system internal energy, between the undissolved SAM chain system prior to heating and the internal energy of the same system after cooling down to the same temperature. The PE of the system decreases as a result of the mixing that occurs in the fully coated gold nanoparticle model during an increase in temperature from 300 K to 600 K.

This mixing may potentially occur completely at 300 K if it were possible to simulate long times (greater than 10 ns) in the MD simulations performed here. On the other hand, it is also possible that the activation energy for this transition to occur may be too high to complete the transition at 300 K. The measured change in system PE is 1000 eV, which is the enthalpy of solution.[51] Denoting this result per unit mass of the fully coated gold nanoparticle results in an enthalpy of solution of 100 kJ/kg. These results show that the SAM has a measurable affect on not only the nanoparticle surface but the entire system.

Conclusions

The alkanthiolate SAM-coated gold nanoparticle has been shown to be mechanically and chemically metastable. The SAM that is adsorbed onto the surface begins to mix with the surface gold atoms at low temperatures because of the weaker gold-gold bonding of the surface atoms to the interior gold atoms. This process is more pronounced as the density of the SAM is increased and as the gold-sulfur binding energy is increased. This is demonstrated by the corrugation numbers computed previously. For full surface coverage, the SAM chains will begin to diffuse into the gold core as the temperature rises from 300 K to 500 K. This mixing process lowers the PE of the system. The diffusion activation energy of the sulfur atoms is measured to be low in this temperature range. Once the temperature has increased to around 600 K, the alkanethiolate chains are completely mixed with the gold atoms, and the activation energy for diffusion greatly increases. The PE and diffusivity results concur with this observation that a phase change in the SAM-coated gold nanoparticle system occurs around 500 K. We have therefore concluded that the melting temperature of the gold is lowered by the adsorption of alkanethiolate SAMs on the nanoparticle surface and creates a metastable system for fully covered nanoparticles of diameters around 50Å. The measured decrease in melting temperature of the nanoparticle is more pronounced as more SAM chains are adsorbed onto the nanoparticle surface.

We have also compared each of the computed results for the low, 126 kJ/mol, gold-sulfur binding energy reported in the literature to the results from the high, 184 kJ/mol binding energy. These comparisons show that the gold-sulfur binding energy does not have a large affect on the meta-stability of the nanoparticle. From these results we are able to conclude that the primary factor in determining if a nanoparticle will become meta-stable is the strength of the chain-chain interactions. Since experiments concur that desorbtion occurs in the 300K to 550K temperature range we have concluded that the 126 kJ/mol binding energy is most likely a better estimate of the true binding energy than the 184 kJ/mol value. Independent of the binding energy is the tilt angle, which is found to be greater for the nanoparticle geometry than for a flat surface.

Acknowledgements

The authors would like to acknowledge the support received by the U.S. Army Major Shared Resource Center (MSRC) at the Aberdeen Proving Ground, MD. Additional support was given by the National Institute for Standards Technology (NIST).

References

1. Manna, A.; Imae, T.; Aoi, K.; Okazaki, M. *Mol. Sim.* **2003**, 29(10–11), 661–665.
2. Hong, R.; Han, G.; Fernández, J. M.; Kim, B.–J.; Forbes, N. S.; Rotello, V. M. *J. Am. Chem. Soc.* **2006**, 128, 1078–1079.
3. Magnusson, M. H.; Duppert, K.; Malm, J.–O.; Bovin, J.–O.; Samuelson, L. *NanoStructured Materials* **1999**, 12, 45–48.
4. Chiang, C.–L. *J. of Colloid and Interface Science* **2000**, 230, 60–66.
5. Kim, J.–U.; Cha, S.–H.; Shin, K.; Jho, J. Y.; Lee, J.–C. *J. Am. Chem. Soc. Communications* **2005**, 127, 9962–9963.
6. Landman, U.; Luedtke, W. D. *Faraday Discussions* **2004**, 125,1–22.
7. Schreiber, F. *Progress in Surface Science* **2000**, 65,151–256.
8. Greer, J. R.; Nix, W. D. *Applied Physics A* **2005**, 80,1625–1629.
9. Cleveland, C. L.; Landman, U.; Schaaff, T. G.; Shafigullin, M. R.; Stephens, P. W.; Whetten, R. L. *Phys. Rev. Lett.* **1997**, 79(10), 1873–1876.
10. Baskes, M. I. *Phys. Rev. B* **1997**, 46(5), 2727–2742.
11. Daw, M. S.; Baskes, M. I. *Phys. Rev. B* **1984**, 29(12), 6443–6453.
12. Lewis, L. J.; Jensen, P.; Barrat, J. –L. *Phys. Rev. B* **1997**, 56(4), 2248–2257.
13. Shevade, A. V.; Zhou, J.; Zin, M. T.; Jiang, S. *Langmuir* **2001**, 17, 7566–7572.
14. Nuzzo, R. G.; Zegarski, B. R.; Dubois, L. H. *J. Am. Chem. Soc.* **1987**, 109, 733–740.
15. Strong, L.; Whitesides, G. M. *Langmuir* **1988**, 4, 546–558.
16. Rosenbaum, A. W.; Freedman, M. A.; Darling, S. B.; Popova, I.; Sibener, S. J. *J. Chem. Phys.* **2004**, 120(8), 3880–3886.

17. Zhang, L.; Goddard III, W. A.; Jiang, S. *Computers in Chem. Eng.* **2002**, 22(10), 1381–1385.
18. Luedtke, W. D.; Landman, U. *J. Phys. Chem.* **1996**, 100(32), 13323–13329.
19. Jiang, S. *Mol. Phys.* **2002**, 100(14), 2261–2275.
20. Bhatia, R.; Garrison, B. J. *Langmuir* **1997**, 13, 765–769.
21. Masens, C.; Ford, M. J.; Cortie, M. B. *Surf. Sci.* **2005**, 580, 19–29.
22. Hawa, T.; Zachariah, M. R. *Phys. Rev. B* **2004**, 69, 035417.
23. Hawa, T.; Zachariah, M. R. *Phys. Rev. B* **2005**, 71, 165434.
24. Hawa, T.; Zachariah, M. R. *J. Chem. Phys.* **2004**, 121(18), 9043–9049.
25. Grönbeck, H.; Curioni, A.; Andreoni, W. *J. Am. Chem. Soc.* **2000**, 122, 3839–3842.
26. Plimpton, .S. J. *J. Comp. Phys.* **1995**, 117, 1–19.
27. Foiles, S. M.; Baskes, M. I.; Daw, M. S. *Phys. Rev. B* **1986**, 33(12), 7983–7991.
28. Mei, J.; Davenport, J.W.; Fernando, G.W. *Phys. Rev. B* **1991**, 43(6), 4653–4658.
29. Lee, B.; Cho, K. *Surf. Sci.* **2006**, 600, 1982–1990.
30. Haftel, M. I. *Phys. Rev. B* **1993**, 48(4), 2611–2622.
31. Jorgensen, W. L.; Madura, J. D.; Swenson, C. J. *J. Am. Chem. Soc.* **1984**, 106, 6638–6646.
32. Fukunaga, H.; Takimoto, J. i.; Doi, M. *J. Chem. Phys.* **2002**, 116(18), 8183–8190.
33. Tries, V.; Paul, W.; Baschnagel, J.; Binder, K. *J. Chem. Phys.* **1997**, 106(2), 738–748.
34. Maiti, P. K.; Lansac, Y.; Glaser, M. A.; Clark, N. A. *Langmuir* **2002**, 18, 1908–1918.
35. Rai, B.; Sathish, P.; Malhotra, C. P.; Pradip; Ayappa, K. G. *Langmuir* **2004**, 20, 3138–3144.
36. Mahaffy, R.; Bhatia, R.; Garrison, B. J. *J. Phys. Chem. B* **1997**, 101, 771–773.
37. Beardmore, K. M.; Kress, J. D.; Gronbech-Jensen, N.; Bishop, A. R. *Chem. Phys. Lett.* **1998**, 286, 40–45.
38. Hautman, J.; Klein, M. L. *J. Chem. Phys.* **1989**, 91(8), 4944–5001.
39. Liu, K. S. S.; Yong, C. W.; Garrison, B. J.; Vickerman, J. C. *J. Phys. Chem. B* **1999**, 103, 3195–3205.
40. Lavrich, D. J.; Wetterer, S. M.; Bernasek, S. L.; Scoles, G. *J. Phys. Chem. B* **1998**, 102, 3456–3465.
41. Jang, S. S.; Jang, Y. H.; Kim, Y.-H.; Goddard III, W. A.; Flood, A. H.; Laursen, B. W.; Tseng, H.-R.; Stoddart, J. F.; Jeppesen, J. O.; Choi, J. W.; Steuerman, D. W.; Delonno, E.; Heath, J. R. *J. Am. Chem. Soc.* **2005**, 127, 1563–1575.
42. Henz, B. J.; Hawa, T.; and Zachariah, M. R. *Langmuir* **2008**, 24(3), 773–783.
43. Allen, M. P.; Tildesley, D. J. *Computer Simulation of Liquids.* Oxford University Press Inc., New York, NY, 1987.
44. Lewis, L. J.; Jensen, P.; Barrat, J. –L. *Phys. Rev. B* **1997**, 56(4), 2248–2257.
45. Badia, A.; Cuccia, L.; Demers, L.; Morin, F.; Lennox, R. B. *J. Am. Chem. Soc.* **1997**, 119, 2682–2692.
46. Dalton, A. S.; Seebauer, E. G. *Surf. Sci.* **2007**, 601, 728–734.
47. Suni, I. I. *Surf. Sci.* **1995**, 309, L179–L183.
48. Sellers, H.; Ulman, A.; Shnidman, Y.; Eilers, J. E. *J. Am. Chem. Soc.* **1993**, 115, 530–542.
49. Thompson, S. M.; Gubbins, K. E.; Walton, J. P. R. B.; Chantry, R. A. R.; Rowlinson, J. S. *J. Chem. Phys.* **1984**, 81(1), 530–542.
50. Dieter, G. E. *Mechanical Metallurgy.* McGraw-Hillm Inc., New York, NY, 1986.
51. Winn, J. S. *Physical Chemistry.* Addison Wesley, Boston, MA, 1995.

220d Quantum Confinement In Core-Shell Particles Via Atomic Layer Deposited Zinc Oxide Films

David M. King, Xinhua Liang, Samantha Johnson, and Alan Weimer. Department of Chemical Engineering, University of Colorado, 1111 Engineering Dr., Campus Box 424, Boulder, CO 80309

ZnO films were coated onto silica nanospheres using atomic layer deposition (ALD) in a fluidized bed reactor. The temperature-dependent difficulties of ZnO ALD on a large quantity of particles are discussed, specifically when an aqueous-based oxidizer is used. An optimal deposition temperature was observed to balance the propensity toward liquid bridging and particle agglomeration at low temperatures and precursor deposition leading to Zn-rich films at high temperatures. Using the optimal deposition temperature, a statistical experimental plan was carried out to quantify the thickness-dependent bandgap shift of ZnO semiconductor films deposited using ALD. Thermal annealing steps were also used to modulate crystallite size for further investigation of quantum confinement. Finally, similar studies were carried out using a non-aqueous oxidizing precursor for the low-temperature ALD of ZnO on particles.

220e In-Situ Coating of Flame-Made TiO2 Nanoparticles by SiO2

Alexandra Teleki, Department of Mechanical and Process Engineering, Swiss Federal Institute of Technology Zurich, Particle Technology Laboratory, Sonneggstrasse 3, Zurich, CH-8092, Switzerland, Martin C. Heine, Particle Technology Laboratory, Swiss Federal Institution of Technology Zurich, Sonneggstrasse 3, Zurich, 8092, Switzerland, Frank Krumeich, Laboratory for Inorganic Chemistry, ETH Zurich, Wolfgang-Pauli-Str. 10, Zürich, CH-8093, Switzerland, Kamal M. Akhtar, Millennium Inorganic Chemicals (a Cristal Company), 6752 Baymeadow Drive, Glen Burnie, MD 21060, and Sotiris E. Pratsinis, Particle Technology Laboratory, Institute of Process Engineering, Department of Mechanical and Process Engineering, ETH Zurich, Sonneggstrasse 3, Zurich, CH-8092, Switzerland.

Mostly rutile titania nanoparticles (doped with Al) about 40 nm in diameter made by flame spray pyrolysis (FSP) were coated at considerable production rates, 30 g/h, in-situ with 2 – 4 nm thick SiO_2 films by mixing the freshly-formed TiO_2 aerosol with hexamethyldisiloxane (HMDSO) vapor just downstream of the flame. A hollow ring delivering HMDSO vapor (precursor to SiO_2) through 16 jets was judiciously positioned in cross-flow above the enclosed flame coating freshly-made Al-doped rutile TiO_2 particles. The as-prepared powders were characterized by (scanning) transmission electron microscopy (STEM and TEM), energy dispersive X-ray analysis, X-ray diffraction, DC plasma optical emission spectroscopy (DCP-OES) and nitrogen adsorption. The coating quality was assessed further in the photocatalytic oxidation of isopropanol to acetone. The coating thickness was controlled by the silica content. The particles were encapsulated by smooth and homogeneous coatings at above 10 wt% SiO_2 contents. Premature or early HMDSO vapor injection resulted in segregated SiO_2 and patchy or poorly-coated TiO_2 particles. The anatase promotion of silica in TiO_2 was significantly reduced as the distance of HMDSO vapor injection point from the FSP burner was increased. Well-coated TiO_2 particles exhibited limited IPA photooxidation. In contrast, spraying and combusting equivalent mixtures of Si/Al/Ti precursors in the enclosed FSP reactor (without delivering HMDSO through the hollow ring) resulted in particles segregated in amorphous (SiO_2) and crystalline (TiO_2) domains and thus high photocatalytic activity.

220f Vesicle Templating of Hollow Oxide Nanoparticles

Spyros Monastiriotis, Dept. of Chemical Engineering, The Graduate Center and The City College of the City University of New York, 140th Street at Convent Avenue, New York, NY 10031 and Alexander Couzis, City College of the City University of New York, Dept. of Chemical Engineering, 140th Street at Convent Avenue, New York, NY 10031.

This research focuses on the development of facile scalable methods for the fabrication of porous hollow inorganic spheres using equilibrium catanionic vesicles as templates. Hollow inorganic nanospheres exhibit catalytic activities and properties from their solid counterparts with the advantages of high surface area and low density. As part of the overall project we have studied how to control size and shell thickness of the inorganic hollow particle. In effect, the structure depends on two length scales that are decoupled from each other. The first scale determines the diameter of the spheres, which is in the range of 50 to 200 nm and can be tuned by controlling the size of the templating vesicle and the second length scale determines the thickness of the inorganic deposited layer. The vesicle size is tuned by the composition ratio of the two surfactants Cetyltrimethylammonium Bromide (CTAB) and Sodium Perfluorooctanate (FC7) and characterized by using Cryo-Transmission Electron Microscopy (Cryo-TEM), Small Angle Neutron Scattering (SANS), and Dynamic Light Scattering. The inorganic shell's thickness was controlled by the vesicle's concentration in the solution versus precursor's concentration, and solution properties as pH and temperature. Multiple inorganic layers were fabricated by surface functionalization of the previous inorganic deposition on the vesicular template. Inorganic shells were characterized with Cryogenic Transmission Electron Microscopy (Cryo-TEM), Transmission Electron Microscopy (TEM), Scanning Electron Microscopy (SEM), Scanning Transmission Electron Microscopy (STEM), Atomic Force Microscopy (AFM), Dynamic Light Scattering (DLS)

226a Application of Coarse Grained Drag Law In Computational Fluid Dynamics Simulations of Fluidized Beds

Alvin U. Chen and E. Nicholas Jones. ExxonMobil Research and Engineering Company, Fairfax, VA 22037

Scale-up of fluidized beds for industrial applications has typically been carried out using dimensional analysis similar to single-phase fluid flow. However, it is impossible to keep all dimensionless groups constant, and therefore impossible to accurately scale up industrial fluidized solids processes using traditional scaling arguments. Because of this inability to fully understand factors affecting scale-up, designs based on small cold-flow experiments or even larger pilot units may overlook critical factors influencing operability of commercial units.

Computational fluid dynamics (CFD) has the potential to directly simulate behavior of reactive commercial scale units, thus overcoming gaps in scale-up methodology. However, CFD technology itself is not without limitations. The popular Eulerian-Eulerian approach (which treats gas and solids phases as interpenetrating continua) suffers from the dependence of predicted dynamics on the computational mesh resolution. Because of this grid dependence, scale-up of the CFD simulations themselves can be difficult or impractical since the grid resolution must be held constant between a typically small validation experiment and a large commercial scale simulation.

The dependence of predicted dynamics on grid resolution arises from ignoring small-scale flow structures which are not adequately resolved with coarse grids. The use of highly resolved computational grids which fully resolve small-scale flow structures can overcome this hurdle (analogous to DNS for single-phase flows), but such fine grid resolutions are impractical for commercial scale systems. Another more computationally efficient method uses so-called coarse grained models which adjust parameters such as interphase drag with changing grid size to capture effects of the small-scale flow structures on overall bed dynamics. Thus, a commercial gas fluidized solids unit can be simulated at relatively low mesh resolution while capturing most features of the overall dynamics.

In this paper, we use such a coarse grained Eulerian-Eulerian model to simulate dynamics of a small-scale cold-flow model of a fluid coker. By comparing our CFD results with experimental measurements, we demonstrate the utility of coarse grain models for fluid-solids flow and also the accuracy of our simulations. We are able to accurately predict gross flow structures in the fluidized bed as well as local voidage and velocity profiles, thus building confidence in the simulation of commercial scale cokers.

CFD Modeling of Cold-Flow Fluidized Beds and Validation with X-ray Computed Tomography

Jian Min[1], Joshua Drake[2], Rodney Fox[1], and Theodore J. Heindel[2]
[1]Department of Chemical and Biological Engineering
[2]Department of Mechanical Engineering
Iowa State University
Ames, Iowa 50011-2161

The hydrodynamics in a 152 mm ID bubbling fluidized bed filled with 500-600 μm glass beads have been simulated using the computational fluid dynamics (CFD) code Fluent 6.3, and the results are compared to experimental data obtained using pressure measurements and 3D X-ray computed tomography. The grids in a Cartesian coordinate system used in simulations are made by 2D (8×8 mm, 4×4 mm, 2×2 mm, and 1×1 mm) and 3D (4×4×4~6 mm). 4 mm grid resolution is sufficient and excellent to use on this scale reactor study to analyze of the time averaged, time and spatial averaged local gas holdup throughout the glass bead bed in 2D and 3D simulations. Throughout the whole fluidized bed, there are two larger symmetric circles of glass beads, in which the gas holdup about 0.46-0.52 is larger than in the middle of the fluidized bed about 0.44-0.46. The 3D simulations with Syamlal-O'Brien and Gidaspow drag models better predict the local gas holdup variation throughout the whole fluidized bed, comparing with the experimental data. The results of simulations with Wen-Yu drag model are generally over predicted. The injector air path extends further into the bed along the axial height from quarter to half of radial long, where the maximum gas holdup is decreasing from 0.9 to 0.6 in the Y-Z slice throughout the fluidized bed with side air injection. But the simulations could not exactly predict the gas path. Because the side air has little effect on the X-Z slice, there remain somewhat symmetric.

Introduction

Fluidized beds are commonly used by the chemical, mineral, pharmaceutical, and energy industries because of low pressure drops, uniform temperature distributions, and the high heat transfer rates. In the biofuel industry, fluidized bed reactors are central components of combustion, pyrolysis, and gasification. The local gas holdup (i.e. solid voidage) distributing in the whole fluidized bed is an extremely important hydrodynamics parameter determining the uniform mass and energy distributions and gasification efficiency. Time-averaged data is more meaningful than transient in the design of commercial-scale fluidized beds.

Because the bed material (solid phase) is typically opaque in gas-solid systems it is more difficult to get data in detail for the whole bed. Since fluidization is a dynamic process, invasive monitoring methods can influence the internal flow, thereby reducing the reliability of the measurements. Currently, there are only a few instances of noninvasing monitoring techniques used with fluidized beds. The techniques for multiphase flows include electrical capacitance tomography, ultrasonic computed tomography, gamma densitometry tomography, X-ray fluoroscopy (radiography/stereography), and X-ray computed tomography, detailed by Heindel et al. (2008). Design of commercial-scale fluidized beds traditionally depends on pilot-scale experiments and too expensive, tough and take time. There has been an increasing usage of computational fluid dynamics (CFD) to complement industrial design of fluidized applications. The situation is complex when multiphase flows are involved, because transport equations for mass, momentum, and turbulence properties have to be solved for each individual phase. A few researchers have attempted to study process characteristics in the fluidized bed using CFD methods and some progress has been made, focusing on terminal settling velocities, minimum fluidization and fluidizability, and residence time of biomass particles. Many numerical simulations of fluidized beds in the literature are tow-dimensional (2D) simulations, mostly due to a lack of computer resources, i.e. memory and processor speeds. To date, even with improved computational facilities, three-dimensional (3D) simulations are still expensive as the equations have to be integrated using small time-steps over long periods of time to compute and average the inherently chaotic and transient fluidization processes; however, parallel computing can speed up the simulational process.

The goal of this research is to find a systemic method for the design of commercial-scale fluidized beds from experimental and simulation techniques. The use X-ray computed tomography and radiography to quantitatively obtain the local time-averaged gas holdup in the fluidized bed was discussed. Experimental validation of local gas-holdup values were done with simulations using Fluent 6.3 to model a comparable bubbling fluidized bed in 2D and 3D. Both quantitative and qualitative predicted local gas holdup distributions calculated using the 3D X-ray flow visualization measurements were compared to the model. The different drag models, Syamlal-O'Brien, Gidaspow and Wen-Yu, were compared in detail. The results of operations with or without side air injection in experiment and simulation were analyzed. The study is useful in understanding how gas holdup is distributed throughout a fluidized bed using different research methods. It is meaningful to the design of commercial-scale fluidized beds.

Experimental Setup

Fluidized Bed Reactor

The fluidized-bed reactor was fabricated from three 152 mm ID acrylic tube sections with 12.7 mm thick acrylic flange plates attached to each end. The 3 sections measure 150 mm, 300 mm, and 640 mm in length, respectively, for the plenum, reactor, and freeboard chambers. Between the plenum and bed chamber is a 1.2 mm thick stainless steel perforated aeration plate containing 130 1 mm diameter holes that are approximately equally space in concentric rings to ensure homogeneous gas distribution and a low pressure drop. The aeration plate has an open area ratio of 0.57%. Air is introduced into the system by an inlet in the bottom of the plenum. The air travels through the marble-filled plenum and is dispersed evenly over the bottom of the aeration plate. Rubber gaskets are placed between each flange, sealing the various components and forcing all gas to flow directly through the bed and out the top. A pressure tap in the bottom of the plenum holds a transducer connected to a data acquisition card to record inlet pressure. A second tap on the side of the fluidized bed 25 mm above the aeration plate is used for side air injection. Data regarding side air injection will not be included in this report however.

500-600 μm diameter glass beads, Geldart Group B, are used as the bed material due to its excellent fluidization and heat transfer properties. The key features of the region are exampled as deaerate quickly when gas is shut off, large bubbles, and intermediate solids mixing. Gas and solid properties of the fluidized bed used in the experiments and simulations are listed on Table 1.

X-ray System

Iowa State University's XFloViz facility was used to image the fluidized bed and has been described in-detail in the literature (Heindel et al., 2005, and Heindel et al., 2008). Only a brief outline will be presented here. Two LORAD LPX200 portable X-ray tubes provide the X-rays. Current and voltage can be adjusted from 0.1 to 10.0 mA and 10 to 200 kV, respectively, with a maximum power of 900 W. Low energy radiation is suppressed by 2 1 mm thick copper and 1 aluminum filters placed directly in front of the X-ray sources. Located opposite each X-ray source is an X-ray detector/CCD camera pair. The CCD camera with the image intensifier has a temporal resolution ranging from 10 frames per second (fps) to 60 fps, depending on binning options, and is primarily used for radiographic imaging. The image intensifier is a 406 mm diameter Precise Optics PS164X screen detector with a 35.0 mm output image diameter. A DVC-1412 monochrome digital camera captures the image from the intensifier. Generally, 2×2 binning (640×512 active pixels) at 20 fps is used for radiographic movies in order to maximize picture quality while maintaining good temporal resolution. A different detector/CCD camera pair is primarily used for CT imaging because of its high spatial resolution. This camera is usually located opposite the second source and is connected to a square 440×440 mm cesium-iodide scintillator screen which transforms radiation into visible light. A 50 mm Nikon lens captures images which are digitized by an Apogee Alta U9 system. This system has 3072×2048 pixels and is thermoelectrically cooled to allow long exposure times. Usually, an exposure time of 1 second with 4×4 binning is chosen to minimize acquisition time while maintaining the signal strength. Both cameras and sources are located on a 1.0 m ID rotation ring that can rotate 360° around the fluidized bed. Data from both CCD cameras are acquired by software developed by Iowa State University's Center for Nondestructive Evaluation (CNDE) and a personal computer with 4 GB of RAM. The software allows for control of both cameras, as well as motion control for the rotation ring. Images acquired by the CCD camera with the image intensifier are normalized for pixel non-uniformity (linearization) and for warping effects produced by the image intensifier. Images acquired by the CCD camera with the CsI screen are also normalized for pixel non-uniformity. These corrections are performed using algorithms within the CNDE software. Volumetric reconstruction of the CT images is performed using CNDE's 64-node LINUX cluster.

CFD Modelling

CFD uses numerical methods to solve partial differential equations that model fluid dynamics. In general, CFD can be applied to reactive, multiphase, laminar and turbulent flows. In this work, governing equations for mass, momentum, and energy for each phase (gas and solid) are solved within a geometrical domain that corresponds to the fluidized-bed reactor used in experiments.

Grid

A Cartesian coordinate system is used for grid generation, where hexahedral elements are formed for the 3D geometry and quadrilateral elements for the 2D geometry. For the 2D simulations we have carried out a grid-refinement study by varying grid size by factors of two for four different grids, 8×8 mm, 4×4 mm, 2×2 mm, and 1×1 mm, respectively, in which a square mesh was used. When side air injection is simulated, the side injector diameter is set to 11 mm with an open area 95 mm^2. The distance between the side air injector centerline and the aeration plate is 30 mm. In order to obtain better meshes, the side air injector area is represented by a rectangle in the 3D grid. The same 3D mesh is used in cases with and without side air injection.

Multiphase Model

There are two approaches for CFD modeling of gas-solid flows: the Lagrangian-Eulerian model and the Eulerian-Eulerian. Using the Lagrangian-Eulerian model, trajectories of each particle are tracked by solving individual equations of motion, whereas the gas phase is modeled using an Eulerian framework.

As a consequence, the Lagrangian-Eulerian model requires large computational resources for large systems of particles. With the Eulerian-Eulerian, the base assumption is that gas and solid phases are interpenetration continua. Therefore, the Eulerian-Eulerian model for gas-solid flows is the more commonly used CFD model to predict the dynamic behavior of dense phase fluidized bed reactors.

Numerical Methodology

CFD simulations of the experimental fluidized bed reactor are carried out with Fluent 6.3 on the High Performance Computing (HPC) machines at Iowa State University. The time step for the simulations is calculated based on the convective Courant-Friedrichs-Levy condition, which depends on the ratio of the grid size and maximum velocity in the flow domain. The time step is set in the range of 10^{-3}-10^{-4} s, depending on simulation conditions. The convergence criterion is set to 10^{-4}. Second-order discretization schemes are used for the spatial and temporal derivatives. Simulations were performed from t = 0 to 70 s. Statistics are not saved from the calculations for t < 10 s to make sure the transients die down. Since the experimental data are time-averaged, all comparisons will be based on time-averaged simulations that span 10 to 70 s.

Results

Grid Resolution Study

The velocity distribution through the gas inlet was uniform throughout all CFD simulations. The overall bulk time-averaged gas holdup distributions show no some substantial changes among the different 2D grids. There are three regions divided by +/- 38 mm, the core region and two wall regions. Experimental values are obtained along mutually perpendicular rays at specified Z heights. There are three zones, the aeration zone 0-38 mm, the fluidized zone 38-152 mm, and the surface zone in the upper portion of the bed. The local fluctuations in the experimental data are due, primarily, to noise in the data. For all 2D simulations, gas holdup is generally over predicted slightly in the core region while being under predicted in the wall region. The condition is not to become better by decreasing the grid size. The experimental gas holdup just above the distributor is smaller due to the effect of the plate, where gas jetting is observed and not all holes are active. Because of the uniform inlet profile used in the simulations, no jetting phenomena are predicted with CFD analysis. The 2D-1 case gives the best overall agreement with the experimental data, while the 2D-4 and 2D-2 cases of give better results in the upper portion of the bed. Based on the analysis of the time averaged, time and spatial averaged local gas holdup throughout the glass bead bed, it is concluded that the 4 mm grid resolution is sufficient and excellent to use on this scale reactor study. The grid size in 3D simulations is also chosen as 4 mm in the following study.

Drag Model Study

Throughout the fluidized bed, there are two larger circulation zones of glass beads both experiments and simulations. Those are almost symmetric in simulations. The glass beads are drawn by pressured air to move upwards and concentrated in the middle of the fluidized bed, where there is a relatively small quantity of gas phase. Therefore the gas holdup is smaller in this region about 0.44-0.46. After the gas carries the glass beads to the top of fluidized bed, it jets out and the glass beads are then circulated back down the reactor along the walls. Some bubbling gas is formed in the two circulation zones; therefore, gas holdup is larger in these regions about 0.46-0.52, where the values by Wen-Yu drag model are larger than the results from Syamlal-O'Brien and Gidaspow drag models qualitatively. And its abrupt change in gas holdup near the top of the bed is higher than that of experiment and the others simulations.

With Side Air Injection

With side air injection, a path of high gas holdup extends from the injector to the bed surface. This air path gradually expands into the bed as it rises up the bed, indicating that horizontal dispersion increases with axial height. But the simulations could not exactly predict the gas path. The radial velocity is under predicted. The side air injection could not break the bulk flux regions throughout the fluidized bed, just following to go through along the wall. Side air injection also produces a non-uniform average bed height above the injector. The highest average bed height occurs near the wall, directly above the injector. The horizontal level Z = 0.25D is just above the side air injection position, where the maximum gas holdup is above 0.9. And simulations can capture the effect and quantitatively agree with the experimental data. The injector air path extends further into the bed along the axial height from quarter to half of radial long, where the maximum gas holdup is decreasing to about 0.6-0.7. The gas holdup is arranged almost from 0.44 to 0.52 throughout the whole fluidized bed in the X-Z slice, because the side air has little effect on the regions, which remain somewhat symmetric.

Conclusions

The Eulerian-Eulerian CFD simulations based on Fluent 6.3 software for a 152 mm ID bubbling fluidized bed containing 500-600 μm glass beads give results in 2D (8×8 mm, 4×4 mm, 2×2 mm, and 1×1 mm), and 3D (4×4×4~6 mm) simulations using three different drag models with or without side air injection. The study is useful in understanding how the gas holdup is distributing in the whole fluidized bed by different research methods including with X-Rays experiments and CFD modelling. It is meaningful to the design of commercial-scale fluidized beds in the continuing process.

The method of the X-ray CT is the good noninvasing monitoring technique to get the whole gas holdup in the opaque fluidized bed. In the whole fluidized bed, there are two larger symmetric circles of glass beads, in which the gas holdup about 0.46-0.52 is larger than in the middle of the fluidized bed about 0.44-0.46.

4 mm grid resolution is sufficient and excellent to use on this scale reactor study to analyze of the time averaged, time and spatial averaged local gas holdup throughout the glass bead bed in 2D and 3D simulations.

The 3D simulations with Syamlal-O'Brien and Gidaspow drag models better predict the time, time and spatial averaged gas holdup variation throughout the whole fluidized bed, comparing with the experimental data. The results of simulations with Wen-Yu drag model are generally over predicted.

The injector air path extends further into the bed along the axial height from quarter to half of radial long, where the maximum gas holdup is decreasing from 0.9 to 0.6 in the Y-Z slice throughout the fluidized bed with side air injection. But the simulations could not exactly predict the gas path. Because the side air has little effect on the X-Z slice, there remain somewhat symmetric.

Extended Abstact

For Session

Industrial Application of Computational and Numerical Approaches to Particle Flow (03B07)

"Industrial Application of CPFD to Biomass Gasification"

Charles Churchman, P.E.,
AIChE Fellow
Senior Director of Technology
SilvaGas Corporation
Atlanta, Georgia 30339

September 23, 2008

AIChE Extended Abstract

SilvaGas Corporation converts wood chips, corn stover and other biomass feedstocks into a clean, medium BTU fuel which can be used directly in powerplant-sized gas turbine engines to produce electricity. The exhaust from the gas turbine can be used to generate more steam for additional electricity production, offering a combined cycle powerplant opportunity for electrical production from biomass feedstocks. The fuel gas is also clean enough to be used downstream in catalytic processes to make liquid fuels, such as the Fischer-Tropsch process, among others.

The SilvaGas process is an indirect gasification process, which pyrolyze the biomass in the absence of air or oxygen. The process was developed by Battelle in the 1980's, to take particular advantage of the high pyrolysis reactivity of biomass. The pyrolysis reaction is accomplished using two fluidized beds of hot sand in separate vessels.

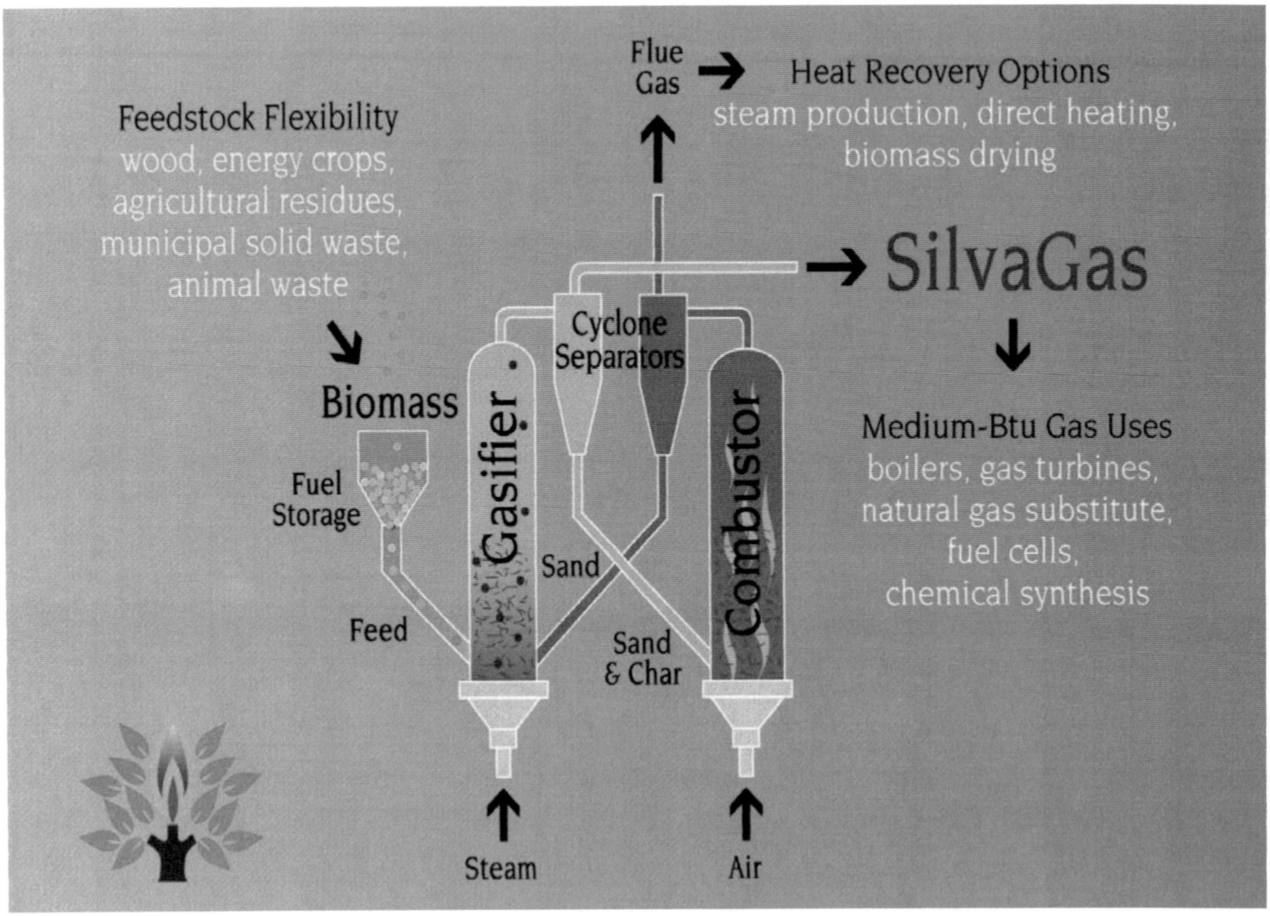

Figure 1
SilvaGas Process Schematic

Sand circulated through the system in a Figure 8 laid-on-its-side pattern. Wood enters the gasifier, and is pyrolyzed, releasing large amounts of gas. This entrains the 1700 deg F sand used to heat the wood upwards to a cyclone, where the sand and some char left over is separated from the fuel gas. The sand and char fall into a separate combustor, where the char is burned with air, reheating the sand, and entraining it overhead to another cyclone. The sand is separated from the ash and flue gas, which go overhead to downstream cleanup. The reheated sand goes back into the gasifier, repeating the process.

In many ways, the SilvaGas process is similar to petroleum FCC units.

A 400+ DTPD commercially-scaled pilot plant was operated from 1998-2001 in Burlington, Vermont, in a plant partially funded by the DOE.

Figure 2
SilvaGas 400 Dry Ton per Day Commercial Scale Facility

This plant accumulated a large amount of computer-based operating data on a wide variety of woody feedstocks. The data was used to quantifiably verify the accuracy of Barracuda prior to it being used for predictive purposes.

Once Barracuda was verified against our Burlington plant data, work was performed on improvements and scaleup studies for future, much larger plants.

The major process equipment which Barracuda has been used on for studies includes:

- The gasifier
- The combustor
- The gasifier cyclones
- The combustor seal pot

The Gasifier

Fluidized solids circulation in the gasifier is complex, with three separate identifiable zones. These are:

- The quiescent unfluidized zone of olivine at the bottom of the gasifier, below the steam fluidizing distributor
- The bubbling bed of olivine which receives the wood
- The entrained bed where the pyrolysis gases lift the sand, char and wood up into the top part of the gasifier.

Not much happens in the quiescent zone except the slow settling of dense, mostly iron contaminants to the bottom, such as nails, wire, and screws. These are drawn off in a batch approach, with the sand cooling to around 300F at the bottom drawoff.

The bubbling bed in the middle is short, at most a few feet in depth, with extremely high heat and mass transfer rates. The signature high biomass reactivity results in the explosive decomposition of the biomass, at rates from 10 to 30 times that of conventional coal fluidized bed combustors.

The fluidizing steam distributor was easy to model in Barracuda; what was harder to model was the decomposition of the wood, which releases roughly six times the volume of gases produced by the steam fluidization. This was modeling initially by adding an additional distributor above the steam distributor, to in effect act like the pyrolyzing biomass.

Barracuda was particularly effective in predicting the circulating patterns in the gasifier in the third section, the entrained bed section. It matched the pressure profile in the gasifier against operating data from the Burlington plant.

The combustor

The main effort here was to evaluate segregation of the char from the sand in the incoming feedline to better understand how to optimize feed distribution. Feed introduction in mature combustion technologies, like fluidized bed coal units, have as many as six feed points. The Burlington combustor was significantly larger than what was required, and is the most expensive vessel in the plant. Reducing its capacity to match that of the gasifier means requires better understanding of its reaction kinetics, and also, feed distribution.

The Barracuda software illustrated that the char in the char/olivine mix rapidly segregates in the feed line, with even more segregation happening once the char enters the combustor.

This is best seen in a movie, in the attached Powerpoint presentation. (This will be performed at the presentation.)

Cyclones

Cyclone design is a mature science, with lots of operational data. Barracuda runs on cyclones were also very time consuming, so less effort was expended on the cyclones with one exception.

Some texts have claimed that the gas/solids mixture flow exiting the top of a cylindrical vessel has a lot of kinetic energy, and that taking advantage of the kinetic energy in a vessel discharge flow can provide a useful separation strategy.

For instance, the high flow velocity of gases leaving the gasfier could be swirled to move the solids to the outer periphery and separated, without the need for a cyclone.

Such an approach was tried using Barracuda for the top head of the gasifier, to determine the relative efficiency of this approach. The approach worked, achieving 90% separation of the solids, but was rejected because the required pressure drop was about 4x that of an exterior cyclone, which would have achieved better than a 99.9% separation efficiency.

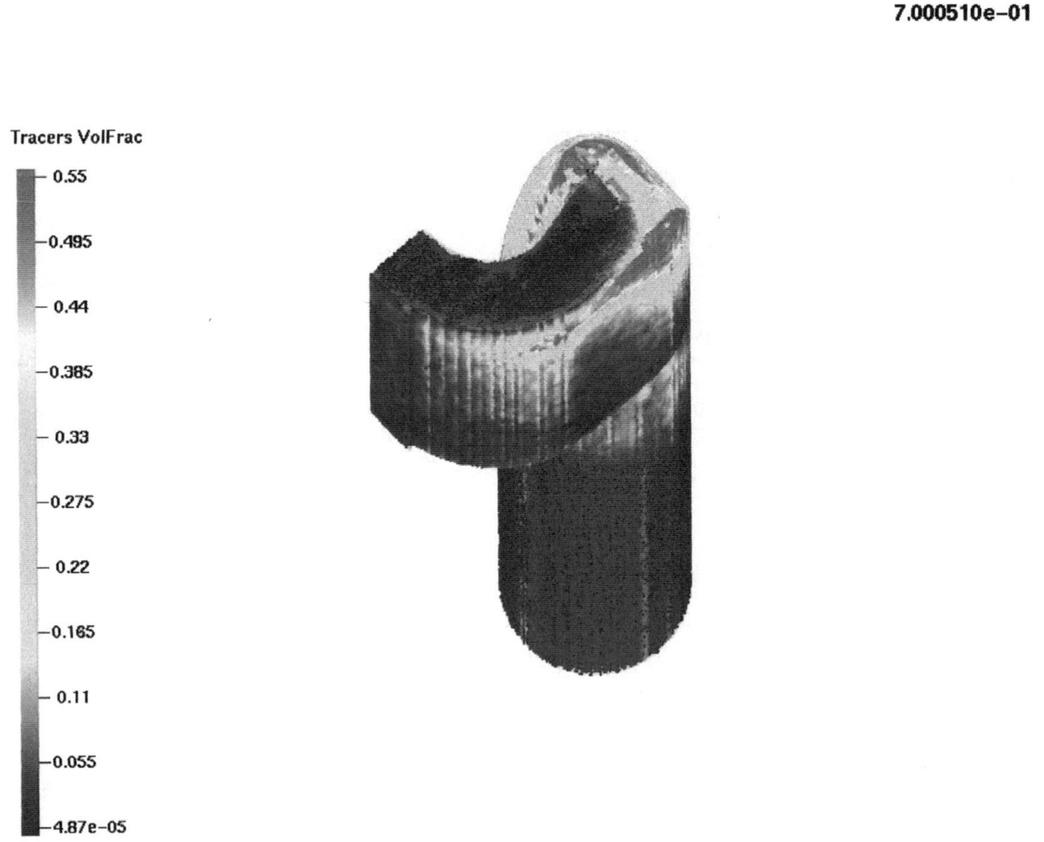

Figure 3
Color coded solids concentration in a swirled head gasifier

The above figure illustrates the variation in the solids volume fraction 0.75 seconds after starup. The red indicates approaching close pack of the olivine along the outer wall of the curved head.

Seal Pots

The gasifier operates at a higher pressure than the combustor. Thus, the transfer lines between the two vessels must be sealed, so that flammable gases from the gasifier train of equipment does not flow into the combustor train of equipment. This is accomplished by using fluidized bed seal pots.

There are distinct advantages to increased pressure operation of the gasifier train , while maintaining the combustor train at a minimum pressure above ambient. This requires a greater seal between the gasifier train and the combustor train, and this affects the design of the seal pots.

Barracuda has been a great help in the evolution of the seal pots used in SilvaGas design efforts.

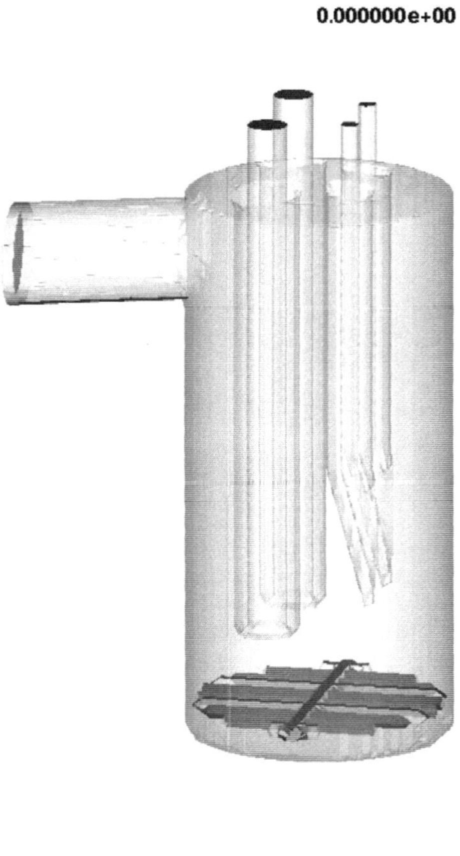

Figure 4
Standpipe configuration in a fluidized seal pot

This representation shows a typical seal pot when it is empty. Four drop legs from cyclones act as standpipes in a fluidized bed. The fluidizing gas used, be it air or steam, is introduced at the bottom of the vessel, and along the bottom of the overflow pipe.

Barracuda has been used to size the standpipes, the overflow, and the sparging rate of the fluidizing gas for both the vessel and the overflow pipe.

Scaleup

Battelle started with a 1 ton pilot plant in the early 1980's, which was scaled up to a 10 ton per day pilot plant in the late 1980's. The Burlington plant was sized for 200 dry tons per day, but was able to run at over 400 dry tons per day. It was limited by the feed system, and may have been able to run at much higher rates.

There is no need to oversize a scaleup of the Burlington plant to 1,000 tons per day or larger with the modern simulation tools we now have available. Better understanding of the circulation flow patterns, pyrolysis reaction kinetics, and fluidization criteria allow for significantly better understanding of how to design these plants economically and accurately.

For example, the entrained flow patterns inside a 6" diameter reactor are far different than what one sees in a 56' diameter reactor, primarily due to Reynolds number differences. Residence times for reacting solids thus are very different.

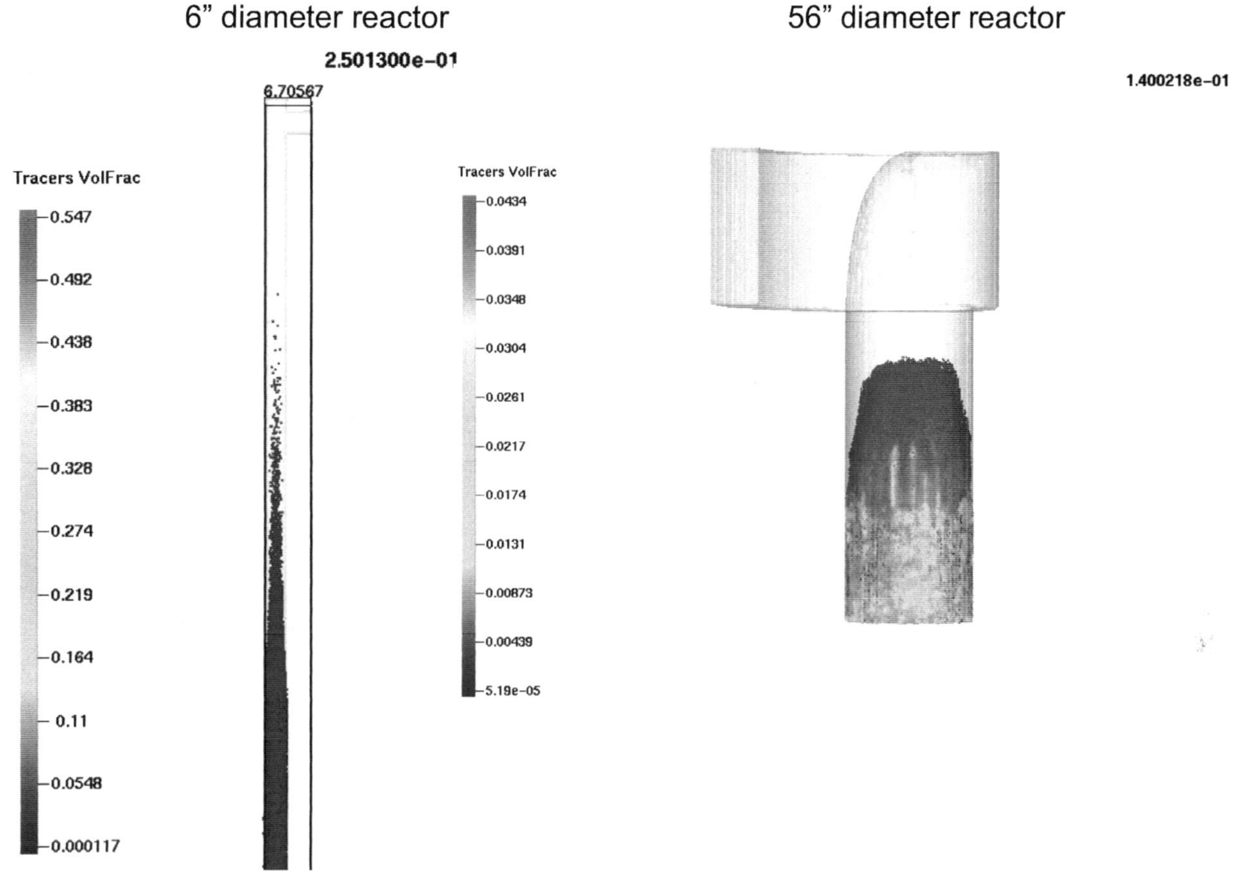

Figure 5
Reynolds and Froude number effects on scaleup of solids flows in vertical pyrolysis gasifiers

This results in limitations on the dependable accuracy of any scaleup factor one might develop from a 6" diameter reactor. In fact, the situation becomes even more different when one begins to look at the downward-flowing solids which occur along the walls of the larger vessel, that do not occur with the smaller vessel.

The problem becomes even more dissimilar, when one considers that business economics drive feedstocks to be used at as large a unit piece as possible, to reduce hogging, grinding and screening costs. How do you model a +2" piece of wood accurately in a 6" diameter reactor? Can 1/16" diameter feedstock particles for a pilot plant yield good scaleup information for a six foot diameter gasifier? The above suggest no, and SilvaGas' experience at Burlington confirms this. Barracuda has offered SilvaGas another route to accurate answers.

Barracuda offers the capability to define the residence time of particles in the system, which facilitates more accurate evaluation of pyrolysis kinetics.

Future Work

Enhancements to the Barracuda software currently in the works include the inclusion of reaction kinetics. Another is the time-dependent size reduction of particles as they react and are consumed.

These sorts of enhancements offer the potential to significantly reduce the early iteration overdesign built into biomass pyrolysis plants such as the SilvaGas process, which should eventually reduce their capital costs significantly. In the 18 months Barracuda has been currently used, SilvaGas has identified capital cost savings of at least $ 1-2 MM in our plants that can be directly attributed to the use of Barracuda. Further work with Barracuda is anticipated to offer even more savings. We figure right now that every dollar spent on Barracuda has saved us $100 in project costs.

226d Large Scale Multiphase Modeling: Continuum Vs. Discrete Approach

Chris Guenther, National Energy Technology Center, US DOE, 3610 Collins Ferry Road, Morgantown, WV 26507

In this investigation leading multiphase CFD models (MFIX, FLUENT, and Barracuda) are used to investigate the behavior of large scale cold flow circulating fluidized beds (CFB's). Experimental data from the CFB's is used to compare computational predictions of the different multiphase models. In addition, computational cost, model set-up, and post-processing advantages and disadvantages of the different models, with respect to large scale CFB's, is presented.

THREE-DIMENSIONAL GAS-SOLID FLUIDIZED BED SIMULATION APPLYING THE CFD TECHNIQUE

Fábio Marini[1]*, Maximilian J. Hodapp*[1]*, Maria G. E. Silva*[1]*, Milton Mori*[1*]
[1]*University of Campinas, Campinas-SP, Brazil*

Abstract

In this study, a three-dimensional two-phase flow model based on the kinetic theory of granular flow (KTGF) was used to predict the behavior of a gas-solid fluidized bed. The model is based on a Eulerian description of the two phases, gas and particles, and is composed of a set of mass conservation and momentum equations for each phase. In this model, the k-epsilon turbulence model and multiphase mixture are used. In order to describe the behavior of several particles in a continuum, the kinetic theory of granular flow was used.

The geometry and numerical mesh were generated using ANSYS ICEM CFD software and the set of partial differential equations was discretized and solved using CFX. Simulation data were verified against data found in Samuelsberg and Hjertager (1996), who used a cold flow laboratory circulating fluidized bed and simulated a two-dimensional model. Results using this model show that the model agrees with the experimental data and predicts a flow behavior similar to that found experimentally. It predicts the core annulus flow, which is known from the literature.
Keywords: Fluidized bed, CFD, Multiphase flow, Kinetic theory

Introduction

Fluidized beds are widely used in many operations in chemical, metallurgical, energy generation and especially petrochemical industries. Major applications are fluid catalytic cracking (FCC) risers and CFB combustor systems. Although fuidized beds are successfully and widely used in commercial industrial operations, much remains to be done due to the complexity of the gas-solid flow. No analytical tools that describe the influences of complex geometries, chemical reaction, internal reflux and heat transfer on the flow pattern in fluidized beds are available. With the increased availability of computers, mathematical models have been applied to predict the behavior of a fluidized bed and several models have been proposed. Gidaspow *et al.* (2004) carried out an extensive review of the models developed for the fluidized bed reactors and Peirano and Leckner (1998) reviewed turbulent gas-solid flow modeling in circulating fluidized beds.

Gas-particle two-fluid models, which treated the particle phase as a continuous fluid based on the Eulerian method, were widely employed in modeling gas-solid flow in the past several decades. A fundamental problem encountered in modeling the hydrodynamics of a gas-solid fluidized bed by the two-fluid method is how to include the stress of the particle phase in the particulate momentum equation. In the more recent two-fluid models, for the particle phase in dense gas-particle flow, the kinetic theory of granular flow has received attention and its use in Eulerian simulations has increased

[*] Corresponding author. Phone + 55-19-3521-3963. E-mail:mori@feq.unicamp.br

(Jenkins and Savage, 1983; Ding and Gidaspow, 1990; Gidaspow, 1994; Hrenya and Sinclair, 1997; Samuelsberg and Hjertager, 1996). The kinetic theory of granular flow is based on the kinetic theory of gases, first developed by Chapman and Cowling (1970). In this theory, the inter particle interactions are taken into account by adding the contribution of collisions between particles (Jenkins and Savage, 1983), which are the main mechanism of transport due to particulate phase properties.

The present study aims to simulate a three-dimensional two-phase flow model, based on the kinetic theory of granular flow (KTGF), to predict the behavior of a gas-solid fluidized bed, using a computational fluid dynamics technique.

Mathematical Model

Based on Eulerian description of the phases, a multiphase computational fluid dynamics model for turbulent gas-solid flow is presented. The Eulerian approach considers the two phases, gas and solid, as a continuum. The conservation equations for the solid phase are based on the kinetic theory for granular flow (Gidaspow, 1994).

The governing equations for transient three-dimensional gas-solid flow are presented as follows:

The continuity equation for phase i (= g,s):

$$\frac{\partial}{\partial t}(\varepsilon_i \rho_i) + \nabla \cdot (\varepsilon_i \rho_i \vec{\upsilon}_i) = S_i^{\rho} \tag{1}$$

where the subscript i can denote gas or solid phase, ε is the volume fraction, ρ is the density, t is the time, $\vec{\upsilon}$ is the velocity vector and S^{ρ} is the source term for mass transfer between the phases. Hence, no mass transfer is allowed between the phases.

The gas phase momentum equation:

$$\frac{\partial}{\partial t}(\varepsilon_g \rho_g \vec{\upsilon}_g) + \nabla \cdot (\varepsilon_g \rho_g \vec{\upsilon}_g \vec{\upsilon}_g) = -\varepsilon_g \nabla p_g + \nabla \vec{\tau}_g + \varepsilon_g \rho_g \vec{g} + \beta_{gs}^{m}(\vec{\upsilon}_s - \vec{\upsilon}_g) \tag{2}$$

where $\vec{g}$ is the gravity vector, p is the pressure and β_{gs}^{m} is the two-phase drag coefficient.

The stress tensor of the gas phase can be written as

$$\vec{\tau}_g = \varepsilon_g \mu_g \left(\left(\nabla \vec{\upsilon}_g + (\nabla \vec{\upsilon}_g)^T \right) - \frac{2}{3}(\nabla \cdot \vec{\upsilon}_g)\vec{I} \right) \tag{3}$$

Here, μ is the shear viscosity.

The gas phase turbulence is modeled with the k-epsilon model.

The solid phase momentum equation:

$$\frac{\partial}{\partial t}(\varepsilon_s \rho_s \vec{\upsilon}_s) + \nabla \cdot (\varepsilon_s \rho_s \vec{\upsilon}_s \vec{\upsilon}_s) = \nabla \vec{T}_s + \varepsilon_s \rho_s \vec{g} + \beta_{gs}^{m}(\vec{\upsilon}_g - \vec{\upsilon}_s) \tag{4}$$

where the solid stress tensor, $\vec{T}_s$, is the sum of a collisional part and a kinetic part and may be formulated as

$$\vec{T}_s = (-p_s + \lambda_s \nabla \cdot \vec{\upsilon}_s)\vec{I} + \mu_s\left(\left(\nabla \vec{\upsilon}_s + (\nabla \vec{\upsilon}_s)^T\right) - \frac{2}{3}(\nabla \cdot \vec{\upsilon}_s)\vec{I}\right) \tag{5}$$

where λ_s is the solid phase bulk viscosity.

The total solid phase pressure, p_s, consists of two terms; one of them, p_s^{k+t}, is determined from Lun *et al.* (1984) and includes both kinetic and collisional pressures, as follows:

$$p_s^{k+t} = \rho_s \varepsilon_s [1 + 2(1+e)\varepsilon_s g_o]\theta \tag{6}$$

where e is the coefficient of restitution and g_o is the radial distribution function.

The other term, p_s^f, is caused by friction and collisions in the solid phase due to highly dense flow, as proposed by Johnson and Jackson (1987), as follows:

$$P_s^f = Fr \frac{(\varepsilon_s - \varepsilon_{s,min})^n}{(\varepsilon_{s,max} - \varepsilon_s)^p} \tag{7}$$

where the values of the constants in Equation (7) are Fr = 0.05, n = 2, p = 3 and $\varepsilon_{s,min}$ = 0.55, as suggested by Van Wachen (1998).

The total solid phase pressure is then given by

$$P_p = P_p^{k+t} + P_p^f \tag{8}$$

The radial distribution function, adopted by Lun and Savage (1986), is

$$g_o = \left(1 - \frac{\varepsilon_s}{\varepsilon_{s,max}}\right)^{-2.5\varepsilon_{s,max}} \tag{9}$$

where $\varepsilon_{s,max}$, the maximum particle packing chosen is 0.62.

Equation (9) is appropriate for dilute flow.

In Equation (6), θ is the granular temperature related to the kinetic turbulent energy of the particle motion. Ding and Gidaspow (1990) proposed the following equation for granular temperature:

$$\theta = \frac{1}{15(1-e)} d_p^2 \left(\frac{1}{2}\left(\nabla \vec{\upsilon}_s + (\nabla \vec{\upsilon}_s)^T\right)\right) \tag{10}$$

where d_p is the particle diameter.

Solid phase bulk viscosity, λ_s, due to particle collisions (Lun *et al.*, 1984) and solid phase shear viscosity, μ_s, for dense and dilute flow (Hrenya and Sinclair, 1997) are given by the following equations:

$$\lambda_s = \frac{4}{3}\varepsilon_s^2 \rho_s d_s g_o (1+e)\sqrt{\frac{\theta}{\pi}} \tag{11}$$

$$\mu_s = \frac{5\sqrt{\pi\theta}}{96}\rho_s d_s \left[\left(\frac{1}{1+\frac{\lambda_{mfs}}{R}}\frac{1}{\eta g_o} + \frac{8\varepsilon_s}{5}\right)\left(\frac{1+\frac{8}{5}\eta(3\eta-2)\varepsilon_s g_o}{2-\eta}\right) + \frac{768}{25\pi}\eta\varepsilon_s^2 g_o\right] \tag{12}$$

where λ_{mfs} and η are given by

$$\lambda_{mfs} = \frac{d_p}{6\sqrt{2}\varepsilon_s} \tag{13}$$

$$\eta = \frac{1}{2(1+e)} \tag{14}$$

The gas-solid drag coefficient when $\varepsilon_g \le 0.8$ is based on the Ergun equation and is given by

$$\beta_{gs}^{m} = 150\frac{\varepsilon_s^2 \mu_g}{\varepsilon_g d_p^2} + 1.75\frac{\left|\vec{\upsilon}_s - \vec{\upsilon}_g\right| \varepsilon_s \rho_g}{d_p} \tag{15}$$

And when $\varepsilon_g > 0.8$, based on the Wen and Yu equation:

$$\beta_{gs}^{m} = \frac{3}{4} C_D \frac{\left|\vec{\upsilon}_s - \vec{\upsilon}_g\right| \varepsilon_s \rho_g}{d_p} f(\varepsilon_g) \tag{16}$$

where $f(\varepsilon_g) = \varepsilon_g^{-2.65}$, as used by Gidaspow and Ettehadieh (1983).

The drag coefficient, C_D, is related to the Reynolds number as follows:

$$C_D = \varepsilon_g^{-2.65} \max\left[\frac{24}{Re_p\, \varepsilon_g}\left(1 + 0.15 (Re_p\, \varepsilon_g)^{0.687}\right)\right] \text{ for } Re_p < 1000 \tag{17}$$

$$C_D = 0.44 \text{ for } Re_p \ge 1000 \tag{18}$$

$$Re_p = \frac{\left|\vec{\upsilon}_s - \vec{\upsilon}_g\right| \rho_g d_p}{\mu_g} \tag{19}$$

Results and Discussion

In order to validate the present numerical results, we compared them with the experimental results of Samuelsberg and Hjertager (1996). Then the simulations were carried out for the riser section of a circulating fluidized bed similar to that used by Samuelsberg and Hjertager (1996) and Mathiesen *et al.* (2000), shown in Figure 1 (a). In this study, just the riser section was simulated. The riser section is 1m high with an inner diameter of d = 0.032m. A secondary air supplier, positioned 0.05m above the gas inlet, feeds the solid back into the riser. The secondary gas inlet and the gas outlet have a diameter of 0.008m (Figure 1 (b)).The initial bed height is 0.05m and the initial solid volume fraction is 0.61. The system properties are defined in **Table 1**.

The riser section, as shown in Figure 1 (b), is modeled and simulated in a three-dimensional Cartesian coordinate system. Numerical predictions were obtained using the finite volume approach through the CFX-10 code, a CFD software developed by ANSYS, and the ICEM CFD code to generate the numerical grid, shown in Figure 1 (b). Numerical tests were conducted in order to optimize the computational grid. A grid of 300,000 control volumes was considered satisfactory considering the precision of the results (10^{-4}) and the computation time consumed. The time step used in all simulations was 10^{-3}s.

At the primary gas inlet, the superficial gas velocity, V_{sup}, was 0.71m/s. At the secondary gas inlet, the gas velocity and volume fraction were 0.05m/s and 0.6, respectively, and the particle volume fraction was 0.4. In order to permit all particles at the outlet back into the riser at the secondary gas inlet, a subroutine was implemented in the CFX code. At the outlet, atmospheric pressure was prescribed. At the walls, no-slip conditions were used for the solid phase as well as the gas phase.

The simulation results were computed through time-averaged distribution of flow variables. All the simulations were run for 7s of real time. To compute the granular temperature, solid-phase pressure and solid-phase bulk and shear viscosity must be programmed into the CFD codes due to the fact that the models used in this study were not presents in the CFX-10.

Figure 2 show a comparison of experimental results due to Samuelsberg and Herjtager (1996) and the present numerical results to the solid radial velocity profiles for heights of 0.16, 0.32 and 0.48m in the riser section and superficial gas velocity, V_{sup}, of 0.71m/s and 1.42m/s in the primary gas inlet.

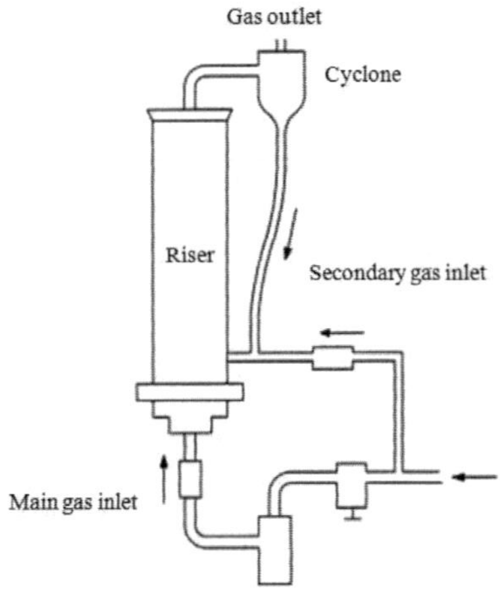

(a)

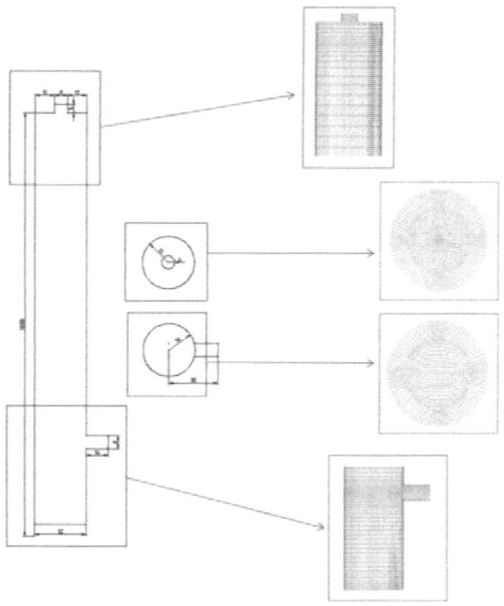

(b)

Figure 1 – (a) Geometry of Mathiesen *et al.* (2000). (b) Calculation domain and grid nodes

Table 1 – System properties

Gas density	1.185kg/m^3
Particle density	1600kg/m^3
Particle size	60μm
Gas viscosity	1.831x10^{-5}kg/m s
Maximum solids volume fraction	0.62
Particle-particle restitution coefficient	0.95

As can been seen in Figure 2 (b), (c), (e) and (f), the agreement is very good in the central region for all heights. The particles move upward in the central part of the riser and downward near the walls, a behavior well known from the literature, which is typical of core-annulus flow. However, in Figure 2 (a) and (d), for the height of 0.16m, the agreement between the curves differs significantly. These differences can probably be attributed to the region, both for the height of 0.16m, being so close to the secondary inlet, causing large perturbations in the flow. In Figure 2, a significant difference can be observed in the wall region for all results compared. This result may indicate that the no-slip wall boundary condition for both phases is not correctly used when employing KTGF.

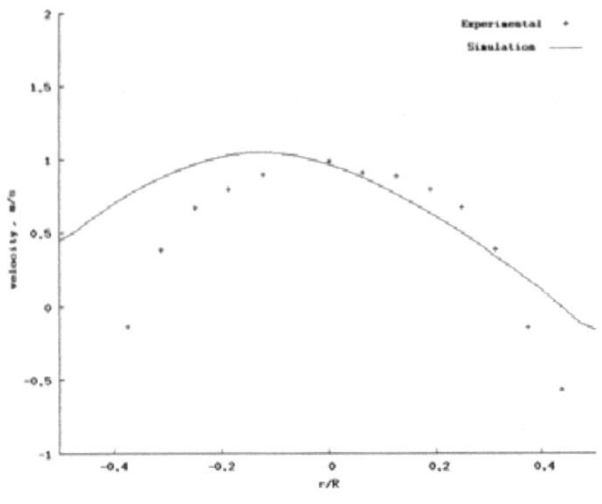

(a) Height of 0.16m and V_{sup} = 0.71m/s

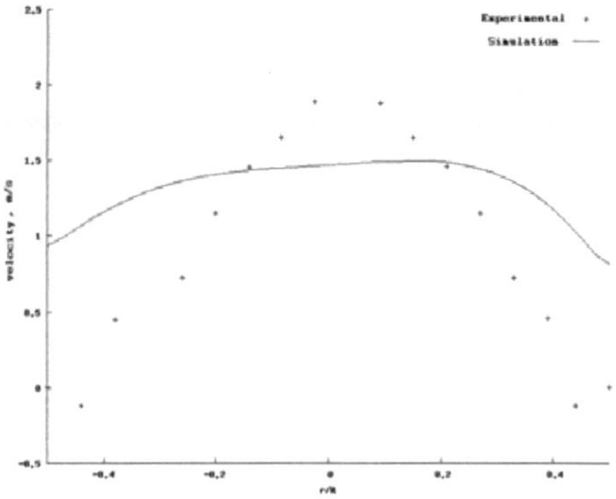

(d) Height of 0.16m and V_{sup} = 1.42m/s

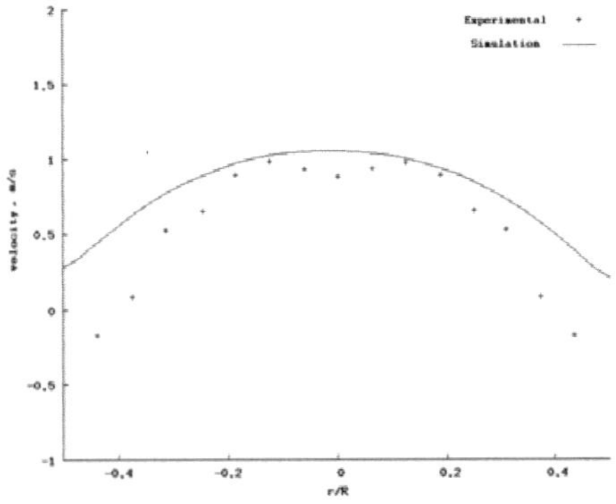

(b) Height of 0.32m and V_{sup} = 0.71m/s

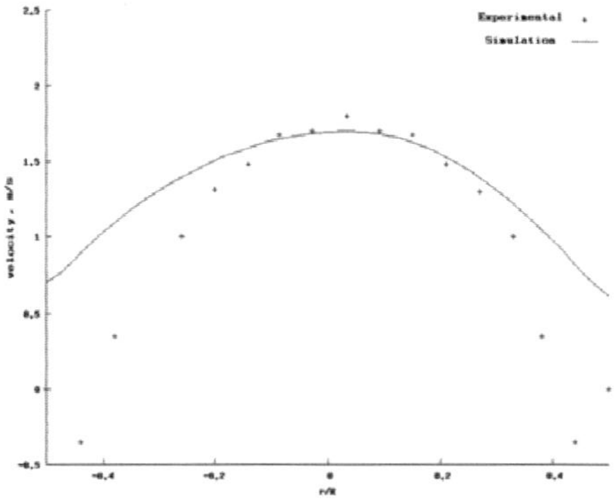

(e) Height of 0.32m and V_{sup} = 1.42m/s

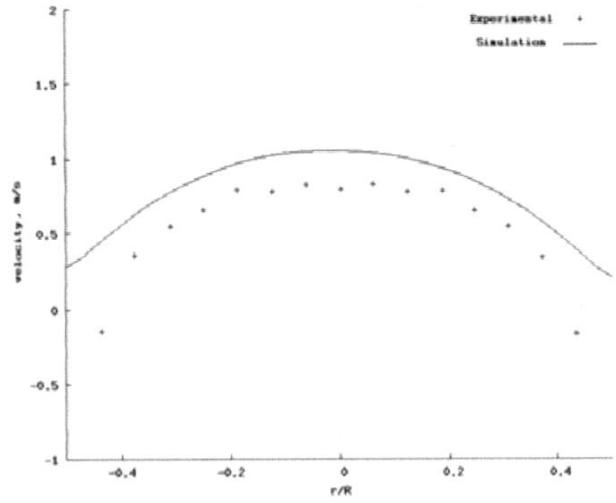

(c) Height of 0.48m and V_{sup} = 0.71m/s

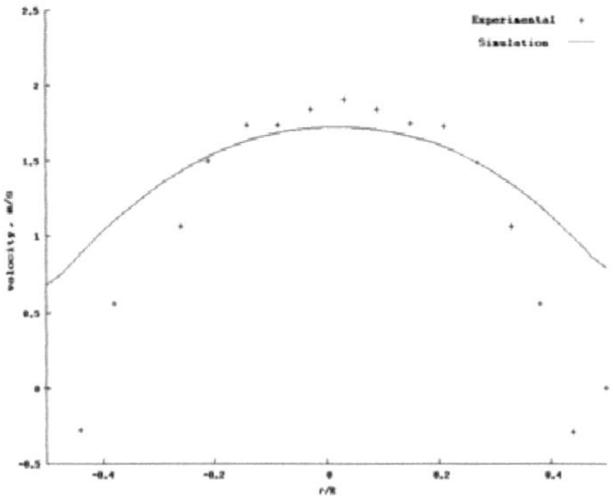

(f) Height of 0.48m and V_{sup} = 1.42m/s

Figure 2 – A comparison of experimental data due to Samuelsberg and Herjtager (1996) and computed solid velocity profiles at different heights in the riser section and superficial gas velocity, V_{sup}, of 0.71m/s and 1.42m/s.

Conclusions

A three-dimensional two-phase flow model based on the kinetic theory of granular flow (KTGF) was used to predict the behavior of a gas-solid fluidized bed. The numerical results are compared against the experimental results of Samuelsberg and Hjertager (1996). The agreement is found to be satisfactory in the riser central region. However, in the wall region, the numerical results indicated a significant difference between all results compared.

References

1. Chapman, S., Cowling, T.G. (1970). *The Mathematical Theory of Non-Uniform Gases*, Cambridge University Press, London.
2. Ding, J., Gidaspow, D. (1990). A bubbling fluidization model using kinetic theory of granular flow. *AICHE J.*, 36, 523-538.
3. Gidaspow, D. (1994). *Multiphase Flow and Fluidization*, Academic Press, London.
4. Gidaspow, D., Ettehadieh, B. (1983). Fluidization in two-dimensional beds with a jet. 2. hydrodynamic modeling. *Industrial and Engineering Chemistry Fundamentals*, 22, 193-201.
5. Gidaspow, D., Jung, J., Singh, R.K. (2004). Hydrodynamics of fluidization using kinetic theory: an emergin paradigm 2002 flour-daniel lecture. *Powder Technology*, 148, 123-141.
6. Hrenya, C.M., Sinclair, J.L. (1997). Effects of particle-phase turbulence in gas-solids flows. *AIChE Journal*, 43, 853-869.
7. Jenkins, J.T., Savage, S.B. (1983). A Theory for the rapid flow of identical, smooth, nearly elastic, spherical particles. *AIChE Journal*, 42, 1536-1546.
8. Johnson, P.C., Jackson, R. (1987). Frictional-collisional constitutive relations for granular materials, with application to plane shearing. *Journal of Fluid Mechanics*, 176, 67-93.
9. Lun, C.K.K., Savage, S.B. (1986). The effects of an impact velocity dependent coefficient of restitution on stresses developed by sheared granular materials. *Acta Mechanica*, 63, 15-44.
10. Lun, C.K.K., Savage, S.B., Jeffrey, D.J., Chepurniy, N. (1984). Kinetic theories for granular flow: inelastic particles in couette flow and slightly inelastic particles in a general flowfield. *Journal of Fluid Mechanics*, 140, 223-256.
11. Mathiesen, V., Solberg, T., Hjertager, B.H. (2000). An experimental and computational study of multiphase flow behavior in a circulating fluidized bed. *International Journal of Multiphase Flow*, 26, 387-419.
12. Peirano, E., Leckner, B. (1998). Fundamentals of turbulent gas-solid flows applied to circulating fluidized bed combustion. *Progress Energy Combustion Science*, 24, 259-296.
13. Samuelsberg, A., Hjertager, B.H. (1996) Computational modeling of gas/particle flow in a riser. *AIChE Journal*, 42, 1536-1546.
14. VanWachen, B.G.M. (1998). Derivation, Implementation and Validation of Computer Simulation Models for Gas-Solid Fluidized Beds, PhD Thesis - Delft University.15.
15. Wen, C.-Y., Yu, Y.H. (1966). Mechanics of Fluidization. *Chem. Eng. Prog. Symp. Series*, 62,100-111.

Dr. George Klinzing

It was the mid 1980's when I was asked if I would be willing to help teach a course in Pneumatic Conveying for the AIChE. I was informed that my portion of the presentation was to be practical as another person would cover the theoretic portion. Although I was told the other person was Dr. George Klinzing, from the University of Pittsburgh, that meant nothing to me.

After only brief exchanges on the details of this course, we met the morning of the first session, and had our first chance to not only meet each other but to look at the materials the other was going to present. When I walked into the class room, I saw George preparing some materials, and I looked at him and wondered who is this smiling energetic person? It certainly could not be a university professor, and even more a dean. I must be in the wrong room!

But, no, it was George, and at that moment a friendship was born which has continued for over 20 years. It is impossible to not like George. His radiant smile and effervescent willingness to help, sets him aside from the normal characters we meet in life and with whom we have to associate. George is different, but I need not take your time to tell that to you.

Indecently, the plans we had individually made for that first course dove tailed together like we had been working together for years. We alternated teaching times, intermingled with coffee and lunch breaks, and I know that at the end of the first day we were amazed how well our subjects worked together.

I have been trying to think of times we spent together, that I know George would like to forget. But, if you know me at all, you know that those are the things I can't keep quiet about.

On one occasion George and I were to give an in-plant school in Oklahoma. We arranged out flights to arrive at about the some time at a convenient airport. Since my flight arrived first, I would rent a car and we would go together to the course location. It was one of those nights that you only see in horror movies. It was dark, a heavy rain was falling, wind was blowing, and occasional flashes of lightening lit the sky, and our pathway. Since time on the road was taking away our sleep time, I drove, perhaps, a little faster than one would recommend. All I can remember of that night was the strong cross winds (I believe a tornado type storm) and when the lightening would flash all I could see was George holding on as tightly as possible and his white knuckles were almost glowing.

On many occasions my wife would travel with me on business trips, and so frequently the two of us would spend time with George. I have some beautiful pictures of George, the Olympic swimming star, at Miami Beach, but my wife wouldn't allow me to show them. George, you owe Myrt a thank you.

Occasionally, I would find a company that was going to surplus some equipment. At that time George was actively running tests in the lower floors of the Engineering building at the University of Pittsburgh. On one occasion I found a pressure tank that could be used to feed fluidizable material into a pressure conveying system. Without much detail, George accepted the offer. I wasn't there to see it, but I hear it went something like this.

The flatbed truck arrived at the engineering building with this 42" diameter, about 10' long pressure coded vessel, which had an integral support. Now, the truck is in the busy street, and now they would figure out what to do. I don't know if this is true, but it makes a good story, George went to one of the engineering classes and brought out all the students. They lined up on both sides of the truck and lifted the vessel off, and carried it into the engineering build, much as one would handle a coffin.

But that was just the start. They wanted the vessel in the basement, but when they tried to put it on the elevator, it was longer that what the elevator could handle. Well, you get the picture.

On a subsequent visit to University of Pittsburgh, we looked at the installation. The tank was in a room off the hall, along with the compressor and receiver. But the conveying line went out into the hallway and proceeded around the building back to the original room, and to a dust collector. Now what would you expect George would be conveying in this set up? Why not? There is plenty of coal in Pittsburgh, and so pulverized coal would be an ideal, inexpensive product to test. I was sure we would read that the engineering build was raised a few stories, as the coal conveying system exploded.

During those "good old days" George and the graduate school were always looking into pneumatic conveying theory. One of the students was none other than, Shrikant. In order to keep my reputation as a heckler, I had to give George a rough time. So, whenever I would see the research work they were doing, such as how long does it take for a gnat to wink his eye, I would tell George it was interesting, but not practical. It wouldn't solve any of the worlds conveying problems.

George was very receptive to my criticism, (at least that is what he led me to think) and for some years, as a new research project would be started, I would meet with the graduate students, discuss the work they were doing, and suggest taking a little bit of additional data which would then make their test work useful to the practical world of conveying. This resulted in several contributions to the pneumatic conveying world that have had a strong influence in the practical application of conveying.

For example, Shrikant, while working on his project and with my prodding, took extra time to computerize several subjects. There were three computer programs that were based on artificial intelligence, that would guide a novice through selection of a material handling system. A second program guided in the best way to feed the material into a conveying system, and the third program was very useful in troubleshooting a conveying system that was operating as desired.

To let George rest for a moment, let's turn our attention to Shrikant. He spent several days with me asking questions for these programs. He would ask "Do you ever do something in pneumatic conveying?" And I might answer "No." Then he would come back "Could you?" I would say "Maybe" Then it would be "What percentage of the time would you do it?" etc. Shrikant got even with me for all the troubles I ever gave George.

Incidentally, the three programs were the property of University of Pittsburgh, and through George's involvement, they were made available for sale to the conveying industry, and many dollars were raised for the university through the sale of these programs. Thank you George and Shrikant.

Every since my first meeting with George, I recognized that he was a man dedicated to helping and working with others. Two times that I can recall, I dropped in on George with a technical question on which I needed his help. The first was a simple question. "George, in our equations to get Mass we divide by G (gravity) What would I do if I were on the moon?" At that time the space administration was considering having people live on the moon, and one of the requirements was to generate Oxygen. This required the pneumatic conveying of, what we shall call moon dust, for lack of a better term. George dropped everything. This was exciting. He had never been asked a question so "out of this world" before, and he was excited. After a bit of research, checking various references, George gave me the guidance needed. And, incidentally, he was right. But who would have guessed otherwise? How do I know he was right? Well conveying tests done under reduced and zero G conditions, proved the theory correct. Thanks George.

On a similar challenge, we had to calculate pneumatic conveying with a gas other than air or nitrogen. I called George for a quick answer, and that is what I got, but I was also surprised in that it was accompanied with a 7 page email with all the details I needed (and maybe a few more).

In these 20+ years of association with George, I have encouraged anyone that approached me with a technical question to "Contact George". To the best of my knowledge, in spite of how busy George may have been with routine work, there wasn't anything he liked more than to dig into the theory of Pneumatic conveying. I have heard back from many of the people who have contacted George, that he was very helpful and friendly. In his present position with the University, one might think that delving into mundane details of science would be too time consuming or beneath their level, but George has always been ready to help anyone.

George, there are a few things you haven't accomplished in this life. You still have that research facility that was with Gulf Oil Company. I'm still waiting for you to turn that into a profitable independent resear5ch facility dedicated to the field of pneumatic conveying.

I am sure I can say for all your friends and colleagues here today, that you have been a great friend, an inspiration, and a great credit to your vocation and the University of Pittsburgh. Thanks for being my friend.

UNLOADING PRESSURE DISCHARGE TRUCKS

Francisco J. Cabrejos, Jenike and Johanson Chile S. A., Viña del Mar, Chile
Thomas G. Troxel, Jenike & Johanson Inc., San Luis Obispo CA, USA
<www.jenike.com>

Abstract

In many industrial processes where raw materials are required in bulk, pressure discharge trucks (usually referred to as PD trucks) are commonly used to transport them from the supplier's warehouse or mine to the processing plant or final client. As the name implies, pressure discharge trucks unload their contents by means of a pneumatic conveying system and transport it through a pipeline into a storage silo. This type of material handling system does not employ bags or containers to transport the materials, it is enclosed and easy to operate, and hence, it is less expensive overall when large quantities of materials are to be transported in bulk on a regular basis.

In this article, attention is centered on the design and operation of PD trucks, from a pneumatic conveying viewpoint, including a description of several problems encountered in the field. Examples of different applications and bulk materials are also presented.

1. INTRODUCTION

The pneumatic transport of bulk materials, or more precisely the flow of a two-phase gas-solids suspension inside a pipe, has been practiced for over a century and has been defined as "the art of transporting bulk materials through a pipeline by either a negative or a positive pressure gas stream" [1]. Pneumatic conveying represents an important topic in engineering and is encountered in a variety of industrial processes. It is not a simple matter because of the large number of variables involved, the interaction between the phases, and the complex dynamic developments occurring in the pipeline. Although gas-solids flows obey all of the basic laws of fluid mechanics, there is not yet a thorough understanding of them and theoretical predictions from first principles have yet to be developed.

Several methods to model the flow of gas-solids suspensions can be found in the literature. Most of these models are based on empirical correlations, discrete-particle simulations, on averaging techniques, and mixture theory [2]. But experimental testing remains the most reliable method for understanding and designing gas-solids flow systems, usually using pilot-scale units. Perhaps the most important constraint to model gas-solids suspensions is the turbulent nature of the flow and the fact that each bulk material has its own unique flow characteristics, making the design of pneumatic conveying systems still "an art rather than a science".

Pneumatic conveying is widely used in industry to handle and transport dry and free-flowing powdered and granular materials because it is suitable for a variety of processes. Fine particles of less than 1 micron as well as 15 mm rocks can be conveyed vertically and/or horizontally from distances of a few meters to a few kilometers at rates of hundreds of tons per hour. Although pneumatic conveying requires a greater power supply and more technology than mechanical conveying, it has a lower initial capital investment, requires less control and maintenance, takes up little floor space (pipelines are easily routed), protects the material from the environment by enclosing it, and it is cleaner and easier to automate. Systems are totally enclosed, which means that potentially hazardous materials can be safely conveyed, and dust generation is minimized. Disadvantages of pneumatic conveying include pipe erosion and potential solids degradation when the system has not been properly designed.

Pneumatic conveying systems can be found in many diverse industries: chemical, pharmaceutical, food, glass, cement, plastics, mining, ports and other industries. Typical applications include: delivery of materials through pipelines to remote plant areas; filling and unloading of silos, rail cars, trucks and ships; feeding and metering of materials into reactors, furnaces and converters; vacuum cleaning and fugitive dust handling systems; etc.

A relatively new industrial application is the pneumatic unloading of pressure discharge trucks, the main focus of this article. For example, lime (calcium oxide) represents a primary ingredient used in many concentrators to prepare 'milk of lime' for milling and flotation processes and for neutralization of acid waters in mines. To satisfy this demand at one mine, several 28-ton trucks are unloaded daily, some of them after traveling days from the supplier's source (see Figure 1). The cement industry also uses this type of transport system to deliver cement in bulk from a cement plant to batch plants or construction sites. Other materials transported in PD trucks include: suggar, flour, plastic pellets, coke, coal, gypsum, flyash, dust, etc.

Figure 1: Examples of two Chilean applications with PD trucks ready to unload lime. [3]

2. BASIC COMPONENTS

The main purpose of pneumatic conveying is to move materials at a controlled rate from one point (i.e. the feed point) to another (i.e. the delivery point) by means of a gas stream flowing inside a pipe. To achieve it, a pneumatic conveying system consists basically of four distinct components as shown in Figure 2 for a positive pressure system: the prime mover, the silo / feeder arrangement, the pipeline, and the separator, each of them with its own characteristics. Controls, safety equipment and instrumentation are also required. A challenge for every designer is to match all the components and to "combine" the different equipment so that the system operates efficiently at the design conditions.

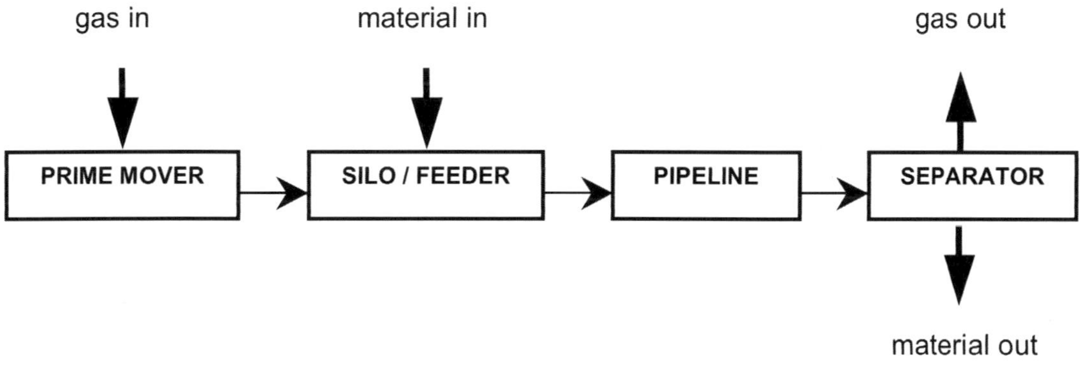

Figure 2: Basic components of a positive-pressure pneumatic conveying system.

The prime mover provides the proper flow rate of gas required for the transport at the right pressure. A wide variety of positive-displacement Roots type blowers and compressors available in the market could be used to unload the material contained in a truck silo. In some cases, the blower is mounted on the truck and is driven by the truck's engine, i.e. a self-unloading PD truck. In other cases, the blower or compressor air supply is fixed at the plant and the driver connects a rubber hose from the plant air supply to a manifold on the truck. Usually, the air is divided to pressurize the top of the stored material in the truck silo, to fluidise the bottom discharge point, and the main stream is directed to transport it through the pipeline.

The material stored in the truck silo is introduced into the pipeline by gravity, usually fluidizing the bottom discharge of the stored material, mixed with the conveying gas, and accelerated to some "steady" transport velocity. This is a crucial zone in all pneumatic conveying systems, and special care must be taken when designing and operating PD truck unloading systems. Reliable flow of material from the truck silo and into the pipeline is an absolute necessity as a starting point. Properly designed silos with mass flow hoppers and aerated bin bottom discharge should be based on the material flow characteristics [4].

The pipeline corresponds to the conveying zone itself, including a connecting rubber hose at the truck's feed point and the pipeline up to the delivery point on top of the storage silo in the plant, using horizontal and vertical pipes, bends, couplings and/or diverters. The most common pipes used in these type of unloading systems are standard 4" or 6" in diameter, ASTM

Sch. 40 or 80 carbon steel pipes. Systems handling ligther, non-abrasive material often use 4″ or 5″ O.D. tubing.

In the separator, the solids are recovered from the gas stream in which they have been transported and then stored in a silo. This is the final point of the transport process where the solid particles are decelerated and the gas is typically released to the environment. Environmental regulations should be considered for the proper release of the conveying gas into the environment. The following most common types of separators may be used: cyclones, settling chambers, bag filters, reverse-jet filters and Hepa absolute filters, depending mainly on material characteristics.

While the conveying gas is usually air at ambient conditions, other gases may be used such as nitrogen and carbon dioxide in applications where there exists a risk of explosion or health and fire hazards.

It is interesting to point out in the case of PD trucks the so-called 'feed point' is not fixed but travels with the truck, from plant to plant, and during discharge, the truck silo itself becomes a "blow tank", as shown in Figure 3, i.e. a positive, low pressure pneumatic conveying system operating in dilute-phase with a Roots type blower.

In general, truck unloading systems are laid out so that the truck can park as close as possible to the storage silo. When the truck can park next to the silo, the pipeline has a vertical run of 20 to 30 meters and may include only one 90 deg. bend, or a single 180 deg. bend. This minimizes the conveying distance which in turn minimizes the unloading time. For truck silos containing roughly 28 to 30 tons of material, the usual unloading time is 1 to 2 hours, depending mainly on conveying gas density (i.e. altitude of the plant) and material characteristics such as particle size and particle density. But in cases where the truck has to park up to hundred meters away from the silo, the unloading time could be several hours. The truck silo (or just silo tank) can be made of carbon steel or stainless steel, with a maximum working pressure rated at 1 barg.

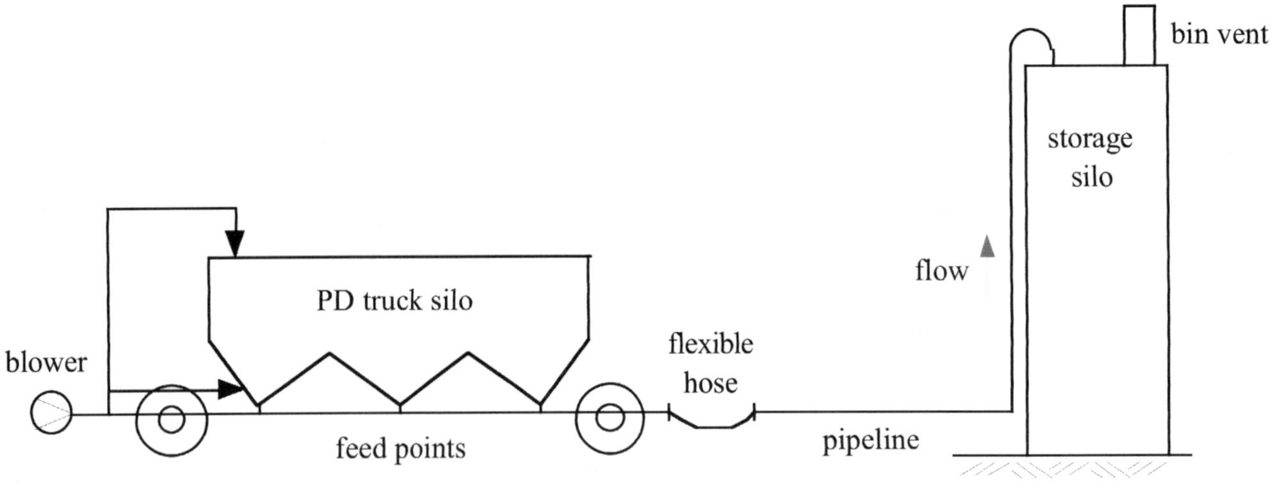

Figure 3: Schematic representation of the pneumatic unloading of a PD truck.

3. GENERAL STATE DIAGRAM

Perhaps the most common way to describe the characteristics of pneumatic conveyance of bulk materials is using the so-called "general state diagram." Here, the pressure drop is plotted as a function of mean gas velocity, with constant solids mass flow rates as a parameter. Figure 4 shows a typical "general state diagram" in which the curve "gas alone" represents the friction loss for the gas-phase alone (Ws=0). Point A represents the saturation carrying capacity at solids flow rate Ws1, usually referred to as "saltation." This point may represent the limit of conveying, where the line plugs and conveying at lower velocities is impossible, or it may represent the point at which conveying becomes unstable and a transition is possible to lower velocity conveying in a dense phase mode. Some materials can be conveyed successfully with a bed of material sliding along the bottom while other materials cannot. Point B represents the optimal operating condition in terms of energy consumption.

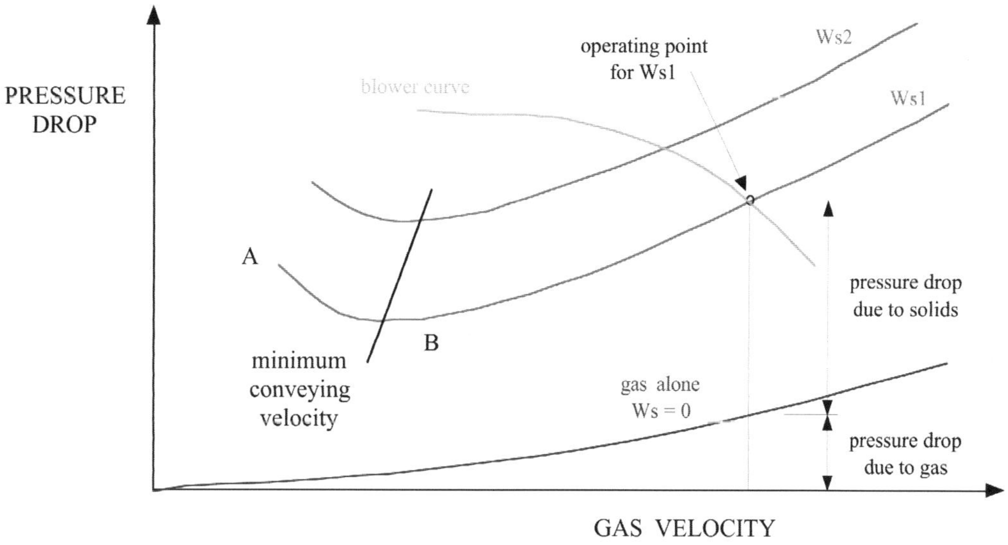

Figure 4: General state diagram for the pneumatic conveyance of bulk materials.

3.1 Minimum conveying velocity

The conveying velocity is one of the most important parameters in the design and operation of pneumatic transport systems, affecting the pressure drop and thus the energy requirements for the transport, pipe erosion, particle attrition, and the flow patterns of the gas-solids suspension. Keeping the mean gas velocity above a minimum value in all horizontal sections of a pipeline ensures no deposition or accumulation of solids in the system. This minimum conveying velocity can be defined as the safe gas velocity for the steady horizontal transport of solids. If this velocity is set at the beginning of a piping system, the velocity will increase along the pipeline due to compressibility effects (density decrease) so the rest of the pipeline should be well above this lower velocity bound. A slightly lower gas velocity results in the deposition of solids on the bottom of the pipe, which can lead to blockage of the system.

There are many terms wrongly used to refer to minimum conveying velocity: saltation velocity, pickup velocity, suspension velocity, deposition velocity, rolling or sliding velocity, critical velocity, etc. Definitions of these terms are based on visual observations and pressure drop measurements, and they are often applied to indicate some transition in the way in which the particles are moving or begin to move. In fact, the terms pickup and saltation represent two completely different mechanisms, but they are often used interchangeable in pneumatic transport. Extensive work has been carried out on the prediction of minimum conveying velocity and its relationship with both pickup and saltation mechanisms of solid particles for different materials [5].

3.2 Flow regimes

Two main distinct regimes exist for horizontal gas-solids conveying: flow at gas velocities above and below the saltation velocity, as illustrated in Figure 5, sometimes referred to as dilute and dense-phase conveying. Above saltation velocity, at high gas velocities, the solids are dispersed and homogeneously suspended in the gas-phase, moving in the same direction as the gas. In this regime, the pressure drop increases with gas velocity. As the gas velocity is decreased, however, the suspension becomes more stratified and a non-uniform solids flow distribution can be obtained. With further decreases in gas velocity, the suspension's velocity for carrying solids is eventually exceeded and solids begin to deposit on the bottom of the pipeline.

Figure 5: Flow regimes in horizontal pneumatic transport: (A) mean gas velocity above saltation, (B) mean gas velocity below saltation.

3.3 Pressure drop calculation

The pressure drop (ΔP) for dilute-phase, fully developed flow inside a straight and horizontal pipe is usually splitted in two contributions, the pressure drop due to the gas-phase (ΔPg) and the additional pressure drop due to the solids (ΔPs):

$$\Delta P = \Delta Pg + \Delta Ps \tag{1}$$

Gasterstadt [6] defined the specific pressure drop (α) dividing equation (1) by the pressure drop due to the gas-phase (ΔPg):

$$\alpha_{HOR} = \Delta P / \Delta Pg = 1 + \Delta Ps / \Delta Pg = 1 + Ks\,\mu \tag{2}$$

Yang [7] proposed a similar model for dilute-phase, fully developed flow inside a straight vertical pipe, including the static head contribution of the solids:

$$\Delta P = \Delta Pg + \Delta Ps + \rho p (1 - \varepsilon) g H \quad (3)$$

The voidage ε is defined as follow: $\varepsilon = 1 - (\rho g / \rho p)(Ug / Up)\mu$ (4)

where ρg and ρp represent the gas and particle density, respectively, and Ug and Up represent the gas and particle velocity, respectively. There are numerous correlations available in the literature for predicting the particle velocity but the one proposed by Konno and Saito [8] in terms of the particle terminal velocity (Ut) is normally selected by simplicity:

$$Up = Ug - Ut \quad (5)$$

Dividing equation (3) by the pressure drop due to the gas-phase (ΔPg) yields:

$$\alpha_{VER} = \{\Delta P - \rho p (1 - \varepsilon) g H\} / \Delta Pg = 1 + Ks\,\mu \quad (6)$$

In equations (2) and (6) Ks is a solids friction factor and represents the slope of a straight line when plotting α as a function of the solids loading ratio μ = Ws /Wg.

These models have been studied in details, in both horizontal and vertical pipelines, and the results obtained for the homogeneous and stable flow of different materials above minimum conveying velocity show that these models work very well in reducing the experimental data obtained [9]. It is interesting to point out that the solids friction factor Ks for vertical flow was found to be slightly lower as compared to the solids friction factor for horizontal flow, approx. 83% on average for all the materials tested. Also, the solids friction factor increases with particle size and particle density, and it is independent of solids flow rate.

Finally, to estimate the total pressure drop required in a pneumatic unloading system like the one shown schematically in Figure 3, other terms should be added to account for the aerated bin bottoms of the truck silo, acceleration length, bends, air disengagers and filter receivers.

4. FLOW PROBLEMS

PD trucks represent a unique class of pneumatic conveying systems because of the mobile nature of the "feed system." Self unloading trucks arrive with their own blower and operator and connect to a fixed system at the plant. Since each plant system that an operator visits is slightly (or, in some cases significantly) different, they have to adapt their operation to each system. Operators may change frequently and may not be knowledgeable in the finer points of pneumatic conveying.

Truck equipment can also vary so that some trucks will perform differently while unloading in the same system with the same material. An operators motivation for choosing operating conditions may also depend on factors other than maximum efficiency. Typically an operator will wish to unload the cargo as quickly as possible in order to complete his shift, but many other factors can affect this decision. Also, an operators perception of the "best" conditions for unloading may not always lead to the most efficient operation. One common misconception about pneumatic conveying systems in general is that more air flow leads to faster transfer rates. In systems constrained by an upper pressure limit and fixed piping, this logic often leads to less efficient conveying and problems such as:

- Excessive wear of the pipeline and/or bends, when handling abrasive materials,
- Excessive particle attrition of the material, when handling friable materials,
- Pipeline pluggage,
- Excessive power consumption,
- Material build-up inside the truck silo,
- Arching problems in the aerated bin bottoms of the truck silo,
- Limited unloading capacity (low tonnage), which results in high unloading times.

By knowing the material flow properties of the bulk solid handled and the general state diagram of the system, these flow problems can be avoided in properly designed new facilities, and also corrected in existing applications.

5. CONCLUSIONS

Pressure discharge trucks (usually referred to as PD trucks) are commonly used to transport materials in bulk such as lime, cement, suggar, flour, plastic pellets, coke, coal, gypsum, flyash, dust, etc. from the supplier's warehouse or mine to the processing plant or final client. By means of a pneumatic conveying system the content of the truck silo can be unloaded and transported through a pipeline into a storage silo. This type of material handling system does not employ any bags and/or containers to transport the materials, it is enclosed and easy to operate, and hence, it is less expensive overall when large quantities of materials are to be transported in bulk on a regular basis. However it is not free of potential flow problems, as described in this article. To avoid them, a proper design must be done.

6. REFERENCES

[1] Stoess H., Pneumatic Conveying, (New York: John Wiley & Sons, 1970), p.1.

[2] Marcus R., Leung L., Klinzing G. and Rizk F., Pneumatic Conveying of Solids - A theoretical and practical approach (London: Chapman and Hall, 1990).

[3] Cabrejos F., del Campo A. y Concha J. "Manejo y Almacenamiento de Cal Viva en Plantas Mineras Mediante Silos con Descargadores Aeroasistidos", Minería Global, Año 5 No. 2, Agosto 2006, pp.46-57.

[4] Purutyan H., Troxel T. and Cabrejos F., "Propel your Pneumatic Conveying System to Higher Efficiency", Chemical Engineering Progress, April 2001, pp. 42-55.

[5] Cabrejos F. and Klinzing G., Bulk Solids Handling, Vol. 14, No. 3 (1994), pp. 541-550.

[6] Gasterstadt J., "Die experimentelle Untersuchung des pneumatischen Fordervorganges," VDI Zeitschrift, Vol.68, No. 24 (1924), pp. 617-624.

[7] Yang W., "A Correlation for Solid Friction Factor in Vertical Pneumatic Conveying Lines," AIChE Journal, Vol. 24, No. 3 (1978), pp. 548-551.

[8] Konno H. and Saito S., "Pneumatic Conveying of Solids Through Straight Pipes," J. Chem. Eng. Japan, Vol. 2, No. 2 (1969), pp. 211.

[9] Cabrejos F. "Recent Developments in Pneumatic Conveying at the Universidad Técnica Federico Santa María", 5th World Congress on Particle Technology (WCPT5), April 24-27, 2006, Orlando, U.S.A.

Acoustic Measurement and Monitoring in Dry Particulate Systems: Inspiration, Education, Application, Appreciation

Clive E. Davies[1], Stephen J. Tallon[2] and Emile Webster[1]

[1] School of Engineering and Advanced Technology, Massey University,
Private Bag 11 222, Palmerston North, New Zealand
C.Davies@massey.ac.nz

[2] Industrial Research Limited, Private Bag 31-310
Lower Hutt 5040, New Zealand

ABSTRACT

Techniques for the use of audio-frequency sound waves in the measurement of flow rate and particle size and the estimation of the volume of material contained in partially closed vessels are described and discussed.

INTRODUCTION

Non-invasive interrogation of systems and processes involving dry particulate materials is particularly challenging. X-rays and γ-rays, for example, offer the prospect of truly non-invasive monitoring of material inside metal pipe work or metal containment as typically found in industrial practice, but require extensive calibration and are usually subject to licensing regulations. Exploitation of inductance and capacitance is limited by constraints on the properties of the pipe work or process vessel holding the material which must be modified to be partly, and usually significantly, non-metallic.

Interrogation by acoustic waves is not a universal panacea to these difficulties, but while especially attractive and effective in dilute flow systems, can also be used in some situations with bulk solids.

In this paper we revisit earlier work on measurement and monitoring of flow rate in both dilute and dense flow systems, particle size measurement, the effect of frequency on acoustic propagation velocity in dense systems, and also report on work in progress on the use of Helmholtz resonance to determine the volume fraction of bulk solids in a partially closed vessel.

DILUTE PNEUMATIC CONVEYING

Natural pressure waves are generated in pneumatic conveying systems, particularly by reciprocating air movers. These waves propagate as sound waves along the pipe and in a dilute system such as pneumatic conveying lines can travel for some distance. Changes in the velocity and magnitude of these waves along the pipe contain information of the flow velocity and concentration of the gas and solids phases and can be used for flow measurement[1-3]. Approximately logarithmic decay in pressure wave intensity was observed with increasing particle concentration, and the conveying gas velocity correlated well with changes in the propagation velocity of the waves, as in figures 1,2. Measurements can be made using a single pressure sensor located downstream of the solids feed point, although it is necessary to minimise the dead volume between the pipeline and the sensing surface of the pressure transducer, and depends on the existence of a constant and reliable source of sound.

To remove this latter limitation it was demonstrated[4] that an introduced wave of a constant frequency could also be used. The use of an introduced sound wave also allows a frequency to be picked that minimises interference effects from other sources of sound waves and from sound wave reflections related to the pipeline geometry. The preferred sound frequency is one which is low enough for the waves to propagate as a one dimensional wave. The frequency limit for this to occur decreases with increasing pipe diameter[4], and is in the mid to low audible frequency range for typical pneumatic conveying pipe sizes. Tallon and Davies[4] report that the measurement method can be applied to either vertical or horizontal conveying. Measurement of the pressure waves at two consecutive points along the pipeline is also reported, so that a relative attenuation between the two points can be measured. This allows a localised measurement to be made, preferably in a region of fully developed flow.

To further improve the measurement reliability it is possible to measure sound waves propagating in both an upstream and downstream direction from a centrally located sound source (figure 3), or between two axially spaced sound sources (figure 4)[5]. The velocity of the sound waves in the downstream direction is faster than the upstream velocity because of the effect of the convective motion of the gas phase that the waves are passing through. The convective velocity is calculated simply from the difference between the upstream and downstream velocities, i.e. the sum of the measured upstream velocity and downstream velocities divided by two. For turbulent and largely plug flow conveying conditions with small particles, the convective velocity is essentially equal to the mean gas velocity.

To calculate a solids mass flow rate from the solids concentration, it is assumed that the solids velocity is related to the gas velocity by a small but constant difference, or slip velocity. The estimated solids velocity is then multiplied by the measured solids concentration to give a mass flow rate. The slip velocity varies under different flow conditions, and at different points in the pipeline where fully developed flow may not have developed[6], but it has been shown that the assumption of a constant slip velocity has little effect on the accuracy of solids flow measurement if the system is calibrated against known solids flow rates[4]; see figure 5. This calibration can be achieved, for

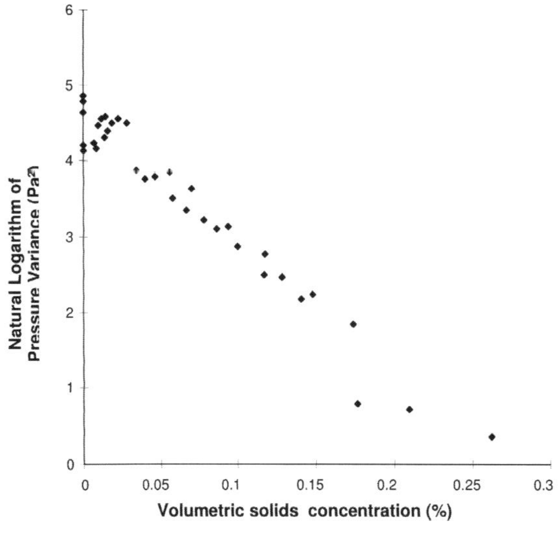

Figure 1. Change in intensity of natural sound waves in a pneumatic conveying line with changing solids concentration. Measured at a single point downstream of the solids feeder.

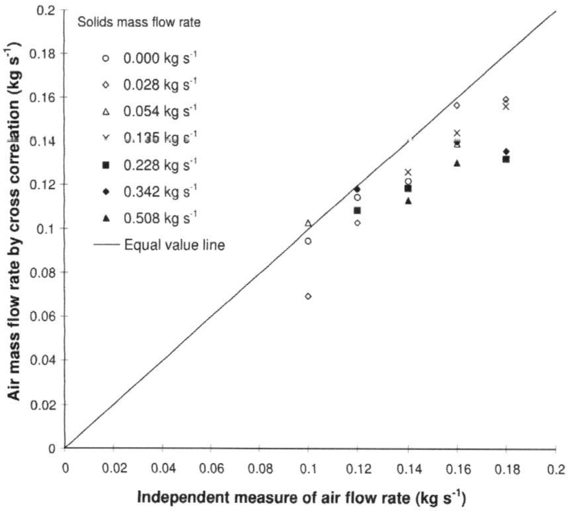

Figure 2. Air mass flow rate calculated from the increase in the sound wave propagation speed due to the convective flow of the gas phase. Sound velocity measured between two points 3m apart. The measurement is influenced by the solids flow rate.

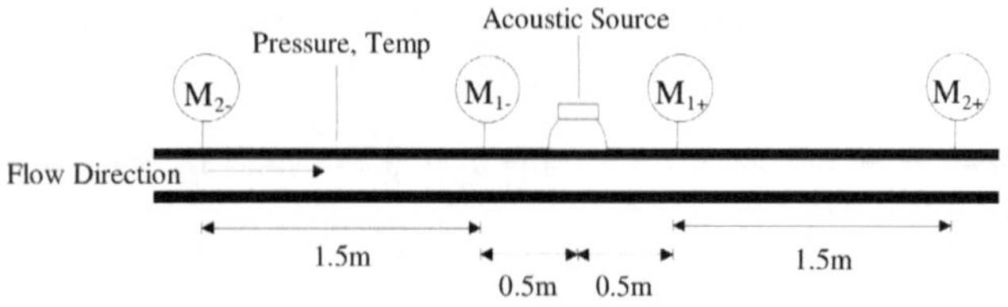

Figure 3. Flow measurement layout using a single acoustic source. Dimensions are indicative for use in a 3" diameter conveying line.

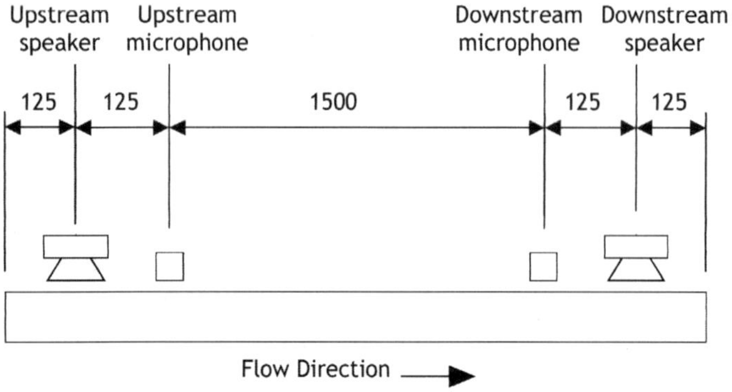

Figure 4. Flow measurement layout using two acoustic sources. Dimensions in mm are indicative for use in a 3" diameter conveying line.

example, by injecting a known mass of solids into the system and integrating the response over time[5].

Solids flow rate measurement has been demonstrated in dilute conveying systems up to 8" in diameter[7] (see figure 6), for conveying velocities up to 35m s^{-1}, and solids concentration up to 0.2% by volume. Better results are achieved if the measurement is made a sufficient distance downstream from the solids feed point or from pipe bends so that the flow is more fully developed.

Theory for the propagation of sound waves through a dilute suspension is described by Tallon,[8] based on theory by Gregor and Rumpf.[9] Accurate calibration of the relationship between attenuation and solids concentration is difficult to achieve theoretically, but good correlation between measured and calculated acoustic velocities was observed. Changes in the propagation velocity due to changes in the particle size could also be predicted and it is possible to use this to give an in-line measurement of particle size, and to compensate for particle size effects on the mass flow rate measurement[10].

DENSE PHASE FLOW SYSTEMS

Sound waves will also propagate through dense particle beds and can be used for measurement of a number of flow and state properties in dense flow systems such as downcomers, fluidised beds, dense phase pneumatic conveying and granulation systems.

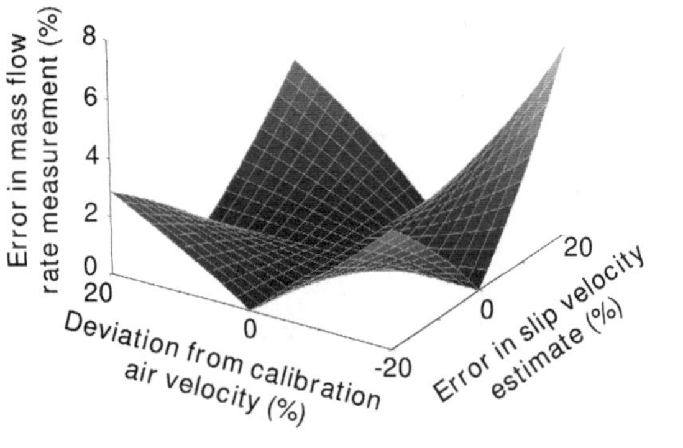

Figure 5. Error in solids mass flow rate measurement when the system is calibrated against a known mass of material. Errors in the slip velocity estimate do not have a large effect on measurement accuracy.

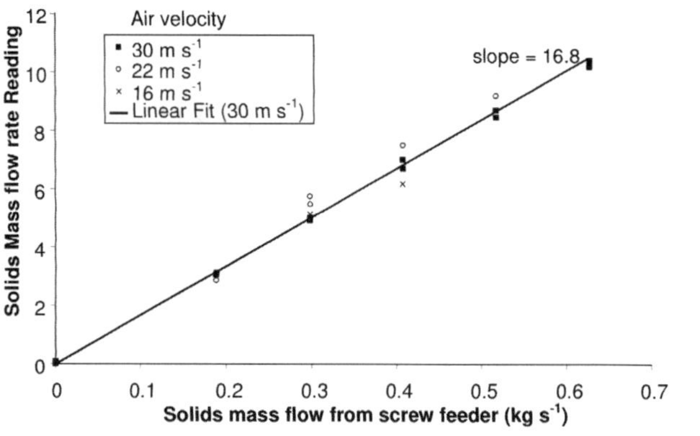

Figure 6. Mass flow rate of coal measured in an 8" diameter conveying line.

Sound waves will propagate through the interstitial gas phase, and also through the solid matrix if sufficient particle to particle contact exists. Interstitial sound waves are rapidly attenuated in a dense system but can still be used for measurement over small distances. Natural pressure waves generated by a gas pulse in an L-valve feeder, for example, have been used to give an indication of linear flow velocity in the conduit above the feeder[11] by measuring changes in the wave propagation velocity.

A more accurate method for flow measurement in a vertical conduit was achieved[12] by propagating introduced sound waves across the diameter of the conduit. The presence of natural variations in the packing voidage created measurable variations in the signal strength which could be tracked between two axial positions to give a flow velocity, as in figures 7,8. This measurement approach was demonstrated[12] in column diameters up to 200mm, and for mass flow rates up to 20 tonnes per hour. Best results were achieved for systems with a particle size greater than 1 mm and materials with plug flow characteristics.

Acoustic propagation velocity through the interstitial gas phase in dense particulate systems has been modelled by Tallon[8] by assuming that the solids particles oscillate in response to the motion of the gas phase. At low frequencies the particles follow the motion of the gas phase well and the propagation velocity is reduced by the greater inertia of the solids phase. At higher frequencies the particles become more stationary and the gas phase propagation velocity is only reduced by the tortuous path length through the particle bed. A comparison with experimental measurement is given in Figure 9.

The sound wave velocity through a packed or fluidised bed is a strong function of voidage and particle size. Velocity measurements in packed beds of a range of different materials are shown in figure 10 plotted against the Carman-Kozeny[13] flow factor. It is also possible to monitor changes in

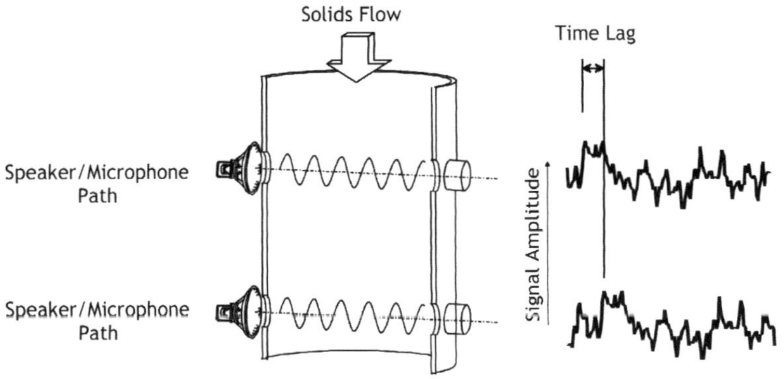

Figure 7. Principle for solids velocity measurement in bulk gravity flow in a conduit.

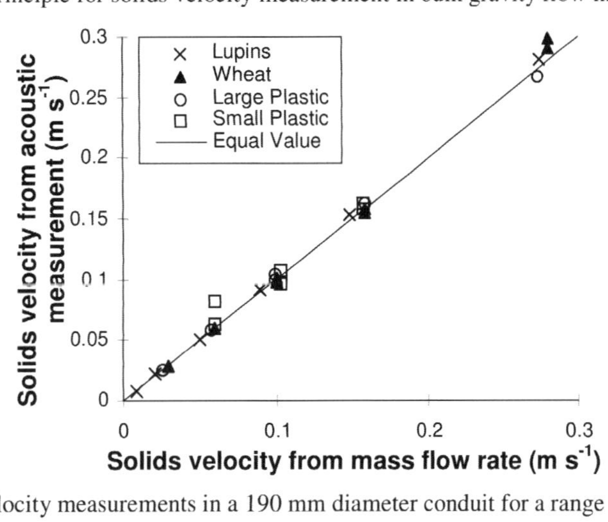

Figure 8. Velocity measurements in a 190 mm diameter conduit for a range of materials.

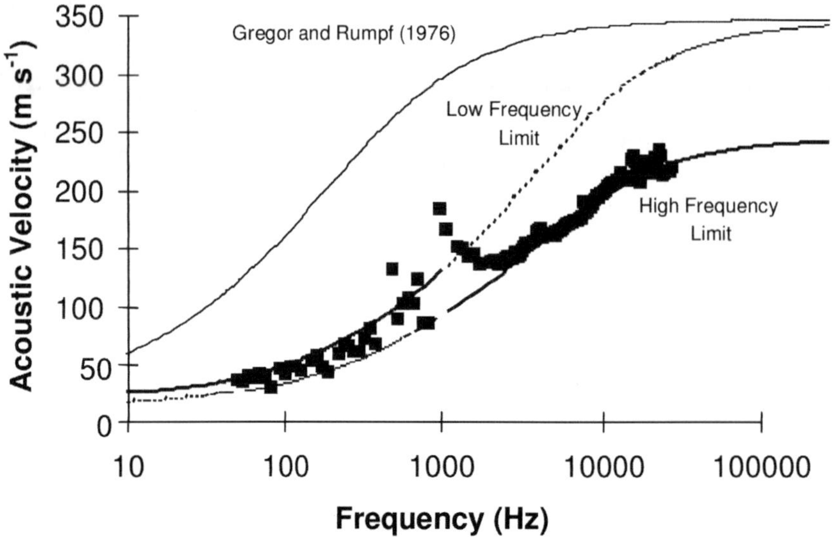

Figure 9. Acoustic velocity measurements through a packed bed of casein. Predicted acoustic velocity described by Tallon,[8] modified from theory by Gregor and Rumpf.[9]

particle size using acoustic velocity measurements.[14]

Acoustic measurements in a dense phase pneumatic conveying system have been used to measure conveying regimes, slug velocity and length, and an approximate solids mass flow[15].

MATERIAL VOLUME AND FILL FACTOR

When a chamber which is connected to the surrounding atmosphere by an open port or ports is excited by an acoustic stimulus, it resonates at a frequency that is determined mainly by the volume of the chamber and the length and diameter of the ports; this is known as Helmholtz resonance. If an object – which may be a massive solid, a powdered or granular material, or a liquid - is placed in the chamber, the volume of the chamber is reduced and the resonant frequency changes, permitting the

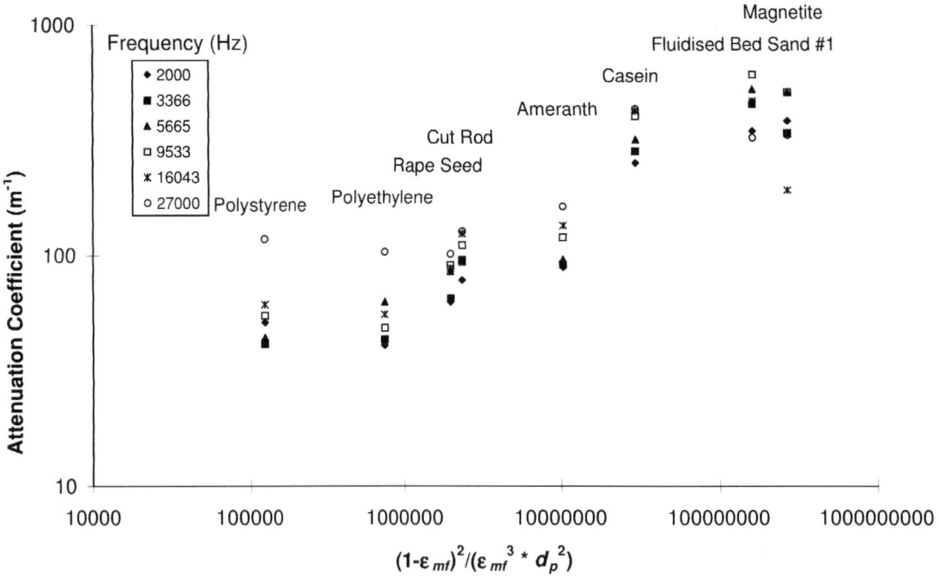

Figure10. Acoustic attenuation in a static bed of particles, plotted against the Carman-Kozeny flow resistance factor.

volume of the introduced object or material to be estimated.

The resonators used in this work were made from Perspex tube with a nominal inside diameter of 140 mm. Figure 11 is a photograph of a 3L resonator with a port 170 mm long and an inside diameter of 22 mm. Excitation was provided by an 8 inch infinite baffle loudspeaker which gave a linear response over the frequency range of 80 Hz to 500 Hz expected in the experimental programme; the loudspeaker was driven by a 100 W stereo audio amplifier. Two PCB103A sound pressure microphones were used to sense the sound signals; one was located centrally in the base of the chamber and the other at the top of the 170 mm port.

The materials used were water; marbles with a mean diameter of 24 mm, termed large marbles; marbles with a diameter of 15 mm termed small marbles; glass ballotinti with a Sauter mean diameter of 261 μm; glass ballotini with a Sauter mean diameter of 713 μm; and silica sand with a Sauter mean diameter of 166 μm. In all experiments the mass of the test material in the chamber was recorded, and its volume calculated using its actual density, *i.e.* the density of water, and the particle density in the case of the ballotini and sand. Full experimental details are given by Webster and Davies.[16]

Figure 12 is a comparison of measured and predicted resonant frequencies for a calibration run using water, and figure 13 is a plot of actual volume of the material in the chamber, *viz* the mass multiplied by the particle density, and predicted volume.

It is apparent that for the water calibration the resonant frequencies measured experimentally and the values predicted by theory are in excellent accord. The results for materials placed in the chamber fall into two broad groups. For large marbles, small marbles, and water, the agreement between theory and experiment is very good. For the sand and ballotini, the relationship between measured and predicted volume is non-linear over virtually the whole of the range investigated. The relationship between bulk volume of the sand and ballotini, as opposed to the actual volume as used here, and predicted volume, will be discussed elsewhere.

CONCLUSIONS

Audio-frequency sound waves are a versatile tool which can be used to interrogate particulate systems for information on flow rate, particle size, and the extent to which a partly closed chamber is occupied by a granular solid. Work to date has demonstrated the potential of acoustic diagnostic techniques for extracting useful and meaningful information from particulate flows in dilute phase pneumatic conveying pipelines, from flows in dense phase pneumatic conveying, from dense gravity flows of bulk solids and from static systems where bulk solids are contained in partially closed chambers.

Figure 11. Photograph showing experimental equipment: loudspeaker, left; resonator, right; signal generator on top of loud speaker.

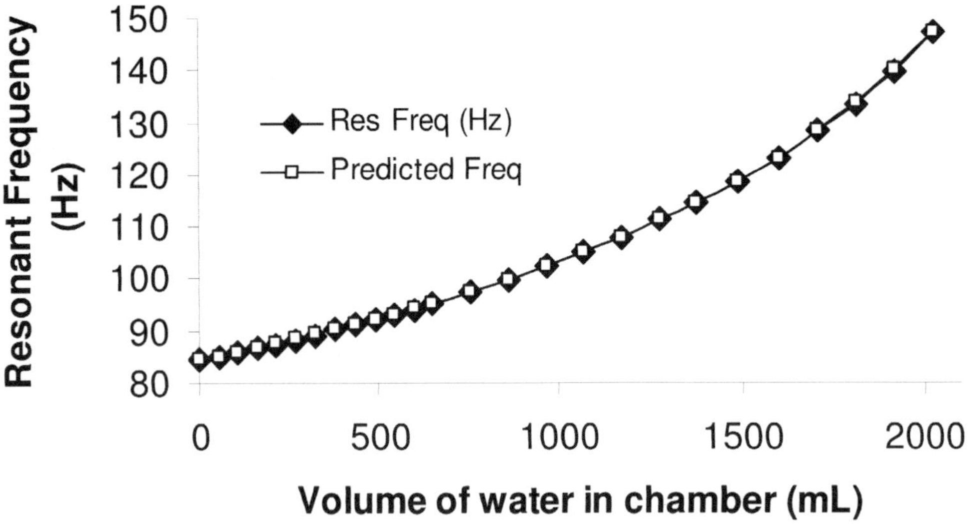

Figure 12 Comparison of resonant and predicted frequency in calibration test with water.

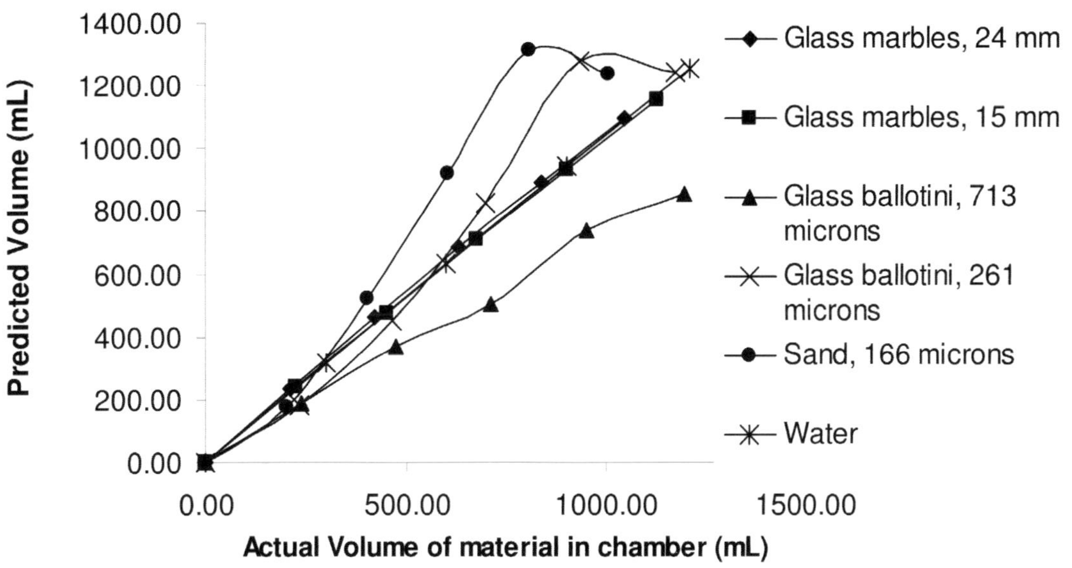

Figure 13. Comparison of actual and predicted volumes of material in 3 L resonator.

ACKNOWLEDGEMENT

In the late 1980s, CED visited Dr George Klinzing at the University of Pittsburg in connection with his interests in bulk solids handling. During the visit, he became aware of the relevance of sound to some diagnostic techniques in dilute pneumatic conveying. That was the beginning of a long and enduring interest in the use of sound as a diagnostic tool in a number of different physical systems. All the work discussed here was spawned by that first visit to Dr Klinzing's laboratories. Over the intervening years, CED and SJT have had many discussions with Dr Klinzing and we gratefully acknowledge the inspiration he has provided to us both.

REFERENCES

1. Tallon, Steve, and Davies, Clive E, 1994; The Use of Pressure Fluctuations to Characterize Gas and Solids Flow Rates in Dilute Phase Pneumatic Conveying, Proceedings of Chemeca 94, Chemical Engineering Conference, Perth, Australia, 25 to 28 September.

2. Tallon, S., 1995, On the characterisation and flow measurement of pneumatic conveying systems using static wall pressure measurements, Masters Thesis, Chemical and Process Engineering, University of Canterbury, Christchurch, New Zealand.

3. Tallon, Steve, and Davies, Clive E; 1995. Flow rate monitoring and measurement in dilute phase pneumatic conveying using pressure fluctuations, *Developments in Fluidization and Fluid Particle Systems*, AIChE Symposium Series, 91, (308), pp137-146.

4. Tallon, Stephen and Clive E Davies, 1996. Velocity and attenuation of sound: Application to flow rate estimation in pneumatic conveying, Powder Handling and Processing, *The International Journal of Storing, Handling and Processing Powder*, 8, (4), pp329-336.

5. Tallon, Stephen and Clive E Davies, 1997, Use of the attenuation of acoustic pulsed pressure waves for concentration measurement of gas-solid pipe flow, *Advances in Fluidization and Fluid-Particle Systems*, AIChE Symposium Series 317, 93, 1997, pp136-140.

6. Tallon, S.J., Davies, C.E., (2000), "The effect of pipeline location on acoustic measurement of gas-solid pipeline flow", Flow Measurement and Instrumentation, Volume 11, Number 3, pp 165-169.

7. Tallon, S., Davies, C.E., Levy, E.K., Yilmaz, A., (2001), "Measurement of solids flow rate in pneumatic conveying pipelines", Proceedings of the 2001 joint conference of SCENZ/FEANZ/EMG, Auckland, New Zealand, April 9th to 10th, pp 49-51.

8. Tallon, S.J., The application of acoustics to the monitoring of gas-solid processing systems, PhD thesis, Chemical and Process Engineering, University of Auckland, Auckland, New Zealand, 2001.

9. Gregor, W., Rumpf, H., (1976), "The attenuation of sound in gas-solid suspensions", Powder Technology, Volume 15, pp 43-51.

10. Stephen Tallon and Clive E Davies, 1997. In-line Monitoring of Particle Size Changes in a Pneumatic Conveying Pipeline using Pulsed Sound Waves, Chemeca 97, Rotorua, New Zealand, 29 September to 1 October.

11. Clive E Davies and Stephen J Tallon , (2004) "A pressure pulse method for monitoring flow rate in dense down flows of Geldart Group B particles", Powder Technology, Volume 148, 2004, pp 102-105

12. Tallon, Stephen, Davies, Clive E and Bernard Barry,1998. Non-intrusive velocity measurement in a downcomer, Fluidization and Fluid-Particle Systems: Recent Research and Development, Ed. Hamid Arastoopour, AIChE Symposium Series, 94, (318), pp42-47.

13. Carman, P.C., (1937), "Fluid flow through granular beds", Transactions of the Institution of Chemical Engineers, Volume 15, pp 150-166.

14. Tallon, S., Davies, C.E., Kirkegaard, M., (2003), "An acoustic method for in-line particle size measurement in flowing bulk solids", Journal of Process mechanical engineering, ImechE, Vol 217, Part E, pp 33-39.

15. Tallon, S., Davies, C.E., Wypych, O., Hastie, D., (2001), "Flow rate measurement in pneumatic conveying pipelines with an audio frequency acoustic meter", Proceedings of the seventh international conference on bulk materials storage, handling and transportation, Newcastle, Australia, October 3rd to 5th, pp 761-771, Volume 2, Ed., Jones, M., The institution of engineers, Australia.

16. Webster, E. S., Davies, C. E., "Determination of Bulk Density and Fill Fraction of Dry Particulate solids Using Helmholtz Resonance", SCENZ Conference, Hamilton, New Zealand, 20 October, 2008.

Pneumatic conveying of blended plastic pellets in dilute phase

Néstor Vásquez [a], Karl Jacob [a], Shrikant Dhodapkar [b], Joerg Theuerkauf [b]

[a] The Dow Chemical Company, Midland, MI 48667, USA
[b] The Dow Chemical Company, Freeport, TX 77541, USA

ABSTRACT

The pneumatic conveying of plastic pellets is a process widely used in the chemical industry. Different products can be transported using the same conveying system thanks to the great flexibility that this technology offers. However, there are also challenges due to the significant variation on pressure drop produced by different products. This has a direct impact on the conveying capacity due to the limited pressure that the air mover can provide. Among the different types of plastic pellets, it is known that products with a higher modulus of elasticity tend to produce higher pressure drop than harder materials.

The present work explores the effect of blending hard and soft pellets, on the pressure drop during dilute phase pneumatic conveying. With this purpose, a series of conveying tests were performed with blends of hard and soft pellets in different proportions. The component pellets had similar size, shape and density but different modulus of elasticity. The blends were conveyed through a 52mm diameter aluminum pipe and approximately 47 m in length. Each blend as well as the neat materials were conveyed at 23 m/s and 30 m/s with three repeats for each setting. The solids rate was controlled by a rotary valve, while the airflow was controlled to maintain a constant average gas velocity along the conveying system. Along the line, three pressure transmitters measured the pressure differential (ΔP) in horizontal and vertical direction as well as total pressure drop. The airflow was measured by a Vortex type flow meter located downstream the filter receiver. In each test, data from the pressure transmitters was recorded for two minutes during steady state condition. Due to segregation at the receiver, the blends were remixed after each trial so a homogeneous blend was assured for each trial.

Using the data form the pressure transmitters, the corresponding friction factors were calculated for each blend and for the neat materials. The results showed how the friction factors are affected by the composite elasticity of the blend, which in its turn is determined by the mixture ratio between the soft and the hard pellets. Additional tests (currently in progress) and analysis allowed the development of an experimental equation that defines the relation between friction factors of soft and hard materials as a function of the ratio of their elasticities.

247e New Perspectives on Fluid-Phase Turbulence Modulation by Particles

Jennifer S. Curtis, University of Florida, Chemical Engineering Department, P.O. Box 116005, Gainesville, FL 32611-6005 and Kunn Hadinoto-Ong, Division of Chemical and Biomolecular Engineering, Nanyang Technological University, 62 Nanyang Drive Blk N1.2 Level B2-31, Singapore.

This presentation will review existing 'rules of thumb' for fluid-phase turbulence modulation in the presence of a dilute concentration of particles, such as found in pneumatic conveying systems. These 'rules of thumb' relate turbulence modulation behavior to turbulent eddy length scale or particle Reynolds number. However, experimental data indicate that turbulence modulation is dependent on additional variables such as the macroscale Reynolds number. For example, it has been shown that gas-phase turbulence intensity which is attenuated with respect to the single phase flow at low Reynolds number, for a given particle loading, is then enhanced at higher Reynolds number. Hence, turbulence modulation behavior is much more complex than described by the simple rules of thumb. This presentation will review some of the new data on turbulence modulation and discuss some new possible mechanisms for the physical origin of fluid turbulence modulation behavior.

LOCAL FRICTION FORCES BETWEEN PLUG AND PIPE WALL IN DENSE PHASE PNEUMATIC CONVEYING

Jae Bum Pahk, University of Pittsburgh Pittsburgh, PA USA

George E. Klinzing, University of Pittsburgh Pittsburgh, PA USA

Summary

Dense phase pneumatic conveying is widely used to transport granular materials or powder due to its efficiencies gained on cost and abrasion and erosion. To date these systems have been designed and analyzed using average frictional representation for the conveyed plug. This study explores the effect of this friction force by measuring strains on the pipe wall when the plug is moving passed the region where strain gauges installed on the outer pipe wall. Strains on axial direction of plug flow measured from four different locations, top, bottom, front side, and backside of outer pipe wall. Dynamics strains with the sampling frequency of 10 KHz were measured by using data acquisition card with related software. With this strain, the frictional force can be determined. In this study, comparison the magnitude of the strain under different plug length, and different material transported will be presented. It was found that the magnitude of the axial direction of strain due to the frictional force was affected on the length of the plug. In addition, it was found that the magnitude of strain was different with its measuring location. The plug can now be models with realistic frictional representation of the conveyed plug. Finally, the voidage variation within the plug was calculated based on the Ergun's Equation.

Experiment

Experiment was performed in the 9.3 m long aluminum pipe. Figure 1 illustrates the schematic of the experimental set up. The entire test rig was 9.3 m long with T6061 schedule 40 aluminum pipe. The materials used for this experiment were polystyrene with mean diameter of 5 mm, cubical shape and polyolefin with mean diameter of 4.8 mm, ellipsoidal shape. The material is delivered from a cyclone, and the pneumatic valve between pipe and cyclone enables the materials go into the pipe. Compressed air is supplied to the pipe to transport materials. Two air pressure transducers were installed from 6 m and 8 m away from the pipe entrance and four strain gauges were installed at the strain measurement section. Three rollers support the pipe after the test section to make pipe as a "free end". Since the pipe can move freely to the flow direction when the plug are moving after the test section, the friction force gives tension to the strain gauge, thus strain signal can be determined.

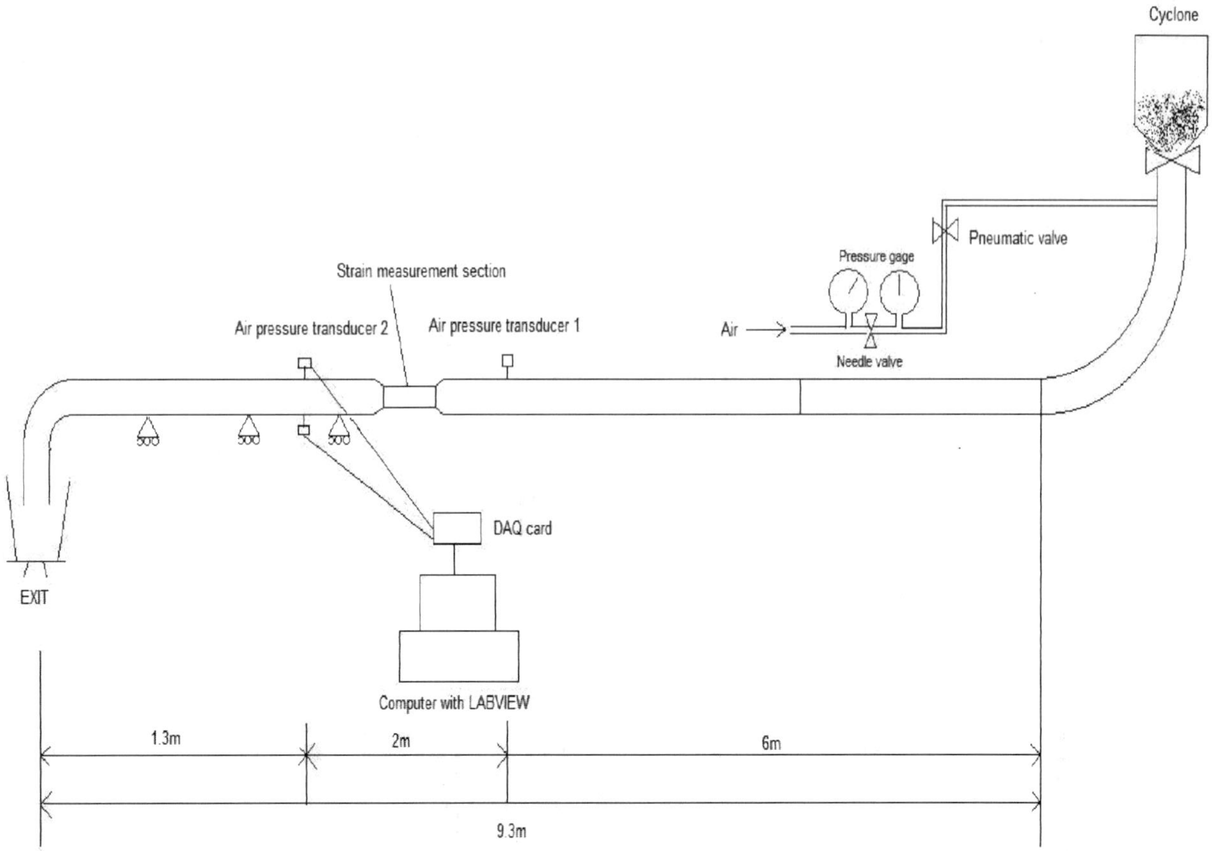

Figure 1 Schematic of test rig

The test section originally designed and assembled with an aluminum block (See Figure 2). This device was not easy to detect the pure axial strain due to the friction force only since the effect of the air pressure was too high. Because the aluminum plate was very thin, it expanded very easily under the air pressure. Next, instead of gluing the strain gauge on the thin aluminum plate, it was glued directly on the inner wall of the pipe, but due to the large and complicate cross-sectional area of the test section, the signal is complicated to be analyzed properly. In addition, there might be a possibility that the strain signal due to the friction was not only between the plug and aluminum block, but also between plug and the strain gauge itself. Thus the whole new strain gauge system design was implemented. The new test section was shown in Figure 2.

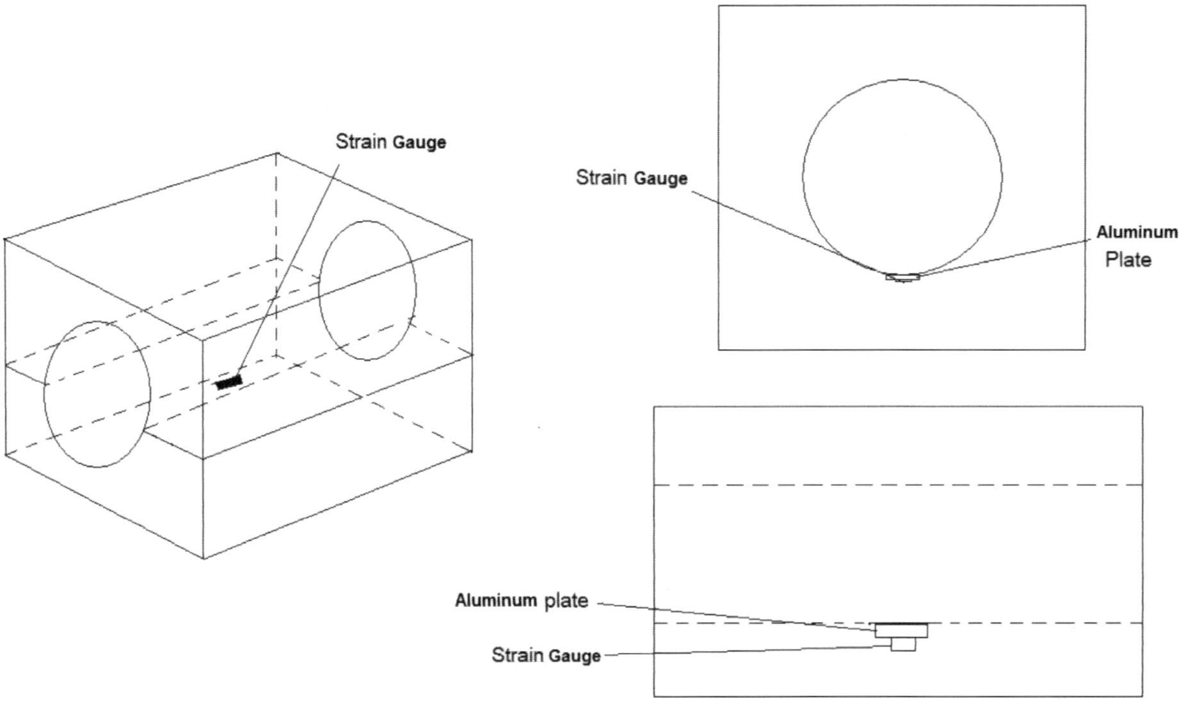

Figure 2 - Strain test section (Original design)

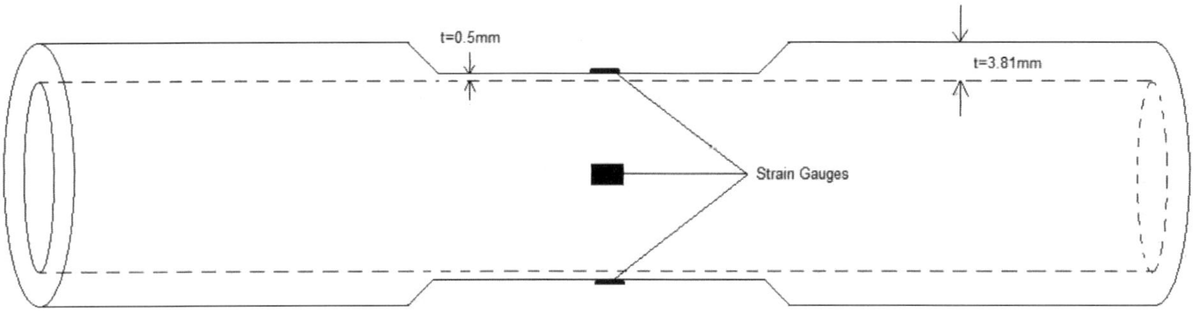

Figure 3 - New strain test section

The new test section where the strain gauges were installed has been milled to 0.5 mm thickness which was much thinner than the original pipe thickness. (3.18 mm) The thinner test section thickness provides larger magnitude strain measurement due to its smaller cross-sectional area. The signal obtained from the strain gauges were amplified for 100 times by AD621 instrumental amplifiers. For each strain gauge, a half bridge configuration was adopted for the purpose of temperature compensation and other signal noise reduction. The sampling frequency for collecting data was 10 kHz (10000 samples / second) and total data acquisition time for each case was 20 seconds. A NI6210 data acquisition cards with LABVIEW from national instrument were used to collect data. For analyzing data for strains and pressures, a MATLAB from Mathworks has used.

The strain signal in this experiment were affected both frictional force and the pressure in the pipe. To convert strain signal to the actual force due to the friction, two different calibrations were performed. The first calibration was to relate force to the strain. To do this, various known weights (4.9N to 29.43N) were hung with pulley on the pipe to generate tensile force on the pipe. Figure 4 is the result of this calibration.

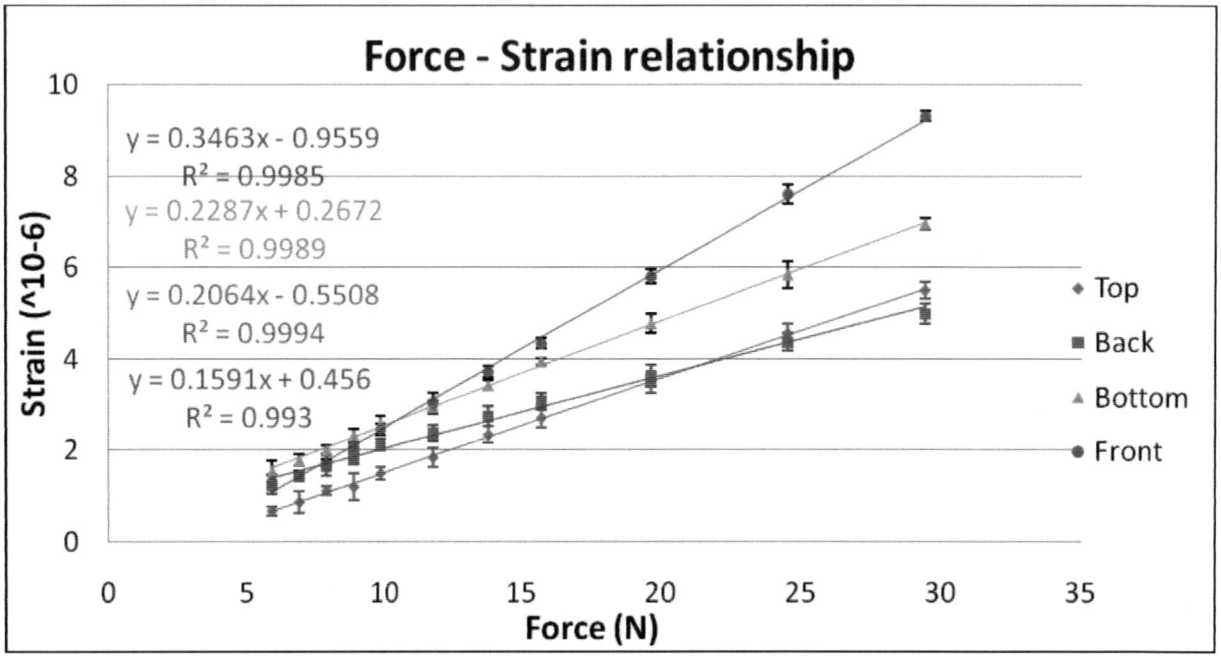

Figure 4 Strain calibration with known weights

The second calibration was to relate strain and the pressure in the pipe. To achieve this, the pipe exit was sealed so that the pipe can be pressurized. The pressure and strain data were collected while pipe was pressurized and related them. From the result on this calibration showed that the effect of the air pressure on the strain measurement was less than 10% thus this study neglect this effect.

To determine the voidage variation along the plug, Ergun's equation was used. The Ergun's equation is defined as follows

$$\frac{\Delta P}{\Delta L} = \frac{150\left(1-\varepsilon\right)^2 \eta U_{slip}}{\varepsilon^3 d_p^2} + \frac{1.75\left(1-\varepsilon\right)\rho_f U_{slip}^2}{\varepsilon^3 d_p^2} \tag{1}$$

Where $\Delta P/\Delta L$ is a pressure drop per unit length, ε is voidage, η is dynamic viscosity of the fluid (air in this study), U_{slip} is particle slip velocity, d_p is mean diameter of particle, and ρ_f is fluid (air in this study) density. Rearrange equation (1) with respect to the voidage, ε yields following third order equation

$$\frac{\Delta P}{\Delta L} d_p \varepsilon^3 - 150\eta U_{slip} \varepsilon^2 + \left(300\eta U_{slip} + 1.75\rho_f U_{slip}^2 d_p\right)\varepsilon$$

$$+\left(-150\eta U_{slip} - 1.75\rho_f U_{slip}^2 d_p\right) = 0 \qquad (2)$$

The pressure data collected from transducers varying with time. The pressure signals from two air pressure transducers enable one to determine the plug length and its velocity by knowing the distance between transducers, time to take when the plug is passing through a pressure transducer and time to take the plug from one transducer to another. After determining the velocity and length of the plug, the pressure gradient and U_{slip} can be determined and the voidage along the plug can be calculated from equation (2).

Result and discussion

1. Strain result

   Test were performed under the air velocity of 3 m/s with 600 g, 800 g, 1000 g, 1200 g of polystyrene and 400 g, 600 g, and 750 g of polyolefin as material was transported. Figure 5 shows the force variations along the plug length based on the strains measured at the four different locations. (Top, front, bottom and back of the test section)

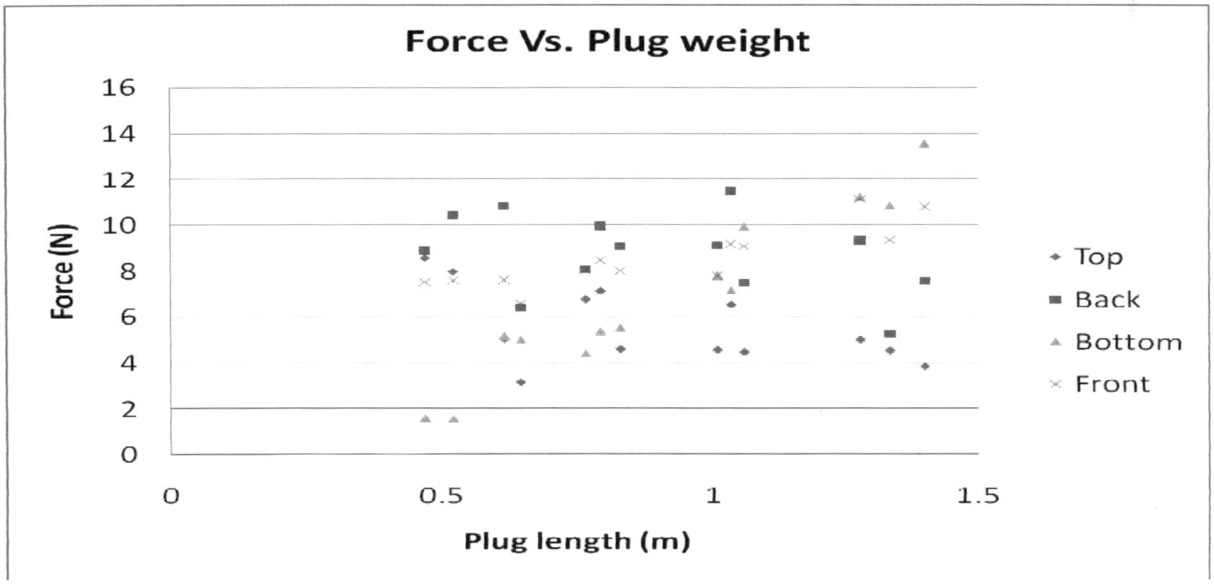

Figure 5 - Force variations with different plug weights (Polystyrene)

From Figure 5, it was observed that the force at the both sides (front and back) were usually higher than the force measured at the top and bottom. At the top of the plug, the particles can be more fluidized than any other sides thus the number of particles that contact the pipe wall would be smaller than other sides yields smaller friction force generated. At the bottom of the pipe, the friction force has a linear relationship to the plug length. Because at this region, increasing plug amount made the particles pack well, thus more gravitational force exists on the bottom. This effect may increase the frictional force. The average frictional force, that is the average of the frictional force applied over the whole cross-sections of the pipe increased when the plug length increased. See Figure 6.

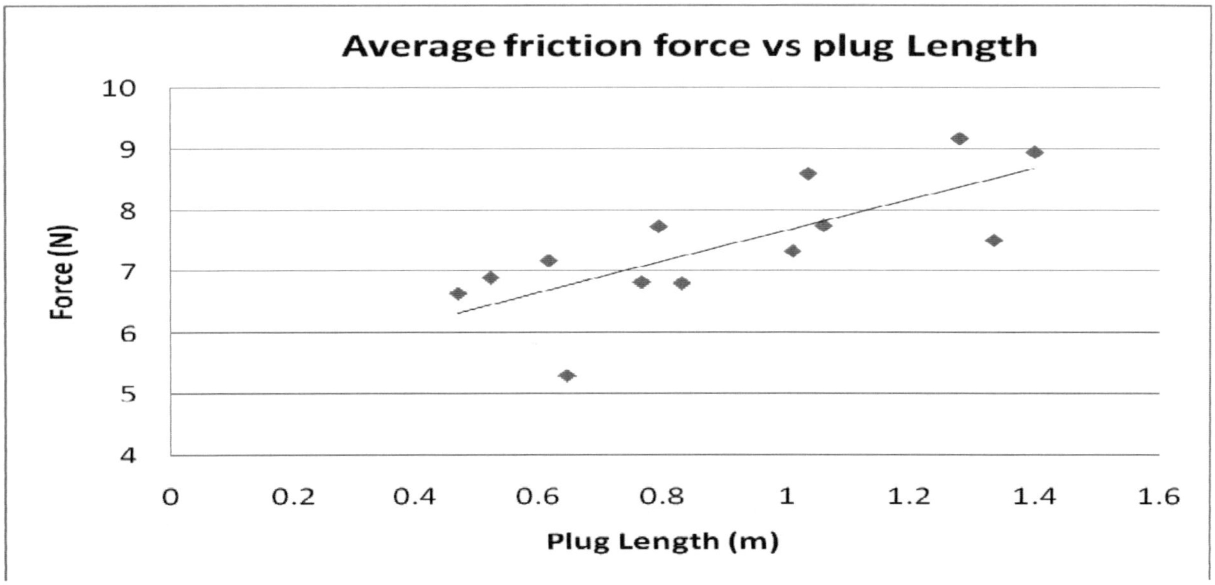

Figure 6 - Average friction force vs plug length, polystyrene.

For polyolefin material, with large amount of plug weight (750 g), the plug length was too long to determine the strain thus it was not able to determine the frictional force. This problem can be solved with longer free end pipe line installed, but this time it will be remained as a future work. Also, the force determined for this particle is very tiny (maximum force measured was 4 Newton) and the number of measurement was small, thus more experiment needed to verify the result. But the behavior of the frictional force with respect to the plug length in each side of the inner wall was similar to the polystyrene.

The limitations of the experiments due to the current system was as follows;

- Air velocity : 3 to 4.8 m/s, if the air velocity is slower than 3 m/s, the plug cannot be conveyed, if it is faster than 4.8 m/s, it is hard to form a plug.
- Plug length : up to 1.8 m, if the plug length is longer, the plug will reach the exit of the system before measure the strain.

Voidage along the plug

The voidage variation along the plug length gave an interesting result. For the polystyrene pellets, in general, higher voidage values were found at the plug front and tail sections and lower at the middle section of the plug. But for the longer plug length, the voidage value was dispersed more than the shorter plug. In addition, the smaller voidage values were fouind for the longer plug. See Figure 7.

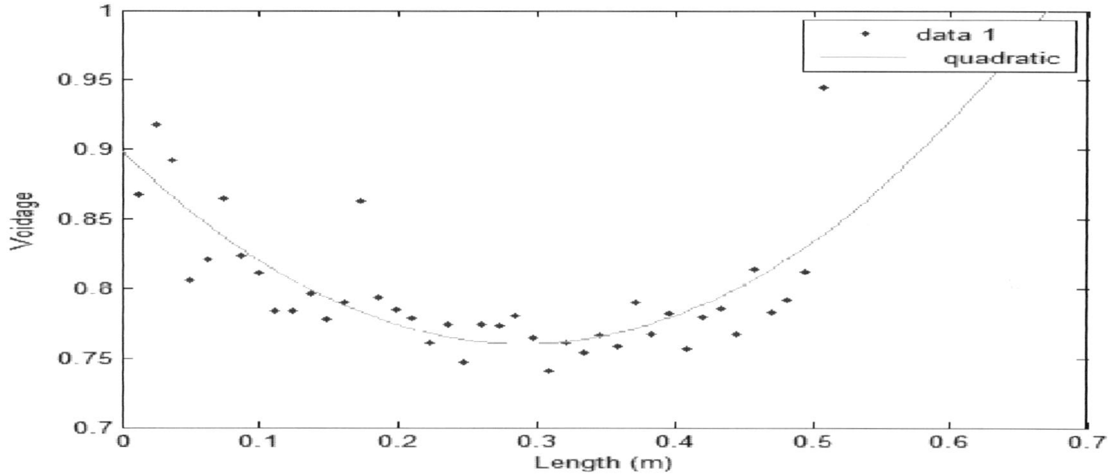

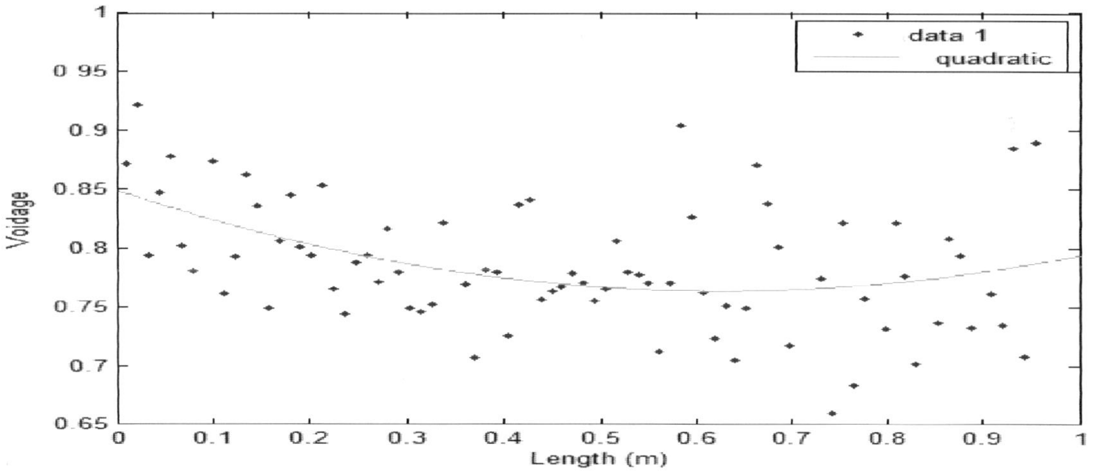

Figure 7 Voidage variation along a single plug, polystyrene.
Plug length = 0.52m (top), 0.97m (bottom)

For polyolefin pellets (Figure 8), in general higher values of the voidage were observed than polystyrene pellets were. No significant voidage difference has been observed with different plug length. In addition, the voidage didnot changed much within a single plug. The particles were found to be well fluidized in the plug for this material.

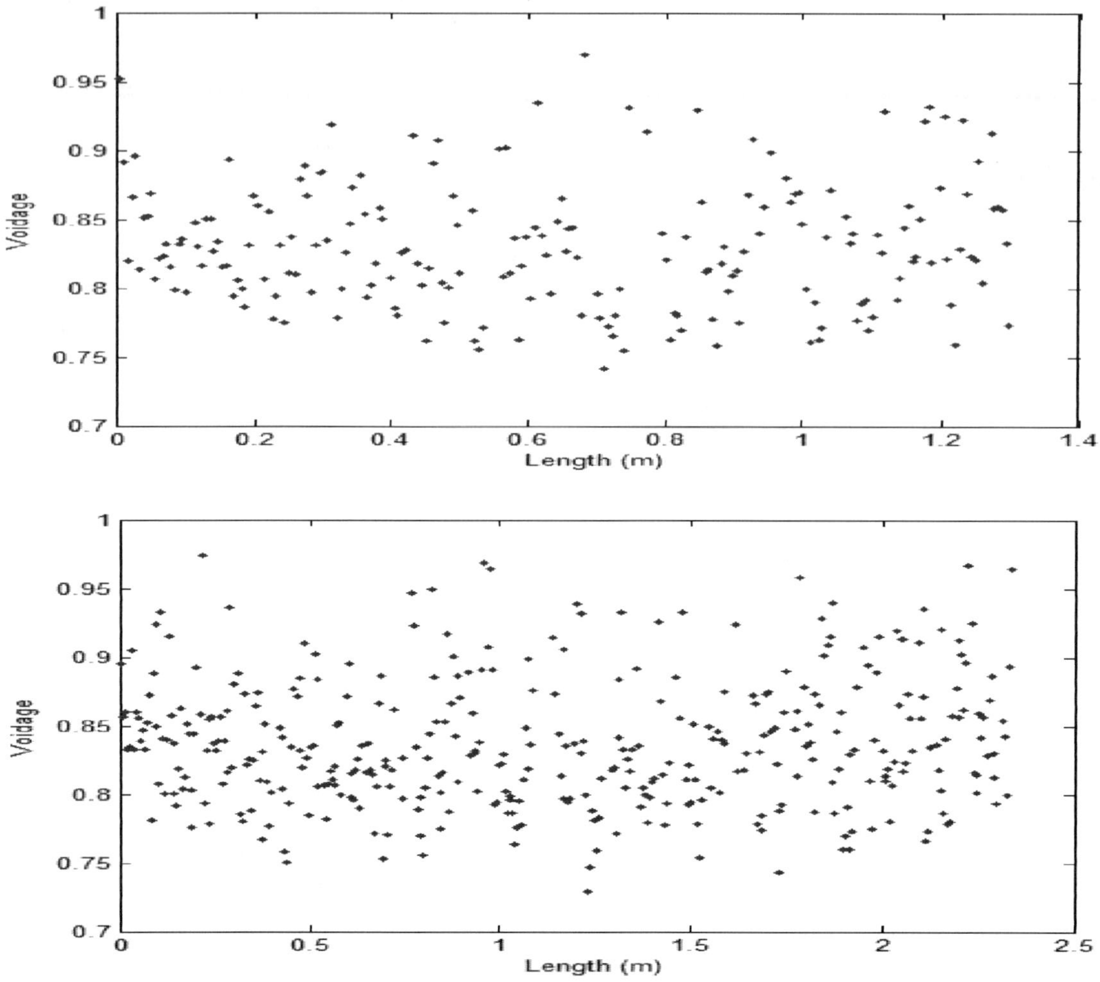

Figure 8 - Voidage variation along a single plug, polyolefin.
Plug length = 1.32m (Top), 2.35m (Bottom)

Conclusions

During this study so far several conclusions can be drawn:.

- The friction force between plug and pipe wall vaires over the cross sectional area of the pipe.
- The friction force at the bottom of the pipe wall has a linear relationship with a plug length.
- The average friction force has a linear relationship with a plug length.
- The voidage determination whithin the single plug was found that the voidage at the plug front and tail was higher than the middle section of the plug for polystyrene, but no significant change was observed for the polyolefin material.

- With the same amount weight of the conveying material, polyolefin plug were much longer than ploystyrene and higher voidage values were observed on this material.

Future work

To develop this study, some suggestions are proposed for future work

- Increasing the free end length will help to investigage frictional force for longer plug.
- Additional data with varying conditions is essential.
- Variation in thetransport air velocity will provide a broader scope..
- Explore different materials and particle sizes of the transported material, such as powders.
- Investigation multi-plug conveying to more realistically address the industrial situation.

Transformative Research in Pneumatic Conveying

George E. Klinzing
University of Pittsburgh

Abstract

This paper address the need to be more risky and exploratory in our research endeavors when studying pneumatic conveying. The term transformative research is the strategy that has been adopted by many of the funding agencies. Significant resources have been spent to provide new research but mostly for projects that provide incremental advancement in our knowledge. This analysis reviews what knowledge that can be considered as transformative advances in the field of pneumatic conveying over the past sixty + years and projects what avenues and areas we can explore to make great future advancements.

1. Transformative Research

In order to advance and improve the operations of any solids processing plant or a particular piece of solid handling equipment knowing the basic factors and physics that contribute to the unit's operation is required. The more involved the process and the equipment the less one seems to know how the unit will operate when material types and sizes vary. As always, knowing the operating characteristics and performance of the materials before the operation is implemented and constructed is most desirable and essential. What is needed to make a significant impact on process and product improvement is to employ research that yields transformative results. Knowing the basics is one step towards this objective. Transformative results are results that give breakthroughs in the field that affect the entire system not just providing small incremental changes. This may sound like a very high risk endeavor and the results may lead to difficulty in obtaining adoption in normal plant operations. The research community is being encouraged to take this risk with transformative research and make the leap. It is understood that not all such transformative research explorations will be successful and there will be failures along the way.

In the U.S. Federal government and other funding agencies transformative research is now being stressed. Incremental research that yields small advances has been assailed by the funding community as not being acceptable. This is probably driven by reviewing the large amount of resources that have been expended to make small advances in science and engineering and the desire to drive the agenda faster and more effectively. The research community has learned how to write proposals that are fundable by suggesting small changes but which are not generally transformative in nature. The researchers please the reviewers by not being too innovative and not too exploratory. It has been a real danger to be transformative in a proposal because reviewers will not write favorable reviews saying that these ideas of too high risk. The Science Board Chair of the National Science Foundation, Steve Bering, who is the former President of Purdue University has been driving the National Science Foundation agenda with the transformative message. The National Institutes of Health are also giving the same message to the research community on the kind of transformative research they want and will fund.

What is proposed here is to review developments in the area of pneumatic conveying and assess that may be considered as transformative research that has taken place in the past to help advance new research in the field. The assessment of these advancements is one in which the author takes full responsibility for choosing the transformative topics and wishes to apologize if other developments which the reader may think as transformative are not considered. It is hoped by reviewing the past we can project into the future with an agenda that will permit us to employ transformative research in an effective manner.

2. Transformative Research Concepts from the Past

- Zenz Diagram – One of the very first people to represent the whole of the pneumatic conveying scope was suggested by Zenz (1962) in his proposing the state diagrams that now often bear his name. The pressure loss per unit length is found to vary with the transport gas velocity at constant solids flow rates. Figure 1 is very informative in providing an understanding of the operational limits of pneumatic conveying. These plots give the designer some insight on where to operate in order not to have an unsteady operation or to have a line plug in their operations. The inherent instability of denser phase flows can be shown to exist at crucial points in these diagrams. It is noteworthy that Zenz performed many of his first experiments in very small equipment located in his garage and using a vacuum cleaner as the air source.

- Barth Analysis - Barth (1954) was also one of the pioneers of pneumatic conveying who reduced the analysis of complex flows into basic physics trying to ascertain the parameters needed in order to develop a predictive model for design. He focused on additional pressure drop due to the presence of solids in addition the standard contributions from the gas flow friction and the gravitational forces. This additional pressure drop is written as

$$\Delta p_z = \mu \lambda_z \frac{\rho}{2} v^2 \frac{\Delta L}{D}$$

Over the years the evasive term (λ_z) has been investigated by many researchers. For the dilute phase operations we have an adequate representation for this value.

- Saltation Analysis – Rizk (1973) working with Barth in his first foray into pneumatic conveying understood that a predictive tool was needed to establish when particles would salt out of the flow. His basic, simple experiments proved to be easily applied to the industrial scene. Further refinement have taken place but Rizk was the first to give us a tool. This work was one of the first that Rizk employed in the career in pneumatic conveying studies and design. The basic salutation equation is given as

$$\mu = \frac{1}{10^{\delta}}\left(\frac{v_s}{\sqrt{gD}}\right)^{\chi}$$

$\delta = 1.44d + 1.96$

$\chi = 1.1d + 2.5$

- Konrad Dense Phase Analysis – In his Ph.D. dissertation Konrad opened the door for a basic analysis of dense phase flow incorporating much of the field of solid mechanics to these conditions. Using the moving packed bed analogy he provided a basic understanding of dense phase conveying. This blend of fluid and solid mechanics was a step forward in attempting to required that one had to test each and every dense phase system in order to have systems that would function. Konrad (1984) incorporated some of the basic findings of Jenike in his analysis. One can note that the common terms developed by Jenike for bin and hopper are part of the models.

- Jenike Shear Cell – As mentioned above Konrad incorporated the finding of Jenike (1964) in his dense phase model development. This provided significant advancement to the understanding of dense phase conveying and is employed extensively today. The Jenike Shear Cell (2008) has been improved continually and a photo of a recent unit offered by the firm, Jenike and Johanson is seen in Figure 2.

- Muschelknautz Dense Phase Analysis – Muschelknautz (1969) was one of the first who addressed dense phase conveying and carried out considerable testing and analysis in this field. His models tend to follow an overall system approach which has had some success in applying to industrial situations. The model he and Krambrock suggested is given as

$$P_1 = P_2 e^{\frac{g \cdot \mu \cdot \beta_o \cdot L}{R \cdot T \cdot c / v}}$$

One noted the term (β_0) term as the product friction coefficient unique to the product conveyed.

As mentioned, Muschelknatuz was one of the pioneers in pneumatic conveying and his thinking about the basic frictional term in the overall analysis and modeling of pneumatic conveying led him to project that the gravitational forces had a significant effect on the basic friction. He was able to be first in carrying out pneumatic conveying in outer space aboard the space shuttle in order to remove the gravitation forces from the basic expression. This was a very ambitious project and while the overall results never did not completely materialize the studies and data obtained pointed to the direction to follow.

- Geldart Diagram – Another development that can be considered transformative that developed in another field of solids processing is the Geldart Diagram conceived with the idea of understanding how particles with different characteristics behaved in the fluidization operation. I believe that Geldart (1973) while working dominating in fluidization had an impact on the whole solids processing field including pneumatic conveying. Through his classification of the various type of solid powder and their flowability conditions and results in pneumatic conveying can often be interpreted. Figure 3 presented here was recently proposed by Yang (2007) in his attempt to expand the analysis to high temperature and high pressure fluidization operations.

- Tsuji Simulation of Dense Phase Flow - Tsuji began his efforts in pneumatic conveying under the tutelage of Morikawa with very classic experimental findings especially in the use of hot wire anemometry. This data has set the standard for many modeling and simulation analyses that followed. Tsuji (1992) branched out into the simulation and modeling arena and began to carry out very basic simulations of complex pneumatic conveying conditions. When his result started to be similar to those found in the laboratory, attention was drawn to him and the field which holds such promise. This is a transformative result. One notes in Figure 4 the similarity to actual experiments.

  Following the modeling and simulation approach Sinclair (1995) has employed detailed fluid and solid mechanics behaviors to provide amazing predictions of actual flow situations.

- Large Scale Systems – Marcus et al. (1986) and Wypych (1999) both took the realm of pneumatic conveying research from what some industrial practitioners call tools to reality by implementing large scale testing with industrial size pipeline both in diameter and length. Industry had to listen to these researchers and could not just pass off University work as cute and that is a nice toy. These studies have been transformative in providing more respect for University scale research because they did show that the smaller scales could provide interesting and useful information for industry. Understanding the scaling parameters can be credited to their transformative work.

- Pressure Fluctuations – Solt was the first to my knowledge that suggested the idea that the noise generated by pneumatic conveying could be used in diagnostics of the problems which such systems experienced. Dhodapkar and Klinzing (1993) took this idea and applied signal analysis of the pressure fluctuation to such flows in order to predict the flow regimes and operations. This type of analysis could be done on-line in the plant. The results point to very distinct signals of the pressure fluctuations depending if the flow is homogenous in nature of is a moving wavy bed. Figure 5 shows one such signal for stratified flow.

  Utilizing pressure measures and other noise analysis techniques has been pioneering by Davies (1987) applying many of his ideas and invention to actual industrial usage.

- Wavelet Analysis – Building on the foundations of signal analysis in electrical engineering Pahk and Klinzing (2008) tried to analyze the signal of the pressure fluctuation in order to what actually made up these fluctuations and could can we indeed read this signal in a more intelligent manner. Detailed analysis using the wavelet approach indicates that the various contributions of the pressure fluctuations can be filtered out of the signal to provide a more detailed analysis of the system. Figure 6 is one result of this analysis which distinctly picks out the blower, feeder, turbulence and solids interactions terms.

- Particle Interactions – In a fundamental analysis of the friction generated in pneumatic conveying it naturally comes down to what the particles are doing between themselves and in their impacts with the wall of the pipe. These questions were first addressed by Louge et al. (1989) who looked at impacts of large particles to gain some insight into how to model and quantify these interactions. Borzone and Klinzing (1990) have shown that the condition of the wall surface can intimately affect the particle impact behavior. One can note in Figure 7 below the unique impact and reflection angles can occur when the surface roughness is varied. In more recent work Vasquez et al. (2008) showed the unusual behavior of flexible particles and their interactions with the wall and themselves. These particles produce considerable rotation as well as unique interactions with the wall even bouncing backward against the main flow stream. The cartoon of this behavior is seen in Figure 8. Are these particle behaviors transformative? I believe the observations challenge us to think of new ideas and models and present to us opportunities to explore the basics more carefully.

- Electrostatics – Richardson and McLeman (1960) explored pneumatic conveying with particles that produced considerable electrostatic forces that presented themselves as an increased in the pressure drop in a recirculation system. See Figure 9. This unique observation focused the community at exploring the cause of such behaviors. Klinzing probed electrostatics first in trying to eliminate the effect but the fascination of the phenomenon led to several experiments trying to establish its amount, direction and conditions for occurrence. Use of the phenomenon in a positive manner was also explored in the metering of flow. Figures 10 shows how the electrostatic effect can give completely different flow behaviors and patterns as such causing considerably more energy to be expended in conveying as well as being dangerous in operation.

- Pickup – Following on the original work of Rizk on saltation Cabrejos and Klinzing (1994) worked in order to determine if there was a broader scope of pick-up and saltation expressions that could be used for a wide cadre of materials. Some interesting results emerged in these studies to show a hysteresis seen between the two phenomena and how difficult determining the exact values for fine particle behaviors could be. Kalman (2008) and his students have explored even further into these phenomena and have developed new generalized correlations for us all to use. Figure 11 shows the Cabrejos experimental findings.

- Tomography - In order to understand the details of the internal structure of pneumatic conveying Dyakowski, Williams and their colleagues employed tomography to obtain a three dimensional view of pneumatic transport including dense phase conveying. They were able to show that the internal structure of this type of flow is complex and unsteady pointing to the reasons correlations are difficult to employ and duplicate. Dyakowski and Williams in 1993 published a paper in Powder Technology which was the beginning of the use of tomography to explore the details of pneumatic conveying with the measurement of the particle velocity distributions in a vertical channel. This technique has blossomed under their direction and attracted other researcher to the field to explore the details of the flows often showing the non-homogeniety of the flow structure. This work has been transformative to the field. See Figure 12.

- Ice, Space and Chickens – Transformative or Just Different

I believe that the next three examples are most unusual and point to the creativity of humans to apply principles completely outside the normal applications and lead to new technologies. In my estimation this type of thinking is transformative. Sheer (1995) met a challenge which came from the need to cool the air in deep mining operation which led to the suggestion to conveying ice from ground level down about 1 kilometer to the mining where the latent heat of melting would be released and cool the air for workman to function in the deep mines. In this application success was achieved by using plastic piping in the operation since metal piping formed a strong bond between the ice and the metal so that plugs of ice would block the pipeline. This points to the strong dependence of the inner piping surfaces to operations including friction. Figure 13 shows the ice plug in the application.

NASA has had a keen interest in moon explorations and as such has carried out a number of tests already on how particles would behave under reduced gravity conditions. The use of a KC135 jet airplane to simulated reduced gravity during the diving operations has been employed to assess the effect of reduced gravity of saltation a pickup and well as fluidization and pressure drop in pneumatic conveying. NASA also has the idea of mining the moon in order to generate oxygen to sustain life. One such process is shown in Figure 14. Actual NASA data on choking under reduced gravity conditions is seen in Figure 15.

It is not uncommon to see pneumatic conveying applied in the food industry. Handling these material pneumatically provides many advantages to the over handling problems of bulk food moving. One lesser known application in the food related industry involved the conveying of live animals as seen in Figure 16. Again, are these transformation developments?

3. Conclusions and a Proposed Future Strategy

One can see that there have been a number of developments over the years that could be called transformative in the field of pneumatic conveying. Sometimes these have occurred within the community itself and sometimes information and discoveries in other fields have enriched the field and brought welcome information and change.

I believe the call is to take a chance and be little riskier is our future approaches.

Some of the topics that appear to be ripe for consideration are:

- Friction – The friction experience between the solids and the wall of the conveying enclosure along with the inter-particle behaviors and the gas turbulence and solid interaction all lead to an energy loss seen in pneumatic conveying. There have been attempts to describe each of these forces and losses (Molerus, 1996) but little success in quantifying these losses in a most general way as to applicable to all conveyed particles . This all point to a more thorough understanding of the basic physics of the process. Because pneumatic conveying operations occur in turbulent flow regimes we are even more challenged to probe turbulence and all its nonlinear behaviors in the presence of particles. We have resorted to many correlations for specific materials and tried to generalize with limited success. Trying to isolate the contributions are being explored and hopefully with new experimental methods and new analyses borrowed from other branches of science and engineering friction will be less of a mystery to us.

- Other Forces – As mentioned, friction does involve several different components that can be broken down into linear forces but also one must anticipate nonlinearities in these considerations. The forces that contribute to the overall energy loss must be considered in trying to understand and analyze the pneumatic conveying. There are the particle-particle, particle wall impacts, rotational spins, turbulence contributions both amplifying and damping components and of course the ever present electrostatics which is often termed "black magic." Our models must determine if these forces have a significant contribution to the overall models both analytical and numerical and fortunately in engineering analysis we can often times reduce or eliminate the significance of some of these forces in our analyses.

- Simulation and Modeling - There has been particularly impressive progress that has taken place in the simulation and modeling arena. It just seems like a few years ago when this approach was being conceived several people would say under their breath, "This will never work." These same people are changing their minds as the field has become more sophisticated and comprehensive in the approaches. We will see more success in this field in the future when we can insert the correct physics into the models

- Industry Perspective – Industry has taken an approach to research in pneumatic transport that is a blend of incremental advances and transformative research. The transformative part has come about when industry must act because of a serious upset to a process that can be pointed at pneumatic conveying. Chances can be taken because if action is not taken even more serious financial losses will occur.

- Films Records – With the advances in recording the physics processes of particle flow with cine and fast videos we have at our disposal the ability to probe the mechanism and slow down their actions that we can address a more realistic model. There are fast video cameras and other photographic techniques available to us at prices that we can afford that hold the potential to unlock some basic observations and information. Use of tracers in the flow can permit possible quantification of particle behaviors.

- Data in the Files – In conversations with various industrial practitioners of pneumatic conveying the topic of having carried out tests before and not really having a chance to analyze these data with newer techniques and concepts often emerges. There is a feeling that there is a wealth of information still sitting in filing cabinets and if we could only go through this data systematically we could provide new incite to existing problems. While in concept this may sound easy, implementing this process is a challenge but I do believe it would be a worthwhile adventure.

4. References

Barth, W., Chem. Ing. Tech., (1954); 20, No. 1, 29-32

Borzone L.A., Klinzing, G.E., Yang, W.-C., Powder Technology (1990); 62: 277-290

Cabrejos, F.J., and Klinzing, G.E., Powder Technology 79 (1994) 173-86

Davies, C.S. and Spedding, N.B., Powder Technology 53 (1987) 131-136

Dhodapkar, S.V. and Klinzing, G.E., Powder Technology 74 (1993) 179-195

Geldart, D., Powder Technology 7 (1973), 285-90

Jenike, A.W., Bulletin 123, (1964), University of Utah, USA

Jenike and Johanson, http://www.jenike.com/ (2008)

Kalman, H., and Rabinovich, H. Powder Technology, 183, (2008), pp. 304-313

Klinzing, G.E., Rizk, F., Marcus, R.D. and Leung, L.S., "Pneumatic Conveying of Solids", Second Edition (1996) Chapman & Hall

Louge, M.Y., Chang, H., Subramanyam, A., Giannelis, E.P., AIChE J (1989) 35, 473-. 1486

Konrad, K. and Davidson, J.F., Powder Technology (1984) 39, 191-8

Marcus, R.D., Rizk, F., and Chambers, J., Second Intl. Conference on Bulk Materials Storage, Handling and Transportation, (1986) 10-14

Molerus, O., Powder Technology, 88 (1996), p. 309–321

Muschelknautz, E. and Krambrock, W., Chem. Ing. Technol. (1969) 41, 1164-72

Pahk, J.B. and Klinzing, G.E., Particulate Science and Technology, 26, 1-10 (2008)

Richardson, J.F. and McLeman, N. (1960) Trans. Inst. Chem. Engrs., 38, 257-66

Rizk, F.,Dissertation, (1973) Universität Karlsruhe

Sheer, T. J., Powder Technology (1995) 85, 203-219

Sullivan. T.A., Koenig, E., Knudsen, C.W., Gibson, M.A., Journal of AIAA (1992) p.1667

Tsuji, Y., T. Tanaka, and T. Ishida, Powder Technology. (1992) 71, 239

Vasquez, N, Jacob, K., Cocco, R., Dhodapkar, S.V., Klinzing, G.E., Powder Technology (2008), 179, 170-175

Wypych P.W., Powder Technology, (1999) 104, pp. 278-286

Yang, W.-C., Powder Technology, (2007) 17, 69-74

Yasuna, J.A., Moyer, H.R., Elliott, S., Sinclair, J.L., Powder Technology, 84, (1995), 23-34

Zenz, F.A., Othmer, D.F., "Fluidization and Fluid Particle Systems," (1962) Reinhold, New York

5. Nomenclature

ΔP : *Pressure difference, (Pa)*
L : *Pipe length, (m)*
H : *Pipe height, (m)*
F : *The stress on the front end of a plug*
D : *Pipe Diameter, (m)*
ρ_B : *Bulk density*

$$\rho_B = \rho_S(1-\varepsilon)$$

ρ_S : *Solid Particle density*
ε : *Voidage, (-)*
g : *Gravitational force, (9.8 m/s²)*
P_2 : Downstream Pressure, (Pa)
P_1 : Upstream Pressure, (Pa)
g : Gravitational force, (9.8 m/s^2)
c/v: Velocity Ratio (Particle/gas), (Between 0-1.3)
μ : Mass flow ratio, (Solid mass flow rate (m_s)/Gas mass flow rate of pipe (m_g)) (Between 50-300)

$$\mu = \frac{m_s}{m_g}$$

T : Temperature

R : Gas Constant

β_D : The value of β_D Weber derived from experiments. Values of β_D=0.6 can be assumed for horizontal dense phase conveying. [16]

$$\beta_D = \sin\delta + \mu_R \cos\delta$$

where:

δ : Angle of inclination

μ_R : Coefficient of sliding friction

β_o : The product frictional coefficient β_o proposed by Muschelknautz and Krambrock can be determined by the previous conveying test from the expression:

Recommended estimating β_o:

β_o = 3 to 5 for plastic pellets

β_o = 4 to 6 for plastic powders at terminal velocities of about 6 to 10 m/s.

Figure 1

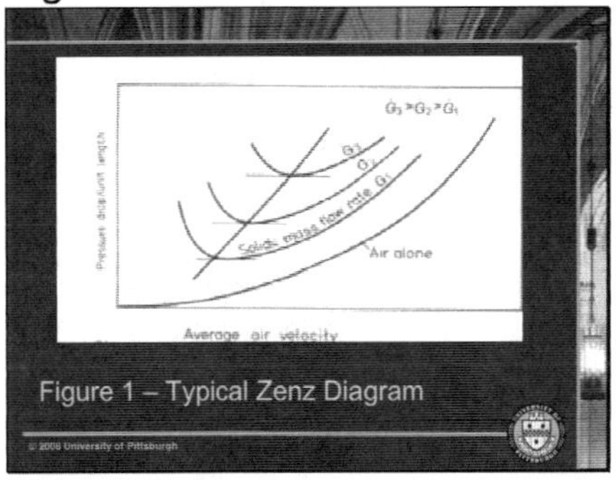

Figure 2

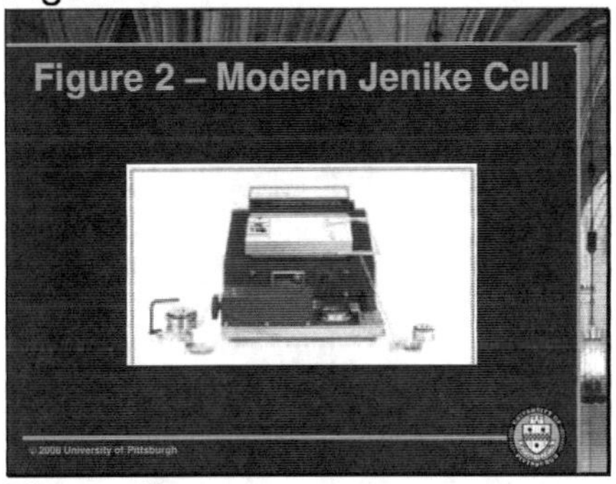

Figure 3

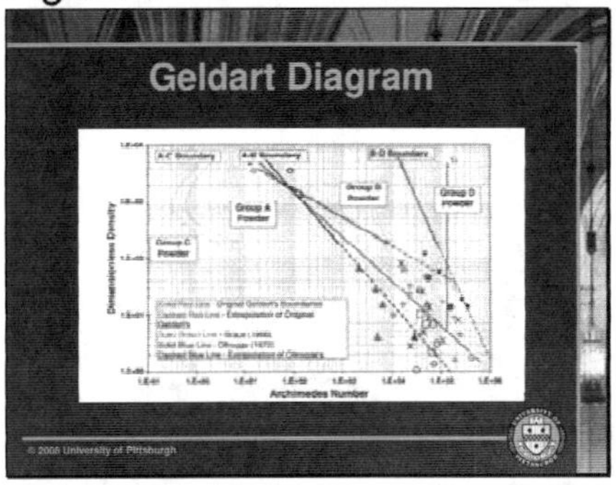

Figure 4

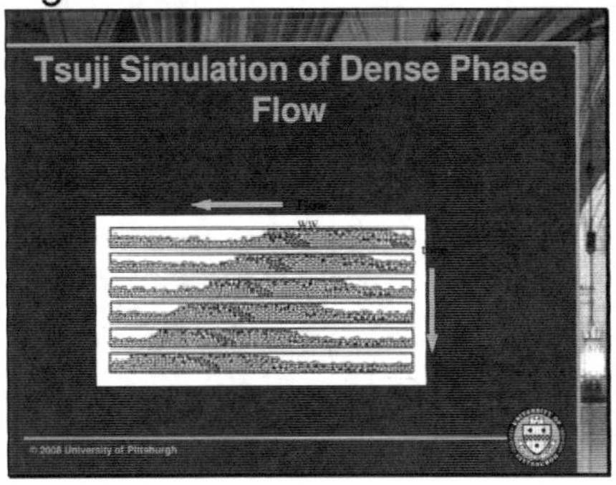

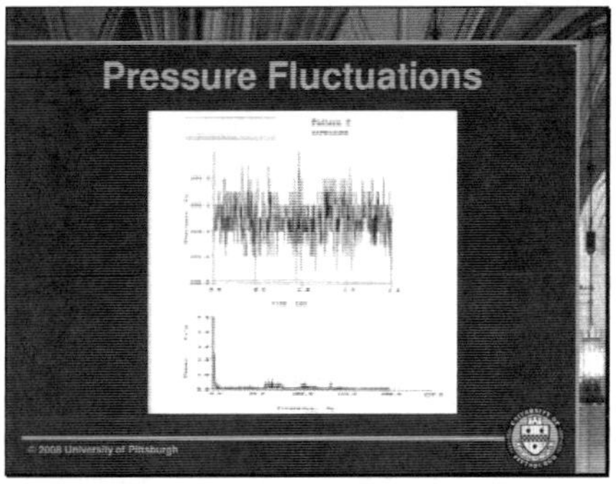

Figure 5

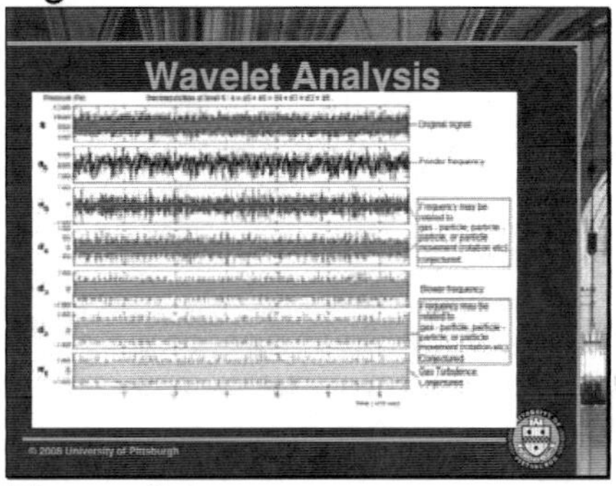

Figure 6-a

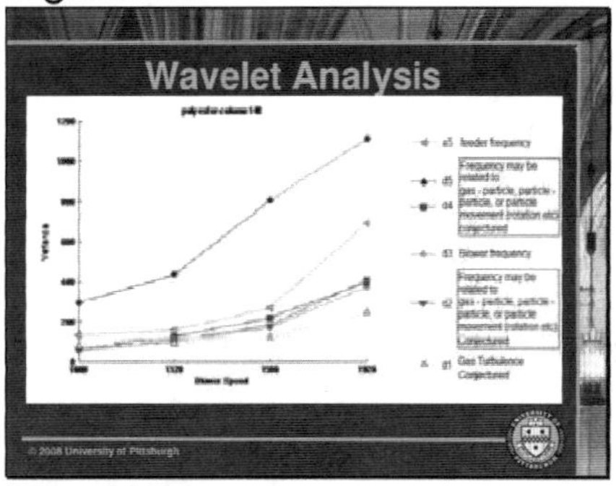

Figure 6-b

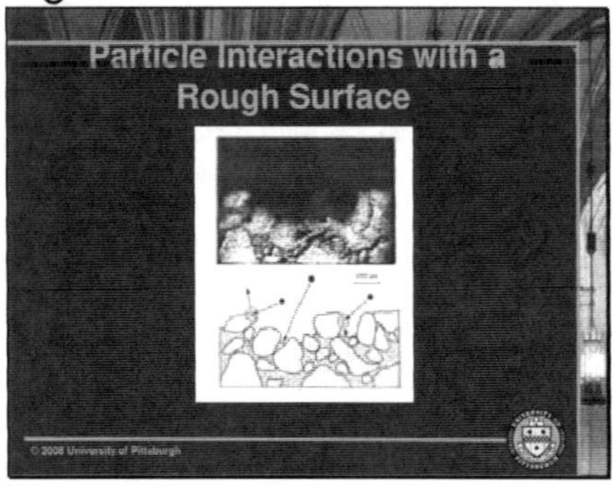

Figure 7-a

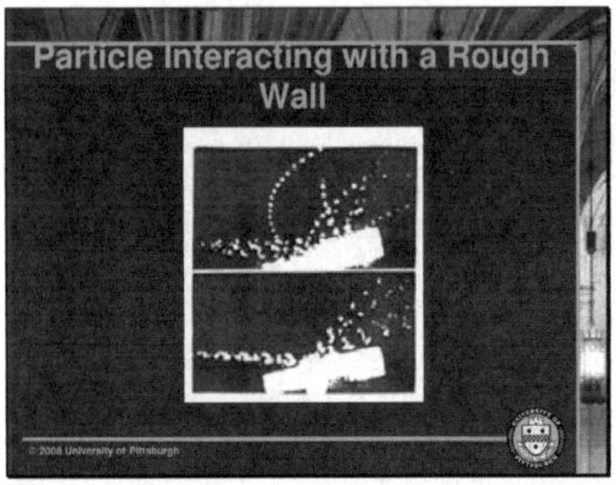

Figure 7-b

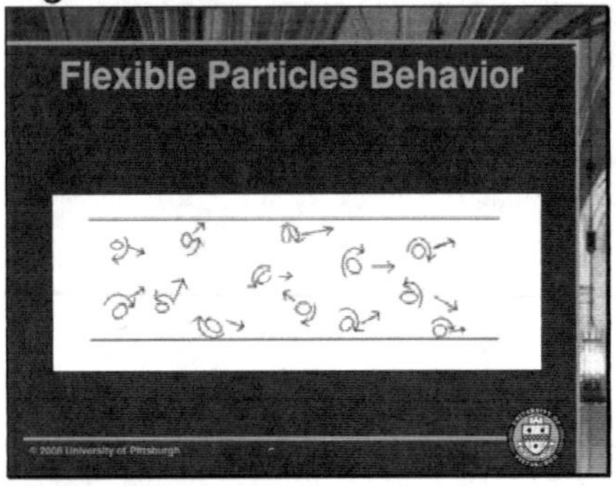

Figure 8

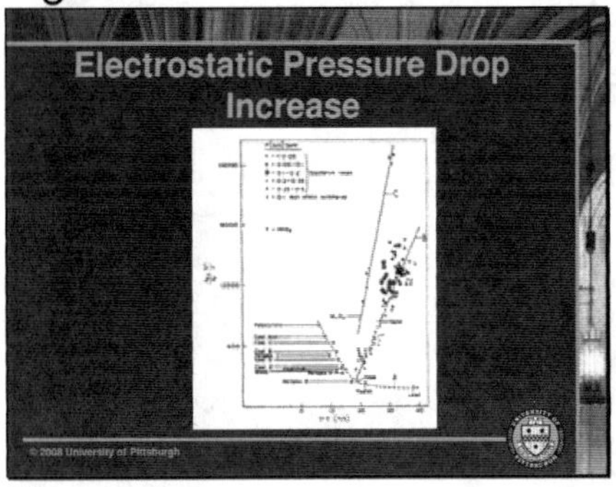

Figure 9

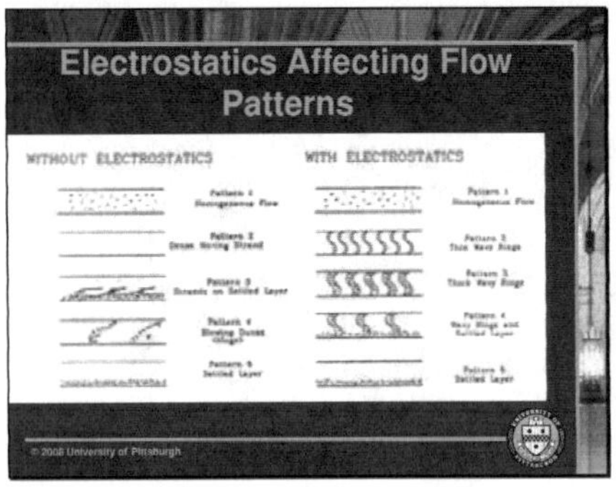

Figure 10

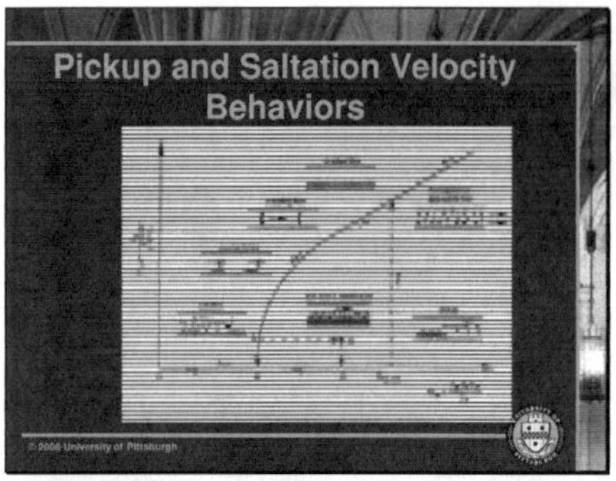

Figure 11

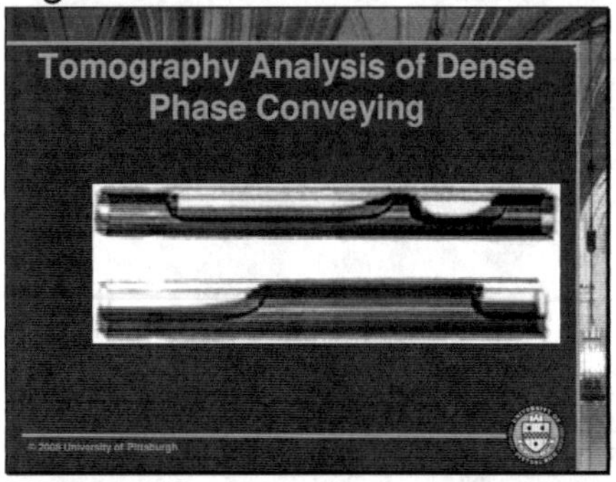

Figure 12

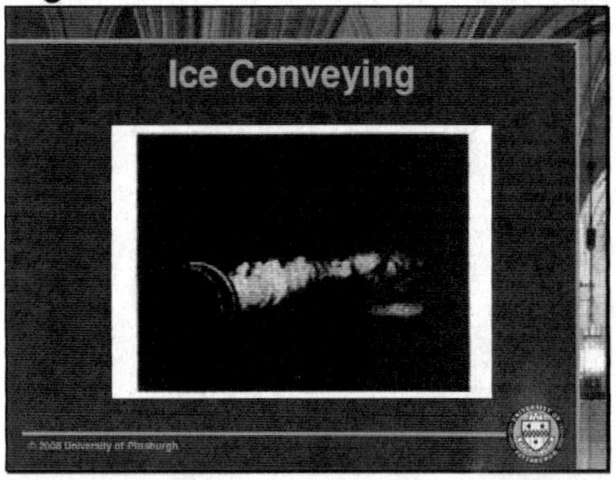

Figure 13

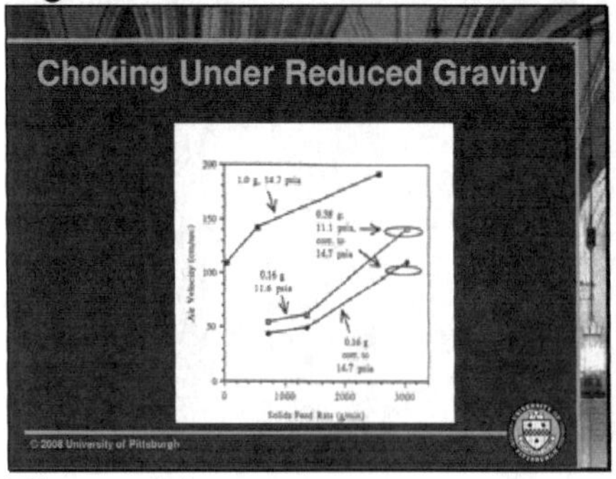

Figure 14

Figure 15

268a Development of Ion-Selective Thermo-Reversible Polymers

Sjoerd F. G. M. van Nispen[1], Leo J.P. van den Broeke[1], Johan P. A. Custers[1], and J.T.F. Keurentjes[2]. (1) Chemical Engineering and Chemistry, Eindhoven University of Technology, PO Box 513, STW0.34, 5600 MB, Eindhoven, Netherlands, (2) Department of Chemical Engineering and Chemistry, Eindhoven University of Technology, P.O.Box 513, Eindhoven, 5600MB, Netherlands

Hardness ions like calcium and barium cause problems in many processes by forming unwanted salt depositions. For different applications, ranging from household applications like laundry and dish washing to drinking water processes, softening of water streams is essentially. The current technology to bind multivalent cations commonly is based on the use of ion exchange resins. Regeneration of these resins typically involves consecutive washing steps with acid or caustic, leading to a large amount of waste salts.

This research project focuses on the development of a new softening technology where regeneration of hardness ions occurs by a change in temperature. In this project Pluronics, triblock copolymer surfactants of PEO-PPO-PEO, are use. The main feature of Pluronic surfactants is the ability to show temperature dependent aggregation behavior.

Pluronic surfactants are available with different critical micellization temperatures (CMTs). In this project Pluronic surfactant P-85 is used. To allow for the reversible binding of multivalent cations, P-85 has been modified to yield a triblock copolymer with carboxylic acid end standing groups (CAE-85) [1] (Fig. 1).

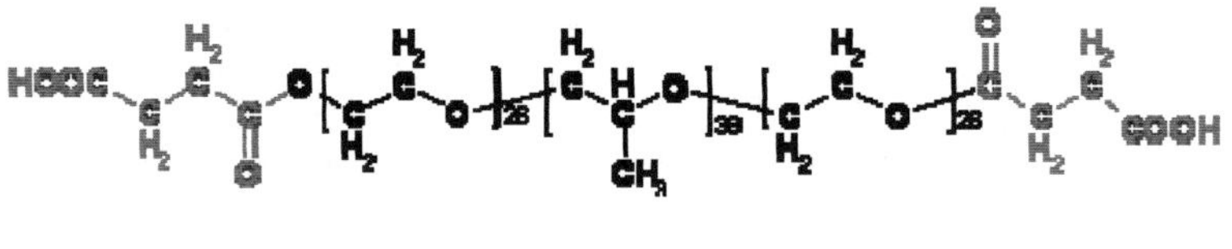

Fig. 1: Chemical structure of CAE-85 surfactant

At a temperature below the CMT (below 30°C), the CAE-85 surfactant and the multivalent cations are present in solution as free unimers. The affinity of isolated carboxylic acid groups for multivalent cations is very low. Increasing the temperature above the CMT brings the charged groups together, which enables the micelle to bind multivalent ions (Fig. 2).

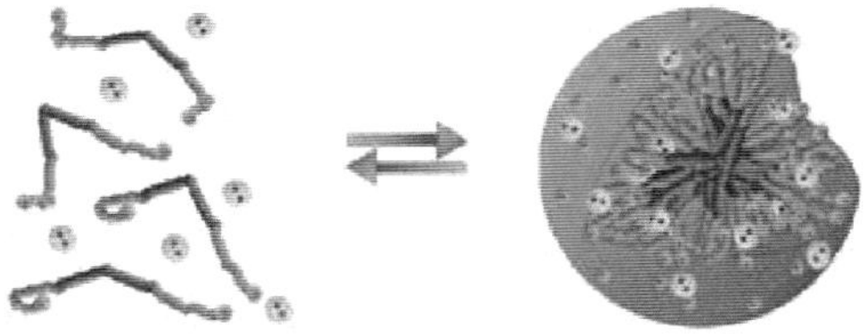

Fig. 2: Mechanism of thermo-reversible cation binding for CAE surfactants

Isothermal titration calorimetry (ITC) measurements have been performed with various mono- and multivalent metalchlorides to gain more insight in the binding mechanism of CAE-85. In these experiments, the metalchlorides were added dropwise to a solution of surfactant in water. While adding the metalchloride, the heat effect was measured.

To evaluate the influence of the addition of divalent cations on the behavior of the monovalent ions Eisenberg plots have been constructed. In these graphs the molar conductivity is plotted as a function of the monovalent molar equivalent conductivity of Li+, Na+ and K+.

References

[1] Custers, J.P.A., Kelemen, P., Van den Broeke, L.J.P., Cohen-Stuart, M.A., Keurentjes, J.T.F. *J. Am. Chem. Soc.* **127**, 1594-1595 (2005).

268b Co-Formulation of API with Polymer Using Pca Process

Ranjit Thakur, Novartis Pharma AG, CHBS, WSJ-145, Lichtstrasse 35, CH-4056 Basel, Switzerland and Maike Lohrmann, Novartis Animal Health Inc., Schwarzwaldallee 215, CH-4058, Basel, Switzerland.

Co-formulation of drug and excipient is one of the new concept coming in Pharma industry. Co-formulation can lead either to coating/encapsulation/solid solution or co-precipitation in different physical forms. Coating/encapsulation can be used for controlled release or taste masking of drugs. Spray drying or melt extrusion are two of the processes which are used in industry commonly to achieve co-formulation. These processes typically suffer from poor reproducibility and process control. At the same time these processes also yield high residual solvent content, high energy consumption and longer processing time. To overcome these challenges supercritical fluid (SCF) based antisolvent method is proposed and studied in this work.

Novartis drug "A" is used as an API and two polymers PLGA and Eudragit EPO are used in this work for co-formulation. Effect of drug to polymer ratio was studied by maintaining same thermodynamic conditions in a SCF based antisolvent system. With Eudragit EPO, it was observed that at higher drug loading, drug was precipitated into crystalline form with polymer coating. Maximum coating was observed at 70 to 30 drug to polymer ratio. But as polymer ratio increased, solid solution of drug and polymer was observed. With PLGA, one trial was conducted which showed good coating over the crystalline drug. Product was analyzed using SEM, Confocal Raman Microscopy, DSC, XRD and dissolution rate. On basis of these analysis, it was concluded that SCF process provides good approach for co-formulation and quality of co-formulation product can be tuned in this process.

268c Engineering Porous Drug Particles by Emulsification-Diffusion for Propellant-Driven Oral Inhalation Formulations

Libo Wu and Sandro R. P. da Rocha. Chemical Engineering and Materials Science, Wayne State University, Detroit, MI 48201

Aerosol inhalation has been traditionally used for the local delivery of therapeutics to the lungs. More recently, it has also been recognized as a potential route for the systemic delivery of both small molecules and large therapeutic biomolecules. Inhalation therapy is non-invasive, provides for fast drug uptake (second only to i.v.) and drugs delivered through the have good bioavailability. Pressurized metered dose inhalers (pMDIs) are the least expensive and most widely used oral inhalation devices. However, there are many challenges associated with the development of hydrofluoroalkane (HFA)-based dispersion pMDIs.

In this work we present a new, facile methodology for engineering porous drug particles with enhanced physical stability and aerosol characteristics in HFA-based pMDIs. The approach consists in preparing highly porous drug particles using a modified emulsification diffusion technique. The methodology allows for the control of the porosity and size of the particles (essential for controlling the aerosol characteristics), at very low concentrations of non-active excipients – typically used in FDA-approved pMDIs. Results for the short-acting β2-adrenergic receptor agonists salbutamol sulfate (SS) and terbutaline hemisulfate (THS) are discussed. The physical stability and the aerosol characteristics of SS formulations are shown to be superior to the commercial formulation prepared with micronized crystals. This study is relevant in the development of formulations for the local and systemic delivery of small polar drugs to and through the lungs using the inexpensive and widely available pMDIs.

Keywords: pulmonary drug delivery; pressurized metered dose inhalers; porous particles, salbutamol sulfate; terbutaline hemisulfate.

268d Effect of Poly-(N-vinyl pyrrolidone) Coating on the Synthesis of Magnetic Particles for the Formulation of Nanoscale Magnetorheological Fluids

Kaustav Sinha and Olivia A. Graeve. Chemical And Metallurgical Engineering, University of Nevada Reno, 1664 N Virginia ST/MS 388, Reno, NV 89557

This work is primarily focused on the synthesis of magnetic iron nanoparticles using poly-(N-vinyl pyrrolidone) as a protective agent. The nanoparticles were synthesized using a modified polyol route using iron chloride as precursor and sodium borohydride as the reducing agent in ethylene glycol. The resulting solutions, consisting of ethylene glycol and PVP-coated particles, were very stable. In order to extract the powders, the solutions were centrifuged and the resulting powders were cleaned and dried. The particle size was controlled by varying the $FeCl_3$: PVP ratio. Electron microscopy results indicate a median primary particle size of ~20 nm. Dynamic light scattering studies were performed on the stable solutions, as well as the redispersed particles in 1-butanol. The bimodal distribution of the particle size in the solution indicates the formation of coated magnetic nanoparticles and salts of NaCl. The average particle sizes after redispersion in 1-butanol was measured at ~300 nm.

268e Optimal SiO_2 Doping for Enhanced SnO_2 Sensitivity and Thermal Stability

Antonio Tricoli[1], Markus Graf[2], and **Sotiris E. Pratsinis**[1]. (1) Particle Technology Laboratory, Institute of Process Engineering, Department of Mechanical and Process Engineering, ETH Zurich, Sonneggstrasse 3, Zurich, CH-8092, Switzerland, (2) Sensirion AG, Laubisruetistrasse 51, Staefa, CH-8712, Switzerland

In metal-oxide gas sensors the relationship between nano-size and thermal stability is very challenging. Miniaturization of the sensitive material down to nano-scale and doping with noble metals has led to successful detection of analytes in ppb concentrations, but long term stability, which may be related to crystal growth, is still unresolved. The sensitivity of SnO_2, the most widely used metal-oxide for gas sensors, increases drastically with decreasing grain size from 20 to 5 nm. Preserving these sizes during sensor fabrication and operation is difficult since high process and operation temperatures (200 – 900 ° C) are applied. Here one-step synthesis of highly sensitive films of SnO_2-SiO_2 nanostructures is explored by FSP direct deposition and in-situ annealing resulting thermally stable and, well-adhering cauliflower-like deposits (1) and transparent sensors (2). Control and stabilization of the SnO_2 particle and neck size is obtained by varying the SiO_2 content during synthesis and direct deposition of these SnO_2-based films. Detailed film characterization is carried out to understand the role of SiO_2 for inhibiting SnO_2 grain and crystal growth as well as the role of primary particle and neck size on film sensitivity of EtOH, a common analyte in sensor development (3).

1. A. Tricoli, M. Graf, F. Mayer, S. Kühne, A. Hierlemann, S.E. Pratsinis "Micropatterning Layers by Flame Aerosol Deposition - Annealing", *Adv. Mater.*, in press, (2008).

2. S. Kühne, M. Graf, A. Tricoli, F. Mayer, S.E. Pratsinis, A. Hierlemann, "Wafer-level flame-spray-pyrolysis deposition of gas-sensitive layers on microsensors", *J. Micromechanics and Microengineering*, **18**, 035040 (2008).

3. A. Tricoli, M. Graf, S.E. Pratsinis "Optimal doping for enhanced SnO_2 Sensitivity and Thermal Stability", *Adv. Funct. Mater.*, in press, (2008).

268f Polymer Microsphere Synthesis by Membrane Emulsification

Jo M. M. Simons[1], Jan Meuldijk[1], Maartje F. Kemmere[2], and Jos T.F. Keurentjes[2]. (1) Chemical engineering and chemistry, Eindhoven University of Technology, P.O.Box 513, Einhoven, 5600 MB, Netherlands, (2) Department of Chemical Engineering and Chemistry, Process Development Group, Eindhoven University of Technology, P.O. Box 513, Eindhoven, 5600 MB, Netherlands

A major driver for innovation in the application of polymer products can be found in the ability to produce well-defined monodisperse polymer particles, allowing the controlled release or dissolution patterns in drug formulations or specific particle sizes for printer toners. Especially in the range between 1 and 10 μ m no current technology is able to produce this type of particles. Additionally, particles consisting of polymer blends or containing functional (solid) additives are often required. This is virtually impossible with existing techniques. It is expected that emulsification using micro-engineered membranes allows for the production of polymer particles with a narrow particle size distribution by choosing the appropriate membrane specifications, temperature and surfactant type and concentration. [1,2]

In this work, available emulsification systems are adapted for use with complex dispersed phases. The final aim is to produce dispersed solid particles. This is done by emulsification of monomer droplets, followed by polymerization or by droplet formation of concentrated polymer solutions or melts. Emulsification experiments are conducted in a cross-flow membrane setup, see Figure 1.

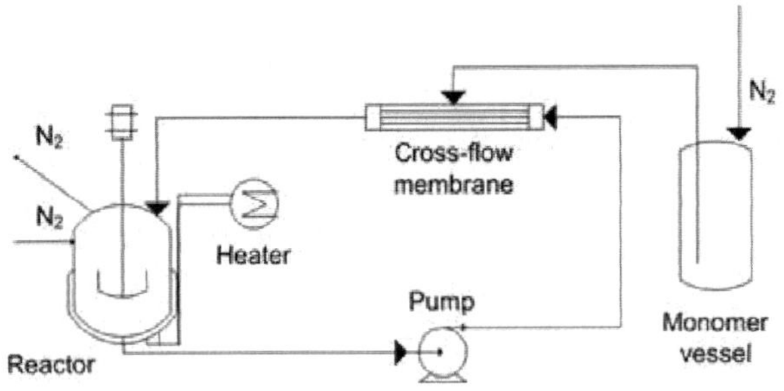

Figure 1. Schematic illustration of the membrane emulsification setup

The droplets formed are polymerized during or after the droplet formation process. Polymerization should be very fast, to avoid coalescence and to avoid the use of substantial amounts of surfactants. However, it should not be so fast that the microchannels become fouled or plugged during emulsification. The time constants of emulsification and polymerization should be closely linked. It is expected that initiating the polymerization during or perhaps even before emulsification can lead to a dramatic change in the emulsification process. Thus, the interplay between loaded dispersed phase, polymerization and emulsification processes are central here.

References

[1] A.J.Abrahamse, R.van Lierop, R.G.M.van der Sman, A.van der Padt, and R.M.Boom, "Analysis of droplet formation and interactions during cross-flow membrane emulsification," *Journal of membrane science*, vol. 204 pp. 125-137, 2002.

[2] S. van der Graaf, C. G. P. H. Schroen, R. G. M. van der Sman, and R. M. Boom, "Influence of dynamic interfacial tension on droplet formation during membrane emulsification," *Journal of Colloid and Interface Science*, vol. 277, no. 2, pp. 456-463, 2004.

287a Preparation and Characterization of NOVEL Polymer-MAGNETIC COLLOIDS

Marco Furlan, Marco Lattuada, and Massimo Morbidelli. Institute for Chemical and Bioengineering, Department of Chemistry and Applied Biosciences, ETH Zurich, Wolfgang Pauli Strasse 10, Zurich, 8093, Switzerland

Magnetic nanocrystals are used in many fields, such as biomedicine, separation technology and material science. For many applications, the nanocrystals need to be encapsulated into a polymeric matrix, thus generating composite nanoparticles. These particles are composed of a core of magnetite nanocrystals and a shell of polymer. It is of utmost importance to control both the amount of nanocrystals encapsulated and the final particle size and size distribution.

In this work, we have prepared novel magnetic colloids having small size (<100nm) and uniform size distribution trough miniemulsion polymerization and have characterized their size distribution and the extent of magnetite incorporation via light scattering and TEM analysis. Magnetite nanocrystals have been prepared through coprecipitation reaction of mixtures of iron salts in presence of ammonium hydroxide and a ligand. Solubility of the nanocrystals in various organic solvents can be tuned by changing the ligands that are anchored on their surface. It was found that the optimization of magnetite encapsulation is strongly dependent upon the chemical nature of ligands, and this motivated us to test different ligands. Our results indicate that, if the ligand is not compatible with the polymer matrix, phase separation occurs and the magnetite accumulates at the surface of the polymer leading to colloidal instability of the dispersion. Ligands carrying reactive polymerizable moieties turned out be particularly effective to achieve a uniform dispersion of magnetite into the polymer matrix.

In order to increase the amount of magnetite incorporated in the magnetic colloid we have added a low boiling point solvent to the monomer before the emulsification. The solvent facilitates the emulsification by reducing the fraction of magnetic material in the droplets, and is subsequently evaporated during and/or after the polymerization reaction. In this manner, it is possible to fine tune the particle size of the colloids by varying the amount of the solvent added.

287b Functionalization of Monodisperse Magnetite Nanoparticles with Fluorescent Amino-Silane-Carboxymethyl Dextran for Cancer Treatment

Adriana P. Herrera, Carola Barrera, Vanessa Ayala, Magda Latorre, Madeline Torres-Lugo, and Carlos Rinaldi. Chemical Engineering, University of Puerto Rico Mayguez Campus, Mayaguez, PR 00680

Monodisperse magnetite nanoparticles coated with oleic acid were synthesized by the thermodecomposition method. These nanoparticles were modified via ligand exchange with amine-silane molecules, which allowed the expression of amine (NH2) end groups onto the nanoparticle surface. Carbodiimide chemistry was used to react these amine groups with carboxylic groups (-COOH) present in carboxy-methyl dextran (CMDx) molecules. We used CMDx with degrees of substitution ranging from 0.35 to 1 to study the colloidal stability of the functionalized nanoparticles by zeta potential measurements. Fluorescent nanoparticles were also obtained by incorporation of fluorescein-amine groups during the synthesis of magnetite-amino-silane-CMDx nanoparticles. These fluorescent nanoparticles were used to visualize the internalization in CaCo-2 and MCF-7 cancer cell monolayers through confocal laser scanning microscopy (CLSM). The cytotoxic effects of these nanoparticles were study to determine their potential application for cancer treatment.

287c Silica-Coated Quantum Dot-Iron Phosphide Nano-Composites

Shuang Deng, Department of Chemical and Biomolecular Engineering, The Ohio State University, Columbus, OH 43210 and Jessica O. Winter, Department of Chemical and Biomolecular Engineering, Department of Biomedical Engineering, The Ohio State University, Columbus, OH 43210.

Nanoparticles have made significant contributions to the field of biological and biomedical imaging. Quantum dots have been extensively studied for their unique optical and electronic properties, sharp emission band with broad excitation, and strong resistance to photobleaching. In addition to these properties, their dimensional similarities to biomacromolecules (e. g., DNA and protein) and facile bioconjugation make them excellent agents for biological imaging. Similarly, magnetic nanoparticles (MNPs) have also been studied for biomedical applications, including MRI contrast enhancement, magnetic immobilization, and drug targeting. Recently, hybrid materials that contain both magnetic nano-particles and quantum dots (QDs) have been synthesized. These materials, which have been initially explored as multimodal imaging agents, may also allow for direct mechanical manipulation and simultaneous observation of sub-cellular features.

Here, we present the synthesis and characterization of a new nano-composite, CdTe-FeP magnetic quantum dot nanoparticles. These materials, which exhibit a dumbbell shape morphology are further encapsulated with silica to produce a core-shell structure. The magnetic quantum dots, CdTe-FeP, were prepared through a one-pot chemical method based on high temperature precursor decomposition of both materials in trioctyl phosphine oxide (TOPO). The silica shell was formed using a reverse micro-emulsion technique.

To characterize this material, we investigated the effect of the synthesis conditions and CdTe: FeP mole ratios on magnetic and fluorescent properties as measured by TEM, XRD, UV, XPS, fluorescence spectroscopy, and superconducting quantum interference device (SQUID). Results indicate that nano-composites exhibit magnetic and fluorescent properties; however, the magnetism exhibited by FeP appeared to change from ferromagnetic to paramagnetic after conjugation with CdTe. Also, the quantum yield of CdTe decreased after conjugation. These data suggest an interaction between the magnetic and fluorescent features of the individual particles. These core-shell nanocomposites have potential applications in diagnostics, circulating drug-delivery systems, and treating cancer at its earliest stages.

287d Mimicking Selectin-Mediated Cell Motion with a Synthetic Receptor System

Maria M. Santore[1], Surachate Kalasin[2], Jeffrey,M. Davis[3], and Ranojoy D. Duffadar[3]. (1) University of Massachusetts Amherst, Conte Polymer Laboratory, 120 Governor s Drive, Amherst, MA 01003, (2) Department of Physics, University of Massachusetts, Amherst, MA 01003, (3) Chemical Engineering, University of Massachusetts, Amherst, 259B, Goesmann lab, North Pleasant Street, Amherst, MA 01002

The rolling of neutrophils over the injury-stimulated endothelium of blood vessels results from the coupling of selectin – sialyl Lewisx bonding and disbonding kinetics to hydrodynamic forces on the far surface of the cell. Much is known about this system, from studies with neutrophils themselves or from studies employing particles and surfaces functionalized with the relevant biomolecules. This contribution explores how an artificial receptor can be designed so that biomimetic particles exhibit similar dynamic motion signatures to neutrophils. We compare the firm adhesion, rolling, and non-adhesive behaviors of neutrophil rolling to a system that relies largely on electrostatic attractions, focusing on the state space diagrams of the Hammer and Schwarz groups. With the artificial receptor we experimentally access regions of parameter space not easily studies with selectins, including systematic variations in the density of the receptor, its distance from the underlying substrate, and the range of repulsion provided by the background surface or passivating layer. These variables are found to be as important as those more typically studied in cell-based and cell-free systems.

287e Functional Nanoparticles for Enantioselective Separations

Nisha Shukla, Institute for Complex Engineered Systems, Carnegie Mellon Univeristy, 5000 Forbes Ave., Pittsburgh, PA 15213, Andrew J. Gellman, Chemical Engineering, Carnegie Mellon University, 5000 Forbes Ave, Pittsburgh, PA 15213, and Melissa Bartel, Chemical Engineering, Carnegie Mellon Univeristy, 5000 Forbes Ave., Pittsburgh, PA 15213.

Layers of organic ligands on the surfaces of nanoparticles dictate many of their properties. Chiral Au nanoparticles have been synthesized by coating with chiral ligands. These ligands render the surface of the particles chiral and active for enantioselective adsorption and other enantioselective processes.

The chirality of the Au nanoparticles has been probed by exposing them to racemic mixtures of propylene oxide. Optical polarization has been used to detect enantiospecific interactions between the propylene oxide and the chiral Au nanoparticles. These measurements show that the chiral Au nanoparticles do selectively adsorb one of the two enantiomers form the racemic mixture, resulting in a net polarization of light passing through the solution of chiral Au nanoparticles and racemic propylene oxide. The polarization depends on the concentration of racemic propylene oxide in solution.

287f Preparation, Assembly and Application of Surface Anisotropic (Janus) Particles

Jingqin Cui, Department of Chemical Engineering, City College, City University of New York, Steinman Hall, 140th ST. & Convent Ave., New York, NY 10031 and Ilona Kretzschmar, Chemical Engineering, City College, City University of New York, Steinman Hall, 140th ST. & Convent Ave., New York, NY 10031.

The chemical and physical properties of materials rely on their composition, shape and size.[1] Micro and sub-micron scale composites containing nano-structure such as core-shell particles have received attention because of their special properties attributed to both their size and their variation in composition.[2] Anisotropic composites, e.g., core-half shell particles, are of interest because of their directional property in addition to their size and composition.[3] Partial surface decoration of symmetric particles is one of the approaches used to fabricate Janus particles employing either templating or vapor deposition methods. The vapor deposition techniques are widely applied taking advantage of their line-of-sight directional deposition, however, at the cost of high temperature, vacuum, and expensive equipment requirements.

In this talk, a liquid phase templating process, electroless deposition (ELD), is introduced because of its moderate conditions, material flexibility, and low cost. A solid template preparation procedure and the asymmetric deposition of various metals such as silver, gold, and platinum on microspheres are discussed.[4] We will show that ELD modified spheres posses a much rougher surface compared to spheres modified using vapor deposition, effecting the self- and directed assembly of these Janus particles. Preliminary results on the combination of both vapor and electroless deposition resulting in dual-patch Janus microspheres will be reported. In addition, Janus particles are loaded into an electroactive polymer and their effect on the elctromechanical deformation of the polymer actuator will be discussed as one application of Janus particles.[5]

1. N. L. Rosi, C. A. Mirkin, Chem. Rev. 2005, 105, 1547-1562

2. A. S. Edelstein, R. C. Cammarata, Nanomaterials: Synthesis, Properties and Applications, Institute of Physics Publishing, Bristol an Philadelphia, 1996

3. A. Perro, S. Reculusa, S. Ravaine, E. Bourgeat-Lami and E. Duguet, J. Mater. Chem., 2005, 15, 3745–3760

4. J.-Q. Cui and I. Kretzschmar, Langmuir, 2006, 22, 8281-8284

5. J.-Q. Cui, A. Maliakal and I. Kretzschmar, "TiO2 nanoparticles/poly(acrylate) elastomer composite films as actuators", 2008 MRS Spring Meeting, March 2008, San Francisco (Oral presentation J8.4)

292a Cfd Modeling of Polymerization Fluidization Reactor

Ram G. Rokkam[1], R. O. Fox[1], and Michael E. Muhle[2]. (1) Department of Chemical & Biological Engineering, Iowa State University, 2114 Sweeney Hall, Ames, IA 50011, (2) Michael E, 5200 Bayway Drive/Baytown, Houston, TX 77520

Polyethylene is one of the most widely used plastics, and over 60 million tons are produced worldwide every year. Polyethylene is obtained by the catalytic polymerization of ethylene in gas and liquid phase reactors. The gas phase processes are more advantageous, and use fluidized bed reactors for production of polyethylene. Since they operate so close to the melting point of the polymer, agglomeration is an operational concern in all slurry and gas polymerization processes. Electrostatics and hot spot formation are the main factors that contribute to agglomeration in gas-phase processes. Electrostatic charges in gas phase polymerization fluidized bed reactors are known to influence the bed hydrodynamics, elutriation, bubble size etc. Accumulation of electrostatic charges in the fluidized bed can lead to operational issues. The objective of this work is to develop a multi-fluid model which can explain the electrostatic behavior in gas-solid fluidized beds. The particles are assumed to carry a prescribed charge. A Poisson equation is solved for the electric field which develops due to presence of charged particles. A multi-fluid model solves mass and momentum equation for one gas and three solid phases. The diameter and volume fraction of the solid phases is obtained from the reaction engineering model and quadrature method of moments. Kinetic theory of granular flow is used for calculating particle pressure and viscosities. Time dependent three dimensional simulations have been performed on a pilot plant reactor using Fluent 6.3.

CFD simulation of the turbulent flow of pulp fibre suspensions

Carla A.F. Ventura, F.A.P. Garcia, P.J. Ferreira, M.G. Rasteiro
Chemical Engineering Department, University of Coimbra, Portugal;
Corresponding author: mgr@eq.uc.pt, phone: +351 239 798 700

Abstract:

Pulp fibre suspensions are complex solid-liquid systems, since their components (fibre, flocs, additives…) present singular and complex interactions between them. As a consequence of this complexity, the understanding of the suspensions flow dynamics remains poor and incomplete, what usually leads to a conservative industrial equipment design and, hence, equipment oversizing.

The purpose of this study is to obtain further knowledge about the dynamic mechanisms of the turbulent flow regime of industrial pulp suspensions, by producing and testing the viability of a model for the turbulent flow of fibre suspensions in conveying pipes, based on fundamental principles of conservation, and also on some experimental information (namely the rheological information) to adjust, empirically, unknown parameters as the fluid viscosity.

With this purpose a numerical model was developed, based on a CFD code, using the Chemical Engineering module of COMSOL Multiphysics Software version 3.4. The $K-\varepsilon$ Turbulence Model was chosen to simulate the pulp suspensions flow. Additionally, the final model was tested and validated using four different industrial pulp suspensions, which were previously fully studied experimentally (flow and rheological tests).

The obtained results demonstrated that the pressure drop profiles obtained using COMSOL Multiphysics Software, for the turbulent flow regime, agree very well with the experimental results obtained in a pilot rig. Additionally, the use of the $k-\varepsilon$ Turbulence Model for the simulation of pulp fibre suspensions flow, associated with the rheological data acquired experimentally, revealed to be a prompt and accurate strategy to attain good prediction of pressure drop values for fibre suspensions flow.

Moreover, the adjustment of the turbulence parameters confirmed previous studies' results, where it was concluded that the existence of particles, such as fibres, in a fluid flow, induces a turbulence damping.

Keywords:
Pulp fibre flow, CFD simulation, turbulence, K-ε model, COMSOL Multiphysics.

1. Introduction

It is known that, contrarily to what happens in the usual suspension systems, the components of the pulp fibre suspensions (fibres, fines, …) are able to develop new "suspension structures" (floccettes, flocs and networks), which modify the suspension nature and consequently the fluid mechanics characteristics.

The flow mechanisms of pulp fibre suspension in pipes have been divided in three different regimes; each regime can be further separated into sub-regimes with well defined shear mechanisms, as can be seen in Figure 1, where pressure drop $\left(\Delta P/L\right)$ is plotted against velocity [1]:

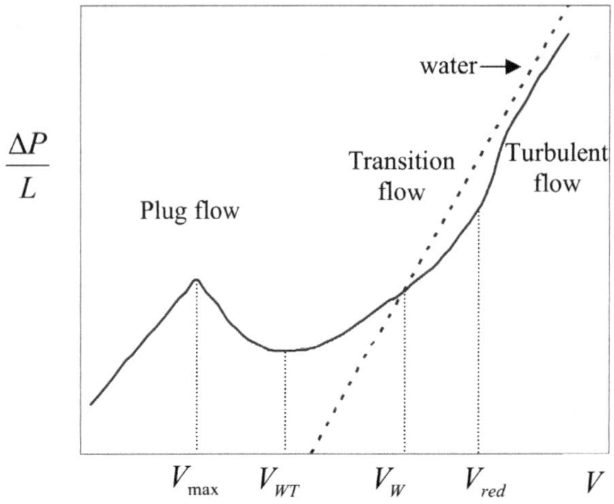

Figure 1 – Pressure drop curve (adapted from [2-6]).

At low velocities the suspension flows as a plug of fibres and water, and the entire shear occurs in a thin layer adjacent to the pipe wall. This induces larger values for pressure drop in pipes than those of water. At intermediate velocities, there is a transition regime where a central and intact plug is surrounded by a turbulent fibre-water annulus. This regime is termed as the transition flow regime, and it starts at the onset of the drag reduction effect, corresponding to a velocity V_W. At high velocities, all the suspension components are in complex turbulent motion, and the pressure drop values are now smaller than the ones expected for water [2-6].

The construction of a flow model able to predict the flow behaviour of pulp fibre suspensions represents an important step in this area, since the numerical modelling can reduce the costs of experiments and prototype equipment usually required to design conveying systems.

The intention of this work is not only to model the turbulent flow of pulp fibre suspensions in pipes, by the use of a finite elements method (FEM) considering a computational fluid dynamics code (CFD), but also to compare the obtained pressure drop results with experimental data.

The $k-\varepsilon$ Turbulence Model is one of the simplest and most used turbulence models for industrial applications, and it was the selected model in this research. Despite of its large application, this model is based on some restrictions, the most important of which are that the Reynolds number is high enough and that turbulence is uniform within boundary layers, which means that production equals dissipation [7]. These assumptions limit the model's accuracy, since they are not always completely correct.

In the present work the suspension is described as a pseudo-homogeneous one phase fluid, which means that the model does not account either for the existence of particles/fibre structures, or for the consistency variation in the flow field. However, it is believed that this restriction does not affect the model's robustness, since for the regime to be modelled all the suspension components are in complete turbulent motion.

This work describes the CFD modelling strategy adopted, making use of the Chemical Engineering module of COMSOL Multiphysics Software version 3.4.

2. Governing Equations

The $k-\varepsilon$ Model is fundamentally based on the Navier-Stokes equations, presented bellow, considering that the fluid is incompressible (ρ=constant) [7-8]:

Continuity equation – represents the conservation of mass:

$$\nabla \cdot u = 0 \tag{1}$$

Vector equation – represents the conservation of momentum:

$$\rho \frac{\partial u}{\partial t} + \rho (u \cdot \nabla) u = -\nabla p + \nabla \cdot \eta \left(\nabla u + (\nabla u)^T \right) + F \tag{2}$$

Energy equation – represents the conservation of energy:

$$\rho C_p \left(\frac{\partial T}{\partial t} + (u \cdot \nabla) T \right) = -(\nabla \cdot q) + \tau : S - \frac{T}{\rho} \frac{\partial \rho}{\partial T}\bigg|_p \left(\frac{\partial p}{\partial t} + (u \cdot \nabla) p \right) + Q \tag{3}$$

where u is the velocity vector ($m.s^{-1}$), ρ is the density ($kg.m^{-3}$), p is pressure (Pa), τ is the viscous stress tensor (Pa), F is the body force vector ($N.m^{-3}$), C_p is the specific heat capacity at a constant pressure ($J.kg^{-1}.K^{-1}$), T is absolute temperature (K), q is the heat flux vector ($W.m^{-2}$), Q contains the heat sources ($W.m^{-3}$) and S is the strain rate tensor: $S = \frac{1}{2}\left(\nabla u + (\nabla u)^T \right)$.

The $k-\varepsilon$ Turbulence Model, which is extended to non-Newtonian fluids by allowing the dynamic viscosity to be a function of the velocity field, introduces two additional transport equations and two dependent variables: the turbulence kinetic energy, k, and the turbulence dissipation rate, ε. Turbulent viscosity is modelled by:

$$\eta_T = \rho C_\mu \frac{k^2}{\varepsilon} \tag{4}$$

where C_μ is a model constant.

The transport equation for k can be described by analogy with the equations for Reynolds stresses:

$$\rho \frac{\partial k}{\partial t} - \nabla \left[\left(\eta + \frac{\eta_T}{\sigma_k} \right) \nabla k \right] + \rho U \cdot \nabla k = \frac{1}{2} \eta_T \left(\nabla U + (\nabla U)^T \right)^2 - \rho \varepsilon \tag{5}$$

An equation for ε can be derived in a similar manner. Such equation is, however, impossible to model on a term-by-term basis. Instead, all the terms that do not have an equivalent term in the k equation are discarded.

$$\rho \frac{\partial \varepsilon}{\partial t} - \nabla \left[\left(\eta + \frac{\eta_T}{\sigma_\varepsilon} \right) \nabla \varepsilon \right] + \rho U \cdot \nabla \varepsilon = \frac{1}{2} C_{\varepsilon 1} \frac{\varepsilon}{k} \eta_T \left(\nabla U + (\nabla U)^T \right)^2 - \rho C_{\varepsilon 2} \frac{\varepsilon^2}{k} \tag{6}$$

The model constants in the aforementioned equations are empirical, and their values can be seen in Table 1.

Table 1 – Model Constants[7-8].

Constant	C_μ	$C_{\varepsilon 1}$	$C_{\varepsilon 2}$	σ_k	σ_ε
Value	0.09	1.44	1.92	1.0	1.3

3. Numerical Implementation

A system with the characteristics described in the following section has been implemented.

3.1 Real System

The system to be modelled is basically a linear pipe, wherein a pulp fibre suspension is flowing. Experimental pressure drop trials with four industrial types of pulp fibre suspensions (recycled pulp, eucalypt bleached kraft pulp, pine unbleached kraft pulp and eucalypt (90%) + pine (10%) bleached kraft pulp) were previously performed in a pipe 4 m long and with 3 in of diameter. This is the geometry to model, so that a comparison of experimental versus predicted values can be made.

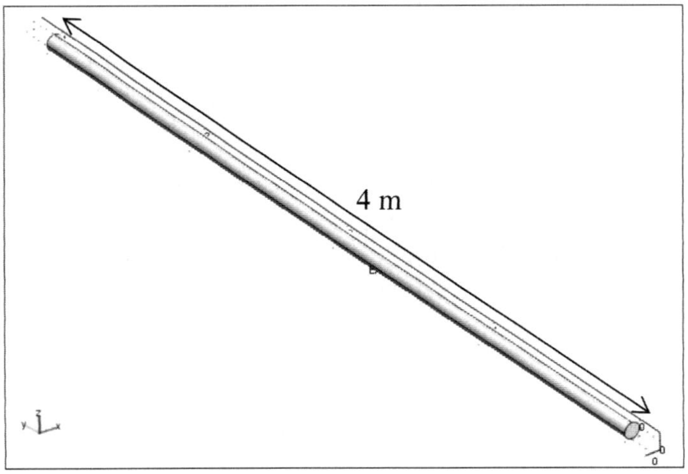

Figure 2 – System geometry.

3.2 Simplification of the geometry

The model size was reduced in order to minimise the calculation time, without loosing accuracy by making use of 2D axial symmetry. Consequently, the modelled domain is simply a rectangle, whose length represents the 1 m pipe length and width corresponds to the 1.5 in (0.0381 m) radius, as presented in Figure 3.

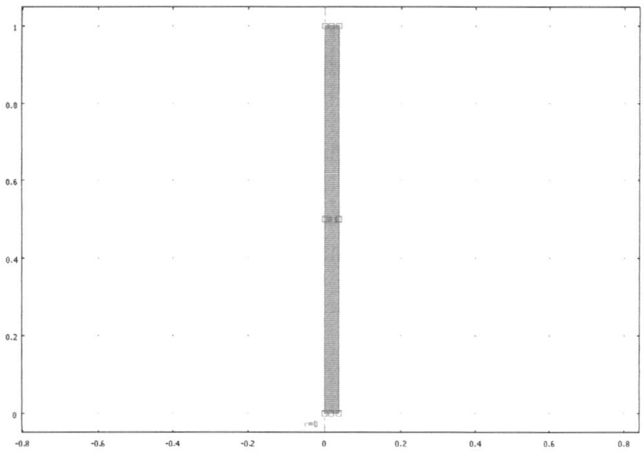

Figure 3 – Simplified geometry.

3.3 Meshing

In order to reach accurate results for the pressure drop, the mesh selected was a mapped mesh consisting of quadrilateral elements. This kind of mesh is structured in its pattern, and is recommended for geometries such that domains are fairly regular in shape and do not contain holes, as is the case. As can be seen in Figure 4, the mesh is more refined near the wall to resolve the so-called viscous sublayer, so that more precise modelling results can be obtained.

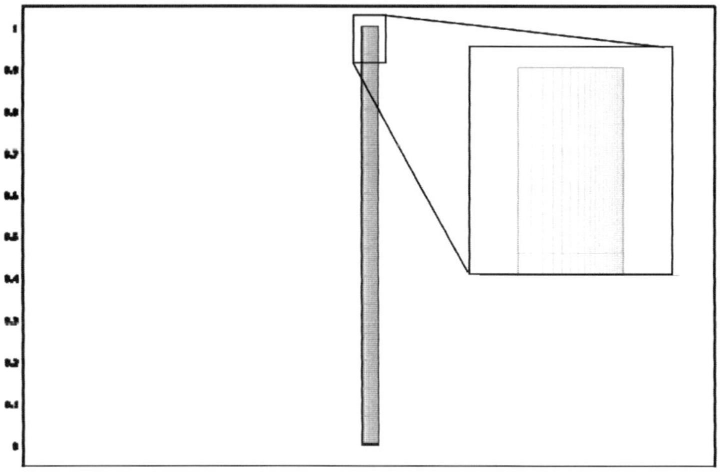

Figure 4 – Mesh mode.

3.4 Physics and boundaries

Subdomain settings	Physical characterization of the system regarding density and viscosity values: ▪ Density: collected experimentally by picnometry; ▪ Viscosity: collected experimentally by the use of a new viscometer specially designed to study multiphase systems. The viscosity values were introduced on the model as a velocity function, since the pulp fibre suspensions are clearly non-Newtonian fluids. The introduced viscosity models were constructed and fully discussed in a previous work [9].

Inlet boundary	The inlet fluid velocity was taken as a boundary condition, considering a uniform cross section velocity profile. Additionally, specific values for the two turbulent quantities, k and ε were required. Alternatively, a turbulent length scale, L_T, and a turbulent intensity scale, I_T, could be specified, which are related with the turbulent parameters as follows: $k=\frac{3}{2}\left(\lvert U\rvert\, I_T\right)^2$ (7) $\varepsilon = C_\mu^{3/4}\frac{k^{3/2}}{L_T}$ (8) These turbulence properties are more intuitive to understand and can more easily be related to the physical characteristics of the problem. The values of I_T and L_T are not exactly known, but they can be estimated following some guidelines from the literature [7]. $I_T = I\ \mathrm{Re}_{D_h}^{-1/8}$ (9) $L_T = l\ D_h$ (10) With no further information, generally, the I and l parameters values are frequently assumed to be 0.16 and 0.07, respectively. It is well referenced in previous studies that, the existence of particles, such as fibres, in a fluid flow, induces a turbulence damping [1, 10], which is reflected on the turbulence parameters decrease, thus the I and l values should be smaller than usually assumed for homogeneous fluids. However, the turbulence damping range and, hence, the I and l decrease are not exactly known; for pulp fibre suspensions it is expectable that the damping effect will depend on fibre type and on the consistency value. Taking into account the aforementioned discussion, and in order to simulate the turbulence damping, for both the I and l values of the modelling equations 9 and 10, smaller values than usually used for homogeneous fluids were introduced in the model. Since the turbulent length scale is mentioned to be mainly dependent on the system geometry, the l value was assumed to be constant for all the fibre types and all the consistencies. The I parameter was adjusted according to the pulp fibre type and concentration.
Outlet boundary	The condition "Normal Stress, Normal Flow" was chosen to describe the outlet condition of the domain, since this condition determines that there must be no tangential velocities on the boundary.
Wall	The wall boundary selected for the model must consider that turbulence close to a solid wall is very different from isotropic free-stream turbulence. To overcome this fact, an approach considering an empirical relation between the value of velocity and wall friction was introduced in the model. This relation known as "wall function" is accurate for high Reynolds numbers and in situations where pressure variations along the wall are not very large [7], which can be assumed in the case reported here. The wall boundary was modelled with a logarithm wall function; this wall function translated into finite elements assumes that the computation domain begins at a distance δ_W from the real wall. The δ_W value considered in this research was 0.0025m, 6% of the pipe radius, in agreement with the the recommendations of COMSOL for this kind of situation. The logarithm wall function also assumes that the velocity is parallel to the wall.

Symmetry boundary	The "Axial Symmetry" condition should be used on all boundaries with coordinate r=0.

3.5 Solver

Different solvers were tested but the best results were achieved through the use of a "Parametric Segregated" solver, which actually is indicated for parameterized sets of stationary multiphysics PDE problems (linear or nonlinear). In the present study, the mentioned solver was used to find the solution for an increasing sequence of velocities.

4. Results and Discussion

Firstly, the numerical implementation was validated with water, considering the water physical characteristics. Then, the pulp's physical characteristics were introduced in the model. It was considered that the suspensions start to flow in fully developed turbulence at velocities above the maximum of the drag reduction effect - V_{red} (see Figure 1).

As an example of an output result, Figure 5 shows the simulated pressure drop along 1 m of pipe for the specific case of the recycled pulp suspension with 2.7% (w/w) consistency, at a velocity of 4.8 m/s.

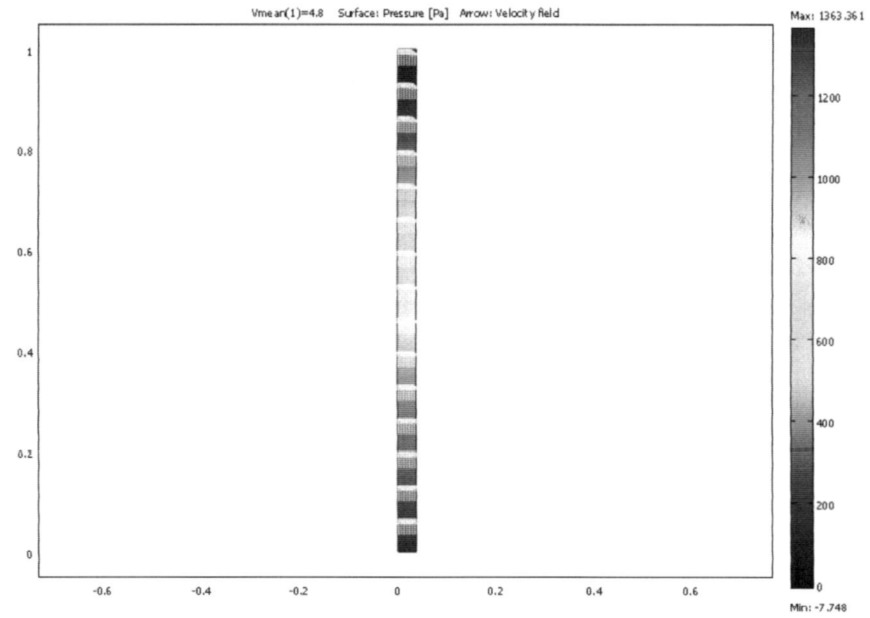

Figure 5 – Simulated pressure drop for the flow of the recycled pulp suspension with 2.7% (w/w) consistency, at a velocity of 4.8 m/s.

It can be seen the pressure drop profile along 1 m of pipe represented by the colour variation and the velocity profile along the pipe length represented by the white arrows. As expected, the simulation results confirm that the pressure decreases along the pipe and, since the system is highly turbulent, that the radial velocity profile remains stable along the pipe.

The model's pressure drop results for each consistency and pulp type were compared with the experimental pressure drop results, and the model's turbulence parameters were adjusted until a good fit with the experimental results was obtained. An example is shown in Figure 6.

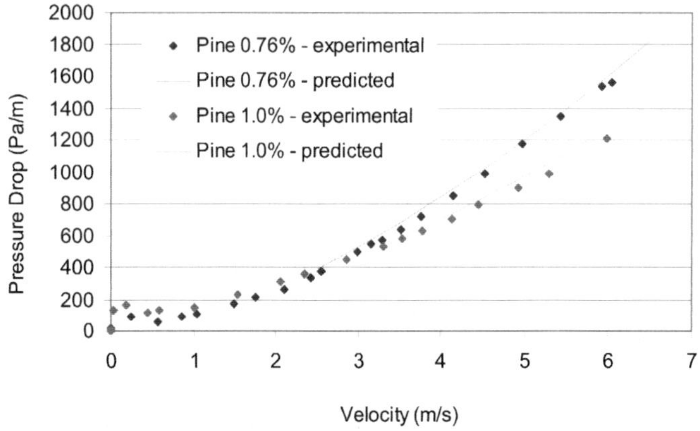

$$C = 0.76\% \left| \begin{array}{l} I = 0.01 \\ l = 0.005 \end{array} \right.$$

$$C = 1\% \left| \begin{array}{l} I = 0.0003 \\ l = 0.005 \end{array} \right.$$

Figure 6 – Experimental versus predicted fit for the pressure drop of pine pulp suspensions at 0.76% and 1.0% of consistency.

Figure 6 shows that the turbulence parameters used to simulate the flow of the pine suspension for the specified conditions led to a good fit between the experimental and predicted values. From the many simulations performed for the different pulps and consistencies, the turbulence parameters values reported in Table 1 were found to be adequate for use at the several flow conditions.

Table 2 - Turbulence parameters values.

Pulp type	Turbulence Parameters	very low consistencies			low consistencies			
Recycled	Consistency (%)	0.72	0.60	0.61	1.40	1.80	2.30	2.70
	I	0.01	0.01	0.01	0.009	0.008	0.007	0.005
	l	0.005	0.005	0.005	0.005	0.005	0.005	0.005
Eucalypt	Consistency (%)	0.77		0.91	1.4		1.5	
	I	0.09		0.09	0.007		0.003	
	l	0.005		0.005	0.005		0.005	
Eucalypt + pine	Consistency (%)	0.71		0.77	0.9	1.2		1.3
	I	0.07		0.05	0.01	0.005		0.003
	l	0.005		0.005	0.005	0.005		0.005
Pine	Consistency (%)	0.66		0.76	0.8		1	
	I	0.01		0.01	0.0005		0.0003	
	l	0.005		0.005	0.005		0.005	

From Table 1 it can be easily concluded that for very low consistencies the I value is minimally influenced by the consistency increase, this means that a slight variation in consistency for these consistencies range does not modify the turbulent flow mechanisms.

Contrarily, for relatively high values of consistency it is observed that as the consistency increases, the I values decreases for all the pulps. This trend was expected, since the presence of solids has an increasing damping effect on the longitudinal turbulence intensity, in agreement with previous studies [10]. Considering the unique

characteristics of pulp fibre suspensions, whose components are able to develop new "suspension structures", such as floccettes and flocs, it can be easily understood the larger decrease of the I value, since these "new" structures, which are not only bigger, but also of a different nature, modify the suspension turbulence characteristics.
The highest values of turbulence damping were observed for the pine fibre suspension, followed by the eucalypt+pine and the eucalypt pulp suspensions. This fact is certainly related with both the fibres length and the fibres morphological characteristics. Pine fibres are the longest and stiffest ones.
Regarding the recycled pulp suspension the effect of consistency was not so important. In fact, this different behaviour has already been recognised in other situations, such as the rheological and flow behaviours [9, 11,12]: this is certainly related to the intrinsic pulp suspension characteristics, such as the presence of other components including adhesives, ink and waste particles, which modify not only its fundamental nature, but also its visual appearance. Actually, this was the pastiest suspension and for this reason it looked more homogenous.
In Figure 7 a comparison between the experimental data and the modelling results for the flow conditions tested (different pulps and consistencies) is presented:

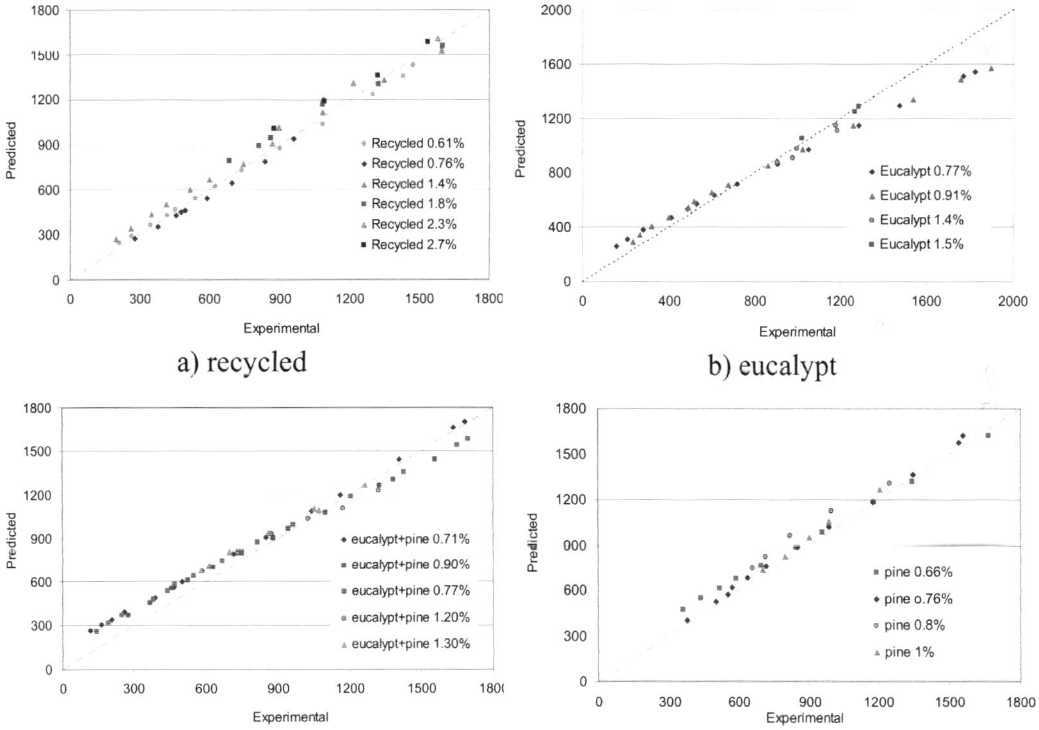

a) recycled b) eucalypt

c) eucalypt+pine d) pine

Figure 7 - Comparison between the experimental pressure drop data and the modelling results for the flow conditions tested.

The predicted values for the pressure drop in pipes are very similar to the experimental ones. Regarding the lowest values of pressure drop for the fresh pulp fibre suspensions, a more careful observation shows that the simulation results slightly over-predict. These values correspond to the lowest flow velocities, so this difference can be due to the fact that for these velocities the turbulent flow regime may not be completely established.
Considering the eucalypt fibre suspension (Figure 9 b)) the CFD model is not completely able to predict pressure drop for the highest velocities: the calculated values are smaller than the experimental ones. These values correspond to the lowest

consistencies at the highest velocities, and the observed trend may be due to the experimental rheological values used. In fact, due to limitations of the rheometer, very high shear velocities were impossible to reach and these values had to be extrapolated from the rheogram.

5. Conclusions

The pressure drop profiles obtained using COMSOL Multiphysics Software agree very well with the experimental results obtained in a pilot rig for the turbulent flow regime. Moreover, the use of the $k-\varepsilon$ Turbulence Model for the simulation of pulp fibre suspensions flow, associated with the rheological data acquired in a viscometer especially designed to study multiphase systems, revealed to be an interesting and accurate strategy to attain good prediction of pressure drop values for fibre suspension flow.

The adjustment of the turbulence parameters confirms previous studies, where it was concluded that the existence of particles, such as fibres, in a fluid flow, induces a turbulence damping. The turbulence intensity scale, I_T (as a function of I), decreases with pulp consistency increase.

6 Acknowledgments

The financial support of NODESZELOSS Project is gratefully acknowledged. The authors would like to thank Prado Karton, GOPACA and CELTEJO for their valuable help in pulp supplying. The authors would like to express their thanks to COMSOL support staff team, for their availability and help.

7. References

1. Duffy, G.G and Cheng, D., "New Insights into the flow of pulp suspensions", *Tappi Peer-Reviewed Paper*, 1 (1): (2001).
2. Moller, K., "General correlations of pipe friction data for pulp suspensions", *Tappi*, 59 (8): 111 (1976).
3. Duffy, G. G. and Titchener, A.L, "Design procedures for obtaining friction loss for chemical pulps", *Tappi Journal*, 57 (5): 162 (1974).
4. Duffy, G.G., Titchener, A.L. and Moller, K., "The mechanisms of pulp suspensions in pipes", *Appita Journal*, 29 (5), 363 (1976).
5. Moller, K., "A correlation of pipe friction data for paper suspensions", *Ind. Eng. Chem., Process Des. Dev.*, 15 (1): 16 (1976).
6. Duffy, G. G., "Flow of medium consistency wood pulp suspensions", *Appita Journal*, 48 (1): 51 (1995).
7. Chemical Engineering Module User's Guide, COMSOL AB, 2007.
8. Wilcox, D.C., "Turbulence Modeling for CFD", DWC Industries Inc., 1998.
9. Ventura, C.A., Garcia, F., Ferreira, P., Rasteiro, M., "Dynamic Factor - A parameter to characterize pulp rheology", submitted to Chemical Engineering & Technology Journal.
10. Zisselmar, R. and Molerus, O., "Investigation of solid-liquid pipe flow with regards to Turbulence Modification", *The Chemical Engineering Journal*, 18: 233 (1979).
11. Ventura, C.A., Blano, A., Negro, C., Ferreira, P., Garcia, F., Rasteiro, M., "Modeling pulp fiber suspension rheology", *Tappi J.,* 6 (7) (2007).

12. Ventura, C.A., Garcia, F., Ferreira, P., Rasteiro, M., "Flow Dynamics of Pulp Fibre Suspensions", *Tappi J.,* (Aug 2008).

Heterogeneous Gas Chemistry in the CPFD Eulerian-Lagrangian Numerical Scheme (Ozone Decomposition)

Dale Snider, CPFD Software, 10899 Montgomery NE, LLC, Albuquerque, NM 87111
Sibashis Banerjee, Cristal Global, 6752 Baymeadow Drive, Gelne Burnie, MD 21060

Abstract

Heterogeneous catalytic chemistry is used throughout the chemical and petro-chemical industry. In predicting the performance of a reactor, knowing the gases and solids flow dynamics is as important as having a good chemical rate expressions. This paper gives the solution of ozone decomposition in a bubbling bed using the CPFD numerical scheme which is an Eulerian-Lagrangian solution method for fluid-solids flow. The ozone decomposition can be described by a single stoichiometric equation and has a first order reaction rate. The ozone decomposition is a standard problem for chemical analysis and has been used to characterize gas-solids contact in fluidized beds. The accuracy of predicting the ozone decomposition comes from correctly predicting the bed dynamics. The solution in this study is three-dimensional and predicts the coupled motion of both solids and gas. The chemical rate equation uses solids volume fraction, but the numerical method could calculate chemistry on the discrete catalyst, including a variation in size (surface area) if such a rate equation was available. The numerical results compare well with an analytic solution of the decomposition rate, and calculated results compare well with the experiment by Fryer and Potter (1976).

Introduction

Ozone decomposition is a standard problem for chemical analysis. It has been used to characterize the gas-solids contact in a fluidized bed. The ozone decomposition can be described by

$$2O_3 \rightarrow 3O_2 \tag{1}$$

and the ozone rate equation is

$$\frac{d[O_3]}{dt} = -c_s [O_3] \tag{2}$$

where [] denotes mole concentration. The rate coefficient is

$$c_s = k\,\theta_S \tag{3}$$

where θ_S is the solids volume fraction and k is a constant based on catalyst activity. The catalyst volume fraction in the chemical rate coefficient equation Eq. (3) does not account for the the effect of catalyst surface area (catalyst sites) from a particle size distribution (PSD). The solid-catalyst size distribution was not available, and a constant size was used in calculations.

The chemical process is simple, and at low concentration, the reaction is essentially irreversible and isothermal. Calculations were made for two stagnant beds and a fluidized bed. One static bed calculation is in a closed vessel where the ozone decomposition reaction increases the pressure. The second static bed calculation is in an open vessel which allows excess volume from the ozone decomposition to exit the vessel and leads to a decrease in mixture density. The ozone decomposition in the static beds has a known analytic decomposition rate and known pressurization in a closed vessel and known density in an open vessel. Because the full set of momentum and mass equations are solved, the static bed calculations provide a check on correctness of the predicted chemistry. The fluidized bed calculation is compared with measured data.

The calculations are made using the CPFD numerical scheme. The CPFD solves the fluid and particle momentum, mass and energy. The particle momentum description is based on the multi-phase particle in cell (MP-PIC) method and the fluid phase is solution of the three dimensional Navier-Stokes equation. The calculations are made with the Barracuda™ software.

CPFD governing equations

The CPFD method solves the fluid and particle momentum equations in three dimensions The fluid is described by the Navier-Stokes equation with strong coupling with the discrete particles. The particle momentum follows the particle-in-cell (MP-PIC) numerical description from P. O'Rourke (Andrews, 1996; Snider, 1998; Snider, 2001) which is a Lagrangian description of particle motion described by ordinary differential equations with coupling with the fluid. The CPFD solution as applied in Arena-flow® and Barracuda-CPFD® is aimed at solving commercial problems, which are generally physically large systems. In the CPFD scheme, a numerical-particle is defined where particles are grouped with the same properties (species, size, density, etc.). The numerical-particle is a numerical approximation similar to the numerical control volume where a spatial region has a single property for the fluid. Using numerical particles, large commercial systems containing billions of particles can be analyzed using millions of numerical-particles.

The volume average two-phase incompressible continuity equation for the fluid with no interphase mass transfer is

$$\frac{\partial \theta_f}{\partial t} + \nabla \cdot \left(\theta_f \boldsymbol{u}_f\right) = 0\,, \tag{4}$$

where $\mathbf{u}_f$ is the fluid velocity and θ_f is the fluid volume fraction. The volume average two-phase incompressible momentum equation for the fluid is

$$\frac{\partial\left(\theta_f \boldsymbol{u}_f\right)}{\partial t} + \nabla \cdot \left(\theta_f \boldsymbol{u}_f \boldsymbol{u}_f\right) = \frac{-1}{\rho_f} \nabla p - \frac{1}{\rho_f} \boldsymbol{F} + \theta_f \boldsymbol{g} + \frac{1}{\rho_f} \nabla \cdot \boldsymbol{\tau}\,, \tag{5}$$

where ρ_f is fluid density, p is fluid pressure, $\boldsymbol{\tau}$ is the macroscopic fluid stress tensor, and $\boldsymbol{g}$ is the gravitational acceleration. $\boldsymbol{F}$ is the rate of momentum exchange per volume between the fluid and particles phases. The fluid phase is compressible or incompressible (incompressible equations shown), and fluid and particle phases are isothermal.

The particle acceleration is

$$\frac{d\,\boldsymbol{u}}{dt} = D_p\left(\boldsymbol{u}_f - \boldsymbol{u}_p\right) - \frac{1}{\rho_p} \nabla p + \boldsymbol{g} - \frac{1}{\theta_p \rho_p} \nabla \tau_p \tag{6}$$

where $\boldsymbol{u}_p$ is the particle velocity, ρ_p is the particle density, $\boldsymbol{g}$ is gravity and τ_p is the particle normal stress. The terms represent acceleration due to aerodynamic drag, pressure gradient, gravity and gradient in the interparticle normal stress, τ_p.

Particle properties are mapped to and from the Eulerian grid. The interpolation operator is the product of interpolation operators in the three orthogonal directions. For a particle located at $\boldsymbol{x}_p$, where $\boldsymbol{x}_p=(x_p,y_p,z_p)$, the x-directional component of the interpolation operator to grid cell i, is an even function, independent of the y and z coordinates, and has the properties.

$$S_i^x(x_p) = \begin{matrix} 0 & x_{i-1} \geq x_p \geq x_{i+1} \\ 1 & x_p = x_i \end{matrix} \tag{7}$$

The x and y interpolation operators have a similar form. The particle volume fraction at cell ξ from mapping particle volume to the grid is

$$\theta_{p\xi}=\frac{1}{\Omega_\xi}\sum_{\iota}^{N_p}\Omega_p n_p S_{p\xi} \tag{8}$$

where Ω_ξ is a grid volume, Ω_p is particle volume, n_p is the number of particles in a numerical particle, and the summation is over all numerical particles N_p. From conservation of volume, the sum of fluid and particle volume fractions equals unity, $\theta_p+\theta_f=1$.

The implicit numerical integration of the particle velocity equation is

$$\boldsymbol{u}_p^{n+1}=\frac{\boldsymbol{u}_p^n+\Delta t\left[D_p\boldsymbol{u}_{f,p}^{n+1}-\frac{1}{\rho_p}\nabla p_p^{n+1}-\frac{1}{\rho_p\theta_p}\nabla\tau_p^{n+1}+\boldsymbol{g}\right]}{1+\Delta t D_p} \tag{9}$$

where $\boldsymbol{u}_{f,p}^{n+1}$ is the interpolated fluid velocity at the particle location, ∇p_p^{n+1} is the interpolated pressure gradient at the particle location, $\nabla\tau_p^{n+1}$ is the interpolated particle stress gradient at the particle location, D_p is the drag coefficient at a particle location. Particles are grouped into numerical-particles each containing n_p particles with identical properties located at position, $\boldsymbol{x}_p$. The numerical new-time particle location is

$$\boldsymbol{x}_p^{n+1}=\boldsymbol{x}_p^n+\Delta t\,\boldsymbol{u}_p^{n+1} \tag{10}$$

The interphase drag coefficient is

$$D_p=C_d\frac{3}{8}\frac{\rho_g}{\rho_p}\frac{\left|\mathbf{u}_f-\mathbf{u}_p\right|}{r}, \tag{11}$$

and the Wun and Yu (1966)drag correlation is used

$$\begin{aligned}&\text{Re}<1000\quad C_d=\frac{24}{\text{Re}}\left(1+0.15\,\text{Re}^{0.687}\right)\theta_f^{-2.65}\\&\text{Re}\geq 1000\quad C_d=0.44\,\theta_f^{-2.65}\end{aligned} \tag{12}$$

$$\text{Re}=\frac{2\rho_f\left|\boldsymbol{u}_f-\boldsymbol{u}_p\right|r}{\mu_f}, \tag{13}$$

where μ_f is the gas viscosity and the particle radius is

$$r=\left(\frac{3\Omega_p}{4\pi}\right)^{1/3} \tag{14}$$

Particle-to-particle collisions are modeled by the particle normal stress. The particle stress is derived from the particle volume fraction which is in turn calculated from particle volume mapped to the grid. The particle normal stress model used here is (Snider, 2001)

$$\tau=\frac{P_s\theta_p^\beta}{max\left[\left(\theta_{CP}-\theta_p\right),\epsilon\left(1-\theta_p\right)\right]}, \tag{15}$$

where ε is a small number on the order of 10^{-7} to remove the singularity. The close-pack limit is somewhat arbitrary and depends on the size, shape and ordering of the particles. Therefore the solution method allows the particle volume fraction, at times, to slightly exceed close-pack which is physically possible considering that shifting or rearranging of granular materials may occur. The particle normal stress is mapped and applied to discrete particles. Because particles have subgrid (no grid) behavior, the application of the normal stress gradient to a discrete particle is modeled and accounts for the particle properties and whether the particle is moving with or against the stress gradient (Snider, 2001).

The conservation equations are approximated and solved by finite volumes with staggered scalar and momentum nodes. The fluid momentum equation implicitly couples fluids and particles through the interphase momentum transfer. The interphase momentum transfer at momentum cell ξ is

$$F_{\xi}^{n+1}=\frac{1}{\Omega_{\xi}}\sum_{p} S_{\xi}\left[D_{p}\left(\boldsymbol{u}_{f,p}^{n+1}-\boldsymbol{u}_{p}^{n+1}\right)-\frac{1}{\rho_{p}}\nabla p_{p}^{n+1}\right] n_{p} m_{p} \quad , \tag{16}$$

Particle drag and properties are interpolated to the grid. If a top hat interpolation is used, the fluid velocity at particle position $\boldsymbol{x}_p$ is either zero or the node velocity. If a trilinear or other interpolation is used, the fluid velocity at particle position $\boldsymbol{x}_p$ includes the node velocities in support of the interpolation.

Mass continuity is from the material motion of particles. The mixture fluid mass is calculated from volume continuity, $\theta_f=1- \theta_p$ A SIMPLE solution scheme with a coupled solids momentum is used to adjust pressure, density and fluid velocity to satisfy fluid continuity. Velocity, pressure and density correction dependence, estimated from the momentum equation are entered into the fluid continuity equation, giving a semi-implicit coupling of mass and momentum in a pressure correction equation. The pressure equation is solved and fluid pressure, density and velocities are corrected to satisfy continuity.

The gas mixture properties are from the mass fractions of the individual gas species properties. An ideal gas equation of state is used for the pressure-density dependency. Gas species transport is calculated from the bulk mixture flow. The Livermore National Laboratory CVODE ordinary differential equation (ODE) solver (Hindermarsh and Serban, 2006), is used to calculate the chemistry rate. For the simple ozone decomposition in this work, the CVODE solver , which is for sparse/stiff ODE equation sets, is over-kill.

Analytic solution for uniform packed bed

If a fluidized bed has an ideal uniform solid distribution, there is an analytic solution for the decomposition of ozone. The one dimensional transport of ozone through a bed is described by

$$\rho_{mix}\theta_f u\frac{dY_{O_3}}{dz}=\theta_f\rho_{mix}\frac{dY_{O3}}{dt} \tag{17}$$

where ρ_{mix} is the gas mixture density, u is the fluid velocity, z is the axial distance, Y_{O3} is the mass fraction of O_3 and θ_f is the fluid volume fraction. The first order reaction, given by Eq. (2), is substituted into Eq. (17) giving the rate of production of ozone

$$\rho_{mix} u\frac{dY_{O_3}}{dz}=-\rho_{mix} k\theta_p Y_{O_3} \tag{18}$$

The mass fraction is

$$Y_{O_3}=Y_{O_3}^{o} e^{-\frac{k\theta_p}{u}z} \tag{19}$$

where Y^o is the initial value (at t=0). The superficial velocity is $U=\theta_f\ u,$ which is substituted into Eq (18) giving

$$Y_{O_3} = Y^o_{O_3} e^{-\frac{k\theta_p\theta_f}{U}z} \tag{20}$$

Figure 3 shows that the Barracuda calculation for a uniform bed of solids compares well to the analytic solution Eq. (20). Because the gas velocity was well above the minimum fluidization velocity, the ideal bed would not be stationary and therefore, the solids were forced to be stationary. While this is not physically correct for solid dynamics, it gives a constant uniform distribution of solids from which the chemistry could be tested.

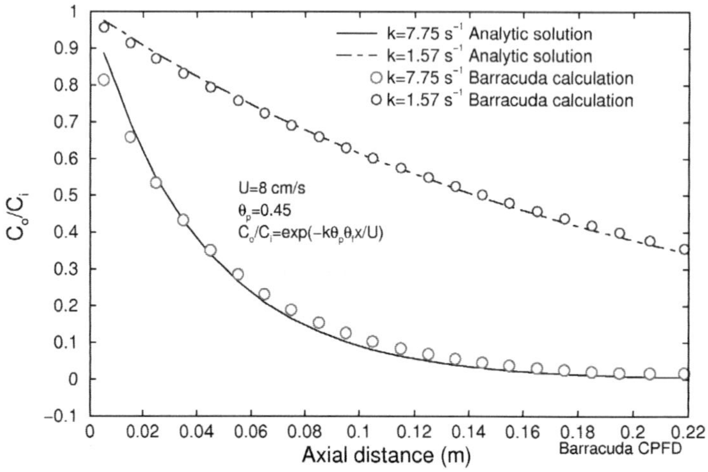

Figure 1. Barracuda calculation compared with analytic solution. Analytic solution uses k=1.57 s^{-1} and k=7.75 s^{-1}, θ_p=0.45, and x=11.5 cm.

Fluidized bed ozone decomposition

The ozone decomposition is calculated in a fluidized bed. The experiment by Fryer and Potter (1976) is modeled. The experiment vessel was 22.9 cm diameter and 200 cm tall. The packed bed contains solid catalyst of silica sand impregnated with Fe_2O_3 .The bed is fluidized with dry air with O_3 added. The parameters are listed in Table 2. All calculations were run for the full three-dimensional bed.

Figure 2 shows the solids in the fluidized bed colored by volume fraction. The volume fraction spatial parameter is mapped to the discrete particle locations for visualization. The bed is bubbling, and like the experiment, the inner details of the bed are not distinct. Figure 3 shows the solid volume fraction by two other graphic methods in a cut section of the bed. Figure 3a shows the volume fraction of solids on the Eulerian grid, which is similar to calculated data using an Eulerian representation of solids. The light areas are void structures or "bubbles", and the dark areas are more dense packed regions of solids. Figure 3b shows the solid volume fraction greater than 0.3. The holes are the void structures of gas rising through the bed.

Figure 4 shows the mass fraction of O_3 next to the solids in the fluidized bed. The view is with the front half of the bed removed. The ozone enters the bed at 0.1 mass fraction and starts decomposing to oxygen in the presence of the catalyst. At the bottom of the bed, the catalyst is relatively uniformly mixed compared to higher in the bed where multiple "bubbles" have formed. This uniform catalyst-gas mixture gives a near uniform O_3 decomposition for the first few centimeters. Further up in the bed, plumes of O_3 rise in the bed in gas voids ("bubbles") while O_3 in denser packed regions of catalyst decompose to oxygen. Above the bed, O_3 which did not decompose to O_2 drifts in the bulk flow and flows out of the bed.

The prediction of the ozone decomposition in a fluidized bed requires accurate prediction of the bed dynamics. In calculating fluidized beds, it is common to adjust the drag coefficient based on a measured fluidization velocity. This allows calculations which use a single particle size to adjust to a best-fit particle size for the calculation. This adjustment also allows a correlation to better represent a particular system. In this study, there were no adjustments in calculation parameters to fit or tune the calculation for the problem. The performance of the drag model is deduced from the predicted pressure, as the inlet velocity is ramped with time. Fryer and Potter (1976) reported the measured minimum fluidization velocity (U_{mf}) was 1.7 cm/s and Fig. 5 shows from the calculated U_{mf} is approximately 1.6 cm/s. A solid size distribution was not available for this study, and possible inaccuracies in calculation parameters with respect to experimental conditions, including inaccuracy in the drag correlation are possible sources for the small discrepancy between measured and calculated U_{mf}. From the relatively good comparison between measured and calculated minimum fluidization velocity, the standard Win and Yu (1966) is judged sufficient without adjustment.

Table 2. Simulation conditions for a fluidized bed

Vessel diameter	22.9 cm
Vessel height	30 cm
Top pressure boundary for open vessel	101 kPa
Temperature	300 K
Inlet mixture N_2	0.702 mass fraction
Inlet mixture O_2	0.198 mass fraction
Inlet mixture O_3	0.1 mass fraction
Mixture flow rate	ranges from 2 to14 cm/s
Solid-catalyst	Fe_2O_3 with a silica base
Solid-catalyst density	2655 kg/m^3
Solid-catalyst radius	60 μm [1]
Solid-catalyst sphericity	1
Reaction rate constant, k	0.33 to 1.57 s^{-1} [2]
Initial solid-catalyst volume fraction	0.6

1. No information was given on the particle size distribution
2. Units from reaction rate equation, (s^{-1}).

The fluidized bed calculations were run sufficiently long at a fixed inlet flow rate to get a quasi steady rate of ozone decomposition. Figure 6 shows the ozone at the top of the vessel for an inlet flow rate of 7 cm/s and a chemical rate coefficient k= 1.57 s$^{-1.}$ At 24 s, the inlet velocity is increased to 11 cm/s and the calculation continues to run beyond the time period shown in the figure. Figure 7 shows the calculated and measured ozone decomposition as a function of inlet flow rate. The comparison is excellent. As the inlet flow rate increases, less ozone decomposes to oxygen. In the bubbling fluidized bed, ozone moving in bubbles bypasses solid catalyst contact. The prediction from the analytic solution for a uniform, stationary packed bed also compares well with the measured ozone decomposition. This is partly due to choosing the solids volume fraction for the analytic solution from the CPFD calculation. In general, an a priori effective solids volume fraction can not be chosen for a fluidized bed.

Figures 8 and 9 show the calculated ozone decomposition for k=0886 s^{-1} and k=0.33 s^{-1}, respectively. The comparison between Barracuda and experiment is also good. The calculation is shifted a little above or a little below the measured decomposition data. The variation could be from slightly different calculation parameters from those used in the experiments. There was a range of measured rate coefficients that where combined to get the average coefficient which is used in the Barracuda calculation. For the average k=7.75, Fryer and Potter reported rate constants from 7.05 to 8.05. The reaction rate from this range of measured coefficients can produce a variation in decomposition of ozone from -105% to 26% relative to the average reaction constant (analysis based on analytic solution). The ozone enters the bed at 1.5 cm from the bottom, while Barracuda calculations have ozone entering at the bottom. As noted by Fryer and Potter (1976) this gives an effectively shorter ozone decomposition zone. A 11.5 cm deep bed has a 13% smaller effective reaction zone. For a 24 cm deep bed, the effective reaction zone is 6% smaller. Again, the the analytic solution for a uniform packed bed compares well with the measured data.

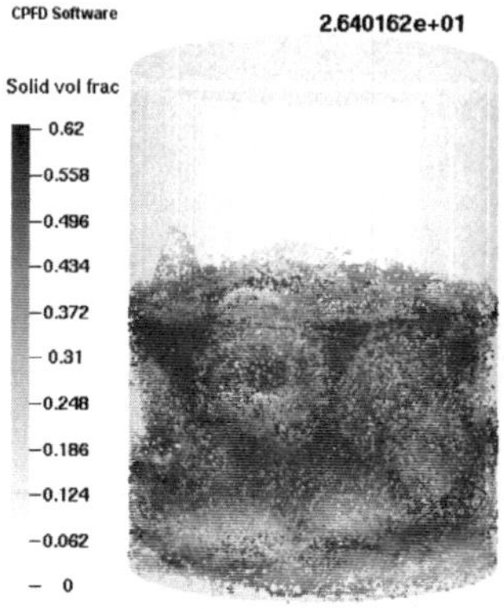

Figure 2. Calculated solids distribution. Discrete solids are shown colored by volume fraction mapped from the grid to particle locations.

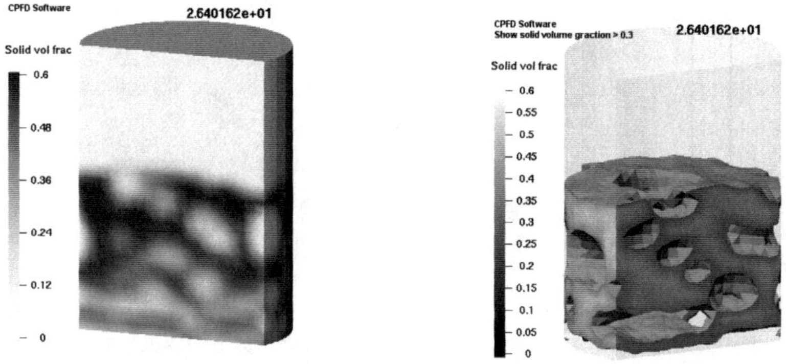

a

b

Figure 3. Calculated solids field, colored by volume fraction. U_{inlet}=11 cm/s, static bed height is 11.5 cm and k=1.57 s^{-1}. Figure b shows volume fraction greater than 0.3.

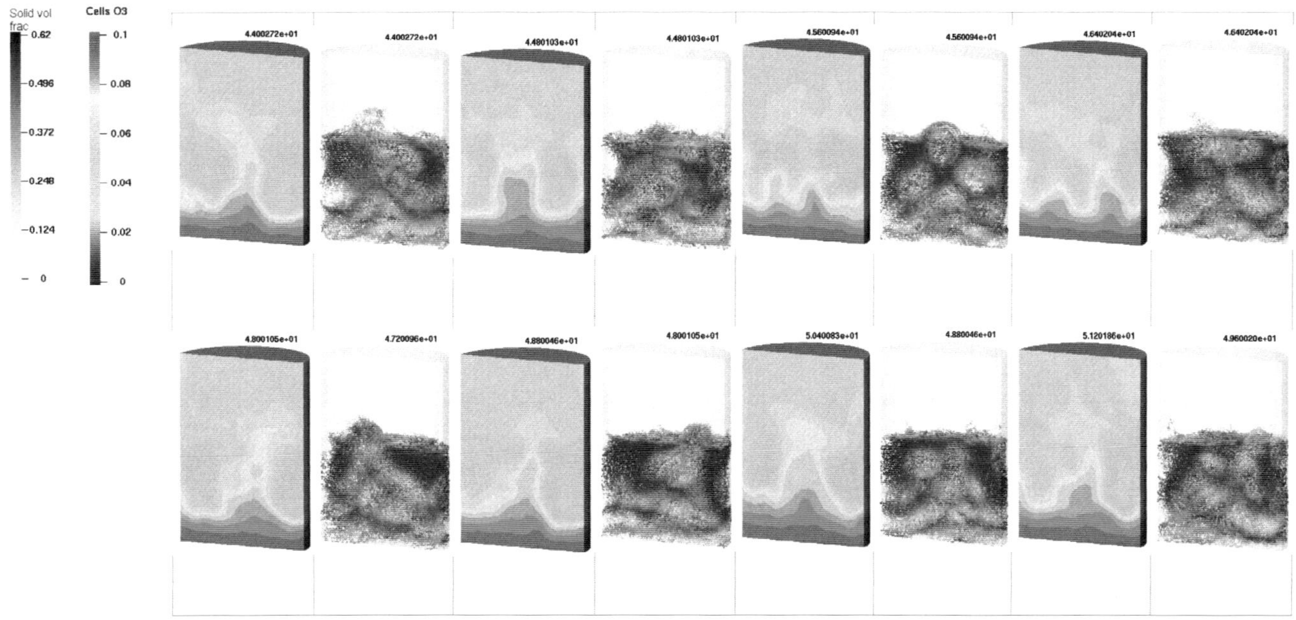

Figure 4. Calculated O_3 mass fraction and catalyst-solids colored by volume fraction. U_{inlet}=11 cm/s, static bed height is 11.5 cm and $k=1.57\ s^{-1}$.

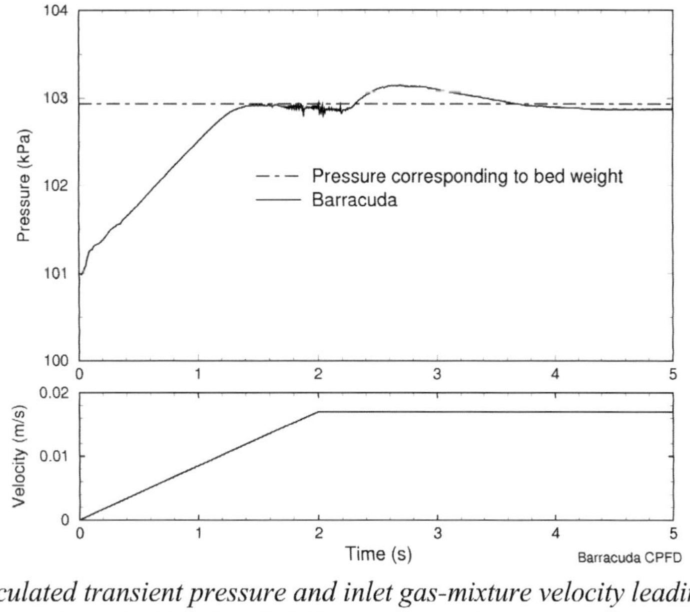

Figure 5. Calculated transient pressure and inlet gas-mixture velocity leading to minimum fluidization. Static bed height of 11.5 cm and reaction rate coefficient $k=1.57\ s^{-1}$

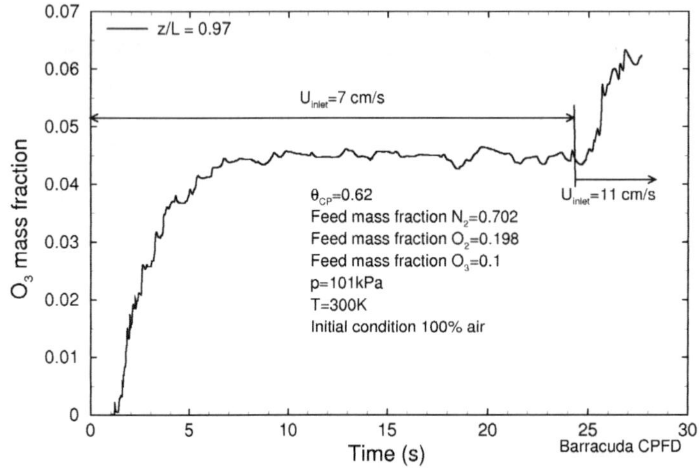

Figure 6. Calculated transient O_3 mass fraction at inlet mixture velocity of 7 cm/s, static bed height of 11.5 cm and reaction rate coefficient $k=1.57\ s^{-1}$

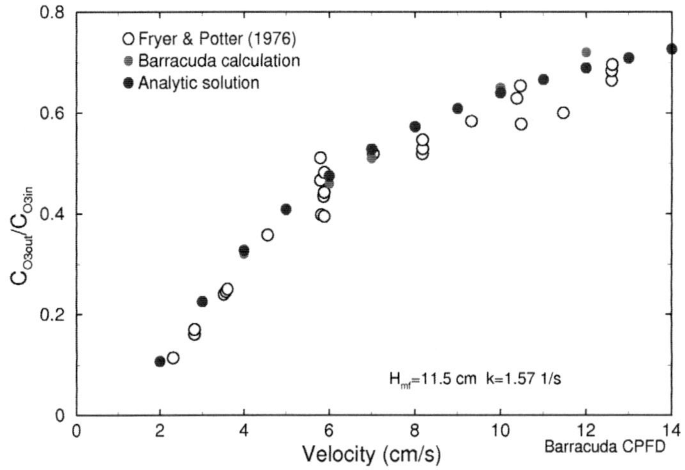

Figure 7. Measured and calculated O_3 normalized mass fraction as function of inlet velocity. Static bed height is 11.5 cm and chemical rate constant $1.57\ s^{-1}$. Analytic solution uses $k=1.57\ s^{-1}$, $\rho_{f_0}=1.2\ kg/m^3$ $\theta_p=0.45$, and $x=11.5$ cm.

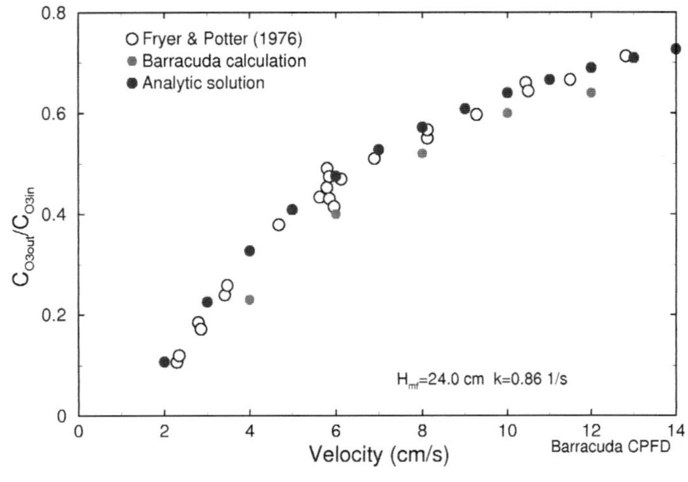

Figure 8. Measured and calculated O_3 normalized mass fraction as function of inlet velocity. Static bed height is 24 cm and chemical rate constant 0.86 s^{-1}. Analytic solution uses $k=0.86\ s^{-1}$, ρ_f,=1.2 kg/m^3 $\theta_p=0.45$, and x=24 cm.

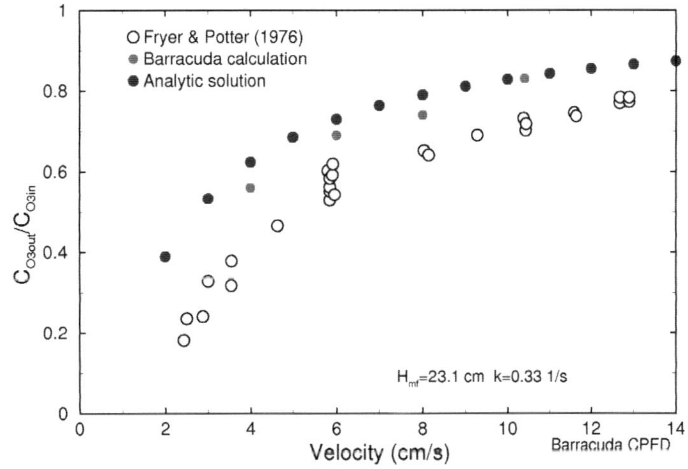

Figure 9. Measured and calculated O_3 normalized mass fraction as function of inlet velocity. Static bed height is 23.1 cm and chemical rate constant 0.33 s^{-1}. Analytic solution uses $k=0.33\ s^{-1}$, ρ_f,=1.2 kg/m^3 $\theta_p=0.45$, and x=23.1 cm.

Concluding remarks

The implicit solution of the ozone decomposition and explicit calculation of oxygen generation was calculated well by the CPFD numerical method. The fluidized bed calculation predicted the "usual" void structures or "bubbles"; however, there were instantaneous complex solids and gas flow patterns. The ozone chemistry is a simple and well behaved reaction rate. The correct calculation of the chemistry depends on correct calculation of the bed motion. The bubbles transport gas with little gas-catalyst contact which reduces the chemistry yield. In this study, the complete set of fluid and solid mass and momentum equations, and the chemistry equations were solved. All calculations were made in three dimensions, and no parameters were tuned or adjusted to fit the experiment.

The analytic ozone decomposition solution also compared well to the experimental data. The good agreement is not surprising considering that the catalyst concentration in the analytic solution was from the CPFD bubbling bed calculation. In general, the bed dynamics and solids concentration are unknown, and if the bed has streamers or various degrees of mixing, a suitable average solids concentration cannot be simply chosen.

Reference

Andrews, M. J., and O'Rourke, P. J. (1996) The Multiphase Particle-in-Cell Method (MP-PIC) Method for Dense Particle Flow, Int. J. Multiphase Flow, 22, 379-402 .

Fryer, C. and Potter, O.E, (1976), "Experimental investigation of models for fluidized bed catalytic reactors," AIChE J., **22**.

Hindermarsh, A. C., and Serban, R., (2006), *User documentation for CVODE v2.5.0*, Center for Applied Scientific Computing, Lawrence Livermore National Laboratory, UCRL-SM-208108.

Snider, D. M., et al. (1998), Sediment Flow in Inclined Vessels Calculated Using Multiphase Particle-in-Cell Model for Dense Particle Flows, Int. J. Multiphase Flow, 24, 1359-1382.

Snider, D. M. (2001) An Incompressible Three-Dimensional Multiphase Particle-in-Cell Model for Dense Particle Flows," J. Comput. Phys., 170, 523-549

Wen, C.Y. and Yu, Y.H., (1966) "Mechanics of fluidization," Chem. Eng. Progr. Symp., Ser. 62, 100-110.

292d Numerical Simulation of Pneumatic Drying Using Computational Fluid Dynamics

Tarek Jamaleddine and Madhumita Ray. Chemical and Biochemical Engineering, University of Western Ontario, Thompson Engineering Building, London, ON N6A5B9, Canada

Despite the industrial importance of drying processes, specifically in drying of slurries, the design and operation of dryers are still mostly based on empirical rules. Although research in this field is progressively growing, it is still in its early stages and the physical phenomena involved in these processes are not yet fully understood. Commercial slurry dryers encounter high manufacturing and operating costs. The design and optimization of these dryers for different slurries and different operating conditions are conducted based on expensive pilot scale studies. With the advent of sophisticated computational fluid dynamics (CFD) codes and high speed computing, CFD can be used as an effective tool for design and optimization of such devices. The objective of this research is to study and evaluate industrial drying processes, in particular, those conducted in pneumatic dryers, using state of the art commercial CFD packages. This approach also enables the prediction of physical properties at discrete points inside the computational domain, which lead to a better understanding of the heat and mass transfer phenomena occurring in slurry dryers. The hydrodynamics and drying mechanisms of a two-dimensional, dilute-phase, gas-solid flow have been predicted in a laboratory- and commercial (large)-scale vertical riser. The CFD method is based on the Eulerian-Eulerian two-fluid multiphase flow model. A control-volume based technique using commercial CFD package and the kinetic theory of granular flow (KTGF) were applied to simulate the flow pattern and heat- and mass-transfer processes for porous particles (sand & PVC) in the presence of hot air. The time-averaged theoretical drying rates in the constant- and falling-rate drying periods agree reasonably well with the published experimental data for similar models. The model also predicts a core-annulus flow in the riser with maximum velocity of both phases in the core center and denser down-flow pattern close to the walls. User-defined subroutines were implemented to extend the CFD capability to account for the mixture properties and to simulate the two-drying stages using mass-transfer models. The effects of different operating conditions and dryer design on the drying kinetics are evaluated.

Key Words: Pneumatic dryer, Pneumatic drying, Two-fluid model, CFD, Modeling, Heat and Mass transfer.

299a Calibration of DEM Material Models

John Favier, DEM Solutions Ltd, 49 Queen Street, Edinburgh, EH1 3NH, United Kingdom

As industry increases its use of DEM for simulation of particulate handling and processing operations, attention is increasingly focused on model calibration and methods for dealing with scaling and multi-scale modeling issues. Here we examine the basis for choosing a DEM model relative to the bulk particulate material and the process being simulated, and outline strategies for calibrating DEM models against measured particle and bulk properties and behavior.

A DEM model is a combination of the particle inertial properties (size, shape, density) and mechanical properties (stiffness, elasticity, plasticity) and interaction terms between particles and other objects – particles or boundary surfaces. The latter depends on the choice of contact models which may include parameters such as friction, restitution coefficient and cohesion for example, and the body force model, usually gravity, but also could include fluid drag, electrostatic force, and magnetic force, and so on. Together these properties and parameter values form what we describe as the DEM material model.

The choice of DEM material model depends on the focus and objectives of the simulation. In bulk transport problems where the objective is to predict local mass flow rates and load transfer to boundary surfaces, the choice and calibration of the material model is dependent only on bulk flow characteristics. In applications where the individual particle dynamics and interactions, for example, particle orientation, number and area of contacting surfaces, load distribution through the bulk are important, then the particle inertial properties and contact physics will need to be closer to the actual system. The third category of DEM simulation is one where the objective is to relate correlate (particle-scale) interactions with process-scale conditions, usually by simulating a sub-region of the system. In that case the focus is usually on the interactions between particles, with equipment surfaces or with surrounding media, which influences the choice of material model to ensure that governing particle properties and interaction terms, such as specific mass, surface area and contact area and deformation are more closely matched. The approach to material model calibration is likely to differ for each of these categories of DEM simulation.

This paper will discuss some of the methods currently employed for calibrating DEM material models and look at developments in computational, experimental and optimization techniques to advance this important aspect of DEM simulation.

299b Evaluation of Contact Models for Discrete Element Method Simulations

Joerg Theuerkauf, Solids Processing, The Dow Chemical Company, Freeport, TX 77566 and D.V.N Prasad, R&D-IMEA Engineering &Process, The Dow Chemical Company, 6th Floor, Tower C/ Panchshil Tech Park, Pune, India.

Discrete element method simulations have increasingly gained importance to solve many complex granular flow systems. The outcome of the simulation depends on the input parameters and the contact models that are used. To gain insight into the flow of granular material as a function of different contact models two different systems have been studied.

The paper focuses on a biaxial shear tester as well as on a rotating drum to show the influence of different contact laws for low as well as high shear environments.

The variation in the flow behavior of granular material is shown when applying four contact models for particle-particle interactions i.e. linear, Hertzian, Luding's and Burger's contact models to particles in the two testers.

299c The Effect of Particle Shape on Bulk Powder Friction In a Schulze Ring Shear Tester Using the Discrete Element Method

William R. Ketterhagen[1], Jennifer S. Curtis[2], Carl R. Wassgren[3], and Bruno C. Hancock[1]. (1) Pfizer Global Research and Development, Pfizer Inc., Groton, CT 06340, (2) University of Florida, Chemical Engineering Department, P.O. Box 116005, Gainesville, FL 32611-6005, (3) Mechanical Engineering, Purdue University, 575 Stadium Mall Drive, West Lafayette, IN 47907-2088

The bulk flow behavior of particulate systems is known to be greatly affected by the morphology of the particles from which it is composed. Materials with angular, blocky, or elongated particle shapes will tend to have greater yield strengths and internal friction angles in addition to reduced flow rates through process equipment such as hoppers. While such quantities are readily measured experimentally for a given material, it is difficult to quantitatively assess the specific role of particle shape. In contrast, by using a computational approach such as the discrete element method (DEM), particle characteristics such as shape can be systematically studied and the effect on the bulk system can then be assessed.

In this work, a Schulze ring shear tester is modeled using DEM and the effects of particle shape and particle-particle friction on the bulk friction angles are observed. Monodisperse systems of frictional, non-cohesive particles are modeled with the commercial DEM package EDEM (DEM Solutions Ltd.), and the particle shape is systematically varied from spheres to elongated rods. The bulk friction measurements consisting of the internal friction angle and the wall friction angle are made for each particle shape and a range of input friction coefficients (particle-particle and particle-wall coefficients). For spherical particles, results show that the internal friction angle reaches a plateau for increasing particle-particle friction coefficients. However, if particle rotation is prevented, the internal friction angle increases significantly to large values. Particles of larger aspect ratios also show an increase in the internal friction angle. Movies from the simulations as well as the translational and rotational velocity data give insight into the physics within the system.

299d Force Model Considerations for Glued Sphere Dem Simulations

Madhusudhan Kodam[1], Rahul Bharadwaj[2], Jennifer Sinclair Curtis[3], Bruno C. Hancock[2], and Carl R. Wassgren[1]. (1) Mechanical Engineering, Purdue University, 500 Central Drive, Room # 209, Potter Building, West Lafayette, IN 47907, (2) Pfizer Inc., MS 8156-007, Eastern Point Road, Groton, CT 06340, (3) University of Florida, Chemical Engineering Department, P.O. Box 116005, Gainesville, FL 32611-6005

Despite knowing that particle shape plays a significant role in the dynamics of powder flow, most discrete element method (DEM) simulations utilize spherical particles. The reasons for using spherical particles are that (a) the contact detection scheme for spherical particles is simple, and (b) the contact force models for contacting spheres are well known (e.g. a Hertzian contact).

Several schemes for modeling non-spherical particles have been proposed including those that concern polyhedra, ellipsoids, sphero-cylinders, and superquadrics. Perhaps the most common approach for modeling non-spherical particles, however, is using "glued spheres," in which irregular particle shapes are produced by rigidly connecting individual, and possibly overlapping, spheres. The advantage of the glued spheres approach is that even for complex particle shapes the simple spherical contact detection algorithm may be retained.

Recent publications have focused on how approximating a given particle shape using a glued sphere geometry affects the rebound of colliding particles (e.g. Price et al., 2007 and Kruggel-Emden et al., 2008). These investigations have focused on the errors introduced by approximating the geometry of the true particle shape. What has not been investigated, however, is how the spherical particle derived force models used in glued sphere particle geometries influence the response of particle collisions. This paper demonstrates that in instances where more than a single component sphere in a glued sphere model is involved in a contact, a modified force model must be used to produce an accurate force-deflection response.

299e Effect of Dem Parameters on the Collision Distribution of a Small Shearing Region beneath the Surface of a Rotating Drum

Ben Freireich, School of Mechanical Engineering, Purdue University, 500 Central Drive, Rm. 209a Potter Building, West Lafayette, IN 47907, James D. Litster, Chemical Engineering and Industrial Pharmacy, Purdue University, Forney Hall of Chemical Engineering, Room 2154, 480 Stadium Mall Drive, West Lafayette, IN 47907-2100, and Carl R. Wassgren, Purdue University, 585 Purdue Mall, West Lafayette, IN 47907.

Traditionally, DEM simulations are validated via particle trajectories and velocity fields using experimental techniques such as PIV or PEPT. However, some researchers use DEM simulations to extract particle-scale measurements such as collision frequency. It is unclear whether a validation of a macroscopic velocity field also implies a validation of particle-scale measurements. In this study, measurements of solid fraction, shear rate, and collision frequency and impact velocity are made in a small region just below the free surface in a rotating drum. The effects of periodic drum length, particle stiffness, coefficient of restitution, and particle size are investigated. It is found that the solid fraction and shear rate within the studied region are unaffected by variations in particle stiffness and coefficient of restitution over the range of values studied. However, the rate of collisions increases for both increasing particle stiffness and coefficient of restitution. Also, the mean collision speed decreases with increasing stiffness, and decreasing coefficient of restitution. The shear rate within the region decreases as particle size increases, while the mean collision velocity remains unaffected. These findings suggest that care must be taken in choosing DEM force models and model parameters in order to accurately model collision distributions, while particle size need not be matched exactly.

361a Controlled Patterning of Nanostructured Surfaces Using Self-Assembly

Chetana Singh, Department of Chemical Engineering, University of Michigan, Ann Arbor, MI 48109-2136 and Sharon C. Glotzer, Chemical Engineering and Materials Science and Engineering, University of Michigan, 2300 Hayward Street, Ann Arbor, MI 48109.

Surface patterns such as stripes or dots have potential applications in several areas including ceramics, photonics, data storage, catalysis, biosensing and bottom-up assembly of novel materials. When the substrate has a high degree of curvature as in the case of nanoparticles and nanotubes, the self-assembled patterns affect its functionalization into a nano-building block and subsequent self-assembly into more complex structures [1,2]. For obtaining self-assembled patterns, block copolymer phase separation has traditionally been used. We present a simulational study of an alternative method to design nanopatterned surfaces by exploiting the immiscibility of a binary mixture of surfactants that form a self-assembled monolayer on the surface [1]. We show how the feature sizes and shapes of patterns obtained by this method can be tailored by varying the length difference or bulkiness of the molecules' tails, surface coverage, relative composition of the two molecules, their immiscibility and substrate curvature [3]. We examine both binary and ternary mixtures on both spherical (nanoparticles) and cylindrical surfaces (nanotubes and nanowires) [4]. A comparison with experimental findings is also provided [2, 3].

1. AM Jackson, JW Myerson, F Stellacci, Nature Materials, 3, 330-336 (2004).

2. RP Carney, GA DeVries, C Dubois, H Kim, JY Kim, C Singh, PK Ghorai, JB Tracy, RL Stiles, RW Murray, SC Glotzer and F Stellacci, Journal of American Chemical Society, (Communication), 130(3), 798-799 (2008).

3. C Singh, PK Ghorai, MA Horsch, AM Jackson, RG Larson, F Stellacci and SC Glotzer, Physical Review Letters 99, 226106 (2007).

4. C Singh and SC Glotzer, preprint.

MOLECULARLY ENGINEERED ORGANIC PARTICULATE FILMS: PREPARATION, STRUCTURE AND PERFORMANCE

Ozge Guney-Altay, Dmitry Pestov and Gary Tepper

Microsensor and Radiation Detector Research Laboratory
Department of Mechanical Engineering, Virginia Commonwealth University, Richmond, VA 23284

Diethyl *p*-phenylenediacrylate (EPA) particles with tunable molecular adsorption characteristics were produced by a novel molecular imprinting procedure applied during the solid-state photo-polymerization reaction. EPA monomer is micronized by Rapid Expansion of Supercritical Solutions (RESS) and deposited directly on Surface Acoustic Wave (SAW) resonators. The templated photo-polymerization of EPA monomer particles was performed on the SAW devices. During the UV initiated solid-state polymerization EPA forms a linear polymer[1,2].

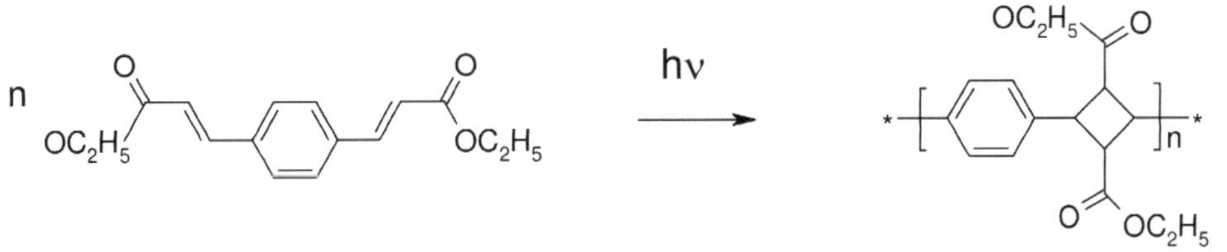

The molecular sieving capability of EPA polymer particles is demonstrated by monitoring the SAW resonance frequency during the sorption of alkane analytes of different sizes. The mass changes due to vapor uptake resulting from the selective sorption into the particulate coatings can be measured accurately due to the nanogram sensitivity of the piezoelectric crystal. Although the molecular porosity was created using template molecules from a different chemical family, these coatings exhibited reversible sorption properties when tested against alkane analytes[3]. A gradual exclusion of smaller analyte molecules from the molecularly imprinted particulate coatings is observed as the size of the template molecule is reduced as shown in Figure 1.

The focus of this presentation will be on the identification of the mechanisms controlling the self-organization and morphological transformations observed during the photopolymerization reaction in the presence of specific template vapors. The structure-property relationship as well as applications of these unique coatings will be discussed.

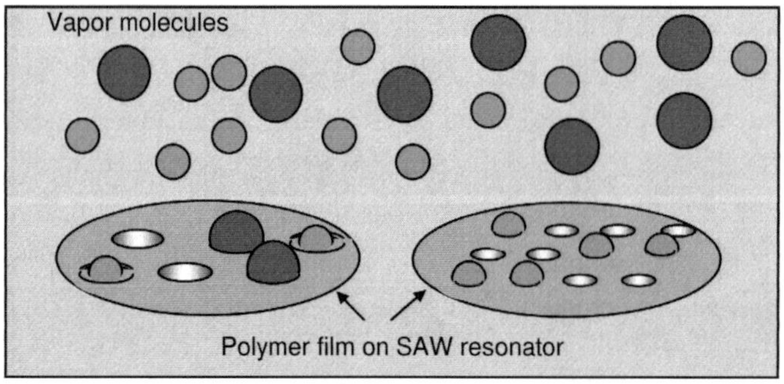

Figure 1. Schematic representation of the exclusion of analyte molecules observed in the molecularly imprinted particulate coatings as a function of template molecule size.

References

1. F. Suzuki, Y. Suzuki, H. Nakanishi, M. Hasegawa, "Four-center type photopolymerization in the solid state. III. Polymerization of Phenylenediacrylic Acids and Their Derivatives" *Journal of Polymer Science, Part A-1-Polymer Chemistry*, **7**, 2319-2332, 1969.

2. M. Hasegawa. "Four-center Photopolymerization in the Crystalline State" *Advances in Polymer Science*, **42**, 1-49, 1982.

3. Guney-Altay, O., Pestov, D., Tepper, G., "Organic Zeolites from a Diolefinic Monomer", *Journal of the American Chemical Society*, **129**, 13957-13962, 2007.

361c Nanostructured Coatings by Combined Flame Spray Synthesis of Nanoparticles and Cluster Beam Deposition

Simone Vinati[1], Alessandro Antonini[1], Paolo Piseri[1], Paolo Milani[1], Luca Belforte[2], Nello Li Pira[2], and **Karsten Wegner**[3]. (1) Tethis S.r.l., Via Russoli, 3, I-20143 Milan, Italy, (2) Advanced Manufacturing & Materials, Centro Ricerche Fiat, Strada Torino, 50, I-10043 Orbassano, Italy, (3) Swiss Federal Institute of Technology (ETH), Sonneggstrasse 3, CH-8092 Zurich, Switzerland

A novel one-step deposition method for nanostructured coatings combining gas-phase flame synthesis of nanoparticles and supersonic cluster beam deposition is presented. Metal-oxide and noble metal nanoparticles are first produced by high temperature flame spray pyrolysis. The combustion aerosol is then expanded through a critical flow nozzle into the first vacuum chamber of the deposition unit. Thereby, the aerosol is rapidly quenched freezing nanoparticle growth. A series of aerodynamic lenses in the vacuum deposition apparatus filters out large particles and focuses the remaining aerosol into a beam. The particle beam is then directed at a substrate generating porous nanoparticulate films, for application in catalysis, gas sensing, or biotechnology.

361d Factors Affecting Strength of Porous Films Made by Gas to Particle Conversion/ Thermophoretic Deposition of Nanoparticle Agglomerates

Oluwatosin Ogunsola[1], Ratna Oetama[2], Tridib Bhowmick[1], Seonmin Kim[1], and **Sheryl Ehrman**[1]. (1) Chemical and Biomolecular Engineering, University of Maryland, Building 90, College Park, MD 20010, (2) Fischell Department of Bioengineering, University of Maryland, Building 90, College Park, MD 20010

High surface area porous films are of interest for applications ranging from catalysts to solar light harvesting for photovoltaic energy conversion. In this work, porous films of titania were synthesized via a hybrid process based upon gas-to-particle conversion and thermophoretic deposition. In the final product of this process, agglomerates of titania nanoparticles made at high temperature are held together by van der Waals forces to form the nanoporous film with a porosity of greater than 95%. These films can be grown rapidly, greater than a micron per minute, but the films are fragile as prepared, with a modulus on the order of 0.50 Mpa as measured using atomic force microscopy. Using this AFM technique, the effects of primary particle size as well as the effects of infiltration and capillary condensation and low temperature reaction of a second species on the Young's modulus of the porous films were explored. The goal of the capillary condensation step was to strengthen the films by condensing material at the neck between particles, increasing the interconnectivity of the film by replacing van der Waals bonds by chemical bonds. Condensation of a second species into the structure increased the Young's modulus of the films, although this increase did not correspond to the saturation ratio of the condensing species as expected. Additional results for effects of primary particle size will be presented and discussed in the context of simulation results for this system.

369a Psri Lectureship in Fluidization Award: Fluidization of Nanopowders

Robert Pfeffer, Chemical Engineering, Arizona State University, Tempe, AZ 85287-6006

Text Not Available.

369b Thomas Baron Award: Rheology and Processing of Particle-Loaded Complex Fluids at Particle Concentrations That Approach the Maximum Packing Fraction

Dilhan M. Kalyon, Stevens Institute of Technology, Castle Point St., Hoboken, NJ 07030

Text Not Available.

AUTHOR INDEX

Aarons, Lee R. 14
Abbas, Ali 47
Abu-Nahar, A. 157
Adschiri, Tadafumi 250
Akhtar, Kamal M. 97, 173, 267
Akhuemonkhan, Osifo 90
Akkisetty, Pavan Kumar 448
Aksomaitytė, Gabrielė 202
Allen, Todd Michael 587
Allendorf, Mark D. 232
Almeida-Rivera, Cristhian P. 752
Alonso, Esther 202
Amundsen, P. A. 492
Anand, Anshu 467
Anderson, Kelly 145
Angelov, Ivan 5
Annapragada, Sriram Kiran 627
Antonini, Alessandro 387
Antunes, Elisabete 483
Arastoopour, Hamid 479
Armenante, Piero M. 465
Aromaa, Mikko 89
Arrington, Yeook 33
Atashfaraz, Mehrnoosh 250
Athanassiou, Evagelos K. 539
Avalos, Josep Bonet 688
Ayala, Vanessa 349
Azurdia, Jose 88
Badiola, Carlo 517
Balani, Prashant N. 438
Bang, Youkyung 556
Barghi, Shahzad 430
Barrera, Carola 349
Bartel, Melissa 352
Bastos, Jaci C. S. C. 204
Bastos, Jaci Carlo Schramm Camara 191
Beach, Lauren 98, 99
Beaucage, Gregory 92, 248, 763
Belforte, Luca 387
Beloni, Ervin 569, 699
Benyahia, Sofiane 235
Bharadwaj, Rahul 188, 382
Bhatia, Surita R. 752
Bhatt, Divesh 736
Bhattacharya, Tathagata 134
Bhowmick, Tridib 388
Bichay, Magdy 587
Birkmire, Andrew 36
Boddu, Veera M. 737
Bongers, Peter 752
Bonnett, Peter 252
Borchert, Christian 3
Bothun, Geoffrey D. 775
Breault, Ronald W. 58
Brennan, John K. 688
Bucalá, Verónica 159
Buesser, Beat 170, 176, 254
Bulian, Chris J. 579
Cabrejos, Francisco J. 297
Cameron, I. T. 6
Caram, Hugo S. 470, 471
Caramellino, Micaela 465
Carpen, Ileana C. 233
Carroll, Nick 555
Castellanos, Antonio 229
Castner, David G. 96
Castro, Miguel 150
Chakka, Siri 737
Chalermsinsuwan, Benjapon 58
Chan, Sui Yung 438
Chandross, Michael E. 145
Chang, Jane P. 152

AUTHOR INDEX

Chaudhuri, Bodhisattwa 136

Chen, Alvin U. 269

Chen, Jiaqi 559

Chen, Julie 586

Chen, Yanjing 775

Chen, Yuhua 98, 99

Cheng, Chungyin 759

Cheng, Yi 559

Chester, Arthur W. 25

Cho, Dong-Lyun 171

Cho, Sung-Yong 172

Choi, Wonjoon 537

Chopra, Sachit 92, 248

Chowdhury, Snehaunshu 725

Chun, Byung Soon 100

Chung, Moon 567

Chung, Sang-Ho 175

Churchman, Charles 275

Cocco, Ray A. 33, 622

Cocero, Maria José 202, 624

Conway, Stephen L. 35

Cooke, Michael 124

Coronella, Charles J. 610, 623

Cotabarren, Ivana 159

Couzis, Alexander 268

Cranford, Bruce 643

Cui, Jingqin 353

Cui, Qingzhou 586

Cuitino, Alberto 55

Curtis, Jennifer S. 188, 316, 381, 382, 467

Custers, Johan P. A. 340

Da Rocha, Sandro R. P. 343

Da Rosa, Leonardo Machado 191

Dalvi, Sameer 626

Damavarapu, Reddy 737

Dang, Hongyi 148

Data, Biplab 492

Dave, Rajesh 98, 138, 229, 528, 562, 626

Davies, Clive E. 40, 306

Davis, Jeffrey M. 351

Davis, Zenda 230

De Broqueville, Axel 563, 611

De Wilde, Juray 83, 611

Decker, Rodrigo K. 204

Deng, Shuang 350

Dhodapkar, Shrikant 315

Dias, Daniel De Brito 218

Diemer, R. Bertrum 449

Ding, S. 26, 244, 492

Ding, Yulong 449

Domingues, Maria Cristina 443

Donahue, Carly M. 44

Doorenbos, Zac 529

Doyle, Francis J. 755

Drake, Joshua 270

Dreizin, Edward L. 516, 517, 528, 569, 594, 699

Du, Bing 56, 57

Du, Juan 560

Dubey, Atul 137

Duffadar, Ranojoy D. 351

Duvvuri, Subbarao 411, 499

Dycus, Eric 36

Dyrøy, Are 244

Eggers, Jan 1

Eggersdorfer, Max L. 176

Ehrl, Lyonel 7, 156

Ehrman, Sheryl 90, 388

Ekatpure, Rahul P. 563

Elsner, Martin Peter 5

Eom, Hyoung-Choon 247

Epstein, Norman 100

Ermoline, Alexandre 516

AUTHOR INDEX

Fan, L. -S. 76, 391, 615
Favier, John 379
Ferreira, Paulo 355, 483
Figueroa, Isabel 16
Fogler, H. Scott 46, 754
Follansbee, David M. 409
Ford, Kenneth J. 470, 471
Fox, R. O. 270, 354, 473, 474, 478, 481, 759
Francis, Todd M. 561
Franka, Nathan P. 75
Freireich, Ben 189, 383
Fricke, Michael 146
Fried, Laurence E. 736
Fritsching, Udo 204
Furlan, Marco 348
Fushimi, Chihiro 625
Gamble, Lara J. 96
Gandhi, Devkant S. 442
Garcia, Fernando A. P. 355, 483
Garg, Rahul 474
Gauer, Cornelius 190, 558, 760, 762
Gellman, Andrew J. 352
Gentzler, Michael 35
George, Steven M. 253
Gevgilili, Halil 666, 667, 687
Ghadiri, Mojtaba 449
Gidaspow, Dimitri 58
Gilchrist, James F. 470, 471
Glasser, Benjamin J. 10, 135, 187, 234, 468
Glasser, David 187, 468
Glotzer, Sharon C. 384
Gnichwitz, Jan-Frederik 541
Gogos, Costas G. 251, 252
Gollamandala, Deepika R 233
Gorb, Leonid 747
Gorrepati, Elizabeth A. 46
Grace, John R. 100
Graeve, Olivia A. 344
Graf, Markus 345
Grass, Robert N. 539
Grest, Gary S. 145
Grinshpun, Sergey A. 594
Gu, Zhiyong 586
Guenther, Chris 285
Guldi, Dirk M. 541
Guney-Altay, Ozge 385
Guo, Bing 93
Gupta, Ram B. 144
Hadinoto-Ong, Kunn 316
Haines, Chris 529, 567
Hajra, Suman K. 24, 134
Hall, Talesha T. 482
Halvorsen, Britt M. 66
Han, Ting(Tim) 56, 57, 153, 498
Hancock, Bruno C. 188, 381, 382, 467
Harris, Michael T. 482
Harshe, Yogesh M. 45
Hausberger, Brendon 187, 468
Hawa, Takumi 255, 689
Heckler, Tracy 780
Heindel, Theodore 75, 270
Heine, Martin C. 173, 267
Heitsch, Andrew T. 774
Hennart, Stephen 443
Henson, Michael A. 752, 753
Henthorn, Kimberly H. 469
Henz, Brian J. 255, 689
Herrera, Adriana P. 349
Heynderickx, Geraldine J. 563
Hildebrandt, Diane 187, 468
Hill, Anita J. 782
Hill, Boyd 664

AUTHOR INDEX

Hill, Priscilla J. 442
Hirlinger, John M. 587
Hirsch, Andreas 541
Hodapp, Maximilian J 178, 286
Horio, Masayuki 11
Howes, Tony 6
Hrenya, Christine M. 44, 473, 478
Huang, Bingyuan 88
Huang, Dale 174, 538
Huang, Yi 782
Igci, Yesim 476
Ikeda, Masahiro 625
Immanuel, Charles D. 755
Irizarry, Roberto 150, 151
Isayev, Olexandr 747
Ismail, Ahmed E. 145
Issangya, Allan 111, 566, 622
Jacob, Karl 315
Jallo, Laila 98, 99
Jamaleddine, Tarek 378
Jang, Hee Dong 246
Janka, Kauko 89
Jayarathna, Chameera K. 66
Jayarathna, Sanoja A. 66
Jiang, Xianjin 528, 569
Johnson, Samantha 266
Johnston, Jack L. 595
Jones, E. Nicholas 269
Jonnavittula, Divya 39
Kalasin, Surachate 351
Kalyon, Dilhan 390, 665, 666, 667, 687
Kang, Choon-Hyoung 171, 172
Kang, Dong-Yuk 171
Kapoor, Deepak 529, 567
Karimipour, Shayan 87
Karlsen, Morten 244
Karri, Reddy S. B. 111, 566
Karri, S. B. Reddy 33, 622
Kashyap, Mayank 58
Kasper, Gerhard 463
Kawajiri, Yoshiaki 5
Kemmere, Maartje F. 346
Kempkes, Michel 1
Keskinen, Jorma 89
Ketko, Marybeth H. 735
Ketterhagen, William R. 381, 467
Keurentjes, J. T. F. 340
Keurentjes, Jos T. F. 346
Khanna, Rajesh 499
Khinast, Johannes G. 10, 135, 234
Khumalo, Ngangezwe 187, 468
Kim, Bok-Min 172
Kim, Chul-Joo 245, 246, 247
Kim, Huiyong 556
Kim, Hyung Rae 391
Kim, Jong-Ho 155
Kim, Min 88
Kim, Sang Done 100
Kim, Seonmin 388
Kim, Seung-Jai 172
Kim, Sung-Don 245, 246
Kim, Young-Chul 155
King, David M. 253, 266
Kirchhof, Martin J. 463
Kitchens, Christopher L. 94
Klinzing, George E. 317, 326
Knieke, Catharina 450, 454
Knowlton, Ted 33, 111, 566, 622
Kodam, Madhusudhan 188, 382
Koenig, Jonathon 601
Kong, Jin-Hwa 155
Korgel, Brian 774

AUTHOR INDEX

Koynov, Athanas A. 55
Kreider, Peter 561
Kretzschmar, Ilona 353
Krinke, Thomas J. 89
Kruis, Einar 475
Krumeich, Frank 173, 267
Kumar, Sandeep 781
Kumar, Varun 759
Laine, Richard M. 88
Lall, Anshuman 95, 174, 228, 538
Lane, J. Matthew D. 145
Langroudi, Mehrdad Kheiripour 466
Latorre, Magda 349
Lattuada, Marco 8, 45, 156, 348, 758, 762
Lee, Dae-Won 175
Lee, Dong Hyun 100
Lee, Kwang Soon 556
Lee, Kwan-Young 175
Lee, Yoong 100
Leon, Madeline 150
Lepek, Daniel 229
Leszczynski, Jerzy 747
Leu, Lii-Ping 249
Lewin, John L. 736
Li, Fanxing 391, 615
Liang, Xinhua 253, 266
Lichty, Paul 561
Lim, Tau Yee 628
Lin, Candy S. 629
Lindberg, William 627
Lindsay, Mitchell 601
Liong, Yuh Inn 630
Lisal, Martin 688
Litster, James D. 6, 189, 383, 755
Littman, Howard 409
Liu, Sha 91
Lohrmann, Maike 342
Lorenz, Chris 145
Luss, Dan 568
Ma, X. 95, 174, 228, 511, 538
Maetani, Toshio 11
Mäkelä, Jyrki M. 89
Malik, Moinuddin 665
Malone, Michael F. 752, 753
Maloney, Stephen W. 737
Mann, Reginald 124
Mao, Yuanbing 152
Maqbool, Tabish 754
Marashdeh, Qussai 76
Marczak, Renata 541, 764
Marin, Guy B. 563
Marini, Fabio 286
Marjamäki, Marko 89
Martignoni, Waldir Pedro 191, 218
Martin, Darold 529, 567
Martin, Lealon L. 409
Martínez, Ángel Martín 624
Martirosyan, Karen S. 568
Mazumder, Jahirul I. 419
Mazzotti, Marco 1
McCann, Ryan 448
McCarthy, J. J. 16, 24, 134
Meesters, G. M. H. 6, 443
Merrill, Chris 601
Meterc, Darja 624
Metzger, Matthew J. 187, 468
Meuldijk, Jan 346
Milani, Paolo 387
Min, Byoung-Jun 172
Min, Jian 270
Monastiriotis, Spyros 268
Moon, Dong-Ju 155

AUTHOR INDEX

Morbidelli, Massimo 7, 8, 45, 156, 190, 348, 558, 758, 760, 762

Mori, Milton 178, 191, 204, 218, 286

Morris, Kenneth 448

Mountziaris, T. J. 780

Mudde, Robert F. 125

Muhle, Michael E. 354

Mujumdar, Ajit 11, 98, 99, 138

Mujumdar, Kaustubh S. 9

Mukasyan, Alexander S. 527

Mulholland, G. M. 95, 228

Muzzio, Fernando J. 34, 137

Nagendra, Basavaraju Agasanapur 610

Nair, Sachin 230

Nakagawa, Kenshiro 44

Nakamura, Naomichi 11

Nere, Nandkishor 3, 448

Ng, Ka M. 629

Ng, Wai Kiong 438, 628

Nirschl, Hermann 761

Ocone, R. 472

Oetama, Ratna 388

Ogunsola, Oluwatosin 388

Olivo, Carlos 762

Olson, Jon H. 449

Olusanmi, Dolapo 449

Ostrowski, Peter P. 587

Otsuji, Yuichiro 625

Paap, Scott M. 462

Paccione, John 409

Pahk, Jae Bum 317

Palmer, Glenn 601

Panau, Freddie 630

Park, Ah-Hyung Alissa 100

Park, Hyungkyu 245, 247

Park, In-Seon 175

Park, Nam-Cook 155

Park, Young-Moo 175

Passalacqua, Alberto 473, 474

Patel, Rajen 567, 667

Pati, Ranjan 90

Penn, R. Lee 781

Pestov, Dmitry 385

Petsev, Dimiter N. 555

Peukert, Wolfgang 450, 454, 541, 764

Pfeffer, Robert 110, 229, 389

Piekiel, N. 511, 725

Piña, Juliana 159

Pingali, Kalyana 34

Pira, Nello Li 387

Piseri, Paolo 387

Piumsomboon, Pornpote 58

Plawsky, Joel L. 409

Pomfret, Michael B. 90

Potoff, Jeffrey J. 735

Prasad, D. V. N 380

Pratsinis, Sotiris E. 97, 170, 173, 176, 254, 267, 345, 761

Prickett, Suzanne 666

Procopio, Adam 35

Prud'Homme, Robert K. 759

Prulliere, Frederic 36

Pugsley, Todd S. 87

Puszynski, Jan 529, 579, 587

Qasim, Mo 747

Qiu, Qi (Grace) 780

Quevedo, Jose A. 110

Quintanilla, Miguel Angel Sanchez 229

Radeke, Charles 25, 39

Radl, Stefan 135

Raha, Sasanka 754

Rai, A. 511

Rai, Neeraj 736

Raikar, Neha B. 752, 753

AUTHOR INDEX

Rajesh, Dave 99
Rajniak, Pavol 453
Ramachandra, Srikanth Bellur 610, 623
Ramachandran, Rohit 755
Raman, Venkatramanan 481
Ramkrishna, Doraiswami 3, 448
Rasteiro, Maria Da Graca 355, 483
Ray, Ajay K. 419
Ray, Madhumita 378
Razzak, Shaikh 430
Redner, Paul 567, 667
Reklaitis, Gintaras V 448
Remy, Brenda 10, 234
Rho, Seon-Gyun 172
Rinaldi, Carlos 349
Roberts, Christopher 149
Rodgers, Thomas L. 124
Rokkam, Ram G. 354
Romagnoli, Jose A. 47
Romanski, Frank 557
Rosenberg, Jason D. 753
Rossit, José 159
Rostedt, Antti 89
Rothenbacher, Sonja 463
Roy, Papiya 499
Saasen, A. 492
Sadoff, Hillary 90
Sahni, Ekneet 136
Samavedham, L 37
Samokhvalov, Alexander 230
Samyudia, Yudi 630
Sanganwar, Ganesh 144
Santore, Maria M. 351
Sarkar, Avik 143, 169
Saunders, Steven R. 149
Scheffe, Jonathan R. 232
Schmid, Hans-Joachim 463
Schoenitz, Mirko 516, 517, 528, 594
Scicolone, James V. 138, 562
Segets, Doris 764
Seidel-Morgenstern, Andreas 5
Seipenbusch, Martin 463
Shariaty-Niassar, Mojtaba 250
Sharma, M. P. 627
Shen, Zheng 101
Shende, Rajesh 529
Shi, Deliang 24
Shin, Jae-Soon 155
Shukla, Nisha 352
Siegel, Nathan P. 232
Siepmann, J. Ilja 736
Silva, Maria Das Graças E. 286
Silva, Maria G. E. 178
Simons, Jo M. M. 346
Singh, Chetana 384
Singh, Dhananjay 626
Singh, Jagdeep 92, 248
Sinha, Kaustav 344
Smith, Benjamin 36
Smith, Danielle K. 774
Smith, Marty J. 35
Snider, Dale 366
Sohn, Jung-Soo 245
Sokkalingam, Nandhini 735
Solt, Paul 294
Soos, Miroslav 7, 156, 190, 558, 760
Spicka, Peter 453
Sridhar, Deepak 391, 615
Srinivasan, Rajagopalan 37
Stamatis, Demitrios 569, 594
Stark, Wendelin J. 539
Stepanek, Frantisek 755

AUTHOR INDEX

Stephenson, David 124
Storti, Giuseppe 762
Strano, Michael S. 537
Strumendo, Matteo 479
Subramaniam, Shankar 474
Sun, Jin 14
Sundaresan, Sankaran 14, 476
Sundmacher, K. 146
Sundmacher, Kai 3
Sung, Yonduck "james" 481
Swiatkiewicz, Jacek J. 579
Swihart, Mark T. 91, 148
Tallon, Steve J. 306
Tan, Reginald B. H. 438, 628
Tang, Hansong 666, 687
Tardos, Gabriel I. 466
Tatarchuk, Bruce 230
Taylor, Robin Klupp 764
Techane, Sirnegeda D. 96
Teleki, Alexandra 97, 173, 267, 761
Tepper, Gary 385
Tester, Jefferson W. 462
Thakur, Ranjit 342
Theuerkauf, Joerg 9, 56, 57, 153, 315, 380, 498
Threlfall-Holmes, P. 472
Tomassone, M. Silvina 39, 557
Tomassone, M. Silvlna 25
Tomassone, Maria. S. 137
Tong, Andrew 615
Torres-Lugo, Madeline 349
Tricoli, Antonio 345
Troxel, Tom G. 297
Trunov, Mikhaylo A. 516, 528
Tsapatsis, Michael 781
Tsutsumi, Atsushi 625
Tungapindi, Navina D. 469
Tüzün, U 157
Valverde, Jose Manuel 229
Van Bokhoven, Jeroen 92, 248
Van Den Broeke, Leo J. P. 340
Van Der Hoeven, M. J. 6
Van Nispen, Sjoerd F. G. M. 340
Vargas, Watson L. 24
Vasilenko, Alisa 34
Vasquez, Nestor A. 56, 57, 315
Vasquez, Victor R. 623
Vats, Alok 529
Vedula, Prakash 473
Venkatasubramanian, Venkat 448
Ventura, Carla A. F. 355
Vetter, Thomas 1
Vicent, Arol 568
Vinati, Simone 387
Viswanathan, Harish 478
Voigt, Andreas 3, 146
Voigt, Michael 764
Von White, Gregory 94
Walker, Robert A. 90
Walter, Kim T. 36
Wang, Dawei 101, 401
Wang, Fei 76, 391, 615
Wang, H. G. 112
Wang, Huanting 782
Wang, Jun 780
Wang, Leizheng 568
Wasito, W. 76
Wassgren, Carl R. 143, 169, 188, 189, 381, 382, 383, 467
Weber, Alfred P. 463
Webster, Emile S. 40, 306
Wegner, Karsten 387
Wei, J. 475
Weidner, Eckhard 624

AUTHOR INDEX

Weimer, Alan W. 232, 253, 266, 561

Weng, Peng 251, 252

Wengeler, Robert 761

Werner, Fabian 541

White, Jeremiah D. E. 527

Wibowo, Christianto 629

Widenski, David 47

Wildeboer, W. J. 6, 443

Winter, Jessica O. 350

Wolfe, Michael S. 462

Wong, Vivian S. Y. 438

Wu, Hua 190, 558, 758, 760

Wu, Libo 343

Wu, Wenjun 66

Xiao, Guoxian 26

Xu, Delong 26

Yang, W. Q. 112

Yau, Robert 136

Ydstie, B. Erik 560

Yim, Hoon 93

Yoon, Ho-Sung 245, 246, 247

York, Trevor A. 124

You, Jun 401

Young, Ming-Wan 251, 252

Yu, Yung-Sz 249

Zaccone, Alessio 758

Zachariah, Michael 95, 174, 228, 255, 511, 538, 689, 725

Zelinskaya, Alexandra 44

Zeng, Liang 391

Zhang, Hongwang 91

Zhou, Lei 95, 228, 511, 725

Zhou, Ying 37

Zhu, Chao 101, 401

Zhu, Jesse 419, 430

Zhu, Linjie 251, 252

Zhu, Wusheng 557

Zhu, Xiaoying 517

Ziomek, Grzegorz 5

American Institute of Chemical Engineers
3 Park Avenue
New York, New York 10016-5991

CURRAN ASSOCIATES INC
proceedings.com

Particle Technology Forum

Presentations at the 2008 AIChE Annual Meeting

Philadelphia, Pennsylvania
16 - 21 November 2008

Volume 2 of 2

ISBN: 978-1-61567-218-9

Particle Technology Forum

Presentations at the 2008 AIChE Annual Meeting

Philadelphia, Pennsylvania
16 – 21 November 2008

Volume 2 of 2

ISBN: 978-1-61567-218-9

Printed from e-media with permission by:

Curran Associates, Inc.
57 Morehouse Lane
Red Hook, NY 12571

Some format issues inherent in the e-media version may also appear in this print version.

Printed by Curran Associates, Inc. (2009)

AIChE
3 Park Avenue
New York, NY 10016-5991

Phone: (203) 702-7660
Fax: (203) 775-5177

www.aiche.org

Additional copies of this publication are available from:

Curran Associates, Inc.
57 Morehouse Lane
Red Hook, NY 12571 USA
Phone: 845-758-0400
Fax: 845-758-2634
Email: curran@proceedings.com
Web: www.proceedings.com

TABLE OF CONTENTS

VOLUME 1

Characterization of Crystal Size and Shape 1
Michel Kempkes, Jan Eggers, Thomas Vetter, Marco Mazzotti

Modeling the Manipulation of Crystal Morphology Distributions 3
Christian Borchert, Nandkishor Nere, Andreas Voigt, Kai Sundmacher, Doraiswami Ramkrishna

Parameter Estimation of Preferential Crystallization Using a Detailed Mathematical Model 5
Yoshiaki Kawajiri, Ivan Angelov, Grzegorz Ziomek, Martin Peter Elsner, Andreas Seidel-Morgenstern

Dynamic Modelling of Particle-Droplet Collisions 6
M.J. van der Hoeven, Tony Howes, James D. Litster, W.J. Wildeboer, G.M.H. Meesters, I.T. Cameron

Shear-Rate Dependence of the Initial Cluster Aggregation Kinetics of Polystyrene Latices Under Turbulent Conditions in Stirred Tank 7
Lyonel Ehrl, Miroslav Soos, Massimo Morbidelli

Shear Induced Aggregation Rate of Colloidal Nanoparticles and Clusters In the Presence of Repulsive Interactions 8
Marco Lattuada, Massimo Morbidelli

Insights in Scaling and Optimization of Ball Mill Operations - A Discrete Element Method Based Approach 9
Joerg Theuerkauf, Kaustubh S. Mujumdar

Characterization of Granular Flow In a Bladed Mixer Using the Discrete Element Method 10
Brenda Remy, Johannes G. Khinast, Benjamin J. Glasser

Mixing Analysis of Binary System of Particles in An Agitated Bed by Discrete Element Method 11
Ajit Mujumdar, Masayuki Horio, Toshio Maetani, Naomichi Nakamura

Rheological Behavior of Quasi-Static Granular Flows Under Unsteady Simple Shear 14
Jin Sun, Lee R. Aarons, Sankaran Sundaresan

Using Janus Particles to Control Mixing/Segregating of Adhesive Granular Materials in Rotating Tumblers 16
Isabel Figueroa, Joseph J. McCarthy

Granular Mixing and Segregation In Zig-Zag Chute Flow 24
Suman K. Hajra, Deliang Shi, Watson L. Vargas, J. J. McCarthy

Discrete Element Method Simulations of Dry Impregnation Processes in Catalysts 25
M. Silvlna Tomassone, Charles Radeke, Arthur W. ChEster

A DE Numerical Study upon Particulate Solids Flow from Model Silos 26
Guoxian Xiao, Delong Xu, S. Ding

Measuring Particle Attrition In a Jet Cup 33
Ray A. Cocco, S.B. Reddy Karri, Yeook Arrington, Ted Knowlton

Shear and Composition Effects on the Flow and Electrostatic Properties of Pharmaceutical Powders 34
Alisa Vasilenko, Kalyana Pingali, Fernando J. Muzzio

Experimental and Numerical Analysis of Stress Generation In Non-Circular Pharmaceutical Tablets - A Manufacturing Case Study ... 35
Stephen L. Conway, Adam Procopio, Marty J. Smith, Michael Gentzler

Scale up and Optimization of Fluid Bed Coating and Layering Processes Using In-Situ Particle Characterization ... 36
Andrew Birkmire, Kim T. Walter, Eric Dycus, Benjamin Smith, Frederic Prulliere

Quantitative Large-Scale Validation of Image-Based Sensors for Online Particle Size Characterization during Crystallization ... 37
Ying Zhou, Rajagopalan Srinivasan, L Samavedham

Characterization of Density of Cohesive Granular Materials Using An X-Ray CT Scanner ... 39
M. Silvina Tomassone, Charles Radeke, Divya Jonnavittula

Application of Helmholz Resonance to Determination of Bulk Density and Fill Fraction of Dry Particulate Solids ... 40
Emile S. Webster, Clive E. Davies

Newton's Cradle Undone: Experiments and Collision Models for the Normal Collision of Three Solid Spheres ... 44
Carly M. Donahue, Christine M. Hrenya, Kenshiro Nakagawa, Alexandra Zelinskaya

Stokesian Dynamics for Rigid Body Motion of Particle Aggregates ... 45
Yogesh M. Harshe, Marco Lattuada, Massimo Morbidelli

Silica Precipitation during Analcime Dissolution ... 46
Elizabeth A. Gorrepati, H. Scott Fogler

Comparison of the Use of Different Solubility Models for the Cooling Crystallization of Acetaminophen ... 47
David Widenski, Ali Abbas, Jose A. Romagnoli

Development and Validation of History-Dependent Inter-Particle Bonding Terms In Compacted Powder Beds ... 55
Athanas A. Koynov, Alberto Cuitino

Discrete Element Method Simulation of a Dense Phase Conveying System ... 56
Bing Du, Ting(Tim) Han, Nestor A. Vasquez, Joerg Theuerkauf

Discrete Element Method Simulation of a Dilute Conveying System ... 57
Bing Du, Ting(Tim) Han, Nestor A. Vasquez, Joerg Theuerkauf

Computation of Turbulence and Low Sherwood Numbers in Fluidized Beds ... 58
Benjapon Chalermsinsuwan, Pornpote Piumsomboon, Dimitri Gidaspow, Mayank Kashyap, Ronald W. Breault

Influence from Particle Size Distributions on the Cfd Simulations and Experiments of Bubbling Fluidized Beds ... 66
Sanoja A. Jayarathna, WenJun Wu, Chameera K. Jayarathna, Britt M. Halvorsen

Jetting Observations In a Cold-Flow Fluidized Bed ... 75
Nathan P. Franka, Theodore J. Heindel

Imaging Gas/Solid Jet Penetration in a Gas-Solid Fluidized Bed Using Electrical Capacitance Volume Tomography ... 76
Fei Wang, Qussai Marashdeh, W. Wasito, L. -S. Fan

A Rotating Chimney to Increase the Flexibility of Rotating Fluidized Beds ... 83
Juray De Wilde

The Effect of Bed Depth on the Hydrodynamics of a Fluidized Bed of Fcc Catalyst ... 87
Todd S. Pugsley, Shayan Karimipour

New Nanopowders and Nanostructured Materials Derived from Liquid-Feed Flame Spray Pyrolysis (LF-FSP) 88
Richard M. Laine, Min Kim, Jose Azurdia, Bingyuan Huang

Generating Nanoparticle Test Aerosol Using Liquid Flame Spray 89
Mikko Aromaa, Jyrki M. Mäkelä, Antti Rostedt, Thomas J. Krinke, Kauko Janka, Marko Marjamäki, Jorma Keskinen

Flame Synthesis of Nano-Scale Nio/ysz and NiO/CeO2/YSZ Powders for Solid Oxide Fuel Cell Anode Catalyst Applications 90
Ranjan Pati, Osifo Akhuemonkhan, Michael B. Pomfret, Hillary Sadoff, Robert A. Walker, Sheryl Ehrman

Synthesis of ZnS Nanoparticles by Spray Pyrolysis 91
Sha Liu, Hongwang Zhang, Mark T. Swihart

Catalysis by Flame Synthesized Supported Gold Nano-Particles 92
Sachit Chopra, Jeroen van Bokhoven, Jagdeep Singh, Gregory Beaucage

Flame Aerosol Synthesis of Functional Composite Nanoparticles 93
Bing Guo, Hoon Yim

Investigation of Ligand Solvation Effects for Alkyl Thiol Stabilized Metal Nanoparticles In CO2 Expanded Liquids Using Small Angle Neutron Scattering 94
Gregory Von White II, Christopher L. Kitchens

A New Ion-Mobility Method for Probing Reaction Kinetics at the Nanoscale 95
Lei Zhou, X. Ma, Anshuman Lall, G. M. Mulholland, Michael Zachariah

Functionalization and Characterization of Gold Nanoparticles 96
Sirnegeda D. Techane, Lara J. Gamble, David G. Castner

The Quality of SiO2 Coatings on Flame-Made TiO2-Based Nanoparticles 97
Alexandra Teleki, Kamal M. Akhtar, Sotiris E. Pratsinis

Properties and Characterization of Pharmaceutical Powders Modified by Dry Coating 98
Lauren Beach, Yuhua ChEn, Laila Jallo, Ajit Mujumdar, Rajesh Dave

Prediction of Inter-Particle Adhesion Force from Surface Energy and Roughness 99
Laila Jai Jallo, Yuhua ChEn, Lauren Beach, Ajit Mujumdar, Dave Rajesh

Axial Particle Concentration Distribution In Two-Phase and Three-Phase Fluidized Beds with Binary Solid Mixtures 100
Dong Hyun Lee, Byung Soon Chun, Yoong Lee, Ah-Hyung Alissa Park, Sang Done Kim, John R. Grace, Norman Epstein

Phase Transfer in a Droplet-Solid Sphere Collision 101
Dawei Wang, Zheng Shen, Chao Zhu

Enhanced Fluidization of Agglomerates of Nanoparticles 110
Jose A. Quevedo, Robert Pfeffer

Emulsion Phase Density In Deep Fluidized Beds of Geldart Group a Materials 111
Allan Issangya, Reddy S. B. Karri, Ted Knowlton

Investigation of Fluidised Bed Dryer by Frequency-Depend Measurement and Different Normalisation Methods with Electrical Capacitance Tomography 112
H. G. Wang, W. Q. Yang

Tomographic Imaging during Semi-Batch Reactive Precipitation of Barium Sulphate In a Stirred Vessel 124
Thomas L. Rodgers, David Stephenson, Michael Cooke, Trevor A. York, Reginald Mann

Time-Resolved X-Ray Tomography of a Fluidized Bed 125
Robert F. Mudde

Improvement of Granular Mixing In Rotating Cylinders 134
Suman K. Hajra, Tathagata Bhattacharya, J. J. McCarthy

Using Dem to Identify Mixing Mechanisms In Bladed Mixers 135
Stefan Radl, Benjamin J. Glasser, Johannes G. Khinast

Understanding Mixing to Enhance Coating Performance In a Pancoater 136
Ekneet Sahni, Robert Yau, Bodhisattwa Chaudhuri

Using Dem Models to Understand Segregation In Mixtures with Continuous Grain Size Distribution In Long Cylinders 137
Atul Dubey, Fernando J. Muzzio, Maria. S Tomassone

Mixing of Nanosized Particles by Magnetically Assisted Impaction Mixing in Dry and Fluid Suspension 138
James V. Scicolone, Ajit Mujumdar, Rajesh Dave

Simulation of a Continuous Granular Mixer: Effect of Operating Conditions on Flow and Mixing 143
Avik Sarkar, Carl R. Wassgren

Nano-Mixing of Dipyridamole Drug and Excipient Nanoparticles by Sonication in Liquid CO2 144
Ganesh Sanganwar, Ram B. Gupta

Forces Between Functionalized Silica Nanoparticles 145
Ahmed E. Ismail, J. Matthew D. Lane, Michael E. Chandross, Kelly Anderson, Chris Lorenz, Gary S. Grest

Molecular Modeling of Particle Nucleation In Emulsion Droplets: Towards An Improved Understanding of Nanoparticle Production Processes 146
Andreas Voigt, Michael Fricke, K. Sundmacher

Computational Modeling of Silicon Nanoparticle Synthesis In a Laser-Driven Reactor 148
Hongyi Dang, Mark T. Swihart

Scale-up of a Nanoparticle Size-Selective Fractionation Process Using CO2-Expanded Liquids 149
Steven R. Saunders, Christopher Roberts

Kinetic Study and Modeling of Metal Particle Formation 150
Roberto Irizarry, Madeline Leon, Miguel Castro

Efficient Optimization of Particulate Systems Described by Monte Carlo Method: Determination of Nucleation Parameters from Kinetic Data 151
Roberto Irizarry

Large Scale Synthesis of Luminescent Rare-Earth Doped Oxide Nanoparticles 152
Yuanbing Mao, Jane P. Chang

Solids Drying Process Simulation by Solidsim - New Tool for Flow Sheeting of Solids Processes 153
Ting(Tim) Han, Joerg Theuerkauf

Characterization of Nano-Sized Ni/ce(x)Zr(1-x)O2 Catalyst for the Auto Thermal Reforming 154
No Author Available

Characterization of Ni/Ce(x)Zr(1-x)O2 Catalyst for Autothermal Reforming 155
Jin-Hwa Kong, Dong-Ju Moon, Jong-Ho Kim, Nam-Cook Park, Jae-Soon Shin, Young-Chul Kim

Generation and Geometrical Analysis of Dense Clusters with Variable Fractal Dimension Providing a Basis for the Interpretation of Data from Optical Particle Characterication Techniques 156
Lyonel Ehrl, Marco Lattuada, Miroslav Soos, Massimo Morbidelli

The Breakage Matrix Approach In Evaluating Inadvertent Breakage of Dense Particulate Mixtures 157
A. Abu-Nahar, U Tüzün

Modeling of An Industrial Vibrating Double-Deck Screen of a Urea Granulation Circuit 159
Ivana Cotabarren, José Rossit, Juliana Piña, Verónica Bucalá

Simulation of Continuous Granular Mixing 169
Avik Sarkar, Carl R. Wassgren

Coagulation of Highly Concentrated Aerosols 170
Beat Buesser, Sotiris E. Pratsinis

Preparation of Dextran Particles by Using Sas Process 171
Dong-Yuk Kang, Dong-Lyun Cho, Choon-Hyoung Kang

Effects for TiO2 Structures In Dye Sensitized Solar Cell 172
Byoung-Jun Min, Bok-Min Kim, Seon-Gyun Rho, Sung-Yong Cho, Seung-Jai Kim, Choon-Hyoung Kang

The Role of Gas-Aerosol Mixing during In-Situ Coating of Flame-Made Titania Particles 173
Alexandra Teleki, Martin C. Heine, Frank Krumeich, Kamal M. Akhtar, Sotiris E. Pratsinis

New Flame Pyrolysis Reactor for Carbon Black Synthesis: Soot Surface Growth and Oxidation 174
X. Ma, Anshuman Lall, Dale Huang, Michael Zachariah

Preparation of CeO2/SiO2 Nano-Abrasive Particles 175
Sang-Ho Chung, In-Seon Park, Young-Moo Park, Dae-Won Lee, Kwan-Young Lee

Shear-Induced and Brownian Coagulation 176
Max L. Eggersdorfer, Beat Buesser, Sotiris E. Pratsinis

Three-Dimensional Gas-Solid Fluidized Bed Cfd Simulation Using a Lattice Boltzmann Based Drag Correlation 178
Milton Mori, Maximilian J. Hodapp, Maria G. E. Silva

Optimization of Particle Agglomeration and Breakage Operations with the Attainable Region (AR) Theory 187
Matthew J. Metzger, Benjamin J. Glasser, Ngangezwe Khumalo, Brendon Hausberger, Diane Hildebrandt, David Glasser

Force Model Concerns for Glued Sphere Dem Simulations 188
Madhusudhan Kodam, Rahul Bharadwaj, Jennifer Sinclair Curtis, Bruno C. Hancock, Carl R. Wassgren

Effect of Dem Parameters on the Collision Distribution of a Small Shearing Region below the Surface of a Rotating Drum 189
Ben Freireich, James D. Litster, Carl R. Wassgren

Behavior of the Elastomer Particles with Different Surface Charge Groups during Aggregation Under Static and Flow Conditions 190
Miroslav Soos, Cornelius Gauer, Hua Wu, Massimo Morbidelli

Simulation of a High-Flux Riser-Reactor Flow Using Cfd Techniques 191
Leonardo Machado da Rosa, Jaci Carlo Schramm Camara Bastos, Milton Mori, Waldir Pedro Martignoni

Synthesis of Doped TiO2 Metal Oxide Nanoparticles in Supercritical CO2 202
Gabrielė Aksomaitytė, Esther Alonso, Maria José Cocero

Simulation of a Confined Turbulent Two Phase Jet Flow 204
Milton Mori, Jaci C. S. C. Bastos, Rodrigo K. Decker, Udo Fritsching

Study of Different Approaches for Modeling Cyclones Using Cfd 218
Daniel de Brito Dias, Milton Mori, Waldir Pedro Martignoni

Investigating Size-Resolved Nanoparticle Reaction Kinetics: A Novel Tandem Ion-Mobility Method 228
Lei Zhou, X. Ma, Anshuman Lall, G. M. Mulholland, Michael Zachariah

Nanofluidization as Affected by Vibration and Electrostatic Fields 229
Daniel Lepek, Robert Pfeffer, Rajesh Dave, Miguel Angel Sanchez Quintanilla, Jose Manuel Valverde, Antonio Castellanos

Preparation and Characterization of Nanodispersed Ag-Based Sorbent for Removal of Sulfur from Logistic Fuels 230
Alexander Samokhvalov, Sachin Nair, Zenda Davis, Bruce Tatarchuk

Synthesis of Nanoscale Cobalt-Iron Spinel Oxides Via Atomic Layer Deposition and Their Applications for Solar Thermochemical Water Splitting 232
Jonathan R. Scheffe, Nathan P. Siegel, Mark D. Allendorf, Alan W. Weimer

Effect of Interactions on Nanoparticle Dispersion In a Polymer Solution 233
Deepika R. Gollamandala, Ileana C. Carpen

Numerical Simulation of Granular Flow In a Bladed Mixer 234
Brenda Remy, Johannes G. Khinast, Benjamin J. Glasser

Verification and Validation Study of Continuum Granular Frictional Flow Theories Using a Discrete Technique and Experimental Data 235
Sofiane Benyahia

Development of An Automatic Fluidization Rig 244
S. Ding, Are Dyrøy, Morten Karlsen

Effect of Process Variables in the Preparation of Bohemite Sol on the Microstructure of Gamma-Alumina Particles 245
Chul-Joo Kim, Ho-Sung Yoon, Sung-Don Kim, Hyungkyu Park, Jung-Soo Sohn

Preparation and Growth of Silica Sol from Sodium Silicate Solution 246
Ho-Sung Yoon, Chul-Joo Kim, Sung-Don Kim, Hee Dong Jang

Preparation of Anhydrous Magnesium Chloride by Dehydration of Ammonium Magnesium Chloride Hexahydrate 247
Hyoung-Choon Eom, Hyungkyu Park, Chul-Joo Kim, Ho-Sung Yoon

Supported Gold Nano-Catalysts by Spray Synthesis and Solution Route. A Comparative Study 248
Sachit Chopra, Jagdeep Singh, Jeroen van Bokhoven, Gregory Beaucage

Characterization of Turbulent Fluidization Regime with Geldart Group B Particles by Wavelet Analysis of Pressure Fluctuation Signals 249
Lii-ping Leu, Yung-Sz Yu

Ex-Situ Surface Modification of BaTiO3 Nanoparticles Prepared by Hydrothermal Synthesis 250
Mehrnoosh Atashfaraz, Mojtaba Shariaty-Niassar, Tadafumi Adschiri

Simulation of Dry Coating Process Using a DEM Method 251
Peng Weng, Costas G. Gogos, Linjie Zhu, Ming-Wan Young

A Novel Process for Simultaneous Milling and Coating Particulates 252
Peng Weng, Costas G. Gogos, Linjie Zhu, Ming-Wan Young, Peter Bonnett

Molecular Layer Deposition of Ultra-Thin Polymer Films on Large Quantities of Cohesive Nanoparticles 253
Xinhua Liang, David M. King, Steven M. George, Alan W. Weimer

Design of High Temperature, In-Situ Coating of Aerosols: Efficiency & Texture 254
Beat Buesser, Sotiris E. Pratsinis

Mechano-ChEmical Stability of Gold Nanoparticles Coated with Alkanethiolate SAMs 255
Brian J. Henz, Takumi Hawa, Michael Zachariah

Quantum Confinement In Core-Shell Particles Via Atomic Layer Deposited Zinc Oxide Films 266
David M. King, Xinhua Liang, Samantha Johnson, Alan Weimer

In-Situ Coating of Flame-Made TiO2 Nanoparticles by SiO2 267
Alexandra Teleki, Martin C. Heine, Frank Krumeich, Kamal M. Akhtar, Sotiris E. Pratsinis

Vesicle Templating of Hollow Oxide Nanoparticles 268
Spyros Monastiriotis, Alexander Couzis

Application of Coarse Grained Drag Law In Computational Fluid Dynamics Simulations of Fluidized Beds 269
Alvin U. ChEn, E. Nicholas Jones

Cfd Modeling of Cold-Flow Fluidized Beds and Validation with X-Ray Computed Tomography 270
Jian Min, Joshua Drake, R. O. Fox, Theodore Heindel

Industrial Application of Cpfd to Biomass Gasification 275
Charles Churchman

Large Scale Multiphase Modeling: Continuum Vs. Discrete Approach 285
Chris Guenther

Three-Dimensional Gas-Solid Fluidized Bed Simulation Applying the Cfd Technique 286
Milton Mori, Fabio Marini, Maria das Graças E. Silva, Maximilian J Hodapp

George Klinzing - Educator, Author, Researcher, Friend 294
Paul Solt

Unloading Pressure Discharge Trucks 297
Francisco J. Cabrejos, Tom G. Troxel

Acoustic Measurement and Monitoring in Dry Particulate Systems: Inspiration, Education, Application, Appreciation 306
Clive E. Davies, Steve J. Tallon, Emile S. Webster

Dilute Phase Pneumatic Conveying of Blended Plastic Pellets 315
Nestor A. Vasquez, Shrikant Dhodapkar, Karl Jacob, Joerg Theuerkauf

New Perspectives on Fluid-Phase Turbulence Modulation by Particles 316
Jennifer S. Curtis, Kunn Hadinoto-Ong

Local Friction Forces Between Plug and Pipe Wall in Dense Phase Pneumatic Conveying 317
Jae Bum Pahk, George E. Klinzing

Transformative Research in Pneumatic Conveying 326
George E. Klinzing

Development of Ion-Selective Thermo-Reversible Polymers 340
Sjoerd F. G. M. van Nispen, Leo J.P. van den Broeke, Johan P. A. Custers, J.T.F. Keurentjes

Co-Formulation of API with Polymer Using Pca Process 342
Ranjit Thakur, Maike Lohrmann

Engineering Porous Drug Particles by Emulsification-Diffusion for Propellant-Driven Oral Inhalation Formulations 343
Libo Wu, Sandro R. P. da Rocha

Effect of Poly-(N-vinyl pyrrolidone) Coating on the Synthesis of Magnetic Particles for the Formulation of Nanoscale Magnetorheological Fluids 344
Kaustav Sinha, Olivia A. Graeve

Optimal SiO2 Doping for Enhanced SnO2 Sensitivity and Thermal Stability 345
Antonio Tricoli, Markus Graf, Sotiris E. Pratsinis

Polymer Microsphere Synthesis by Membrane Emulsification 346
Jo M. M. Simons, Jan Meuldijk, Maartje F. Kemmere, Jos T.F. Keurentjes

Preparation and Characterization of NOVEL Polymer-MAGNETIC COLLOIDS 348
Marco Furlan, Marco Lattuada, Massimo Morbidelli

Functionalization of Monodisperse Magnetite Nanoparticles with Fluorescent Amino-Silane-Carboxymethyl Dextran for Cancer Treatment 349
Adriana P. Herrera, Carola Barrera, Vanessa Ayala, Magda Latorre, Madeline Torres-Lugo, Carlos Rinaldi

Silica-Coated Quantum Dot-Iron Phosphide Nano-Composites 350
Shuang Deng, Jessica O. Winter

Mimicking Selectin-Mediated Cell Motion with a Synthetic Receptor System 351
Maria M. Santore, Surachate Kalasin, Jeffrey,M. Davis, Ranojoy D. Duffadar

Functional Nanoparticles for Enantioselective Separations 352
Nisha Shukla, Andrew J. Gellman, Melissa Bartel

Preparation, Assembly and Application of Surface Anisotropic (Janus) Particles 353
Jingqin Cui, Ilona Kretzschmar

Cfd Modeling of Polymerization Fluidization Reactor 354
Ram G. Rokkam, R. O. Fox, Michael E. Muhle

Cfd Simulation of the Turbulent Flow of Pulp Fibre Suspensions 355
Maria da Graca Rasteiro, Carla A.F Ventura, Fernando A. P. Garcia, Paulo J. Ferreira

Heterogeneous Gas ChEmistry in the Cpfd - Eulerian-Lagrangian Numerical Scheme (Ozone Decomposition) 366
Dale Snider

Numerical Simulation of Pneumatic Drying Using Computational Fluid Dynamics 378
Tarek Jamaleddine, Madhumita Ray

Calibration of DEM Material Models 379
John Favier

Evaluation of Contact Models for Discrete Element Method Simulations 380
Joerg Theuerkauf, D.V.N Prasad

The Effect of Particle Shape on Bulk Powder Friction In a Schulze Ring Shear Tester Using the Discrete Element Method 381
William R. Ketterhagen, Jennifer S. Curtis, Carl R. Wassgren, Bruno C. Hancock

Force Model Considerations for Glued Sphere Dem Simulations 382
Madhusudhan Kodam, Rahul Bharadwaj, Jennifer Sinclair Curtis, Bruno C. Hancock, Carl R. Wassgren

Effect of Dem Parameters on the Collision Distribution of a Small Shearing Region beneath the Surface of a Rotating Drum ... 383
Ben Freireich, James D. Litster, Carl R. Wassgren

Controlled Patterning of Nanostructured Surfaces Using Self-Assembly ... 384
ChEtana Singh, Sharon C. Glotzer

Molecularly Engineered Organic Particulate Films: Preparation, Structure and Performance ... 385
Ozge Guney-Altay, Dmitry Pestov, Gary Tepper

Nanostructured Coatings by Combined Flame Spray Synthesis of Nanoparticles and Cluster Beam Deposition ... 387
Simone Vinati, Alessandro Antonini, Paolo Piseri, Paolo Milani, Luca Belforte, Nello Li Pira, Karsten Wegner

Factors Affecting Strength of Porous Films Made by Gas to Particle Conversion/ Thermophoretic Deposition of Nanoparticle Agglomerates ... 388
Oluwatosin Ogunsola, Ratna Oetama, Tridib Bhowmick, Seonmin Kim, Sheryl Ehrman

Psri Lectureship in Fluidization Award: Fluidization of Nanopowders ... 389
Robert Pfeffer

Thomas Baron Award: Rheology and Processing of Particle-Loaded Complex Fluids at Particle Concentrations That Approach the Maximum Packing Fraction ... 390
Dilhan M. Kalyon

VOLUME 2

Syngas ChEmical Looping Process ... 391
Fanxing Li, Deepak Sridhar, Hyung Rae Kim, Liang Zeng, Fei Wang, L. -S. Fan

Heterogeneous Flow Structure in Gas-Solid Risers: Continuous MODELING ... 401
Dawei Wang, Jun You, Chao Zhu

Design, Development and Operation of a Novel Recirculating Liquid/solids Reactor Based on a Type 2 Draft Tube Spout Fluid Bed ... 409
John Paccione, Howard Littman, Lealon L. Martin, Joel L. Plawsky, David M. Follansbee

Mass Transfer Resistance in Downer Reactors ... 411
Subbarao Duvvuri

Modeling and Multi-Objective Optimization of Continuous Protein Recovery in Liquid-Solid Circulating Fluidized Bed Ion Exchange System ... 419
Jahirul I. Mazumder, Jesse Zhu, Ajay K. Ray

Local Holdups and Phase Propagation Velocity Measurement in Glscfb Riser Using Electrical Resistance Tomography and Optical Fibre Probe ... 430
Shaikh Razzak, Shahzad Barghi, Jesse Zhu

Use of Co-Milling to Improve Physical Stability of Amorphous Salbutamol Sulphate ... 438
Wai Kiong Ng, Vivian S.Y. Wong, Prashant N. Balani, Sui Yung Chan, Reginald B. H. Tan

Modeling of Changes In Particle Size-Shape Distribution Due to Breakage ... 442
Priscilla J. Hill, Devkant S. Gandhi

Investigation of Sub-Micron Milling Process of Poorly Water Soluble Products Using Factorial Design ... 443
Stephen Hennart, G.M.H. Meesters, W.J. Wildeboer, Maria Cristina Domingues

Milling of Roller Compacted API with Excipients: Model Identification and Verification 448
Pavan Kumar Akkisetty, NandKishor Nere, Ryan McCann, Kenneth Morris, Doraiswami Ramkrishna, Gintaras V Reklaitis, Venkat Venkatasubramanian

Solving Inverse Problems In Fragmentation / Breakage 449
R. Bertrum Diemer Jr., Jon H. Olson, Yulong Ding, Mojtaba Ghadiri, Dolapo Olusanmi

Delamination of Graphite Sheets in Stirred Media Mills 450
Catharina Knieke, Wolfgang Peukert

Theoretical Study of Granule Friability during Pneumatic Conveying 453
Peter Spicka, Pavol Rajniak

Particle Breakage in the Nanometer Range 454
Catharina Knieke, Wolfgang Peukert

Nanoparticle Milling and Encapsulation in Supercritical Carbon Dioxide Mixtures 462
Scott M. Paap, Michael S. Wolfe, Jefferson W. Tester

Stability of Silica Nanoparticle Agglomerates 463
Martin Seipenbusch, Sonja Rothenbacher, Martin J. Kirchhof, Hans-Joachim Schmid, Alfred P. Weber, Gerhard Kasper

High Shear De-Agglomeration of Nanoparticle Agglomerates in Stirred Tanks 465
Micaela Caramellino, Piero M. Armenante

Development of a Visco-Plastic Constitutive Law for Dense Flow of Granular Matter 466
Mehrdad Kheiripour Langroudi, Gabriel I. Tardos

Predicting Discharge Dynamics from a Rectangular Hopper Using the Discrete Element Method (DEM) 467
Anshu Anand, Jennifer S. Curtis, Carl R. Wassgren, Bruno C. Hancock, William R. Ketterhagen

Use of the Attainable Region (AR) Analysis to Optimize Particle Agglomeration and Breakage Operations 468
Matthew J. Metzger, Benjamin J. Glasser, David Glasser, Brendon Hausberger, Diane Hildebrandt, Ngangezwe Khumalo

Classification of Fine Particles Using a Taylor-Couette Device 469
Navina D. Tungapindi, Kimberly H. Henthorn

Power Consumption and Granular Flow In a Vibro-Fluidized Stirred Granular Bed 470
Kenneth J. Ford, James F. Gilchrist, Hugo S. Caram

Density Measurements In a Vibro-Fluidized Deep Granular Bed 471
Kenneth J. Ford, James F. Gilchrist, Hugo S. Caram

Spray Drying of Large Particles of Film Forming Materials 472
P. Threlfall-Holmes, R. Ocone

A Kinetic-Based Model for a Non-Isothermal Granular Gas with Mono- and Bi-Dispersed Particles 473
Alberto Passalacqua, Prakash Vedula, Christine M. Hrenya, R. O. Fox

Comparison of Euler-Euler and Euler-Lagrange Simulations of Finite-Stokes-Numbers Gas-Particle Flows in a Lid-Driven Cavity 474
Rahul Garg, Alberto Passalacqua, Shankar Subramaniam, R. O. Fox

Combination of Monte Carlo Simulation of Population Balances and Cfd for Modeling of Nanoparticle Formation 475
Einar Kruis, J. Wei

Coarse-Grid Simulation of Fluidized Gas-Particle Flows 476
Yesim Igci, Sankaran Sundaresan

Representation of a Continuous Particle Size Distribution Using a New Polydisperse Kinetic Theory 478
Harish Viswanathan, Christine M. Hrenya, R. O. Fox

Solution of the Boltzmann Equation by the Fcmom 479
Matteo Strumendo, Hamid Arastoopour

Direct Numerical Simulation of Nanoparticle Evolution In Turbulent Flames 481
Yonduck "James" Sung, Venkatramanan Raman, R. O. Fox

Determining the Density and Moisture Content of Powders Using Microwave Dielectric Spectroscopy 482
Talesha T. Hall, Michael T. Harris

Correlating Retention and Drainage in Papermaking with Flocculation: Effect of Polyelectrolyte Branching 483
Elisabete Antunes, Fernando A.P. Garcia, Paulo Ferreira, Maria da Graca Rasteiro

Experimental Investigation into Shielding Effect of Magnetic Powder on Drilling Fluid 492
S. Ding, Biplab Data, a. Saasen, P.a. Amundsen

Solids Drying Process Simulation by Solidsim --- a New Tool for Flow Sheeting of Solids Processes 498
Ting(Tim) Han, Joerg Theuerkauf

Granulation Time in Fluidized Bed Granulators 499
Papiya Roy, Rajesh Khanna, Subbarao Duvvuri

Size Resolved Kinetics of Nickel Nanoparticle Oxidation 511
Lei Zhou, A. Rai, N. Piekiel, X. Ma, Michael Zachariah

Preparation and Characterization of Metal-Bound Hybrid Reactive Particles 516
Alexandre Ermoline, Mirko Schoenitz, Mikhaylo A. Trunov, Edward L. Dreizin

Reactive Nanocomposite Materials Prepared by Cryogenic Milling 517
Carlo Badiola, Edward L. Dreizin, Mirko Schoenitz, Xiaoying Zhu

Effect of Mechanical Activation on Reaction Kinetics In Gasless Heterogeneous Systems 527
Jeremiah D.E. White, Alexander S. Mukasyan

Mechanical Alloying and Reactive Milling In a High Energy Planetary Mill 528
Xianjin Jiang, Mikhaylo A. Trunov, Mirko Schoenitz, Rajesh Dave, Edward L. Dreizin

Pyrophoric Films Based on Nano-Sized Iron 529
Zac Doorenbos, Alok Vats, Jan Puszynski, Rajesh Shende, Deepak Kapoor, Darold Martin, Chris Haines

Directed Release of ChEmical Energy In Single Walled Carbon Nanotube Films 537
Wonjoon Choi, Michael S. Strano

A Flame-Pyrolysis Reactor for Carbon Black Production: Yield from Thermal Decomposition of Heavy Hydrocarbon Fuels 538
X. Ma, Anshuman Lall, Dale Huang, Michael Zachariah

Large-Scale Gas Phase Synthesis of Zinc Sulfide Nanoparticles 539
Evagelos K. Athanassiou, Robert N. Grass, Wendelin J. Stark

Synthesis and Characterization of Zinc Oxide Nanoparticles for Dye-Sensitized Solar Cells 541
Renata Marczak, Fabian Werner, Jan-Frederik Gnichwitz, Andreas Hirsch, Dirk M. Guldi, Wolfgang Peukert

Simultaneous Molecular and Microemulsion Templating of Inorganic Oxide Particles with Hierarchal Porosity 555
Nick Carroll, Dimiter N. Petsev

Use of Calorimetry Model and Batch Control Technique for Scale-up of Unseeded Batch Cooling Crystallization of Polyhydroxybenzophenon 556
Kwang Soon Lee, Huiyong Kim, Youkyung Bang

Atomistic Simulation Studies of Stability Enhancement of Griseofulvin Nanocrystal Aqueous System by Adding Surfactant, Polymer, and Surfactant-Polymer Mixtures 557
M. Silvina Tomassone, Wusheng Zhu, Frank Romanski

Characterization of Elastomer Particle Aggregation and Implications for Industrial Processing 558
Cornelius Gauer, Miroslav Soos, Hua Wu, Massimo Morbidelli

Reactor Analysis of Coal Pyrolysis to Acetylene In Hydrogen Plasma Reactor 559
Jiaqi ChEn, Yi ChEng

Multi-Scale Modeling and Control of Particulate Cvd for Production of Solar Grade Silicon In Fluidized Beds 560
B. Erik Ydstie, Juan Du

A Study of the Effect of Feeding Device Choice on Conversion In An Aerosol Flow Reactor 561
Todd M. Francis, Peter Kreider, Paul Lichty, Alan W. Weimer

Fluidized Bed Mixing of Nanosized Particles 562
James V. Scicolone, Rajesh Dave

Experimental Investigation of a Rotating Fluidized Bed in a Static Geometry 563
Rahul P. Ekatpure, Geraldine J. Heynderickx, Axel De Broqueville, Guy B. Marin

Effect of Baffles on Gas Bypassing In a 0.9-M-Diameter Unit 566
Allan Issangya, Reddy S. B. Karri, Ted M. Knowlton

Advanced Materials Development at Rdecom-Ardec 567
Paul Redner, Deepak Kapoor, Rajen Patel, Chris Haines, Darold Martin, Moon Chung

Development of High-Efficient Nanoenergetic Gas-Generators 568
Karen S. Martirosyan, Leizheng Wang, Arol Vicent, Dan Luss

Aluminum Burn Rate Modifiers Based on Reactive Nanocomposite Powders 569
Demitrios Stamatis, Xianjin Jiang, Ervin Beloni, Edward L. Dreizin

Tunability of Nanoenergetic Materials 579
Chris J. Bulian, Jan A. Puszynski, Jacek J. Swiatkiewicz

Fabrication and Characterization of Bimetallic Al-Ni Reactive Nanostructures as Nanoheaters 586
Qingzhou Cui, Julie ChEn, Zhiyong Gu

Metastable Intermolecular Composites (MIC) for Small Caliber Cartridges and Cartridge Actuated Devices 587
Peter P. Ostrowski, Magdy Bichay, John M. Hirlinger, Jan Puszynski, Todd Michael Allen

Filled Nanocomposite Materials for Biocidal Applications 594
Mirko Schoenitz, Demitrios Stamatis, Edward L. Dreizin, Sergey A. Grinshpun

Design for Demilitarization - Be Creative Now or Pay Me Later 595
Jack L. Johnston

Characterizing and Modeling of Particle Size Distributions from Propellant Open Burning or Open Detonation (OB/OD) 601
Mitchell Lindsay, Jonathon Koenig, Chris Merrill, Glenn Palmer

Analysis of Transport Phenomena & Modeling the Behavior of Single Particle during Drying Process 610
Srikanth Bellur Ramachandra, Basavaraju Agasanapur Nagendra, Charles J. Coronella

Gas-Solid Heat Transfer In Rotating Fluidized Beds In a Static Geometry 611
Axel de Broqueville, Juray De Wilde

Particle Size Optimization for Syngas ChEmical Looping Process 615
Fei Wang, Fanxing Li, Andrew Tong, Deepak Sridhar, L. -S. Fan

Computational Fluid Dynamics of Gas Bypassing In Fluidized Beds with Imposed Solids Flux 622
Ray A. Cocco, Allan Issangya, S.B. Reddy Karri, Ted Knowlton

Equilibrium Moisture Content and Drying Analysis of Biosolids 623
Srikanth Bellur Ramachandra, Charles J. Coronella, Victor R. Vasquez

Precipitation of Green Tea Polyphenols Using a Supercritical Anti-Solvent Process 624
Darja Meterc, Maria Jose Cocero, Ángel Martín Martínez, Eckhard Weidner

Nanoencapusulation Using Water-in-Supercritical CO2 Microemulsion Technology 625
Yuichiro Otsuji, Masahiro Ikeda, Chihiro Fushimi, Atsushi Tsutsumi

Precipitation of Ultra-Fine Particles of Fenofibrate with Controlled Size Distribution by Resolv: Rapid Expansion Vs Stabilizer Concentrations 626
Sameer Dalvi, Dhananjay Singh, Rajesh Davé

A Study of Production of Micronized Powders Using Supercritical Fluid Atomization Techniques 627
Sriram Kiran Annapragada, M. P. Sharma, William Lindberg

Effect of Co-Precipitation on Crystallinity of Pharmaceutical Compounds Using Supercritical Anti-Solvent Process 628
Tau Yee Lim, Wai Kiong Ng, Reginald B. H. Tan

Re-Engineering Particle Sizes by Precipitation with Compressed Antisolvent (PCA) Process 629
Candy S. Lin, Ka M. Ng, Christianto Wibowo

Hydrodynamics Modeling and Analysis of Rapid Expansion Systems of Supercritical Solutions (RESS) 630
Yudi Samyudia, Freddie Panau, Yuh Inn Liong

Energetic Materials, What Are They? 643
Bruce Cranford Jr., PE

History of Energetic Materials 664
Boyd Hill

Incorporation of Pressure Dependent Wall Slip to the Simulation of the Continuous Processing of Energetic Suspensions 665
Dilhan Kalyon, Moinuddin Malik

Prediction of Conditions Under Which Flow Instabilities Are Onset to Give Rise to Extrudate Distortions during the Extrusion of Highly Filled Energetic Suspensions 666
Suzanne Prickett, Hansong Tang, Halil Gevgilili, Dilhan Kalyon

Twin Screw Extrusion Processing of a Nanoalumina Based Formulation in Conjunction with a Molding Powder Preparation Methodology 667
Rajen Patel, Paul Redner, Halil Gevgilili, Dilhan Kalyon

Inverse Problem Solution of Squeeze, Rectangular Slit, Steady Torsional and Capillary Flows to Enable the Determination of Parameters of Wall Slip and Shear Viscosity for Energetic Suspensions 687
Hansong Tang, Halil Gevgilili, Dilhan Kalyon

Mesoscale Modeling of Energy Release Processes: Dissipative Particle Dynamics Simulations of Shear Initiated Behavior 688
John K. Brennan, Martin Lisal, Josep Bonet Avalos

Molecular Dynamics Simulation of the Kinetic Reaction of Ni and Al Nanoparticles 689
Brian J. Henz, Takumi Hawa, Michael Zachariah

Experimental Study of Ignition of Magnesium Powder by Electro-Static Discharge 699
Ervin Beloni, Edward L. Dreizin

T-Jump Mass Spectrometry to Probe Condensed State Reactive Processes 725
Lei Zhou, Nicholas Piekiel, Snehaunshu Chowdhury, Michael Zachariah

Thermophysical Property Prediction of Energetic Materials with Atomistic Computer Simulations 735
Nandhini Sokkalingam, Marybeth H. Ketko, Jeffrey J. Potoff

Monte Carlo Simulations of 1,3,5-Triamino-2,4,6-Trinitrobenzene (TATB): Pvt Data, Phase Equilibria, and Solubility 736
J. Ilja Siepmann, Neeraj Rai, John L. Lewin, Divesh Bhatt, Laurence E. Fried

Prediction of Physicochemical Properties of Energetic Materials 737
Siri Chakka, Veera M. Boddu, Stephen W. Maloney, Reddy Damavarapu

Mechanism and Kinetics of Thermal Decomposition of Cl-20: 747
Olexandr Isayev, Leonid Gorb, Mo Qasim, Jerzy Leszczynski

Identifying Emulsion Drop Breakage Mechanisms for Population Balance Equation Models of High Pressure Homogenization 752
Neha B. Raikar, Surita R. Bhatia, Michael F. Malone, Cristhian P. Almeida-Rivera, Peter Bongers, Michael A. Henson

Population Balance Equation Modeling of Emulsion Drop Coalescence in High Pressure Homogenization 753
Jason D. Rosenberg, Neha B. Raikar, Michael F. Malone, Michael A. Henson

A Generalized Geometric Population Balance Model for Asphaltene Aggregation 754
Tabish Maqbool, Sasanka Raha, H. Scott Fogler

A Combined Mechanistic Model for Nucleation, Aggregation and Breakage In a Population Balance Model of Granulation 755
Rohit Ramachandran, Charles D. Immanuel, James D. Litster, Francis J. Doyle III, Frantisek Stepanek

The Elasticity and Breakage of Colloidal Aggregates In Shear and Turbulent Flows 758
Alessio Zaccone, Hua Wu, Marco Lattuada, Massimo Morbidelli

Study on Aggregation of Gold Nanoparticles by Tuning the Surface Properties 759
Varun Kumar, Chungyin ChEng, R. O. Fox, Robert K. Prud'homme

Effect of Particle Surface Charge Groups on the Aggregation of Elastomer Particles Under Static and Flow Conditions 760
Miroslav Soos, Cornelius Gauer, Hua Wu, Massimo Morbidelli

Distinguishing Between TiO2 Aggregates and Agglomerates by High Pressure Dispersion 761
Alexandra Teleki, Robert Wengeler, Hermann Nirschl, Sotiris E. Pratsinis

Use of Asymmetric Flow-Field Flow Fractionation to Characterize Aggregating Colloidal Dispersions 762
Marco Lattuada, Carlos Olivo, Cornelius Gauer, Giuseppe Storti, Massimo Morbidelli

A Structural Model for Aggregation and Its Application to Small-Angle Scattering 763
Gregory Beaucage

Size and Shape Engineering of ZnO Nanoparticles 764
Renata Marczak, Robin Klupp Taylor, Michael Voigt, Doris Segets, Wolfgang Peukert

Multifunctional Particles: Magnetic Nanocrystals and Gold Nanorods Coated with Fluorescent Dye-Doped Silica Shells 774
Danielle K. Smith, Andrew T. Heitsch, Brian Korgel

Phospholipid-Assisted Formation and Dispersion of Aqueous Nano-C60 775
Yanjing ChEn, Geoffrey D. Bothun

Synthesis of Doped Zinc Selenide Quantum Dots Using Microemulsions as Templates 780
Tracy Heckler, Qi (Grace) Qiu, Jun Wang, T. J. Mountziaris

Zeolite Crystal Growth: A High-Resolution Cryo-TEM Study 781
Sandeep Kumar, R. Lee Penn, Michael Tsapatsis

Synthesis of Size Distribution Controllable Zeolite Nanocrystals Via a Novel Confined Space Strategy 782
Yi Huang, Anita J. Hill, Huanting Wang

Author Index

Syngas Chemical Looping Process for Hydrogen Production: Bench Scale Demonstration and Process Simulation

Fanxing Li, Deepak Sridhar, Hyung Kim, Liang Zeng, Fei Wang, L. -S. Fan[1]
Department of Chemical and Biomolecular Engineering
The Ohio State University
Columbus, OH, USA

Abstract

The Syngas Chemical Looping (SCL) process converts gaseous fuels such as syngas and hydrocarbons into hydrogen and electricity while capturing 100% of the CO_2 generated [1, 2]. The SCL process simplifies the overall fuel conversion scheme while reducing pollutants in an indirect but highly effective manner. This is achieved via a specially tailored Fe_2O_3 based composite particle. As a result, hydrogen production cost is substantially reduced. Extensive demonstrations of the SCL process have been carried out under a bench scale (2.5 KWt) unit using coal-derived syngas, natural gas, and other light hydrocarbons as feedstock. Hydrogen with an average purity higher than 99.95% was produced.

In this paper, the SCL process demonstrations are discussed. ASPEN® simulation based on the bench scale demonstration results was used to analyze the SCL process.

1. Introduction

The sustainable development of the modern society demands efficient and environmentally friendly energy conversion techniques. Through the aid of an iron based reaction medium, the syngas chemical looping (SCL) process converts syngas into hydrogen and/or electricity while generating a sequestrable CO_2 stream.

In this paper, comprehensive ASPEN simulation models are developed to simulate both the traditional coal to hydrogen process and the SCL process based on common assumptions. The ASPEN simulation on the SCL process, which is supported by the bench scale demonstration results on the major chemical looping units, show that the SCL process is 5 – 7% more efficient when compared to the traditional coal-to-hydrogen process. The results from the bench scale demonstration and the process simulation suggest that the SCL process can be a technically attractive process to efficiently produce hydrogen and/or electricity from coal.

---

[1] Corresponding Author

2. Traditional Coal-to-Hydrogen Process and the Syngas Chemical Looping Process

Hydrogen is mainly produced from steam reforming of methane. With soaring crude and natural gas prices, coal is emerging as a viable choice for hydrogen generation. Coal is used for hydrogen production through the well known gasification process. Figure 1 shows the schematic of a specific configuration of a conventional coal to hydrogen process.

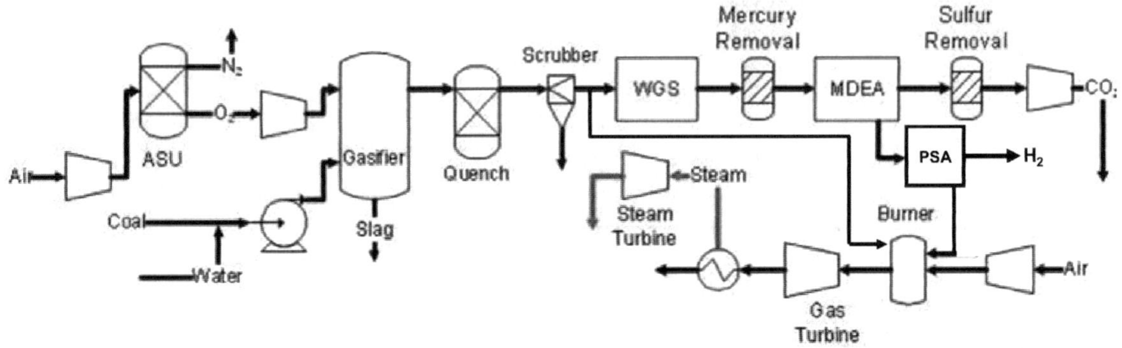

Figure 1. Simplified schematic of the traditional Coal-to-Hydrogen process

In this process, coal is first pulverized and mixed with water to form coal slurry. The coal slurry is then pressurized and introduced to the gasifier to be partially oxidized at 1500 °C and 30 atm. The high temperature raw syngas after gasification is then quenched to 250 °C with water. The quenching step solidifies the ash. Moreover, most of the NH_3 and HCl in the syngas are removed during this step. After quenching, the syngas is sent to a Venturi scrubber for further particulate removal. The particulate-free syngas, saturated with steam, is then introduced to the raw syngas water gas shift (WGS) unit. The syngas exiting the WGS unit contains mainly of H_2 and CO_2 with small amount of CO, H_2S, and mercury. This gas stream is then cooled down to around 40 °C for mercury removal using an activated carbon bed. The CO_2 and H_2S in the syngas are then removed using an MDEA unit, resulting in a concentrated hydrogen stream with small amounts of CO_2 and unconverted CO. The hydrogen rich gas stream is then passed through a pressure swing adsorption unit (PSA) to obtain a purified hydrogen stream with a purity of above 99.99%. The tail gas from the PSA, which is a mixture of H_2, CO, and CO_2, will be re-compressed and sent to a combined cycle system for power generation to compensate the parasitic energy requirements. The combined cycle system consists of a gas turbine with an inlet firing temperature of 1430 °C and a two stage steam turbine working at 550 °C and 35 atm. The CO_2 obtained from the MDEA unit is compressed to 150 atm for sequestration.

As can be seen in Figure 1, the conventional coal to hydrogen process involves elaborate unit operations including air separation, coal gasification, syngas cleanup, conditioning, and separation and CO_2 compression. As a result, the capital investment for conventional coal to hydrogen process is much higher than a steam methane reforming plant with a similar hydrogen production capacity. In order for the gasification based coal to hydrogen process to be economically attractive, process intensification that can reduce the capital investment while improving the energy conversion efficiency of the traditional coal to hydrogen process is required. The SCL process simplifies the syngas conversion and product gas separation scheme through the assistance of an iron oxide based oxygen carrier particle. The simplified schematic flow diagram of the SCL process is shown in Figure 2.

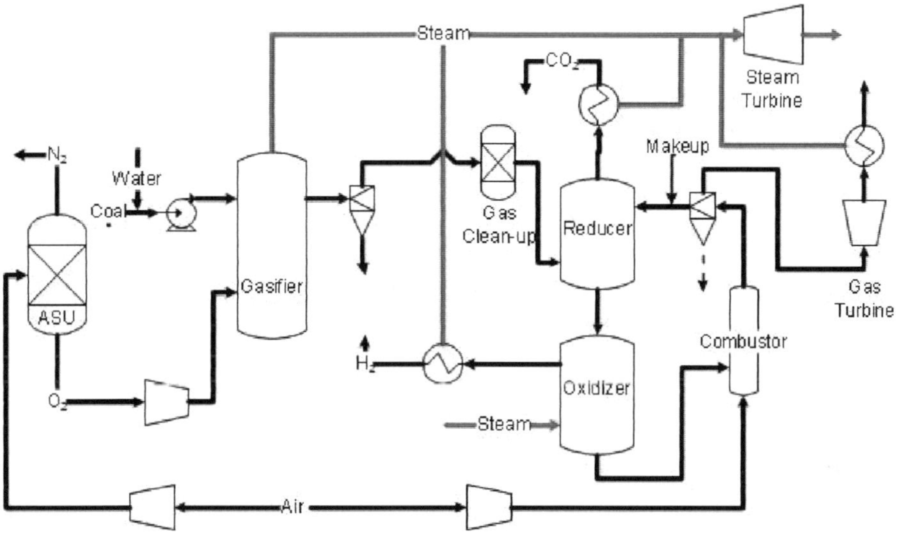

Figure 2. Simplified schematic of the Syngas Chemical Looping process

The SCL process has air separation and coal gasification units similar to that in the conventional coal to hydrogen process. The difference lies in the syngas cleaning, conversion, and separation steps. In the SCL process, a candle filter is used to remove the particulates in the syngas stream. This is followed by a hot gas cleaning step at 400 °C for bulk sulfur removal. The particulate free raw syngas after the hot gas clean up unit, which has a H2S concentration of lower than 20 ppm, is then introduced to a reducer operated at 900 °C and 30 atm. The reducer is a countercurrent moving bed reactor with Fe_2O_3 fed from the top and syngas fed from the bottom. The main reactions in the reducer include:

$$Fe_2O_3 + CO/H_2 \rightarrow 2FeO + CO_2/H_2O$$
$$FeO + CO/H_2 \rightarrow Fe + CO_2/H_2O$$

The impurities in the syngas stream such as sulfur compounds and elemental mercury will not react with the iron oxide. Therefore, these pollutants will leave the reducer along with the CO_2 and steam resulted from the oxidation of the syngas stream. After condensing out steam, the concentrated CO_2 stream can be further compressed and sequestrated.

The reducer is slightly endothermic. The heat required in the reducer is fully compensated by the sensible heat carried by the hot iron oxide particles introduced from the combustor.

The iron oxide exiting from the bottom of the reducer is comprised mainly of FeO and Fe. Such a solid stream is then introduced to the oxidizer, which is also a countercurrent moving bed operated at 900 °C and 30 atm. The oxidizer converts steam into hydrogen while regenerates the FeO and Fe introduced from the reducer into Fe_3O_4. The main reactions in the oxidizer are:

$$3Fe + 4H_2O \rightarrow Fe_3O_4 + 4H_2$$
$$3FeO + H_2O \rightarrow Fe_3O_4 + H_2$$

The above reactions are slightly exothermic. The reducer is adjusted to be heat neutral by introducing low temperature steam. The gaseous product from the reducer is a mixture of hydrogen and steam. Steam can be easily condensed out to obtain a concentrated hydrogen stream.

The partially regenerated Fe_3O_4 from the oxidizer is further combusted in the combustor using air. The combustor is an entrained bed reactor operated at adiabatic conditions. The hot oxygen depleted air generated in the combustor can be used for power generation while the hot Fe_2O_3 particle is recycled back to the reducer.

As can be seen, the hydrogen product is generated from the oxidizer whereas CO_2 and other contaminants are generated in the reducer. Therefore, the energy required for CO_2 separation and contaminant removal is reduced. In the following sections, moving bed test results on the SCL reducer is reported. A detailed ASPEN simulation is then provided.

3. Bench Scale Tests of the SCL Process

3.1 Reactor Setup

A bench scale moving bed reactor was used to demonstrate the SCL process[3]. The bench scale reactor setup is illustrated in Figure 3. In a typical demonstration, solid reactants such as oxygen carrier particle and/or coal char are first loaded to the top funnel and then moved downwards steadily by the screw conveyor system. The bed height is maintained by the bed height sensor system. Reactant gas is introduced from the gas inlet located at the bottom of the moving bed to react with the solids in a counter current manner. The gas composition along the reactor is constantly monitored using a Varian CP-4900 micro GC. The solid samples can be taken from the solid sampling ports after the experiment for further characterization. The moving bed setup mimics the actual gas-solid contacting pattern in the proposed reducer and oxidizer design. More detailed information regarding the reactor setup can be found in Gupta et al. [3].

a b

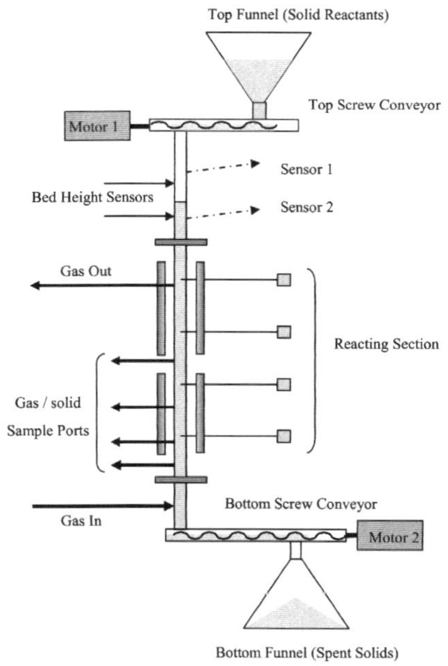

Figure 3. Bench Scale Demonstration Unit for SCL Process a. Schematic flow diagram of the unit; b. picture of the unit.

3.2 Reducer/Oxidizer Test Procedure and Results

The experiment to demonstrate continuous reducer operation was carried out first. Before the experiment, around 12 kg composite pellets with 60% Fe_2O_3 and 40% inert support was loaded to the top funnel. The reactor was then sealed and flushed overnight using N_2 at a flow rate of 500 ml/min.

Prior to the experiment, the solid transport system was activated to continuously move the particles through the reacting section. Meanwhile, the reactor was gradually heated up to around 900 °C. Once the desired reacting temperature was achieved, simulated syngas that consists of CO, H_2, and CO_2, was introduced into the bottom of the reactor. N_2 was also introduced along with the simulated syngas as an internal standard.

After the injection of the reactant gas, the gas compositions at the various positions of the reactor were constantly monitored and sampled using the MicroGC. The steady flow of both solid and gas was maintained for at least three solid residence time to ensure the achievement of steady state operation. After the experiment was finished, the reactant gas, the solid transport system, and the heating elements were simultaneously switched off. Nitrogen was introduced to the system to prevent re-oxidation of the solid samples. The solid samples at various positions of the reactor were taken from the reactor once it is cooled down to room temperature. These samples were than characterized using a CM-120 total carbon analyzer (UIC Inc.) for carbon content. The same samples were oxidized in a TGA using air. The solid conversions were then determined using the weight change obtained in the TGA and the carbon content characterized in the carbon analyzer.

After the experiment, it was found that nearly 99% syngas was converted to CO_2 and H_2O. Meanwhile, Fe_2O_3 was reduced by 49.5%. The carbon content in the reduced metal oxide was found to be lower than 0.1% (by weight).

Test on the oxidizer operation was carried out using the reduced composite Fe_2O_3 particle obtained from the previous test. The reduced particles were made to flow through the reactor in a manner similar to the previous test. The reacting gas comprised of steam balanced with inert gas (N_2). The same analysis method was used to monitor the H_2 production and extent of solid oxidation. It was observed that the reduced particles can be completely converted at 500 °C or even lower. The average hydrogen product purity was found to be 99.95%.

4. Process Simulation

4.1 summary of interested system

The ASPEN flow sheets of the traditional coal to hydrogen process and the SCL process are shown in Figure 4 and Figure 5. The process configurations are identical to that described in Section 2.

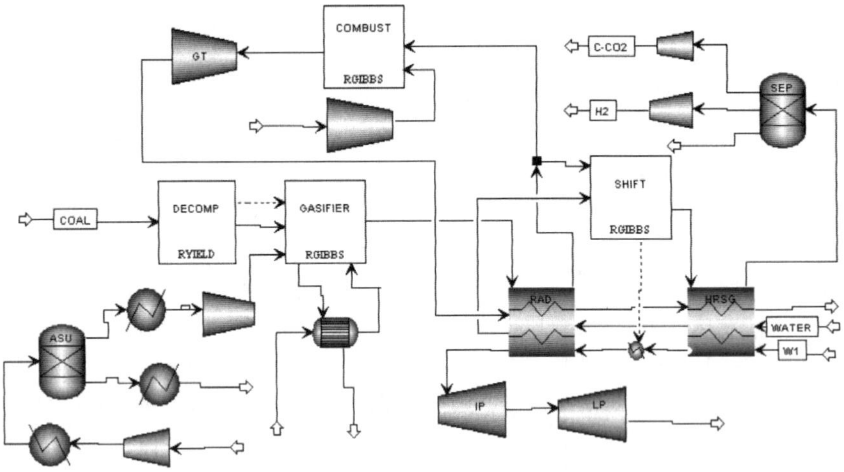

Figure 4 ASPEN Flowsheet of Conventional Coal to Hydrogen Process

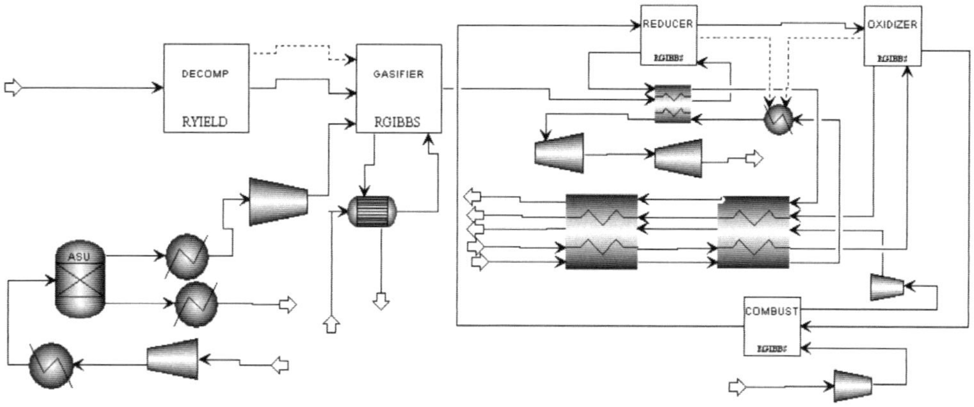

Figure 5 ASPEN Flowsheet of the SCL Process

4.2 Assumptions

To compare the differences between the SCL process and conventional process for hydrogen production from coal, two detailed models have been set up on ASPEN plus under the same assumptions, which are:

- 132.9 tonne/hr of Illinois #6 coal is fed into the system (1000 MW in HHV), the composition of the Illinois #6 coal is shown in Table 1;
- Energy consumed for units such as acid gas removal are simulated based on performance data of the commercial units;
- The GE slurry feed gasifier has a carbon conversion of 99%, heat loss in the gasifier is 0.6% of the HHV of coal, heat loss in all the other units are negligible [4];
- A GE 7H gas turbine combined cycle system is used, all the exhaust gas is cooled down to 130 °C before exiting the Heat Recovery Steam Generator (HRSG);
- At least 90% of the CO_2 generated needs to be captured and compressed to 15 Mpa for sequestration;
- The mechanical efficiency of pressure changers is 1, whereas the isentropic efficiency is 0.8~0.9;
- Solids circulating in SCL consist of 60% Fe_2O_3, and 40% inert support, by weight;
- The pressure of H_2 coming out of the system should be 3 Mpa ;
- steam is superheated to 680 °C, 30 atm to recover heat and then drive steam turbine;
- Pressure drop in the combustor is 0.2 MPa; pressure drop both the reducer and the oxidizer is 0.1 MPa;
- The reactions in high temperature reactors such as gasifier can reach their thermodynamic equilibriums;
- Experimental results obtained from the SCL reducer test is used to simulate reducer performance;
- The CO_2 capture cost is about 100.6 kWh per tonne CO_2 for traditional case[5];

- All the compressors are designed as 4-stage type with intercooler outlet temperature at 40 °C;
- The PSA unit can recover 77% hydrogen with 0.2 MPa pressure drop for the product gas and 2.5 MPa pressure drop for the tail gas;
- The air separation unit (ASU) consumes 367 kWh per ton of oxygen, mainly calculated based on gas compressions.

Table 1: Physical and Chemical Properties of Illinois #6 coal [6]

Proximate Analysis	Wt% (As-Received)	Wt%, dry	**Ultimate**	Wt% (As-Received)	Wt%, dry
Moisture	11.12		Moisture	11.12	
Fixed Carbon	44.19	49.72	ASH	9.7	10.91
Volatiles	34.99	39.37	CARBON	63.75	71.72
Ash	9.7	10.91	HYDROGEN	4.5	5.06
	100	100	NITROGEN	1.25	1.41
			CHLORINE	0.29	0.33
HHV (MJ/kg)	**27.13511**	**30.53107**	SULFUR	2.51	2.82
			OXYGEN	6.88	7.75

5. Results and Discussions

In order to co-produce of H_2 and electricity from coal, 551.4 tonne/hr of air is compressed into ASU to obtain a stream of 96% pure O_2 for the gasifier. The entrained bed GE/Texaco slurry feed gasifier can covert 132.9 ton/hr of coal into high temperature, high pressure raw syngas. The high temperature syngas from the gasifier is then cooled down using a radiant cooler. The syngas composition of this gasifier is shown in Table 2.

Table 2 Syngas from GE gasifier obtained from the ASPEN model

Gas Species in the Syngas Stream	Percentage (%)
H2O	25.32845
N2	1.46731
O2	6.61E-08
S	1.86E-04
SO2	3.05E-03
H2	23.59942
CO	39.31033
CO2	9.576
H3N	1.04E-03
COS	3.85E-02
H2S	0.676
CH4	1.75E-04
Total Flow kmol/hr	14483
Total Flow tonne/hr	303
Temperature °C	1631
Pressure atm	30

In the conventional case, syngas is split into two streams, 10% of the syngas is directly used for power generation and the rest of the syngas is sent to the hydrogen production units for low temperature (220 °C) water gas shift reaction, CO_2 separation, and product upgradation in PSA. The tail gas including part of hydrogen product from PSA is also used for power generation. In this configuration, 90% of the feed carbon is ready for sequestration. Hydrogen production rate is estimated to be 13.30 tonne/hr along with 71.6 MW net electricity generation.

In the SCL case, all the syngas, after bulk sulfur removal, is introduced to the reducer. The syngas is fully oxidized in the reducer with 49.5% iron oxide reduction, which is consistent to the experimental results. Therefore, the solid product from the reducer is a mixture of metallic iron and Wuestite. The reduced iron oxide is then introduced to the Oxidizer (800 °C, 30 atm) to perform the iron-steam reaction while being partially regenerated to Fe_3O_4. The combustor will further combust Fe_3O_4 (1128 °C, 30 atm) from the oxidizer, forming Fe_2O_3. The high temperature exhaust gas streams from all three looping units are used for steam/power generation. After the compensation of parasitic power requirements, 74.8 MW of net power and 14.67 tonne/hr of H_2 are produced with 100% CO_2 captured and compressed for sequestration. Table 3 compares the results between SCL process and conventional process. As can be seen in Table 3, significant efficiency improvement can be potentially realized through the utilization of the chemical looping strategy proposed in the SCL process.

Table 3: Summary of Hydrogen from Coal Cases

	Conventional Process	SCL Process
Coal feed(tonne/hr)	132.9	132.9
Carbon Capture(%)	90	100
Hydrogen(tonne/hr)	13.30	14.67
Net Power(MW)	71.6	74.8
Efficiency(%HHV)	59.7	65.3

6. Conclusions

A novel Syngas Chemical Looping process that converts coal derived syngas into hydrogen and electricity is discussed in this paper. Experimental results obtained from a bench scale unit indicated that the process is technically feasible. ASPEN simulation further confirmed that the SCL process has the potential to improve the energy conversion efficiency of the conventional coal to hydrogen process by over 6% while capturing all the carbon in coal.

7. Acknowledgement

This work has been supported by the Ohio Coal Development Office of the Ohio Air Quality Development Authority, U. S. Department of Energy, Ohio State University and an industrial consortium.

8. References

1. Gupta P, Velazquez-Vargas LG, Fan LS. Syngas redox (SGR) process to produce hydrogen from coal derived syngas. *Energy & Fuels.* Sep-Oct 2007;21(5):2900-2908.
2. Velazquez-Vargas LG, Thomas T, Gupta P, Fan L-S. Hydrogen production via redox reaction of syngas with metal oxide composite particles. Paper presented at: Proceedings of AIChE Annual Technical Meeting, 2004.
3. Gupta P, Velazquez-Vargas LG, Valentine C, Fan LS. Moving bed reactor setup to study complex gas-solid reactions. *Review of Scientific Instruments.* Aug 2007;78(8):-.
4. Higman C. *Gasification.* 2ND ED. ed. Boston: GULF PROFESSIONAL; 2008.
5. Shah VA, Huurdeman TL. Synthesis gas treating with physical solvent process using Selexol process technology. *Ammonia Plant Saf. FIELD Full Journal Title:Ammonia Plant Safety & Related Facilities.* 1990;30:216-224.
6. NETL. Shell Gasifier IGCC Base Cases; 2000.

HETEROGENEOUS FLOW STRUCTURE IN GAS-SOLID RISERS: CONTINUOUS MODELING

Dawei Wang[1],Jun You[1], Chao Zhu[1]*

1 New Jersey Institute of Technology, Newark, NJ 07102, USA

Abstract - Gas-solid riser flow exhibits strong heterogeneous structure in both axial direction and radial directions. Recent experimental studies reveal that the general radial solids concentration profiles present a double ring structure and the formation of a solid core region which have a relative higher concentration than the annulus region. This paper is focused on a comprehensive modeling of continuous gas-solids flow structure both in radial and axial directions. The specific transport mechanism due to collisional diffusive mass transfer and turbulent mass transfer are modeled. The radial heterogeneous flow structure of solids and gas at the different stage of the riser are investigated in detail. This mechanistic model, implemented with a detailed axial flow structure model, consists of a set of coupled ordinary-differential equations developed from conservation laws of mass, momentum and kinetic energy of both gas and solids phases. The solving algorithm is based on the Runge-Kutta method. The proposed model predicts the phase transport profiles such as the solids concentration, phase velocities and pressure drops in different regions along the riser. The model also yields the critical information of flow structure characteristics such as back flow, wall frictions and choking.

INTRODUCTION

Recent measurement of circulating fluidized bed riser with Electrical Capacitance Tomography (ECT) reveals a double ring structure of solids concentration and a solids core region flow structure in radial direction (Du et al. (2004)). This double ring structure appears to be stable along the riser, and could be observed under a wide range of CFB operation conditions.

Most models on heterogeneous flow structure are based on the core-annulus (wall) flows (Bolton and Davidson, (1988); Rhodes and Geldart, (1987); Horio et al, (1988); Senior and Brereton, (1992)), which typically consider a dilute uniform core flow, and a dense wall flow along the riser. Most of these models ignore the detailed mechanisms in the bottom region of riser where the flows can be very dense and complex. A primitive model was lately proposed to interpret the reported core-annulus-wall structure (Zhu et al., (2005)), using a simplified kinetic theory model to account for the solids acceleration in collision dominated dense flow regimes near the bottom of riser. It is realized, however, that most traditional momentum-based models with the assistance of kinetic theory modeling approach may be insufficient to describe some basic physics of collision-induced energy dissipation in fluidization, such as energy dissipations from tangential slip and rotational slip. This deficiency may be represented by the inability of correctly predicting the pressure distribution in the dense flow regime near the bottom of a CFB riser. The importance of correct account of energy transport and dissipation in the momentum equation may be analogous to that of k-ε model in the turbulent momentum transport equations in turbulence flows. Hence an additional term due to energy dissipation should be introduced in solid momentum transport equation in the collision and acceleration dominated regime (Zhu and You, (2006)).

A 3-zone model is recently presented to simulate the heterogeneous structure of the riser flow (Zhu et al. (2007)). The model yields a reasonable explanation not only for "core-annulus (wall)" flow structure but also the "core-annulus-wall" flow structure in riser flows. While, the model artificially divides the riser into 3 different zones and uses averaged values to describe the characteristics of the flow in each zone, thus it can not reveal the intrinsic mechanism and relationship of solids mass transport of each place. To make the problem to be closed, the model uses the pre-defined wall zone conditions to simulate the structure in core and annulus zones.

This paper is aimed to present a continuous modeling approach with intrinsic mass transfer mechanism to characterize the formation mechanisms of heterogeneous structure in a CFB riser.

* corresponding author: zhu@adm.njit.edu

GENERAL GOVERNING EQUATIONS

Basically, the flow in a riser could be described with following governing equations based on the mass and momentum conservation of each phase.

$$\frac{d}{dz}(\int_A \alpha_g \rho_g U_g dA) = 0 \quad (1)$$

$$\frac{d}{dz}(\int_A \alpha_s \rho_s U_s dA) = 0 \quad (2)$$

$$-\frac{d(\int_A PdA)}{dz} = \tau_w l_w + \int_A \alpha_g \rho_g g dA + \frac{d(\int_A \alpha_g \rho_g U_g^2 dA)}{dz} + \int_A F_D dA \quad (3)$$

$$\frac{d(\int_A \alpha_s \rho_s U_s^2 dA)}{dz} = \int_A F_D dA - \tau_{sw} l_{sw} - \int_A \alpha_s \rho_s g dA - \int_A F_C dA \quad (4)$$

With the volumetric fraction relations of gas and solids phase and the equation of state of gas phase: $\alpha_g + \alpha_s = 1$ and $\rho_g = \frac{P}{RT}$.

Where, F_D is the drag force which is given by Richard-Zaki equation (Richardson and Zaki (1954)),

$$F_D = \frac{18\mu}{d_s^2} \frac{\alpha_s}{(1-\alpha_s)^L} \cdot (U_g - U_s) \quad (5)$$

F_C is the solid momentum transport due to energy dissipation in the dense and acceleration region (Zhu and You (2006)) which is

$$F_C = -\frac{\Gamma - (U_g - U_s)(-\frac{dP}{dz})}{U_s} \quad (6)$$

Where, Γ represents the energy dissipations due to inter-phase frictional and inter-particle collision.

$$\Gamma \approx \sum_{j=0}^{N} k_j \alpha_s^j \quad (7)$$

τ_w and τ_{sw} are respectively the friction stresses between the wall and the gas phase and the solids phase.

For radically uniform flow, the integrals in above equation could be replaced by an averaged value; the closure of the problem for axially heterogeneous flow structure is fulfilled

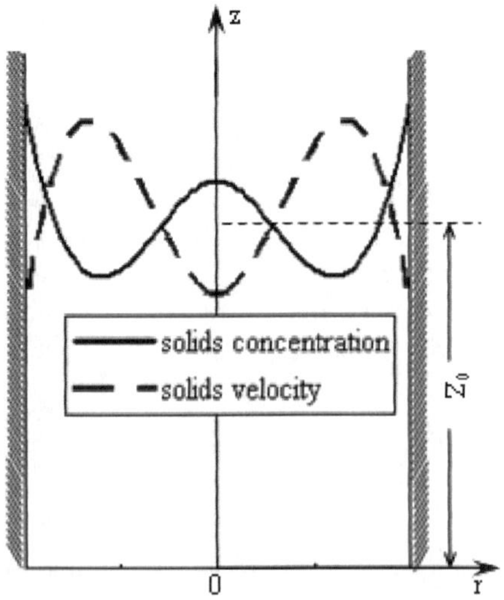

Figure 1. Representative flow structure by term of 3rd order polynomials

For the heterogeneous flow structure in both radial and axial directions, we assume that the transport parameters ϕ could be expressed as polynomials with following form:

$$\phi(r,z) = \sum_{i=0}^{n} c_{\phi i}(z) \cdot r^i .$$

We use 3rd order polynomials to approximate the distribution,

$$\phi(r,z) = c_{\phi 3}(z) r^3 + c_{\phi 2}(z) r^2 + c_{\phi 1}(z) r + c_{\phi 0}(z) \quad (8)$$

Here, ϕ could be $\alpha_s(r,z)$, $U_s(r,z)$ or $U_g(r,z)$.

It is noted that the pressure distribution at an arbitrary cross-section is uniform. That means the gas density is constant at a cross-section.

The coefficients of $C_{\Phi i}$ could be given by

$c_{\phi 0} = \phi_0(z)$ is the value of $\phi(r,z)$ at the center line;

$c_{\phi 1} = 0$ from axi-symmetricity of the riser;

$$c_{\phi 2} = \frac{3(\phi_w(z) - \phi_0(z)) - \delta_{\phi w}(z)R}{R^2}; \qquad c_{\phi 3} = \frac{\delta_{\phi w}(z)R - 2(\phi_w(z) - \phi_0(z))}{R^3}$$

Where, $\phi_w(z)$ is the value of $\phi(r,z)$ at the wall, and $\delta_{\phi w}(z)$ is the radial gradient of $\phi(r,z)$ at the wall.

Where, $\phi_w(z)$ is the value of $\phi(r,z)$ at the wall, and $\delta_{\phi w}(z)$ is the radial gradient of $\phi(r,z)$ at the wall.

Put the polynomials of solids concentration, solids velocity and gas velocity with above form into governing equation (1)-(4), and with the truth that the pressure across any cross-section of the riser is uniform, we could have

$$\begin{bmatrix} C_{1,1} & C_{1,2} & C_{1,3} & C_{1,4} & C_{1,5} & C_{1,6} & C_{1,7} & C_{1,8} & C_{1,9} & C_{1,10} \\ C_{2,1} & C_{2,2} & C_{2,3} & C_{2,4} & C_{2,5} & C_{2,6} & C_{2,7} & C_{2,8} & C_{2,9} & C_{2,10} \\ C_{3,1} & C_{3,2} & C_{3,3} & C_{3,4} & C_{3,5} & C_{3,6} & C_{3,7} & C_{3,8} & C_{3,9} & C_{3,10} \\ C_{4,1} & C_{4,2} & C_{4,3} & C_{4,4} & C_{4,5} & C_{4,6} & C_{4,7} & C_{4,8} & C_{4,9} & C_{4,10} \end{bmatrix} \begin{pmatrix} \frac{d\alpha_{s0}}{dz} \\ \frac{d\alpha_{sw}}{dz} \\ \frac{d\delta_{\alpha sw}}{dz} \\ \frac{dU_{g0}}{dz} \\ \frac{dU_{gw}}{dz} \\ \frac{d\delta_{ugw}}{dz} \\ \frac{dU_{s0}}{dz} \\ \frac{dU_{sw}}{dz} \\ \frac{d\delta_{usw}}{dz} \\ \frac{dP}{dz} \end{pmatrix} = \begin{pmatrix} 0 \\ 0 \\ -\frac{\int F_D dA + \tau_w l_w}{A} \\ \frac{\int_A (1-\lambda_c) F_D dA - \tau_{ws} l_{ws}}{A} \end{pmatrix} \quad \text{(Group A)}$$

Where $C_{i,j}$ are coefficients explicitly expressed by terms of α_{s0}, α_{sw}, $\delta_{\alpha sw}$, U_{g0}, U_{gw}, δ_{ugw}, U_{s0}, U_{sw} and δ_{usw}. (see appendix A)

we have ten unknowns (α_{s0}, α_{sw}, $\delta_{\alpha sw}$, U_{s0}, U_{sw}, δ_{usw}, U_{g0}, U_{gw}, δ_{gw}, P), and we have only 4 governing equations ((8)-(11)). These 4 governing equations describe the axial heterogeneous flow structure along the riser and could only be used to represent the change of averaged flow structure parameters $\bar{\alpha}_s, \bar{U}_g, \bar{U}_s$.

To closure the problem of heterogeneous flow structure (in both radial and axial directions), 6 additional intrinsic correlations (in radial direction) are needed. These 6 correlations will quantify the radial transport of phases in terms of $\phi_0(z), \phi_w(z), \delta_{w\phi}(z)$, which link the transport properties of centerline to those at the wall and shall cover the flow structure of concentrations and velocities of both gas and solids phases across any cross-section of the riser. They shall reveal the radial mass transport caused by the turbulence and collisional diffusion as well as the effect of wall boundary constraint on the flow structure.

INTRINSIC MECHANISM AND PROBLEM CLOSURE

Based on the non-slip condition of gas phase at the wall boundary, the gas velocity at the wall shall be zero. We have $U_{gw}(z) = 0$. We also assume that the friction between the gas phase and the wall is small enough and could be neglected, thus the radial change rate of gas velocity at the wall could be assumed to be zero.

We could write $\delta_{Ugw}(z)=0$. The above two equations along with the averaged gas velocity given by the previous governing equation could give the gas velocity distribution across the cross-section of riser which is characterized with three characteristic values $U_{g0}(z)$, $U_{gw}(z)$, $\delta_{ugw}(z)$.

For the solids volume fraction, we could also assume it is packed on the wall and thus is the constant value, which gives

$$\alpha_{sw}(z)=\alpha_{packed}=\text{constant} \tag{9}$$

Based on the force balance of solids phase at the wall, we could have

$$F_{Dw}-\frac{\partial\tau_{sw}}{\partial r}l_{sw}-\rho_s\alpha_s g=\alpha_s\rho_s\frac{dU_{sw}}{dz} \tag{10}$$

Where, F_{Dw} is the drag force of gas phase on the solids phase at the wall.

The formation mechanism of radial heterogeneous flow structure is the radial solids transport due to the turbulent mass transfer and collisional diffusive mass transfer of solids particles. The intensity of turbulent mass transfer is dependent on the local turbulent intensity and the velocity gradient of solids and is from high velocity to low velocity. The intensity of mass transfer is dependent on the local solids concentration and the concentration gradient of solids phase, the direction is from high concentration to low concentration. The net value of these two mass transfers gives the local solids transport. $J_{net}=J_{T.C.}+J_{C.D.}$

Based on the mass conservation of solids phase, the total mass flow of solids phase across the cross-section shall keep unchanged although the solids may transfer inwards or outwards locally, and we could write:

$$\int\frac{\partial J_{net}}{\partial r}dA=0 \tag{11}$$

Where, For the simplicity of the problem, we assume the coefficients of turbulent convective and collisional diffusive mass transfers are respectively proportional to the local velocity and solids concentration.

$$J_{TC}=-D_{turb}U_s\rho_s\frac{\partial U_s}{\partial r};\qquad J_{CD}=-D_{diff}\alpha_s\rho_s\frac{\partial\alpha_s}{\partial r}.$$

The radial transport of solids also causes the radial solids momentum shift, which is balanced by the radial stress of the wall boundary and represented as the form of solids pressue. We assume that the solids pressure is equal to the gravity of the solids phase at each height.

$$\int J_{net}U_s dA=\sigma_{sw} \tag{12}$$

With above equations of intrinsic mechanism and simplifications, we could obtain 7 independent equations (equation group A and 10-12) for 7 variables (α_{s0}, $\delta_{\alpha sw}$, U_{s0}, U_{sw}, δ_{usw}, U_{g0}, P) to describe the heterogeneous flow structure in a riser. They are given as:

$$\begin{bmatrix} D_{1,1} & D_{1,2} & D_{1,3} & D_{1,4} & D_{1,5} & D_{1,6} & D_{1,7}\\ D_{2,1} & D_{2,2} & D_{2,3} & D_{2,4} & D_{2,5} & D_{2,6} & D_{2,7}\\ D_{3,1} & D_{3,2} & D_{3,3} & D_{3,4} & D_{3,5} & D_{3,6} & D_{3,7}\\ D_{4,1} & D_{4,2} & D_{4,3} & D_{4,4} & D_{4,5} & D_{4,6} & D_{4,7}\\ D_{5,1} & D_{5,2} & D_{5,3} & D_{5,4} & D_{5,5} & D_{5,6} & D_{5,7}\\ D_{6,1} & D_{6,2} & D_{6,3} & D_{6,4} & D_{6,5} & D_{6,6} & D_{6,7}\\ D_{7,1} & D_{7,2} & D_{7,3} & D_{7,4} & D_{7,5} & D_{7,6} & D_{7,7} \end{bmatrix} \begin{pmatrix} \frac{d\alpha_{s0}}{dz}\\ \frac{d\delta_{\alpha sw}}{dz}\\ \frac{dU_{g0}}{dz}\\ \frac{dU_{s0}}{dz}\\ \frac{dU_{sw}}{dz}\\ \frac{d\delta_{usw}}{dz}\\ \frac{dP}{dz} \end{pmatrix} = \begin{pmatrix} 0\\ 0\\ -\frac{\int F_D dA+\tau_w l_w}{A}\\ \frac{\int_A(1-\lambda_c)F_D dA-\tau_{ws}l_{ws}}{A} \end{pmatrix} \qquad \text{(Group B)}$$

Where $D_{i,j}$ are coefficients explicitly expressed by terms of α_{s0}, $\delta_{\alpha sw}$, U_{g0}, U_{s0}, U_{sw} and δ_{usw}. (see appendix B).

The closure of the problem is obtained.

EXPECTED RESULTS AND DISCUSSIONS

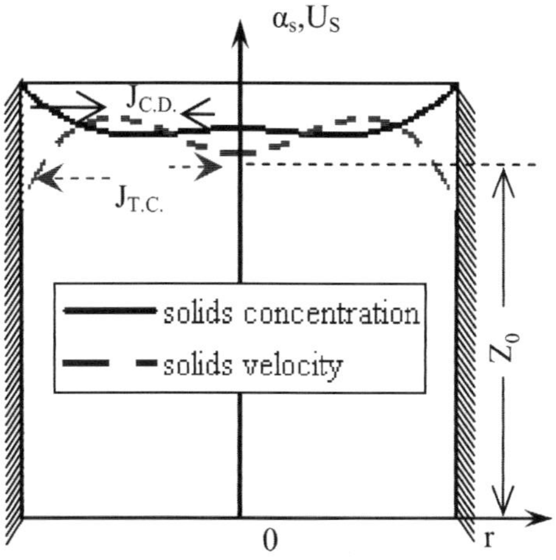

(a) solids concentration and velocity distribution at riser inlet

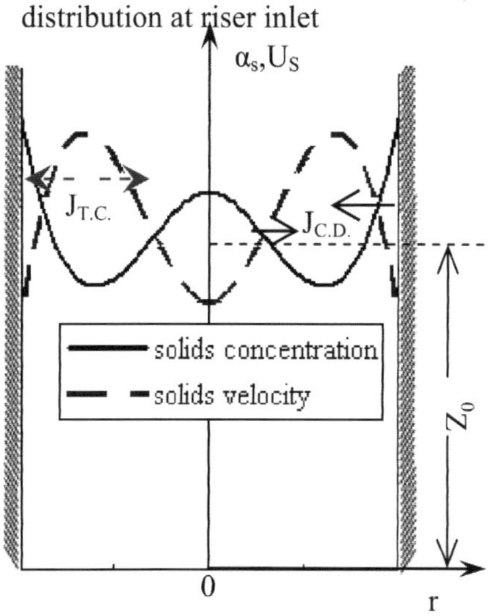

(a) solids concentration and velocity distribution at middle of riser height

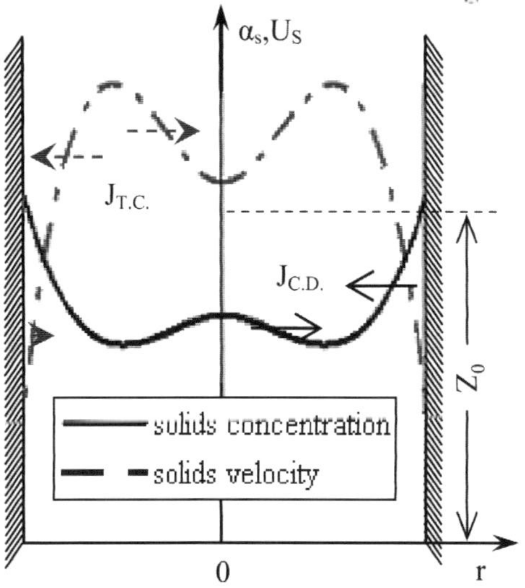

(a) solids concentration and velocity distribution at middle of riser height

To preliminarily evaluate the correctness of our model, some analysis and discussion are made. We assume that at the inlet of the riser, the solids concentration and solids velocity are slightly non-uniform across the cross-section as shown in figure 2(a). The directions of the solids mass transfers due to turbulence and collisional diffusion are also shown. Figure 2(b) shows the net mass transport of solids phase. The solids always transport outwards from the inner of the riser across the cross-section, which could make the flow structure more non-uniform. As the transport amount in the middle part of the riser is higher than that in the center part, the middle part shall become dilute much faster than the center part.

After the process of solids acceleration, the solids velocities in the middle and center regions increase while the velocity at the wall boundary shall decrease due to the wall fraction. The non-uniformity of the radial flow structure becomes higher (figure 3(a)). Now, the direction of net mass transport would be from the outer side into the inner side of the riser (figure 3(b)).

With further radial mass transport of solids and development of flow structure, the solids velocity and concentration distributions change further as shown in figure 4(a). The solids concentrations of core and annulus zone are very dilute, while the concentration in the core zone is still higher than that of annulus zone. Due to the friction of the wall, the solids velocity at the wall boundary decreases further and becomes to move downwards. Now, the solids transport radially from inner side to outer side of the riser.

CONCLUSIONS

This paper presents a complete mechanistic model to interpret the heterogeneous flow structure in riser flows. The governing equations of mass and momentum conservation of gas and solids phases are given to describe the axial flow structure. The intrinsic mechanism of radial mass transport is presented to reveal the formation process of the radial heterogeneous flow structure. The proposed mechanism yields a reasonable explanation for the heterogeneous riser flow structure in both radial and axial directions. Typical radial distributions of solids velocity and concentration at different stage of riser flow are described and analyzed. More quantitative comparison and investigation of parametric study could be conducted in future studies.

NOTATION

α_g	gas volume fraction	p	pressure, Pa
U_g	gas velocity, m/s	L	length, m
A	area, m^2	µm	
z	riser height, m	p_{ref}	reference pressure at cyclone exit, Pa
ρ_g	gas density, kg/m^3		overall riser pressure drop, Pa
α_s	solids volume fraction	R	diameter of the riser, m
U_s	solids velocity, m/s	t	time, s
ρ_s	solids density, kg/m^3	u	superficial gas velocity, m/s
F_D	drag force, N/m^3	U	signal of fiber-optical probes, V
g	gravity constant, m/s^2	U_0	signal of fiber-optical probes for $c_v = 0$, V
H_b	height of the bottom zone, m	ε	voidage, -
j	exponent defined by eq. 2, -	$\overline{\varepsilon}$	cross-sectional average voidage, -
k	exponent defined by eq. 2, -	P	Pressure, Pa
r	radial position, m		

REFERENCES

Du, B., W. Warsito and L.-S. Fan, "ECT studies of the choking phenomenon in a gas-solid circulating fluidized bed", AIChE J., 50 (7), 1386 (2004).

Bolton, L.W. and Davidson, J.F.(1988). Recirculation of Particles in Fast Fluidized Risers. In Circulation Fluidized Bed Technology _. Ed. Basu and Large. Toronto: Pergamon Press

Rhodes, M.J. and Geldart, D. (1987). A Model for the Circulating Fluidized Bed. Powder Tech., 53, 155

Horio, M., Morishita, K., Tachibana, O. and Murata, N. (1988). Solid Distribution and Movement in Circulating Fluidized Beds. In Circulating Fluidized Bed Technology _. Ed. Basu and Large. Toronto: Pergamon Press

Senior, R.C. and Brereton, C.M.M. (1992). Modeling of Circulating Fluidized Bed Solids Flow and Distribution. Chem. Eng. Sci., 47, 281.

Zhu,C., Yu, Q., and L.S.Fan, Modeling on Core-annulus-wall Structure in Circulating Fluidized Bed Riser, Circulating Fluidized Bed Technology _, 354, 2005

Zhu, C. and You, J. An Energy-Based Model of Gas-Solids Transport in a Riser, Powder Tech., 2006

Chao Zhu, Jun You, Dawei Wang, Liang-Shih Fan, Modeling on Heterogeneous Structure in Accelearation Regime of Gas-Solid Riser Flows, Fluidization XII, 119

Rhodes, M.J., M. Sollaart and X.S. Wang, "Flow structure in a fast fluid bed", Powder Technology, 99, 194 (1998).

Richardson, J.F., Zaki, W.N., Sedimentation and fluidization, Trans. Inst. Chem. Eng., vol. 32: p. 35-53. (1954)

APPENDIX A

Where

$$\begin{cases} D_1 = \frac{\delta_{\alpha gw} R^2}{168} - \frac{\alpha_{gw} R}{28} - \frac{\alpha_{g0} R}{28} \\ D_2 = \frac{\delta_{ugw} R^2}{168} - \frac{U_{gw} R}{28} - \frac{U_{g0} R}{28} \\ D_3 = \frac{2\alpha_{gw}}{7} + \frac{9\alpha_{g0}}{140} - \frac{\delta_{\alpha gw} R}{28} \\ D_4 = \frac{2U_{gw}}{7} + \frac{9U_{g0}}{140} - \frac{\delta_{ugw} R}{28} \\ D_5 = \frac{9\alpha_{gw}}{140} + \frac{3\alpha_{g0}}{35} - \frac{\delta_{\alpha gw} R}{70} \\ D_6 = \frac{9U_{gw}}{140} + \frac{3U_{g0}}{35} - \frac{\delta_{ugw} R}{70} \end{cases}$$

and

$$\begin{cases} D_7 = \frac{\delta_{\alpha sw} R^2}{168} - \frac{\alpha_{sw} R}{28} - \frac{\alpha_{s0} R}{28} \\ D_8 = \frac{\delta_{usw} R^2}{168} - \frac{U_{sw} R}{28} - \frac{U_{s0} R}{28} \\ D_9 = \frac{2\alpha_{sw}}{7} + \frac{9\alpha_{s0}}{140} - \frac{\delta_{\alpha sw} R}{28} \\ D_{10} = \frac{2U_{sw}}{7} + \frac{9U_{s0}}{140} - \frac{\delta_{usw} R}{28} \\ D_{11} = \frac{9\alpha_{sw}}{140} + \frac{3\alpha_{s0}}{35} - \frac{\delta_{\alpha sw} R}{70} \\ D_{12} = \frac{9U_{sw}}{140} + \frac{3U_{s0}}{35} - \frac{\delta_{usw} R}{70} \end{cases}$$

and

$$\begin{cases} D_{\delta 1w} = I_1\delta_{2w}^2R^3 + I_2\phi_{2w}\delta_{2w}R^2 + I_3\phi_{20}\delta_{2w}R^2 + I_4\phi_{2w}^2R + I_5\phi_{20}\phi_{2w}R + I_6\phi_{20}^2R \\ D_{\phi 1w} = I_7\delta_{2w}^2R^2 + I_8\phi_{2w}\delta_{2w}R + I_9\phi_{20}\delta_{2w}R + I_{10}\phi_{2w}^2 + I_{11}\phi_{20}\phi_{2w} + I_{12}\phi_{20}^2 \\ D_{\phi 10} = I_{13}\delta_{2w}^2R^2 + I_{14}\phi_{2w}\delta_{2w}R + I_{15}\phi_{20}\delta_{2w}R + I_{16}\phi_{2w}^2 + I_{17}\phi_{20}\phi_{2w} + I_{18}\phi_{20}^2 \\ D_{\delta 2w} = 2I_1\delta_{1w}\delta_{2w}R^3 + I_2\delta_{1w}\phi_{2w}R^2 + I_3\delta_{1w}\phi_{20}R^2 + 2I_7\phi_{1w}\delta_{2w}R^2 + I_8\phi_{1w}\phi_{2w}R + I_9\phi_{1w}\phi_{20}R + 2I_{13}\phi_{10}\delta_{2w}R^2 + I_{14}\phi_{1w}\phi_{2w}R + I_{15}\phi_{10}\phi_{20}R \\ D_{\phi 2w} = I_2\delta_{1w}\delta_{2w}R^2 + 2I_4\delta_{1w}\phi_{2w}R + I_5\delta_{1w}\phi_{20}R + I_8\phi_{1w}\delta_{2w}R + 2I_{10}\phi_{1w}\phi_{2w} + I_{11}\phi_{1w}\phi_{20} + I_{14}\phi_{10}\delta_{2w}R + 2I_{16}\phi_{10}\phi_{2w} + I_{17}\phi_{10}\phi_{20} \\ D_{\phi 20} = I_3\delta_{1w}\delta_{2w}R^2 + I_5\delta_{1w}\phi_{2w}R + 2I_6\delta_{1w}\phi_{20}R + I_9\phi_{1w}\delta_{2w}R + I_{11}\phi_{1w}\phi_{2w} + 2I_{12}\phi_{1w}\phi_{20} + I_{15}\phi_{10}\delta_{2w}R + I_{17}\phi_{10}\phi_{2w} + 2I_{18}\phi_{10}\phi_{20} \end{cases}$$

APPENDIX B

$$D_1 \frac{d\delta_{ugw}}{dz} + D_2 \frac{d\delta_{\alpha g}}{dz} + D_3 \frac{dU_{gw}}{dz} + D_4 \frac{d\alpha_{gw}}{dz} + D_5 \frac{dU_{g0}}{dz} + D_6 \frac{d\alpha_{g0}}{dz} = 0 \qquad (9)$$

$$D_7 \frac{d\delta_{usw}}{dz} + D_8 \frac{d\delta_{\alpha s}}{dz} + D_9 \frac{dU_{sw}}{dz} + D_{10} \frac{d\alpha_{sw}}{dz} + D_{11} \frac{dU_{s0}}{dz} + D_{12} \frac{d\alpha_{s0}}{dz} = 0 \qquad (10)$$

$$-\frac{dP}{dz} = \frac{\tau_w l_w}{A} + \frac{\int F_D dA}{A} + (2D_{\delta\alpha g} - \frac{\rho_g gR}{10})\frac{d\delta_{\alpha gw}}{dz} + (2D_{\alpha gw} + \frac{7\rho_g g}{10})\frac{d\alpha_{gw}}{dz} + (2D_{\alpha g0} + \frac{3\rho_g g}{10})\frac{d\alpha_{g0}}{dz} + 2D_{\delta ug}\frac{d\delta_{ugw}}{dz} + 2D_{ugw}\frac{dU_{gw}}{dz} + 2D_{ug0}\frac{dU_{g0}}{dz} \qquad (11)$$

$$(2D_{\delta\alpha s} - \frac{\rho_s gR}{10})\frac{d\delta_{\alpha sw}}{dz} + (2D_{\alpha sw} + \frac{7\rho_s g}{10})\frac{d\alpha_{sw}}{dz} + (2D_{\alpha s0} + \frac{3\rho_s g}{10})\frac{d\alpha_{s0}}{dz} + 2D_{\delta us}\frac{d\delta_{usw}}{dz} + 2D_{usw}\frac{dU_{sw}}{dz} + 2D_{us0}\frac{dU_{s0}}{dz} = \frac{\int_A (1-\lambda_c)F_D dA - \tau_{ws} l_{ws}}{A} \qquad (12)$$

Where
$$\begin{cases} D_1 = \frac{\delta_{\alpha gw}R^2}{168} - \frac{\alpha_{gw}R}{28} - \frac{\alpha_{g0}R}{28} \\ D_2 = \frac{\delta_{ugw}R^2}{168} - \frac{U_{gw}R}{28} - \frac{U_{g0}R}{28} \\ D_3 = \frac{2\alpha_{gw}}{7} + \frac{9\alpha_{g0}}{140} - \frac{\delta_{\alpha gw}R}{28} \\ D_4 = \frac{2U_{gw}}{7} + \frac{9U_{g0}}{140} - \frac{\delta_{ugw}R}{28} \\ D_5 = \frac{9\alpha_{gw}}{140} + \frac{3\alpha_{g0}}{35} - \frac{\delta_{\alpha gw}R}{70} \\ D_6 = \frac{9U_{gw}}{140} + \frac{3U_{g0}}{35} - \frac{\delta_{ugw}R}{70} \end{cases}$$
and
$$\begin{cases} D_7 = \frac{\delta_{\alpha sw}R^2}{168} - \frac{\alpha_{sw}R}{28} - \frac{\alpha_{s0}R}{28} \\ D_8 = \frac{\delta_{usw}R^2}{168} - \frac{U_{sw}R}{28} - \frac{U_{s0}R}{28} \\ D_9 = \frac{2\alpha_{sw}}{7} + \frac{9\alpha_{s0}}{140} - \frac{\delta_{\alpha sw}R}{28} \\ D_{10} = \frac{2U_{sw}}{7} + \frac{9U_{s0}}{140} - \frac{\delta_{usw}R}{28} \\ D_{11} = \frac{9\alpha_{sw}}{140} + \frac{3\alpha_{s0}}{35} - \frac{\delta_{\alpha sw}R}{70} \\ D_{12} = \frac{9U_{sw}}{140} + \frac{3U_{s0}}{35} - \frac{\delta_{usw}R}{70} \end{cases}$$

and
$$\begin{cases} D_{\delta 1w} = I_1\delta_{2w}^2R^3 + I_2\phi_{2w}\delta_{2w}R^2 + I_3\phi_{20}\delta_{2w}R^2 + I_4\phi_{2w}^2R + I_5\phi_{20}\phi_{2w}R + I_6\phi_{20}^2R \\ D_{\phi 1w} = I_7\delta_{2w}^2R^2 + I_8\phi_{2w}\delta_{2w}R + I_9\phi_{20}\delta_{2w}R + I_{10}\phi_{2w}^2 + I_{11}\phi_{20}\phi_{2w} + I_{12}\phi_{20}^2 \\ D_{\phi 10} = I_{13}\delta_{2w}^2R^2 + I_{14}\phi_{2w}\delta_{2w}R + I_{15}\phi_{20}\delta_{2w}R + I_{16}\phi_{2w}^2 + I_{17}\phi_{20}\phi_{2w} + I_{18}\phi_{20}^2 \\ D_{\delta 2w} = 2I_1\delta_{1w}\delta_{2w}R^3 + I_2\delta_{1w}\phi_{2w}R^2 + I_3\delta_{1w}\phi_{20}R^2 + 2I_7\phi_{1w}\delta_{2w}R^2 + I_8\phi_{1w}\phi_{2w}R + I_9\phi_{1w}\phi_{20}R + 2I_{13}\phi_{10}\delta_{2w}R^2 + I_{14}\phi_{1w}\phi_{2w}R + I_{15}\phi_{10}\phi_{20}R \\ D_{\phi 2w} = I_2\delta_{1w}\delta_{2w}R^2 + 2I_4\delta_{1w}\phi_{2w}R + I_5\delta_{1w}\phi_{20}R + I_8\phi_{1w}\delta_{2w}R + 2I_{10}\phi_{1w}\phi_{2w} + I_{11}\phi_{1w}\phi_{20} + I_{14}\phi_{10}\delta_{2w}R + 2I_{16}\phi_{10}\phi_{2w} + I_{17}\phi_{10}\phi_{20} \\ D_{\phi 20} = I_3\delta_{1w}\delta_{2w}R^2 + I_5\delta_{1w}\phi_{2w}R + 2I_6\delta_{1w}\phi_{20}R + I_9\phi_{1w}\delta_{2w}R + I_{11}\phi_{1w}\phi_{2w} + 2I_{12}\phi_{1w}\phi_{20} + I_{15}\phi_{10}\delta_{2w}R + I_{17}\phi_{10}\phi_{2w} + 2I_{18}\phi_{10}\phi_{20} \end{cases}$$

399c Design, Development and Operation of a Novel Recirculating Liquid/solids Reactor Based on a Type 2 Draft Tube Spout Fluid Bed

John Paccione[1], Howard Littman[2], Lealon L. Martin[2], Joel L. Plawsky[3], and David M. Follansbee[2]. (1) Environmental Health Sciences, SUNY at Albany, School of Public Health, 547 River Street, Flanigan Square, Troy, NY 12180, (2) Department of Chemical and Biological Engineering, Rensselaer Polytechnic Institute, 110 8th St., Troy, NY 12180, (3) The Isermann Department of Chemical and Biological Engineering, Rensselaer Polytechnic Institute, Ricketts Building, 110, 8th Street, Troy, NY 12180

There is a wide variety of multiphase reactions that require the solid phase to be regenerated. Very often the primary and regeneration reactions must be run under different operating conditions. A novel, next-generation draft tube spout fluid bed has been developed to provide a flexible and quantifiable system to optimize the operating conditions of a given fluid/solids regeneration reaction. Separating the locations of the annulus and draft tube effluent converts a standard DTSFB into a second generation (type 2) DTSFB. By separating the effluents of the annulus and the draft tube, the pressure drops across these two regions and the fluid flows through them can be independently varied. This allows the particle velocity and the solids concentration in the draft tube to be determined independently. Therefore, for a given solids circulation rate, both the solids concentration and particle velocity can be determined to accommodate the conditions at which the regeneration may take place. A type 2 DTSFB has been designed and constructed to run advanced oxidation reactions that use titanium dioxide to generate hydroxyl radicals to destroy organics in water. The reactor has been used to collect hydraulic data that were compared to existing fluid/solid transport correlations. The inlet section to the draft tube was also examined to determine the maximum and minimum operating limits of the inlet length. The maximum limit for solids concentration in the draft tube was also examined and compared to literature data. The type 2 DTSFB has been shown to be an efficient recycle reactor that can be used to run reactions that require the solids to be regenerated. Because the annulus need not be fluidized, the particles that approach the inlet to the draft tube have the longest residence time and are not randomly mixed with those that have been freshly regenerated. Other novel applications and configurations of the type 2 DTSFB are also discussed. The flexibility of the type 2 DTSFB allows them to be cascaded into a series that allows two or more reactions that have differing rates to be run continuously.

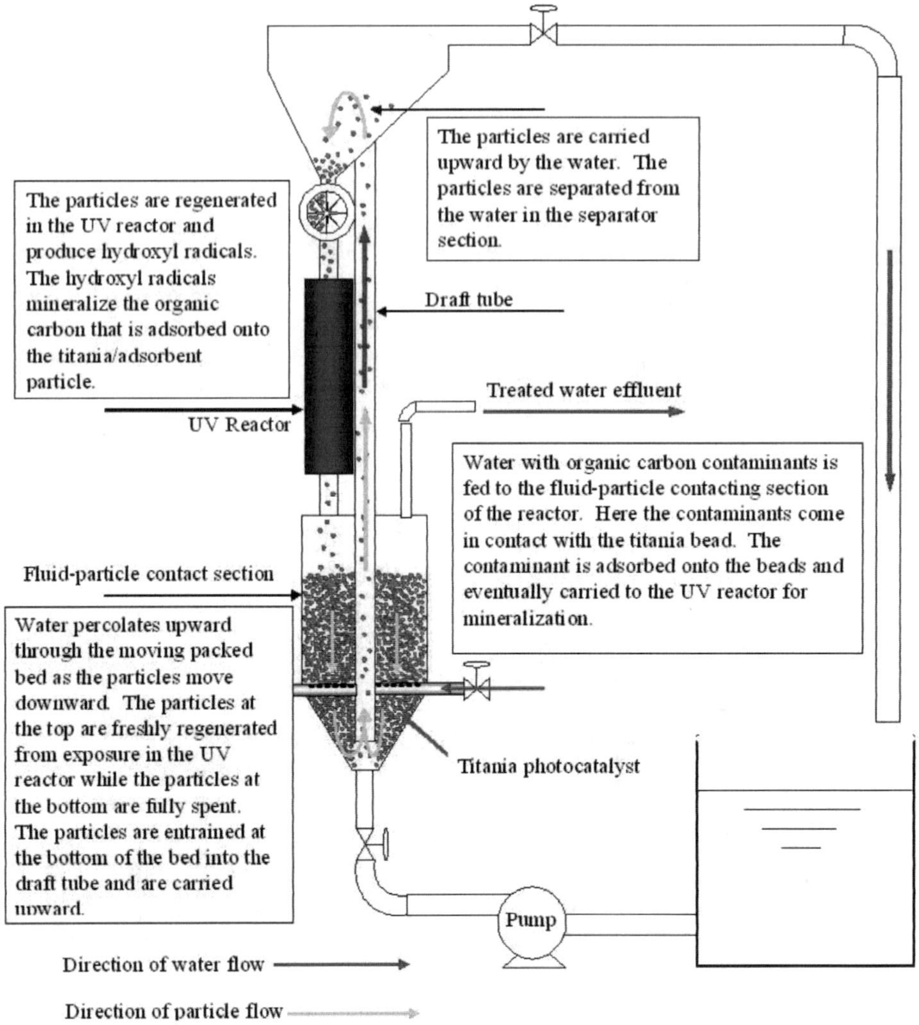

Titanium Dioxide Water Treatment Reactor

Mass Transfer Resistance in Downer Reactors

399d

Duvvuri Subbarao
Department of Chemical Engineering
Universiti Teknologi PETRONAS
Bandar Seri Iskandar, 31750, Tronoh, Perak, Malaysia

Abstract

Conversions in downer reactors are lower than what is expected from a pseudo homogeneous plug flow reactor model. This was attributed to particles moving as clusters, axial gas mixing and gas to particle mass transfer. In the present paper, a two phase tanks in series model is developed for the downer reactors considering gas to particle mass transfer. The model is compared with recent experimental observations on a downer reactor reported in the literature to estimate mass transfer coefficients. Estimated Sherwood Numbers are observed to increase as a function of the ratio of gas to particle velocities. This result is in line with the recently reported direct observations on gas to particle mass transfer in downers.

Introduction

Downers are receiving attention as a possible efficient gas solid contactors [1]. Talman et al [2] and Talman and Reh [3] reported experimental observations on the gas-solid catalytic reactions in downers reactors. Conversions were observed to be lower than the expected from a pseudo homogeneous plug flow model at higher solid loadings. This was attributed to particles moving as clusters. Fan et al [4] investigated downer reactor performance by catalytic ozone decomposition and observed that gas to particle mass transfer may be affecting the conversions.

In the present paper, a two phase tanks in series model is developed for the downer reactor performance considering gas to particle mass transfer. The model parameters are evaluated in the light of the experimental observations of Fan et al [4].

The Model

In a downer reactor, reactant gas reacts at the surface of catalyst particles as gas and particles flow concurrently downward through a vertical pipe. Along the length of flow axially, the reactor is visualized to consist "n tanks" in series through which gas and particles flow down. Each tank consists of gas phase and particle phase, each well mixed with in itself, with mass transfer of reactants and products between the two phases. All the tanks are assumed to be of equal size with a height of Δz.

A schematic of the structure of the model is shown in fig.1. Processes taking place in "ith compartment" are shown in fig.2. The model is applied to develop equations for decomposition of ozone by 1st order irreversible catalytic reaction. Component material balance for each phase in "i"th compartment can be written as

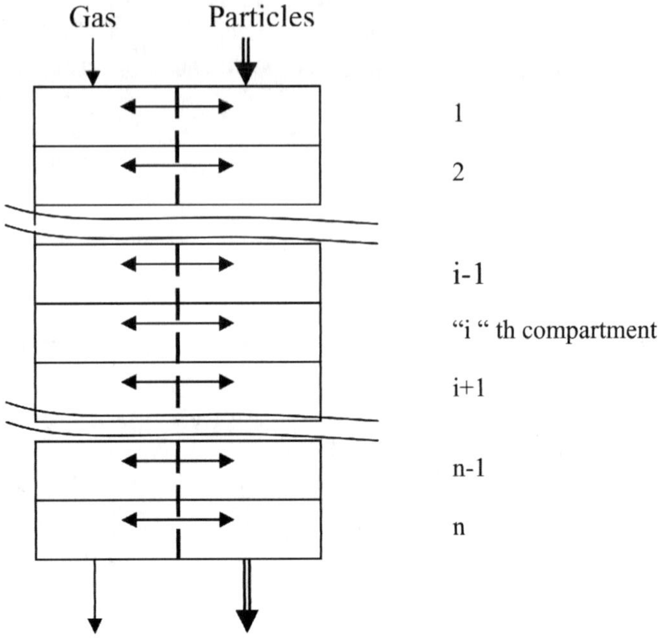

Fig.1. Structure of the two phase tanks-in-series model for a downer reactor.

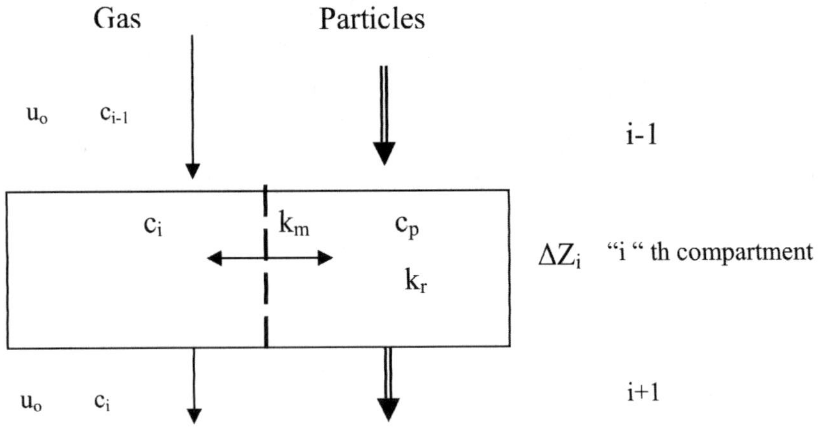

Fig. 2. Mass Transfer and reaction processes between gas phase and particle phase in ith compartment

For gas phase

input rate = output rate + mass transfer rate

$$u_0 c_{i-1} = u_0 c_i + k_m \Delta z (1-\varepsilon) \frac{s_p}{v_p} (c_i - c_p)$$

(1)

for particle phase

mass transfer rate = reaction rate

$$k_m \Delta z (1-\varepsilon) \frac{s_p}{v_p} (c_i - c_p) = k_r \Delta z (1-\varepsilon) \rho_p c_p$$

Eliminating particle phase concentration between these equations

$$\frac{c_i}{c_{i-1}} = \left[1 + \frac{\Delta Z(1-\varepsilon)}{u_o} \frac{1}{\dfrac{1}{k_r \rho_p} + \dfrac{1}{k_m \, {}^{s_p}\!/_{v_p}}} \right]^{-1} \tag{2}$$

Number of tanks in the downer can be obtained as

$$n = \frac{L}{\Delta z} \tag{3}$$

Then, fraction of reactant not converted in the reactor can be obtained as

$$(1-X) = \frac{c_{out}}{c_{in}} = \left[1 + \frac{L(1-\varepsilon)}{n\, u_o} \frac{1}{\dfrac{1}{k_r \rho_p} + \dfrac{1}{k_m \, {}^{s_p}\!/_{v_p}}} \right]^{-n} \tag{4}$$

This formulation explicitly brings out the importance of mass transfer and degree of axial mixing in heterogeneous catalytic reactors. With zero mass transfer coefficient no conversion is possible; with infinitely large mass transfer coefficient, the reaction can be considered as homogeneous in nature. The reactor behaves as a CSTR for n=1 and as a PFR for extremely high values of n.

Model Evaluation

Fan et al [4] measured Ozone conversion (X) by catalytic decomposition along the reactor length (L) in a downer reactor (column height of 8.5 m and diameter of 0.09 m) as a function of gas velocity (2.2 m/s to 3.7 m/s) and particle flux (8.4 to 26.8 kg/m^2/s. FCC catalyst particle of 62 micron size with density of 1747 kg/m^3 impregnated with Ferric Oxide were used for the study. The equations developed for two phase tanks in series model for the downer reactor can be used to evaluate this data.

Equation (4) can be rearranged as

$$\frac{L(1-\varepsilon)}{n\,u_o\left\{(1-X)^{-1/n}-1\right\}} = \frac{1}{k_r\rho_p} + \frac{1}{k_m\,{}^{s_p}/_{v_p}} \qquad (5)$$

Downer reactors consist of three zones along the axial length; in the first zone particles accelerate due to gas drag force and gravity; in the second zone particles accelerate due to gravity while drag force retards the acceleration; in the third zone particles will be moving at terminal velocity as the gravity and drag force balance each other. Particle holdup in the particle accelerating zone can be slightly more than the terminal zone. Particle holdup is estimated for each operating gas velocity and particle flux by the equation proposed by Xiao-Bo Qi et al [5] for the terminal zone

$$\left(1-\varepsilon\right) = 0.125\left(\frac{W}{\rho_p\left(u_o+u_t\right)}\right)\left(\frac{u_o}{\sqrt{gD_p}}\right)^{0.25} Ar^{0.15} \qquad (6)$$

and is assumed to represent the acceleration zone as well.

Zhang and Zhu [6] observed that there can be particle mixing accompanied by gas mixing in downers at low gas velocities. Radial non uniformity in particle holdup and clustering of particles can create mixing of continuous (gas) phase similar to liquid mixing in bubble columns operating in churn turbulent regime. Based on the experimental observations on axial concentration profiles reported by Fan et al (2008), there could be a lot of longitudinal backmixing. In view of this, the parameter (LHS of the equation)

$$\frac{L(1-\varepsilon)}{n\,u_o\left\{(1-X)^{-1/n}-1\right\}}$$

was evaluated for various values of n ranging between 1 to 10000. This parameter is observed to vary as a function of gas velocity and particle flux. This could be due to gas to particle mass transfer coefficient term in the RHS of equation (5).

Luo et al [7] investigated mass transfer between gas to porous particles in downers and observed that mass transfer coefficients are a function of the parameter ($u_o\rho_p/W$). Raghuramulu et al [8] investigated mass transfer between gas to non-porous coarse sand particles in downers and proposed

$$Sh_{pd} = 8\ .10^{-6}\left(\frac{u_o\rho_p}{W}\right)^{1.43} \qquad \text{for} \quad D_t=2.5\text{ cm}$$

(7)

$$Sh_{pd} = 5\ .10^{-5}\left(\frac{u_o\rho_p}{W}\right)^{0.94} \qquad \text{for} \quad D_t=5\text{ cm}$$

For gas to particle mass transfer in risers, assuming particles move as clusters, Subbarao [9] proposed an equation of the form

$$\frac{Sh_{pr}}{Sh_{sp}} = c_1\left(\frac{u_o\rho_p}{W}\right)^m \tag{8}$$

From eqs (5) and (8)

$$\frac{L(1\text{-}\varepsilon)}{n\,u_o\left\{(1\text{-}X)^{-1/n}-1\right\}} = \frac{1}{k_r\rho_p} + \frac{D_p^{\,2}}{6D_m Sh_{sp}c_1}\left(\frac{W}{u_o\rho_p}\right)^m \tag{9}$$

This equation suggests a linear relation between

$$\frac{L(1\text{-}\varepsilon)}{n\,u_o\left\{(1\text{-}X)^{-1/n}-1\right\}} \quad \text{and} \quad \frac{1}{Sh_{sp}}\left(\frac{W}{u_o\rho_p}\right)^m$$

To proceed with the calculations, values for n and m have to be assigned; values 1, 2, 3, 4, 5, 10, 100, 1000 and 10000 are explored for n; m values in the range of 0.9 to 1.4 are explored. The intercept is ($1/k_r\rho_p$) and has to be positive. Fan et al [4] reported k_r value as 0.098 ml/g cat/s and density ρ_p is 1.747 g/cm^3. With these values, the intercept is expected to be 5.84 s. However the intercept is much lower indicating the rate constant, most probably, to be much higher. Also, experimentally observed conversions in downer reactor are much higher than what can be expected in a plug flow reactor with reaction rate constant of 0.098 ml/gcat/s. In view of this, m value for possible intercept values of 0, 0.006, 0.01 and 0.02 as a function of n value are estimated and shown in fig.3. Intercept value of 0.006 corresponds to 0.098 m^3/kg cat/s and is used for further discussion. Axial concentration profiles suggest that n value is closer to 1. For a value of n equal to 1, value of m is 1.22.

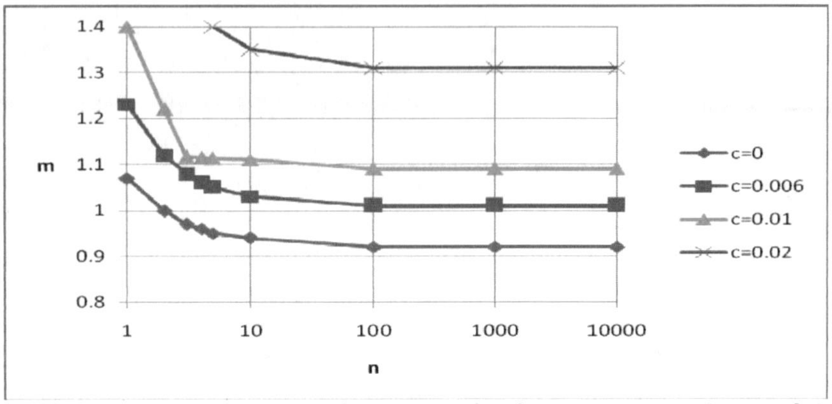

Fig3. Parameter m as a function of n for various values of the intercept.

Data Fan et al [4] for n equal to 1 and m equal to 1.22 are presented in fig 4.
With this mass transfer correlation for this data is obtained as

$$\frac{Sh_{pd}}{Sh_{sp}} = 2.19x10^{-7}\left(\frac{u_o\rho_p}{W}\right)^{1.22} \qquad (10)$$

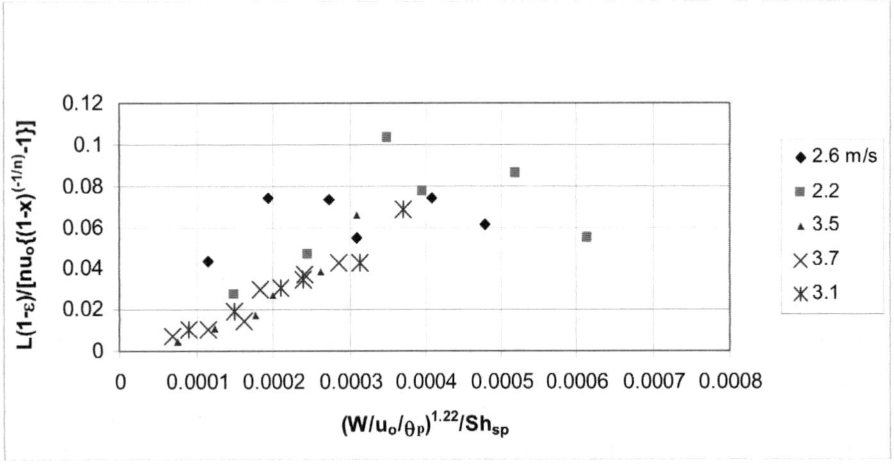

Fig.4 Data Fan et al [4] for n equal to 1 and m equal to 1.22

In fig.5, data of Fan et al [4] for n equal to 10000 and m equal to 1 are presented for comparison with fig.4.
With this mass transfer correlation for this data is obtained as

$$\frac{Sh_{pd}}{Sh_{sp}} = 4.2\ .10^{-7}\left(\frac{u_o\rho_p}{W}\right) \qquad (11)$$

The difference in the estimated mass transfer correlation is not very significant. However,

Axial concentration profiles will be significantly different. As the experimentally observed axial concentration profiles indicate significant axial mixing, one tank appears to be more representative.

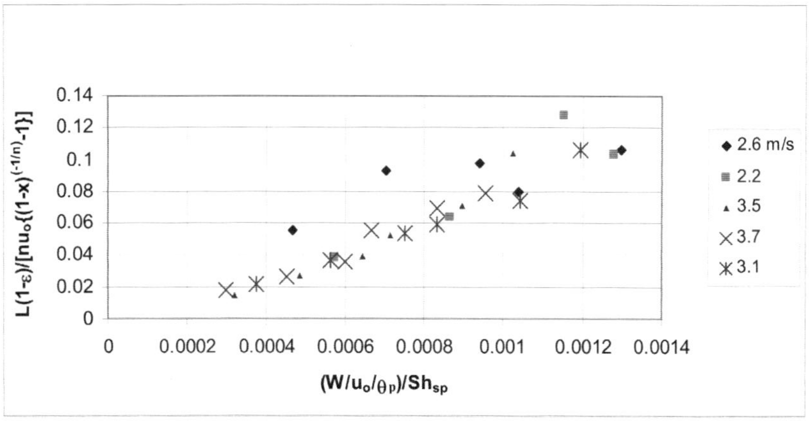

Fig.5 Data of Fan et al [4] for n equal to 10000 and m equal to 1

Concluding Remarks

This value of c_1 (2.19×10^{-7} or 4.2×10^{-7}) is much lower than the value reported by Raghuramulu et al [8] for sand particles. In risers, for fine particles, model of Subbarao [9] brought out the effect of particle size by the following equation

$$\frac{Sh_{pr}}{Sh_{sp}} \propto \left(\frac{D_p}{D_v}\right)^2 \left(\frac{\rho_P u_o}{W}\right)^m \qquad (12)$$

This equation brings out the importance of particle size. Even this equation predicts an order of magnitude higher Sherwood numbers than the ones estimated from the downer reactor data of Fan et al [4]. Thus, it appears that Sherwood numbers in downers are smaller compared to risers. Further experimental study on the effect of mass transfer on downer reactor performance is needed.

Acknowledgements

The author wishes to record his deep sense of appreciation for the financial support by Universiti Teknologi PETRONAS.

Nomenclature

Ar	Archimedes Number
c_1	Constant
c_{i-1}	Gas phase concentration in the (i-1)th compartment
c_i	Gas phase concentration in the ith compartment
C_{in}	Reactor inlet gas phase concentration
C_{out}	Reactor outlet gas phase concentration

C_p	Particle phase concentration in the ith compartment
Dm	Diffusion Coefficient
D_p	Diameter of particles
D_t	Diameter of the Downer column
D_v	Diameter of the void
g	Acceleration due to gravity
i	Compartment number
K_m	Mass transfer coefficient
K_r	Reaction Rate Constant
L	Length of the downer
m	index, constant
n	Number of compartments in series
s_p	External surface area of catalyst particle
Sh_{pd}	Sherwood Number based on particle size in downers
Sh_{pr}	Sherwood Number based on particle size in risers
Sh_{sp}	Sherwood Number based on particle size for single particle
u_o	Gas velocity
u_t	Particle terminal velocity
v_p	Volume of particle
W	Particle mass flux
X	Conversion of Ozone

References

1. Cheng , Changning Wu, Jingxu Zhu, Fei Wei, Yong Jin, Downer reactor: From fundamental study to industrial application, Powder Technology 183 (2008) 364–384.

2. Talman, J.A., R. Geier, L. Reh, Development of a downer reactor for fluid catalytic cracking, Chemical Engineering Science 54 (1999) 2123-2130.

3. Talman.J.A. and L. Reh, An experimental study of fluid catalytic cracking in a downer reactor, Chemical Engineering Journal 84 (2001) 517–523.

4. Chuigang Fan, Yong Zhang, Xiaotao Bi, Wenli Song, Weigang Lin and Ling'ai Luo, Evaluation of downer reactor performance by catalytic ozone decomposition, Chemical Engineering Journal (2008) (in press).

5 Xiao-Bo Qi, Hui Zhang, Jesse Zhu, Solids concentration in the fully developed region of circulating fluidized bed downers, Powder Technology 183 (2008) 417–425.

6 H. Zhang and J.-X. Zhu, Gas-solids flow behaviour with a gas velocity close to zero, China Particuology, 4, (2006), 167-177.

7 Baolin Luo, Dong Yan, Ying L. Ma, Shahzad Barghi, Jesse Zhu, Characteristics of gas–solid mass transfer in a cocurrent downflow circulating fluidized bed reactor, Chemical Engineering Journal 132 (2007) 9–15.

8. Raghuramulu, Shantanu Roy , D.Subbarao, Gas to particle mass transfer in Downers, CFB 9, Germany (2008).

9. Subbarao. D, A Cluster Model for Mass Transfer in Risers, Journal of Engineering Science and Technology, (2008) (in press).

Modeling and Multi-Objective Optimization of Continuous Protein Recovery in Liquid-Solid Circulating Fluidized Bed Ion Exchange System

Jahirul I. Mazumder, Jesse Zhu and Ajay K. Ray
Department of Chemical & Biochemical Engineering, The University of Western Ontario, 1151 Richmond St., London, ON N6A 5B9, Canada

Introduction

Advancement in biotechnology, petrochemical and metallurgical industries leads to the requirement for new types of liquid-solids reactor system which is capable of simultaneous reaction and regeneration in a continuous mode. This is particularly true for cases where the production rate is limited by the activity of the catalysts/capacity of the ion exchange particles. If catalyst/ion exchange particles can be continuously regenerated, higher productivity/throughput is achievable even using catalysts/ion exchanger with relatively low kinetics. Liquid-solid circulating fluidized bed (LSCFB) is an integrated two column system which can accommodate two separate processes (simultaneous reaction and regeneration) in the same system unit with continuous circulation of the solid particles between the two columns. LSCFB reactors provide high liquid-solid contact efficiency, favorable mass and heat transfer and reduced back mixing of phases, and are capable of treating streams containing suspended particulates (Zhu et al. 2000, Lan et al., 2000). Having these advantages, LSCFB emerges as promising liquid-solid contacting equipment in chemical and biochemical processes for both the production and the removal of contaminants from products. Lan et al. (2000; 2002) developed a LSCFB ion-exchange system for the continuous recovery of proteins from biological broths.

Comprehensive modeling and multi-objective optimization of the LSCFB system for continuous protein recovery are carried out in this study. Detailed modeling of the hydrodynamics, mass transfer and kinetics of adsorption and desorption of protein in the LSCFB ion-exchange system is fundamental and crucial for better understanding of the adsorption and desorption behaviors, the design and scale up of the LSCFB system, and optimization of the operating parameters. The only model for continuous protein recovery in LSCFB ion-exchange system was developed by Lan et al. (2000) assuming the process is limited by surface adsorption/desorption; but ion-exchange processes are generally controlled by slow diffusion. Again, detailed hydrodynamics of the LSCFB was not included in their model.

Two separate operations are carried out simultaneously in the two columns of LSCFB system and the performance of these operations are mutually dependent. Therefore, optimum design and operation of the LSCFB system for the better overall performance is very critical. Moreover, the design and operation of LSCFB system for continuous protein recovery are associated with several important objectives such as production rate, recovery, and ion exchange resin requirements, which need to be optimized. Optimization of all individual objectives simultaneously is the best way to approach such multi-objective problems. The results of multi-objective optimization are conceptually different from single objective optimization problems. Solution of multi-objective optimization problems give an entire set of equally good solutions known as Pareto-optimal set. The choice of one solution for better performance requires additional information on the system which is often non-quantifiable. Srinivas and Deb (1995) developed non-dominated sorting genetic algorithm (NSGA) to solve multi-objective optimization problems. Later on Deb et al. (2002) incorporated elitism, a method in preserving good solution, in NSGA, which is referred as elitist non-dominated sorting genetic algorithm (NSGA-II). NSGA-II provides better convergence and better spread of Pareto-optimal solutions. Subsequently, the performance of NSGA-II has been further improved by incorporating the concept of jumping gene (JG) (portion of a chromosome string replaced by new, same-sized, randomly generated binary string) and several JG adaptations are made available (Kasat and Gupta, 2003; Agarwal and Gupta, 2008). In this study, a binary-coded NSGA-II-aJG was used for the multi-objective optimization of LSCFB ion-exchange system for continuous protein recovery.

Modeling of LSCFB Ion Exchange System

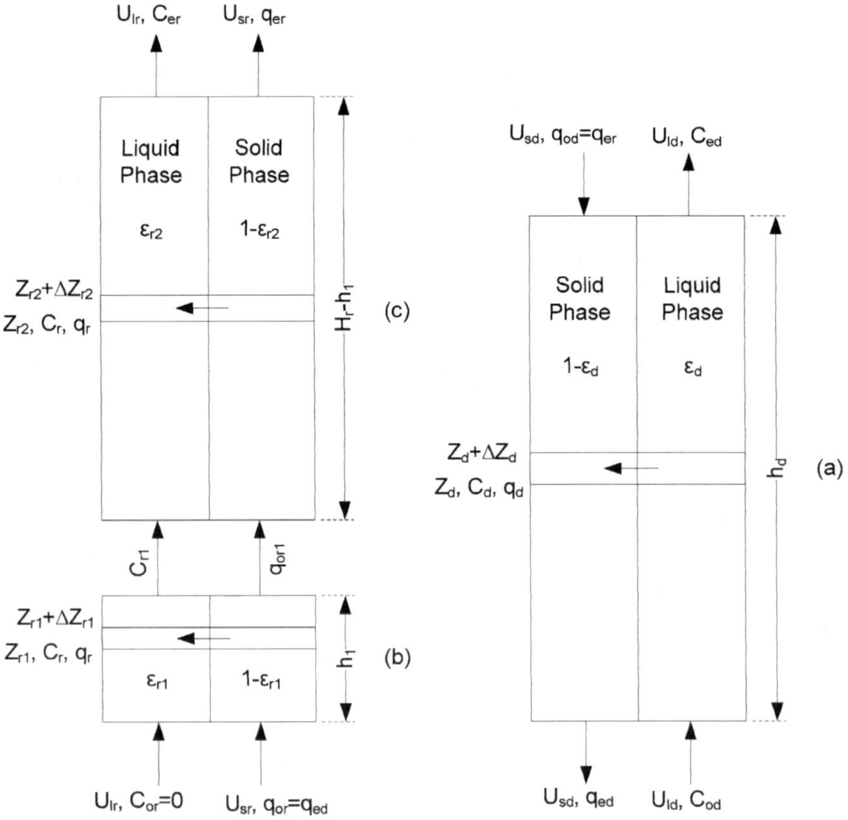

Figure 2: Flow patterns in (a) the downcomer, and (b) distributor zone and (c) upper dilute zone of the riser.

Design equation for the downcomer:

Lan et al. (2002) observed three different operating zones in the downcomer which differ in solids holdup: the dense phase zone, the dilute phase zone and the freeboard zone. The protein concentration is very low in the dilute phase and the freeboard zone. Solids hold up in the dense phase zone is much higher than that of the other zones and contains most of the ion-exchange particles. Hence, the dense phase zone is considered as effective bed for adsorption.

$$U_{ld}\frac{\partial C_d}{\partial Z_d} + \psi k_f a(C_d - C_s)(1-\varepsilon_d) = 0 \tag{1}$$

where, C_d is the protein concentration in the bulk liquid phase of the downcomer; U_{ld} is the superficial liquid velocity in the downcomer; Z_d is the axial distance from the bottom of the downcomer; ε_d is the voidage in the downcomer dense phase; ψ is a constant factor which includes the effects of intra-particle diffusion and liquid phase axial dispersion; a is the specific surface area of the ion-exchange resins. C_s is the equilibrium liquid phase protein concentration at liquid-solid interface predicted using the Langmuir isotherm:

$$C_s = \frac{K_d q_d}{q_m - q_d} \tag{2}$$

where, q_m is the maximum adsorption capacity of the ion-exchange particles, K_d is the dissociation constant, q_d is the solids phase protein concentration in the downcomer.

$$q_d = \frac{U_{ld}}{U_{sd}}(C_d - C_{ed}) + q_{od} \tag{3}$$

k_f is the film mass transfer co-efficient (k_f) in the downcomer dense phase calculated as a function of solids holdup (ε_{sd}) and particles Reynolds number (Re_p) using the correlation reported by Fan et al. (1960):

$$k_f = \frac{D_m}{d_p}(2 + 1.03(\varepsilon_{sd}\,\mathrm{Re}_p)^{1/2}(Sc)^{1/3} \tag{4}$$

where, $$\mathrm{Re}_p = \frac{d_p U_{slip}\rho}{\mu};\; Sc = \frac{\mu}{\rho D_m};\; u_{slip} = \frac{U_{ld}}{\varepsilon_d} + \frac{U_{sd}}{1-\varepsilon_d} \tag{5}$$

The value of D_m for BSA solution was estimated using the following correlation reported by Young et al. (1980):

$$D_m = 8.34\times 10^{-8}\left(\frac{T}{\mu M^{1/3}}\right) \tag{6}$$

□$_d$ is the voidage in the downcomer dense phase and is calculated using the modified Richardson-Zaki equation:

$$U_{ld} + U_{sd}\frac{\varepsilon_d}{1-\varepsilon_d} = U_i \varepsilon_d{}^n \tag{7}$$

The bed expansion index (n) can be determined from the following correlation:

$$n = (4.4 + 18\frac{d_p}{D_c})\mathrm{Re}_t^{-0.01} \tag{8}$$

$$\text{where, } \mathrm{Re}_t = \frac{U_t d_p \rho}{\mu} \tag{9}$$

U_i, the superficial liquid velocity at $\varepsilon = 1$, which can be determined using following empirical equation (Khan and Richardson, 1989):

$$\frac{U_i}{U_t} = 1 - 1.15(\frac{d_p}{D_c})^{0.6} \tag{10}$$

Design equation for the riser:

The riser distributor separates the liquid flow into two portions: primary liquid flow and auxiliary liquid flow. Because of the arrangement of the riser distributor, two distinct zones, namely a distributor zone and a upper dilute zone, are observed along the riser based on their solids holdup.

Distributor zone of the riser, $$U_{sr}\frac{\partial q_r}{\partial Z_{r1}} + k_{r1} q_r \varepsilon_{sr1} = 0 \tag{11}$$

Boundary Conditions:

At $Z_{r1} = 0$ $\quad C_r = 0\ ;\ q_r = q_{or}$ (12a)

At $Z_{r1} = h_1$ $\quad C_r = C_{r1}\ ;\ q_r = q_{or1}$ (12b)

Upper dilute zone of the riser, $$U_{sr}\frac{\partial q_r}{\partial Z_{r2}} + k_{r2} q_r \varepsilon_{sr2} = 0 \tag{13}$$

Boundary Conditions:

At $Z_{r2} = h_1$ $\quad C_r = C_{r1}\ ;\ q_r = q_{or1}$ (14a)

At $Z_{r2} = 3$ $\quad C_r = C_{er}\ ;\ q_r = q_{er} = q_{od}$ (14b)

where, k_{r1} and k_{r2} are desorption rate constants in the distributor zone and the upper dilute zone respectively. ε_{sr1} and ε_{sr2} are the solid hold up in the in the distributor zone and in the upper dilute zone of the riser, and can be calculated using the following equations:

$$U_{lr} - U_{sr}\frac{\varepsilon_{r1}}{1-\varepsilon_{r1}} = U_i \varepsilon_{r1}^{\ n} \tag{15}$$

$$\varepsilon_{sr2} = 2.64 \times 10^{-14} U_{lr}^{-5.343} + 2.57 \times 10^{-5} G_s U_{lr}^{-1.578} \tag{16}$$

n and U_i can be calculated using Eqs. (8)-(10).

Formulation of Multi-objective Optimization

Protein adsorption and desorption are carried out in two separate columns of a LSCFB ion-exchange system with continuous solids circulation between the two columns. As the operation of one column also influence the performance of the other column, optimization of the entire system concurrently is of crucial importance to maximize the overall performance. From the view point of recovery of protein from broths, the most pertinent objectives are to maximize the fraction of protein recovered (*R*) and the production rate (*P*) (throughput or capacity). In terms of the operating costs another important objective is to minimize the amount of solids (resins) required (*S*). *P*, *R* and *S* are defined as follows:

$$P = \text{(flow rate of the extract stream) x (protein concentration in the extract)}$$
$$= U_{lr} A_r C_{er} \quad (17)$$

$$R = \frac{\text{amount of protein in the extract}}{\text{amount of protein in the feed}} = \frac{U_{lr} A_r C_{er}}{U_{ld} A_d C_{od}} \quad (18)$$

$$S = \sum \left(\text{Amount solids in different sections of the LSCFB} \right)$$

$$= (h_d A_d \varepsilon_d + h_1 A_r \varepsilon_{r1} + (H_r - h_1) A_r \varepsilon_{r1} + Vp\varepsilon_p)\rho_p \quad (19)$$

Because of the conflicting influence of decision variables on these performance parameters as seen in the sensitivity analysis, it is not possible to maximize the recovery and the production rate, and/or to minimize the amount of solids required simultaneously. One must perform a systematic multi-objective optimization study to determine the optimal operating conditions of the LSCFB ion-exchanger.

Two and three objectives optimization studies were performed to find out the optimal conditions for the operation of existing setup. Objective functions used in this study are:

$$\text{Maximize } J_1 = R \quad (20)$$
$$\text{Maximize } J_2 = P \quad (21)$$
$$\text{Minimize } J_3 = S \quad (22)$$

For optimizing the operation of the existing LSCFB unit, five adjustable (manipulative) parameters, which significantly affect the performance of the system, were chosen as the decision variables. Following are the decision variables and their bounds used for the operating stage optimization:

$$0.4 \leq U_{ld} \leq 1.2 \text{ mm/s} \quad (23)$$

$$0.6 \leq G_s \leq 1.5 \text{ kg/m}^2\text{s} \quad (24)$$

$$9.0 \leq U_{lr} \leq 21 \text{ mm/s} \quad (25)$$

$$0.5 \leq C_{od} \leq 2.0 \text{ kg/m}^3 \quad (26)$$

$$0.4 \leq h_d \leq 1.0 \text{ m} \quad (27)$$

Upper and lower bound of these operating parameters were chosen based on the stability of the system as observed by Lan et al. (2002) and mathematical feasibility.

Optimization programs used in this work are for the maximization of objective functions and hence minimization of a function (*J*) was converted to maximization function (*I*) by using the transformation of $I = 1/(1+J)$.

Results and Discussions

Maximization of Recovery (*R*) and Production rate (*P*)

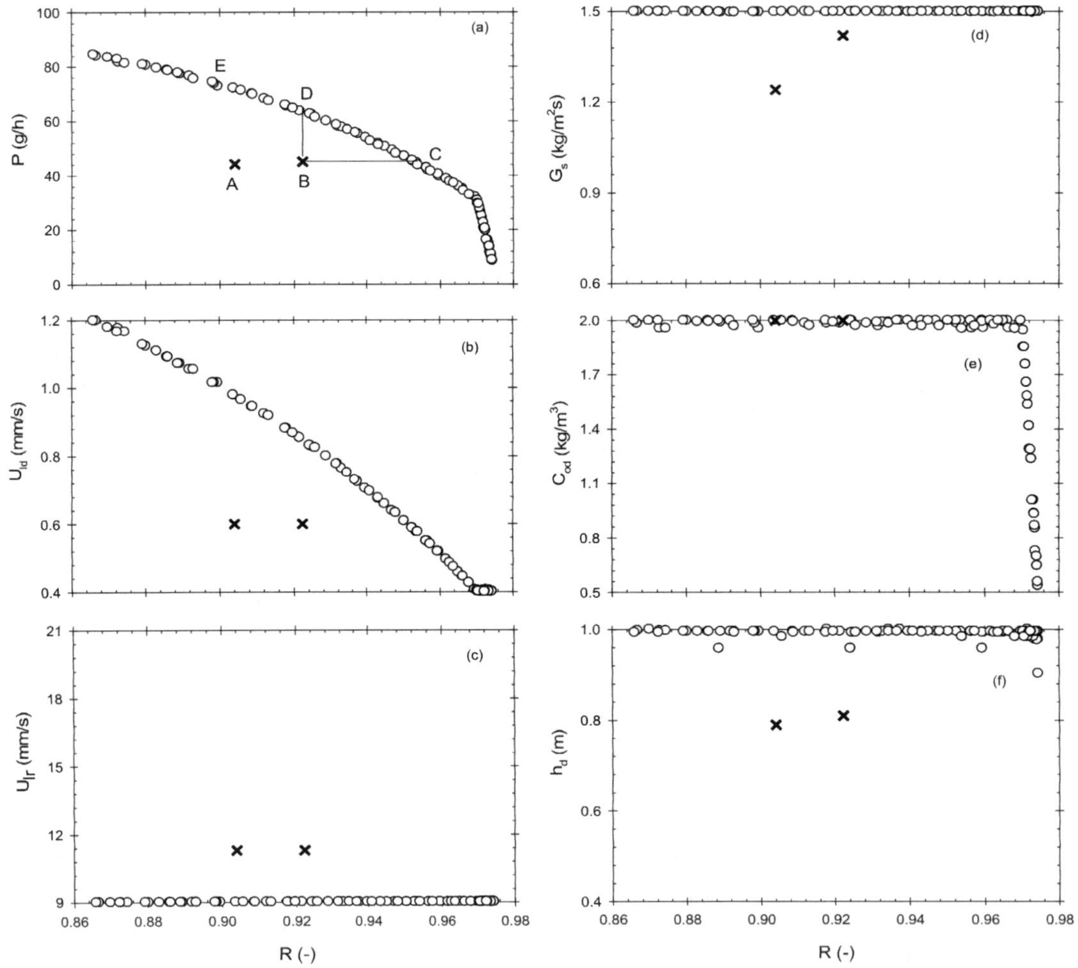

Figure 1: Results for two-objective optimization: maximization of production rate and recovery) (a) Pareto-optimal set; (b-f) values of decision variables corresponding to the points of Pareto-optimal set shown in part (a); x indicates current experimental operating points.

Figure 1a shows the Pareto-optimal set obtained for the simultaneous maximization of the protein recovery (*R*) and the production rate (*P*). It shows a contradictory behavior between the two objectives, i.e., moving from the left to the right (for example, from point E to C) the protein recovery is increased at the cost of reduced production rate. The maximum possible recovery is very close to 1.0 while the maximum possible production rate equal to about 86 g/hr. Each point on the Pareto-optimal front corresponds to a set of decision variables, which are plotted in *Figure 1b-f* against the *R*. The maximum and minimum values of y-axis in the *Figure 1b-f* corresponds to the upper and lower bounds of the respective decision variables. Among the decision variables superficial liquid velocity in the downcomer (U_{ld}) is the most sensitive one with respect to *R* and *P*, and the Pareto was mainly formed because of the conflicting behaviour of the U_{ld} on the two objective functions. *Figure 1b* shows with the decease in U_{ld}, the recovery is increased while production rate is decreased, which was expected from the sensitivity analysis results discussed earlier. With the decrease in U_{ld}, the residence time is increased, but the protein loading rate decreased. U_{ld} reached its lower bound when *R* is equal to 0.97. After that sharp decrease in C_{od} contribute to the increase in *R*. It also resulted in sharp decreased in *P*. The optimum values of rest of the decision variables (G_s, U_{lr}, h_d) obtained are constant as expected. G_s and h_d hit the upper bound as the increase in G_s or h_d, resulted in higher values of both recovery and production. On the other hand U_{lr} hit the lower bound.

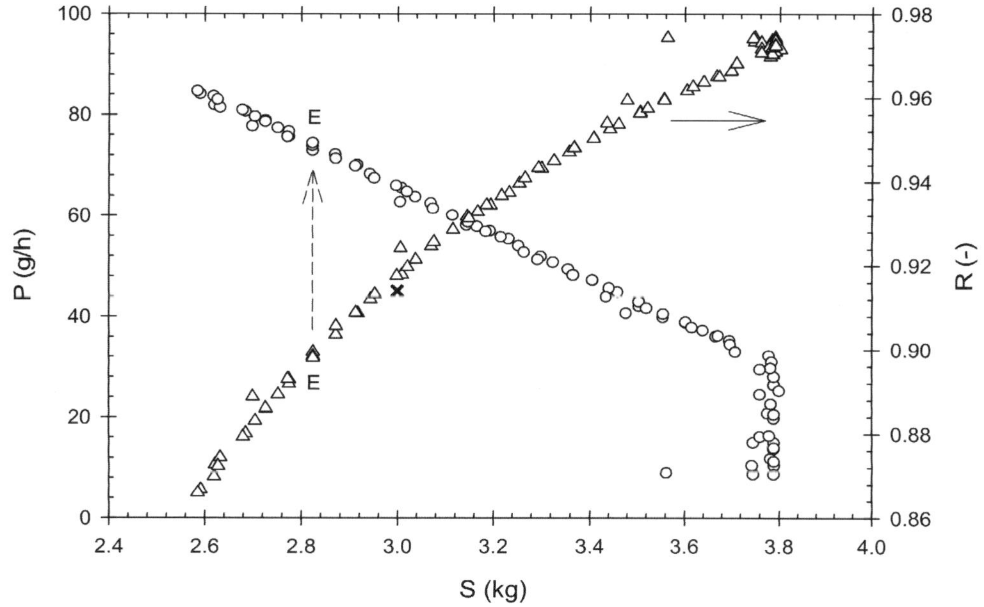

Figure 2: Protein production rate (○) and recovery (Δ) against amount of the solids required (S) for Case A (maximization of production rate and recovery). Production rate versus solids required of the current operation point B (*Figure 1*) is shown by x. E indicates points corresponding to the same point on the Pareto in *Figure 1*.

Figure 1 also shows the current experimental operating points. Point B is better than point A in terms of both the recovery and the production rate. However, all the points on the Pareto set are much better than the current operating points. It shows that the production rate can be increased by ~ 43% for a fixed recovery same (point D over point B) or the recovery can be improved by ~ 4% without decreasing the current production rate (point C over D). Selection of a point from the entire Pareto sets for the efficient operation depends on other factors such as cost of the feed, downstream processing, operating costs, etc., which are site and time specific. Of the two objectives which one is most important depends on these factors. To see the amount of particles required corresponding to the points on the Pareto set, the production rate and the recovery are plotted against the calculated values of the solids required (S) in *Figure 2*. In case of protein recovery from the biological broth 90% recovery is acceptable. *Figure 1* and *Figure 2* show that for *R* equal to 0.9, the value of *P* increased to 74.2 g/h (point E) while the amount of solids required is less than the current experimental operating point. With a small sacrifice in the recovery, one can achieve much higher production rate using less ion-exchange resins. So, for the case of protein recovery from cheese whey, the point E is a reasonable selection. On the other hand, one also needs to remember that moving from right to left in the Pareto front will increase the pumping costs because of the increase in U_{ld}. However, *Figure 1* and *Figure 2* provide a wide range of competing options for the improvement over the current operation.

Maximization of Production Rate (*P*) and Recovery (*R*), and minimization of Solids Required (*S*)

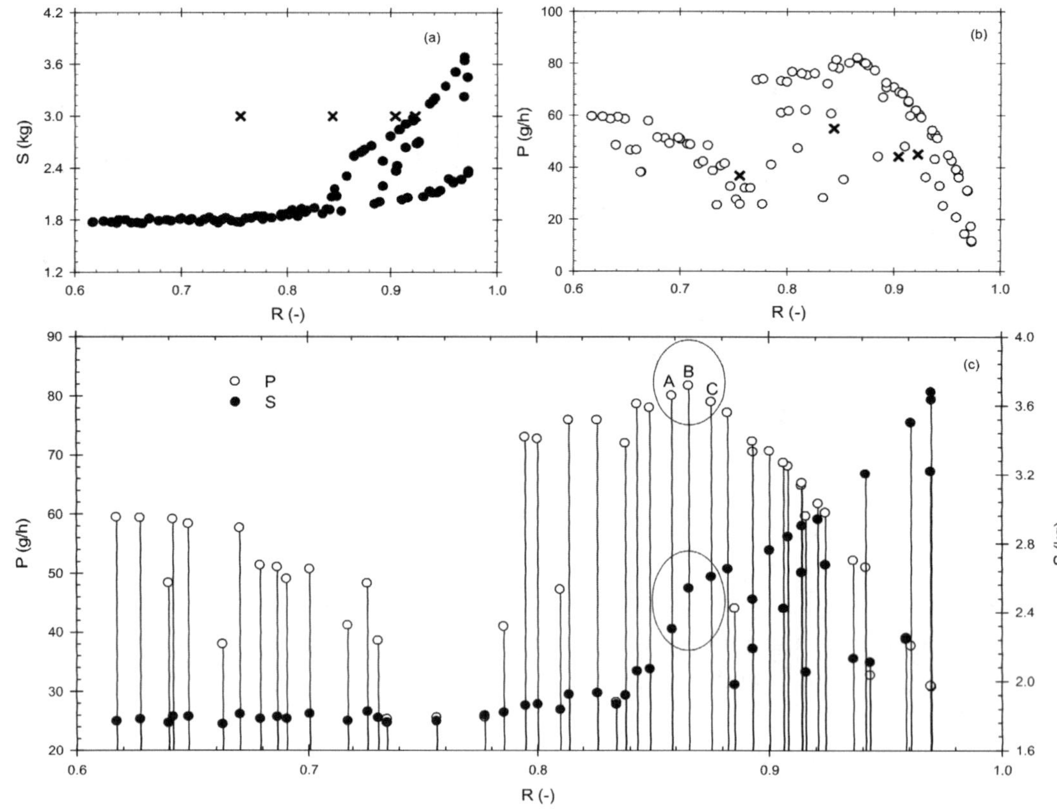

Figure 3: Results for three-objective optimization: (maximization of production rate and recovery, and minimization of solids required); x indicates current experimental operating points.

To obtain a clear understanding of the system performance in terms of production rate (*P*), recovery (*R*) and amount of ion-exchange particles required (*S*), we carried out a three objective optimization: maximization of *P* and *R*, and minimization of *S*, simultaneously. *Figure 3* represents the optimal Pareto set generated in this case. Solids required (*Figure 3a*) and production rate (*Figure 3b*) are plotted against the recovery. The figures show that the Pareto is wide spread and the common trend is with increasing *R*, *S* is increased (not desired), while *P* is gradually increased up to a certain point and then decreased sharply. Although the plots are visually scattered, all points are non-dominating i.e., when we move one point to another in the plot at least one objective function improves and at least another deteriorates. To establish the fidelity of the non-domination, *S* and *P* are plotted against increasing *R* in the same figure (*Figure 3c*). To make it clearly visible, only alternative chromosomes are plotted. It should be noted that in case of three-objective optimization, a Pareto (non-dominating) consists of points where either two objective functions improve with one deteriorating or only one objective function improving with two deteriorating. For example, in *Figure 3c*, from point A to B, both the *P* and *R* increased, but *S* is also increased (two improving, one deteriorating) whereas from point B to C, both *P* and *S* deteriorated, while *R* improved (one improving, two deteriorating). Similarly, if we compare any two chromosomes in *Figure 3c*, one would not find any pair where all three objectives improved or deteriorated. Therefore all points in *Figure 3c* are equally good (non-dominated). Once again, compare to the current experimental points, the optimal Pareto provides lot of options for improvement.

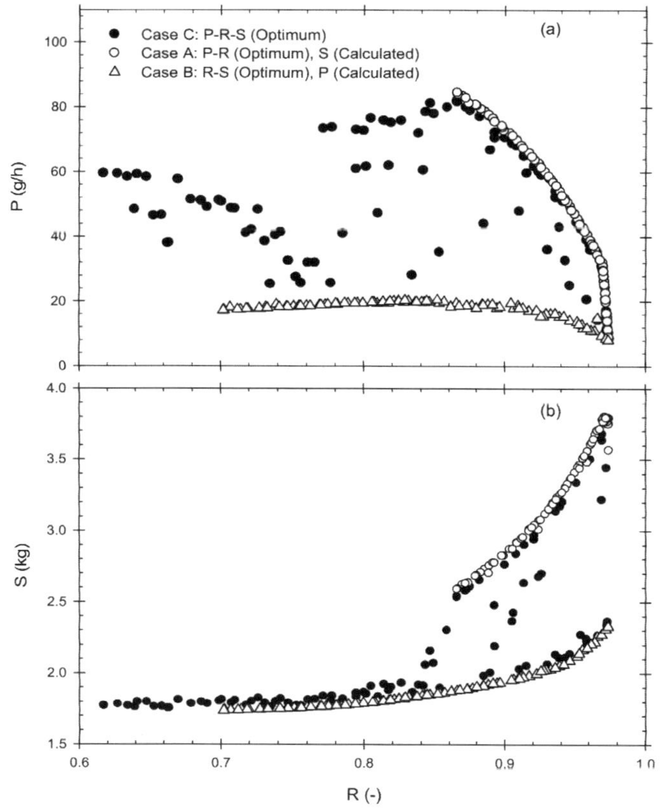

Figure 4: Comparison of optimization results plotted as (a) production rate vs recovery and (b) solids required vs recovery, obtained from three-objective optimization Case C (●) and two-objective optimization Case A (○) and Case B (Δ). (Case A: maximization of *P* and *R*; Case B: maximization of *R* and minimization of *S*; Case C: maximization of *P* and *R*, and minimization of *S*)

Three-objective optimization results are compared with the results obtained from two-objective optimizations. For both the three and two-objective optimizations, production rate (*P*) and solids required (*S*) are plotted against the recovery (*R*) in *Figure 4*. Note that, for the Case A (maximization of *P* and *R*) the values of *S* and for the Case B (maximization of R and minimization of *S*) the values of *P* corresponding to their Pareto front are calculated and shown in the figure. *Figure 4* shows that the *P-R* trade-off is slightly better for the two-objective optimization (Case A) than the three-objective optimization (Case C). However, the calculated values of the *S* corresponding to the Pareto front of the Case A are much higher than that in Case C where S was simultaneously minimized. On the other hand, Case B provides options to recover a particular fraction of proteins using less solids. Whereas in Case C one may achieve much higher production rate than that in Case B. Therefore, three-objective optimization provides better solution to achieve high recovery with improved production rate using less solids. Furthermore, three-objective optimization provides wider range of options to select the best possible solutions.

Conclusions

A comprehensive mathematical model of the Liquid-Solid Circulating Fluidized Bed (LSCFB) ion exchange system for continuous protein recovery was developed considering the hydrodynamics of LSCFB and the ion exchange kinetics, in order to understand clearly the adsorption and desorption behaviour of the system. Subsequently, this model was used for multi-objective optimization of the system at both the operation and the design stages.

Three important performance parameters of the system were used as objective functions for multi-objective optimization study: maximization of the protein production rate, maximization of the fraction of protein recovered, and minimization of the amount of ion exchange resins required. Five operating parameters which have significant influence on the objective functions were considered as decision variables. Two- as well as three-objective function optimization were carried out to find out the optimal values of decision variables. The optimization results for all problems showed significant improvements in the system performance over the current experimental operating points. Optimization results obtained for simultaneous maximization of the production rate and the recovery of the existing LSCFB units showed that the production rate can be increased about 43% over the current experimental results using only 1% excess ion exchange resins for the same recovery level (92.2%) while for the fixed production rate (of 45 g/hr) the recovery increased by 4% over the experimental results. Whereas when all the three objectives were optimized simultaneously, for fixed recovery (of 92.2%), the production rate increased circa 33% over the current experimental results using only 89% of the existing ion exchange resins used. In design stage optimization, design parameters are also considered as decision variable in addition to operating parameters. Hence, the design stage optimization provides far more flexibility compared to the optimization of the performance of the existing unit.

References

Agrawal, A., Gupta, S.K., "Jumping Gene Adaptations of NSGA-II and their Use in the Multi-objective Optimal Design of Shell and Tube Heat Exchangers" *Chem. Eng. Res. & Design* 86, 123-139 (2008)

Chang, Y.K., Chase, H.A., "Development of Operating Conditions for Protein Purification Using Expanded Bed Techniques: The Effect of the Degree of Bed Expansion on Adsorption performance" *Biotechnol. Bioeng.* 49, 512–526 (1996)

Deb, K., Pratap, A., Agarwal, A., Meyarivan, T., "A Fast and Elitist Multiobjective Genetic Algorithm: NSGA-II" *IEEE Transactions on Evolutionary Computation* 6, 182–197 (2002)

Fan, L-T., Yang, Y-C., Wen, C-Y., "Mass Transfer in Semifluidized Beds for Solid-Liquid System" *AICHE J.* 6(3), 482-487 (1960)

Kasat, R.B., Gupta, S.K., "Multi-Objective Optimization of An Industrial Fluidized Bed Catalytic Cracking Unit (FCCU) Using Genetic Algorithm with the Jumping Genes Operator" *Comp. Chem. Eng.* 27, 1785–1800 (2003)

Lan, Q.D., Zhu, J.X., Bassi, A.S., Margaritis, A., Zheng, Y., Rowe, G.E., "Continuous Protein Recovery Using a Liquid-Solid Circulating Fluidized Bed Ion Exchange System: Modeling and Experimental Studies" *Can. J. Chem. Eng.* 78, 858–866 (2000)

Lan, Q.D., Zhu, J.X., Bassi, A.S., Margaritis, A., "Continuous Protein Recovery with a Liquid-Solid Circulating Fluidized Bed Ion Exchanger" *AICHE J.* 48, 252-261 (2002)

Srinivas, N., Deb, K., "Multi-Objective Function Optimization Using Non-Dominated Sorting Genetic Algorithms" *J. Evolutionary Computation* 2, 221- 248 (1995)

Zhu, J-X., Zheng, Y., Karmanev, D., Bassi A.S., "(Gas-) Liquid – Solid Circulating Fluidized Beds and their Potential Applications to Bioreactor Engineering" *Can. J. Chem. Eng.* 78, 82-94 (2000)

Local Holdups and Phase Propagation Velocity Measurement in GLSCFB Riser Using Electrical Resistance Tomography and Optical Fibre Probe

S. A. Razzak, S. Barghi, J-X. Zhu

Department of Chemical and Biochemical Engineering,
The University of Western Ontario,
London, ON, Canada N6A 5B9

Abstract

The application of Gas-Liquid-solid circulating fluidized beds (GLSCBE) has increased recently in chemical, petrochemical and biochemical industries. Electrical Resistance Tomography (ERT) and optical fibre probe were applied to investigate local phase holdups distribution. Since ERT is applicable only to conductive phase(s), e.g. liquid phase in this study, optical fibre probe was employed simultaneously to quantify all three phases. Water was used as continuous and conductive phase, air as the gas phase and glass beads as solid non-conductive phases. The local conductivity measured by a number of electrodes located at the periphery of the plane, was then further converted into a local phase concentration distribution based on Maxwell's relation. Fibre optic probe was also employed to measure gas holdup independently. A new model was developed to exploit the fibre optic data in differentiating gas bubbles from solid particles in the riser. Gas holdup was higher in the central region and decreased radially, while opposite trend was observed with solid holdup due to the drag forces imposed on solid particles by the gas and liquid upward flow in the riser. By applying cross correlation between the data obtained at two different levels in the riser, the propagation velocity of the nonconductive phase was obtained. Propagation velocity was higher in the central region compared to the wall region and increased with increasing liquid superficial velocity

Keywords: Three Phase Circulating Fluidized bed, Electrical Resistance Tomography, fibre optic, phase hold-up, Propagation velocity.

1. Introduction

Gas-liquid-solid circulating fluidized beds (GLSCFB) have been widely used in chemical, petrochemical and biochemical and environmental processes, such as hydrogenation, desulfurization, fermentation, due to its efficient mixing, heat and mass transfer capabilities. Most of the studies on gas-liquid-solid fluidization systems have mainly focused on conventional expanded bed regime in the past decades [1]. Different measurement techniques have been developed to measure phase holdups such as; optical fiber technique, ultrasound technique, electric conductive probe technique, process tomographic technique. Most of the techniques are suitable for two phase systems.

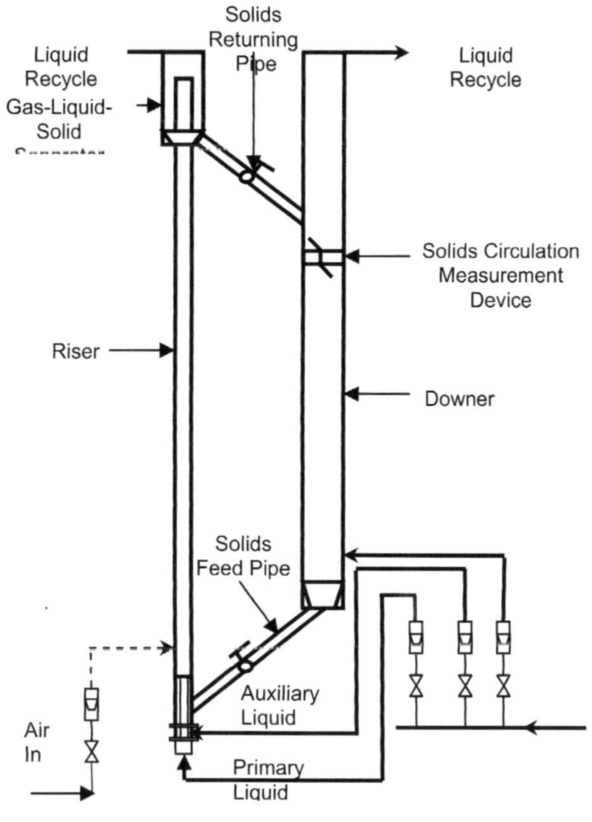

Figure 1: Schematic Diagram of GLSCFB system

However, there are no imaging techniques available for the study of three phase systems in real time. George et al.[2] developed a combined system of Electrical Impedance Tomography (EIT) and gamma-densitometry tomography (GDT) to measure distribution of phases in a vertical three-phase flow system simultaneously. Razzak et al.[3] measured phase holdups and velocities in a GLSCFB system by combining ERT and Pressure Transducers (PT). In this study, Electrical

Resistance Tomography (ERT), a newly developed method for the phase holdups measurement, is presented. However ERT cannot measure phase holdups for all the phases, therefore an optical fiber probe and pressure transducers are used simultaneously to measure phase holdups for all three phases. In the experiments, water was used as the liquid (continuous and conductive) phase, air as the gas phase and glass beads with 500 microns range as the solids phase.

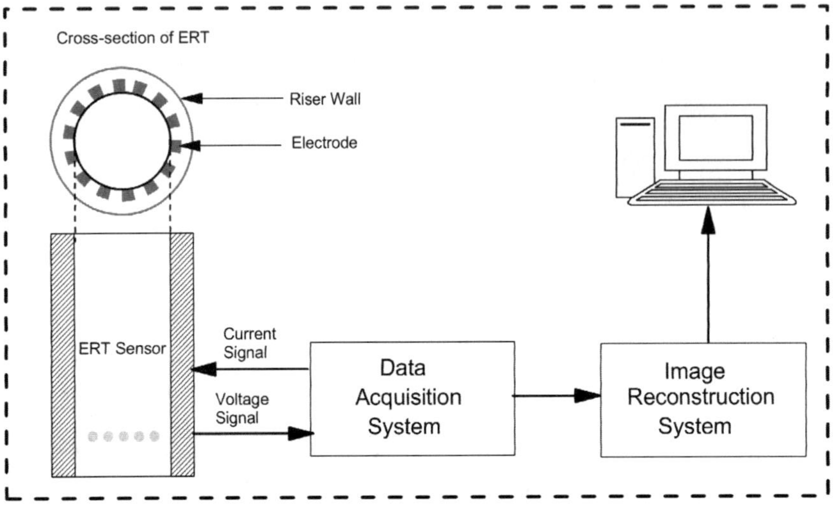

Figure 2: Schematic diagram of the measurement prinicple of ERT.

A schematic diagram of the typical ERT setup is shown in Figure 2. The ERT consists of a sensor section, an electronic circuit and a PC-based data acquisition system. The inner diameter of the sensor section is built equal to the inner diameter of the riser so that the sensors can be lined up with the riser.The optical fiber probes used in the present study were model PV- 5, produced by the Institute of Process Engineering, Chinese Academy of Sciences, are capable of measuring solids concentration in three phase fluidized beds. Details of the optical fiber probes system are shown schematically in Figure 3. All experiments were conducted with two different types of particles which were similar in size but different in density and shape. Particle density of glass beads was 2500 kg/m^3. Schematic diagram of the experimental set-up of GLSCFB is shown in the Figure 1. The GLSCFB consists of two main sections, riser and downer, both made of Plexiglas.

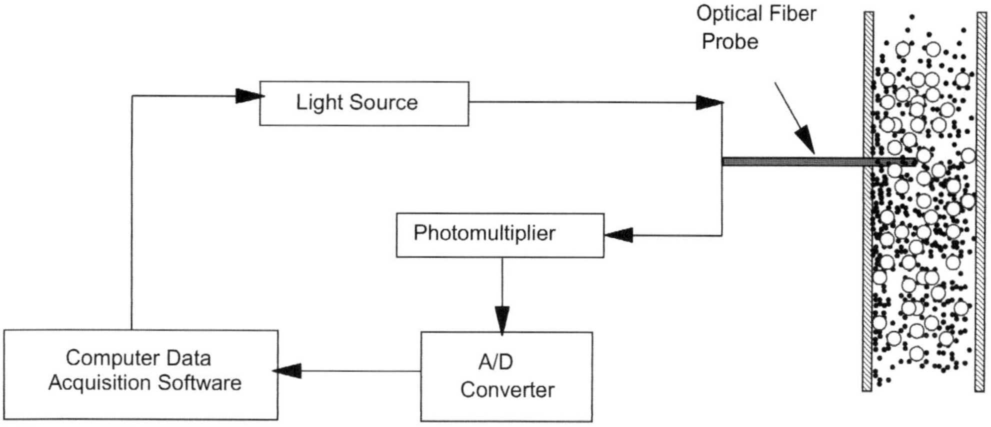

Figure 3: Optical Fibre Probe system for gas holdup measurement in GLSCFB riser.

Combination of these measurement techniques provided valuable information about the hydrodynamics of GLSCFB riser. Average phase holdups, and local distribution of phases were obtained. The phase propagation velocity was estimated by applying cross correlation technique to the two sets of data obtained at upstream and downstream planes in ERT experiments. All measurements in this work were carried out at seven radial positions (distributed radially, centered at r/R = 0, 0.2034, 0.492, 0.6396, 0.7615, 0.8641, 0.9518).

2. Method

A schematic diagram of the typical ERT setup and Optical Probe system have shown in Figure 1b and c. For each driving current, the ERT measures the electrical potential distribution through the electrodes flush mounted on the pipe wall. With input values of the electrical potentials and currents, the local conductivity (or resistivity) of the mixture can be reconstructed through a state-of-the-art optimization algorithm. The conductivity distribution is then further converted into a local phase concentration distribution based on Maxwell's relation.

Figure 4 shows the signals produced in the gas-liquid-solid system for a particular operating case, U_g = 4.88 cm/s. It can be observed that bubble produces distinctively bigger signals than those of solid particles. Signals associated with large bubbles can be easily distinguished and differentiated for the signal produced for liquid-solid system. The amplitudes of the small bubbles signals were quite small and quite close to that of solid particles, which made it difficult to differentiate this bubbles. By combining both ERT and Optical probe techniques gas, liquid and solids holdup distinguished and measured.

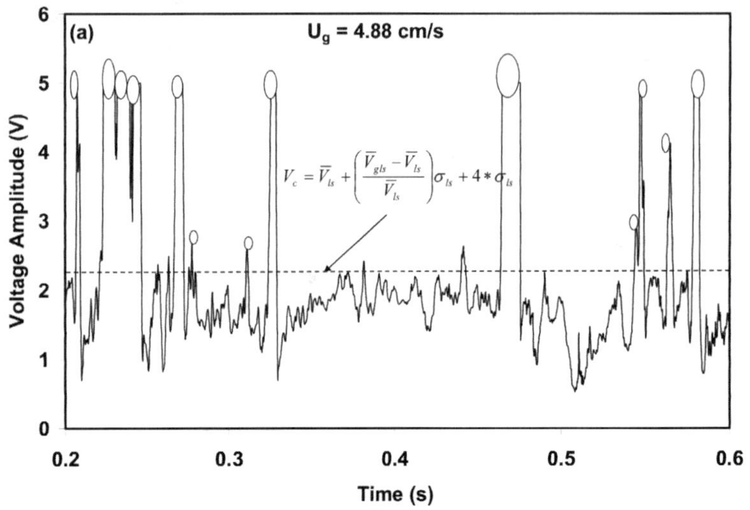

Figure 4: Gas holdup measurement method using optical probe.

The interfacial propagation velocities of the non-conductive phase/phases are obtained by cross-correlation technique. Two imaging planes are placed in short distances in the ERT sensors. The image reconstructed by the ERT was divided into a number of finite elements, each having a value which was indicative of the resistivity/conductivity of the region occupied. Propagation velocity was estimated by applying cross-correlation analysis to phase distributions at the two levels. The total number of finite elements used to measure propagation velocities was 256 per plane. If it takes a time lag, τ, for a void wave to propagate from upstream plane to downstream plane, the wave velocity is

$$C_k = \frac{d}{\tau} \tag{1}$$

where d and τ are the distance between the two sets of electrodes and the time lag of the interfacial wave propagation, respectively. The distance is set by the ERT manufacturer and the time lag is obtained from cross-correlation analysis. Basically, a cross-correlation function can be defined as

$$S_{AB} = \lim_{T\to\infty}\frac{1}{T}\int_{-T/2}^{T/2}\sigma_A(t)\cdot\sigma_B(t+\tau)dt \qquad (2)$$

for a certain period of time, T. The function describes the general dependence between the upstream, $\sigma_A(t)$, and the downstream, $\sigma_B(t)$, conductivities.

3. Results and Discussions

Figures 5 have shown the solids, gas and liquid holdup measured by using newly invented combined methodology. With this method local phase holdup successfully measured in situ for the first time in GLSCFB system. Figure 2 b, and c show the radial distributions of gas and solids phase holdups at different superficial gas velocities where superficial liquid velocity U_l=5.6 cm/s in the GLSCFB riser. All measurements were carried out at an axial location of H=2.02 m above the distributor. Solids holdup initially remains constant at the central location and increases radially toward the wall. Opposite trend was observed for the gas holdup as it sharply decreased in wall region. On the other hand with the increase of gas superficial velocity liquid holdups (not showing in the figure) are decrease. And holdup found lower at the central region and gradually increases radially towards the wall.

Phase propagation velocity under different superficial liquid and solid velocities for glass beads is shown in Figure 6. The nonconductive phase propagation velocity was higher at the central region and decreased radially towards the wall region for all superficial liquid velocities applied. Gas bubbles have a tendency to move upward in the central region. The liquid velocity was higher at the central region and imposed higher drag force on particles in that region. Liquid superficial velocity had also a profound effect on the phase propagation velocity. The rate of change in propagation velocity was much higher at higher liquid superficial velocities.

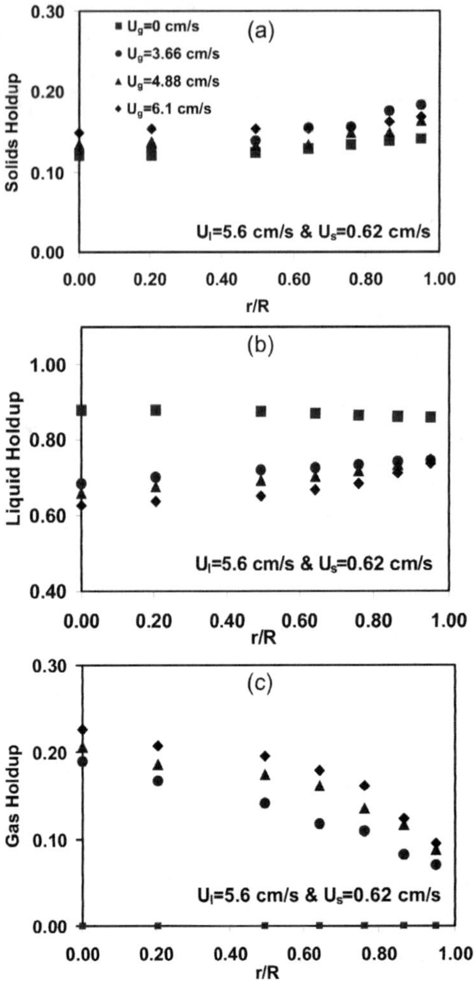

Figure 5: Radial distribution of (a) solids (b) liquid and (c) Gas phase holdups under different gas superficial velocity at Ul = 5.6 cm/s & Us = 0.62 cm/s using glass beads.

4. Conclusions

Radial distribution of phase hold-ups and phase propagation velocities were determined successfully by applying Electrical Resistance Tomography and fibre optic methods. ERT signals are influenced by conductive phase(s) only, therefore the application of fibre optic method was necessary to determine phase hold-ups for all three phases. An empirical method was developed for quantification of the fibre optic data. The fibre optic signals obtained for glass beads and lave rocks were substantially different in amplitude and average values, which required minor modifications in the model. Although the fibre optic was not sensitive to the small gas bubbles the model provided reasonably good results. Different results were obtained

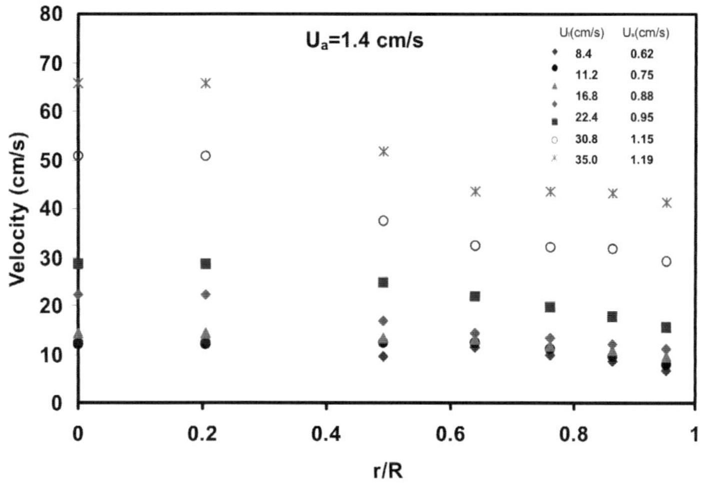

Figure 6: Radial distribution of combined gas-solid phase propagation velocity profile under different superficial liquid and solid velocities for glass beads.

for glass beads and lava rock particles but both particles showed similar trends in radial distribution of phase holdups and propagation velocities. Gas holdup was always higher in the central region and decreased radially towards the wall while solid holdup showed opposite trend. Solid propagation velocity was also higher in the central region compared to the wall region due to higher velocities of the gas and liquid phases which imposed larger drag force on particles in that region.

4. References

[1] Muroyama, K., Fan, L.S., Fundamental of gas-liquid-solid fluidization. *AIChE J.* (1985) 40, 83-92.

[2] George, D.L., shollenberger, K.A., Torczynski, J.R., O' Hern, T.J., Ceccio, S.L., Three-phas material distribution measurement in a vertical flow using gamma-densiometry tomography and electrical-impedance tomography. *Int. J. of Multiphase Flow.* (2001) 27, 1903-1930.

[3] Razzak, S. A., Barghi, S., Zhu, J-X., Eectrical Resistance Tomography for flow characterization of a Gas-Liquid-Solid Three-Phase Circulating Fluidized Bed. *Chem. Eng. Sc.* (2007) 62, 7253-7263.

Use of Co-milling to Improve Physical Stability of Amorphous Salbutamol Sulphate

In recent years, co-milling of active pharmaceutical ingredients (APIs) with excipients is being investigated as a promising technique to improve desired properties such as solubility and dissolution profiles[1], bioavailability[2], aerosolization[3] and physical stability[4]. Traditionally, excipients are simply blended into a drug formulation to improve the manufacturability or product quality, e.g. to improve the ease of tablet compaction, to increase solubility or bioavailability[5]. As milling is an energy intensive process, it tends to induce structural disorders leading to the formation of amorphous regions, which are located particularly at the surfaces of milled particles[6]. This amorphous phase possesses high molecular mobility and tends to revert back to the more stable crystalline state upon storage. Stabilization of this amorphous form is important because this solid-state change can be detrimental to the drug performance as it affects critical particle properties such as morphology, particle size distribution, specific surface area, chemical and physical reactivity[7] and dissolution rates[8].

In this study, salbutamol sulphate (SS) was selected as a model drug as it typically transforms from the crystalline to the amorphous form during milling. As the glass transition temperature (T_g) of SS is close to typical storage temperatures at elevated relative humidity conditions, milled SS particles are physically unstable during normal storage and tend to undergo agglomeration[9]. Pfeiffer and co-workers[9,10] suggested that the increase in particle size of micronized SS was caused by re-crystallization of the amorphous regions. It was also shown that storage of partially amorphous SS at accelerated stress test conditions of 40°C and 75% RH resulted in complete re-crystallization in 5 hrs[9]. In our previous work, a new co-milling method using crystalline excipients was devised to minimize the amorphization of SS below the detection limits of dynamic vapor sorption (DVS) and differential scanning calorimetry (DSC) [11,12]. However, although the amorphous form is unstable, it can sometimes be highly desired because of its higher reactivity and better dissolution properties. As polyvinyl pyrrolidone (PVP) was shown to inhibit the crystallization in amorphous solid dispersions and promote stabilization of the drug in metastable state[13], the present study aims to stabilize the amorphous form of SS by co-milling with PVP.

Different mass ratios (1:1, 1:2, 1:3, 1:5, 2:1, 3:1, 5:1) of SS and PVP were co-milled at 300 rpm in a planetary ball mill for 1 h. The co-milled (CM) mixtures along with their individual components and physically blended (PB) mixtures were subjected to storage conditions of 22°C/75%RH and 22°C/15%RH for 7 days. Powder X-Ray Diffraction (PXRD) and Scanning Electron Microscopy (SEM) analyses prior and after storage were carried out to evaluate changes in crystallinity and morphology. Before storage, PXRD patterns of all CM and PB mixtures showed an amorphous halo, which indicated that SS was present in the amorphous state. After storage at 22°C/75%RH (Figure 1), only CM mixture at SS:PVP (1:5) ratio was able to retain the fully X-ray amorphous character. All other mixtures including PB mixtures were found to be unstable and showed re-crystallization behavior upon storage at 22°C/75%RH, which was similarly observed in milled SS. The results showed that co-milling of SS with PVP at (1:5) ratio was effective in stabilizing the amorphous form of SS. As a control, all CM and PB mixtures remained

X-ray amorphous when stored at 22°C/15%RH as the humidity conditions were not sufficient to induce re-crystallization (Figure 2). The PXRD results were further supported by SEM analysis, as no agglomeration behavior was observed in CM mixtures having mass ratios higher than SS:PVP (1:2). This indicated that CM mixtures at ratios higher than SS:PVP (1:2) were able to minimize the agglomerative tendency of milled SS and improved the physical stability of amorphous SS. The thermal properties of these CM mixtures were also analyzed using Differential Scanning Calorimetry (DSC). Preliminary DSC studies indicated the absence of a recrystallization exotherm in the thermograms of CM mixtures, which was observed in the thermogram of milled SS. All PB mixtures were found to re-crystallize upon heating as indicated by the presence of a re-crystallization exotherm. This suggested that in CM, the amorphous SS has fused with PVP into a solid dispersion.

Figure 1

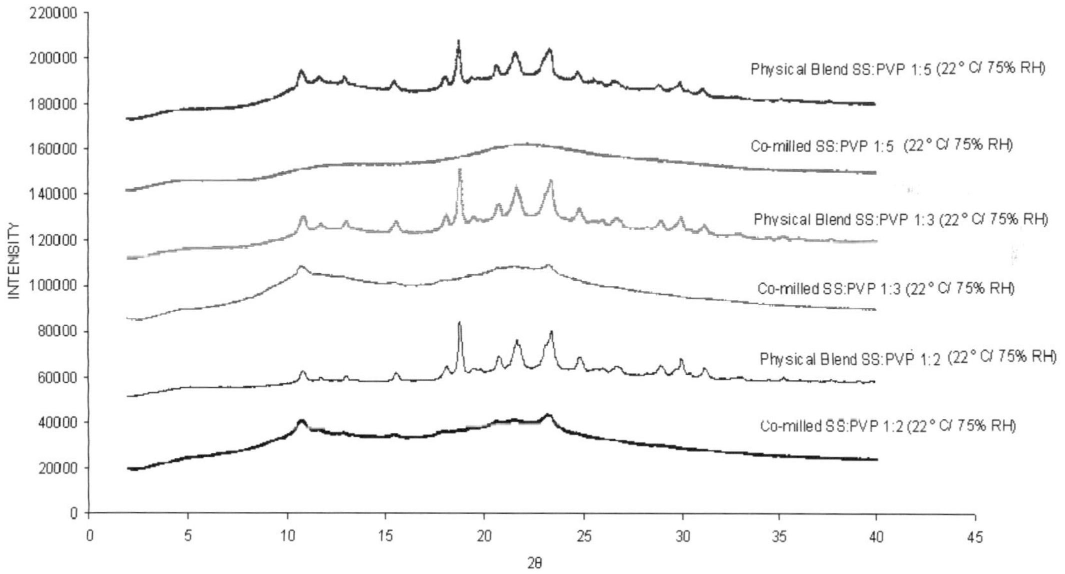

Figure 2

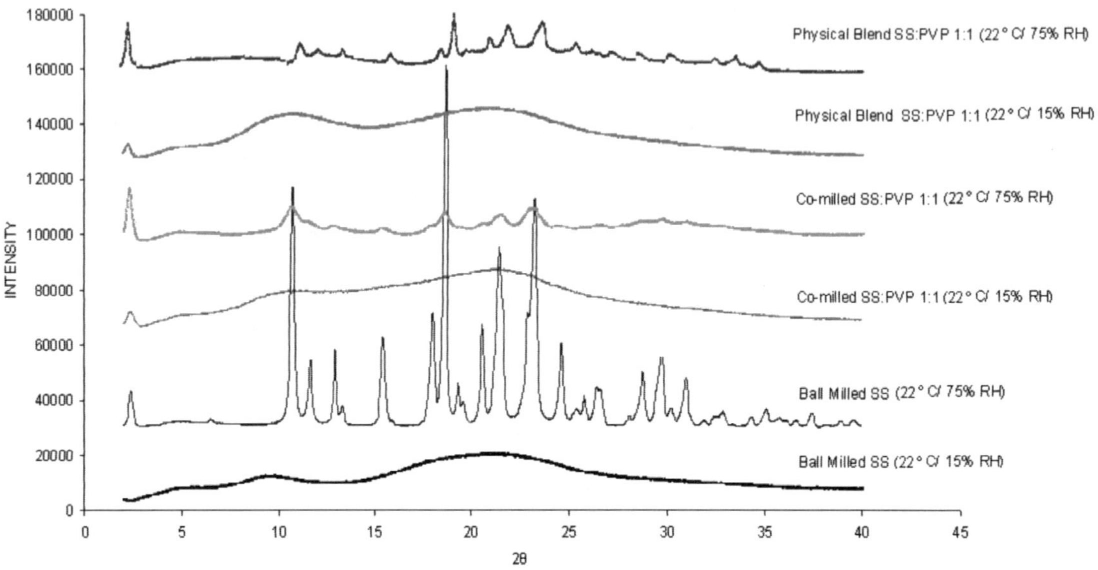

References:

1. Friedrich, H., A. Nada and R. Bodmeier. Solid state and dissolution rate characterization of co-ground mixtures of nifedipine and hydrophilic carriers. Drug Dev. Ind. Pharm. 31: 719-728 (2005).
2. Miro, A., F. Quaglia, U. Sorrentino, M.I.La. Rotunda, R.D.E.D.V. Bianca and R. Sorrentino. Improvement of gliduidone hypoglycemic effect in rats by cyclodextrin formulations. Eur. J. Pharm. Sci. 23: 57-64 (2005).
3. Giry, K., J.M. Pèan, L. Giraud, S. Marsas, H.Rolland and P. Wüthrich. Drug/lactose co-micronization by jet milling to improve aerosolization properties of a powder for inhalation. Int. J. Pharm. 321: 162-166 (2006).
4. Bahl, D and R.H. Bogner. Amorphization of indomethacin by co-grinding with Neusilin US2: amorphization kinetics, physical stability and mechanism. Pharm. Res. 23: 1-9 (2006).
5. Shakhtshneider, T.P., F. Danède, F. Capet, J.F. Willart, M. Descamps, L. Paccou, E.V. Surov, E.V. Boldyreva and V.V Boldyrev. Grinding of drugs with pharmaceutical excipients at cryogenic temperatures Part II. Cryogenic grinding of indomethacin-polyvinylpyrrodione mixtures. J. Therm. Anal. Chem. 89(3): 709-715 (2007).
6. Chikalia, V., R.T. Forbes, R.A. Storey and M. Ticehurst. The effect of crystal morphology and mill type on milling induced crystal disorder. Eur. J. Pharm. Sci. 27: 19-26 (2006).

7. Crowley, K and G, Zografi. Cryogenic grinding of indomethacin polymorphs and solvates: assessment of amorphous phase formation and amorphous phase physical stability. J. Pharm. Sci. 91:492-507 (2002).
8. Guinot, S. and F. Leveiller. The use of MTDSC to assess the amorphous content of a micronized drug substance. Int. J. Pharm. 192: 63-75 (1999).
9. Pfeiffer, K.B., H. Haeusler, P. Grass and P. Langguth. Conditioning following powder micronization: Influence on particle growth of salbutamol sulfate. Drug Dev. Ind. Pharm., 29: 1077-1084 (2003).
10. Pfeiffer, K.B., P. Langguth, P. Grass and H. Haeusler. Influence of mechanical activation on the physical stability of salbutamol sulphate. Eur. J. Pharm. Biopharm. 56: 393-400 (2003).
11. Balani, P.N., W.K. Ng, S.Y. Chan and R.B.H.Tan. Stabilizing crystalline salbutamol sulphate by co-milling. AAPS annual exposition and meeting. San Diego, CA (2007).
12. Balani, P.N., W.K. Ng, S.Y. Chan and R.B.H.Tan. Effects of co-milling on crystallinity of salbutamol sulphate using adipic acid. AIChE annual meeting. Salt lake City, UT (2007).
13. Shakhtshneider, T.P., F. Danède, F. Capet, J.F. Willart, M. Descamps, L. Paccou, E.V. Surov, E.V. Boldyreva and V.V Boldyrev. Grinding of drugs with pharmaceutical excipients at cryogenic temperatures Part II. Cryogenic grinding of piroxicam-polyvinylpyrrodione mixtures. J. Therm. Anal. Chem. 89(3): 709-715 (2007).

402b Modeling of Changes In Particle Size-Shape Distribution Due to Breakage

Priscilla J. Hill, Mississippi State University, Box 9595, Swalm School of Chemical Engineering, Mississippi State, MS 39762 and Devkant S. Gandhi, Dave C. Swalm School of Chemical Engineering, Mississippi State University, Box 9595, Mississippi State, MS 39762.

Fragmentation and attrition of particles are of significant industrial and scientific interest. They affect many chemical engineering processes ranging from crushing and grinding to breakage of particles in agitated slurries. To better design such processes, it is necessary to model changes in particle shape as well as in particle size. In addition, it is essential to determine whether or not the models are able to represent real systems.

This paper presents experimental data for the breakage of crystals in a stirred vessel and compares the data with simulation results. Although a stirred vessel is used for experiments, the models are general and can be applied to comminution. Particles are analyzed before and after agitation in a stirred vessel to determine changes in the size-shape distribution. The approach is to model the breakage distribution function as a bivariate function of the particle volume and the shape factor. A comparison of models with recent experimental results for several aqueous systems is presented. This work provides a basis for incorporating particle shape into the modeling of particulate processes.

INVESITGATION OF A SUB-MICRON MILLING PROCESS OF POORLY WATER SOLUBLE PRODUCTS USING FACTORIAL DESIGN.

S.L.A. Hennart[1,2], M.C. Domingues[3], W.J. Wildeboer[1] and G.M.H. Meesters[1,2]

1. DSM Food Specialties, R&D Recovery and Formulation, Delft, THE NETHERLANDS
2. Delft University of Technology, Nano Structured Material Group, Delft, THE NETHERLANDS.
3. University of Coimbra, Faculty of Sciences and Technology, Coimbra, PORTUGAL.

Abstract – The objective of this work was to investigate the mechanisms of very fine grinding in a wet ball milling set-up as a function of the product properties and process parameters. The ball mill used was a Dynomill and the grinding media consisted of zirconium oxide beads. The grinded powder was a poorly water-soluble product. Laser diffraction techniques were used to analyze the particle size distributions.

During grinding the average particle diameter of a particulate product is reduced to a minimum size. The grinding experiments showed that for a specific product this minimum size is a fixed constant within the range of tested operating conditions.

The grinding parameters (rotation speed of the mill, the grinding media size) are being further investigated to be able to control the grinding process (and optimize the grinding time) and to reach the desired particle attributes (minimizing size).

The grinding time was strongly dependent on the grinding media size and on the rotation speed in a second order of importance. The grinding process becomes faster when the rotation speed increases and the grinding media decreases.

1. INTRODUCTION

Poorly water soluble products are challenging for formulation scientists. The impossibility to dissolve compounds in water based matrixes requires innovative research for the optimization of the application of the product. In the field of pharmaceuticals for example, target release and bio availability are strongly affected by the solubility properties (solubility and solubilisation rate) of the product. One way to increase the efficiency of the product is often to tune the particle size distribution of the powder material. Decreasing the particle size increases the surface area and thus increases the availability of the product.

High shear wet ball milling enables to reach extra fine particle sizes and thus to improve the product efficiency[1]. In order to be capable to reach desired particle size distributions, the grinding process needs to be well understood. Not only the particle size, the time of grinding is an important variable as well.

A grinding process in a ball mill can be influenced by several parameters. Molls, H.H. and Hornle, R. described 44 of them[2]. Even if not all parameters have the same degree of importance, and some of them could be considered as not influent; the complexity of the system is evident. The focus in this research was to study the more important parameters: the rotation speed of the stirrers and the grinding media size. All other possible parameters (flows, charge of the grinding chamber, concentration of product, etc) are kept constant. The present work aims to understand the importance of those main parameters in order to optimize the milling process and reach the product specifications in terms of particle size and time of grinding.

A factorial design allows the simultaneous analysis of the effect of the parameters as independent variables and their degree of interaction. This tool is especially efficient for complex system with multiple variables and responses, such as ball mills. The statistical computation by software enables to get with a restricted number of experiments a quantitative overview of the different process parameters studied. In the present study, the variables are the rotation speed of the stirrer in the mill, the grinding media size. The response variables is the time of grinding.

2. EXPERIMENTAL WORK DICTED BY FACTORIAL DESIGN

2.1. Grinding process

The grinding has been performed on a ball mill (Dynomill, Bachofen AG, Switzerland). Particle size distributions have been measured with a laser diffraction size analyzer (LS230 equipment from Beckman Coulter).

The ball mill was operated in a recirculation mode. Figure 1 shows how the suspension of product is pumped in the grinding chamber which is filled with a grinding media and how the product is drawn out through a filter preventing the grinding media to exit the milling chamber. The system is cooled to keep constant controlled temperature in the entire system.

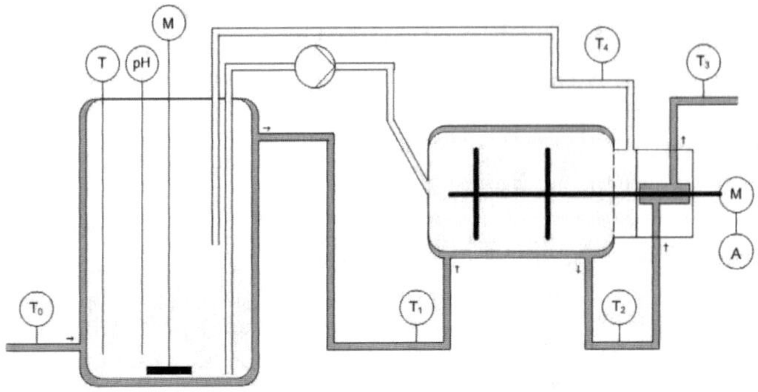

Figure 1 . Grinding set-up and controls, where T is a temperature measurement and pH is a pH control point.

The particle size distributions of the samples taken from the stirred vessel were measured at different times during grinding. Figure 2 is an example of the data obtained from the experimental work.

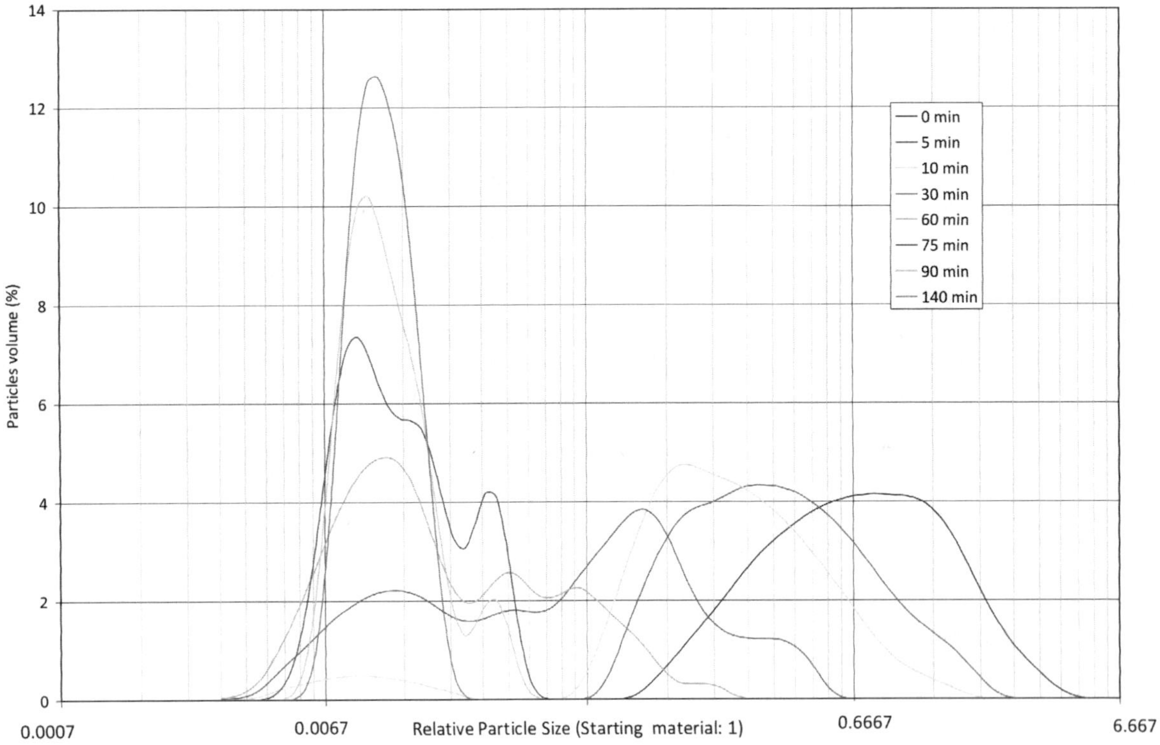

Figure 2 . Evolution of the particle size distribution at different grinding times.

2.2. The factorial design.

In the following study, the number of level for the different parameters is chosen:

The rotation speed is studied at 2 levels: the maximum and the minimum of the operating range: 2000rpm and 6000rpm.

The grinding media size is studied at three levels: 0.3, 0.5 and 0.8mm beads are tested. In the study of the contamination, a 2-level factorial design is drawn using 0.3 and 0.8mm beads.

The factorial design software used, Statraphics Plus 5.1 (StatPoint, Inc), defines a work plan to produce the required output data.
The objectives are to determine the influence of the rotation speed and the grinding media on the required time of grinding to reach a given particle size. Several response variables are therefore created: the time to reach a product particle size of 5µm, 2µm, 1µm, 0.5µm and 0.2µm.
For each response variable, the statistical program runs an analysis of variance giving the effect of each process parameter (grinding media and rotation speed) on the studied response.

3. COMPUTATION OF EXPERIMENTAL RESUTLS

3.1. The effect of individual parameters and of combined parameters on the time of grinding

To analyze the effect of the parameters on the time of grinding, factorial design are carried out for different product particle sizes. The calculations are done according to the method used by Statgraphics 5.1 software called ANOVA (ANalysis Of VAriance). The effect of the parameters is measured as the average response at the high level of the factor minus the average of the response at the low level of the factor. Therefore, the higher the effect, the more influence the parameter has. As shown in Figure 3 the pure effect of the parameters is increasing exponentially when considering smaller particle sizes. Thus the influence of the parameters is very strong on the time of grinding, especially when extra fine particle size as sub-micron particles are desired.

The effect of the parameters rotation speed and the combination rotation speed and grinding media size is negative. Thus, the higher the parameter, the less grinding time is required. Opposite to this the grinding media size has a positive effect on the time: the smaller the grinding media, the less time is required.

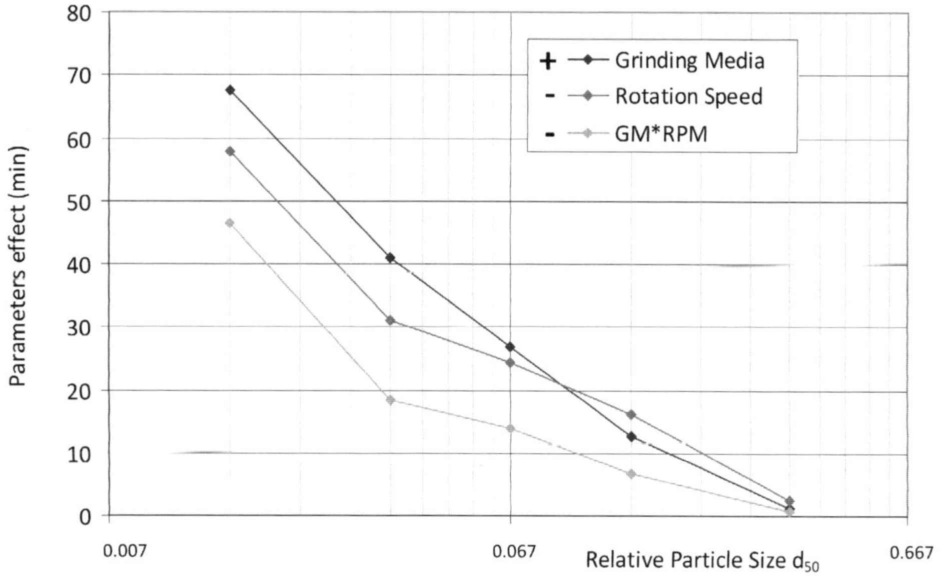

Figure 3 . Parameters effect on the time of grinding to reach different product d_{50} particle size.

In previous studies using population balances to model the ball milling process (to be published[3]), the time of grinding has been defined as function of the process parameters.
The time required to reach 100.P percent of particles below a given size x_i is a function of the initial quantity of particles ($R_i°$) above x_i and the selection function Q_i. (Eq. 1)

$$t(P\% < x_i) = \frac{-1}{Q_i} \cdot \ln\left(\frac{1-P}{R_i^0}\right) \tag{1}$$

In the factorial design, the target particle sizes are below the initial particle size distribution of the product particles, therefore, $R_i°$ equals 1. The target particle sizes are median sizes (d_{50}) and P equals 0.5.
According to the model developed and validated, the expression of the grinding time is thus as in Eq. 2.

$$t(50\% < x_i) = \frac{0.69 \cdot d_{GM}}{v_{tip} \cdot x_i^{d_{GM}^3 \cdot v_{tip}^2}} \quad (2)$$

The trends are equal: The bigger the grinding media, and the slower the rotation speed, the longer the grinding time. The time is a function of the inverse of the particle size, but the effect of the parameters are defined as the difference between the average between the maximum and the minimum value of the parameters and thus this difference will increase with time. To look at the real changes in the effect of the parameters and to be able to compare the parameters to each other, the sum of the effects needs to be normalized.

In order to be able to compare the relative effect of different parameters versus the particle size, the effect have been standardize to 100%. Thus the value of the relative effect represents the contribution of the studied parameter to the grinding time for a desired particle size. Graphs are represented in Figure 4. The rotation speed is the most important parameter for the grinding of coarse particles. To reach sub micron particles, the grinding media is the dominant factor.

In a previous study (Hennart et al.[4]) the grinding mechanism in the same stirred ball mill for the same product was determined. Coarse particles mainly undergo cleavage and the production of extra fine particles is resulting from an abrasion mechanism. The importance of the grinding parameters on the time of grinding is resulting of an influence of the parameters themselves on the grinding mechanism.

Cleavage occurs by the application of low and high forces such as compaction in the centrifuge bed of grinding media. Thus grinding media as individual particles are not relevant, but much more the loading of media in the grinding chamber is key. The loading is kept constant in the series of experiment. In the case of cleavage, the intensity of the applied forces and therefore the speed of centrifugation is much more important than the number of contact and thus the grinding media size.

Abrasion is a result of tangential forces applied on the mother particle. Typically those forces are also present in the centrifuge bed of grinding media. The speed and the energy of the impact is then less important than the number of the potential impact. The number of those impacts depends on the number of grinding media particles that are present in the mill and is at constant load ratio, the number of grinding media particles is increasing when decreasing the media size.

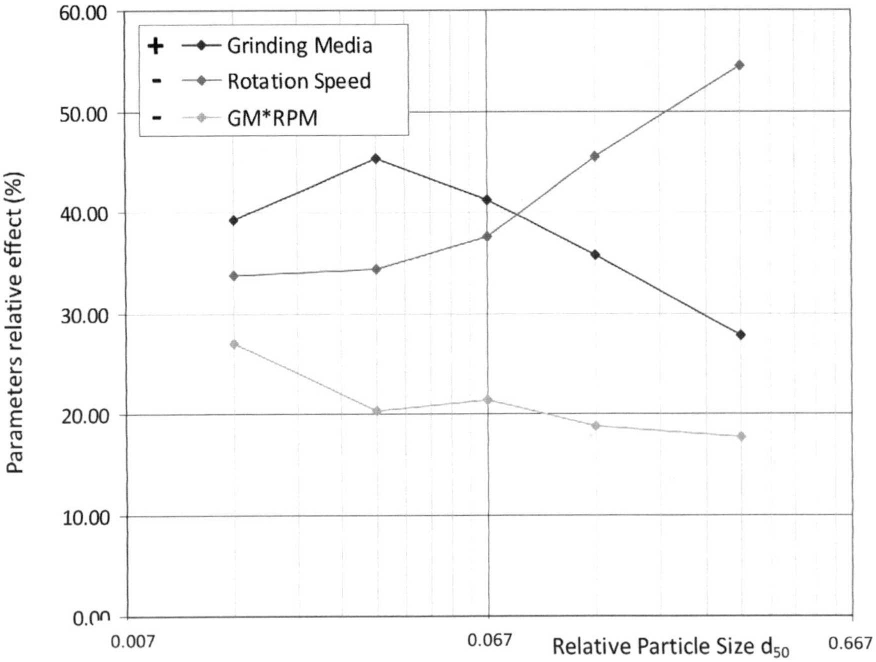

Figure 4 . Relative effect of the different parameters and combination of parameters on the time of grinding to reach different particle sizes.

By taking into consideration the grinding mechanism, it is then clear that the effect of the grinding media size and the rotation speed calculated (Figure 4) is dependent on the size of the particles being broken. The grinding media size is more relevant than the rotation speed in the case of production of fine particles. The number of impacts determined by the number of grinding media particles is dominant. And in case of coarse particles the breakage is initiated by high centrifuge forces of compression in the packed bed of grinding media particles, thus explaining that the rotation speed is a dominant factor.

4. CONCLUSION

The use of a factorial design to determine the effect of the principle process parameters brings the following conclusion:

In the studied wet stirred ball milling process, the grinding media size is the most important factor with regard to the time of grinding. The smaller the grinding media, the quicker the grinding takes place. Higher rotation speed enables as well shorter grinding time.

Optimization of the mill would therefore mean, firstly a reduction of the grinding media size, thus lowering the grinding time. Secondly, faster grinding can be obtained by increasing the rotation speed.

5. NOMENCLATURE

P	: Percentage, dimensionless
X_i	: Product particle size class, µm
t	: Time, min
Q_i	: Selectivity function from the model, dimensionless
R_i^o	: Initial cumulative weight fraction, dimensionless
d_{GM}	: Diameter of the grinding media, mm
v_{tip}	: Tip speed, rpm or m/s
GM	: Grinding media
RPM	: Rotation speed

6. REFERENCES

[1] E. Merisko-Liversidge, G.G. Liversidge, E.R. Cooper; Nanosizing: a formulation approach for poorly-water-soluble compounds; *European Journal of Pharmaceutical Sciences*; **2003**; 18; 113-120.

[2] Molls, H.H. and Hornle, R.; DECHEMA; **1972**; Monography 69 Tl 2; 631-661.

[3] S.L.A. Hennart, V. Drouet, W.J. Wildeboer and G.M.H. Meesters; Modeling of Sub-Micron Milling Processes Limited by re-Agglomeration Phenomena; To be submitted.

[4] S.L.A. Hennart, W.J. Wildeboer and G.M.H. Meesters; Identification of The Grinding Mechanisms and Their Origin in a Stirred Ball Mill Using Population Balances; To be submitted.

402d Milling of Roller Compacted API with Excipients: Model Identification and Verification

Pavan Kumar Akkisetty[1], NandKishor Nere[2], Ryan McCann[2], Kenneth Morris[3], Doraiswami Ramkrishna[2], Gintaras V Reklaitis[2], and Venkat Venkatasubramanian[2]. (1) Chemical Engg, Purdue University, 480 Stadium Mall Drive, West Lafayette, IN 47907-2100, (2) Chemical Engineering, Purdue University, 480 Stadium Mall Drive, West Lafayette, IN 47907-2100, (3) Industrial and Physical Pharmacy, Purdue University, 480 Stadium Mall Drive, West Lafayette, IN 47907-2100

Milling is one of the more common, deceptively simple but inherently complex unit operations in use in the process industry in general and the pharmaceutical industry in particular. The particle size distribution produced in milling is the result of a convoluted interplay between the material properties, the configuration of the milling equipment and mill specific operating parameters, which jointly give rise to specific breakage mechanisms. Consequently, the prediction of breakage behavior becomes difficult. Population balance modeling offers a suitable approach to address some of these problem features. It has been shown that certain forms of linear as well as nonlinear breakage behavior exhibit interesting scaling behavior, which is amenable to model extraction through inverse problem approach proposed by Sathyagal et al (1995). It is this methodology that has been implemented along with the nonlinear milling analysis of Bilgili and Scarlett (2005) and Nere et al. (2008) for model identification and subsequently, its verification.

We present the experimentally characterized dynamic evolution of the particle size distributions for two lab scale quadro comills, namely Quadro U5 and Quadro 197. Three types of roller compacted granules of Acetaminophen, Ibuprofen and Tolmetin with MCC and SiO2 as excipients were used for the milling experiments. Both the mills were operated in batch mode, a requisite to generating the data for inverse problem. The operating parameters such as the impeller speed and the batch sizes were chosen to gather dynamic data. The data thus generated is subjected to the inverse problem approach to extract the models for both the breakage rates and daughter size distributions. These models are validated by comparing the forward simulation of the population balance equation with experimental measurements.

Furthermore, an attempt will be reported to unify the extracted milling models to account for the effects of various operating parameters using an artificial neural network approach. The overall aim of the present work is to develop a general framework integrating population balances with artificial neural networks with the aim of providing a flexible and practical approach for the development of hybrid models for milling model.

Reference

1. Sathyagal, A. N., Ramkrishna, D., and Narsimhan, G. Solutions of Inverse Problems in Population Balances II. Particle Break-Up. Comput. Chem. Eng. 19, (1995), 437.

2. Bilgili, E. and Scarlett, B. Population balance modeling of non-linear effects in milling processes, Powder Technology 153, (2005), 59– 71.

3. Nere, N. K. McCann, R., Morris, K. and D. Ramkrishna. On the Modeling of Milling in Pharmaceutical Industry. AIChE Annual meeting, Philadelphia, PA , November 2008.

402e Solving Inverse Problems In Fragmentation / Breakage

R. Bertrum Diemer Jr.[1], Jon H. Olson[2], Yulong Ding[3], Mojtaba Ghadiri[3], and Dolapo Olusanmi[3]. (1) DuPont Engineering Research & Technology, DuPont Company, 1007 Market Street, DuPont Building 5072, Wilmington, DE 19898, (2) Chemical Engineering, University of Delaware, Colburn Laboratory, Newark, DE 19716, (3) Institute of Particle Science & Engineering, University of Leeds, Clarendon Road, Leeds, LS2 9JT, United Kingdom

This paper will discuss progress in solving the inverse problem for systems amenable to modeling via power-law breakage with self-similar daughters. Application where possible to milling of organic materials such as aspirin will be discussed.

DELAMINATION OF GRAPHITE SHEETS IN STIRRED MEDIA MILLS

C. Knieke, W. Peukert

Institute of Particle Technology, University of Erlangen-Nuremberg, Germany

Abstract

The structural changes of graphite particles during wet grinding in a stirred media mill have been investigated by scanning electron microscopy, atomic force microscopy and X-ray diffraction analysis. A delamination of the graphite sheets occurs in the shear field of the mill. With an appropriate stabilization method thin graphite flakes with a thickness smaller than 10 nm could be produced.

Introduction

Graphite is a well-known allotropic form of carbon. It consists of stacks of parallel two-dimensional graphene sheets. Each carbon atom is covalently bonded to three neighbors in plane with atoms placed in the corners of hexagon. The graphite structure is characterized by its strongly anisotropic behavior due to the strong covalent forces inside the graphene layers and the weak van der Waals forces perpendicular to the sheets. As a consequence the mechanical, thermal and electrical properties of graphite are highly anisotropic. Graphite and especially graphene are known as promising materials for applications in the field of electronics.

Results and Discussion

In this paper we will present results of different grinding experiments in the stirred media mill. For this purpose the initial graphite particles are dispersed in water and afterwards stressed under different milling conditions (rotational speed of the stirrer, milling bead size). Beside the breakage of the particles between the milling beads, a simultaneous stabilization of the new created surfaces against agglomeration is necessary to advance the grinding progress. Especially in the submicron size range large interparticle interactions and high collision rates of the particles can lead to agglomeration or in the case of a multi-layered material to rearrangement of the single sheets [1]. Therefore different stabilization mechanisms have been investigated. Steric stabilization with polymers as well as electrostatic stabilization with ionic surfactants was used in the milling experiments. By means of X-ray diffraction analysis, SEM and AFM measurements we found that a delamination of the graphite layers occurs in the shear field of the mill [2]. Figure 1 shows a graphite particle before stressing in the mill.

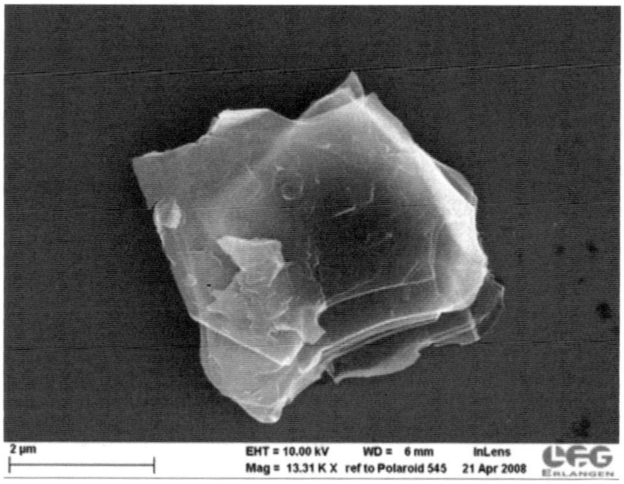

Figure 1: SEM picture of the graphite before milling (feed material)

During the milling process we observe nearly no breakage along the graphene planes, but a good grinding progress perpendicular to the sheets. An optimization of the stabilization mechanism and the milling process, i.e. the use of smaller milling beads, enables the production of extremely thin graphite flakes with a thickness smaller than 10 nm but a mean diameter of about 1 µm. In figure 2 and 3 graphite flakes after different milling times and stabilization methods are shown. One can observe in figure 2 that the graphite sheets are extremely thin and seem to be very flexible. The thickness of the graphite flakes was obtained by AFM measurements. A typical height profile of a very thin graphite sheet is shown in figure 4.

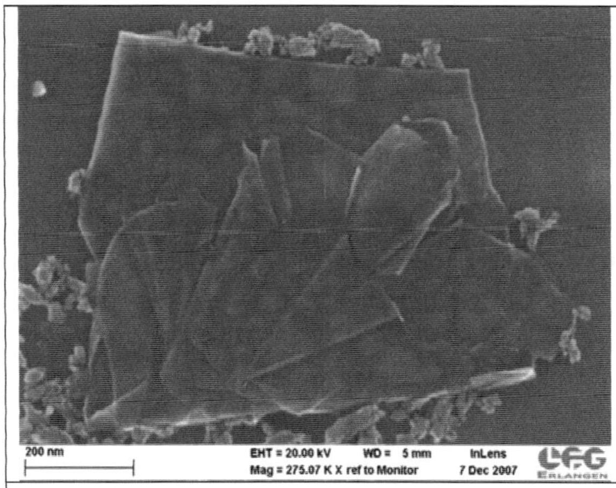

Figure 2: Flexible graphite sheets after 15 h of milling, stabilized with a polymer.

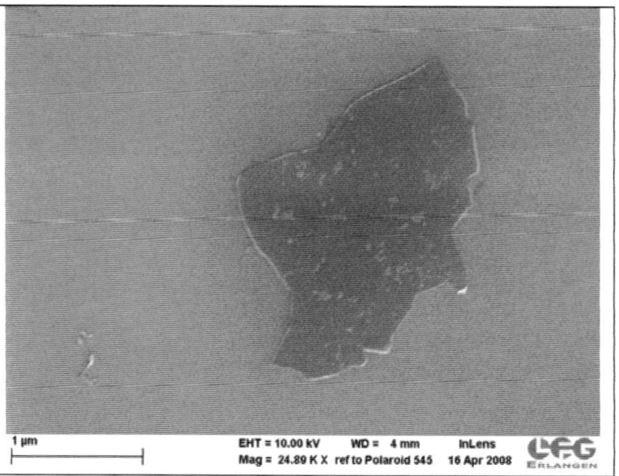

Figure 3: Graphite sheet after 5 h of milling, electrostatically stabilized with ionic surfactants.

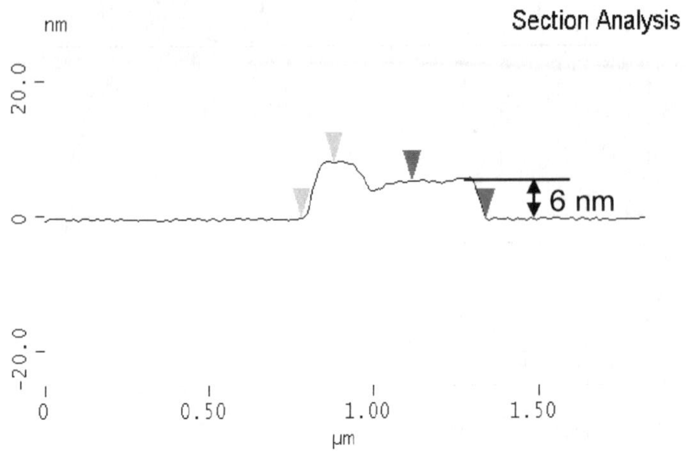

Figure 4: Height-profile of a graphite flake (milled for 5 h, electrostatically stabilized) obtained by AFM measurements.

Under the mild milling conditions with small stress intensities (small milling bead size, low rotational speed of the stirrer) the crystalline microstructure in a layer will barely be affected. Therefore the specific properties, for example the electrical conductivity along a graphene layer, are still present after stressing. In addition we found out that those thin layers seem to be very flexible. This opens totally new applications in the field of printable electronics. In general we showed for graphite particles as an example of a multi-layered material that it is possible to reach a selective and orientation depending comminution which can be controlled by the stabilization mechanism as well as the milling process parameters.

Conclusion

An exfoliation of graphite sheets could be achieved by stressing graphite particles dispersed in water in a stirred media mill. During the grinding process the created graphite-flakes could be stabilized electrostatically as well as sterically to avoid a rearrangement of the exfoliated sheets. Therewith flake like particles with a thickness in the lower nanometer range were produced in the milling process. By changing the stabilization method and the process parameters the morphology of the graphite sheets can be controlled.

References

[1] Stenger, F., Mende, S., Schwedes, J., Peukert, W.; *Nanomilling in stirred media mills*, Chemical Engineering Science (2005), 60(16), 4557-4565

[2] Milev, A., Wilson, M., Kannangara, G. S. K., Tran, N.; *X - ray diffraction line profile analysis of nanocrystalline graphite,* Materials Chemistry and Physics (2008), 111(2-3), 346-350.

402g Theoretical Study of Granule Friability during Pneumatic Conveying

Peter Spicka, Fluent Inc., 10 Cavendish Court, Lebanon, NH 03766 and Pavol Rajniak, Merck & Co, Inc., WP78-110, PO Box 4, West Point, PA 19486.

This work presents an improvement of methodology for predicting the friability of granules as an extension of our recent work [1]. The methodology of collecting experimental data of granule attrition using a laser diffraction particle size analyzer and the model combining CFD Eulerian gas-solid model with QMOM methodology for solution of particle breakage was developed and described in [1]. The combined model allowed us to relate the breakage kernel parameters to particular flow properties. It was found that for a given experimental set-up with diluted gas-solid flow, the breakage rate of granules is proportional to the characteristic particle size and to the square of the impact velocity between a granule and the equipment wall, and can be expressed using the form proposed by Moreno-Atanasio and Ghadiri [2]. Empirical parameters were evaluated by fitting the model to experimental data. The impact velocity was computed in [1, 2] using the CFD code as the solid-phase velocity magnitude in the cells next to the wall. Different components (normal vs. tangential) of the impact velocity vector are employed as in this work and results are compared with previous calculations. Theoretical simulations are also performed for (hypothetical) different curvature of the bend of the particle size analyzer. Simulation results indicate higher sensitivity of the breakage rate (1) to the bend curvature when employing the normal velocity or tangential velocity than for the velocity magnitude. As a conclusion we hypothesize that the velocity components are more relevant for quantification of the attrition and breakage during the dilute phase conveying of granules.

[1] P. Rajniak, K. Dhanasekharan, C. Sinka, N. MacPhail, R. Chern, S. Fitzpatrick, Modeling and measurement of granule attrition during pneumatic conveying in a laboratory scale system, In press, Powder Technology (2008).

[2] R. Moreno-Atanasio, M. Ghadiri, Mechanistic analysis and computer simulation of impact breakage of agglomerates: Effect of surface energy, Chem. Eng. Sci. 61 (2006) 2476 – 2481.

PARTICLE BREAKAGE IN THE NANOMETER RANGE

C. Knieke, W. Peukert

Institute of Particle Technology, University of Erlangen-Nuremberg, Germany

Abstract

Comminution in stirred media mills enables the production of ultrafine particles down to the nanometer range. The grinding results are mainly determined by two contrary mechanisms: particle breakage and agglomeration of the so-created fragments. In this paper the breakage mechanism in the nanometer range and the existence of the overall limit of grinding will be discussed. By means of X-ray diffraction analysis microstructural changes during the grinding process are analyzed. One can observe that lattice imperfections are generated which enhance the total amount of stored energy in the particles. This energy is necessary to reach a breakage of the particles in the nanometer range. At a critical crystallite size defects cannot be stored in the lattice anymore and the limit of grinding is reached.

Introduction

Nanoparticles are increasingly used in many areas of chemical and pharmaceutical as well as in ceramic and microelectronic industry. Besides the direct synthesis of these materials by chemical methods, wet grinding in stirred media mills is a suitable technique for the production of nanoparticles in the liquid phase with high solid concentrations [1], [2]. The manufacturing of fine particles is influenced by particle breakage and interparticle interactions. The interactions become relevant especially for particles smaller than 1 μm because of an increasing collision rate of the particles due to their Brownian motion. In most grinding experiments a limit in particle size is reached where further energy input does apparently not lead to smaller particle sizes. Stenger et al [1] demonstrated that this observed limit is dependent on the stability of the suspension. By means of electrostatic, steric or electrosteric stabilization and shear stress this so called apparent grinding limit can be shifted to smaller particle sizes. Whereas the apparent grinding limit can be controlled by the interparticle interactions, the true grinding progress only refers to the actual breakage of particles. The existence of a true grinding limit as the minimal achievable particle size in grinding experiments, where no further particle breakage occurs even after excessive energy input, is an essential question in nanogrinding. By means of X-ray diffraction analysis the evolution of internal particle microstructural features such as the crystallite size (coherent domain size) and the lattice strain can be examined to understand the breakage mechanism in the nanometer range.

In this paper we present results of long term grinding experiments of different materials in a stirred media mill and discuss the influencing parameters on breakage kinetics and the grinding limit. With the knowledge about the breakage behavior in the nanometer range comminution processes in the field of wet ultrafine grinding can be controlled and optimized in order to avoid excessive energy input which in turn increases the efficiency of milling.

Experimental

Materials

Milling material

The materials used in milling experiments along with their initial particle sizes are summarized in Table 1.

Table 1: Particle size parameters of the inserted materials (measured with laser diffraction)

Material	Trade name	Manufacturer/ Supplier	$x_{10,3}$	$x_{50,3}$	$x_{90,3}$
SiO_2		Carl Roth GmbH & Co. KG	3.4 µm	14.3 µm	39.1 µm
Al_2O_3	CT 1200 SG	Almatis	1.2 µm	2.4 µm	4.4 µm
$CaCO_3$		Carl Roth GmbH & Co. KG	0.8 µm	7.0 µm	13.4 µm
SnO_2		Merck GmbH	0.2 µm	1.6 µm	5.2 µm
ZrO_2		Sigma Aldrich	1.0 µm	1.8 µm	3.3 µm

Solvents

Ethanol (96 % denatured with 1 % MEK, VWR, Germany) and distilled water were used as dispersants.

Grinding media

Wear resistant, commercially available yttrium stabilized zirconia (ZrO_2) milling beads in the size range of 0.4 – 0.63 mm (TOSOH, Japan) were used in the experiments. According to the manufacturer the grinding media has a density of 6065 kg/m³ and a chemical composition of 95 % ZrO_2 and 5.0 % Y_2O_3.

Experimental Set-up

All milling experiments in aqueous suspensions were carried out in a six disc stirred media mill with a content of 0.95 liter. The suspension flows axially through the mill and a sieving cartridge with gaps of 100 µm prevents milling beads from leaving the grinding chamber. The grinding chamber is lined for wear protection reasons with ceramic walls (SiSiC) and the stirrer is equipped with discs of polyurethane (PU). The experimental set-up provides a circuit mode comminution of the product. The suspension is pumped with a hose pump from the grinding chamber into a stirred vessel, where samples can be taken and stabilizers added. From the vessel the suspension flows back into the mill. Further details about the set-up and the construction of the mill are given elsewhere [1].

The commercial laboratory mill LabStar LS 1 (Netzsch, Germany) was used for the milling experiments in ethanol. The mill has an explosion-proof design and is equipped with a centrifugal separating system for the milling beads. This allows the use of small grinding beads down to 0.1 mm. The grinding chamber with a volume of 0.68 liter as well as the stirrer is lined with ZrO_2. The experimental set-up for this mill corresponds to the apparatus described above. Both grinding chambers are equipped with a double wall for cooling, which is connected to an internal cooling water system.

Sample Characterization

Specific Surface Area (BET). BET (Brunauer-Emmett-Teller) measurements to determine the specific surface area of the particles were carried out on a Quantachrome NOVA 2000 gas sorption analyzer. The samples were degassed for 12 h at 150 °C under vacuum to remove adsorbed solvent molecules. The specific surface area was determined using a 7-point method with nitrogen as adsorption gas.

X-ray Diffraction (XRD). The XRD data was collected using a Siemens D 5000 powder diffractometer with graphite secondary monochromator using Cu K_α radiation ($\lambda = 0.15406$ nm). The XRD patterns of the samples were recorded with a step width of 0.02° and a counting time of 3.5 sec/step. A divergence silt and an anti-scatter slit of 0.5° as well as a detector slit of 0.2 mm were used. The samples were measured in a 2Θ range from $20° \leq 2\Theta \leq 80°$.

Results and Discussion

Long term grinding experiments of different materials have been carried out in a 20 wt% suspension in a stirred media mill. Samples were taken during the milling process and the dried powder was measured by the nitrogen adsorption method (BET) and X-ray diffraction. Micropore analysis of the feed materials and the final powders at low relative pressures were accomplished within the BET-measurements to exclude the existence of micropores in the materials which can distort the measured particle size.

Under the assumption that particle breakage occurs at the grain boundaries, i.e. the weakest points in a polycrystalline material, the development of the crystallite size can be used to explain the breakage process. Fig. 1 illustrates the X-ray diffraction patterns of zirconia particles of the initial state and after 50 h of stressing in a stirred media mill. One can clearly observe the loss in crystallinity with milling time. The peaks of the milled zirconia are significantly broadened which may be caused by either decreasing crystallite size and/ or enhanced strain in the lattice of the domains.

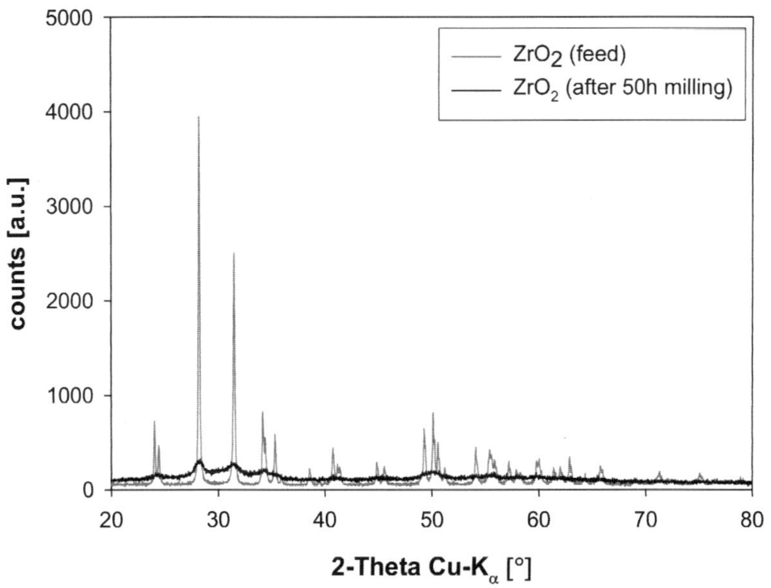

Fig. 1: Diffraction patterns of zirconia before and after 50 h milling

The crystallite sizes and the microstrains of the milled powders were determined from the X-ray patterns with a Rietveld refinement [3] using the commercial software package TOPAS (Bruker AXS). For modelling microstructural effects TOPAS is supported by the Double-Voigt Approach (e.g. Balzar, 1999 [4]). The possibility for a separate size-strain analysis is given by the effect that size broadening is angle independent, whereas strain broadening depends on the diffraction angle [4].

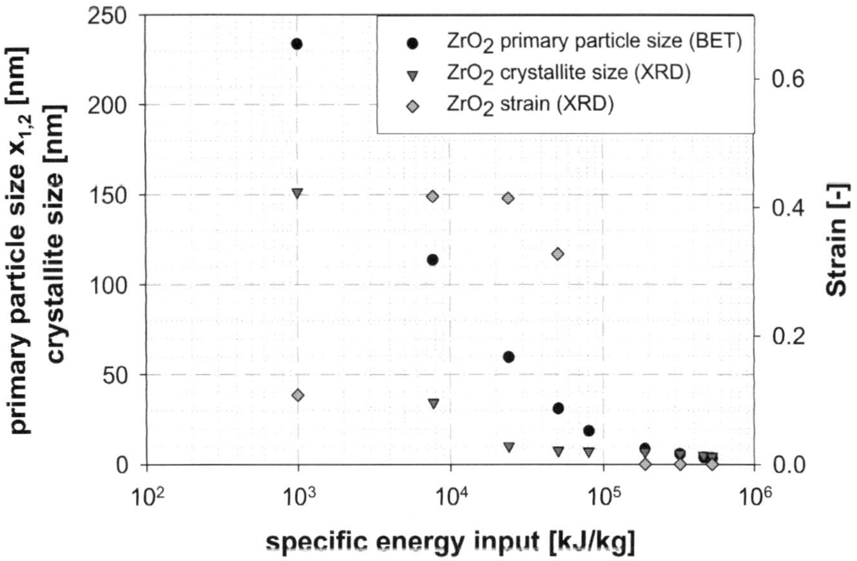

Fig. 2: Evolution of primary particle size, crystallite size and strain of zirconia with specific energy input

In Fig. 2 the evolution of the primary particle size, the crystallite size and the strain of zirconia particles over the specific energy input are shown. One can observe that the crystallite size decreases until a plateau at a critical crystallite size (volume weighted) of about 4.5 nm and a primary particle size (area weighted) of approximately 4 nm are achieved. This plateau defined as true grinding limit is the minimal achievable particle size in grinding experiments and can be reached under stable milling conditions where the particle size is not influenced by an agglomeration process but only through real breakage of particles. Comparing the crystallite sizes with the primary particle sizes one can observe that above a certain energy input the initially polycrystalline zirconia particles become monocrystalline. From this point on particle breakage can only occur following a reduction of the crystallite size. The strain inside the crystallites increases at the beginning of the milling experiment until a maximal value is reached and afterwards decline to nearly zero. Due to the intensive stressing of the particles between the milling beads lattice imperfections are generated, enhancing the total amount of elastic stored energy in a crystallite. After reaching a critical crystallite size it is assumed the domains become so small that it is not possible to store further defects in the lattice [5]. As a result of this mechanism the strain decreases to a minimum. In Fig. 2 the disappearance of lattice strain corresponds to the point where no further reduction of the crystallite size occurs and the true grinding limit is reached. It is supposed that below a certain particle size the transferred stress intensity from the milling beads is no longer sufficient to achieve a brittle fracture with a single impact. Hence, further breakage can only occur if the elastic stored energy provided from lattice imperfections and external mechanical impact is sufficient to fulfill the integral energy balance and enable a fracture to traverse the entire particle [6]. This behavior of a fatigue fracture explains the reduced kinetics in the nanometer range where the available stress intensity becomes insufficient. Accordingly one can conclude that the generation of defects enables the breakage in the nanometer range whereas their absence below a critical size determines the grinding limit.

The kinetics of particle breakage as well as the grinding limit depend on a multitude of influencing variables. These influencing variables can be separated into three groups:

- Material parameters (material properties of the milling powder, properties of the solvent, stability conditions)
- Process parameters (rotational speed of the stirrer, milling bead size and material, etc.)
- Machine parameters (design of the mill, kind of stressing, etc.)

We have investigated the influence of different materials on the breakage behavior of particles in the nanometer range and their grinding limit. For this study five different materials (ZrO_2, Al_2O_3, SnO_2, SiO_2 and $CaCO_3$) were stressed in a stirred media mill. The milling experiments were carried out under the same following conditions: milling bead size 0.4 – 0.5 mm, filling ratio of the grinding chamber with milling beads 80 % and stirred tip speed 8 m/s. Whereas particles of SnO_2 and $CaCO_3$ were milled in water the other materials were milled in ethanol. However, as ethanol and water do not vary much in their viscosity, the influence of the different solvents can be neglected.

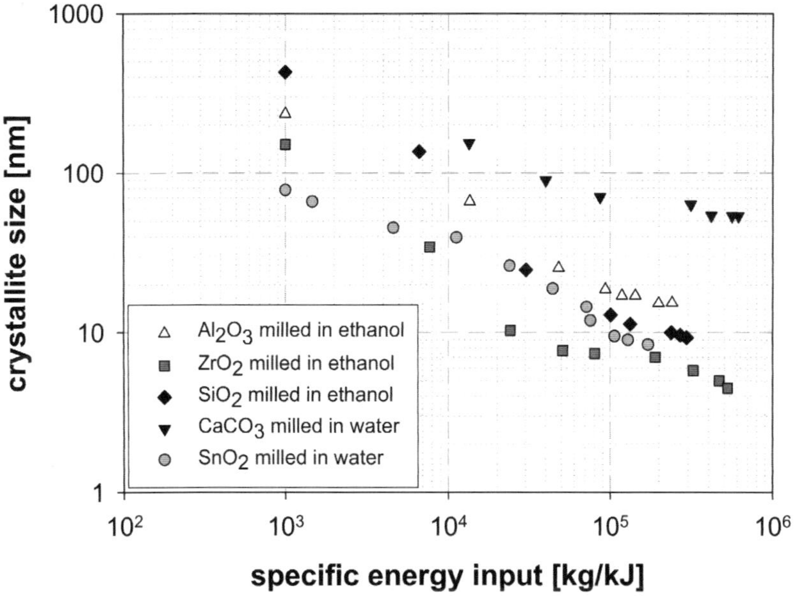

Fig. 3: Development of crystallite sizes for different materials with specific energy input

After stressing the particles for about 50 hours, most of the materials have reached their grinding limit. The continually decreasing final mean crystallite sizes of some materials may be caused by a crystallite size distribution which gets narrower as large crystallites break whilst the smaller ones have already reached their grinding limit. Additionally an amorphization of the particle surface due to intensive mechanical stressing can also lead to slightly smaller crystallite sizes without implying a breakage of the crystallites.

From Fig. 3 one can observe that each material shows a different breakage behavior expressed in different breakage kinetics as well as in a different grinding limit. In general the kinetics of crystallite breakage can be approximated by a power function:

$$x = a \cdot E_m^{-b} \qquad (1)$$

where E_m is the mass specific energy input and a and b are parameters depending on material and process properties.

For ductile materials like calcium carbonate the smallest crystallite size reached in the milling experiment under the chosen conditions was about 55 nm whereas for very brittle materials like zirconia a grinding limit of less than 5 nm was observed. Ductile materials respond to mechanical stressing with distinct plastic deformation whereas in brittle materials the applied energy is mainly used for particle fracture. We have to point out that the grinding limits reached are not fixed values. It should become clear that under different milling conditions, for example by changing the rotational speed of the stirrer other stress intensities and therefore different values of the true grinding limit can be achieved.

A direct correlation of the minimal crystallite size with material properties is difficult for the time being, because of a superposition of material and process parameters both influencing the breakage behavior. To correlate the material properties with the grinding limit and the breakage kinetics a separation of process und material parameters is necessary. Further investigations on this topic are in progress and will be presented in forthcoming papers.

Conclusions

The particle size in an agglomerate-free suspension is only determined by actual particle breakage. A grinding limit is observed in long term experiments for different materials and could be characterized by the crystallite size measured by X-ray diffraction. This true grinding limit is the minimal achievable particle size under the current milling conditions. It is assumed that the grinding limit is reached due to an insufficient intensity of stressing. During the grinding process lattice imperfections are generated which enhance the amount of elastic stored energy in a crystallite. Hence, the necessary energy transferred from the milling beads to reach a particle or crystallite breakage is reduced and can be bring up by the actual stress intensity for larger crystallite sizes. The procedure of generating defects in several stressing events until the stored energy combined with the constant 'external' energy are large enough to lead to a breakage event slows the breakage kinetics in the nanometer range. The grinding limit is reached when the crystallites become so small that nearly no defects can be stored in the lattice anymore. The total available energy below this critical crystallite size is insufficient and could not be increased by further milling. Accordingly the generation of defects enables nanogrinding by increasing the elastic stored energy, but determines with its disappearance below a critical size concurrently the overall limit of grinding.

The kinetics of particle breakage and the true grinding limit depend on material properties as well as on process parameters influencing the stress intensity. A correlation between those parameters and the actual primary particle size especially the critical particle size will to be determined by a separation of the material function and the machine function in future works.

Acknowledgement

This work was financially supported by the German Research Council (DFG) and the International Fine Particle Research Institute (IFPRI). We gratefully acknowledge Prof. Neubauer and Christoph Hesse of the Institute of Mineralogy for their support in analyzing the XRD patterns. Additionally we thank the NETZSCH-Feinmahltechnik GmbH for providing the laboratory mill "LabStar".

References

[1] Stenger, F., Mende, S., Schwedes, J., Peukert, W.; *Nanomilling in stirred media mills*, Chemical Engineering Science (2005), 60(16), 4557-4565

[2] Stenger, F., Mende, S., Schwedes, J., Peukert, W.; *The influence of suspension properties on the grinding behavior of alumina particles in the submicron size range in stirred media mills*, Powder Technology (2005), 156(2-3), 103-110

[3] Rietveld, H. M.; *Profile refinement method for nuclear and magnetic structures.* Journal of Applied Crystallography (1969), 2(2), 65-71.

[4] Balzar, D. *Voigt function model in diffraction-line broadening analysis.* International Union of Crystallography Monographs on Crystallography (1999), 10 (Defect and Microstructure Analysis by Diffraction), 94-126.

[5] V.G. Gryaznov, I.A. Polonsky, A.E. Romanov, L.I. Trusov. *Size effects of dislocation stability in nanocrystals.* Physical Review B (1991), 44(1), 42.

[6] Bernotat, S., Schönert, K.; *Size reduction*, Ullmann's encyclopedia industrial chemistry, B2-5, (1988).

429b Nanoparticle Milling and Encapsulation in Supercritical Carbon Dioxide Mixtures

Scott M. Paap[1], Michael S. Wolfe[2], and Jefferson W. Tester[1]. (1) Chemical Engineering, Massachusetts Institute of Technology, 77 Massachusetts Avenue, Room 66-053, Cambridge, MA 02139, (2) Imaging Technologies, DuPont, P.O. Box 80402, Wilmington, DE 19880-0402

In the conventional process used to produce pigment dispersions, agglomerates of primary pigment particles in an aqueous slurry are broken down to aggregates with a diameter of approximately 100 nm in a media mill. As the milling process proceeds, polymeric dispersant molecules dissolved in the aqueous phase adsorb onto newly exposed particle surfaces, forming a continuous coating which prevents reagglomeration of the particles and enhances the stability of the resulting dispersion. In this work we present a new process to produce polymer-encapsulated carbon black nanoparticles, in which milling and encapsulation with benzyl methacrylate/methacrylic acid random copolymers are carried out simultaneously in high-pressure CO_2-expanded liquids. In the final process step, the solvent is flushed from the system with CO_2 and the encapsulated particles are recovered as a dry powder which can be redispersed in aqueous or solvent-based carriers to yield pigment dispersions. The use of CO_2-expanded liquids enables precise control of the polymer solubility and adsorption behavior, and also allows the use of a wide variety of water-insoluble polymers that are inaccessible with the conventional process. Experimental investigations of the high-pressure phase behavior of polymer-CO_2-solvent systems as well as the adsorption of polymers onto pigment particles from CO_2-expanded solvents provided a basis for analysis and optimization of the new process. Encapsulated particles obtained using the optimized CO_2-based process had a number-average diameter of 65 nm, and were easily redispersed in water to form pigment dispersions. Characterization of the encapsulated particles was performed using TEM and dynamic light scattering techniques.

429c Stability of Silica Nanoparticle Agglomerates

Martin Seipenbusch[1], Sonja Rothenbacher[1], Martin J. Kirchhof[2], Hans-Joachim Schmid[3], Alfred P. Weber[4], and Gerhard Kasper[1]. (1) Institut für Mechanische Verfahrenstechnik und Mechanik, Universität Karlsruhe, Am Forum 8, Karlsruhe, Germany, (2) Institute of Particle Technology, University of Erlangen-Nuremberg, Cauerstr. 4, Erlangen, Germany, (3) Mechanische Verfahrenstechnik und Umweltverfahrenstechnik, Universität Paderborn, Pohlweg 5, Paderborn, Germany, (4) Institut für Mechanische Verfahrenstechnik, Technische Universität Clausthal, Leibnizstr. 19, Clausthal-Zellerfeld, Germany

In the synthesis of nanoparticles the high number concentrations, unavoidable in an even moderate yield process, lead to the formation of agglomerates. This is especially true in the aerosol phase, where stabilization of particles in the unagglomerated state is quite challenging. Upon contact the primary particles of an agglomerate form bonds which can be of different nature, depending on the synthesis process. At sufficiently high temperatures for instance, the viscosity of the particles or the diffusivity allow sintering of the particles in contact, and thus hard aggregates are formed.

Depending on the chemical nature of the particles the formation of chemical bonds between the particle surfaces may also occur. As a baseline of the interparticle energy van der Waals forces are always present, leading to relatively weak agglomerates if no other mechanism contributes to the bond strength. While aggregates are very stable structures that can hardly be broken up, agglomerates can be disintegrated in suitable processes for further use. The distinction between agglomerates and aggregates is significant to estimate the product properties of nanopowders since the strength of interparticle bonds to a large extent determines the physical properties and applicability in subsequent manufacturing steps.

For fumed silica a limit of dispersibility in liquid media is observed, where an increase of the energy input does not lead to further fragmentation. The particles remain at diameters of about 100 nm, while the primary particles are much smaller than this, in the range of 10-20 nm (e.g. Pohl et al. 2005). While electron micrographs show no signs of sintering there are obviously substructures of the agglomerates with high interparticle forces that can not be fragmented. It can be speculated that these substructures are aggregates, formed in regions of high temperature in the synthesis process where minute sintering in the contact regions was possible. These aggregates may then agglomerate in cooler regions to form weak agglomerates.

The aim of this work was the systematic investigation of the interparticle forces in silica agglomerates. To this end silica particles were generated and allowed to agglomerate at room temperature, forming van der Waals contacts. The formation of high energy bonds was then induced by controlled sintering at various temperatures between 1000 and 1500°C at a constant residence time of 30 ms. The bond energies of the primary particles within the agglomerates were determined using the method of impact fragmentation, which was adapted to the nanoscale (Seipenbusch et al., 2002). The method enables fragmentation of agglomerates under variation of the kinetic energy prior to impaction. Analysis of the fragmentation patterns at different initial kinetic energies then yields fragmentation curves that show the distribution of interparticle forces within the agglomerates. In parallel to the fragmentation experiments the evolution of solid state bridges was analysed using electron microscopy.

As a reference Aerosil® 200 was analysed for its fragmentation behaviour. It showed a fragmentability of 30% and a kinetic energy necessary to break 50% of the interparticle contacts of 5.5x1017 J/bond.

The experiments with our synthesized SiO2 showed, that particles agglomerated at room temperature are indeed held together by relatively weak forces and can be deagglomerated almost entirely. At temperatures higher than 1000°C however, the maximum degree of deagglomeration (fragmentability) obtainable in the applied energy range rapidly decreased. When sintering necks eventually became visible at temperatures above 1300°C the fragmentability had already dropped to about 50%. The fragmentation curve of Aerosil® 200 was approximated for temperatures exceeding 1400°C. The interparticle contacts in fumed silica therefore appear to be dominated by solid state necks.

429d High Shear De-Agglomeration of Nanoparticle Agglomerates in Stirred Tanks

Micaela Caramellino, Otto H. York Dept. of Chemical Engineering, New Jersey Institute of Technology, 323 King Blvd., Newark, NJ 07102 and Piero M. Armenante, Otto H. York Department of Chemical Engineering, New Jersey Institute of Technology, 323 King Blvd., Newark, NJ 07102-1982.

The use of nanoparticles in liquid suspensions is becoming increasingly common in the pharmaceutical industry, where applications involving nano-sized active pharmaceutical ingredients offer significant potential benefits. However, processing of nano-particles often results in nanoparticle agglomeration, due to the high surface area and inter-particle forces per unit area acting between these particles. Therefore, de-agglomeration of nanoparticle agglomerates is a critical requirement for the target dosage form, since the size of the agglomerates obtained influences the final properties of the material.

In this work, de-agglomeration of nano-particles agglomerates in stirred tanks was experimentally studied using a stirred tank provided with a high-shear homogenizer. The agitation system consisted of a centrally mounted pitched-blade turbine (PBT) as the primary impeller, and a high-shear homogenizer, mounted in an off-set position, as the secondary impeller. The former impeller was used to suspend the particles, while the latter was used to impart the necessary shear for the breakage of the nano-particle agglomerates. The system examined here consisted of a suspension of silica nanoparticles in water, where the nanoparticles initially formed agglomerates about 30 μm in size. Experiments were conducted in which the system was subjected to high-shear mixing under different operating conditions. Samples were taken at different times and the agglomerate size distribution was measured by laser diffraction.

The results show that breakage of the nano-particle agglomerates is time dependent and is a function of the shear applied by the high-speed homogenizer. For a given set of operating conditions, the size of the agglomerate eventually reaches an asymptotic value, which is primarily a function of the homogenizer speed. Other parameters, such as tank size, primary impeller type and speed, also play a role but they only affect the time required by the system to approach the asymptotic value. The data were regressed accordingly and the results could be interpreted using a correlation relating the agglomerate size to the energy input and the aggregative strength of the particle agglomerates.

445a Development of a Visco-Plastic Constitutive Law for Dense Flow of Granular Matter

Mehrdad Kheiripour Langroudi, Department of Chemical Engineering, The Graduate Center , City University of New York, 140th Street and Convent Avenue, New York, NY 10031 and Gabriel I. Tardos, Department of Chemical Engineering, The City College of the City University of New York, 140th Street & Convent Ave, New York, NY 10031.

The rheological behavior of dense granular materials was studied using a modified annular Couette cell, which allows for porosity changes inside the bed. In these experiments, a small increase in porosity partially breaks the force chains and provides enough space for random collision to occur inside the bed resulting in a significant decrease of shear stresses on the rotating cylinder. This also increases the domain (width) of shear bands and eliminates the existence of points with zero velocity.

We were able to measure normal stresses inside the shearing zone (on the rotating cylinder) and on the stationary (outside) walls of the Couette cell. We found that the coefficient of apparent friction, which is the ratio of shear to normal stress, remains constant (material property) at low shear rates when the regime of flow is quasi-static and plasticity rules applies but it increases exponentially ($n < 1$) by increasing the shear rate to moderately high values when the granular matter behaves somewhat like a liquid, i.e. both collisions and enduring contacts play a role.

The experimental results suggest a visco-plastic constitutive law for dense granular materials. In this model, shear stresses were found to be independent of shear rate so that the quasi-static regime dominates while increasing the shear rate makes the bed acquire a fluid-like behavior and changes the regime of flow to intermediate. We also found that the dependency of the shear stress to shear rate increases by further increases in shear rate because collisions between particles become more important. In order to generalize the idea, we did experiments with different materials like glass beads (different size), acrylic powders, elastomer granules and catalytic powders. We found that material properties like bulk density, size and stiffness have an important effect on the rheology of flow.

445b Predicting Discharge Dynamics from a Rectangular Hopper Using the Discrete Element Method (DEM)

Anshu Anand[1], Jennifer S. Curtis[1], Carl R. Wassgren[2], Bruno C. Hancock[3], and William R. Ketterhagen[4]. (1) Chemical Engineering, University of Florida, Gainesville, FL 32611, (2) Purdue University, 585 Purdue Mall, West Lafayette, IN 47907, (3) Pfizer Inc., Groton, CT 06340, (4) Pfizer Global Research and Development, Pfizer Inc., Groton, CT 06340

Accurate prediction of the discharge rate from hoppers is important in many industrial processes involving the handling of granular materials. The present work investigates the parameters affecting the discharge rate using the Discrete Element Method (DEM). The effects of particle properties (particle size and size distribution) and hopper geometry (hopper width, outlet width, angle and fill height) are studied and compared to previously published experimental correlations. The results indicate that DEM simulations are fully capable of reproducing trends in the discharge rate that are well-known experimentally. For example, particle size and hopper width are shown to have a minimal influence on the discharge rate. In addition, for rectangular hoppers, the discharge rate is shown to vary with the outlet width raised to the 3/2 power as given by the modified Beverloo correlation. The DEM simulations are also used to explore a wider range of parameters that have not been or are not easily explored experimentally. For example, the effects of hopper friction, particle friction, coefficient of restitution are investigated, and particle friction is shown to have a significant influence on the hopper discharge dynamics.

445c Use of the Attainable Region (AR) Analysis to Optimize Particle Agglomeration and Breakage Operations

Matthew J. Metzger[1], Benjamin J. Glasser[2], David Glasser[3], Brendon Hausberger[4], Diane Hildebrandt[3], and Ngangezwe Khumalo[5]. (1) Chemical and Biochemical Engineering, Rutgers University, 98 Brett Rd, Piscataway, NJ 08854, (2) Dept. of Chemical and Biochemical Engineering, Rutgers University, 98 Brett Road, Piscataway, NJ 08854, (3) COMPS, School of Chemical and Metallurgical Engineering, Wits University, Private Bag x3, PO Wits, Johannesburg, 2050, Johannesburg, 2050, South Africa, (4) Centre of Material and Process Synthesis, School of Chemical and Metallurgical Engineering, University of the Witwatersrand, Johannesburg, 1 Jan Smut Avenue, Johannesburg, South Africa, (5) University of the Witwatersrand, School of Process and Materials Engineering, Wits University, P. Bag 3, Wits, Johannesburg, 2050, South Africa

Tight particle size distribution control is vital to the success of any process involving particles. However, even controlling crystal size during crystallization - a process with a relatively well-known mechanism - is a daunting chore. In other operations central to the pharmaceutical and chemical industries, such as agitated drying and granulation, the task of size control is complicated due to the competition between particle agglomeration and breakage. A tool that with the ability to encompass all aspects of particle agglomeration and breakage within one framework is needed to advance the development of solid processes.

Presented here is the application of the Attainable Region (AR) approach to optimizing the simplest case of an industrial sizing process: milling. Previously, the AR approach has been used to optimize complex reactor networks and separation systems. Similarities between species in chemical reaction networks and particle sizes in agglomeration and breakage processes, suggest the AR is a capable tool for determining the optimal control strategy for a milling project. Though the optimization done in this work is for a batch ball mill and a specific material, the concepts can be applied to any system consisting of breakage, agglomeration, and classification.

We have experimentally determined the optimal control strategy for achieving a certain size distribution with the shortest amount of time and most efficient energy usage. The control strategy recommends a variation of speed throughout the process - fast speed at first followed by a lower speed - to achieve the maximum amount of material of a desired size. Benefits of this approach can be realized in any size alteration operation, as a size profile over time can be predicted. Therefore, a design methodology can be constructed without complicated mathematical tools.

445d Classification of Fine Particles Using a Taylor-Couette Device

Navina D. Tungapindi and **Kimberly H. Henthorn**. Chemical Engineering, Missouri University of Science & Technology, 143 Schrenk Hall, 1870 Miner Circle, Rolla, MO 65409

The study of fine particle classification is common, especially since a well-defined particle size distribution is important in most processes and products. Many methods have been developed to classify micron-size particles, and although each has its own advantage, most cannot classify particles to an extremely narrow range. This is partly due to the cohesive nature of fine particles, which can be due to electrostatic attraction, moisture, or van der Waals forces. In order to overcome this problem, a wet method of classification has been chosen here. Taylor vortices, discovered in 1923, develop in the fluid contained in the annulus of two concentric rotating cylinders. It has been shown in recent years that Taylor vortices can be used to classify particles regardless of their size (which may range from a few microns to millimeters) and density difference between the particles and fluid. This method also allows for a semi-continuous throughput of particles. The present work outlines the use of Taylor vortices to classify fine particles to a narrower size distribution. The main objective here is to determine the parameters affecting classification of glass spheres from an initial distribution ranging from 10 microns to 30 microns. Factors such as annulus width, rotational velocity, and particle feed rate have been investigated. Preliminary numerical investigations have shown that up to 90% of particles larger than 25 microns can be removed using a Taylor-Couette device, and experimental work is being carried out to validate the results obtained numerically.

445e Power Consumption and Granular Flow In a Vibro-Fluidized Stirred Granular Bed

Kenneth J. Ford, James F. Gilchrist, and Hugo S. Caram. Lehigh University, Department of Chemical Engineering, Bethlehem, PA 18015

The behavior of static and vibrated granular beds is explored through measurements of the power required to rotate a vane within the granular media and capacitance measurements to evaluate local density. These measurements of "granular rheology" aim to define a quantitative framework for granular bed states and transitions between these states. The effects of bed height, bed diameter, vibration frequency and glass bead size on the observed states and transitions are explored. Deep granular beds are subjected to vibration in the frequency range of 10 Hz to 300 Hz with acceleration in the range 0 to 4.0g. The bed aspect ratio (effective bed height to bed diameter) is varied up to 4.1 with effective bed height range from 0 cm to 23 cm and bed diameter from 5.6 cm to 7.5cm.

Through measurements of vane power and granular density, three granular bed states, the static granular state, the quasi-static state and vibro-fluidized state, along with the transitions between the states are identified. A granular bed state phase diagram is constructed, mapping the effects of vibration frequency and intensity input. Transition to a vibro-fluidized state at a critical dimensionless vibration acceleration is characterized by dramatic decreases in vane power, dynamic zero-shear rate torque and measurable density fluctuations throughout the bed. In the vibro-fluidized state, the vane power and extrapolated zero-shear rate torque are weakly influenced by the vibration force and frequency. The measured vane power becomes invariant with bed height above a critical bed aspect ratio. This phenomenon can be characterized by the Janssen equation, which relates static pressure to the height of the granular media.

445f Density Measurements In a Vibro-Fluidized Deep Granular Bed

Kenneth J. Ford, **James F. Gilchrist**, and Hugo S. Caram. Department of Chemical Engineering, Lehigh University, Bethlehem, PA 18015

The local state of a granular bed is explored through capacitance measurements of granular density fluctuations. The deep granular bed is subjected to vibration in the frequency range of 10 Hz to 200 Hz with acceleration in the range 0 to 2.0g. The bed aspect ratio (effective bed height to bed diameter) is varied up to 3.6 with effective bed height range from 0 cm to 20 cm and bed diameter of 5.6 cm. Three granular bed states, the static granular state, the quasi-static state and vibro-fluidized state, along with the transitions between the states are identified. Lack of density fluctuations characterize the static granular state. Large density fluctuations during the vibration cycle are characteristic of the vibro-fluidized state. The entire height of the bed can reside in a static or vibro-fluidized state at low and high vibration levels, respectively. A quasi-static bed state occurs at intermediate vibration levels and is characterized by vibro-fluidized state density fluctuations in the upper portion of the bed and static state lack of density fluctuations in the lower portion of the bed. The extent of density fluctuation penetration is a function of distance from the top of the bed, rather than total bed height. Density fluctuations during thc vibration cycle scale with frequency at lower frequencies and are not present at higher frequency vibrations. This talk will attempt to characterize the "melting" of granular beds through the transition to vibrofluidization.

445g Spray Drying of Large Particles of Film Forming Materials

P. Threlfall-Holmes, Wilton Applied Research Group, AkzoNobel Research, Development and Innovation, Redcar, United Kingdom and **R. Ocone**, Chemical Engineering, Heriot-Watt University, School of Engineering and Physical Sciences,, Edinburgh, EH14 4AS, United Kingdom.

We have observed that large spray dried particle product with uniform morphology can be created when using a nozzle atomiser working on the principle of Rayleigh instability of varicose jets. In this paper, we report results from experiments conducted at AkzoNobel Applied Research Group, Wilton, UK, drying dispersions of starch in water in a Niro P-6,3 pilot spray dryer, 1.6m diameter, 7.8m fall height, with co-current gas flow, using both rotary and nozzle atomisers. The starch dispersions used in these experiments form films, rather than porous crusts, as they dry out. This behaviour is key to their industrial use as encapsulation matrices. It was found that, under otherwise identical drying conditions with the nozzle atomiser, by varying the composition of the liquid feed, the maximum mean droplet diameter that could be dried varied from 250μm to 450μm. Our experimental data indicate that for these large droplets containing film-forming polymeric materials, the drying kinetics of the droplets may be rate controlling. This is contrast to the normal expectation of equilibrium limitation in spray drying. A different approach to performance modelling and prediction is therefore required. Ideally, drying kinetic data for a material should be measured at realistic industrial spray dryer operating conditions. An apparatus design is proposed to measure the drying rates of small droplets of aqueous liquids, employing an ultramicrobalance to measure mass loss rate from droplets as small as 500μm diameter, in an air stream which velocity and temperature may be as high as 30m/s and 300°C, with humidity from dry to near saturation. In this paper, we demonstrate that a simple scaling model using distributions of droplet size and residence times can qualitatively explain the experimental observations of increased droplet size and uniform dry particle morphology.

468a A Kinetic-Based Model for a Non-Isothermal Granular Gas with Mono- and Bi-Dispersed Particles

Alberto Passalacqua, Department of Chemical and Biological Engineering, Iowa State University, 2114 Sweeney Hall, Ames, IA 50011-2230, Prakash Vedula, School of Aerospace and Mechanical Engineering, University of Oklahoma, Norman, OK 73019-0601, Christine M. Hrenya, Chemical and Biological Engineering, University of Colorado at Boulder, Boulder, CO 80309-0424, and R. O. Fox, Department of Chemical & Biological Engineering, Iowa State University, 2114 Sweeney Hall, Ames, IA 50011-2230.

A granular gas between two stationary Maxwellian walls at different temperatures is studied by means of numerical simulations of the Boltzmann equation with a hard-sphere collision kernel for mono- and bi-dispersed particles.

The behaviour of a granular gas in these conditions is influenced by the thickness of the wall Knudsen layers: if its thickness is comparable to the characteristic length scale of the system, the traditional description based on the hydrodynamic equation is invalid, and it is necessary to account for the presence on the Knudsen layers.

In this work the system is described by solving the full Boltzmann equation using a third-order quadrature-based moment method (Fox, 2008).

These results are compared with the predictions of molecular dynamics simulations of an equivalent system with finite-size particles (Galvin et al., 2007; Hrenya et al., 2008). Results for constitutive quantities such as the heat flux and the stress tensor are provided, and show the capability of the quadrature-based approach to predict them in agreement with the molecular dynamics simulations.

Fox, R. O., A quadrature based third-order moment method for dilute gas-particle flows, Journal of Computational Physics, 227, 6313 – 6350, 2008.

Galvin J. E., Hrenya C. M., Wildman R. D., On the role of the Knudsen layer in rapid granular flows, Journal of Fluid Mechanics, Vol. 585, pp. 73 – 92, 2007.

Hrenya C. M., Galvin J. E., Wildman R. D., Evidence of higher-order effects in thermally-driven, rapid granular flows, Journal of Fluid Mechanics, 598, 429 – 450, 2008.

Vedula, P., Fox, R. O., A quadrature-based method of moments for solution of the collisional Boltzmann equation, Journal of Statistical Physics, Submitted, 2008.

468b Comparison of Euler-Euler and Euler-Lagrange Simulations of Finite-Stokes-Numbers Gas-Particle Flows in a Lid-Driven Cavity

Rahul Garg[1], **Alberto Passalacqua**[2], Shankar Subramaniam[1], and R. O. Fox[3]. (1) Mechanical Engineering, Iowa State University, 95-E Black Engineering, Ames, IA 50011-2161, (2) Department of Chemical and Biological Engineering, Iowa State University, 2114 Sweeney Hall, Ames, IA 50011-2230, (3) Department of Chemical & Biological Engineering, Iowa State University, 2114 Sweeney Hall, Ames, IA 50011-2230

Classical Euler-Euler two-fluid models based on the kinetic theory of the granular flow assume the particle phase to be dominated by collisions, even when the particle volume fraction is low and hence collisions are negligible. This leads to erroneous predictions of the particle-phase flow patterns and to the inability of such models to capture phenomena like particle trajectory crossing for finite Stokes numbers. To correctly predict the behaviour of dilute gas-particle flows a more fundamental approach based on solving the Boltzmann equation is necessary to treat non-zero Knudsen numbers and finite Stokes numbers. In this work a dilute collision-less (i.e. infinite Knudsen number) gas-particle flow with mono-dispersed finite-Stokes particles in a lid-driven cavity is studied by means of numerical simulations. An Eulerian quadrature-based moment method for the direct solution of the Boltzmann equation (Fox, 2008) is adopted to describe the particle phase, and it is fully coupled with an Eulerian fluid solver to account for the two-way coupling between the phases. Various Stokes numbers in the range of 0.03-3 are considered to show the capability of the Euler-Euler approach to correctly capture the particle-phase behaviour and the correct segregation phenomena. Results are compared with the predictions of both a classical two-fluid model, whose limitations are pointed out, and of Euler-Lagrange simulations, showing good agreement with the latter.

Fox, R. O., A quadrature based third-order moment method for dilute gas-particle flows, Journal of Computational Physics, 227, 6313 – 6350, 2008.

468c Combination of Monte Carlo Simulation of Population Balances and Cfd for Modeling of Nanoparticle Formation

Einar Kruis, Faculty of Engineering, University Duisburg-Essen, Bismarckstr 81, Duisburg, Germany and J. Wei, Engineering, University Duisburg-Essen, Bismarckstr 81, Duisburg, Germany.

In this work, we combine a Monte Carlo based method to model the particle dynamics with a transport model based on CFD data. Monte Carlo methods are ideally suited to model multivariate particle properties, such as size, charge and composition.

Transport by convection and diffusion is decribed on a probabilistic basis ,on the basis of CFD data for the carrier gas transport. The particle dynamics includes nucleation, coagulation, growth and thermoionic charging. A twodimensional case study is done using 400 cells. Each cell has a predefined number of simulation particles, which each represent a number of real particles, which is different for each cell. Separate studies are done to check the validity of the transport and particle dynamic modules. The model is applied to nanoparticle generation at high temperatures where thermoionic effects lead to charged particles.

The software is implemented on a LINUX cluster and applies

MPI.

468d Coarse-Grid Simulation of Fluidized Gas-Particle Flows

Yesim Igci and Sankaran Sundaresan. Chemical Engineering, Princeton University, A-214 Equad, Princeton, NJ 08544

Gas-particle flows in fluidized and circulating fluidized beds are inherently unstable, and they manifest fluctuations in velocities and local suspension density over a wide range of length and time scales. In riser flows, these fluctuations are associated with the random motion of the individual particles (typically characterized through the granular temperature) and with the chaotic motion of particle clusters, which are repeatedly formed and broken apart. The two-fluid model equations are able to capture these clusters in a robust manner; however, to resolve the clusters at all length scales, extremely fine spatial grids are necessary [1]. Due to computing limitations, the grid size used in simulating industrial scale gas-particle flows is invariably much larger than the length scales of the finer particle clusters. Such coarse-grid simulations for industrial scale gas-particle flows will clearly not resolve the structures which exist on sub-grid length scales; however, these small-scale unresolved structures affect the resolved flow characteristics. We pursue a filtered equations approach in which the influence of the small scale structures appears as residual correlations for which constitutive models should be constructed. If constructed properly, the filtered equations should produce a solution with the same macroscopic features as the finely resolved kinetic theory model results; however, as the filtered equations place less stringent requirements on the grid resolution than the original two-fluid model equations, they would be easier to solve.

In a recent study [2], we extracted closures for the filtered two-fluid models by filtering the statistical "data" generated through highly resolved MFIX simulations in large periodic domains – employing a kinetic theory based two-fluid model with Wen & Yu drag for uniformly sized particles [3,4]. We found that both the filtered drag coefficient decreased systematically with increasing filter width, whereas the particle-phase stresses increased with increasing filter width.

The objectives of the present study are two-fold:

(a) What boundary conditions should be used with the filtered equations at bounding walls?

(b) How do the solutions of filtered model equations compare with those of the kinetic theory based model which were used to generate the closures for the filtered equations?

We have completed several 2-D simulations with different boundary conditions at the vertical walls and have found that as filter size increases, the boundary conditions which should be employed with filtered equations approach free slip conditions for both phases.

In order to evaluate the filtered two-fluid models, we have run a number of simulations of the filtered two-fluid models and compared the results predicted by such simulations with those from finely resolved simulations of the kinetic theory based two-fluid model. Through these comparisons, we assess the filtered two-fluid and bring forth their strengths and weaknesses.

The details of these results will be described in the presentation.

References:

[1] Agrawal, K., Loezos, P. N., Syamlal, M. & Sundaresan, S. 2001 The role of meso-scale structures in rapid gas-solid flows. J. Fluid Mech. 445, 151 – 185.

[2] Igci Yesim, Andrews, A.T. IV, Sundaresan S., Pannala S. & O'Brien T. 2008 Filtered two-fluid models for fluidized gas particle suspensions. AIChE J. 54, 1431-1448.

[3] Gidaspow, D., 1994 Multiphase Flow and Fluidization, Academic Press, CA. 31-58, 197 – 238.

[4] Syamlal, M., Rogers, W. & O'Brien, T. J., 1993 MFIX Documentation, U.S. Department of Energy, Federal Energy Technology Center, Morgantown, WV.

468e Representation of a Continuous Particle Size Distribution Using a New Polydisperse Kinetic Theory

Harish Viswanathan[1], Christine M. Hrenya[1], and R. O. Fox[2]. (1) Chemical and Biological Engineering, University of Colorado at Boulder, Boulder, CO 80309-0424, (2) Department of Chemical & Biological Engineering, Iowa State University, 2114 Sweeney Hall, Ames, IA 50011-2230

Flows of polydisperse particles are common in nature as well as in industrial applications. In the past, kinetic theories for monodisperse systems have not only been successful in predicting the rapid flow of particles but have also been extended to gas-solid systems. Previous kinetic theories for polydisperse systems have been based on the assumptions of a Maxwellian velocity distribution or an equipartition of energy. Recently, a kinetic theory for an s-component mixture of inelastic, smooth hard disks (2D) and spheres (3D) has been derived that rigorously incorporates the non-Maxwellian and non-equipartition (Garzó, Dufty, and Hrenya, 2007 and Garzó, Hrenya & Dufty, 2007) effects. It is applicable to wide range of restitution coefficients and a wide range of concentrations (dilute and moderately dense). At the center of the current effort is using this new theory to represent a continuous size distribution. Two questions are considered: (i) What technique should be used to choose discrete particle sizes to represent the continuous size distribution, and (ii) How many discrete sizes are appropriate for an accurate representation? Regarding the first question, a representation of a continuous particle size distribution (PSD) using discrete species is attempted via the method of moments. To answer the second question, all transport coefficients are evaluated using an increasing number (s) of discrete species. The results show that s = 3 (three discrete particle sizes chosen according to the method of moments) provides an accurate description of a continuous PSD for both Gaussian and lognormal distributions.

References:

Garzó, V., Dufty, J. & Hrenya, C. M. (2007) Enskog theory for polydisperse granular mixtures. I. Navier-Stokes order transport. Physical Review E, 76, art. no. 031303.

Garzó, V., Hrenya, C. M. and Dufty, J. W. (2007) Enskog theory for polydisperse granular mixtures. II. Sonine polynomial approximation. Physical Review E, 76, art. no. 031304.

468f Solution of the Boltzmann Equation by the Fcmom

Matteo Strumendo and Hamid Arastoopour. Department of Chemical and Biological Engineering, Illinois Institute of Technology, Chicago, IL 60616

The fluid dynamics of the particulate phase in gas-solid flows (in fast and relatively dilute conditions) is described by the Boltzmann equation. Several methods are available to solve the Boltzmann equation, i. e. generalizations of the Chapman-Enskog expansion, method of moments, Monte Carlo simulations, discretization of the Boltzmann equation in the velocity space, Lagrangian methods; in this work, the focus is on the solution of the Boltzmann equation by the method of moments.

The first application of the method of moments to the Boltzmann equation was developed by Grad [1], who solved the Boltzmann equation for simple (elastic) gases using a third order approximation of the particle velocity distribution. Strumendo and Canu [2] generalized the Grad approach and were able to compute the anisotropy of the granular flows, however their technique is not successful when extended to higher order moments.

The Boltzmann equation can be interpreted as a tri-variate population balance equation, in which the internal variables are the three components of the particle velocities. In this perspective, concepts and techniques developed in the solution of population balance equations can be used for the solution of the Boltzmann equation, and vice versa.

Recently, several variations of the method of moments (QMOM, DQMOM, FCMOM) were developed and were capable to provide numerically efficient and accurate solutions of population balance equations. It is interesting to investigate whether and to which extent such techniques can be applied for solving the Boltzmann equation.

Fox [3] explored the possibility of applying quadrature-based closures in the solution of the kinetic equation for dilute gas-particle flows, thus enlarging the spectrum of applications of some of the core ideas of the QMOM/DQMOM techniques.

In this work, the FCMOM technique is extended for solving the kinetic equation representing the fluid dynamics of the particulate phase. The FCMOM (Finite size domain Complete set of trial functions Method Of Moments) was developed by Strumendo and Arastoopour ([4], [5]) to solve mono-variate and bi-variate population balance equations. The fundamental idea is to construct a method of moments on a finite domain of the internal variables (typically the particle size in mono-variate population balance equations, or the particle velocities in the kinetic equation). An advantage of the FCMOM is that it provides both the moments and the reconstructed particle distribution evolution. Further, the domain of the internal variables is always well defined (a property which is relevant in multi-variate applications).

The proposed method is illustrated through applications to the Riemann problem and to the relaxation of Maxwellian molecules.

[1] Grad, H., "On the Kinetic Theory of Rarified Gases", Communications on Pure and Applied Mathematics, 2, 331-407, (1949).

[2] Strumendo, M., Canu P., "Method of Moments for the Dilute Granular flow of Inelastic Spheres", Physical Review E 66, 041304/1-041304/20, (2002).

[3] Fox, R.O., "A quadrature-based third-order moment method for dilute gas-particle flows", Journal of Computational Physics, Volume 227, Issue 12, 6313-6350, (2008).

[4] Strumendo, M., Arastoopour, H., "Solution of PBE by MOM in Finite Size Domains", Chemical Engineering Science", doi: 10.1016/j.ces.2008.02.010, (2008).

[5] Strumendo, M., Arastoopour, H., "Solution of Bivariate Population Balance Equations Using the FCMOM", submitted to Industrial and Engineering Chemistry.

468g Direct Numerical Simulation of Nanoparticle Evolution In Turbulent Flames

Yonduck "James" Sung, Department of Mechanical Engineering, University of Texas at Austin, 210 E. 24th street, Bldg WRW, 305, Austin, TX 78712, Venkatramanan Raman, Department of Aerospace Engineering & Engineering Mechanics, University of Texas at Austin, 210 E. 24th street, Bldg WRW, 305, Austin, TX 78712, and R. O. Fox, Department of Chemical & Biological Engineering, Iowa State University, 2114 Sweeney Hall, Ames, IA 50011-2230.

Gas-phase combustion based particulate synthesis provides an attractive route for large-scale production of practically relevant nanoparticles. Here, precursors are introduced in a pre-existing flame leading to the formation of metal-oxides in the high-temperature environment. However, the size and number distribution of the particles are critically dependent on the turbulent combustion process. If this multiscale interaction can be understood and modeled accurately, viable fast-throughput processes could be designed. As a first step towards this goal, a novel direct numerical simulation approach is developed here. The gas-phase turbulent combustion is fully resolved by the computational grid. The nanoparticle population evolution is tracked using the direct quadrature method of moments (DQMOM) approach. A planar reacting jet configuration is used to study the effect of gas-phase combustion on the particle evolution process. The rate of particle growth, sintering, and agglomeration are very sensitive to the gas-phase temperature and thermochemical composition. In particular, the history of gas-phase conditions encountered by the population ultimately determines the outlet size distribution. In this context, the DNS database is used to identify the important modeling hurdles. Further, specific numerical issues relevant to solving the population balance equation in this framework are discussed.

484a Determining the Density and Moisture Content of Powders Using Microwave Dielectric Spectroscopy

Talesha T. Hall and Michael T. Harris. Chemical Engineering, Purdue University, 480 Stadium Mall Drive, FRNY 3007, West Lafayette, IN 47907

The pharmaceutical industry provides a significant opportunity to derive value by focusing on improvements in development and manufacturing. One improvement that could transform the way in which drugs are manufactured is the development of continuous manufacturing processes, as opposed to the typical batch processes currently used in the industry. The emergence of continuous processes requires the development of new sensors that can be easily integrated into equipment currently being used in industry. These sensors need to have the ability to monitor and control bulk flow properties. Fundamental limitations of most existing sensors reside in the lack of transferability of calibration across instruments and the need for individual calibration for each type of material. Some of the reasons are related to the fact that these calibrations are often established for instrument specific parameters. They are based on statistical analysis and lack any physical foundations. The transformative impact lies in understanding material properties at a fundamental level and using that information to develop new sensors and/or configurations of existing sensors. In this study, a dielectric probe in conjunction with a microwave network analyzer was used to establish fundamental relationships between density, moisture content, and the dielectric properties of common pharmaceutical excipients. This microwave probe method allows the monitoring of both density and moisture content using one device. This is an important asset because both of these physical properties, density and moisture content, were found to be important in controlling the drug dosage during the manufacturing of tablets.

CORRELATING RETENTION AND DRAINAGE IN PAPERMAKING WITH FLOCCULATION: EFFECT OF POLYELECTROLYTE BRANCHING

*E. Antunes, F. A. P. Garcia, P. Ferreira and M.G. Rasteiro**

Research Centre for Chemical Processes Engineering and Forest Products, Chemical Engineering Department, Coimbra University, Pólo II, Pinhal de Marrocos, 3030-290 Coimbra, Portugal

* Corresponding author: Tel: +351239798700; Fax: +351239798703.
E-mail: mgr@eq.uc.pt

Abstract

The effect of the degree of polymer branching on retention and drainage performance of flocculated kraft pulp fiber suspensions containing precipitated calcium carbonate (PCC) was investigated and has been correlated with flocs properties obtained by light diffraction scattering (LDS). The results show that polymers of medium charge density are more adequate to be used as retention aids since lower drainage time and higher filler retention are obtained at low flocculant contact time and low flocculant dosage. The results also demonstrate that it is possible to correlate the flocculation process evaluated by LDS with the flocculant's performance in the drainage test.

Introduction

In papermaking, chemical flocculation is fundamental for achieving high retention and high drainage rate simultaneously (1, 2, 3). The flocculation evaluation is of great importance to control the wet-end stage in paper production since retention, drainage and sheet formation depend on the flocculation mechanisms involved and on flocs characteristics. Additionally, flocculation mechanisms depend on several factors namely on flocculants characteristics and dosage and on contact time, among others (1). For example, flocculant overdosage, which results in large flocs, can reduce drainage rates since it can be more difficult to remove the interstitial water from large flocs (4).

The polymer structure is another parameter that affects the flocculation process. In fact, some studies have shown that the branched polymers have a significant potential as papermaking retention aids on the fast paper machine. Firstly, Nicke and co-workers (5) demonstrated that a branched copolymer, is an attractive flocculant for the paper industry. In the same way, Shin and co-workers (6, 7) have studied the potential of the highly branched polymer as a retention aid for microparticulate systems by performing flocculation, retention and formation tests. They concluded that the branched polymer exhibits better retention efficiency than the linear polyelectrolytes without affecting sheet formation. More recently, Brouillette et al. (8, 9) have also studied the performance, on retention, drainage and sheet formation but in a high turbulent environment, of a branched C-PAM of high molecular weight in a microparticle retention system. They confirmed that this retention aid system improves filler retention comparing with the conventional ones and that this improvement is particularly significant as the turbulence level increases.

We believe it is of great interest to study these new polymers in single component systems to better understand the mechanisms involved since, so far, the few fundamental investigations conducted studied retention and drainage performance of these polymers in microparticulate systems.

In this present study, the effects of the degree of polymer branching on retention and drainage performance were evaluated. Additionally, the effects of flocculant concentration, flocculant charge density and flocculant contact time were investigated. The results were correlated with flocculation kinetics and flocs structure. Several new cationic polyacrylamides (C-PAM) of high molecular weight were used to flocculate the suspension. These C-PAMs differ on their charge density and number of branches. The dynamic drainage analyser (DDA) was used to evaluate retention and drainage of a flocculated kraft pulp fiber suspension containing precipitated calcium carbonate (PCC). Flocculation kinetics were assessed by using a light diffraction scattering (LDS) technique which gives simultaneously information about the flocs size distribution and the flocs structure. The flocs structure was quantified by both the mass fractal dimension and the scattering exponent parameters (10).

Methods and materials

A short fiber bleached kraft pulp was used in this study. The fiber suspension was prepared at a consistency of 1% in distilled water. The PCC suspension was prepared at a concentration of 1% in distilled water (11). The median size of the PCC particles evaluated by LDS was 0.5μm.

The main characteristics of the six new C-PAM emulsions of very high molecular weight, developed and supplied by AQUA+TECH used in this study are summarized in Table 1. Flocculant solutions were prepared daily with distilled water at 0.1% (w/w).

Table 1. Flocculant characteristics.

Alpine Floc™	Molecular Weight	Number of Branches	Charge density
E1	1.2×10^7	linear	50%
E1+	1.3×10^7	1	
E1++++	1.2×10^7	4	
G1	4.6×10^6	linear	20%
G1+	4.7×10^6	1	
G1++++	4.4×10^6	4	

Drainage tests were carried out in the DDA (AB Akribi Kemikonsulter). The DDA measures drainage performance but can simultaneously give information about retention and wet sheet permeability for the same sample.

The kraft pulp fiber suspension was mixed with the PCC suspension in the equipment during two minutes before the addition of the flocculant. In the DDA, the vacuum was maintained at 30 kPa and the stirring speed in the vessel was 800 rpm. For each experiment, the flocculant contact time varied from 30 to 90 seconds and a drainage test without flocculant (blank) was performed daily.

The wet sheets obtained from the drainage tests were dried at 105°C to calculate the total solid retention. Afterwards, the samples were burned at 600°C during 16 hours to determine the PCC retention (12).

The Malvern Mastersizer 2000 (Malvern Instruments) was used to monitor the flocculation of the PCC particles. The flocculation kinetics and the flocs structure were studied for various flocculant concentrations as described in previous papers (11, 13). Figure 1 shows an example of the flocculation kinetics obtained by LDS for E1 and E++++ and for the corresponding optimum dosage (13). For the optimum flocculant dosage, the flocculation rate is faster and larger flocs are reached. The results obtained by LDS (Table 2) will be used in this paper to correlate the drainage time and the retention performance with the flocculation kinetics and the flocs structure.

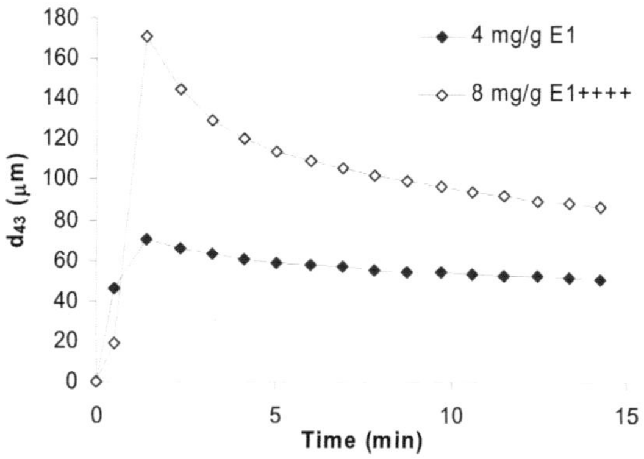

Figure 1. Flocculation kinetics for the optimum flocculant dosage of E1 and E1++++ (13).

Table 2. Main flocculation results obtained by LDS (13).

			Max. kinetics curve	
Alpine Floc™	Optimum dosage	d_{43} (µm)	d_F	SE
E1	4	46	1.33	2.36
E1+	12	31	1.13	1.37
E1++++	8	19	1.46	1.61
G1	10	373	1.52	1.50
G1+	30	436	1.67	1.51
G1++++	30	345	1.63	1.34

Results and Discussion

Drainage and retention results

Drainage tests were performed for the optimum flocculant dosage found by LDS (Table 2) and for a common flocculant concentration, 6 mg/g. Moreover, for G1+ and G1++++, drainage tests were performed for 20 mg/g and for E1 and E1+, flocculant concentrations of 2 mg/g and 16 mg/g respectively were tested. In order to compare drainage results, the drainage times and the PCC retention are normalized relatively to the drainage time and the PCC retention of the blank test respectively. The normalized drainage times for 30 and 90 seconds of contact time are represented in Figure 2 as a function of flocculant concentration for all the flocculants tested. The average drainage time for blank experiments is 5.1 seconds (± 0.5s). In Figure 3, the normalized PCC retention is plotted for the six polymers studied. Here, the total solid retention is not presented since the change in total retention is mainly caused by filler retention (3). The average total solid retention and the average PCC retention of the unflocculated suspension are 84.3% (± 0.5%) and 11.5% (± 1%) respectively.

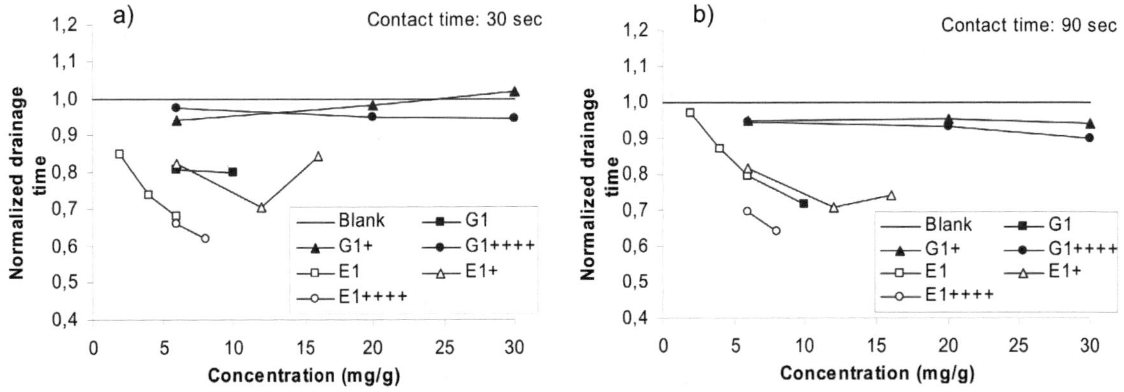

Figure 2. Normalized drainage time as a function of flocculant concentration for a) 30s and b) 90s of contact time.

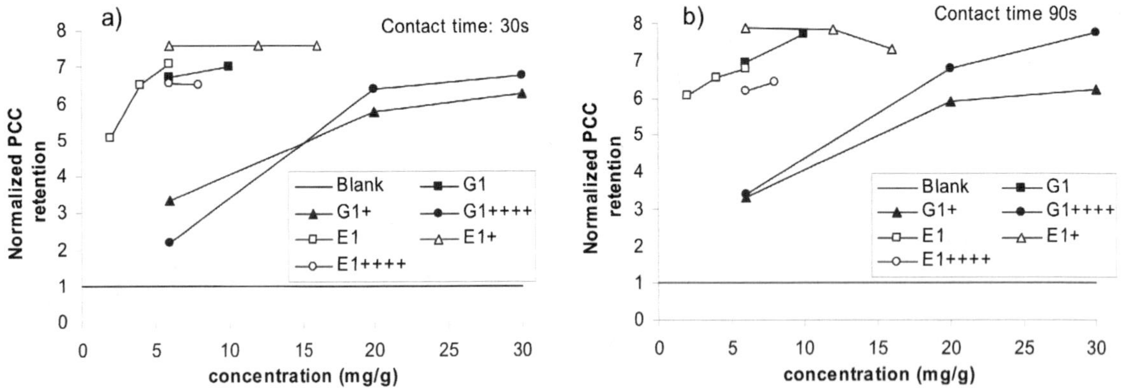

Figure 3. Normalized PCC retention as a function of flocculant concentration for a) 30s and b) 90s of contact time.

The addition of G1+ and G1++++ does not improve the drainage time relatively to the blank. For the high flocculant concentrations for which G1+ and G1++++ reach the optimum flocculant dosage, the amount of flocculant is too high leading to an increase of the suspending medium viscosity. Thus, the increase of the medium viscosity which makes water flow more difficult can explain the increase of the drainage time for these two flocculants. At the lowest flocculant concentration, G1+ and G1++++ not only impair drainage but also present the worst results for PCC retention. A poor flocculation results in a low drainage rate and in a low PCC retention because the flocculated suspension behavior is close to the one observed for the unflocculated suspension. At higher dosages, the PCC retention is similar to the ones observed for the other polymers.

For the other polymers, all the flocculated suspensions exhibit a better drainage time than the unflocculated suspension. As the flocculant concentration is close to the optimum dosage, a lower drainage time is obtained. In addition, PCC retention is, in general, maximum for the optimum dosage. Hence, despite the flocculation results being related only with the flocculation of the PCC and the operating conditions being different in the DDA and in the LDS, it is possible to see a good agreement between flocculation tests and drainage tests. This can be explained by the fact that in a composite furnish containing refined fibers, fines and filler, polymer preferentially flocculates the filler (2). Thus LDS and DDA tests can be complementary techniques to pre-screen flocculants performance in papermaking.

Results obtained for E1+ indicate that if the flocculant concentration increases too much, the drainage time increases again. When the flocculant is in excess, flocculation progresses at a lower rate (13) and, thus, for the flocculant contact time used in these studies the flocs are still too small (poor flocculation) and the sheet structure is relatively closer to the blank and so is the drainage time. In general, as flocculant concentration increases further retention tends to reach a plateau. Hence, it is possible to find a flocculant dosage range where a low drainage time and a high PCC retention can be achieved simultaneously. In this case, this range is 5-10 mg/g for all the polymers except the G1+ and G1++++ for the reasons explained previously.

For the E1 series the increase in the contact time results in an increase in the drainage time and PCC retention while for the G1 series the increase in the contact time results in a decrease in the drainage time. However, the highest drainage time variations with contact time are observed for the linear polymers E1 and G1. The trend of the drainage time with flocculant contact time observed for the E1 series agrees with the work of Forsberg and Ström (14). However, when the polymer E1+ is in excess (16 mg/g), the drainage time decreases as the contact time increases: for such dosage the flocculation degree is higher at 90s than at 30s resulting in the improvement of the drainage time. Nevertheless, the G1 series does not follow this behavior.

These trends can be explained by the LDS results. These results have shown that flocs produced with the G1 series are much larger than those produced with E1 series due to the lower charge density, therefore resulting in overflocculation and producing too large flocs that reduce the drainage performance. In this case, the decrease of the flocs size with flocculation time due to polymer reconformation and degradation (15) reduces the effect of the overflocculation and thus results in drainage time decrease with flocculant contact time increase. Moreover, the same effect of the flocculant contact time was observed with the PCC retention (Figure 3). Thus, for the E1s series, better drainage times and PCC retention are obtained for lower flocculant contact times while for the G1s series, the opposite happens.

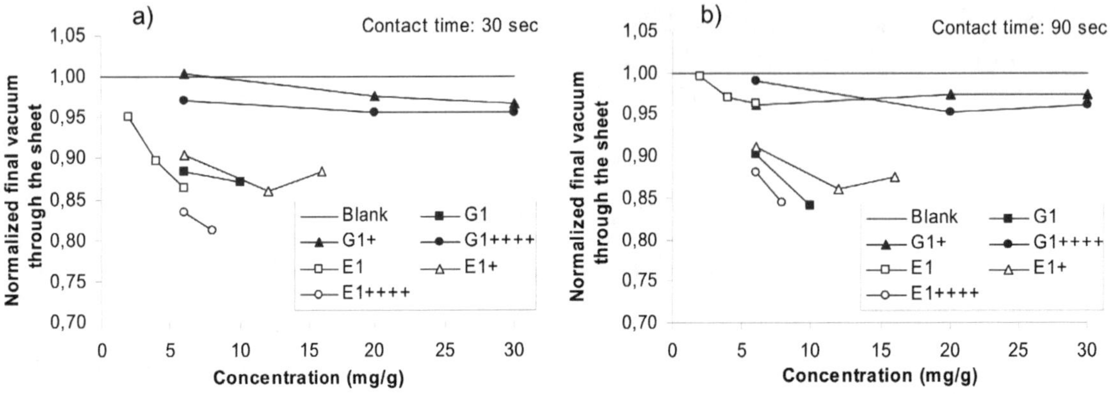

Figure 4. Normalized final vacuum as a function of flocculant concentration for a) 30s and b) 90s of contact time.

Figure 4 summarizes the final vacuum normalized relatively to the final vacuum of the blank test, through the sheet as a function of the flocculant concentration. A low final vacuum corresponds to high sheet permeability, i.e., to high sheet porosity. The final vacuum average through the sheet for the unflocculated suspension is 16.8 kPa (±0.6 kPa). The same trend observed for the drainage time is verified for the sheet permeability when the flocculant dosage varies. In fact, lower drainage times correspond to higher sheet permeability, i.e., higher sheet porosity. Figure 5 confirms the linear correlation between the drainage time and the sheet porosity.

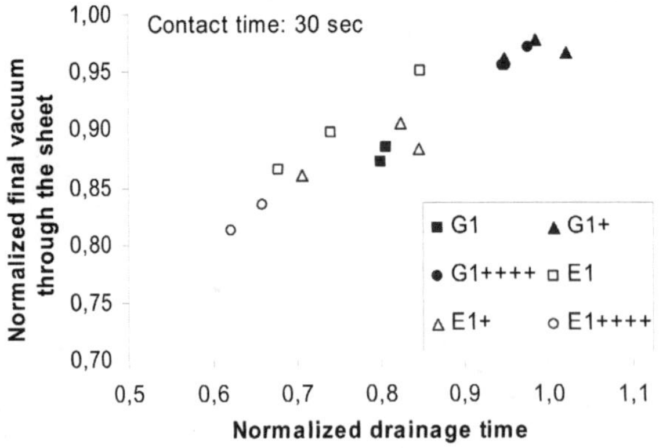

Figure 5. Normalized final vacuum as a function of normalized drainage time for 30s of contact time.

Correlation with flocs properties

The effect of flocs size and flocs structure on the drainage time is investigated for the optimum flocculant dosage obtained by LDS.

In Figure 6a, the normalized drainage time for the optimum flocculant dosage is represented as a function of the mean flocs size. For the E1 series, the results correspond to a flocculation time of 30 seconds while for the G1 series, results refer to a flocculation time of 90 seconds for which both drainage and retention give best results. The drainage time decreases with the decrease in the flocs size and the E1 series produce the smallest flocs. Additionally, E1++++ is the flocculant that produces the smallest flocs and gives the lowest drainage time. Thus, it is possible to have fast dewatering and high filler retention with small flocs. Larger flocs reduce drainage rate as confirmed for the G1 series. This happens because flocs that are too large and have much more interstitial water which is difficult to remove. So, overflocculation (very large flocs) results in low drainage despite retention being not affected.

In Figure 6b, the drainage time is related with the flocs structure quantified by the mass fractal dimension and by the scattering exponent for the optimum flocculant dosage. The mass fractal dimension gives indication about the structure of the primary flocs while the scattering exponent gives information about the structure of secondary flocs that result from the aggregation of the primary ones. The mass fractal dimension and the scattering exponent are calculated for the maximum in the flocculation kinetic curve observed in the previous study as shown, for instance, in Figure 1 for the E1 and E1++++ polymers (13). Primary flocs produced with E1 are open (small d_F) while the secondary flocs are compact (high SE). Both primary and secondary flocs produced with E1+ are open (small d_F and SE) when comparing with E1 flocs. The configuration of flocs produced with E1++++ seems to be the most adequate to remove easily the water from the flocs. Primary flocs are slightly more compact while secondary flocs are open comparing with E1 and E1+. This contributes to a lower drainage time as shown in Figure 6b.

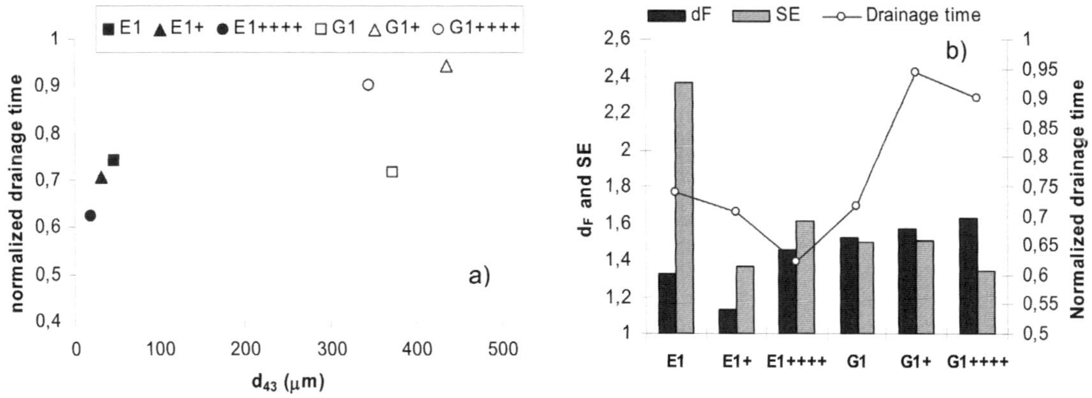

Figure 6. Normalized drainage time as function of a) mean floc size and b) flocs structure for the optimum flocculant dosage.

However, the structure of the flocs produced with the G1 polymers is similar to the structure of the E1++++ flocs but drainage time is much higher. In this case, the drainage time is mainly affected by the larger floc size (overflocculation).

Conclusions

The correlation between retention and drainage results and flocculation behavior (namely the flocs properties) is fundamental to understand how the flocculant characteristics affect the retention and drainage performance in papermaking.

The retention and drainage performance is enhanced, in general, close to the optimum flocculant dosage found by the flocculation results. A low flocculation degree also results in low drainage rate but in poor filler retention.

Flocculants of low charge density do not improve drainage times compared to the unflocculated suspension but offer very high filler retention. An increase of the flocculant contact time can slightly decrease the drainage time and increase filler retention.

On the other hand, flocculants of medium charge density offer simultaneously low drainage times and very high filler retentions at low flocculant dosage and at low flocculant contact time. When a branched structure of these flocculants is used it improves significantly the drainage rate and the filler retention comparing with the linear ones.

Moreover, it was shown that for the branched polymers of medium charge density, the improvement in the drainage time is due to the formation of small flocs sizes with an open structure, mainly at the secondary aggregate level. The increase of the drainage time for the linear polymer is due to the more compact structure of the small flocs formed.

Polymers of medium charge density are more suitable to be used as retention aid because low drainage time and very high filler retention are obtained simultaneously at low flocculant contact time and low flocculant dosage. Moreover, highly branched polymers can be considered an adequate choice because the balance between flocculant dosage, drainage time and filler retention is the best. Thus, these polymers represent a promising additive for papermaking.

Acknowledgments

The authors thank the support of FCT (Fundação para a Ciência e Tecnologia, Portugal) and to AQUA+TECH Specialties SA (La Plaine, Geneva, Switzerland) for supplying the flocculants samples.

References

1. Norell, M.; Johansson, K.; Persson, M. (1999), “Retention and Drainage. Book 4: Papermaking Chemistry”, Papermaking Science and Technology, TAPPI PRESS, Finland.

2. Whipple, W.L.; Maltesh, C. (2002), “Adsorption of cationic flocculants to paper slurries”, *J. Colloid Interf. Sci.*, 256, 33-40.

3. Cadotte, M.; Tellier, M.E.; Blanco, A.; Fuente, E.; van de Ven, T.G.M.; Paris, J. (2007), "Flocculation, retention and drainage in papermaking: a comparative study of polymeric additives", *Can. J. Chem. Eng.*, 85, 240-248.
4. Hubbe, M.A.; Heitmann, J.A. (2007), "Review of factors affecting the release of water from cellulosic fibers during paper manufacture", *Water release, papermaking. BioResources*, 2(3), 500-533.
5. Nicke, R.; Pensold, S.; Hartmann, H-J.; Tappe, M. (1992), "Polydiallyldimethylammonium chloride as a flocculating agent", *Wochenbl. Papierfabr.*, 120(14), 559-564.
6. Shin., J-H.; Han, S.H.; Sohn, C.; Ow, S.K.; Mah, S. (1997), "Highly branched cationic polyelectrolytes: filler flocculation", *Tappi J.*, 80(11), 179-185.
7. Shin., J-H.; Han, S.H.; Sohn, C.; Ow, S.K.; Mah, S. (1997), " Highly branched cationic polyelectrolytes: fines retention", *Tappi J.*, 80(10), 185-189.
8. Brouillette, F.; Morneau, D.; Chabot, B.; Daneault, C. (2005), "A new microparticulate system to improve retention/drainage in fine paper manufacturing", *Appita J.*, 58(1), 47-51.
9. Brouillette, F.; Morneau, D.; Chabot, B.; Daneault, C. (2004), "Paper formation improvement through the use of new structured polymers and microparticle technology", *Pulp Paper Can.*, 105(5), 108-112.
10. Biggs, S.; Habgood, M.; Jameson, G.J.; Yao-de-Yan (2000), "Aggregate structures formed via a bridging flocculation mechanism", *Chem. Eng. J.*, 80, 13-22.
11. Rasteiro, M.G.; Garcia, F.A.P.; Ferreira, P.; Blanco, A.; Negro, C.; Antunes, E. (2008), "The use of LDS as a tool to evaluate flocculation mechanisms", *Chem. Eng. Proc.*, 47, 1329-1338.
12. Ferreira, P.; Velho, J.; Figueiredo, M.; Mendes, A. (2005), "Effect of thermal treatment on the structure of PCC particles", *Tappi J.*, 4(11), 18-22.
13. Antunes, E.; Garcia, F.A.P.; Ferreira, P.; Blanco, A.; Negro, C.; Rasteiro, M.G. (2008), "Effect of water cationic content on flocculation, flocs resistance and reflocculation capacity of PCC induced by polyelectrolytes", *Ind. Eng. Chem. Res.*, 47, 6006-6013.
14. Forsberg, S.; Ström, G. (1994), "The effect of contact time between cationic polymers and furnish on retention and drainage", *J. Pulp Paper Sci.*, 20(3), 71-76.
15. Rasteiro, M.G.; Antunes, E.; Garcia, F.A.P; Ferreira, P.; Wandrey, C.; Hunkeler, D. (2008), "Flocculation by polyelectrolytes: correlating the flocculation process with polymer characteristics", Polyelectrolytes 2008, Coimbra, Portugal.

Experimental investigation into shielding effect of magnetic powders in drilling fluids

S. Ding[1], B. K. Datta[2], A. Saasen[3] and P. A. Amundsen[3]

1. POSTEC dept. av Tel-tek R &D Centre, N-3914 Porsgrunn, Norway
2. System Engineering, Aker Solutions, Oslo, Norway
3. University of Stavanger, N-4036 Stavanger, Norway

Introduction

The drilling fluid is one of the factors that may contribute significantly to errors in directional surveying of wellbores, because the magnetic properties of the fluid shields the Earth's magnetic field measured by the magnetic sensors used in measurement while drilling (MWD) directional tools (Wilson and Brooks, 2001; Amundsen et al. 2006). The magnetic properties of a drilling fluid is known to vary significantly, depending on factors, such as mechanical wear of down-hole and surface equipment, the mineralogical composition and the bulk material transferring procedures of the drilling fluid additives, and the circulation properties of the drilling fluid (Torkildsen et al., 2004; Amundsen et al. 2008). In this contribution we shall show that it can also depend significantly on operational aspects of the drilling and data logging processes.

To understand this shielding effect and its dynamics we have performed laboratory measurements where powders of known magnetic of properties have been added to a well defined non-magnetic model drilling fluid. The time dependent shielding of the Earth's magnetic field was then monitored using a fluxgate magnetometer immersed into the fluid.

Here we present results for shielding caused by the addition of magnetite and soft iron powders. These materials were chosen because they are certain (iron) or likely (magnetite) to be present in a real drilling fluid, and have quite different magnetic properties: Iron is a soft ferromagnetic; magnetite is strongly paramagnetic. As a base fluid which will keep the powders in suspension we have chosen a common xanthan gum solution, widely used as a base fluid for water-based drilling fluids in the petroleum industry, and which is very weakly diamagnetic.

Experimental set up.

To create a situation mimicking the condition of the fluid as in drilling field, an experimental set-up as sketched in Figure 1 was built. The drilling fluid was contained in a parallelepipedal tank, made of transparent Plexiglas plate of a thickness of 12 mm, with internal size of 1200mm x 750mm x 200mm. The magnetic field was measured with a ***Mag*-01**(H) fluxgate magnetometer and probe **Mag B** (see Figure 2), connected to a PX1-1042 data acquisition unit and analysed using a NI-Labview platform. In the present paper we only report results with the probe placed vertically at the geometric centre of the fluid volume. We have used the vertical component of the Earth's field as the external magnetic field.

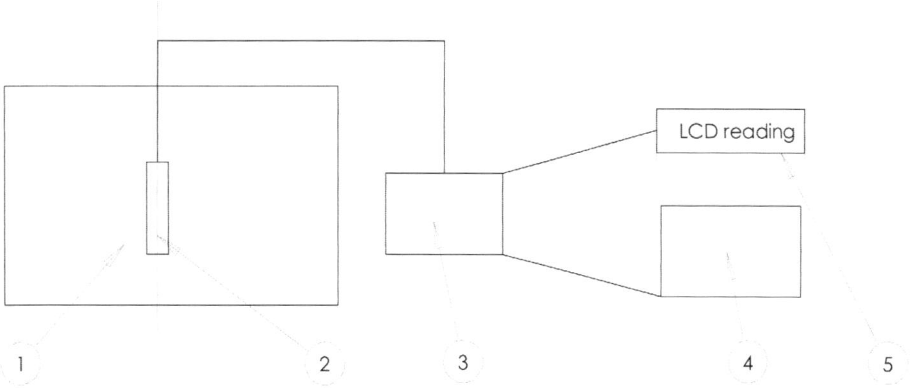

1, Drilling Fluid, 2. Probe, 3. Mag-01(H), 4. Ni-logger+PC, 5. LCD reding.

Figure 1 Schematic of measuring system

The drilling fluid was first prepared as solution of water with 0.3% (by weight) xanthan gum powder, whereafter magnetite or iron powders was added. The mixture was thoroughly stirred. For further details on the experimental set-up and preparation of the fluid, see Amundsen et al. (2008).

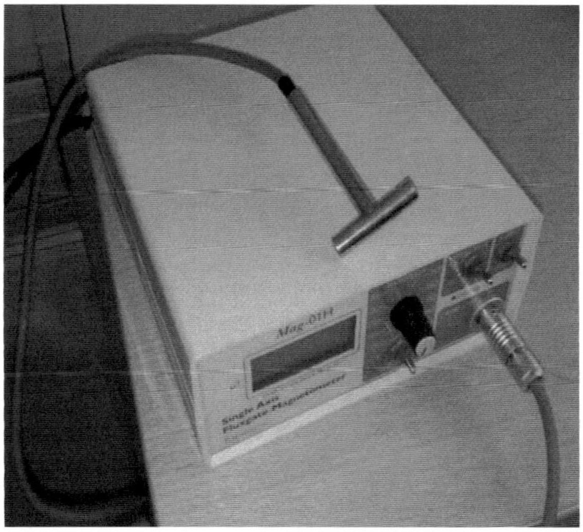

Figure 2 Fluxgate magnetometer Mag-01 (H) and the single-axis probe

For comparison purposes we also recorded measurements of the shielding effect of dry powders by just burying the magnetometer probe in the powders and recording the vertical components of the Earth's magnetic field. The results are given in Table 1, along with the Earth's magnetic strength as measured without shielding.

One can see from Table 1 that the magnetic fields were reduced about to one third of the Earth's field in the media of either powder. The shielding was measured not to vary with time, indicating that the magnetic properties of the powders were stable, as expected.

Table 1 Shielding effects of magnetic powders

	Magnetic field (μT)		
Powders used	No powder	Iron	Magnetite
Measurement 1	46.63	15.65	22.18
Measurement 2	45.76	14.26	18.79

Measurement results

Once the powders were mixed into the xanthan gum fluid, the probe was immediately immersed into the drilling fluid and positioning at the geometric centre of the fluid volume. The vertical component of the Earth's field in the magnetite drilling fluids was measured. There was no significant difference in the measured vertical magnetic field strengths between the empty (air-filled) tank and one filled with the pure xanthan gum solution, 45.3μT.

The measured magnetic field did hardly change when adding iron powder to the fluid. A typical result is shown in Figure 3 for a concentration of iron powder in the fluid of 2.67%. It is seen that the field strength remains at the same value as for an empty tank. This absence of shielding is quite unexpected, considering the strong shielding caused by the dry powder (see table 1), although the total mass of iron surrounding the probe is of course much higher in the latter case.

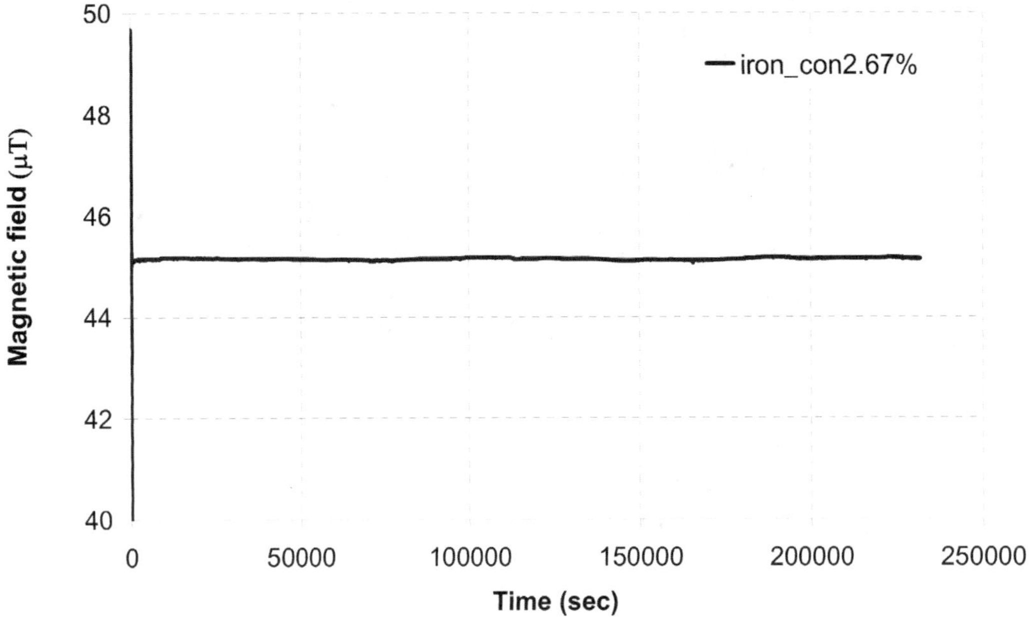

Figure 3 Magnetic field as a function of time in the fluid mixed with iron powders

In contrast to the case of iron powder, for the fluid mixed with the magnetite powder we observed a strong, and strongly time dependent, shielding effect. In Figure 4 is shown the evolution of the measured magnetic field within the fluid as a function of time over a 3 day period for different concentrations of magnetite powders, which were set to be 0.67% 1.33 % 2.67% 5.33% in different rounds of measurements.

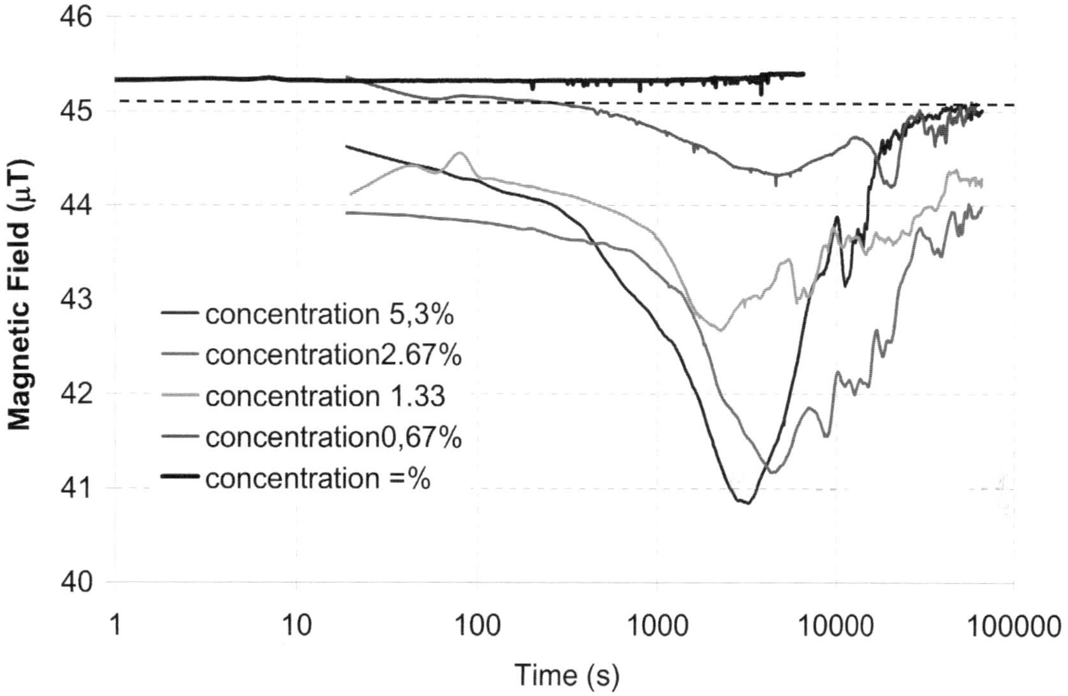

Figure 4. Magnetic field as a function of time for different concentrations of magnetite

The main feature of the results for magnetite is a rich and unexpected dynamical structure of the magnetic shielding as shown in Figure 4. There is an initial strong and rapid decay of the field for an hour, or so, until it reaches a minimum. This maximal shielding is significantly larger than expected based on the magnetic susceptibility of the dry powder. Thereafter the measured field gradually recovers, but with large superimposed slow variations. After some 3 days the field recovered to about the original level as its origin.

For the same fluid composition, the strength of the magnetic field up to the time of the minimum was quite similar from run to run; also the value of the minimum of the field strength was reproducible within 10 %, although the time for reaching it can differ by, say about 1000 s.

It might be expected that there should be an instantaneous initial shielding of the external field when the magnetite was added, but this was not seen, just as there was no shielding at all in the case of iron powder. The variations of the initial values seen in Figure 4 may well be attributable to the operation of the measurements: the probe might not have been positioned at the exact same orientation from one measurement to another.

In Figure 5 we show the maximum shielding effect, which is the relative decrease of the magnetic field strength from its initial value to the minimum value. It is seen that it depends strongly, and nearly linearly on the magnetite concentration.

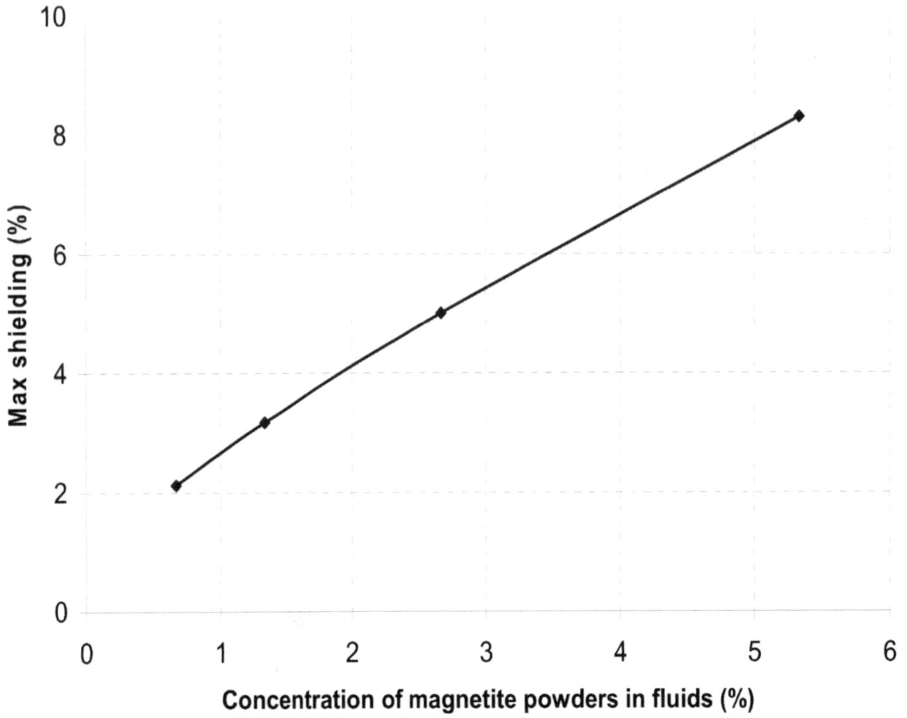

Figure 5 The maximum shielding effects

Discussion

Amundsen et al. (2008) explained the observed behaviour of the shielding effect for magnetite by invoking the fact that the magnetic polarity of individual grain will tend to orient itself to a state in the opposite direction of the external field, and hence cancelling its strength. This reorientation of the individual grain will be hampered by the viscous forces in the fluid. Since magnetic forces in the system are generally weak, the reorientation requires time. Obviously, this process of reorientation depends on the balance between magnetic moments and the hampering viscous forces exerted on the individual grain, which will be determined by the grain's magnet property, size, shape, density, the viscosity of the drilling fluid and the strength of the external magnet field.

The subsequent rise of the measured field can be explained by the fact that the magnetite dispersion is not completely stable with respect to sedimentation. Indeed, it was observed that a substantial fraction of the magnetite powders had settled out onto the bottom of the tank after 3 days, no longer contributed to the shielding of magnetometer probe.

The very different behaviour of iron and magnetite is not readily explained by the above considerations, and the very different screening effect of dry iron powder versus iron powder in solution is indeed remarkable. One major difference between the two powders is that the iron powder is ferromagnetic, so that the induced magnetic moments are much larger than for magnetite, which is paramagnetic. This means that the relaxation times for the dipoles to respond to the external field is much shorter for iron, which explains that we do not see time dependent effects on the time scale of our measurements. It does not explain that we do not see any shielding effects for iron powder in solution at all, however.

Conclusion

Experiment has been carried out with a fluxgate magnetometer to measure magnet shielding effect of the magnetite or iron powders. It was confirmed that both have a very strong shielding effect when the probe was submerged in dry powders. However, the iron powders showed no significant shielding effect when suspended in a drilling fluid, whereas magnetite powder shows a considerable shielding effect with a complex dynamical behaviour.

These results are of importance to understand the shielding caused by magnetic contaminations in drilling fluids used in petroleum drilling, as this can cause errors when performing magnetic surveying of wellbores. The strong time dependence of the shielding effects explains problems encountered in making repeatable measurements of the magnetic susceptibility of drilling fluids using simple laboratory instruments. Even slight variations in the drilling fluid properties and measurement procedures can cause significantly different magnetic responses.

References

Amundsen P. A., Ding S., Datta B. K., Torkildsen T. and A. Saasen, 2008, “Magnetic Shielding during MWD Azimuth Measurements and Wellbore Positioning”, paper SPE 113206, Indian Oil and Gas Technical Conference and Exhibition, Mumbai, India, 4–6 March.

Amundsen, P. A., Torkildsen, T., and Saasen, A, “Shielding of Directional Magnetic Sensor Readings in a Measurement While Drilling Tool for Oil Well Positioning”, Journal of Energy Resources Technology, vol. 128, no. 4, pp. 343-345, 2006.

Torkildsen T., Edvardsen I., Fjogstad A., Saasen A., Amundsen P. A. and Omland T. H., 2004, “Drilling Fluid affects MWD Magnetic Azimuth and Wellbore Position”, paper SPE/IADC 87169 presented at the IADC/SPE Drilling Conference held in Dallas, Texas, 2–4 March.

Wilson H. and Brooks A. G., 2001, “Wellbore Position Errors Caused by Drilling Fluid Contamination”, paper SPE 71400 presented at the SPE Annual Technical Conference and Exhibition, New Orleans, LA, Sept. 30–Oct. 3.

484d Solids Drying Process Simulation by Solidsim --- a New Tool for Flow Sheeting of Solids Processes

Ting(Tim) Han and Joerg Theuerkauf. Solids Processing, Engineering & Process Science, The Dow Chemical Company, Freeport, TX 77566

Solidsim is a new Microsoft WindowsTM based flow sheet simulation software for complex solids processes which can be used to predict the performance of a processing plant and to simulate plant extensions, modifications or designs.

In this paper, Solidsim was used to simulate the processes of two types of convective dryers: the co-current flow dryer and the cross-current flow dryer, respectively. The material and energy balances for each process was calculated by Solidsim. The results were evaluated. In addition, by changing input and output data of the dryers, Solidsim has been further investigated on predicting and optimizing the performance of the convective dryers.

Granulation Time in Fluidized Bed Granulators

Papiya Roy[1], Rajesh Khanna[1], Duvvuri Subbarao[2]*
Department of Chemical Engineering
[1] Indian Institute of Technology Delhi, New Delhi, 110016, India
[2] Universiti Teknologi PETRONAS, 31750,Tronoh, Perak, Malaysia

Abstract

Size enlargement of fine powders by wet granulation involves mixing of fine particles with a binder liquid to form larger wet granules and drying them by evaporation of the binder liquid to form larger dry granules. Identification of the time for completion of granulation process is critical as further mixing of dry granules is providing extra energy for their attrition. In a batch fluidized bed granulator, monitoring the bed pressure drop and bed temperature with time can provide information on time for completion of the granulation process. Experimental observations on granulation time and granulation size for wet granulation of urea powder in a lab-scale batch fluidized bed granulator are presented.

1. Introduction

Granulation is a process of forming coarser particles by combining finer particles. Its purpose is to improve the flow properties of particles, to modify the bulk densities and to control the dissolution rates *etc.* It is widely used in pharmaceutical, food, agrochemical, dyestuffs industries. In wet granulation process, feed particles and a liquid containing solute (binder) are well mixed to form wet granules; the liquid is then evaporated to bind feed particles together to form dry granules. Equipments such as rotary drums, rotating helical ribbons, anchor agitators with mechanical agitation or fluidized beds can be used for wet particle mixing as well as drying of wet granules. Particles in a gas-solid fluidized bed are in a state of good mixing due to the gas flowing through the fluidized bed of particles in the form of bubbles. Spraying a binder solution on to particles in a fluidized state can provide uniform wetting of the particles to generate wet granules. Also fluidized beds offer higher drying rates while maintaining bed isothermality. Process variables that affect fluidized bed granulation include liquid volume fraction, operating gas temperature and gas velocity [1—5]. Liquid volume fraction has to be within an acceptable range for successful granulation [6, 7]. Excess liquid volume fraction can lead to exponential growth of size of wet granules and eventual defluidization. Lesser liquid volume fraction can lead to poor granulation. Operating gas temperature and velocity define the fluidization behavior and rate of drying of wet granules. Identification of the endpoint of drying and hence granulation process is critical. Continuing fluidization process beyond the endpoint of granulation can cause attrition and breakage of dry product granules. Techniques such as sound and vibration analysis [3,8-10] growth kinetics of the particles [11-12] have been investigated for the detection of endpoint of granulation. Present study is an attempt to explore the possibility of utilizing the bed pressure drop and bed temperature for identifying the endpoint of granulation and study the effect of operating variables on the time of granulation and size of granules of urea powder as a model system.

2. Experimental study

Figure 1 presents a schematic diagram of the experimental setup which has been used in the present study.

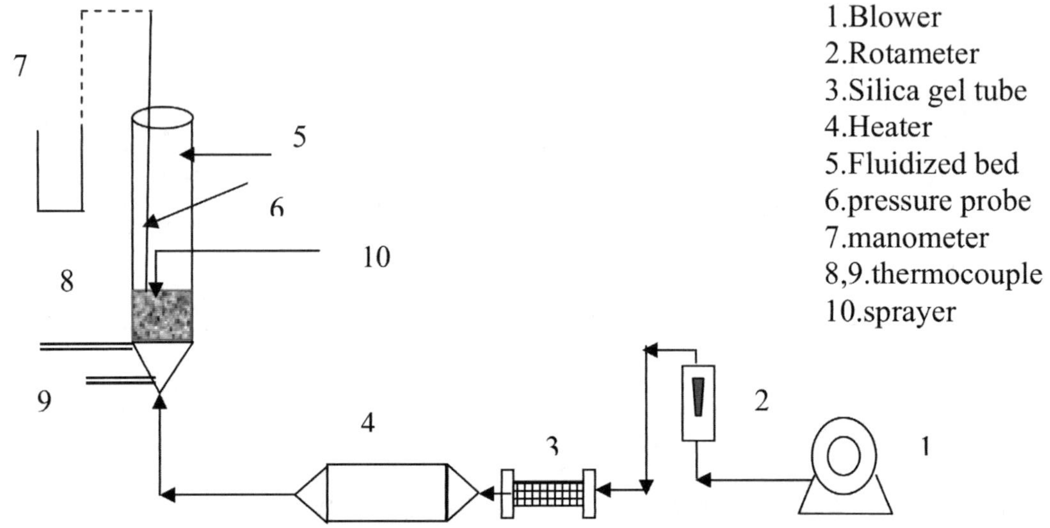

Figure 1: Schematic diagram of the experimental set up

Air flow from the blower (1) is metered (2), dried over silica gel (3) and heated to the required temperature as it flows over an electrical heater assembly (4) before it enters the fluidized bed granulator (5). The fluidized bed granulator is a 2.54 cm diameter glass column of height 15 cm fitted with a porous plate distributor. A pressure probe (6) is connected to water manometer (7) to measure bed pressure drop. Two thermocouples (8) and (9) are provided to monitor temperature of inlet gas and fluidized bed respectfully. Graded urea particles were used for fluidization and granulation experiments. A sprayer (10) is used to spray a predetermined quantity of the binder liquid (saturated solution of urea) on to the bed of particles from the top over a short period. Experiments were performed at various combinations of the following operating parameters

- Initial particle size - 327.5, 390 and 512.5 micron
- Operating velocity - 65.8, 98.8 and 131.7 cm/s
- Inlet air temperature - 30°C, 37°C and 50° C
- Amount of binder - 8.7%, 15.8% and 26.8%

(100*weight of solution /weight of bed particles)

Urea particles of a given cut size were fluidized with hot inlet air. After the bed got heated to the inlet air temperature, binder solution was sprayed onto the top of the bed. Binder solution was added in small quantities at a time for a short period. (There is an upper limit to addition of binder solution per weight of bed at which the bed can get defluidized and was experimentally observed to be around 30.5%). As liquid and the bed particles get well mixed, wet granules form and grow while getting dried. Consequently, inlet air velocity was chosen sufficiently high to keep the bed of wet granules fluidized.

3. Identification of End point of granulation

Addition of binder solution at the start of granulation process increases the bed weight and corresponding bed pressure drop. As liquid evaporates due to drying, bed pressure drop decreases and bed temperature reduces to approach wet bulb temperature. Towards completion of liquid evaporation by drying and granulation, bed pressure drop reduces to the initial value. Bed temperature reduces to a minimum and then increases it equals the inlet gas temperature. Typical bed pressure drop and temperature traces with time during the period of granulation are shown in fig. 2a and fig. 2b for the experiment performed at a gas velocity of 131.7 cm/s with initial particle size of 327.5 micron, inlet gas temperature of 50 °C, moisture content of 8.7% . From fig.2a on bed pressure drop, it can be seen granulation was complete in 20 minutes.

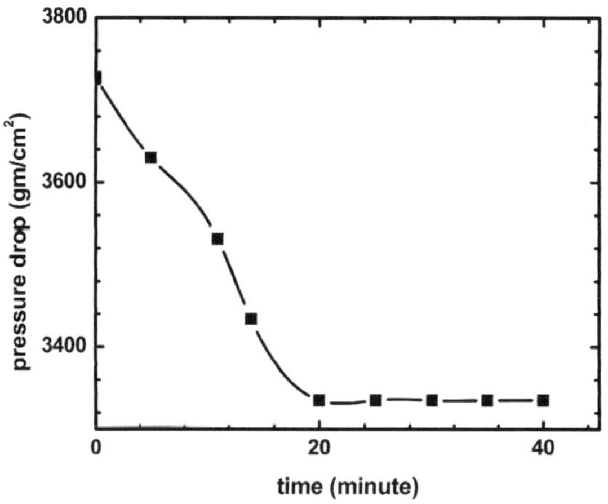

Figure 2a: Pressure drop as a function of time. The parameters are initial particle size: 327.5 micron, operating gas velocity: 131.7cm/s, moisture content: 8.7% and inlet gas temperature: 50°C.

From fig.2b on bed temperature with time, temperature reached a minimum at around 10 minutes and then increased to equal the inlet temperature by 40 minutes. At a time of 20 minutes (corresponding to the completion of granulation time recognized from bed pressure drop measurements), an inflection point can be recognized on the rising section of temperature trace. Beyond the inflection point, evaporation of moisture from the bed is negligible and bed temperature increased due to heating by incoming gas.

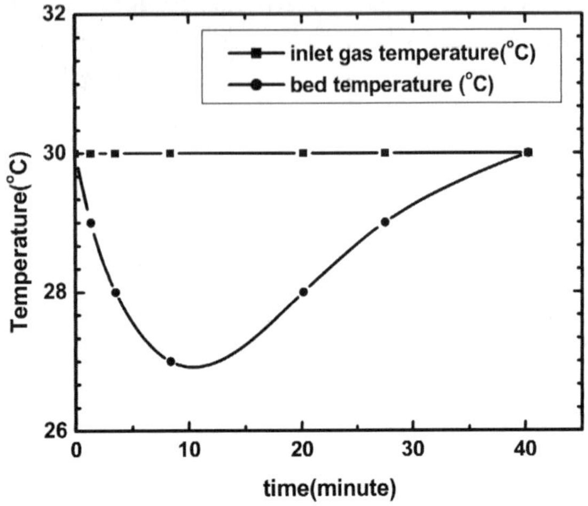

Figure 2b: Bed temperature as a function of time. The parameters are initial particle size: 327.5 micron, operating gas velocity: 131.7cm/s, moisture content: 8.7% and inlet gas temperature: 50°C.

The initial decrease in the bed temperature compared to inlet gas temperature is due to rapid evaporative cooling. As the fraction of moisture in the bed decreases, the effect of evaporative cooling diminishes and the bed temperature rises to reach the inlet gas temperature due to heating by the incoming gas. Slope of the temperature change with time increases during the period of evaporative cooling followed by a decrease during the period of heating. The transition can be identified by locating the inflection point (dT/dt is maximum or d^2T/dt^2 is zero). In the present work, end of granulation process is identified by these two techniques.

Granulation Time:

Typical experimental observations on the effect of moisture content, gas velocity and inlet gas temperature on time of granulation are presented in Fig.3 a,b,c.

Fig. 3a presents the effect of moisture content on granulation time at an operating velocity of 65.8 cm/s and inlet air temperature of 30°C with initial particle size as a parameter. Granulation times are more for higher moisture contents. It is interesting to note that granulation of smaller particles need longer granulation times. Fig. 3b shows the effect of operating gas velocity on granulation time with moisture content as a parameter and inlet gas temperature of 30°C. Higher operating gas velocity will lead to faster evaporation and lesser granulation time. Fig. 3c shows the effect of inlet gas temperature on granulation time at 8.7% moisture content and gas velocity of 65.8 cm/s with initial particle size as a parameter. At higher inlet gas temperatures, moisture evaporates at a higher rate and hence, the time required for granulation is less.

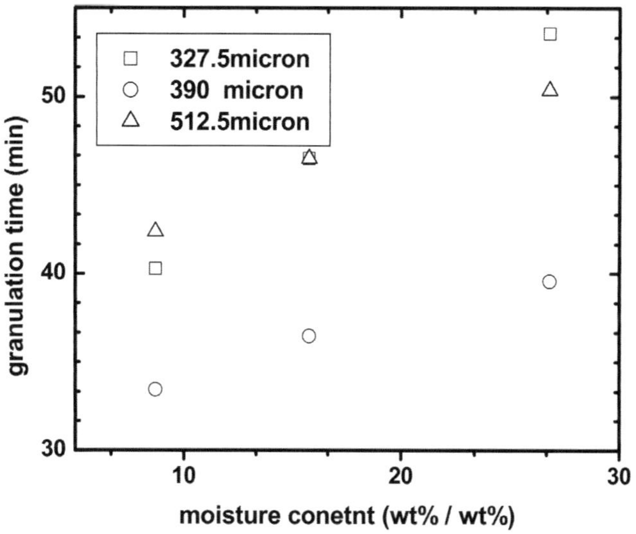

Figure 3.a: Effect of moisture content on granulation time for different initial particle sizes. The parameters are operating gas velocity: 65.8cm/s and inlet air temperature 30°C.

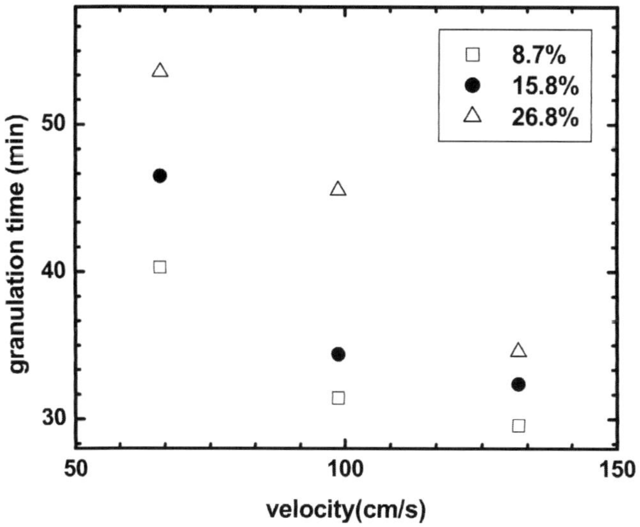

Figure 3.b: Effect of operating gas velocity on granulation time for different moisture contents. The parameters are initial particle size 327.5 micron and inlet air temperature 30°C.

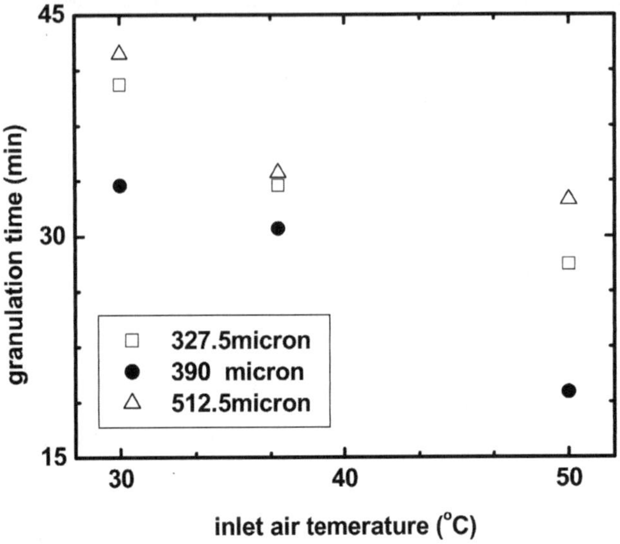

Figure 3.c: Effect of inlet gas temperature on granulation time for different initial particle sizes. The parameters are operating gas velocity 65.8cm/s and moisture content 8.7%.

Model for granulation time:

Wet granulation process involves mixing of liquid binder with particles to form wet granules and their drying to form dry granules. Drying step is much slower than mixing step and, hence time required for granulation is essentially time needed for drying. Drying rate depends on rate of heat transfer between gas and granules. Due to good granule mixing, temperature of bulk of the bed is uniform and heat transfer from gas to granules takes place over a thin layer ΔX very near the grid. Then, time for granulation can be expressed as

$$t_g \cong t_d = \frac{\textit{Heat required to evaporate Moisture Content in the bed}}{\textit{Heat Transfer from gas to bed}}$$

$$= \sum_{t_d} \left[\frac{\Delta\left(\frac{MC}{100}\right) AL(1-\varepsilon)\rho_p \lambda}{A u_o \rho_g c_{pg} \left(T_{gi} - T_b\right)} \right] \qquad \textbf{(1)}$$

Temperature difference between inlet gas temperature and the bed increases from zero initially and then decreases to zero. To estimate time for drying this equation needs to be integrated over small time intervals with appropriate temperature difference. As a first guess, it is proposed that this equation be approximated as

$$t_g \simeq \frac{\left(\frac{MC}{100}\right) AL(1-\varepsilon)\rho_p \lambda}{A u_o \rho_g c_{pg}\left(T_{gi} - \frac{T_{b\min}}{2}\right)} \quad \textbf{(2)}$$

As a first approximation, data on time of granulation is presented as a function of the ratio of moisture content to gas velocity in fig.4.

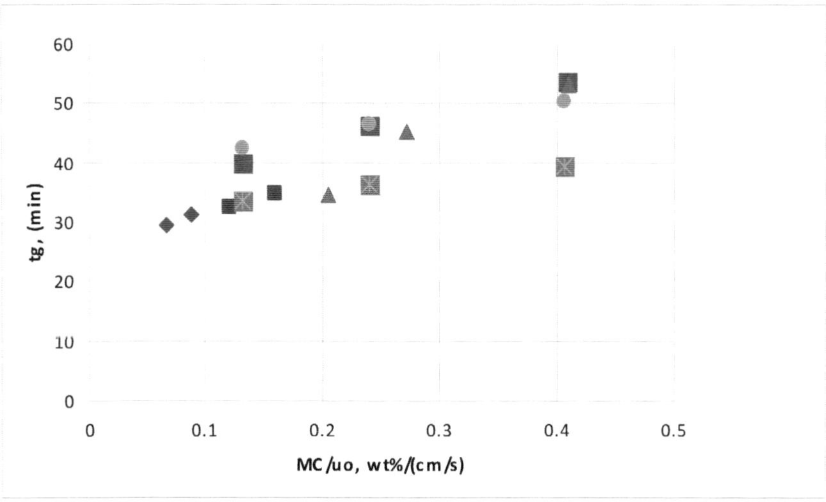

Fig.4. Correlation for time of granulation as a function of the ratio of moisture content to gas velocity.

This model needs to be improved considering the heat transfer process between the gas and the granules to explain the effect of initial particles size on granulation time.

4. Size of Granules

Figure 4 presents the variation of the ratio of the average final granule size to the average initial particle size as a function of operating parameters.

Fig. 4a presents the effect of initial particle size on the ratio of granule size to initial particle size at an operating velocity of 65.8 cm/s and inlet air temperature of 30°C with moisture content as a parameter. The ratio increases with moisture content for all the particle sizes. The ratio is lower for larger initial particle size.

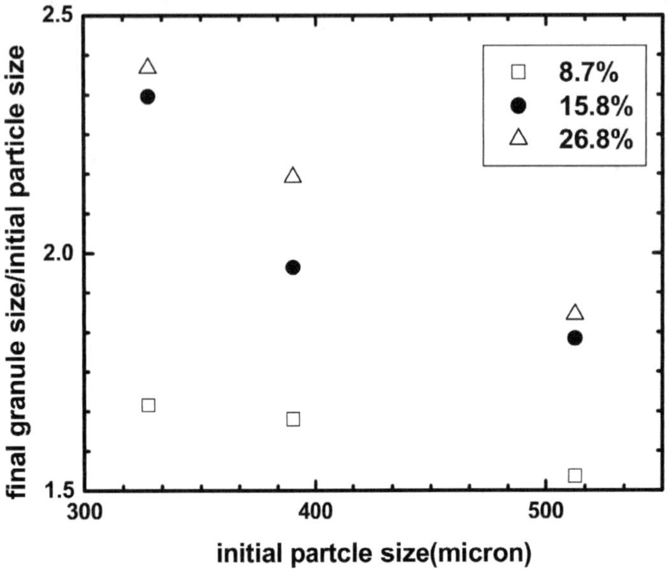

Figure 4a: Effect of initial particle size on the ratio of average final granule size and initial particle size. The parameters are operating velocity: 65.8cm/s and inlet air temperature: 30°C.

Fig. 4b shows the effect of operating gas velocity on the ratio at 8.7% moisture content and inlet gas temperature of 30°C with initial particle size as a parameter. Higher operating gas velocity resulted in lower ratio of granule size to initial particle size. The ratio is higher for smaller initial particles.

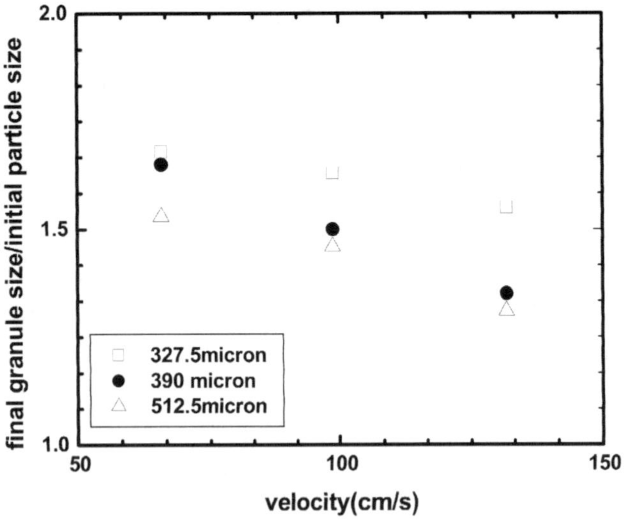

Figure 4b: Effect of operating velocity on the ratio of average final granule size and initial particle size. The parameters are inlet gas temperature: 30 °C and moisture content 8.7%.

Fig. 4c shows the effect of inlet gas temperature on the ratio for moisture content of 26.8% and initial particle size of 327.5 microns with gas velocity as a parameter. The ratio increases with increase in inlet gas temperature.

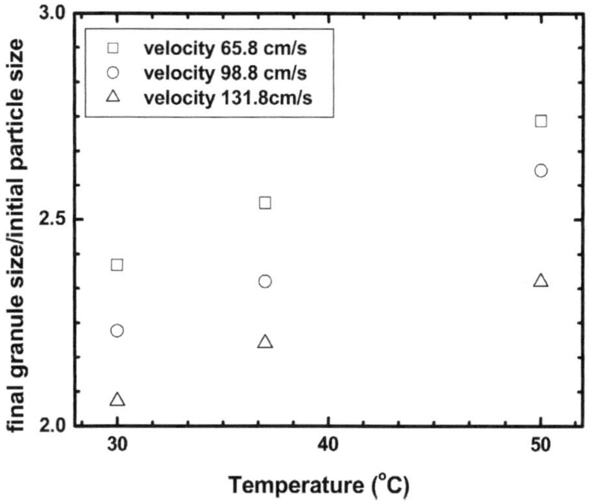

Figure 4c: Effect of inlet gas temperature on the ratio of average final granule size and initial particle size. The parameters are initial particle size = 327.5 micron and moisture content 26.8%.

These observations are in line with the bulk of information in the literature.

Model for size of Granules:

Wet granules form due to particle mixing with binder solution. The particles are held together by cohesive capillary forces due to the presence of liquid in between the particles. In fluidized bed granulators, the wet granules are suspended by gas drag force against gravitational force. The resulting shear determines the size of wet granules. Hydraulic capillary force which keeps the granule cohesive is

$$F_{ca} = \pi D_h \sigma \frac{MC}{100} \quad \textbf{(3)}$$

Diameter of capillaries in the granule, considering that a granule is similar to a packed bed, can be obtained as

$$D_h = 4\frac{\textit{Volume of voids in the granule}}{\textit{surface area of particles in the granule}} = 4\frac{\pi D_{pi}^{3}}{6}\frac{\varepsilon_g}{(1-\varepsilon_g)}\frac{1}{\pi D_{pi}^{2}}$$
$$= \frac{2}{3}\frac{\varepsilon_g}{(1-\varepsilon_g)} D_{pi} \quad \textbf{(4)}$$

This capillary force needs to be in equilibrium with drag force because of gas flow through the bed

$$\frac{2}{3}\frac{\varepsilon_g}{(1-\varepsilon_g)}\pi D_{pi}\sigma\frac{MC}{100} = c_D \frac{\pi D_g^{2}}{4}\frac{\rho u_o^{2}}{2\varepsilon^{2}} \quad \textbf{(5)}$$

From this ratio of wet granule size to initial fine particle size can be obtained as

$$\frac{D_g}{D_{pi}} = \left(\frac{16}{3c_D}\frac{\varepsilon_g}{1-\varepsilon_g}\right)^{1/2}\left[\frac{\sigma(MC/100)\varepsilon^2}{D_{pi}\,\rho u_o^2}\right]^{1/2} \quad \textbf{(6)}$$

Voidage in the granules ε_g may be assumed to be around 0.5. Inter granular movement and wall friction can cause attrition and breakup. These are also manifestations of drag force exerted by gas on granules. Also, for zero moisture content, the ratio has to reduce to a value of one. In view of this eq.(6) can be adopted as

$$\frac{D_g}{D_{pi}} = 1 + k_{gs}\left[\frac{\sigma(MC/100)}{D_{pi}\,\rho u_o^2}\right]^{1/2} \quad \textbf{(7)}$$

where k_{gs} is granular size constant.

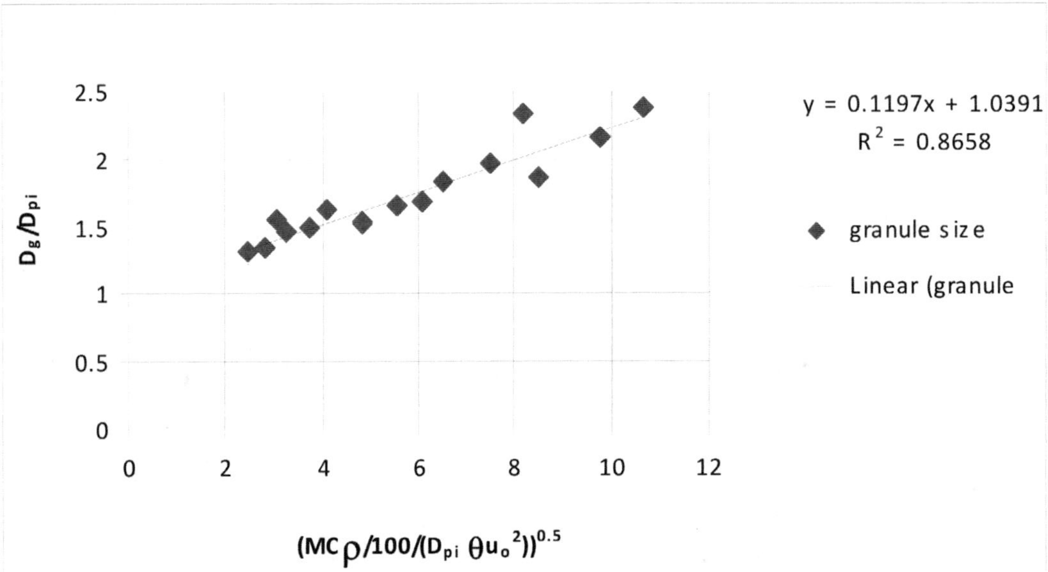

Fig.5 Comparison of the model equation (7) with experimental data.

From a comparison of this equation with experimental data as shown in the fig.5, k_{gs} is obtained as 0.1197. These are preliminary results and need to be further evaluated with a wider data base..

Conclusions.

1. Identification of the endpoint of granulation is necessary to discontinue further fluidization to avoid breakage due to attrition of the dry granules.
2. An easy method to measure the endpoint of wet granulation in fluidized beds is presented based on the bed pressure drop and temperature measurement as a function of time.
3. Bed pressure drop decreases with time to a limiting value as the liquid used for wet granulation evaporates.
4. During the wet granulation process. bed temperature decreases with time in the beginning and increases back as the evaporation of liquid is complete. The

inflexion point on the temperature- time trace in the rising temperature section corresponds to the end of evaporation and time for granulation.

5. Granulation time increases with increase in moisture content of the binder solution, decreases with increase in operating velocity, decreases with increase in inlet temperature of the fluidizing air.

6. A model is presented to estimate ratio of granule size to initial particle size and the ratio is observed to be a function of Weber Number

Acknowledgements

One of the authors (Duvvuri Subbarao) wishes to record his deep sense of appreciation for the financial support by Universiti Teknologi PETRONAS.

Nomenclature

A	Area of fluidized bed
c_1	Constant
D_g	Diameter of the granule
D_h	Hydraulic diameter of voids in the granule
D_{pi}	Initial Particle Size fed to the granulator
F_{ca}	Capillary force
h	Gas to Particle Heat transfer Coefficient
k	Gas thermal conductivity
k_{gs}	Constant for granule size
L	Length of fluidized bed
MC	Moisture content per weight of bed particles in %, wt/wt %
n	Constant
T_b	Fluidized Bed temperature
T_{gi}	Inlet gas Temperature
t_d	Time required for drying the granules
t_g	Time for granulation
s_g	Surface area of a granule
v_g	Volume of a granule
u_o	Superficial gas velocity

Greek Letters

ΔX	Length of inlet gas to fluidized bed heat transfer zone
ε	Bed Voidage
ε_g	Voidage in the granules
λ	Latent heat of evaporation
μ	Viscosity of the gas
ρ	Density of gas
ρ_p	Density of partiles
σ	Surface tension

REFERENCES

1. Nienow.A.W., Smith.P.G.,Particle growth mechanisms in fluidized bed granulation-I, the effect of process variables, Chem.Engg.Sci., 38, (1983), 1223
2.R.D.Becher, E.U.Schlunder. Fluidized bed granulation the importance of drying zone for the particle growth mechanism, Chem Engg. And Proc.,37, (1998), 1-6
3.Mackaplow B.M, Lawrence A.R, Michaels, Effect of primary particle size on granule growth and endpoint determination in high shear wet granulation, Powder Technology, Volume 108, Issue 1, 1 March 2000, Pages 32-45
4. Hemati. M.,Cherif.R,Saleh.K, Pont.V, Fluidized bed coating and granulation: influence of process related variable and physicochemical properties on the growth kinetics, Powder Technology, 130, 1-3, 2003, 18
5. Rajniak P, Mancinelli C, Chern R T, Stepanek F, Farber L, Hill B T, Experimental study of wet granulation in fluidized bed: Impact of the binder properties on the granule morphology, Int. Journal of Pharmaceutics, Volume 334, Issues 1-2, 4 April 2007, 92-102
6. Miwa.A.,Yajima.T,Shigeru.I.,Prediction of suitable amount of water addition for wet granulation Int.J. of Pharm, 195, (2000),81
7. Goldszal.A, Bousquet.J, Wet agglomeration of powders: from physics toward process optimization, Powder Tech, 117, 2001, 221
8. Frake.P, Greenhalgh.D, Grierson. S. M., Hempenstall. J. M., Rudd. D. R. Process control and end-point determination of a fluid bed granulation by application of near infra-red spectroscopy, International Journal of Pharmaceutics, Volume 151, Issue 1, 12,May,1997,75-80.
9. Laicher.A., Profitlich. T, Schwitzer. K, Ahlert. D, A model signal analysis system for endpoint control during granulation, European J of Pharm Sci, 5, 1,1997, 7-14
10. Derek Daniher, Lauren Briens and Andy Tallevi, End-point detection in high-shear granulation using sound and vibration signal analysis Powder Tech, Dec 2006,
11.Simons.S.J.R, Fairbrother R.J, Direct observation of liquid binder –particle attractions: the role of wetting behavior in agglomerate growth, Powder Tech ,110, 200, 44-58
12. Findley W. P. ,Peck. G. R.,Morris K.R., Determination of fluidized bed granulation endpoint using near infrared spectroscopy and phenomenological analysis, 94, 3 , 2005.

AIChE Annual Meeting - November 16-21, 2008 Philadelphia, PA

Size Resolved Kinetics of Nickel Nanoparticle Oxidation by Ion-Mobility Classification

L. Zhou, A. Rai, N. Piekiel, X.F. Ma and M. R. Zachariah

Department of Mechanical Engineering and Department of Chemistry and Biochemistry University of Maryland, College Park, 20742, USA

1. Introduction

Recent advancement on field of so called "nanoenergetic" materials are focused on either enhancing or tuning reactivity[1]. On one level this issues reduces to a length-scale argument, whereby smaller fuel/oxidizer combinations result in smaller diffusion lengths and therefore higher reactivity. On another level, this discussion leads to choices of other materials beyond aluminum as the primary thermite based fuel to Ti, or Ni, or even metal based composites such as Ni/Al. While the oxidation of Ni metal in form of bulk sample or thin films has been studied for over a century, there are only a few studies on the oxidation of nickel nanoparticles and most of the studies were carried out by using conventional dynamic thermal techniques such as thermogravimetry. It is well know that those methods are greatly influenced by heat and mass transfer effects such that the results are biased by experimental artifacts[2, 3]. In this work we employ aerosol based techniques to study the oxidation kinetics of Ni nanoparticles. The basic idea of the experiment approach is to prepare Ni particles of characterized size (i.e. Monodisperse), and monitor changes during oxidation in free-flight (i.e. no substrate). The study consists of two experiments, a Tandem Differential Mobility Analyzer (DMA) system[4-7] is used to measure the size change after oxidation, while the mass change is tracked by a DMA-APM (Aerosol Particle Mass Analyzer) system[8-10]. This technique allow a direct measure of mass and volume change of individual particles and thus enables us to explore the intrinsic reactivity of nanoparticles with minimizing the sampling error introduced by mass and heat transfer. The mass and size changes of nanoparticles are studied from room temperature to 1100C. The average density obtained show the nickel nanoparticle oxidation process can be correlated to the formation of both NiO and Ni_2O_3 and a phase change region where both the oxidation of nickel and thermal decomposition of Ni_2O_3 to NiO occur simultaneously. The reaction kinetics parameters (activation energy and effective diffusion coefficient) were then extracted from the experiment data as a function of particle size.

2. Experimental Approach

In this work, high purity nickel nanoparticles were prepared in an oxygen free environment using gas-phase thermal pyrolysis of nickel carbonyl as shown in Figure 1.

The primary analytical tools employed in the experiments were a tandem differential mobility analyzer system (TDMA)[4-7] and DMA-APM (aerosol particle mass analyzer) systems[8-10]. In the experiments, DMA-1 is used as a band pass-filter to create the monodisperse particle source, by selecting particles with same electrical mobility size. A second DMA was operated in voltage-step mode with a condensation particle counter (CPC) as a particle size distribution measurement tool to track the size change after the oxidation process. In a parallel experiment the change in particle mass after oxidation was measured by an aerosol particle mass analyzer (APM) coupled with a CPC.

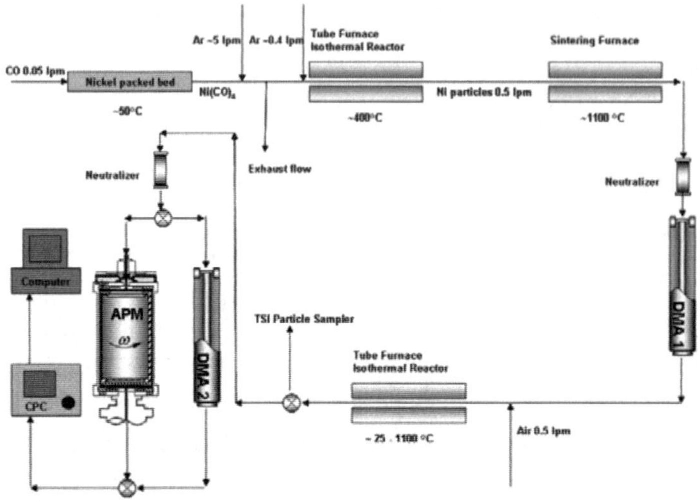

Figure 1: Schematic of Ni generation and oxidation experiment.

3. Size and Mass Measurements of Nickel Nanoparticle Oxidation

In this study, mobility sizes of 40, 62, 81, and 96 nm spherical nickel particles were obtained for oxidation study. Size selected particles mixed with air and oxidized, following which the size or mass of the reacted particles is measured by a second DMA or APM. The TDMA experimental results of size change ΔDp in Figure 2 (a) show that the particles size first increases as we increase the furnace temperature, and further increases in the reaction temperature, result in decreases in particle size.

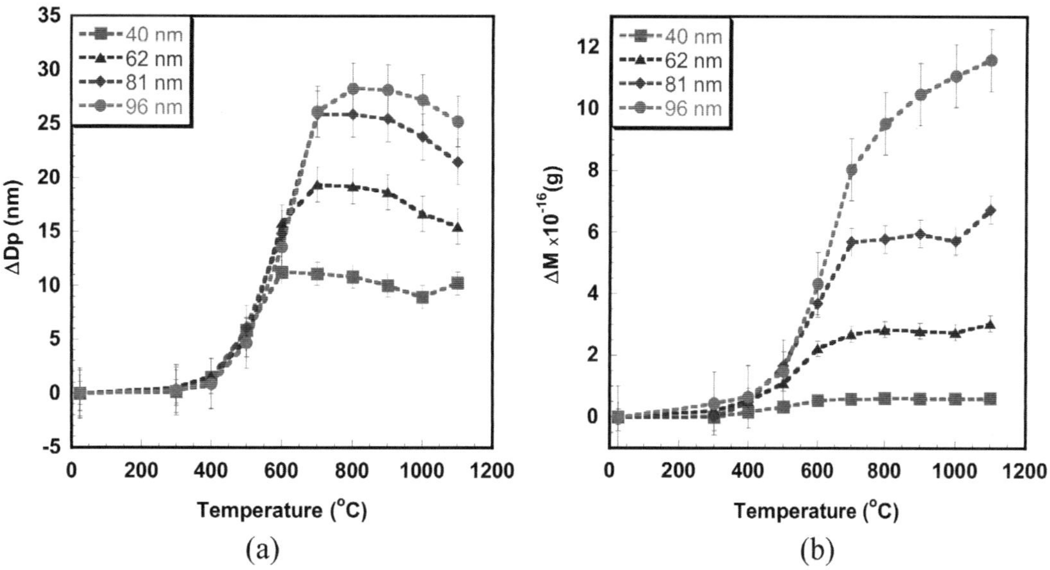

Figure 2 the change of (a) particle size and (b) particle mass as function of oxidation temperatures.

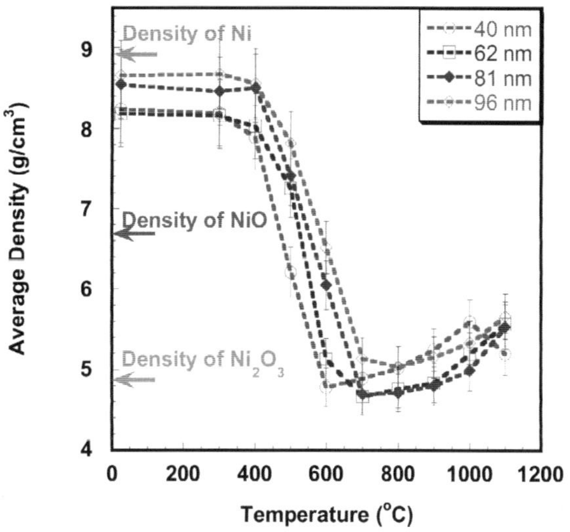

Figure 3 Average Density measured using DMA-APM setup

The APM measured mass change ΔM are shown in figure 2 (b). The average density profiles of reacted particles are calculated and shown in figure 3, We find that as the furnace temperature increases, the average density of the reacted particles decreased monotonically to 4.7-5.0 g/cm3, consistent with the density of Ni_2O_3 (4.84 g/cm^3). At higher temperatures the particle density increases to 5.5-5.7 g/cm^3 and at the highest temperature investigated is roughly at a density half way between NiO (6.67 g/cm^3) and Ni_2O_3 (4.84 g/cm^3). The oxidation to form Ni_2O_3 should be dominant in the low temperature region while the process of formation of the two types of oxides, and the phase transition are coupled at higher temperatures.

4. Size-resolved Oxidation Kinetics of Nickel Nanoparticles

The well known diffusion controlled shrinking core model is employed in this work to extract the reaction rate constant. At steady state the mass change rate for the reacted nickel nanoparticle is

$$\frac{dM}{dt} = 4\pi M_{O_2} DeCo_2 \frac{r_1 r_2}{r_2 - r_1} \quad (1)$$

where r_1, r_2 are the radius of the nickel core, and the reacted particle radius. *De* is the diffusion coefficient for ions diffusion in oxide layer. Co_2 is the oxygen molar concentration in gas. Mo_2 is the molecular weight of oxygen. To simplify the problem we approximate the instantaneous mass changing rate with the average mass changing rate by using the mass change measured from the APM. The size-resolved activation energies are obtained from an Arrehenius plot as shown in figure 4(a) and the effective diffusion coefficients of oxidation process were plotted in figure 4 (b). Two different regions can be distinguished from the Arrehenius plot 4 (a), as the oxidation process transitions to a phase change region. The calculated activation energies at low temperature region decrease from 54 KJ/mol to 35 KJ/mol as the particle mobility size decrease from 96 nm to 40 nm. The size effect on the activation energy observed here may be explained by the enhanced surface area-to-volume ratio with decreasing particle mobility size.

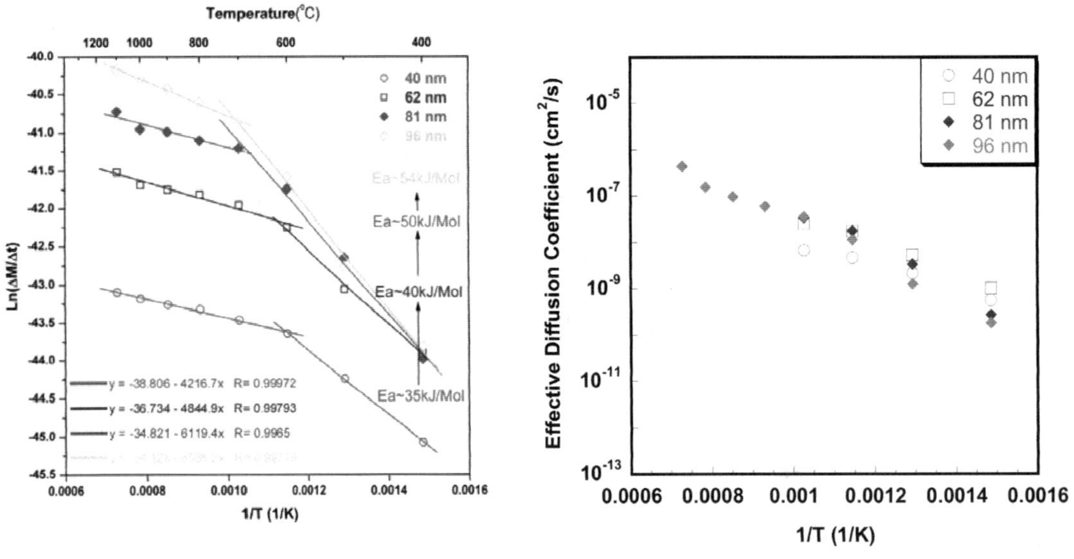

(a) (b)

Figure 4 (a) Arrehenius plots of average mass changing rate as a function of inverse temperature (b) the effective diffusion coefficient of nickel nanoparticles oxidation

5. Concluding Remarks

We applied an online aerosol based TDMA and DMA-APM methods to study oxidation and reactivity of nickel nanoparticles. The experimental data can be divided into an oxidation region and a phase transit region. We found that the activation energy of oxidation process decreased from 54 KJ/mol to 35 KJ/mol as the particle mobility size decrease from 96 nm to 40 nm.

References

(1) A. Prakash, A. M., and M.R. Zachariah *Advanced Materials* **2005**, *17*, 900-903.

(2) Park, K.; Lee, D.; Rai, A.; Mukherjee, D.; Zachariah, M. R. *J. Phys. Chem. B* **2005**, *109*, 7290-7299.

(3) Mahadevan, R.; Lee, D.; Sakurai, H.; Zachariah, M. R. *J. Phys. Chem. A* **2002**, *106*, 11083-11092.

(4) Knutson, E. O.; Whitby, K. T. *Journal of Aerosol Science* **1975**, *6*, 443-451.

(5) Kim, S. H.; Fletcher, R. A.; Zachariah, M. R. *Environmental Science & Technology* **2005**, *39*, 4021-4026.

(6) Jung, H.; Kittelson, D. B.; Zachariah, M. R. *Combustion and Flame* **2005**, *142*, 276-288.

(7) Higgins, K. J.; Jung, H. J.; Kittelson, D. B.; Roberts, J. T.; Zachariah, M. R. *Journal of Physical Chemistry A* **2002**, *106*, 96-103.

(8) Park, K.; Rai, A.; Zachariah, M. R. *Journal of Nanoparticle Research* **2006**, *8*, 455-464.

(9) Park, K.; Kittelson, D. B.; Zachariah, M. R.; McMurry, P. H. *Journal of Nanoparticle Research* **2004**, *6*, 267-272.

(10) Ehara, K.; Hagwood, C.; Coakley, K. J. *Journal of Aerosol Science* **1996**, *27*, 217-234.

491b Preparation and Characterization of Metal-Bound Hybrid Reactive Particles

Alexandre Ermoline[1], Mirko Schoenitz[2], Mikhaylo A. Trunov[1], and Edward L. Dreizin[2]. (1) Reactive Metals, Inc., Edison, NJ 08817, (2) Chemical Engineering, New Jersey Institute of Technology, University Heights, Newark, NJ 07102

This study is directed to establish the feasibility of preparing hybrid reactive materials with increased strength and energy density. The goal of this study is to produce reactive materials with coarse, fully dense, and mechanically strong particles, which, at the same time, are highly reactive. In order to achieve these characteristics, hybrid particles consisting of reactive nanocomposite filler and metallic binder were produced by mechanical milling. In a preliminary step, aluminum-rich nanocomposite thermite particles with CuO and MoO3 serving as oxidizers were prepared by Arrested Reactive Milling. These nanocomposite powders were embedded into Al matrix, also using mechanical milling with properly selected process parameters. The product hybrid material with particle sizes close to 1 mm was obtained. It consisted of aluminum matrix and micron-sized inclusions of the nanocomposite thermite material. Using the same technique, reference material was also prepared. It consisted of only aluminum, while the particle sizes and morphologies were the same as for the hybrid material. The effect of metal – thermite composition and milling time on the hybrid particle size distribution was studied. The compositions and structures of the produced hybrid particles were characterized by X-ray Diffraction and Scanning Electron Microscope. Individual Al-bound hybrid material particles and reference aluminum particles of approximately the same size were ignited with a CO2 laser. It was observed that the ignition of hybrid particles containing nanocomposite material inclusions occurred faster than for the reference pure Al particles. The combustion of hybrid particles occurred with ejection of burning fragments and was completed sooner compared to the nearly steady and slow combustion of the similar size and morphology reference aluminum particles.

Synthesis and Characterization of Aluminum-Rich Nanocomposite Powders at Cryogenic Temperatures

Carlo Badiola, Mirko Schoenitz, Xiaoying Zhu, Edward L. Dreizin
New Jersey Institute of Technology, Newark, NJ 07103

Abstract: Nanocomposite materials with reactive components are of interest for many applications in pyrotechnics, explosives, and propellants. Several such materials have been recently prepared by Arrested Reactive Milling (ARM), a method based on mechanical milling of µm-sized component powders to form µm-sized composite particles in which the components are mixed on a scale of 100 nm or finer. The temperature at which the milling is performed affects significantly both the rate at which the material is refined and the final properties of the product. In the present paper we report on an effort prepare Al-CuO and Al-MoO_3 reactive nanocomposites at cryogenic temperatures. A SPEX Certiprep 6815 Freezer/Mill was used to prepare the nanocomposites with aluminum-rich compositions from µm-sized component powders. The material was processed in steel vials using steel balls of different sizes as milling medium. The number and dimensions of the milling balls as well as the milling time were systematically varied. The prepared powders were characterized by X-ray diffraction, particle size analysis, and scanning electron microscopy. Thermal characteristics were studied using a custom wire-ignition setup and differential scanning calorimetry. Preliminary results show that the uniformity of mixing and reactivity of the nanocomposite powders can be substantially improved using milling at cryogenic temperatures.

I. Introduction

Nanocomposite reactive materials, in particular, thermites, are gaining much attention after recent studies showing that high reaction rates are possible to achieve in addition to the substantial reaction enthalpy when the reactive components are mixed on the scale of about 100 nm, see, for example refs. [1-7]. Among different techniques used to prepare such nanocomposite materials, Arrested Reactive Milling (ARM) method offers an inexpensive and scalable technique that is very versatile and can be applied to a wide range of reactive material compositions [8-13]. The ARM-prepared powders comprise micron-sized particles, where each particle is a fully dense composite with nano-sized inclusions of one component embedded in a matrix of the second component. The powders are prepared by high-energy mechanical milling of micron-sized powders of the individual components. Both a shaker mill and a planetary mill operated at room temperature were used in previous studies to prepare reactive nanocomposite materials by ARM. The milling is interrupted (or arrested) before the highly exothermic and typically self-sustaining reaction between the material components is triggered mechanically. Previous work showed that a slight reduction in the temperature at which the reactive materials are milled results in a substantial change in the product properties[14]. In parallel, substantial efforts have been reported on mechanical alloying performed at cryogenic temperatures[15,16]. The material properties, altered at cryogenic temperatures result in better-refined structures produced during shorter milling times. In some experiments, the phase make up of the mechanically alloyed compound was observed to be different depending on the milling temperature. Mechanical alloying, as ARM, relies on high-energy ball milling to refine the material structure. The key differences of ARM are that the starting components are capable of highly exothermic reaction and that the initiation of such a reaction or its precursors is undesirable. Thus, it is expected that ARM performed at cryogenic temperatures can result in a very attractive combination of intimately refined reactive components for which the low processing temperature suppresses chemical interaction. This paper investigates the feasibility of preparation reactive nanocomposite materials by ARM at a cryogenic temperature.

II. Experimental

The experiments in this study were performed on a SPEX Certiprep 6815 Freezer/Mill. In this vibratory mill, four stainless steel vials are arranged horizontally, while submerged in liquid nitrogen. Two solenoids impose an alternating magnetic field that moves the magnetic milling media back and forth along the axis of the milling vials. Collisions occur among the milling media and between the media and the end caps of the milling vials. Available milling media were single cylindrical impactors, and hardened steel balls of varying diameter.

Initially, a 12Al·MoO_3 nanocomposite powder that had been previously prepared in a planetary mill was cryogenically milled to alter its particle size distribution. Later, aluminum (-325 mesh, <45 μm, 99.5%, Atlantic Equipment Engineering), and CuO (25 μm, 99+%, Sigma Aldrich), or MoO_3 (98%, Alfa Aesar) were placed in the milling vials directly to prepare nanocomposite powders with overall compositions of 12Al·3CuO, and 12Al·MoO_3, respectively. These metal-rich thermite compositions, which are substantially less reactive than stoichiometric compositions, were selected for initial cryogenic milling experiments in order to minimize possible damage to the equipment in case of undesirable thermite reaction occurring during the milling. Milling vials were loaded at room temperature in ambient air. The milling vials consist of stainless steel tubes which are closed off at the ends by loosely-fitting end caps. The end caps do not seal hermetically; therefore loading in a protective atmosphere was not attempted. After loading, the vials were pre-cooled by immersing in liquid nitrogen.

Milling parameters varied in experiments were the milling time, the frequency of the magnetic field, the type, mass, and number of the milling media, and the amount of powder in each vial. Details of the experiments are shown in Table 1.

Table 1: Overview of milling parameters

Composition	Powder Load, g	Milling Media	Milling Time, min	Frequency, Hz
12Al·MoO_3, prepared in planetary mill	1.5 3	Cylindrical steel impactor 3/8" diameter by 2 3/8" length Steel balls: 3×1/2"; 7×3/8"; 58×3/16"	10 20 300	10 15
12Al·3CuO, 12Al·MoO_3, direct milling	3	Steel balls: 3·7/16"; 58·3/16"; 3·7/16" + 10·3/16"	30 - 1200	15

The lack of a tight seal on the milling vials posed a significant problem, as during milling as much as 2/3 of the powder load was lost in some experiments. Likewise, unknown amounts of liquid nitrogen would enter the vials, changing the milling environment in poorly controllable ways. To alleviate this problem, the end caps were dipped in water before inserting into the ends of the vials. The water then filled the small gap between end cap and vial, and on freezing provided an effective seal. Material loss was reduced down to 0.3 % over a 20-h milling period. Later experiments were all conducted using this procedure.

In runs without the water seal, the powder was recovered by simply opening the vial in a protective atmosphere to prevent condensation from ambient humidity. For experiments with the water seal, the vials were first placed in a vacuum chamber to sublimate the ice.

Product powders were characterized using various techniques. X-ray diffraction (XRD) measurements were conducted on a Philips X'pert MRD powder diffractometer using unfiltered Cu radiation (λ=1.5438 Å), operating at 45 kV and 40 mA. XRD provided information about the phase composition of the prepared materials; in particular, identification of any products of partial thermite reaction occurring during the milling was of interest. Particle size distributions (PSD) were measured by light scattering using a Beckman-Coulter LS230 Particle Counter.

To study morphology of the prepared materials, powders were embedded in epoxy and cross-sectioned. After polishing, cross-sectioned samples were examined on a LEO 1530 Field Emission Scanning Electron Microscope (SEM) operated at 10 kV. For the 12Al·3CuO material, at least five representative SEM images of each sample were taken at a fixed magnification and processed to determine the average size distribution of CuO inclusions in the aluminum matrix.

Reactivity of the prepared materials was analyzed using differential scanning calorimetry (DSC) and a custom wire-ignition set up. DSC experiments were performed using a Netzsch Simultaneous Thermal Analyzer STA409PC. Samples were heated using different heating rates ranging from 2 to 20 K/min in argon environment. The custom wire-ignition apparatus is shown in Figure 1. Its operation was presented in detail elsewhere[17] and it is only briefly described below. A slurry is prepared with the powder being tested placed in hexane or another quickly evaporating hydrocarbon liquid. A thin layer of the slurry is coated on a portion of a Nichrome® wire with a paintbrush and dried. The wire is then heated electrically with heating rates varying from 100 to 50,000 K/s. The wire temperature is monitored in real time using an infrared pyrometer. Simultaneously, the emission from the powder coating is measured with a photodiode, and the instant of powder ignition is detected by a spike in the photodiode signal. Thus, ignition temperatures at different heating rates are measured.

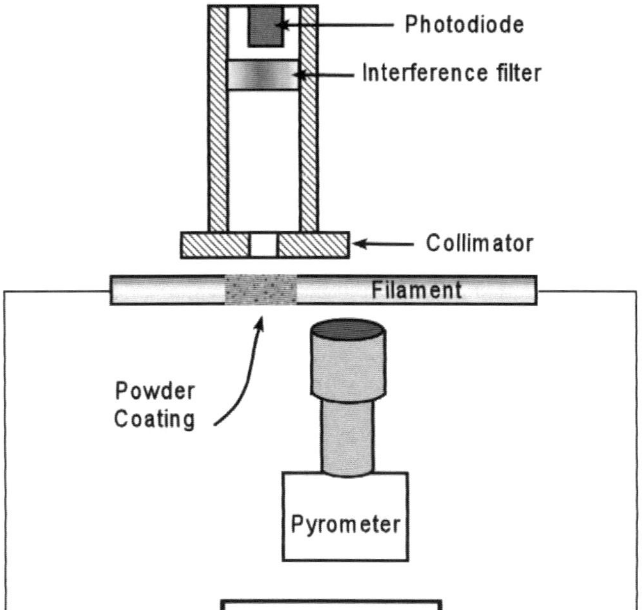

Figure 1. Simplified diagram of the heated filament ignition set up.

III. Preliminary Experiments and Results

Early cryogenic milling experiments used a nanocomposite powder with composition 12Al·MoO_3 prepared using a planetary mill as a starting material. The goal of these experiments was to establish whether measurable changes in the material morphology and particle size can be detected. Four milling vials 7/8" diameter, 3 7/8" length, were used in each milling run. The four milling vials are stacked in an array with two on the bottom and two on the top. It was observed that the powder milling occurs differently, depending on the location of the milling vial. Specifically, the powders were refined faster in the vials placed on the top of the vial array.

Considering powders prepared in the top vials, the change in morphology of the nanocomposite powder was undetectable. However, the particle sizes were found to change as a function of the milling time. Based on these observations, a milling progress function was considered to compare experiments with varied milling conditions. Similarly to the specific milling dose considered elsewhere[18], it was assumed that the milling progress function □ is proportional to the milling time *t*, the operating frequency of the mill *f*, as a measure of the energy of individual collisions, the number of milling media units present, *n*, proportional to the number of collisions, and the inverse of the powder mass *m*:

$$\Phi = n \cdot f \cdot t / m$$

Figure 2 shows the moments of the measured particle size distributions as a function of the progress function for a series of runs with nanocomposite $12Al \cdot MoO_3$ used as a starting material. As mentioned above, for these initial experiments consistent differences in the milling progress were observed between the vials placed in different positions within the mill. In Fig. 2 each run is represented by the 10th and the 90th percentiles of the volume-based size distribution, and by the median. Horizontal lines show respective values for the initial powder. It can be seen that maximum and median particle sizes systematically decrease as a function of the milling progresses.

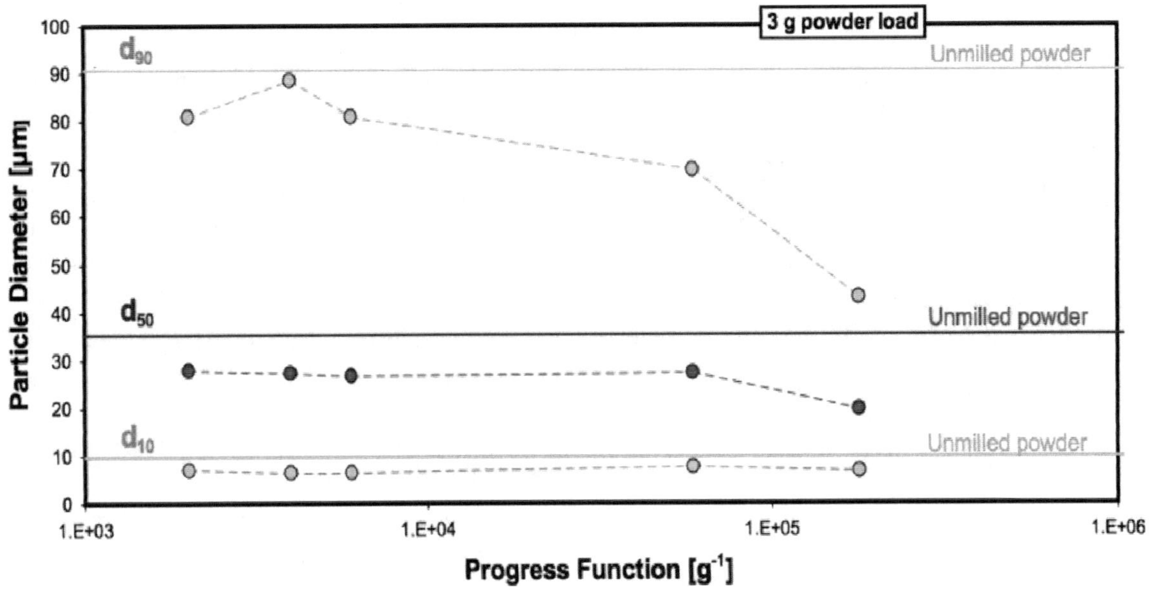

Figure 2. Progress function for preliminary experiments with $12Al \cdot MoO_3$ composite processed in top vials. At increasing progress function values, the particle diameter decreases. The percentiles for the composite powder used as starting material are shown for comparison.

No clear improvement in the rate of powder refinement was observed for using milling balls as compared to the cylindrical impactor. Note that the experimental results were most significantly affected by the location of the milling vial (top vs bottom position), while this position was not considered as a meaningful experimental parameter and thus was not recorded in the initial tests. The difference between the milling in top and the bottom vials is best illustrated in a follow-up set of experiments. The morphologies of powders produced as a result of direct milling of starting Al and CuO powders are shown in Fig. 3. The cross-sections shown in Fig. 3 are for powders prepared under identical conditions, but retrieved from different vials of the same milling run. The sample shown on the left side of Fig. 3 appears to be only at the initial stage of the material refinement. Deformed from its initial shape, flake-like particles of Al are mixed with unattached (brighter) CuO inclusions with only few particles forming composite structure. The sample shown on the right side, however, is seen to contain a uniform nanocomposite material. Systematic investigation of this problem revealed that the temperature variation between the vials was the main factor affecting milling efficiency, although the presence of liquid nitrogen inside the vials due to non-hermetic sealing may also have played a role. The difference in temperature was associated with slow but continuous loss of liquid nitrogen from the milling compartment due to its evaporation. Thus, the top two vials were immersed into the liquid nitrogen only for a portion of the entire milling time and the effective temperature in the top two vials was somewhat higher than that in the bottom two vials, which remained in liquid nitrogen for the entire milling time. The temperature difference was expected to be small because all vials were placed in a thermally insulated compartment of the freezer mill containing liquid nitrogen. In further experiments, the level of liquid nitrogen was maintained constant during the entire milling period. The experiments were stopped periodically, the milling compartment was opened and the liquid nitrogen refilled.

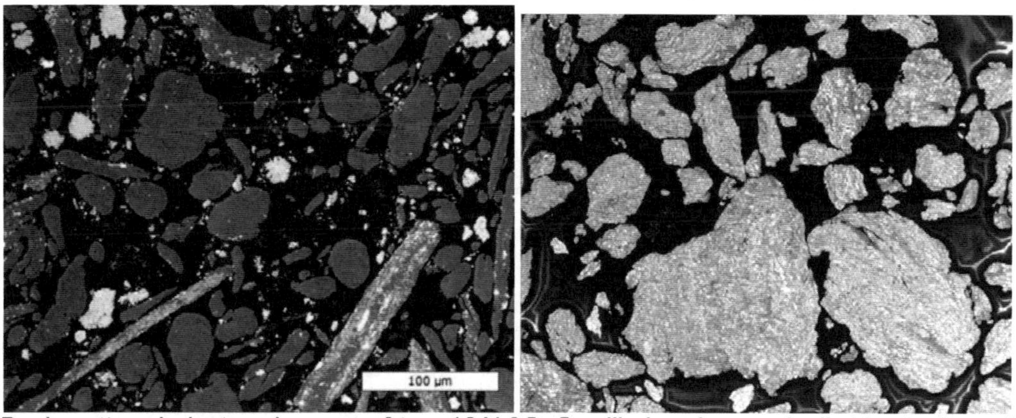

Figure 3. Backscattered electron images of two 12Al·3CuO milled under nominally identical conditions. The images have the same magnification. The sample on the left was retrieved from a lower vial, while the one on the right was retrieved from an upper vial. The powders were milled using a combination of 3 7/16" and 10 3/16" steel balls.

It was further found that when the temperature was maintained constant, manufacturer-provided cylindrical steel impactor operated in a similar fashion for the vials placed in both top and bottom positions. However, the powder refinement with steel balls used as milling tools depended on the vial position even if the liquid nitrogen level was maintained constant during the experiment.

Note that most of the experiments described below were performed before the effects of the milling tools and position of the vials were well understood. As the number of experiments increased it became clear that the milling in the vials placed on top of the vial array results in a more consistent powder processing using both cylindrical impactor and milling balls. Thus, the results described below are for the experiments performed in the top vials only.

IV. Results and Discussion

The results are discussed in the context of comparing the properties of two nanocomposite materials with the same bulk composition of 12Al·3CuO. One of the materials was prepared using milling at cryogenic temperatures and the other was prepared using a planetary mill operated at room temperature. The same starting powders of Al and CuO were used to prepare both materials.

Material Characterization

The size distributions for starting powder blend used in both cryogenic and planetary mills, and for both produced composite powders are shown in Fig. 4. The particle size distribution for the powder prepared in the planetary mill is somewhat narrower than the initial size distribution of the powder blend. Alternatively, the composite powder prepared at the cryogenic temperature appears to include quite a few relatively coarse particles. Note that the volume fractions of the particle size distributions shown in Fig. 4 are heavily affected by larger particles.

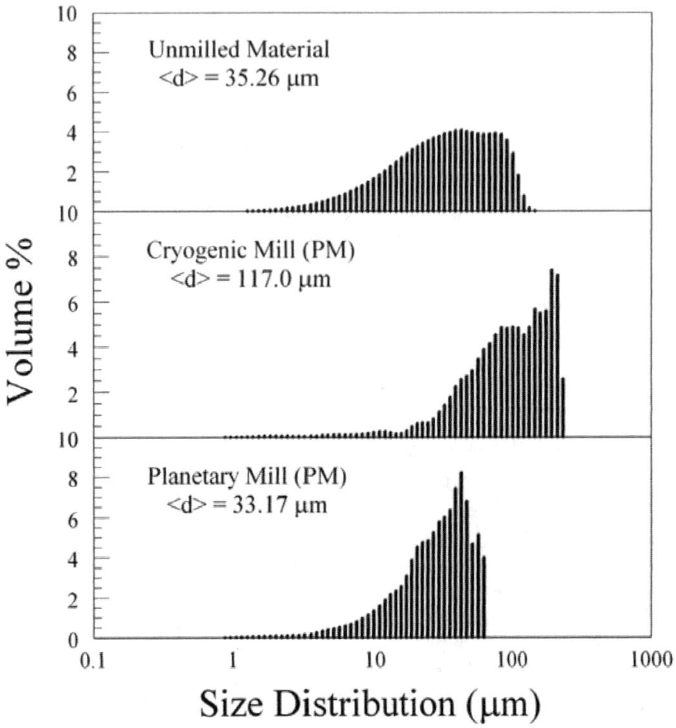

Figure 4. Particle Size Distribution of unmilled 12Al·3CuO in comparison to cryomilled and planetary milled materials. The figure indicates that particle sizes for cryomilled materials were not as fine as the planetary or starting material.

To compare the morphology of two prepared materials, the size distributions of CuO inclusions in Al matrix were measured using processing of the SEM images of the particle cross-sections. Representative SEM images are shown in Fig. 5. Images for the powders produced in both planetary and cryogenic mills showed brighter CuO inclusions within the darker Al matrix. The major apparent difference between the images is that the cryomilled material seems to be more homogeneous in terms of the distribution of CuO inclusions as compared to the material prepared in the planetary mill. The sizes of the CuO inclusions do not appear to be substantially different. This is confirmed by the size distributions shown in Fig. 6. These distributions were obtained by image processing.. The images were thresholded, lighter objects (CuO inclusions) were

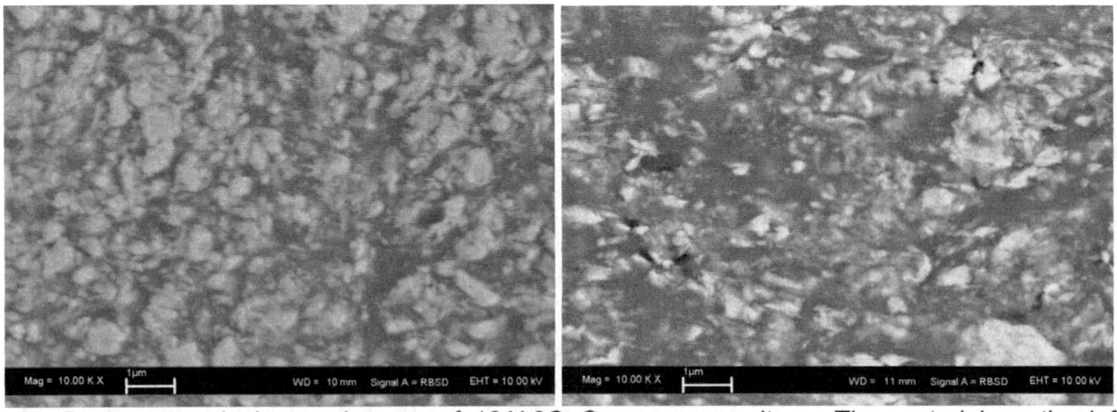

Figure 5. Backscattered electron images of 12Al·3CuO nanocomposites. The material on the left was cryomilled, and the material on the right was milled at room temperature in a conventional planetary mill. In both images, bright inclusions are CuO, while the dark matrix is Al.

selected, and their Feret diameters was automatically determined using the UTHSCSA ImageTool package by S. Brent Dove available as freeware In both cases, the inclusions size distributions peak at around 160 nm. The magnification used to collect the SEM images restricted the capability to resolve finer inclusions, so that the measurements are likely biased to the larger inclusion sizes. This limitation is manifested by an apparent sharp drop off in the count number for inclusions under 80 nm. Thus, the difference in the mean inclusion size identified for the two powders is considered to be insignificant, and the inclusion size distributions are seen as effectively equivalent for both powders.

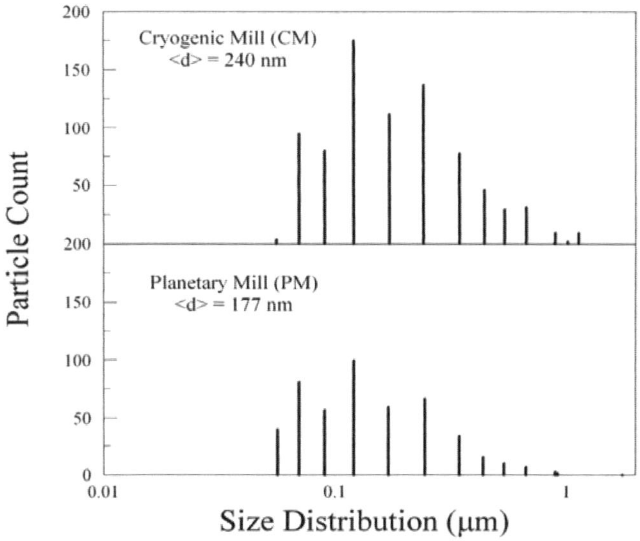

Figure 6. Inclusion Size Distribution for 12Al3CuO cryomilled and planetary milled. Both materials have similar distributions and evidence of analysis limitations by the sharp drop left of the peak.

Figure 7 shows XRD patterns collected for the powders prepared in the cryogenic and planetary mills. The differences between the patterns are subtle. The strongest peaks are, not unexpectedly, produced by the CuO and Al structures for both materials. However, the measured peak intensities are lower for the material prepared in the cryogenic mill. In addition, the peaks are wider for the material prepared at cryogenic temperatures, both indicators pointing out at a finer crystallite size for these materials as compared to their counterparts prepared in the planetary mill. A weak increase in the baseline indicating the presence of minor amounts of poorly crystalline aluminum oxides was occasionally observed for the powders prepared in the planetary mill. Such oxides could have formed as a result of undesirable thermite reaction occurring locally in the vial during the milling. Similar features were never detected for the powders prepared in the cryogenic mill.

Material Reactivity

DSC traces are shown in Figure 8. Traces for both powders were measured at different heating rates in argon. It is observed that for both materials a relatively weak endothermic feature precedes the observed exothermic peak. However, the DSC traces are substantially different for the powders prepared at different milling temperatures. The endothermic peak is substantially stronger for the powder prepared at the cryogenic temperature. This peak occurs around 820 K and its position does not shift to higher temperatures at higher heating rates. Based on its parameters, this peak is assigned to a eutectic point between Al and the $CuAl_2$ alloy in the binary Al-Cu system[19]. In order for this transition to occur, the $CuAl_2$ alloy must have formed prior, indicating that some of the copper was reduced and available to react with aluminum. Thus, the presence of a stronger endothermic feature corresponding to the eutectic reaction for the material prepared at the cryogenic temperature indicates its higher reactivity at lower temperatures, compared to the material prepared in the planetary mill. Indeed a raised baseline observed clearly for the cryomilled material prior to the endothermic peak supports this reaction sequence.

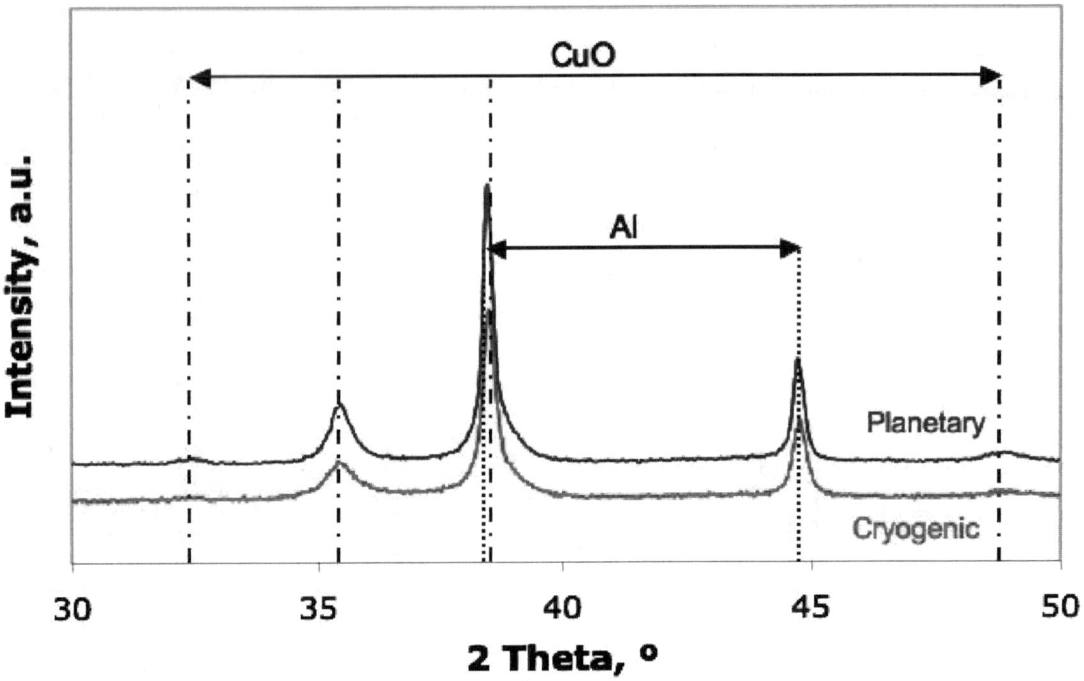

Figure 7. XRD traces for both cryogenic and planetary mill. Both materials are very similar in composition with differences in the peak widths and intensity. This suggests that milling cryogenically will result in more refined components.

The second important difference in the DSC signals for the two materials has to do with the main exothermic peak clearly visible in all DCS traces. For the cryomilled material, this peak occurs at significantly lower temperatures as compared to that for the material prepared in the planetary mill. Furthermore, the kinetic shift to higher temperatures at increased heating rates is smaller for the material prepared at cryogenic temperatures, indicating that the equivalent activation energy for the exothermic reaction is higher as compared to the material prepared in the planetary mill.

The results of the heated filament ignition experiments are shown in Fig. 9. Both materials were heated at different heating rates. In both cases, the measured ignition temperatures are

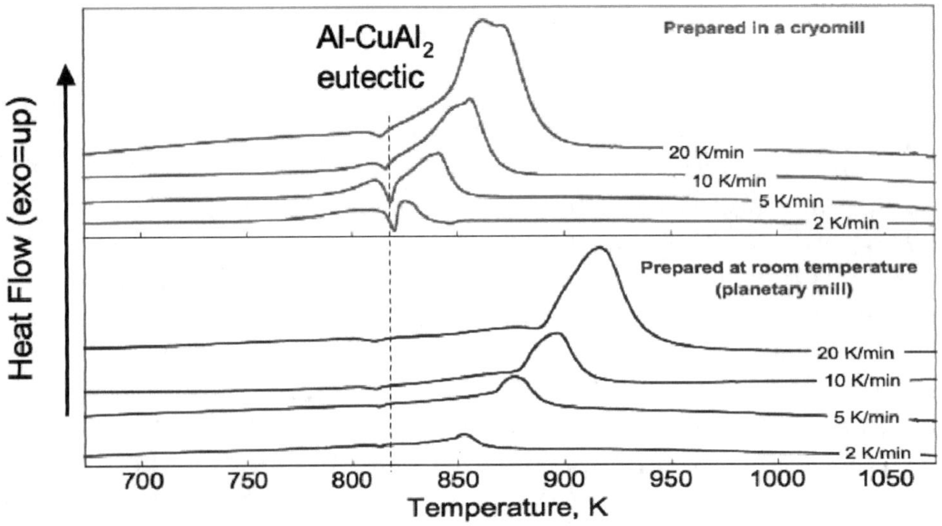

Figure 8. Differential scanning calorimetry traces measured at various heating rates for materials prepared in cryogenic mill and planetary mill.

close to the temperature range in which the eutectic melting of Al and $CuAl_2$ alloy occurs in the binary Al-Cu system. However, the observed effect of the heating rate on the ignition temperature is greater for the material prepared at cryogenic temperature. It can be initially proposed that the ignition in the material prepared in the planetary mill is controlled by the eutectic melting in the sample, while for the material prepared in the cryomill the reaction may be thermally activated and may initiate even before the eutectic melting.

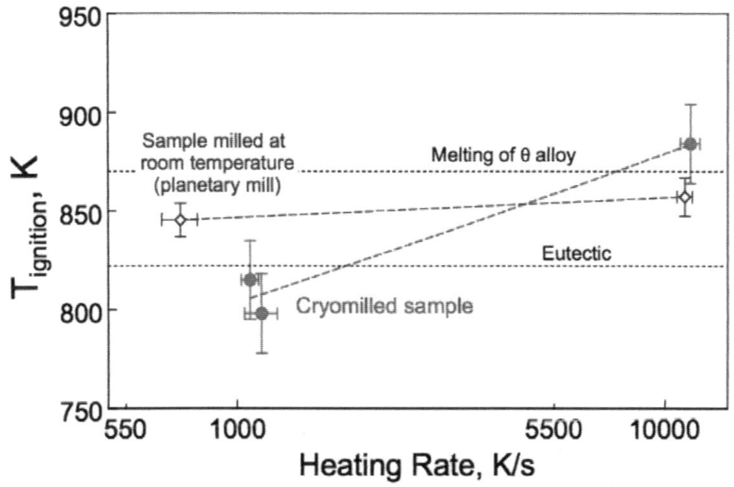

Figure 9. Ignition temperatures measured as a function of the heating rate in the heated filament ignition experiments.

IV. Conclusions

The feasibility of preparing reactive nanocomposite materials by ARM performed at cryogenic temperatures is established. The inclusions of the harder phase (metal oxide) distributed in the matrix of a softer component (aluminum) are approximately of the same size for materials prepared at both cryogenic and room temperatures. However, the inclusions are distributed more uniformly for the materials prepared in the cryomill. X-ray diffraction analysis shows that the crystallite sizes of the components mixed by ARM performed at the cryogenic temperature are finer compared to their counterparts produced by the room temperature milling. It is also observed that effectively no reacted material can be detected in the nanocomposite prepared at the cryogenic temperature. Thermal analysis indicated that the material prepared at cryogenic temperatures is more reactive at low temperatures. It is also observed that the reaction mechanisms and activation energies differ for the materials with the same bulk chemical composition prepared at different temperatures.

V. Acknowledgement

This work was supported in parts by Defense Threat Reduction Agency and by RDECOM-ARDEC, Picatinny.

VI. References

1. Puszynski JA, Bulian CJ, Swiatkiewicz JJ. The Effect of Nanopowder Attributes on Reaction Mechanism and Ignition Sensitivity of Nanothermites. Materials Research Society Symposium Proceedings 2006; 896:147-58.
2. Tillotson TM, Gash AE, Simpson RL, Hrubesh LW, Satcher Jr. JH, Poco JF. Nanostructured energetic materials using sol-gel methodologies. Journal of Non-Crystalline Solids 2001; 285(1-3):338-45.
3. Danen WC, Jorgensen BS, Busse JR, Ferris MJ, Smith BL. Los Alamos nanoenergetic metastable intermolecular composite (super thermite) program. 221st ACS National Meeting, San Diego, CA, United States, Abstracts of Papers, 2001.
4. Perry WL, Tappan BC, Reardon BL, Sanders VE, Son SF. Energy release characteristics of the nanoscale aluminum-tungsten oxide hydrate metastable intermolecular composite. Journal of Applied Physics 2007; 101(6):064313/1-064313/5.

5. Prakash A, McCormick AV, Zachariah MR. Synthesis and reactivity of a super-reactive metastable intermolecular composite formulation of $Al/KMnO_4$. Advanced Materials 2005; 17(7):900-3.
6. Walter KC, Pesiri DR, Wilson DE. Manufacturing and performance of nanometric Al/MoO_3 energetic materials. Journal of Propulsion and Power 2007; 23(4): 645-50.
7. Pantoya ML, Granier JJ. Combustion behavior of highly energetic thermites: Nano versus micron composites. Propellants, Explosives, Pyrotechnics 2005; 30 (1):53-62.
8. Dreizin EL, Schoenitz M. Nano-composite energetic powders prepared by arrested reactive milling. US Patent Application 20060053970; published in March 2006.
9. Schoenitz M, Ward TS, Dreizin EL. Fully Dense Nano-Composite Energetic Powders Prepared By Arrested Reactive Milling. Proceedings of The Combustion Institute 2005; 30:2071-8.
10. Umbrajkar S, Trunov MA, Schoenitz M, Dreizin EL, Broad R. Arrested Reactive Milling Synthesis and Characterization of Sodium-Nitrate Based Reactive Composites. Propellants, Explosives, Pyrotechnics 2007; 32(1):32-41.
11. Dreizin EL, Schoenitz M, Shoshin YL, Trunov MA. Highly-Energetic Nanocomposite Powders Produced by Arrested Reactive Milling. Internaional Annual Conference of ICT 2005, 36th (Energetic Materials), p. 138/1-138/12.
12. Umbrajkar SM, Schoenitz M, Dreizin EL. Control of Structural Refinement and Composition in $Al-MoO_3$ Nanocomposites Prepared by Arrested Reactive Milling. Propellants Explosives and Pyrotechnics 2006: 31(5):382-9.
13. Umbrajkar SM, Seshadri S, Schoenitz M, Hoffmann VK, Dreizin EL. Aluminum-Rich Al-MoO_3 Nanocomposite Powders Prepared by Arrested Reactive Milling. Journal of Propulsion and Power 2008; 24(2)192-8.
14. Umbrajkar SM, Schoenitz M, Jones SR, and Dreizin EL, Effect of Temperature on Synthesis and Properties of Aluminum-Magnesium Mechanical Alloys. Journal of Alloys and Compounds 2005; 40(1-2)70-77
15. Perez, RJ, Huang B, Lavernia EJ. Thermal Stability of Nanocrystalline Fe-10 wt.% Al Produced by Cryogenic Mechanical Alloying. Nanostructured Materials 1996; 7 (5) 565-572
16. Witkin, DB, Lavernia EJ. Synthesis and mechanical behavior of nanostructured materials via cryomilling. Progress in Materials Science 2006; 51 (1)1-60
17. Ward, TS, Trunov, MA, Schoenitz, M, Dreizin, EL, Experimental methodology and heat transfer model for identification of ignition kinetics of powdered fuels. International Journal of Heat and Mass Transfer 2006; 49(25-26) 4943-4954
18. Ward, TS, Chen, W, Schoenitz, M, Dave, RN, Dreizin, EL. A study of mechanical alloying processes using reactive milling and discrete element modeling. Acta Materialia 2005; 53(10), 2909-2918.
19. Massalski, TB., The Al−Cu system. Journal of Phase Equilibria, 1980 1(1) 27-33

491d Effect of Mechanical Activation on Reaction Kinetics In Gasless Heterogeneous Systems

Jeremiah D.E. White and Alexander S. Mukasyan. Department of Chemical and Biomolecular Engineering, University of Notre Dame du Lac, 182 Fitzpatrick Hall, Notre Dame, IN 46556

In this work, we investigated the effects of mechanical activation (MA) on the kinetics of several gasless reaction systems, including nickel-aluminum, hafnium-aluminum, iron oxide--titanium, and others. Reaction kinetics were primarily studied by differential thermal analysis (DTA), while the microstructures of the investigated reaction mixtures were characterized by scanning electron microscopy (SEM), electron dispersive spectroscopy (EDS), and x-ray diffraction analyses (XRD). It was demonstrated that under certain conditions, MA can significantly (>10 times) change the activation energy in such gasless high energy systems. This effect, in turn, influences the self-ignition temperature, as well as the velocity of the reaction wave propagation in these systems. Based on established correlations between the kinetic parameters and microstructural characteristics of heterogeneous reaction media, an explanation of the observed effect is suggested.

491e Mechanical Alloying and Reactive Milling In a High Energy Planetary Mill

Xianjin Jiang[1], Mikhaylo A. Trunov[1], Mirko Schoenitz[1], Rajesh Dave[2], and Edward L. Dreizin[1]. (1) Chemical Engineering, New Jersey Institute of Technology, York Center, NJIT, 138 Warren St, Newark, NJ 07103, (2) New Jersey Center for Engineered Particulates, New Jersey Institute of Technology, 138 Warren Street, YCEES 208, Newark, NJ 07102

Powder refinement in a high energy planetary mill is investigated experimentally and analyzed using discrete element modeling (DEM). Experimental parameter tracking the progress in the material refinement, specific milling dose, is introduced as a product of the charge ratio and milling time. This parameter is determined experimentally for reactive milling of boron and titanium powders for which the time of initiation of a self-sustained reaction is measured for different milling condition. It is assumed that the reaction is always initiated when the same degree of structural refinement is achieved. A parameter analogous to the specific milling dose is considered based on the DEM calculations. It is established that for the milling conditions used for reactive milling in this project, the primary type of ball motion is their rolling along the vial's circumferential surface. The reciprocal of the rate of energy dissipation resulting from the ball rolling is used to create the DEM analog of the specific milling dose which correlates well with experimental observations. The usefulness of the DEM model is established in predicting the milling conditions for varied process parameters for mechanical alloying or reactive milling performed in a planetary mill.

Pyrophoric Films Based on Nano-Sized Iron

Zac Doorenbos[1], Alok Vats[1], Jan Puszynski[1], Rajesh Shende[1], Deepak Kapoor[2], Darold Martin[2], Chris Haines[2]

[1]Department of Chemical and Biological Engineering, South Dakota School of Mines and Technology, Rapid City, SD

[2]Armament Research, Development, and Engineering Center, Picatinny, NJ

Abstract

Pyrophoric α-iron nanostructured porous coatings were prepared using a sol-gel technique. Iron oxide coating deposited on a non-porous metallic substrate was formed from $FeCl_2{\cdot}2H_2O$ dissolved in ethanol in the presence of Brij-76 or Pluronic P123 surfactant. The gelling process was conducted by the addition of propylene oxide. The surfactant was removed, in some experiments, by washing in ethanol and the resulting coating was heat treated to form Fe_2O_3. The Fe_2O_3 coatings were reduced in H_2/N_2 atmosphere at temperatures above 500°C and after cooling tested for their pyrophoricity in air. These coatings have shown a strong affinity to oxygen. The effect of processing conditions as well as thickness of the coating and it morphology on maximum combustion temperature and the duration of the pyrophoric reaction will be discussed. Experimental data will be compared to results obtained from one-dimensional mathematical model.

Introduction

The use of pyrophoric materials in everyday applications is important in both the military and commercial applications. Its uses range from the protection of aircraft to hand and muscles warmers. One area that has had a great deal of focus lately is the use of pyrophoric materials for infrared countermeasures for aircraft.

Of the several different pyrophoric materials that could be used, iron (Fe) in the form of fine particles is highly reactive and pyrophoric [1] and has not received as much attention as the others. Fe nano particles have found a great deal of interest for their use in non-pyrophoric applications such as catalysis [2], magnetic resonance imaging [3], magnetic data storage [4], coatings [5], synthesis of carbon nanotubes [6], synthesis of highly oriented and ordered nanostructures for field emission devices [7] etc. Previously, Gash et al. [8] and Shende et al. [9,10] have published work showing Fe based foils with a pyrophoric response when exposed to air. The work done by Gash et al. was based on a sol-gel technique using a precursors of $Fe(NO_3)_3.9H_2O$, $F_2Cl_3.6H_2O$, or anhydrous $FeCl_3$. The gels were subsequently reduced in H_2 and H_2/Ar environments at temperatures ranging from 350-700°C. The films did not show a pyrophoric response when exposed to air. However, once thermally ignited the films were self-sustaining. Gash et al. suggested that the addition of another metal would aid in the pyrophoric response of the Fe films. Shende et al. showed that it was possible to generate a pyrophoric response from iron obtained by thermal decomposition and reduction of Fe-oxalate/yttria films [9]. The Fe-oxalate/yttria films were made by dip coating a thin stainless steel substrate with a mixture of yttria sol-gel and Fe-oxalate using different weight ratios of starting precursors. The Fe-oxalate films showed a pyrophoric temperature response over 800°C.

Very little previous work has been done on the mathematical modeling of combustion front propagation in foils when exposed to air. In the paper written by Wilharm [11] a mathematical model that predicts the pyrophoric response of Fe based foils when exposed to air was proposed. Because of the small pore diameter in porous iron foils it is assumed that the diffusion of the oxygen is the rate limiting step of the reaction. The author reports the results of the mathematical model but does not compare modeling results to any experimental data. Wilharm [11] concludes that the foils characteristics are dependent on "The pore diameter, specific surface area, and porosity…". The foil characteristics, as mentioned by Wilharm, are dependent on the production method and substrate use for foil generation.

Experimental

The goal of this research was to generate a pyrophoric foil that will generate a pyrophoric response of greater than 800°C in less than 1 second. It is also important that the film adheres to the substrate and that the process of making the films is cheap and non-labor intensive. The first method which uses Fe-oxalate films as was reported by Shende et al. [9], is a labor intensive and time consuming process. A better way of generating these films would be to use a one step Fe_2O_3 sol-gel based process and spin or dip coating of the substrate with a sol-gel.

To accomplish this task the following raw materials were used: $FeCl_2 \cdot 2H_2O$ powder (Alfa Aesar), absolute ethanol (Sigma Aldrich), propylene oxide (PO)(Sigma Aldrich Reagent Grade), Pluronic P123 Surfactant (BASF), Brij-76 Surfactant (Sigma Aldrich), stainless steel foil and alumina felt (McMaster Carr).

The generation of the sol-gel was done using the following steps: 1) Addition of the $FeCl_2 \cdot 2H_2O$ to the absolute ethanol, 2) The addition of the surfactant (if being used), 3) Sonication to dissolve $FeCl_2 \cdot 2H_2O$ and surfactant, 4) addition of PO and mixing. After these steps are completed the gel was formed in 1 to 9 minutes, depending on the amount of PO added. The sol-gel was coated onto the stainless steel substrate by spin coating at 80% of its total gel time. When a porous substrate was used instead, a dip coating was applied at 80% of its total gel time. The substrates were then calcined in air at a heating rate of 0.5°C up to 400°C and allowed to cool to room temperature. The samples were then reduced in a H_2 atmosphere at 500°C for 10-20 minutes. The reduction time is dependent on the amount of Fe_2O_3 that is present in the initial sample. The foils were then exposed to air and their pyrophoric temperature responses are recorded using a Fluke IR pyrometer. Photographs of the FeOOH sol-gels coated onto the two different substrates are shown in Figure 1.

Figure 1: *a) Sol-gel coating on stainless steel substrate after calcination, b) sol-gel coating on porous alumina substrate after calcination.*

Results

The FeOOH sol-gel is converted to Fe_2O_3 after calcination. When the Fe_2O_3 is reduced in a H_2 environment results in the formation of α-Fe. This α-Fe is pyrophoric when exposed to air. The pyrophoric temperature profiles generated when the α-Fe is in powder form, coated on stainless steel and coated onto a porous substrate when exposed to air are shown in Figure 2.

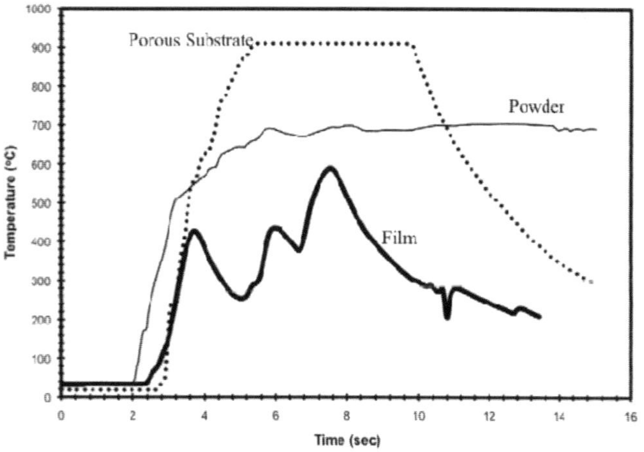

Figure 2: *Pyrophoric temperature response of sol-gel based α-Fe powder, Fe based film coated on stainless steel, and Fe coated porous substrate.*

The pyrophoric responses shown in Figure 2 indicate that the porous substrate has the best pyrophoric response reaching a temperature of over 900°C. The use of a porous substrate allows for an increase in surface area allowing for more Fe to present in the foil and oxygen to diffuse from both sides of the foil.

To investigate the surface morphology of the α-Fe after reduction a small amount of oxygen was introduced into the furnacc, for one sample, passivating the iron allowing for it to be exposed to air without burning. A SEM image of the porous substrate coated with passivated Fe is shown in Figure 3.

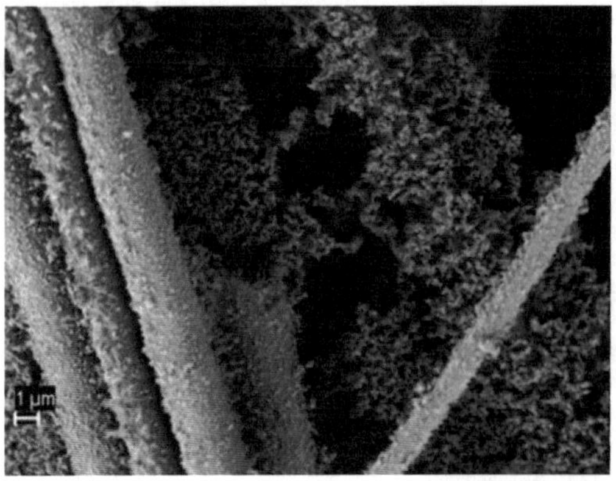

Figure 3: *SEM of the porous alumina substrate coated with passivated iron from the FeOOH sol-gel.*

In the experimental section it was mentioned that a surfactant can be added to the sol-gel. In theory this would increase the surface area of the material formed. Surface area analysis was done using BET. The Fe_2O_3 after calcination actually decreased with increased surfactant concentration. The use of a surfactant did not appear to have an effect on either the rise time or the maximum temperature that could be achieved but instead had an effect on the time length at which the foil burned at its maximum temperature.

In order to obtain a better understanding of the burning characteristics and be able to predict the pyrophoric response when exposed to air a mathematical model has been proposed. It was assumed for the initial model that the Fe foil could be modeled using the particle-pellet model for a gas solid reaction. Sampath et al. [12] proposed the particle-pellet model for the following reaction:

$$A_{(g)} + bB_{(s)} \longrightarrow \text{Products}$$

The use of this type of model would allow for the modeling of either diffusion controlled or reaction controlled reactions. Using Sampath's approach and the kinetic data data from Grosvenor et al. [12] the following mathematical model was derived. When modeling this system there are three equations that must be considered: 1) mass transfer of oxygen into the foil, 2) heat transfer from the reacting foil to the surroundings, and 3) the kinetic equation for the oxidation reaction. The key model equations are shown below (Eqs 1-3) and a schematic of the foil that is being modeled is shown in Figure 4.

$$D_{eff}\frac{\partial^2 C_A}{\partial x^2} - (-r_A) = \varepsilon_2 \frac{\partial C_A}{\partial t} \tag{1}$$

$$k_{eff}\frac{\partial^2 T}{\partial x^2} + (-\Delta H)(-r_A) = \left[(1-\varepsilon_1-\varepsilon_2)\rho_{Fe}C_{p_{Fe}} + \varepsilon_1\rho_i C_{p_i} + \varepsilon_2\rho_g C_{p_g}\right]\frac{\partial T}{\partial t} \tag{2}$$

$$-\frac{dr_c}{dt} = AR^{0.6}T^{0.6}C_A^{0.6}e^{-E/RT}e^{-\gamma(r_0-r_c)/RT} \tag{3}$$

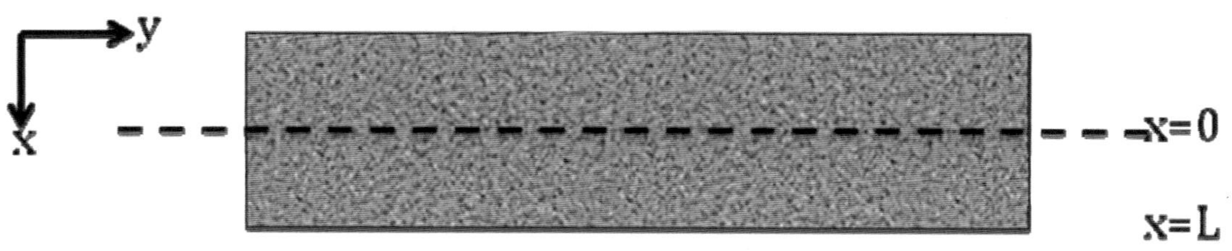

Figure 4: *Schematic of Fe foil for the mathematical model.*

The model equations were converted to a dimensionless form for numerical convience. The dimensionless equations are shown below (Eqs 4-6). The equations were derived with the assumption that diffusion inside a porous structure is the rate limiting step. Knudsen diffusion was considered in this case only.

$$\frac{\partial^2 Y_A}{\partial \xi^2} - K_2 Z^2 \Theta^{0.6} Y_A^{0.6} e^{\frac{-(\beta_1+\beta_2(1-Z))}{\Theta}} = K_1 \frac{\partial Y_A}{\partial \tau} \tag{4}$$

$$\frac{\partial^2 \Theta}{\partial \xi^2} + \Omega K_2 Z^2 \Theta^{0.6} Y_A^{0.6} e^{\frac{-(\beta_1+\beta_2(1-Z))}{\Theta}} = K_3 \frac{\partial \Theta}{\partial \tau} \tag{5}$$

$$\frac{\partial Z}{\partial \tau} = Y_A^{0.6} \Theta^{0.6} e^{\frac{-(\beta_1+\beta_2(1-Z))}{\Theta}} \tag{6}$$

$$\tau = t \, {}^{A}\!/_{r_0} \, R^{0.6} T_g^{0.6} C_{Ag}^{0.6} \tag{7}$$

The dimensionless constants listed in equations 4-6 are defined in equations 8-13.

$$K_1 = \frac{\varepsilon_2 L^2}{D_{eff}} \frac{A}{r_0} R^{0.6} T_g^{0.6} C_{Ag}^{0.6} \tag{8}$$

$$K_2 = \frac{L^2}{C_{Ag} D_{eff}} \frac{3A}{b} \frac{(1-\varepsilon_1-\varepsilon_2)}{r_0 M_{Fe}} \rho_{Fe} R^{0.6} T_g^{0.6} C_{Ag}^{0.6} \tag{9}$$

$$K_3 = \frac{\left[(1-\varepsilon_1-\varepsilon_2)\rho_{Fe} C_{p_{Fe}} + \varepsilon_1 \rho_i C_{p_i} + \varepsilon_2 \rho_g C_{p_g}\right] L^2}{k_{eff} r_0} A R^{0.6} T^{0.6} C_{Ag}^{0.6} \tag{10}$$

$$\Omega = \frac{D_{eff} C_{Ag} (-\Delta H)}{k_{eff} T_g} \tag{11}$$

$$\beta_1 = \frac{E}{R T_g} \tag{12}$$

$$\beta_2 = \frac{\gamma r_0}{RT_g} \tag{13}$$

The initial and boundary conditions are as follows:

Initial conditions:

t = 0

$$\frac{\partial \Theta}{\partial \xi} = 0 \tag{14}$$

$$\frac{\partial Y_A}{\partial \xi} = 0 \tag{15}$$

Boundary conditions:

x = 0

$$\frac{\partial Y_A}{\partial \xi} = 0 \tag{16}$$

$$\frac{\partial \Theta}{\partial \xi} = 0 \tag{17}$$

x = L

$$\frac{\partial Y_A}{\partial \xi} = Sh(1 - Y_A|_L) \tag{18}$$

$$\frac{\partial \Theta}{\partial \xi} = Nu(\Theta - 1) \tag{19}$$

This system of partial differential equations is presently being solved using FlexPDE computational software. The terms, ε_1 and ε_2 account for the SiO_2-Al_2O_3 substrate and iron porosity, respectively. The generation of pyrophoric foils using a Fe based sol-gel method has been shown and the mathematical model is currently being investigated.

Acknowledgements

The authors of this paper acknowledge the financial support by the Army Research Development Engineering Center in Picatinny Arsenal, NJ (Contract No. WI5QKN-06-D-0006).

Definition of Variables

A = pre-exponential factor for Arrhenius equation
B = stoichiometric coefficient of reaction
C_A = concentration of oxygen at any point, (mol/m^3)
C_{Ag} = concentration of oxygen in the bulk, (mol/m^3)
$C_{p,FE}$ = Specific heat capacity, (J/kg K)
$C_{p,g}$ = Specific heat capacity, (J/kg K)
$C_{p,i}$ = Specific heat capacity, (J/kg K)
D_{eff} = effective diffusivity, (m^2/s)
E = activation energy (kJ/mol)
k_{eff} = effect thermal conductivity, (W/m K)
Nu = Nusselt number
R = gas constant, (8.314 J/mol K)
r_c = radius at the reaction front, (nm)
r_o = radius of the particles, (nm)
Sh = Sherwood number
t = time, (sec)
T = Temperature at any point (K)
T_g = Temperature of the surrounding gas (K)
Y_A=dimensionless concentration , C_A/C_{Ag}
Z = dimensionless radius, r_c/r_o

Greek Symbols

ξ = dimensionless distance, x/L
ε_1 = porosity of the inert substrate
ε_2 = porosity of the iron
Θ = dimensionless temperature T/T_g
τ = dimensionless time
ρ_{Fe} = density of Fe, (m^3/kg)
ρ_g = density of gas, (m^3/kg)
ρ_i = density of inert substrate, (m^3/kg)
γ = increase in activation energy with respect to oxide layer, (kJ/mol nm)

References

[1] E.A. Shafranovsky, Yu. I. Petrov, "Aerosol Fe Nanoparticles with the Passivating Oxide Shell", Journal of Nanoparticle Research. 6, 1, 71-90, 2004.

[2] Xuanke Li, Zhongxing Lei, Rongcui Ren, Jing Liu, Xiaohua Zuo, Zhijun Dong, Houzhi Wang, Jianbo Wang, "Characterization of carbon nanohorn encapsulated Fe particles", Carbon, 41, 15, 3068-3072, 2003.

[3] Stefan G. Ruehm, Claire Corot; Peter Vogt, Stefan Kolb, Jörg F. Debatin, "Magnetic Resonance Imaging of Atherosclerotic Plaque With Ultrasmall Superparamagnetic Particles of Iron Oxide in Hyperlipidemic Rabbits" Circulation, 103 , 415-422, 2001.

[4] S. A. Majetich, Y. Jin, "Magnetization Directions of Individual Nanoparticles" Science, 284, 5413, 470 – 473, 1999.

[5] Satoshi Tomita, Masahiro Hikita, Minoru Fujii, Shinji Hayashi, Keiichi Yamamoto, "A new and simple method for thin graphitic coating of magnetic-metal nanoparticles", Chemical Physics Letters, 316(5-6),361-364, 2000

[6] Yoshikazu Homma, Takayuki Yamashita, Paul Finnie, Masato Tomita, Toshio Ogino, "Single-Walled Carbon Nanotube Growth on Silicon Substrates Using Nanoparticle Catalysts", Jpn. J. Appl. Phys., 41, L89-L91, 2002

[7] C. H. Liang, G. W. Meng, L. D. Zhang, Y. C. Wu, Z. Cui, "Large-scale synthesis of β-SiC nanowires by using mesoporous silica embedded with Fe nanoparticles", Chemical Physics Letters, 329, 3-4, 323-328, 2000.

[8] A.E. Gash, J.H. Satcher Jr., R.L. Simpson, "Preparation of porous pyrophoric iron using sol-gel method", US Patent 20060042412, Appl. No. 11/165734.

[9] R. V. Shende, A. Vats, Z.D. Doorenbos, D. Kapoor, D. Martin, J.A. Puszynski, "Formation of pyrophoric iron particles by H_2 reduction of oxalates and oxides", 7th International Symposium on Special Topics in Chemical Propulsion, Kyoto, Japan, Sep. 17-21, 2007 (to be published).

[10] Rajesh Shende, Alok Vats, Zac Doorenbos, Deepak Kapoor, Christopher Haines, Darold Martin, and Jan Puszynski, "Formation of Pyrophoric α-Fe Nanoparticles from Fe(II)-Oxalate," NSTI-Nanotech (ISBN 978-1-4200-8503), 1, 692, 2008.

[11] C.K. Wilharm, "Combustion Model for Pyrophoric Metals Foils", Propellants, Explosives, Pyrotechnics, 6, 28, 296-300, 2003.

[12] B.S. Sampath, P.A. Ramacandran, R. Hughes, "Modeling of Non-Catalytic Gas-Solid Reaction –I. Transient Analysis Of The Particle-Pellet Model", Chemical Engineering Science, 30, 125-134, 1975.

[13] A.P. Grosvenor, B.A. Kobe, N.S. McIntyre, "Activation Energies for Oxidation of Iron by Oxygen Gas and Water Vapour", Surface Science, 574, 317-321, 2005.

491g Directed Release of Chemical Energy In Single Walled Carbon Nanotube Films

Wonjoon Choi, Mechanical engineering, Massachusetts Institute of Technology, Building 66-153, 25 Ames st, Cambridge, MA 02139 and Michael S. Strano, 66-566 Department of Chemical Engineering, Massachusetts Institute of Technology, 77 Massachusetts Avenue, Cambridge, MA 02139.

Single Walled Carbon Nanotubes (SWNT) are promising materials as one-dimensional thermal conduits due to long phonon mean free paths along the axial length of the nanotube. Previous calculations predict that oxidation reactions conducted in an annular region encompassing the nanotube experience an increased reaction velocity in the direction of the nanotube length up to a factor of 10 to the bulk isotropic reaction(1). According to this model, vertically aligned carbon nanotube films should demonstrate even greater enhancement. In this work, we describe the synthesis, characterization and testing of a new class of 1-D energetic materials as chemical actuators and unique materials for propulsion. The porosity of aligned SWNT films was used to disperse mixtures of RDX and sodium azide by a wet impregnation method. SEM of the resulting film cross section shows annular coatings of the energetic crystals encompassing SWNT bundles in the film. While control samples without the carbon scaffold are unreactive, impregnated films demonstrate rapid energetic release that is orientated along the alignment direction of the film. High speed video microscopy is utilized to estimate reaction front velocities and specific impulse. Lastly, we discuss measurements of the temporal force evolution of such reactions. Applications to MEMS actuators, propulsion devices and synthetic platforms are discussed.

References

1. Joel T Abrahamson, Nitish Nair, Michael S Strano, Nanotechnology 19, (2008) 195702

496a A Flame-Pyrolysis Reactor for Carbon Black Production: Yield from Thermal Decomposition of Heavy Hydrocarbon Fuels

X. Ma[1], Anshuman Lall[2], Dale Huang[2], and Michael Zachariah[3]. (1) University of Maryland-College Park, College Park, MD 20742, (2) Department of Mechanical Engineering, University of Maryland-College Park, College Park, MD 20742, (3) University of Maryland and NIST, 2125 Glenn L Martin Hall, College Park, MD 20742

A new flame pyrolysis reactor for carbon black synthesis from liquid hydrocarbon fuels is developed. The reactor system consists of a surface mixed oxygen/methane diffusion burner, downstream of which the heavy hydrocarbon precursor is injected. The hot mixture flows directly into a high temperature isothermal plug-flow reactor, from which samples are directly taken for on-line measurement of aerosol size and yield. The reactor is designed to produce carbon black (CB) at a rate of 100 g/h at overall equivalence ratio varying from 2.5 to 4.0. For these studies Methyl-naphthalene is used as the hydrocarbon source for the CB which is injected downstream of the flame fuel in vapor form. Particles nucleate and subsequently grow by coagulation and surface growth in the isothermal reactor region. Nitrogen injection is used to quench the chemistry and aerosol dynamics and minimize transport losses during sampling. Particle size distributions are obtained on-line, using a differential mobility analyzer, and total carbon black yield is obtained by on-line oxidation of the CB and measurement of the resulting CO/CO2 concentrations. Our preliminary yield measurements with varying overall equivalence ratios indicate that the soot yield increases with the overall equivalence ratio at a given burner equivalence ratio. Also, the temperature independence of soot yield measured at furnace temperature between 1300C-1500C indicates that the soot formation process is not kinetic controlled over the testing temperature range.

496b Large-Scale Gas Phase Synthesis of Zinc Sulfide Nanoparticles

Evagelos K. Athanassiou, Robert N. Grass, and Wendelin J. Stark. Chemistry and Applied Biosciences, ETH Zurich, Wolfgang-Pauli-Str. 10, ETH Hönggerberg, Zurich, 8093, Switzerland

Flame synthesis has been applied for several decades for the large-scale manufacturing of metal oxides such as silica and titania. Recent developments including flame spray pyrolysis have allowed the large scale production of functional nanometerials starting from nano-gypsum[1], calcium-phosphate[2] for bio-applications, to non-noble metal[3-7] nanoparticles, complex alloys[8] and metallic / ceramic[9] nanocomposites for mechanical, magnetic or electronic applications.

Here we present how thermodynamics allowed us to further extend the principle of flame synthesis for the production of sulfide nanoparticles. The oxygen-limiting conditions in collaboration with sufficient amounts of a sulfur source allowed the production of zinc sulfide nanoparticles (10 to 40 nm, Figure 1) at much higher production rates (in a lab-scale reactor: 5 to 10 g h-1) than typical wet-based processes of quantum dots. The zinc sulfide nanoparticles could be further doped with other elements to further enhance its electrical properties such as photoluminescence. We demonstrate how versatile gas phase processes are and how they could offer an interesting alternative for the large-scale production of photoluminescent materials.

Fig. 1: The principle of flame spray pyrolysis has been further extended for the preparation of zinc sulfide nanoparticles. Transmission electron microscopy images of zinc sulfide nanoparticles show that the as prepared particles have an average size of 10 to 40 nm (left) and exhibit high crystallinity (right).

[1] N. Osterwalder, S. Loher, R. N. Grass, T.J. Brunner, L. K. Limbach, S. C. Halim, W. J. Stark, J. Nanopart. Res., 2007, 9, 275.

[2] T.J. Brunner, M. Bohner, C. Dora, C. Gerber, W.J. Stark, J. Biomed. Mater. Res. Part B, 2007, DOI: 10.1002/jbm.b.30809

[3] E. A. Athanassiou, R. N. Grass, W. J. Stark, Nanotechnology, 2006, 17, 1668.

[4] R. N. Grass, W. J. Stark, J. Mater. Chem., 2006, 16, 1825.

[5] R.N. Grass, W.J. Stark, J. Nanopart. Res., 2006, 8(5), 729.

[6] R.N. Grass, M. Dietiker, C. Solenthaler, R. Spolenak, W.J. Stark, Nanotechnology, 2007, 18(3), 035703.

[7] R.N. Grass, E.K. Athanassiou, W.J. Stark, Angew. Chem. Int. Ed., 2007, in print

[8] E.K. Athanassiou, R.N. Grass, N. Osterwalder and W.J. Stark, Chem. Mater, 2007, 19, 4847.

[9] R.N. Grass, T.F. Albrecht, F. Krumeich, W.J. Stark, J. Mater. Chem., 2007, 17, 1485.

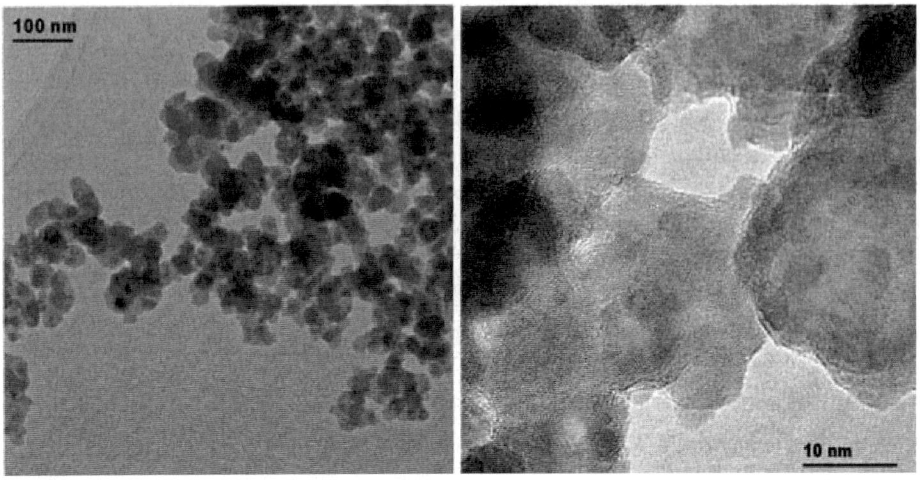

Synthesis and characterization of zinc oxide nanoparticles for dye-sensitized solar cells

Renata Marczak, Wolfgang Peukert
Institute of Particle Technology, Friedrich-Alexander-University Erlangen-Nuremberg, Cauerstr. 4, 91058 Erlangen, Germany

Fabian Werner, Dirk M. Guldi
Department of Chemistry and Pharmacy, Friedrich-Alexander-University Erlangen-Nuremberg, Egerlandstr. 3, 91058 Erlangen, Germany

Jan-Frederik Gnichwitz, Andreas Hirsch
Institute of Organic Chemistry, Friedrich-Alexander-University Erlangen-Nuremberg, Henkestr. 42, 91054 Erlangen, Germany

Abstract

Stable ZnO nanoparticles suitable for further derivatization of their surface were synthesized in the liquid phase from homogeneous ethanolic solutions of the precursors, lithium hydroxide and zinc acetate. The particle growth was monitored by in situ absorption spectroscopy and dynamic light scattering (DLS). It was found that the growth of the particles was governed by temperature as well as the presence of the reaction by-product lithium acetate during the ripening process. In particular, the reaction could be almost completely arrested by removal of this by-product. The "washing" consisted of repeated precipitation of the ZnO particles by addition of alkanes such as heptane, removal of the supernatant, and redispersion in ethanol. The ZnO nanoparticles, before and after "washing", were characterized by absorption spectroscopy, DLS, thermogravimetric analysis (TGA), X-ray powder diffraction (XRD) and high resolution transmission electron microscopy (HRTEM).

The functionalization of the ZnO nanoparticles with photoactive organic compounds, i.e., 5-(N-(3,4-dihydroxyphenethyl)-2-phenoxyacetamide)- 10, 15, 20-(p-tert-butyl-triphenyl)-porphyrinato-zinc and 5-(3,4-dihydroxy-N-phenylbenzamide)-10,15,20-tris(4-tert-butylphenyl) porphyrinato-zinc, was examined. Mesoporous films of the ZnO nanoparticles suitable for use as electrodes in dye-sensitized solar cells (DSSCs) were prepared by the doctor blade technique. Following sintering the films were sensitized by the dye molecules mentioned above. The morphology of the films was characterized by using scanning electron microscopy (SEM). The current generated during photoelectrochemical measurements of these films was ascribed to an efficient photoinduced electron transfer from the photoexcited porphyrin to the ZnO moiety.

Introduction

Dye-sensitized solar cells (DSSCs),[1] based on photoelectrochemical dye-sensitized mesoporous metal-oxide electrodes, have been a subject of intense interest for several years. This is due to the fact that they offer a potentially low cost, medium performance alternative to traditional photovoltaic devices.[2] In most work on DSSCs, titanium dioxide (TiO_2) has been the material of choice and so far has been shown to enable the highest overall light conversion efficiency.[1, 3] However, due to it having a similar size band gap energy, ZnO (~ 3.4 eV) has recently been explored as an alternative material for DSSCs.[4, 5] Although its potential has yet to be fully realized ZnO has significant advantages as a mesoporous electrode for DSSCs when compared to TiO_2, including much lower UV photocatalytic activity,[5] fewer electron-trapping surface states,[5] simpler tailoring of the nanostructure and easier dye-functionalization of the surface.[6] These advantages are thought to provide a promising means for improving the overall solar cell performance by improving the working electrode in dye-sensitized solar cells.

To obtain ZnO nanostructures different synthesis routes have been devised, including solid-vapor phase thermal sublimation,[7] spray pyrolysis,[8] RF plasma synthesis,[9] sonochemical or microwave-assisted synthesis[10] and hydrothermal processing.[11] However, wet-phase synthesis of ZnO is an area of intense interest, as it provides a low-temperature, economical way to produce various ZnO nanostructures.[12, 13] Growth from solution typically starts with nucleation and is followed by growth of the nuclei by addition of dissolved species until the metal cation concentration reaches the solubility of the oxide. Then particle ageing proceeds in the mother liquors. Synthesis conditions such as temperature,[14] the overall concentration of the precursors,[15] water concentration,[16] ions present in solution[17], and solvent[18] have been shown to impact particle nucleation and growth. Therefore, the synthesis and investigation of ZnO nanoparticles suitable for further functionalization of their surface for DSSCs application is of great interest.

Herein, we report about a simple colloidal method to synthesize ZnO nanoparticles suitable for further derivitization of their surface. The particles were prepared from zinc acetate dihydrate in ethanolic solution under basic conditions. To learn more about the growth mechanism, we have employed in situ Uv-vis absorption spectroscopy and dynamic light scattering (DLS) to quantify both particle size and energy band gap as a function of time in order to study influence of the ageing conditions on the growth of the ZnO nanocrystals. Next, we describe the photosensitization of mesoporous films of the ZnO nanoparticles by dye molecules, i.e., 5-(*N*-(3,4-dihydroxyphenethyl)-2-phenoxyacetamide)-10, 15, 20-(p-tert-butyl-triphenyl)-porphyrinato-zinc and 5-(3,4-dihydroxy-N-phenylbenzamide)-10,15,20-tris(4-tert-butylphenyl) porphyrinato-zinc (Scheme 1), for use as electrodes in dye-sensitized solar cells (DSSCs).

Experimental

All chemicals (analytical grade reagents) were purchased from commercial sources and used without further purification. Colloidal ZnO nanoparticles were prepared by hydrolyzing zinc acetate dihydrate in basic ethanol solution. The overall preparation procedure was adapted from Spanhel, Anderson[12] and Meulenkamp[13] with a few modifications. Homogenous solutions containing 0.1 M Zn^{2+} and 0.14 M OH^- were prepared by dissolving separately the reactants in boiling ethanol at atmospheric pressure. Both solutions were cooled to synthesis temperature. A white powder of anhydrous zinc acetate precipitated in a range close to room temperature.[13] The lithium hydroxide solution was added dropwise to the zinc acetate solution under vigorous stirring. A molar ratio of $[Zn^{2+}]/[OH^-]$ ratio was adjusted to 0.71. The reaction mixture became transparent after addition of about 1/3 volume of the lithium hydroxide solution. The ZnO sol was stored at ≤ 10 °C to prevent rapid particle growth. Next, the ZnO sol was washed by repeated flocculation of ZnO induced by addition of n-heptane. The supernatant was removed by centrifugation and decantation. For colloidal characterization the ZnO flocculates were redispersed in ethanol. In order to obtain the powder, the ZnO white precipitate was dried under nitrogen for about 5 minutes.

In order to obtain the 5-(N-(3,4-dihydroxyphenethyl)-2-phenoxyacetamide)- 10, 15, 20-(p-tert-butyl-triphenyl)-porphyrinato-zinc (DOPAZ) a solution of 500 mg of 5-(p-tert-butyl-acetato-phenoxy)-10, 15, 20-(p-tert-butyl-triphenyl)-porphyrin[19] (0.55 mmol) and 200 mL formic acid was stirred for 4 hours to obtain the deprotected acid. The completed reaction was observed via TLC. After that period the solvent was removed on a rotary evaporator, transferred into toluene and evaporated twice. The product was finally dried under reduced pressure. The dried product was dissolved in dichloromethane-DMF 1:1 mixture at 0° C. 315 mg of EDC (1.65 mmol), 222 mg HOBT (1.65 mmol) and 200 mg DMAP (1.65 mmol) were added and the solution was stirred for one hour at 0° C. After that period 155 mg of Dopamin (0.825 mmol) were added to the solution and the mixture was stirred at room temperature for 48 hours. The solvent was removed on a rotary evaporator and the residue was purified with column chromatography on silica gel with a mixture of dichloromethane/methanol (19:1) as eluent. The product was obtained by crystallization in pentane. The yield was 250 mg (45%).110 mg of 5-(N-(3,4-dihydroxyphenethyl)-2-phenoxyacetamide)- 10, 15, 20-(p-tert-butyl-triphenyl)-porphyrin (0.11 mmol) was dissolved in THF and 80 mg of zinc acetate (0.44 mmol) was added. The mixture was heated to reflux over a period of 4 hours. The reaction was monitored by TLC by the disappearance of the free porphyrin. The solvent was removed on a rotary evaporator and the residue was dissolved again in dichloromethane and filtered to remove any unreacted zinc acetate. The dried product was purified with column chromatography on silica gel with a mixture of dichloromethane/methanol (95:5) as eluent. Final crystallization in pentane gave a purple red solid in good yields (40 mg, 35%). To synthesise 5-(3,4-dihydroxy-N-phenylbenzamide)-10,15,20-tris(4-tert-butylphenyl) porphyrinato-zinc (CAMIZ) 100 mg of 3,4-diphenylmethylenedioxyprotocatechuic acid[20] (0.31mmol) was dissolved in 200 mL dichloromethane at 0° C then 130 mg of DCC (0.62 mmol) and 85 mg HOBT (0.62 mmol) were added and the solution was stirred for one hour at 0° C. After that period 250 mg of 5-(4-aminophenyl)-10,15,20-tris(4-tert-butylphenyl)porphyrin[21] (0.62 mmol) were added to the solution and the mixture was stirred at room temperature for 96 hours. The solvent was removed on a rotary evaporator and the residue was purified by column chromatography on silica gel with dichloromethaneas eluent. The yield was 120 mg (35%). 100 mg of the dried residue were dissolved in 100 mL of formic acid. The solution was stirred further for 5 hours. The progress of the reaction can be monitored via TLC. After completion of the reaction, the solvent was removed by rotary evaporation. The mixture was transferred into toluene and evaporated twice. The final product is obtained by recrystallization in toluene over night at 0°C with a yield of 83 mg (98%). 80 mg of 5-(3,4-dihydroxy-N-phenylbenzamide)-10,15,20-tris(4-tert-butylphenyl) porphyrin (0.09 mmol) was dissolved in 50 mL THF and 65 mg of zinc acetate (0.36 mmol) was added. The mixture was heated to reflux over a period of 4 hours. The reaction was monitored by TLC by the disappearance of the free porphyrin. After completion of the reaction water is added till the product is precipitated. The solid is filtered and then dried in vacuum to give the desired product in a nearly quantitative yield (81 mg, 95%).

Particle size distribution, zeta potential and conductivity of ZnO nanoparticle suspensions were determined via dynamic light scattering (DLS) by using a Malvern Nano ZS Instrument with a 633 nm "red" laser. For each sample ten measurements were performed and the average value per sample was calculated. The optical properties of the nanoparticles were determined from Uv-vis absorption spectra recorded with a Cary 100 Scan Spectrometer (Varian). Spectra from colloidal suspensions were obtained with 10 mm cells. Structural analysis of the ZnO nanoparticles was performed in a D8 Advance (Bruker AXS) X-Ray diffractometer (XRD) using Cu-Kα radiation (0.154 nm). The measurement was in the range of $20° \leq 2\theta \leq 70°$. Ultra thin carbon films on a 200 mesh Cu-grid were used to place the samples for high resolution transmission electron microscopy (HRTEM) observations. The HRTEM images were obtained using a Philips CM 300 UltraTwin microscope. The measurements were carried out at an acceleration voltage of 300 kV in the dark-field mode.

Mesoporous films of the ZnO nanoparticles were prepared by the doctor blade technique.[22] For the preparation of the starting coating paste, 0.10 g of starch was solubilised in 0.75 g of ethanol and then 0.15 g of the ZnO nanoparticles were added to the solution. Pieces of F:SnO_2 conducting glass (8Ω/cm^2 TCO) were cleaned subsequently with water, acetone and isopropanol in an ultrasonic bath. A drop of the ZnO paste was then applied to a piece of F:SnO_2 conducting glass on one of the bare edges of the substrate and flattened with a glass rod by sliding over the tape-covered edges. Such prepared films were dried at 80°C for 10 min and heated at 30°C min^{-1} to 450 °C; this temperature was held for 30 min. Next, the films were cooled slowly down to room temperature. The morphology of the films was examined by scaning electron microscope (SEM). The images were obtained by using a Zeiss Gemini Ultra 55 SEM. The measurements were carried out at the acceleration voltage of 20 kV in the dark-field mode.

Photoelectrochemical measurements were carried out in a custom made cell holder with the dye molecule/mesoporous ZnO/F:SnO_2 working electrode and a Platinum coated F:SnO_2 counter electrode. The electrolyte was 0.6 M 1,2-dimethyl-3-propylimidazolium iodide, 0.1 M lithium iodide and 0.05 M iodine in methoxyacetonitrile. Photocurrent measurements were carried out using a Keithley Sourcemeter 2400 during illumination. A collimated light beam from a adjustable 1000 W Xenon arc lamp was used for UV and VIS illumination closely matching AM 1.5 conditions. When white light was used, a 350 nm cut-off filter was used. When recording the photoaction spectrum, a B&M Spektronik grating monochromator was introduced into the path of the excitation beam for selecting the required wavelengths. All measurements were performed after subtracting stable dark current. The incident photon-to-current conversion efficiency (IPCE) defined as the number of electrons collected per incident photon, was determined from short-circuit photocurrent measurements at different wavelengths. All experiments were performed at room temperature.

Results and discussion

ZnO nanoparticles were prepared from zinc acetate dihydrate in ethanolic solution under basic conditions. The zinc acetate and lithium hydroxide solutions in ethanol were mixed at different temperatures under vigorous stirring. In order to follow the progress of the oxide formation, the suspensions of the nanoparticles were aged in their mother liquors. The influence of nanocrystal size on the electronic structure of semiconducting material is represented by the band gap increasing with decreasing of the particle size, which is attributed to the so-called quantum confinement effect. ZnO shows this effect for particles smaller than 8 nm.[23] Hence, Uv-vis absorption spectra provide a convenient way to investigate particle growth. Figure 1 shows the absorption spectra of the ZnO suspension recorded immediately after the preparation (about 2 minutes) and performed in 20 minutes intervals for about 14 h. Just after mixing the zinc acetate with the lithium hydroxide the absorbance spectrum shows a well-defined exciton peak, which suggests a very fast nucleation. Beside, a meaningful red shift in the absorption onset was observed during the early formation stage, which is attributed to the fast crystal growth. Additionally, the absorption measurements provide a qualitative indication of the crystal size distribution. The sharp excitonic peak in the absorption spectra in the case of small nanocrystals (at shorter times) is indeed indicative of the narrow size distribution of the nanoparticles in the sample. For the larger particles, sharp excitionic features are not present in the absorption spectra (at longer times), but only a broad and featureless absorption edge. This is due to the fact that a number of exciton peaks appear at different energies corresponding to different sized nanocrystals which overlap with each other. Therefore, a broadened size distribution of the nanocrystals can be expected. This assumption is confirmed by the number size distributions recorded by using DLS at different times during the aging of the particles (Figure 2). The longer the aging time, the larger particles with broader size distributions were obtained. The influence of the ageing temperature, i.e. 2 °C, 15 °C, 25 °C and 35 °C, on the particle growth was examined. The higher the temperature the larger

ZnO nanocrystals were formed, which is confirmed by the hydrodynamic diameter changes during the ZnO nanoparticle ageing recorded by DLS (Figure 3). These results show that the nanocrystals continued to age after the synthesis, even when stored at low temperatures. The ability to obtain various particle sizes is based on the phenomena that the growth of the particles is governed by temperature and time, but also by the presence of the reaction byproduct lithium acetate during the ageing process. The reaction could be almost completely stopped by removing the byproduct. This process, so-called "washing", consisted of reversible flocculation of the ZnO nanoparticles by addition of n-heptane after about four hours of ageing.[13] The particles precipitated immediately after adding heptane to the ethanolic suspension. After centrifugation and removing of the supernatant, the white ZnO flocculates were either redispersed in ethanol by using ultrasonic bath to obtain a transparent suspension or dried under nitrogen for about 5 minutes to get a white powder.

Structural analysis of the sample was performed by using X-Ray diffraction. In the XRD pattern of the ZnO nanoparticles (Figure 4) no impurity peaks are observed. All the diffraction peaks are well assigned to standard hexagonal phase ZnO with a wurtzite structure reported in JCPDS card (No. 36-1451, a = 3.249 Å, c = 5.206 Å). The broadness of the XRD peaks reveals the nanocrystalline nature of the ZnO powder. The average crystallite size of about 5 nm was estimated by using the Debye-Scherrer's equation.[24] The crystalline nature of the ZnO nanoparticles is evidenced also in the HRTEM image (Figure 5) by the clearly observed lattice features.

The ZnO nanoparticles were used for preparation of mesoporous films of the ZnO nanoparticles by the doctor blade technique. The surface morphology of the prepared nanostructured ZnO films was investigated by SEM. The SEM image shows that the ZnO particles form a porous film (Figure 6) suitable for DSSCs application. Next, the ZnO films were immersed into a 5 mM ethanolic solution of a dye-molecule and characterized as a photoelectrochemical cell with a Pt-counter electrode. Upon preparation of the F:SnO_2/mesoporous ZnO-dye molecule/Iodine-Iodide/Pt cell its photoaction spectra were recorded (Figure 7). An excellent match between the photoaction spectrum and the absorption spectrum of the dye molecule is observed. Both the Soret- (at 424 nm) and *Q*-band (550 and 600 nm) features are discerned. This photoaction spectrum confirms the origin of the photocurrent generation, namely, the chromophore activity of the dye molecule which is mainly responsible for the observed photocurrent. To conclude, the following electron-transfer sequence should be anticipated as a basis for the photocurrent generation: (i) initial photoexcitation of the dye molecules, (ii) charge separation between the photoexcited dye molecules and ZnO, (iii) electron shift from the reduced ZnO to the F:SnO_2 electrode, and (iv) electron transfer from the redox couple in the electrolyte to the oxidized dye molecules.

Conclusions

We have synthesized ZnO nanoparticles in ethanol by using a simple colloidal method. It was shown that the growth of the particles was governed by temperature as well as the presence of the reaction byproduct lithium acetate during the ageing process. This process could be almost completely stopped by removal of this byproduct by repeated flocculation of the ZnO particles by addition of n-heptane. We have shown the preparation of mesoporous films of the ZnO nanoparticles suitable for use as electrodes in dye-sensitized solar cells (DSSCs) and successful functionalization of the surface of the particles by dye molecules, i.e., 5-(N-(3,4-dihydroxyphenethyl)-2-phenoxyacetamide)- 10, 15, 20-(p-tert-butyl-triphenyl)-porphyrinato-zinc and 5-(tris-ethoxy-silane-propyl-amide-acetato-phenoxy)- 10, 15, 20-(p-tert-butyl-triphenyl)-porphyrinato-zinc. The current generated during photoelectrochemical measurements of these films was ascribed to an efficient photoinduced electron transfer from the photoexcited porphyrin to the ZnO moiety.

References

1. O'Regan, B.; Grätzel, M. *Nature* **1991**, *353*, 737.
2. Westermark, K.; Rensmo, H.; Siegbahn, H.; Keis, K.; Hagfeldt, A.; Ojamae, L.; Persson, P. *J. Phys. Chem. B* **2002**, *106*, 10102.
3. a) Green, M. A.; Emery, K.; King, D. L.; Hishikawa, Y.; Warta, W. *Prog. Photovol.: Res. Appl.* **2006**,*14*, 455. b) Hagfeldt, A.; Grätzel, M. *Acc. Chem. Res.* **2000**, *33*, 269. c) Grätzel, M. *J. Photochem. Photobiol. C* **2003,** *4*, 145.
4. a) Quintana, M.; Edvinsson, T.; Hagfeldt, A.; Boschloo, G. *J. Phys. Chem. C* **2007**, *111*, 1035. b) Redmond, G.; Fitzmaurice, D.; Grätzel, M. *Chem. Mater.,* **1994**, *6*, 686.
5. Zeng, L.; Dai, S.; Xu, W.; Wang, K. *Plasma Science & Technology* **2006**, *8*, 172.
6. Kakiuchi, K.; Hosono, E.; Fujihara, S. *J. Photochem. Photobiol. A* **2006**, *179*, 81.
7. Wang, Z. L. *J. Phys. Condens. Matter* **2004**, *16*, R829.
8. a) Park, S. B.; Kang, Y. C. *J. Aeroslo Sci.* **1997**, *28*, S473. b) Okuyama, K.; Lenggoro, I. W. *Chem. Eng. Sci.* **2003**, *58*, 537.
9. Sato, T.; Tanigaki, T.; Suzuki, H.; Saito, Y.; Kido, O.; Kimura, Y.; Kaito, C.; Takeda, A.; Keneko, S. *J. Cryst. Growth* **2003**, *255*, 313.
10. a) Hu, X. L.; Zhu, Y. J.; Wang, S. W. *Mater. Chem. Phys.* **2004**, *88*, 421. b) Hong, R. Y.; Shen, Z. H.; Li, H. Z. *J. Process Eng.* **2005**, *5*, 693.
11. a) Zhang, H.; Yang, D.; Ji, Y.; Ma, X. Y.; Xu, J.; Que, D. L. *J. Phys. Chem. B* **2004**, *108*, 3955., b) Liu, B.; Zeng, H. C. *J. Am. Chem. Soc.* **2003**, *125*, 4430.
12. Spanhel, L; Anderson, M. A. *J. Am. Chem. Soc.* **1991**, *113*, 2826.
13. Meulenkamp, E. A. *J. Phys. Chem. B* **1998**, *102*, 5566.
14. Wong, E. M.; Bonevich, J. E.; Searson, P. C. *J. Phys. Chem. B* **1998**, *102*, 7770.
15. a) Pacholski, C.; Kornowski, A.; Weller, H. *Angew. Chem. Int. Ed.* **2002**, *41*, 1188. b) Verges, M. A. Mifsud., A.; Serna, C. J. *J. Chem. Soc. Faraday Trans.* **1990**, *86*, 959.
16. Hu, Z.; Escamilla Ramirez, D. J.; Heredia Cervera, B. E.; Oskam, G.; Searson, P. C. *J. Phys. Chem. B* **2005**, *109*, 11209.
17. a) Hu, Z.; Oskam, G.; Penn, R. L.; Pesika, N.; Searson, P. C.; *J. Phys. Chem. B.* **2003**, *107*, 3124. b) Hu, Z.; Herrera Santos, J. F.; Oskam, G.; Searson, P. C.; *J. Colloid. Interface Sci.* **2005**, *288*, 313.
18. Hu, Z.; Oskam, G.; Searson, P. C. *J. Colloid Interface Sci.* **2003**, *263*, 454.
19. Wessendorf, F.; Gnichwitz, J.-F.; Sarova, G. H.; Hager, K.; Hartnagel, U.; Guldi, D. M.; Hirsch, A. *J. Am. Chem. Soc.* **2007**, *129*, 16057.
20. Iacazio, G.; Perissol, C.; Faure, B. *Journal of Microbiological Methods* **2000**, *42*, 209.
21. Liu, X.; Liu, J.; Jin, K.; Yang, X.; Peng, Q.; Sun, L. *Tetrahedron* **2005**, *61*, 5655.
22. Ito, S. *J. Mater. Chem.* **2004**, *14*, 385.
23. a) Koch, U.; Fojtik, A.; Weller, H.; Henglein, A. *Chem. Phys. Lett.* **1985**, *122*, 507. b) Wong, E. M.; Bonevich, J. E.; Searson, P. C. *J. Phys. Chem. B* **1998**, *102*, 7770. c) Brus, L. E. *J. Phys. Chem. B* **1986**, *90*, 2555.
24. West, A. R. Solid State Chemistry and Its Applications, John Wiley & Sons: New York, 1984; pp. 174-175.

a)

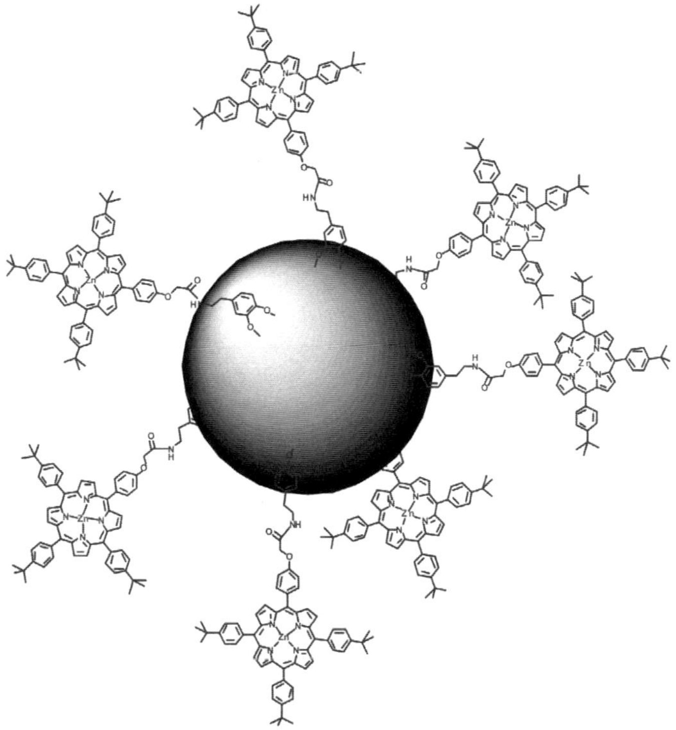

b)

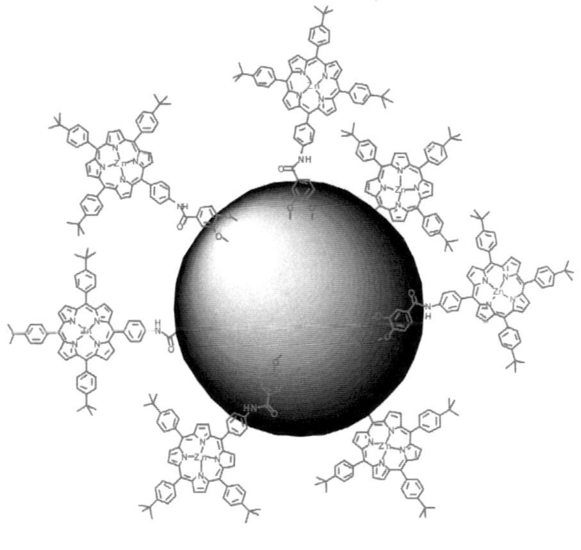

Scheme 1. a) 5-(N-(3,4-dihydroxyphenethyl)-2-phenoxyacetamide)- 10, 15, 20-(p-tert-butyl-triphenyl)-porphyrinato-zinc (DOPAZ) and b) 5-(3,4-dihydroxy-N-phenylbenzamide)-10,15,20-tris(4-tert-butylphenyl) porphyrinato-zinc (CAMIZ) molecules adsorbed on the surface of a ZnO nanoparticle.

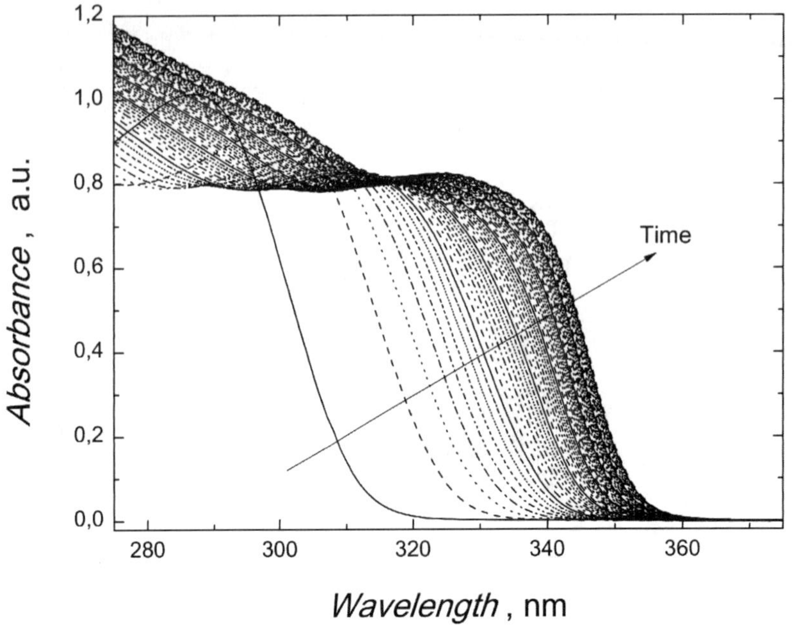

Figure 1. Room temperature absorption spectra recorded at different aging time of ZnO nanocrystals in ethanolic suspension at 35°C.

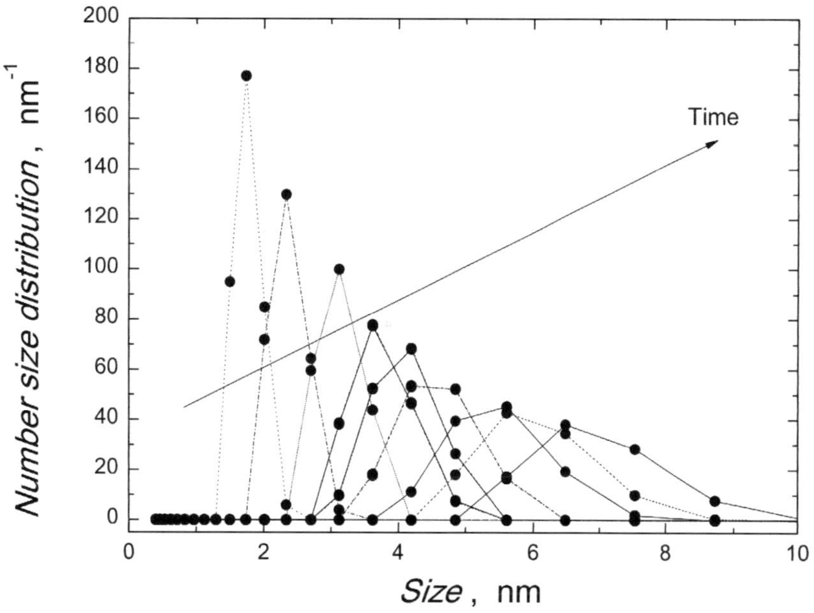

Figure 2. Number size distributions recorded by DLS at different aging time of the ZnO crystals in ethanolic suspension at 35 °C.

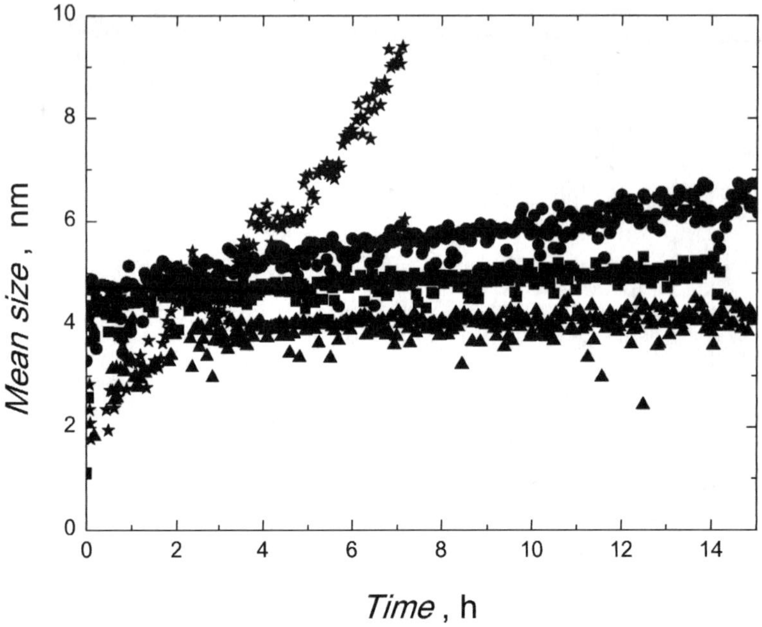

Figure 3. Particle ageing in the ZnO ethanolic suspension at 2 °C (triangles), 15 °C (squares), 25 °C (circles) and 35 °C (stars) monitored by DLS.

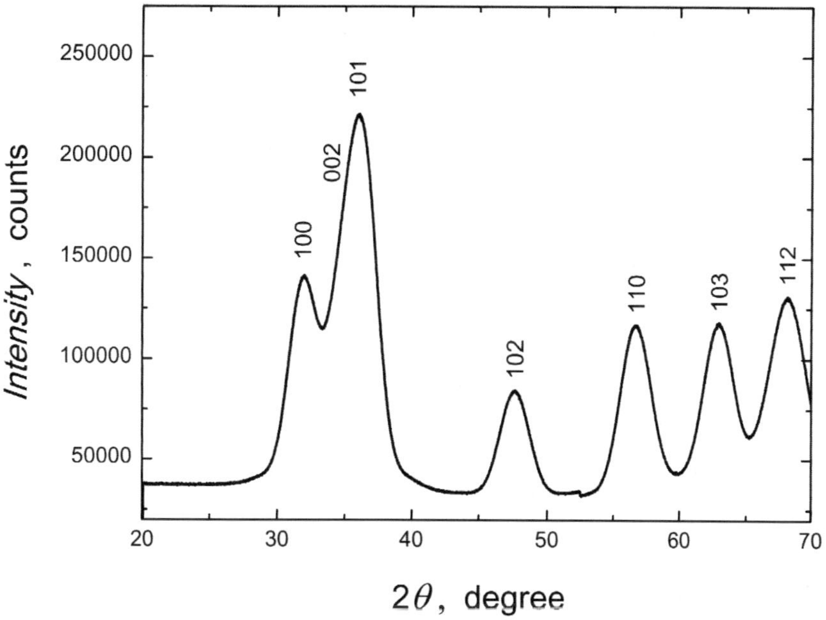

Figure 4. X-ray powder pattern of ZnO nanoparticles.

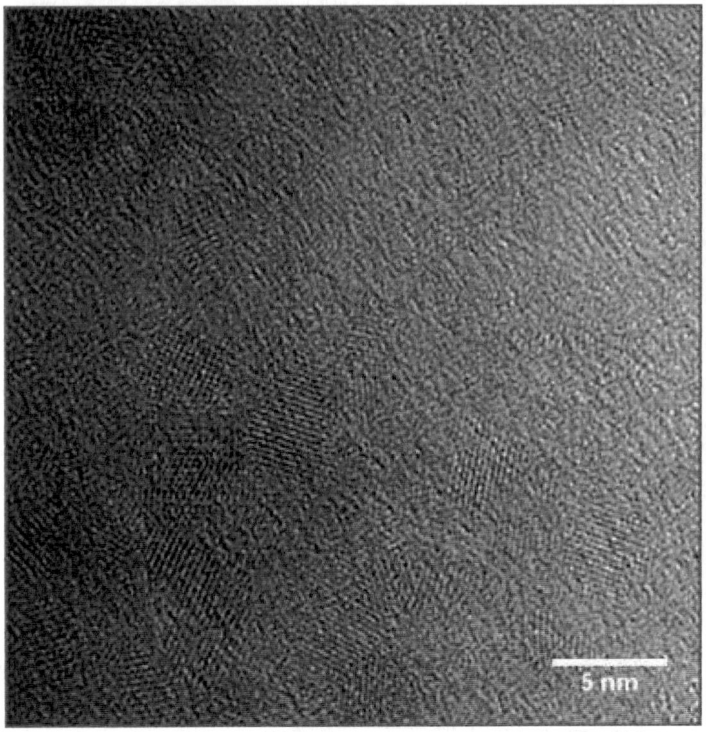

Figure 5. HRTEM image of ZnO nanoparticles. The crystalline nature of the ZnO nanoparticles is evidenced by the clearly observed lattice features.

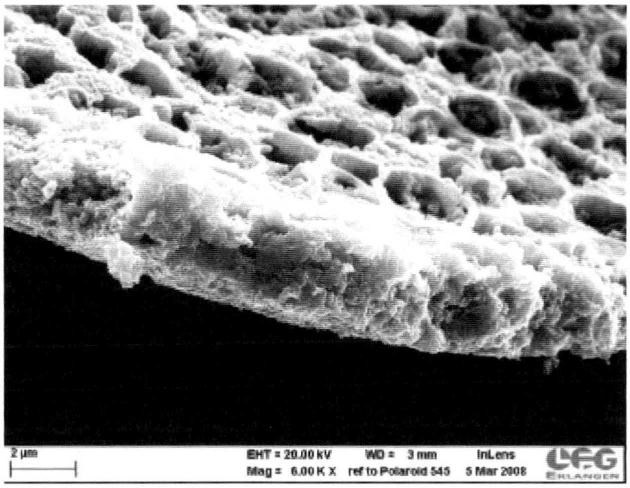

Figure 6. SEM image of a mesoporous ZnO film.

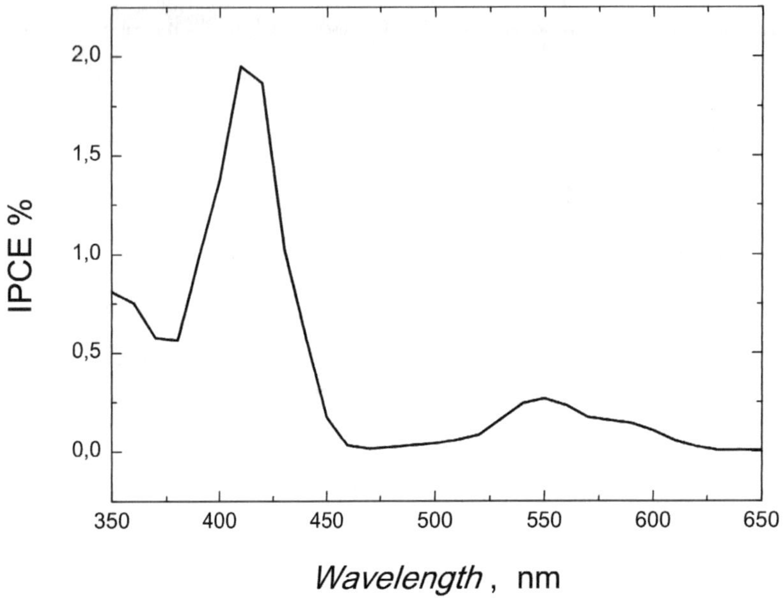

Figure 7. Photoaction spectra of the F:SnO_2/mesoporous ZnO-dye molecule/iodine-iodide/Pt cell filled with 0.6 M DMPII, 0.1 M LiI and 0.05 M Iodine in methoxyacetonitrile supporting electrolyte. Photocurrents were recorded upon monochromatic light illumination with no bias voltage being applied.

496d Simultaneous Molecular and Microemulsion Templating of Inorganic Oxide Particles with Hierarchal Porosity

Nick Carroll, Chemical and Nuclear Engineering, University of New Mexico, Farris Engineering Center MSC01 1120, 1 The University of New Mexico, Albuquerque, NM 87131 and Dimiter N. Petsev, Chemical & Nuclear Engineering, University of New Mexico, 1 University of New Mexico, MSC01 1120, Albuquerque, NM 87131.

Siliceous mesostructured microspheres with well-defined pores ranging in size from a few nanometers to over 20 nanometers have been synthesized using evaporative induced self assembly within emulsion droplets. The highly ordered microspheres were templated using a unique combination of amphiphile surfactants and oil microemulsions. The templated nano-structure of the microspheres were characterized by X-ray scattering, nitrogen sorption, and TEM techniques. BET surface areas reported are between 800 and 1000 square meters per gram. Additionally, this technique has been applied to synthesize other inorganic oxide particles, specifically, niobium oxide powders. The open system with pores of varying size and high surface area makes these particles attractive for applications in fuel cell anode templating, drug delivery, and as catalyst supports.

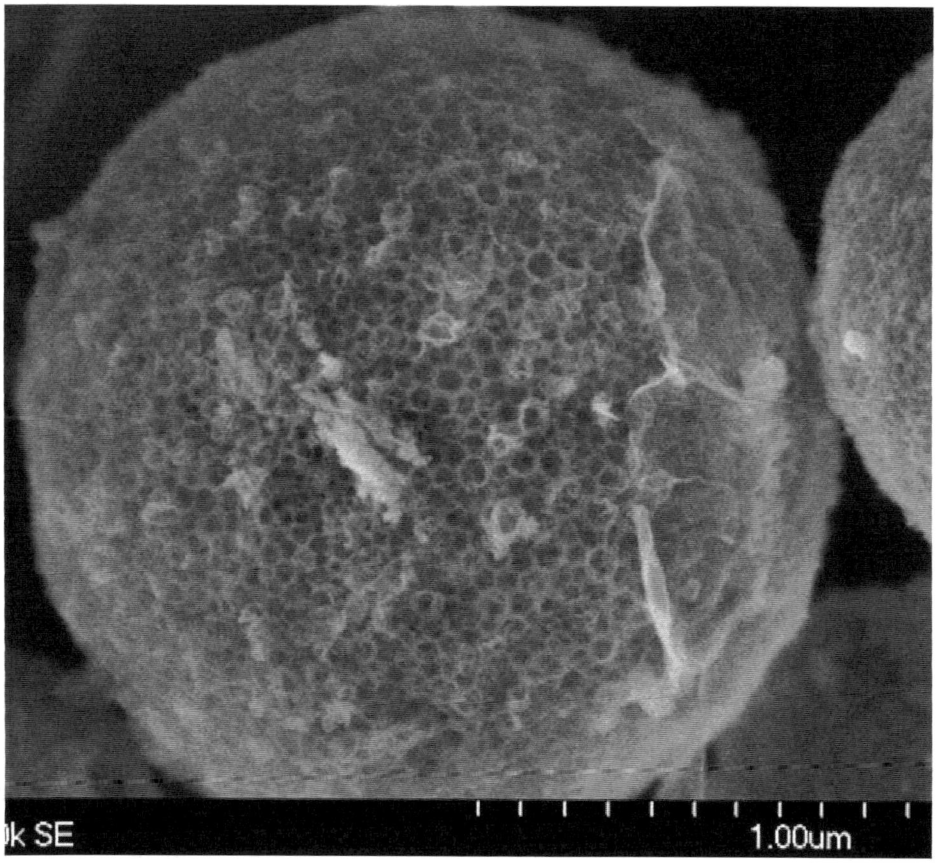

496e Use of Calorimetry Model and Batch Control Technique for Scale-up of Unseeded Batch Cooling Crystallization of Polyhydroxybenzophenon

Kwang Soon Lee[1], Huiyong Kim[2], and Youkyung Bang[2]. (1) Chemical & Biomolecular Engineering, Sogang Univ., 1-Shinsoodong, Mapogu, Seoul, 121-742, South Korea, (2) Chemical and Biomolecular Engineering, Sogang Univ., 1-Shinsoodong , Mapogu, Seoul, 121-742, South Korea

A scale-up method to design and implement a cooling profile for unseeded batch cooling crystallization has been investigated. Due to the reduced heat transfer area to volumn ratio, temperature differnce between the reactor and jacket increases in a large crystallizer, which may induce unwanted nucleation on the crystallizer wall.

Addressing this problem, a method to design a cooling-holding profile that minimizes the undershoot of the jacket temperature has been proposed together with a fine batch control technique that can practice the profile as designed.

Experiments have been conducted for scaling-up the result with a 100 mL crystallizer to a 5 L system for crystalliation of polyhydroxybenzophenon. As a consequence, it has been observed that the proposed method correctly works as anticipated.

496f Atomistic Simulation Studies of Stability Enhancement of Griseofulvin Nanocrystal Aqueous System by Adding Surfactant, Polymer, and Surfactant-Polymer Mixtures

M. Silvina Tomassone, **Wusheng Zhu**, and Frank Romanski. Chemical and Biochemical Engineering, Rutgers University, 98 Brett Road, Piscataway, NJ 08854

In recent years, major pharmaceutical companies have started using high-throughput screening and combinatorial chemistry to select good drug candidates. However, many poorly water soluble drugs (hydrophobic) are screened out far too early. Recent pharmaceutical advances have shown that preparing this type of drugs as nanocrystals can greatly increase their solubility and bioavailability. In order to maintain nanoparticle size, and reduce agglomeration, surfactants and polymers are used in the drug formulation. While experiments have shown that surfactants aid in nanocrystal stabilization, not all are created equally and different surface functionalities determine completely different behaviors. Therefore, surfactant/drug-surface interactions emerges as a relevant topic to apply molecular dynamic simulations. These simulations can validate experiments and even screen potential surfactant candidates prior to conducting experiments. In addition, molecular simulations can also examine the overall stability of the drug-surfactant system. In this work, atomistic simulations via Materials Studio® are employed to study the stability of nanocrystal-surfactant systems. Griseofulvin has been chosen as our poorly water soluble drug candidate. Griseofulvin is an antifugal drug with poor solubility which is used to treat skin or nails infections of both animals and humans. Several polymers and surfactants are tested including Tween 80, Pullulan, and Hydroxylpropylmethylcellulose (HPMC).

In our simulations, the drug is cleaved along the (100), (010), and (001) planes in order to study the interaction between the drug surface and the surfactant. The attachment energies between a specific drug crystal surface, the pure solvent, and the surfactant/polymer are calculated during the simulations. Hartman–Perdok theory suggests that the relative growth rate of crystal faces is proportional to the attachment energy of the faces. A comparison between attachment energies can conclude the overall effect of the stabilizer, or combination of stabilizers. In turn, the results reflect the stability of the nanocrystal aqueous system. Simulation results are obtained for non-ionic surfactant Tween 80, water-soluble polymer Pullulan and polymer HPMC. At this point, all three stabilizers have shown significant effects in stabilizing the nanocrystals in the aqueous system. Overall, HPMC is the most efficient stabilizer. Pullulan and Tween 80, while significant, are not as efficient in terms of stabilization if used alone. However, using a mixture of the two has shown a remarkable synergetic effect specifically on a side-by-side adsorption configuration of the surfactant and the polymer on the drug crystal surface. Other adsorption configurations, such as surfactant on top of the polymer or vice versa, do not show nearly as much improvement. This can imply that experimentally mixing Tween 80 with Pullulan with the nanocrystal suspension will achieve better stabilization than either stabilizer alone. Other comparisons between the three stabilizers have also been conducted, although they have not shown results as effective as the Tween 80/Pullulan system.

496g Characterization of Elastomer Particle Aggregation and Implications for Industrial Processing

Cornelius Gauer[1], Miroslav Soos[1], Hua Wu[1], and Massimo Morbidelli[2]. (1) Institute for Chemical and Bioengineering, Department of Chemistry and Applied Biosciences, ETH Zurich, Wolfgang-Pauli-Str. 10, HCI F 138, Zurich, 8093, Switzerland, (2) Institute for Chemical and Bioengineering, Department of Chemistry and Applied Biosciences, ETH Zurich, Wolfgang-Pauli-Str. 10, HCI F 133, Zurich, 8093, Switzerland

Coagulation is a typical process step to facilitate recovery of solid phase from the carrier liquid in the colloidal domain. Industrial coagulation takes advantage of the fact that particle aggregation is controlled not only by diffusive motion but as well by shear flow which limits particle size by breakage. For a given particle size the effectiveness of breaking aggregates greatly depends on their structure. This has implications on the processability of a coagulated colloid which is relevant for many applications e. g. in coagulation of elastomer particles. Aggregation studies in quiescent fluid can greatly contribute to understanding of particle size evolution and cluster structure. This work addresses the issue of transferability of information gained in static fluid to flow conditions.

Focus of our study is to characterize the aggregation of elastomer particles in static conditions. Simulation of the aggregation kinetics with comprehensive, physical models (population balances) provides useful information about the physics and chemistry governing the aggregation process. Experiments were carried out on partially as well as fully destabilized elastomer colloids at temperatures ranging from 25°C to 70°C. Static and dynamic light scattering were employed to measure particle size. Scanning electron micrographs of frozen colloids provided further information about cluster structure of selected samples.

Comparisons of latexes which differ in material as well as surface properties allow us to draw following general picture. Given a sufficiently low polymer viscosity or high polymer chain mobility clusters of surfactant stabilized elastomer particles always coalesce to form sphere-like clusters. Elastomer particles of the same bulk material but carrying fixed charges can coalesce at high temperatures e.g. 70°C but not at 25°C. This rather surprising finding is discussed in another contribution in terms of a temperature dependent structural barrier within the surface-to-surface contact region. High polymer chain mobility as pre-condition for coalescence seems greatly affected by elastomer bulk properties. Depending on the material composition cluster coalescence was found to be reduced or absent without significant influence of temperature. The results obtained from static aggregation have important consequences for aggregation in shear flow such that cluster breakage is impaired by coalescence resulting in increased steady state particle size. Even though latter quantity is the most important one for design of (industrial) coagulation process, static aggregation experiments are useful to characterize latexes in terms of cluster coalescence. Information about influence of material and surface properties might be therefore already considered in design of a preceding polymerization process.

522a Reactor Analysis of Coal Pyrolysis to Acetylene In Hydrogen Plasma Reactor

Jiaqi Chen and **Yi Cheng**. Department of Chemical Engineering, Tsinghua University, Beijing, 100084, China

The rapid decrease in the world reserves of hydrocarbons and natural gas and the considerable increase in their prices have stimulated the utilization of coal as the source of energy for the logical solution to the long-term energy problem. Meanwhile, coal has also been treated to produce hydrocarbons, pipeline gas and coal-derived liquid fuels, such as gasoline, methanol and olefins (e.g., by MTO process), for the chemical industry. As an example, a widely developed technique currently is coal gasification to syngas at the temperature of 1200-1500 C, which is the first step for coal chemistry to downstream chemical synthesis.

Different from these conventional routes for coal utilization, rapid pyrolysis of coal to gaseous products was ever paid much attention from 1960s to 1980s and seemed hindered for some reason in recent 20 years. When coal is heated to temperatures above 1500 C in milliseconds, a gaseous mixture is produced with the acetylene as the principal hydrocarbon constituent. Since the reaction environment is in hydrogen atmosphere, little CO and almost no CO2 are generated. In addition, the coal powders after the pyrolysis process can either serve as high-valued carbon materials or be burned to generate heat and electricity. Therefore, the coal pyrolysis to acetylene in hydrogen plasma is a rather clean process for coal utilization, which opens up a direct means for the conversion of coal to downstream chemical processes, e.g., to provide the feedstock to PVC industry. This novel process is an alternative to replace the conventional calcium carbide method with high energy cost and heavy pollution.

However, the high-severity operation of the coal pyrolysis process, i.e., high-temperature pyrolysis in a few milliseconds with a very rapid quenching rate, imposes great challenge to the commercialization in industry. This work is to introduce the latest development of a 2-MW plasma reactor for the purpose of coal pyrolysis to acetylene in China. The pyrolysis reactor consists of a V-shaped plasma generator, a downer reactor with unique coal injection nozzle design and a fast quenching operation. The volume concentration of acetylene in the product gas has reached 9.6% under the conditions of 2-MW power plasma and the feeding rate of 1 t/hr coal. The whole system has been scaled up to a 5-MW unit, which is ready for industry test.

To better understand the milliseconds reacting flow process inside the reactor, a CFD approach is established to model the coal pyrolysis in the hot hydrogen gas. The comprehensive model includes the sub-models for the turbulent gas flow, the multiphase interaction, the particle reactions, the gaseous reactions and the radiative energy transport. For gas phase, the Realizable k-≤` model and a mixture fraction approach are chosen to model the turbulent reacting flow. For the solid phase, the discrete phase model (DPM) is employed. The chemical percolation devolatilization (CPD) model is used to describe the devolatilization behavior of rapidly heated coal powders, and a simplified surface reaction is introduced to consider the C-H reactions at extremely high temperatures (> 2000 K). All the numerical simulations are conducted in 3-D domains using FLUENT-6.3 software package.

522b Multi-Scale Modeling and Control of Particulate Cvd for Production of Solar Grade Silicon In Fluidized Beds

B. Erik Ydstie, Department of Chemical Engineering, Carnegie Mellon University, 5000 Forbes Ave, Pittsburgh, PA 15213 and Juan Du, Chemical Engineering, Carnegie Mellon University, 5000 Forbes Ave, Pittsburgh, PA 15213.

In this paper we develop a multi-scale model to describe growth of silicon particles due to Chemical Vapor Deposition (CVD) in a Fluidized Bed Reactor (FBR). The reactor system is designed to make poly-silicon for solar cell applications by thermal decomposition of Silane. It is expected that the new FBR system will reduce the cost of solar grade silicon relative to the energy intensive decomposition of Tri-Chloro-Silane in the Siemens process, which dominates the market at present.

The complex interplay between continuous and disperse phases and CVD onto the Silicon particles in the FBR is modeled using three computational modules. The Computational Fluid Dynamics (CFD) module describes simplified hydrodynamics via momentum exchange, mass and heat transfer between different phases. The CFD system is assumed to have fast dynamics relative to the deposition rate so that we can use the quasi-steady state approximation. The reaction module describes the heterogeneous and homogeneous gas phase reactions which allow Silane gas to decompose and deposit on poly-crystalline silicon particles. Hydrogen escapes the system and carries with it fine silicon particles. The formation of fines represents a yield loss. It should be minimized by optimizing the process conditions. The mechanism for fines production is modeled in the reaction module. Finally, we have developed a novel approach to discretize and solve the population balance equations. This last module is used to represent the dynamic evolution of the particle size distribution function. The reaction and population balance models are implemented and solved in the MATLAB modeling environment, whereas the fluid bed system is solved using FEMLAB multi-physics modules.

The simulation studies show that the dynamics of the open loop model is very sensitive to external perturbations and that very long time (24 hours or more) is needed before the system reaches steady state. Control is therefore needed to stabilize the particle size distribution and optimize yield. The control problem is difficult since it involves a wide range of time-scales, ranging from seconds to several days. The dynamics are highly interacting and nonlinear.

In the paper we show how a simple feedback-feedforward control strategy allows the system dynamics to stabilize at desired operating points by changing system parameters like temperature and seed flow. We furthermore show that the controller maintains the particle mass hold up by adjusting silicon product outlet rate while the temperature inside the reactor is kept constant to maintain expected recovery through manipulating heat input through the wall of the fluid bed reactor. The particle size distribution and average product size is controlled using the seed rate and seed distribution functions.

The paper will review briefly results from an industrial pilot plant system which are used to verify the model system and we will demonstrate the effectiveness of the closed loop control system in simulation studies. The overall stability of the control system has been verified using the passivity-based theory based on thermodynamic storage functions which we have developed in our research group.

522c A Study of the Effect of Feeding Device Choice on Conversion In An Aerosol Flow Reactor

Todd M. Francis, Peter Kreider, Paul Lichty, and Alan W. Weimer. Department of Chemical and Biological Engineering, University of Colorado, 1111 Engineering Drive, Boulder, CO 80309-0424

Feeding a dispersed cloud of powder into a transport tube reactor, particularly on a laboratory scale, presents challenges. A major difficulty is dispersing the powder with minimal gas flow to maintain a long enough residence time in the reactor to achieve high conversions. In this study, two different laboratory feeders were designed and built to be mounted on top of an aerosol flow reactor incorporating different dispersion techniques: (1) a fluid bed feeder uses primarily gas flow to shear and disperse powder and (2) a spinning wheel feeder uses mechanical shear as the primary dispersion force. A factorial design was used to compare particle size distributions from both feeder types along with identifying factors that may influence the collected particle size distributions. Both Mn_2O_3 and ZnO powders were tested, because of their importance to solar-thermochemical based water splitting cycles. The feeders were mounted on an aerosol flow reactor and a 2 factor central composite design was executed to quantify how much the particle size distributions affected thermal reduction of Mn_2O_3 and ZnO. This talk will present the results from this study

522e Fluidized Bed Mixing of Nanosized Particles

James V. Scicolone and Rajesh Dave. New Jersey Center for Engineered Particulates, New Jersey Institute of Technology, Newark, NJ 07102

Mixtures of nanoparticles and nanocomposites, manufactured using nanoparticles as building blocks, have unique properties owing to small particle/grain size, and large interface area between individual nano-sized constituents. These properties have a great potential to improve performance of drugs, biomaterials, catalysts, energetic materials etc. However, to fully employ this potential, nanoparticles should be mixed at nanoscale level, which is a difficult task because they tend to form aggregates on the order of microns. Therefore, a major challenge in utilizing nanoparticles is to achieve their mixing at nano-scale level.

In this project, the mixing of nano-powders in an environmentally benign fluidized bed system was explored. The objective of this study is to explore the homogeneity of nanoparticle mixtures obtained using various types of fluidize bed systems, and determining what variables will improve mixing quality and why. The fluidized bed systems studied were conventional versus magnetically assisted fluidized bed systems. All of the fluidized bed methods are environmentally benign dry mixing techniques. The characterization of the resulting nanoparticle mixtures were conducted using Scanning Electron Microscope (SEM) equipped with Energy Dispersive X-ray Spectroscopy (EDS).

522f Experimental Investigation of a Rotating Fluidized Bed in a Static Geometry

Rahul P. Ekatpure[1], Geraldine J. Heynderickx[1], Axel De Broqueville[2], and Guy B. Marin[1]. (1) Laboratory for Chemical Technology, Ghent University, Krijgslaan 281, S5, Ghent, 9000, Belgium, (2) Grez-Doiceau, 1390, Belgium

The first experimental results obtained with a flexible cold-flow set-up aiming at the validation and the presentation of a novel reactor concept, the Rotating Fluidized Bed reactor in a static geometry (RFB-SG) [1] are presented. The rotational motion of a bed of fluidized particles is realized by feeding the fluidizing gas tangentially via several inlets situated on the circumference of the circumferential reactor. The fluidizing gas is discharged via a centrally located chimney. The presence of a central chimney reduces the fluidizing chamber of the reactor to an annular ring with a given thickness (Figure 1). In a conventional stable fluidized bed, the drag force is balanced by the gravity, whereas, in the RFB-SG, the former is balanced by the centrifugal force. Compared to the conventional Fluidized Bed Reactors (FBR), this set-up is expected to operate more efficiently for high-throughput as well as short residence time operations.

The experimental set-up constructed to perform an analysis of this novel reactor concept has a maximum of 36 gas injection slots, positioned under a 10° angle at an outer diameter of 540 mm. The width of the fluidizing annular chamber of the reactor is fixed at 100 mm. The side walls of the complete reactor are made of thick Plexiglas to visualize the flow behavior in the reactor. The number of inlet slots can be varied evenly from 0 to 36, while each inlet slot thickness can be varied from 0 to 7 mm. Different types of central chimneys with varying outer diameters, of gas exhaust slots and of variable gas exhaust slot thicknesses, can be implemented. Decreasing or increasing the central chimney outer diameter, increases or decreases the fluidizing annular chamber thickness. The fluidizing annular chamber of the experimental set-up can be mounted either vertically or horizontally, to take into account the effects of gravity. Pressure measurements to monitor the pressure variation due to the solids particle bed content are carried out by means of non-intrusive pressure probes located at several positions in the fluidizing annular chamber. They are connected to pressure sensors. A high speed camera is used to monitor the solids particle velocities and the solids particle bed height.

Experiments are performed with variable air mass flow rate (fluidizing gas) and solids particle flow rate at fixed geometrical conditions. Two different types of solids particles (Geldart classification: Type-B, D) are studied. The obtained experimental results show a stable rotating solids bed, with an optimum solids content, without slugging or channeling effects. A sharp inter-phase at the solids particle and freeboard in the fluidizing annular chamber is observed. A vertically positioned reactor shows a difference in the solids bed height at the top-most and the bottom-most position of the fluidizing annular chamber at lower gas mass flow rates, caused by the influence of gravity. Upon increasing the gas mass flow rate, the rotating solids particle bed height tends to be equal in both positions. Furthermore, as the fluidizing gas mass flow rate is increased, an increase in the pressure drop over the solids bed is observed (Figure 2). The pressure drop over the rotating solids bed is observed to be proportional to the square of the gas mass flow rate. Based on the theoretical prediction of the pressure drop due to rotating solids bed and the prediction of the experimentally measured values, a "nearly packed" rotating solids bed is predicted. A detailed comparative analysis of the rotating solids particle velocity and the rotating solids bed thickness as a function of various geometrical conditions and solids particle properties is performed.

Reference:

[1] De Wilde J, Wautier L, Marin G, Heynderickx G, De Broqueville A, *A Rotating Fluidized Bed in a static Geometry: Experimental Proof of Concept*, FLUIDIZATION XII, Vancouver, B.C., Canada, 13-18 May, 2007

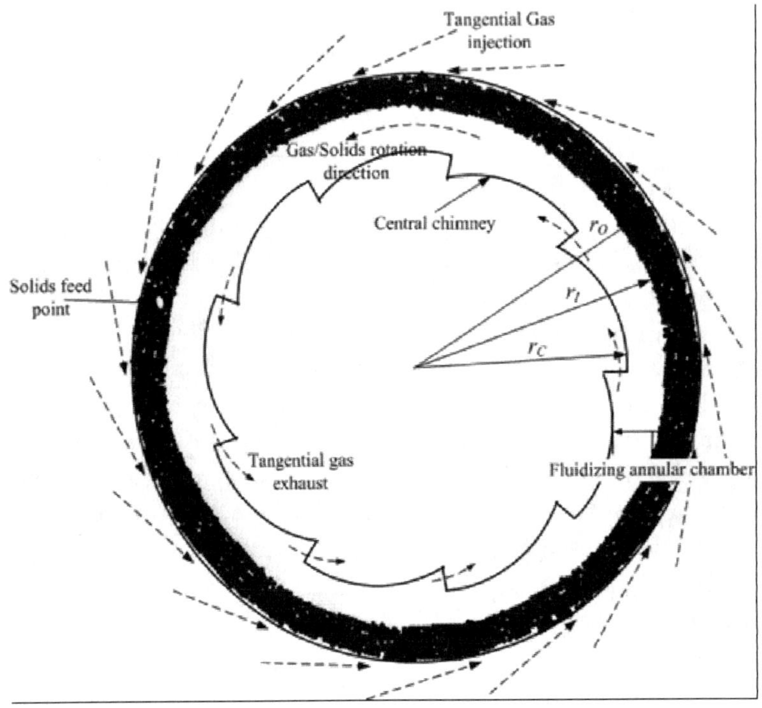

Figure 1: Fluidizing annular chamber with rotating gas/solids, with an outer radius r_O. The central gas discharge chimney with an outer radius of r_C. Tangential feeding the gas via gas inlet slots situated at the circumference of a circular reactor. Solids particle feeding directly in the fluidizing

chamber

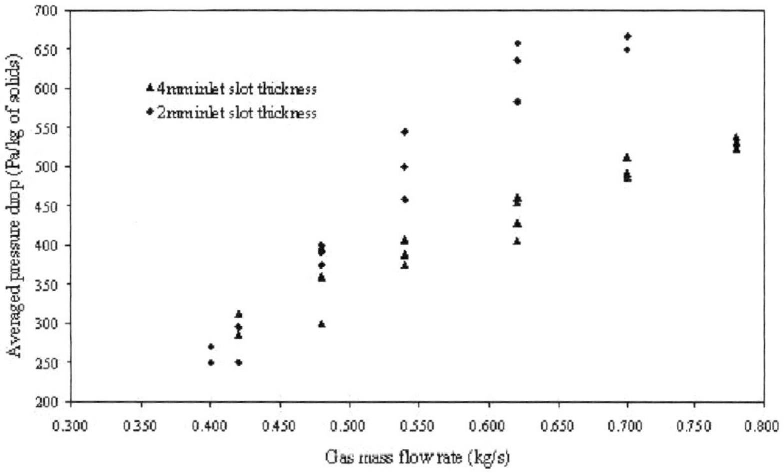

Figure 2: Averaged pressure drop per kg of rotating solids as a function of gas mass flow rate, in the fluidizing annular chamber of 60mm thick and two different tangential gas inlet slot thickness cases

522g Effect of Baffles on Gas Bypassing In a 0.9-M-Diameter Unit

Allan Issangya, Reddy S. B. Karri, and Ted M. Knowlton. Particulate Solid Research, Inc., 4201 West 36th Street, Suite 200, Chicago, IL 60632

Various types of internals have traditionally been used to improve fluidization quality. Horizontal baffles often enhance gas-solid contacting by breaking and redistributing the bubbles over the bed cross section. This is especially important for Group B materials whose bubbles are large and grow continuously with bed height. In Group A particles bubbles are usually small and their size is not affected as significantly by baffles. One area where horizontal baffles can be especially useful is in preventing gas bypassing in deep fluidized beds.

Some recent studies have shown that under some conditions there can be severe gas mal-distribution of gas in deep fluidized beds of Geldart group A materials. The conventional knowledge is that bubbling and near-turbulent beds of these materials are well mixed. Knowlton (2001) presented videos of fluidization tests conducted with 4 and 11% fines (<44 micron) FCC catalyst particles in a transparent 0.3-m-diameter unit. For the 4% fines catalyst particles, a significant gas bypassing was observed in the bed in a form of a high velocity air jet that preferentially flowed up near the column wall while the rest of the bed stayed defluidized or poorly fluidized. The air jet moved around the wall but it occasionally anchored or stayed longer at the cyclone dipleg return location. When the fines content in the bed was increased to 11% < 44 micron the gas bypassing disappeared and the fluidization became quite uniform.

The study of Wells (2001) using FCC catalyst particles in a semi-circular Plexiglas unit for bed heights of 2.4 to 4.9 m found that a "snake" of streaming gas formed a short distance above the grid and passed through the bed, bypassing the mostly stagnant bed. Issangya et al (2007) using FCC catalyst in a 0.3-m-diameter suggested that the standard deviation of differential pressure fluctuations measured across short intervals around the column can be used as a diagnostic tool for gas bypassing in deep fluidized beds.

Gas bypassing can lead to serious problems if it occurs in industrial beds. For example, in FCC regenerators and strippers or in CFB combustors it can result in afterburning, poor gas/solids contacting and poorly fluidized entrances to standpipes, poorly fluidized discharge regions of cyclone diplegs and poor solids flows around loop seals. The paper discusses results of tests conducted for FCC catalyst particles in a 0.9-m-diameter fluid bed unit in which horizontal baffles were employed to eliminate gas bypassing. The unit had a transparent Plexiglas section for visual observation of the fluidization behavior as well as instrumentation to measure differential pressure fluctuations and bubble void fraction profiles. Videos will be presented.

547a Advanced Materials Development at Rdecom-Ardec

Paul Redner[1], Deepak Kapoor[1], Rajen Patel[2], Chris Haines[1], Darold Martin[1], and Moon Chung[1]. (1) Energetics and Warheads Division, US Army ARDEC, Bldg 3022, AMSRD-AAR-MEE-W, Picatinny, NJ 07806, (2) Energetics and Warheads Division, US Army, ARDEC, Bldg 3028, AMSRD-AAR-MEE-W, Picatinny Arsenal, NJ 07806-5000

A number of approaches are being explored in order to fabricate materials that take advantage of the fundamental change in a material's attributes at the nanoscale. Nanomaterials and nanostructured materials exhibit superior strength-to-weight ratios, improved conductivity, improved reactivity and, most importantly, tunability of all of these attributes at the nanoscale. Materials are being developed using both "top-down" and "bottom-up" approaches to meet various technology gaps that have been identified by the Warfighter.

ARDEC has established a Nanomaterials Technology Center to directly address the need for both a consistent source of various nanomaterials – such as metals, metal oxides, non-oxide ceramics, and composites – as well as a central location to both characterize and fabricate these materials into useful parts for end item applications. The NTC provides the DoD with the ability to identify, evaluate, and develop new technologies to meet current and future technology gaps and directly address single-point failures.

ARDEC's internal capabilities are enhanced by teaming with outside organizations that bring their own technologies to the table. Through these partnerships, ARDEC has been applying new technologies with its own expertise to tackle Advanced Materials needs for the Army.

The potential of advanced materials, such as nanomaterials and nanostructured materials, is great. Research will continue with the goal of increasing the speed at which these materials are processed, evaluated, and incorporated. This will be accomplished by expanding on current prototyping capabilities and by evaluating each material through a multidisciplinary approach.

547b Development of High-Efficient Nanoenergetic Gas-Generators

Karen S. Martirosyan, Leizheng Wang, Arol Vicent, and Dan Luss. Chemical and Biomolecular Engineering, University of Houston, 4800 Calhoun, Bldg. 1, S222, Houston, TX 77054

Nanoenergetic thermite materials release energy much faster than conventional energetic materials and have various potential military applications such as rocket propellants, aircraft fuel and explosives. They are likely to become the next-generation explosive materials, as they enable flexibility in energy density and power release through control of particle size distribution, stoichiometry and choice of fuel and oxidizer. The reduction of the reactant powders from micro- to nano-size generates a more intimate contact between the particles. This, in turn, increases the reaction front propagation velocity in some systems by two to three orders of magnitude. Most researchers have investigated the dependence of the temperature evolution and velocity of the thermal front propagation on the thermite reactions. However gas pressure evolution and rate of gas release is not well investigated and understood.

We report here the development of high-efficient Nanoenergetic Gas-Generators (NGG) systems having various potential military applications. We systematically studied micro and nano energetic reactants mixtures such as MoO3-Al, MoO2-Al, WO3-Al, Bi2O3-Al, MnO2-Al, Fe2O3-Al, CuO-Al, some containing boron and carbon additives, to find those generating the highest gas discharge pressure peak. Loose reactant mixtures were loaded (m=0.5 g) into a ceramic boat that was placed inside a cylindrical stainless steel vessel (Parr, V=342 cc) and then ignited. Addition of some boron to the reactant mixture for MoO3/Al and WoO3/Al increased the peak pressure. The maximum combustion temperature during the studied reactions exceeded 2000°C. The thermal front emitted many sparks (hot zones) and small particles. The estimated average rate of temperature rise was ~40,000 C/s. The peak pressure during some nanoenergetic thermite reactions exceeded 1000 psi. Key factors that affect the ignition sensitivity and peak pressure evolution will be discussed.

Aluminum Burn Rate Modifiers Based on Reactive Nanocomposite Powders

Demitrios Stamatis, Xianjin Jiang, Ervin Beloni, Edward L. Dreizin

Abstract

Aluminum powders have long been used in reactive materials for such applications as propellants, pyrotechnics and explosives. Aluminum has a high enthalpy of combustion but relatively low combustion rate. Addition of reactive nanocomposites can increase the burn rate of aluminum and thus the overall reaction rate. Replacing a small fraction of the fuel by a nanocomposite material can enhance the reaction rate with little change to the thermodynamic performance of the energetic formulation. This research showed the feasibility of the above concept using nanocomposite powders prepared by Arrested Reactive Milling (ARM), a scalable "top-down" technique for manufacturing reactive nanocomposite materials. The nanocomposite materials used in this study were 2B+Ti, and Al-rich 8Al+3CuO, and 8Al+MoO_3. The reactive nanocomposites were added to micron sized aluminum powder and the mixture was burned in a constant volume chamber. The combustion atmosphere was varied using oxygen, nitrogen, and methane. The resulting pressure traces were recorded and processed to compare different types and amounts of modifiers.

Introduction

Various nanocomposite materials are currently under development as potential components of different energetic formulations, from propellants to explosives, to pyrotechnics, e.g., [1 – 14]. The advantages anticipated from such materials are primarily due to a very developed reactive surface that facilitates a rapid initiation of the exothermic reaction and results in a nearly adiabatic reaction temperature. At the same time, the overall energy outputs from many exothermic reactions employed in such materials, including thermites, intermetallic, and metal-metalloid compositions, are smaller than the benchmark values for aluminum combustion in air or in other practically important oxidizers (e.g., ammonium perchlorate). Thus, replacement of aluminum as a fuel in most metallized energetic formulations with almost any of the nanocomposite materials currently under development would result in an overall reduction of the theoretical reaction enthalpy. This negative effect may be offset by an increase in the efficiency of metal combustion, so that the overall increase in practical performance is still anticipated. Therefore, the optimized composition would combine the high energetic output with the accelerated reaction rate. The approach discussed in this paper suggests that replacing a fraction of aluminum fuel with a reactive nanocomposite material could result in an acceleration of the ignition kinetics for all metal fuel. Aluminum particles located in vicinity of the igniting reactive nanocomposite particles would be heated more efficiently and ignite sooner. It is anticipated that a relatively small addition of the reactive nanocomposite material would provide a number of localized hot spots distributed in the igniting energetic formulation, which would accelerate ignition of the nearby aluminum particles, which, in turn, will accelerate ignition of their own neighbors. Effectively, the nanocomposite material will serve as a burn rate modifier for an aluminized energetic formulation. The amount of such modifier is expected to be a function of the specific formulation. In this paper, the proposed concept is initially explored for aluminum particles burning in a gaseous oxidizer in presence of products of hydrocarbon combustion. Such environments are relevant for both enhanced blast explosives and metallized solid propellants.

Materials

Reactive nanocomposite powders were prepared by arrested reactive milling (ARM), a high-energy mechanical milling technique [8 – 14]. Samples of three micron-sized, fully dense nanocomposite powders of 2B+Ti, 8Al+3CuO, and 8Al+MoO_3 were produced using a Retsch 400 PM planetary mill. Further details on the material synthesis are available elsewhere [13, 14]. Typically, the nanocomposite materials consist of micron-sized particles whereas each particle is a fully-dense, three-dimensional composite with characteristic dimension of material mixing of about 100 nm. Commonly, the morphology of composite is that of inclusions of one component, e.g., B, CuO, or MoO_3, embedded into a matrix of another component such as Ti or Al. In this study, the nanocomposite materials prepared by ARM were added to a spherical aluminum powder, 10 – 14 μm nominal particle size by Alfa Aesar. The mixing of aluminum and nanocomposite powders was performed using a SPEX Certiprep 8000 shaker mill operated without milling balls for three minutes. The particle size distributions for all powders used in this project were measured using a Coulter LS 230 Enhanced Laser Diffraction particle size analyzer. The size distributions and respective volume mean particle sizes for all powders are shown in Figure 1.

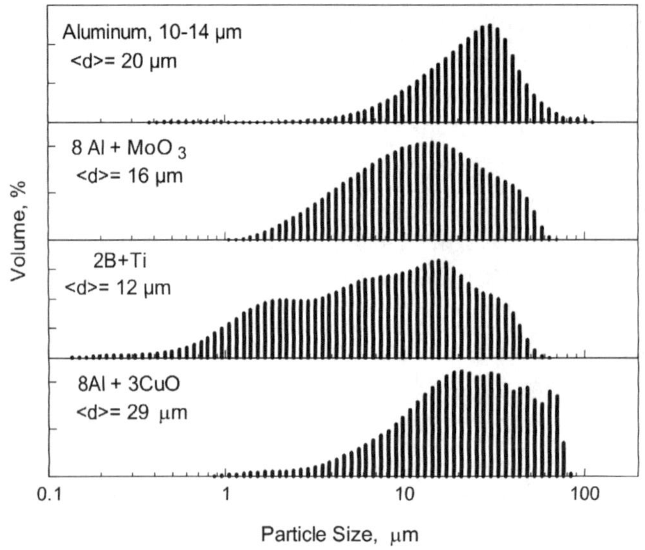

Figure 1 Particle size distributions for the different powders used in this project

Experimental

Constant Volume Explosion (CVE) experiments were performed with a set of materials including aluminum and aluminum mixed with different amounts of added nanocomposite burn rate modifiers. The details of the CVE experimental methodology and setup are described elsewhere [15 – 17]. Fig. 2 shows a simplified drawing of the CVE apparatus.

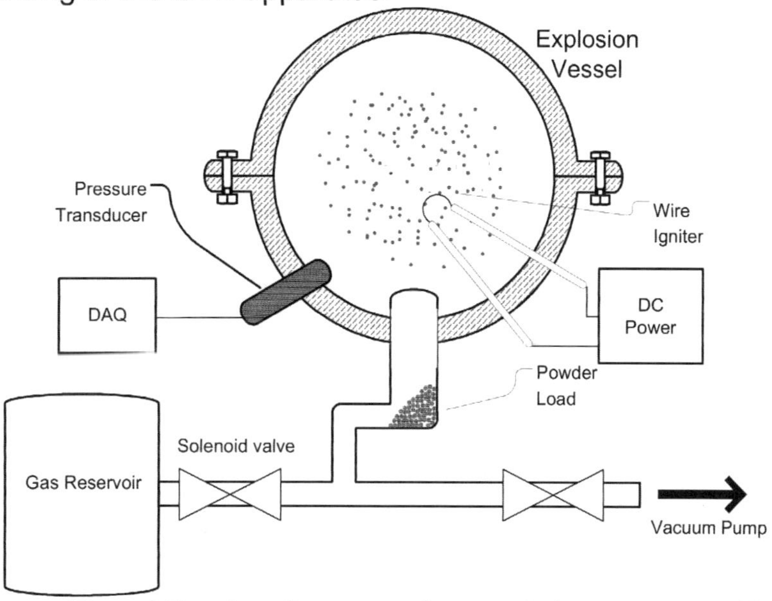

Figure 2 Drawing of constant volume explosion apparatus

In one set of experiments, the oxidizing environment was fixed to nominally include 3% CH_4, 21% O_2, and 76% N_2. The amounts of the nanocomposite powders added to aluminum were 10, 20, and 30 mass %.
The gaseous combustion products include moisture and carbon oxides, imitating the environment in actual propellants better than the previous CVE experiments performed in air [15, 16]. The powders were introduced into a nearly spherical 9.2 l vessel as an aerosol and ignited at the center of the vessel. The powder load was selected based on thermodynamic equilibrium calculations considering combustion of aluminum in the described above gas mixture. It was predicted that the maximum adiabatic flame temperature for the 9.2 l vessel occurs at an aluminum load of 2.64 g. Respectively, all experiments were performed with 2.64 g of powder loaded. Before the powder was introduced to the vessel, it was evacuated to less than 1 torr. The vessel then was filled with the 171 torr of O_2.The powder was introduced into the vessel with a blast of a gas mixture comprising nitrogen and methane produced by opening a solenoid valve connecting the vessel with a 2-gallon gas reservoir filled with nitrogen/methane gas mixture at 4,200 torr. Before each experiment, the gas mixture was prepared by evacuating the gas reservoir and re-filling it with 163 torr of methane and the balance of nitrogen. The duration of the gas blast pulse was 200 ms. At the end of the blast, the pressure in the vessel was close to 1 atm. To reduce the turbulence in produced gas powder mixture, the gas blast was followed by a

300 ms waiting period. Finally, the powder was ignited using an electrically heated tungsten wire placed in the center of the vessel. The combustion pressure traces were measured in real time using an American Sensor Technology AST 4700 transducer. The values and the rates of pressure rise produced by the combustion were compared for different powders.

A second set of experiments was carried out using a constant mass % of additive of each modifier in aluminum powder load and varying the methane concentration between 1.5 and 4.5 % while keeping the oxygen concentration constant.

Results and Discussion

A common concern for reactive metal powder addition is their sensitivity to electro-static discharge (ESD) ignition. All the materials used in this project were tested using a firing test system model 931 by Electro-tech Systems, Inc., according to standard Mil-1751A. Table 2 shows the measured values of the minimum ignition energy (MIE) for each material.

Table 1 Sensitivity of materials to electro-static discharge

Material	MIE (mJ)	Propagation
Spherical Al 10-14 micron	25.7	No
8Al+3CuO nanocomposite	3.8	Yes
8Al+MoO_3 nanocomposite	<0.8	Yes
2B+Ti nanocomposite	1.2	No
Al (10-14 micron) + 20%(8Al+3CuO) blend	13.2	No
Al (10-14 micron) + 20%(8Al+MoO_3) blend	6.9	No
Al (10-14 micron) + 20%(2B+Ti) blend	1.9	No

The data indicates that mixing the nanocomposites with the aluminum results in a powder that is more sensitive than pure aluminum but substantially less sensitive than nanocomposite material itself. In particular, it is worth noting that the flame did not propagate in the powder mixtures, unlike in the individual nanocomposite powders.

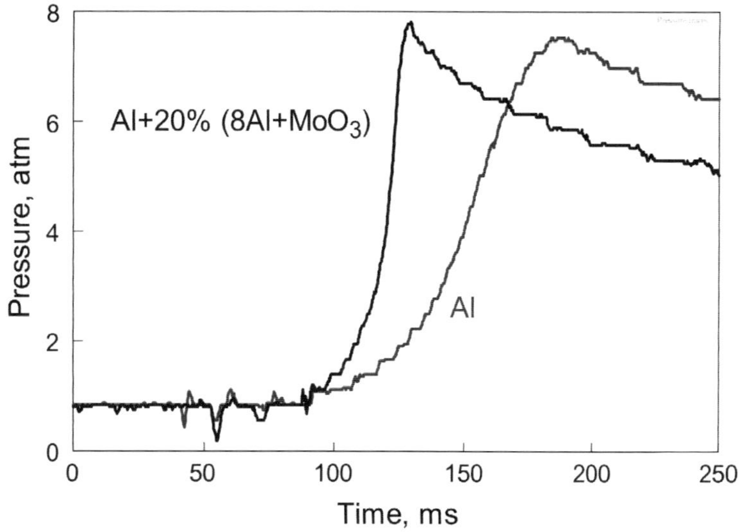

Figure 3 Constant volume explosion results obtained for regular aluminum powder and modified aluminum powder

Shown in Fig. 3 are two pressure traces from the CVE experiment. The modified aluminum powder shows both increased maximum pressure and rate of pressure rise. The accelerated burn rate was indeed observed for all experiments using nanocomposite powders as burn rate modifiers. However, the maximum pressure could be both higher and lower then for pure Al powder. The results are presented in Fig. 4 and 5.

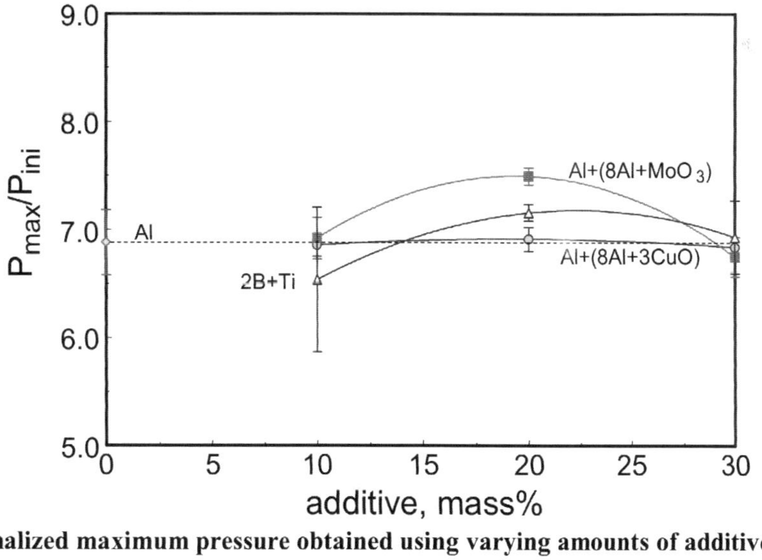

Figure 4 Normalized maximum pressure obtained using varying amounts of additives for 3% CH_4 21% O_2 and 76% N_2 atmosphere

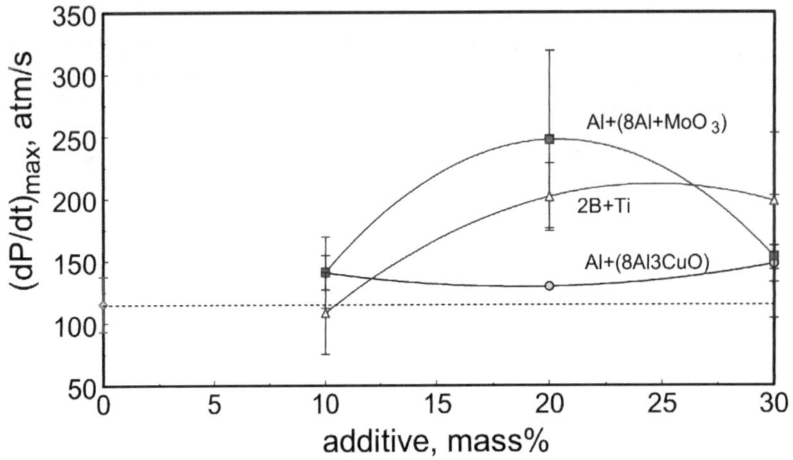

Figure 5 Maximum rate of pressure rise obtained using varying amounts of additives for 3% CH_4 21% O_2 and 76% N_2 atmosphere

Initial experiments were carried out to determine the effect of various amount of modifier. Figure 4 shows ratio of the maximum pressures observed in explosions over respective values of the initial gas pressure in the vessel. Results are shown for different modifiers at different additive mass percents. Pressure ratio shown in Fig. 4, P_{max}/P_{ini}, is proportional to the combustion temperature. Results for the maximum rates of pressure rise, $(dP/dt)_{max}$, are shown in Fig 5. The maximum rate of pressure rise, $(dP/dt)_{max}$, is proportional to the flame speed. The dashed lines in Figs. 4 and 5 indicate the reference values for pure aluminum powder. The data show that the most significant improvement was gained using 20 mass% additives for both 8Al+MoO_3 and 2B+Ti nanocomposites. The effect is consistently small for 8Al+3CuO.

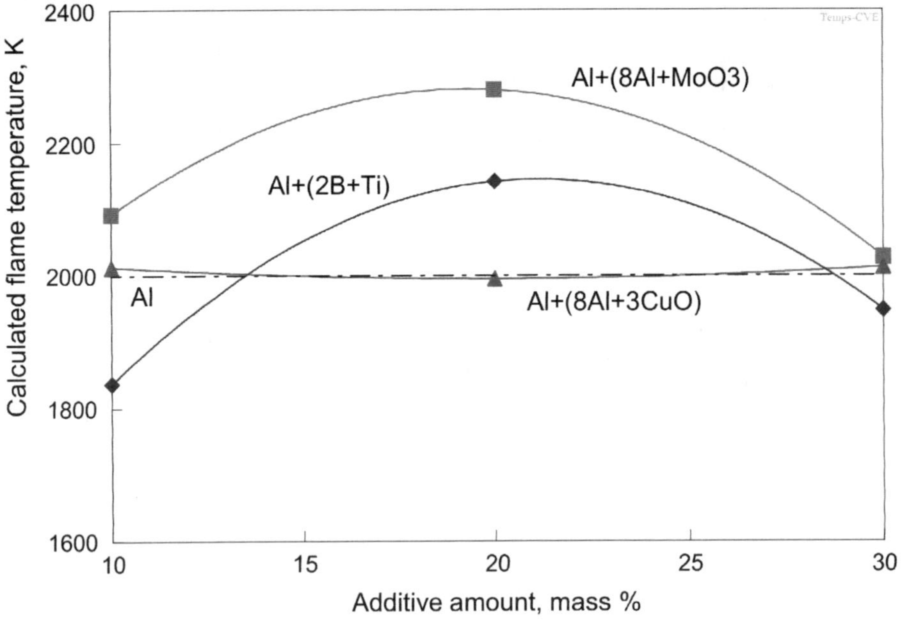

Figure 6 Combustion temperatures for different additives calculated using ideal gas law and maximum pressure recorded from CVE experiment

The combustion temperatures for different experiments were estimated from the measured pressures using the ideal gas law and neglecting the changes in the number of moles of gas in the vessel. The results for these estimates are presented in Figure 6. In agreement with the measured maximum pressures, the temperature increase over the case of pure Al is most significant when 20 % of $8Al+MoO_3$ was added to the aluminum powder.

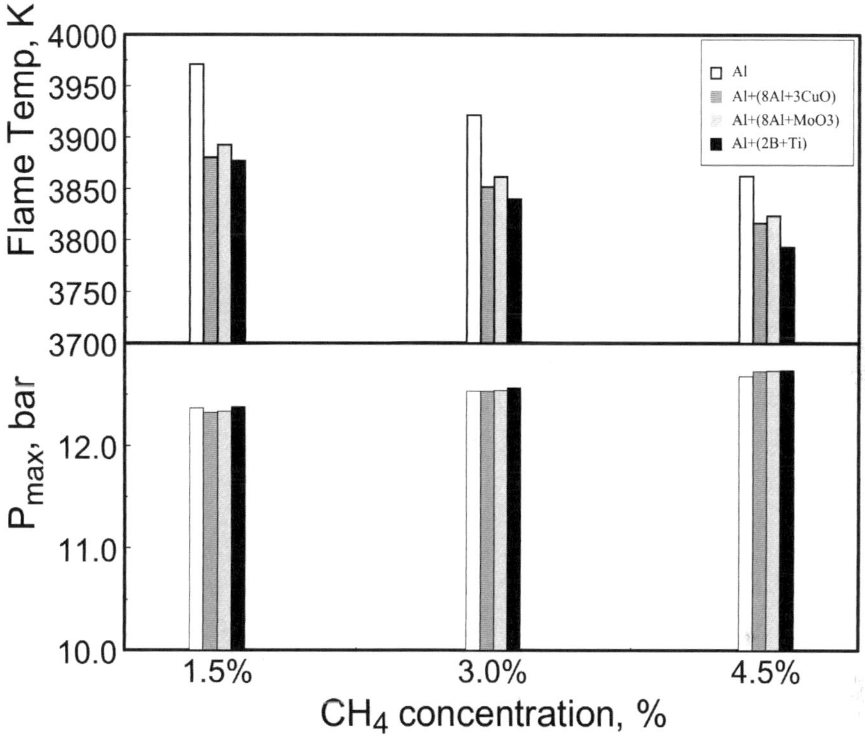

Figure 7 CEA calculations for pressure and flame temperature

Fig. 7 shows the calculated values for pressure and flame temperature using the NASA equilibrium code Chemical Equilibrium and Applications (CEA). The calculations were performed for constant volume combustion. Pure aluminum flames are characterized by substantially higher temperatures. However, the difference in the predicted pressure is much more subtle. The results indicate that there is a small increase in pressure with increasing methane concentration. The flame temperatures decrease with increasing methane concentration indicating a substantial change in the predicted make-up of the equilibrium combustion products. The calculations predict that the highest pressures are obtained by the 2B+Ti modifier followed by pure aluminum then the MoO_3 and finally the CuO.

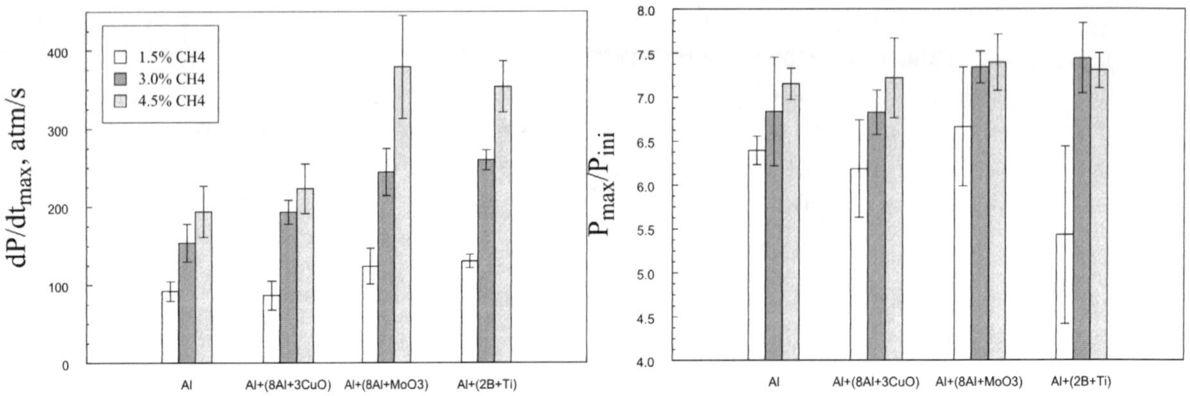

Figure 8 Constant volume explosion results for different atmospheres

Figure 8 shows the results of the CVE test in different environments of methane. There is an increasing trend in the combustion rate for all modifiers as the methane concentration increases. The most effective modifier seems to be the 8Al+MoO_3 or 2B+Ti depending on environment. The 8Al+3CuO on the other hand has little effect on the combustion performance. Similar to the CEA calculations, the 2B+Ti and MoO3 modifiers achieve the highest pressures. The CuO modifier has the lowest pressure of the modifiers.

The combustion products were collected and analyzed by energy-dispersive x-ray spectroscopy to determine the amount of aluminum and oxygen. This ratio was normalized by dividing it by the oxygen to aluminum ratio for an ideal reaction. Therefore, a ratio of 1 would indicate an ideal and complete reaction. The final normalized oxygen to aluminum ratio is shown in Fig. 9. Interestingly, it seems that the most complete reaction was achieved by the CuO modifier although it did not attain the highest pressure. The trend is followed for all three concentrations of methane.

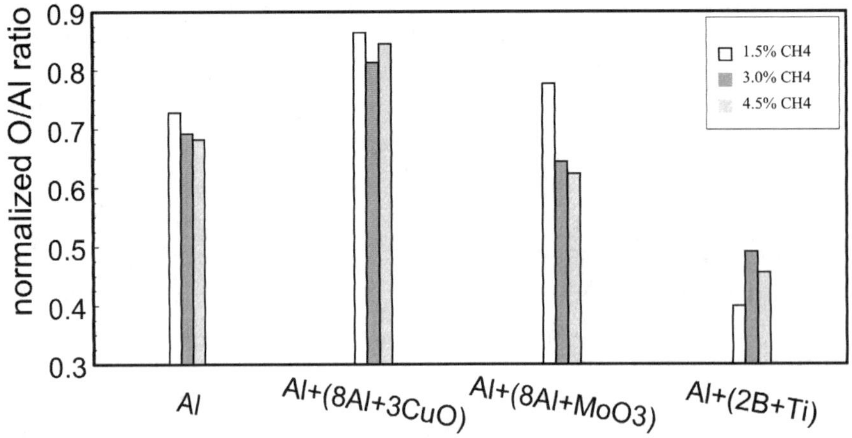

Figure 7 Oxygen to aluminum content ratio from analysis of combustion products

Conclusions

Addition of relatively small amounts of nanocomposite burn rate modifiers to a metal-oxidizer formulation enables substantial increase in the burn rate without a substantial reduction in the overall theoretical combustion enthalpy. In the present experiments, additives of nanocomposite aluminum-rich $8Al+MoO_3$ thermite and B-Ti materials were found effective in improving both the rate and maximum pressure of aluminum combustion. However, additives of a metal-rich nanocomposite 8Al+3CuO thermite did not show appreciable improvements in aluminum combustion.

Acknowledgment

This project was supported in parts by the U.S. Edwards Air Force Base (Mr. Scott Shackelford) and RDECOM-ARDEC Picatinny (Mr. Paul Redner).

References

1. Jouet, R.J., Carney, J.R., Granholm, R.H., Sandusky, H.W., Warren, A.D., "Preparation and Reactivity Analysis of Novel Perfluoroalkyl Coated Aluminium Nanocomposites," ***Materials Science and Technology*** 22 (4), pp. 422-429 (2006)

2. Perry, W.L., Smith, B.L., Bulian, C.J., Busse, J.R., Macomber, C.S., Dye, R.C., Son, S.F., "Nano-Scale Tungsten Oxides for Metastable Intermolecular Composites," ***Propellants, Explosives, Pyrotechnics*** 29 (2), pp. 99-105 (2004)

3. Danen, W.C., Jorgensen, B.S., Busse, J.R., Ferris, M.J., Smith, B.L., "Los Alamos Nanoenergetic Metastable Intermolecular Composite (Super Thermite) Program," Abstracts of Papers, 221st ACS National Meeting, San Diego, CA, United States, April 1-5, 2001, (2001)

4. Perry, W. L., Tappan, B.C., Reardon, B.L., Sanders, V.E.; Son, S.F., "Energy Release Characteristics of the Nanoscale Aluminum-Tungsten Oxide Hydrate Metastable Intermolecular Composite," ***Journal of Applied Physics*** 101(6), 064313/1-064313/5 (2007)

5. Son, S. F., Hiskey, M. A., Naud, D. L., Busse, J. R., Asay, B. W., "Lead-Free Electric Matches," Proceedings of the 29th International Pyrotechnics Seminar, Publisher: Defence Science & Technology Organisation, Pyrotechnics Group pp. 871-877 (2002)

6. Prakash, A., McCormick, A.V., Zachariah, M. R., "Synthesis and Reactivity of a Super-Reactive Metastable Intermolecular Composite Formulation of Al/KMnO4," ***Advanced Materials*** 17(7), 900-903 (2005)

7. Walter, K.C., Pesiri, D.R., Wilson, D.E., "Manufacturing and Performance of Nanometric Al/MoO3 Energetic Materials," ***Journal of Propulsion and Power*** 23(4), 645-650 (2007)

8. Dreizin, E.L., Schoenitz, M., "Nano-Composite Energetic Powders Prepared by Arrested Reactive Milling," *US Patent Application 20060053970*, published in March 2006

9. Schoenitz M., Ward T., Dreizin E.L., "Preparation of Energetic Metastable Nano-Composite Materials by Arrested Reactive Milling," *Materials Research Society Proceedings*, V. 800, pp: AA2.6.1-AA2.6.6 (2004)

10. Schoenitz, M., Ward, T.S., Dreizin, E.L., "Fully Dense Nano-Composite Energetic Powders Prepared By Arrested Reactive Milling," *Proceedings of The Combustion Institute* 30 pp. 2071-2078 (2005)

11. Umbrajkar, S., Trunov, M.A., Schoenitz, M., Dreizin, E.L., Broad, R., "Arrested Reactive Milling Synthesis and Characterization of Sodium-Nitrate Based Reactive Composites," ***Propellants, Explosives, Pyrotechnics*** 32 (1), 32-41 (2007)

12. Dreizin, E.L., Schoenitz, M., Shoshin, Y.L., Trunov, M.A., "Highly-Energetic Nanocomposite Powders Produced by Arrested Reactive Milling," *International Annual Conference of ICT*, 36th (Energetic Materials), 138/1-138/12, (2005)

13. Umbrajkar, S.M., Schoenitz, M., Dreizin, E.L., "Control of Structural Refinement and Composition in Al-MoO_3 Nanocomposites Prepared by Arrested Reactive Milling," ***Propellants, Explosives, Pyrotechnics*** 31(5), pp. 382-389, (2006)

14. Umbrajkar, S.M., Seshadri, S., Schoenitz, M., Hoffmann, V.K., Dreizin E.L., "Aluminum-Rich Al-MoO_3 Nanocomposite Powders Prepared by Arrested Reactive Milling," ***Journal of Propulsion and Power*** 24 (2) pp. 192 – 198 (2008)

15. Schoenitz, M., Dreizin, E.L., Shtessel, E., "Constant Volume Explosions of Aerosols of Metallic Mechanical Alloys and Powder Blends," ***Journal of Propulsion and Power*** 19(3), pp. 405-412 (2003)

16. Eapen B. Z., Hoffmann V. K., Schoenitz M., Dreizin E. L., "Combustion of Aerosolized Spherical Aluminum Powders and Flakes in Air," ***Combustion Science and Technology*** 176(7), pp. 1055-1069 (2004)

17. Trunov, M.A., Hoffmann, V.K., Schoenitz, M., Dreizin E.L., "Combustion of Boron-Titanium Nanocomposite Powders in Different Environments," ***Journal of Propulsion and Power*** 24 (2) pp. 184-191 (2008)

Tunability of Nanoenergetic Materials

Chris J. Bulian, Jan A. Puszynski, and Jacek, J. Swiatkiewicz
Department of Chemical and Biological Engineering
South Dakota School of Mines and Technology
501 East Saint Joseph Street
Rapid City, SD 57701, USA

Abstract

Expansive research in the area of nanoenergetics over the past several years has led to the discovery and development of a wide array of reactive systems. These reactive systems have many varying properties including: ignition sensitivity, reaction rate, adiabatic reaction temperature, pressure generation, and theoretic maximum density. Additional investigations into the factors affecting ignition and reaction characteristics have shown that many physical and chemical properties of the individual reactants, as well as processing methods, can have a significant effect on the overall behavior of the nanoenergetic system. Application of nanoenergetic materials as components of various products requires that the ignition and reaction characteristics be tailored to have specific sensitivities and energy outputs. This research effort addressed how ignition sensitivity thresholds, reaction rate, and pressure generation can be tailored to have a wide range of values. Several different individual reactants are used including: aluminum nanopowder, iron oxide nanopowders, bismuth oxide nanopowders, and additional binders and gasifying agents. The primary system of interest in this research consisted of aluminum with both iron oxide and bismuth oxide incorporated as oxidizing agents. The effect of stepwise increases in iron oxide content on pressure generation is shown, as well as the effect of consolidation on ESD sensitivity. The use of nitrocellulose as a gasifying agent is also presented along with its effect on pressurization rate, peak pressure generation, and ESD sensitivity.

Introduction

There are many different methods by which the ignition and reaction parameters of nanothermite systems can be modified. This paper will focus on certain methods that can be used to modify or 'tune' specific properties including: electrostatic discharge (ESD) sensitivity, ignition delay time, pressure generation, and pressurization rate. ESD sensitivity for nanothermite systems can range from the microjoule level, as exhibited by loose powder Al-Bi_2O_3, to tens of millijoules for consolidated systems[1]. Ignition delay is dependent on the rate at which energy is added to the sample and the form of the energy (i.e. laser, electrostatic discharge, flame) in addition to the physical and chemical properties of the individual nanothermite system[2]. Pressures generated during reaction of the nanothermite are due to two different processes; expansion due to heating of surrounding gases and evolution of gaseous products. The latter of these processes generally produces much higher pressures. Pressurization rate is obviously proportional to the reaction rate as pressures generated by both processes subside after a period of time due to cooling and condensation of gaseous products.

With the wide variety of nanoscale fuels and oxidizers available on the commercial market today, different nanothermite mixtures can be made to provide near limitless combinations of the reaction properties listed above. Individual properties of materials that can affect reaction rates include: particle size, specific surface area, particle morphology, and the physical-chemical nature of the material[3]. It stands to reason that if specific reaction properties for an application are required, one or more of these material properties could be changed to achieve the desired result. This process of developing a mixture with specific properties by changing the material being used, however, is a very tedious process that has several potential complications. These complications include the fact that there are no direct correlations to be made between nanothermite systems containing different types of fuels or oxidizers, and that suppliers do not necessarily produce the exact same quality of product from batch to batch. In a fast growing field like nanotechnology, suppliers of nanoscale reactions also come and go quickly. Because of this, it is not always possible to procure a reactant with the exact same physical properties over a long period of time.

Tuning of nanothermite materials, on the other hand, is a process by which the reaction parameters are controlled by varying concentrations of reactants, methods of mixing, degree of consolidation of the material, use of binders, and incorporation of gas generating agents. These methods are less reliant on having specific reactant properties such as particle size and specific surface area because most of them can be changed on a stepwise level to adjust the reaction parameters. This makes it easier to fit a specific nanoenergetic material to a specific application with less guess and check work as is necessary when using different types and sizes of reactants to do so. The research presented in this paper is mainly focused on nanothermite systems based on nanoscale aluminum as a fuel. Previous studies have investigated different loose powder nanothermite systems including Al-Fe_2O_3, Al-CuO, Al-MoO_3, Al-WO_3, and Al-Bi_2O_3[4-7]. These studies have shown the Al-Fe_2O_3 and Al-Bi_2O_3 systems to have significantly different reaction properties. The system containing iron oxide reacts much slower and produces much lower pressures where as the system containing bismuth oxide has a very fast reaction rate and produces a significant amount of gaseous products. Based on these results, it was hypothesized that including both iron oxide and bismuth oxide in the reactive mixture in various ratios could make a nanothermite material with reaction properties covering the entire spectrum between the two systems. Other methods for tuning reaction parameters to be discussed include method of processing, use and amount of binder to form consolidated material, and coating of material in gasifying agents to increase pressure and decrease sensitivity.

Experimental

Two different methods of mixing nanothermite reactants were used in this research, each resulting in the formation of a different form of nanothermite material. The first method, used to make loose powder nanothermite, involves combining the reactants in isopropyl alcohol and mixing the components ultrasonically in the absence of binders. Upon thorough mixing, the slurry was then poured into a grounded metal drying pan to evaporate the solvent and yield the loose powder material. The second method involves processing the reactants in water with the inclusion of a binder material. This method eliminated the use of flammable solvents but came with a drawback. If untreated, aluminum nanopowders will react with water for form aluminum oxide. To counteract this, 2.5 wt% of ammonium dihydrogen phosphate, with respect to the mass of aluminum nanopowders, was added to the water prior to adding aluminum. A water soluble binder material, such as arabic gum or methyl cellulose, was dissolved into the solution as well at 4-6 wt% with respect to the total nanothermite reactant mass. The metal oxide was then finally added and the thick slurry was ultrasonically mixed. Upon mixing, the slurry was then micro-pipetted into a grounded metal drying tray to form small granular material.

Pressure output of the material was measured using one of two closed volume pressure cells shown in Figure 1. The first pressure cell has an internal volume of 20 cc. The nanothermite material is ignited by a laser-pulse directed through a glass window and the pressure is measured using a piezoelectric pressure transducer from PCB Piezotronics. The second pressure cell has a much smaller internal volume of 2.75 cc to measure higher pressures produce under more confined conditions. This setup using a Ni-Cr wire coil to ignite the nanothermite material. Again, a piezoelectric pressure transducer is used to measure dynamic pressure.

Figure 1: Closed volume-pressure cell for measuring dynamic pressure response of nanothermite reaction. Internal volume of (20cc left, 2.75 cc right).

Another important reaction property, aside from pressure generation, is sensitivity of the material to ignition, especially by electrostatic discharge (ESD). ESD testing of the materials was performed using a Model 930D-FTS Electrostatic Discharge Simulator from Electro-Tech Systems, Inc. The apparatus allowed for the easy exchange of capacitance over the range of 100 pF to 20 nF. The voltage level could also be varied over the range of 0 – 5000 V. In accordance with standard military testing of materials for ESD sensitivity, each sample was subjected to 20 consecutive discharges as a specific energy level. If no ignition occurred, the voltage and/or capacitance were increased and the sample was again subjected to 20 consecutive discharges at the new energy level until ignition occurred. The lowest energy level at which ignition occurred is the ignition threshold.

Results

In order for future development of products containing nanothermite materials, we have focused on systems based on a combination of aluminum nanopowders fuel with various ratios of iron oxide and bismuth oxide nanopowders. Fe_2O_3 and Bi_2O_3 each have different reaction properties when combined with aluminum. Iron oxide reacts relatively slowly (tens of m/s) with aluminum compared to bismuth oxide (hundreds of m/s) and also generates much less gas upon combustion (~8% by weight vs. ~89% by weight for bismuth oxide)[8]. Certain applications require fast combustion rates and high pressures while others require controlled reaction rates with hot particulate reaction products. In order to investigate the effect of iron oxide content on reactivity, the mass % with respect to total oxide was increased incrementally and pressure evolution during reaction was recorded. For each test, ~35 mg of nanothermite mixture was ignited by laser impulse in a closed volume pressure cell with an internal volume of 20 cc. Figure 2 shows the effect of increasing iron oxide content on the peak pressure generated during reaction.

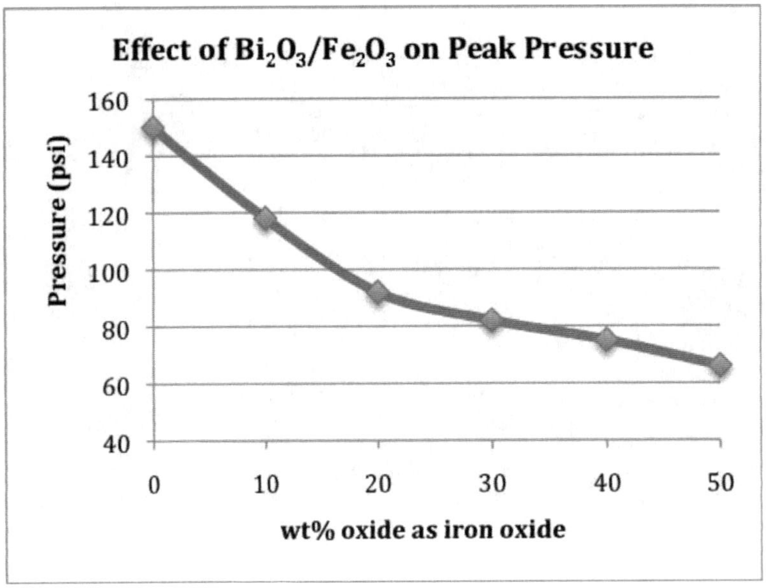

Figure 2: Plot of peak pressure generated as a function of iron oxide mass percentage of total oxide content.

In order to investigate the effect of consolidation, it was determined that a mixture of aluminum nanopowder combined with a 50/50 molar ratio of iron oxide and bismuth oxide would produce significant pressure upon reaction while also yielding molten iron particulates. The inclusion of iron oxide also slows the reaction down enough to allow for longer interaction time with subsequent propellants that may need to be ignited by the nanothermite. Nanoenergetics containing iron oxide are also less sensitive to ignition than those containing only bismuth oxide as the oxidizer. Water-based processing was used to render the resultant energetic material into granular form. A thick slurry containing 18.7 wt% of 80 nm aluminum from Novacentrix, 20.8 wt% of 30 nm iron oxide from Nanophase Materials, and 60.5 wt% of 416 nm bismuth oxide from Accumet Materials, along with 4 wt% arabic gum binder was micropipetted into grounded metal drying trays.

In order to measure the ESD sensitivity of the material in consolidated form, several granules were placed in close proximity to each other and the discharge tip of the ESD gun was lowered into the group of granules. Using this procedure, the minimum energy at which 20 consecutive discharges could not be completed without at least one ignition was 1.6 mJ, a value much higher than the loose powder ignition threshold

Replacing conventional materials in various applications usually requires that certain reaction parameters are either matched or exceeded. The two important factors that needed to be investigated were pressurization rate and peak pressure generated. Burning of propellants requires the containment of the gases produced to drive the reaction faster and produce high pressures. Proper testing of these materials required the construction of a smaller 2.75 cc volume pressure cell. Pressure generation for the combustion of nanothermite granules is shown in Figure 3.

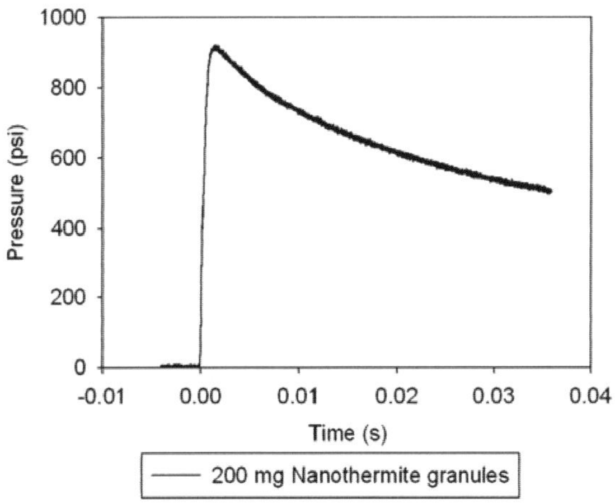

Figure 3: Pressure generation of nanothermite granules

Since it can be seen that the nanothermite reaches its peak pressure quickly and then begins to sharply drop off, it was decided to coat the granular material in a gasifying agent to slow down reaction and produce higher pressures. The first agent considered was nitrocellulose, due to the fact that its byproducts are not environmentally harmful and it is a common component in pyrotechnic materials. One drawback of trying to incorporate nitrocellulose into the granules is that it water insoluble and the granules are formed using a water-based process. To get around this obstacle, it was decided to add the nitrocellulose in the form of a coating on the outside of the granules once they have been formed and dried. A solution of nitrocellulose dissolved in acetone was made. This solution could then be spray onto the dried granules. Since the binder material used in the granules is water soluble, there was minimal penetration of nitrocellulose or acetone into the pores of the granules. Once the acetone is evaporated from the granules, the nitrocellulose remained as a coating on the outer surface. Not only did the coating slow down pressurization rate and increase peak pressure as is shown in Figure 4, but the ESD sensitivity was mitigated to the point that 25 mJ was not sufficient to ignite the granules.

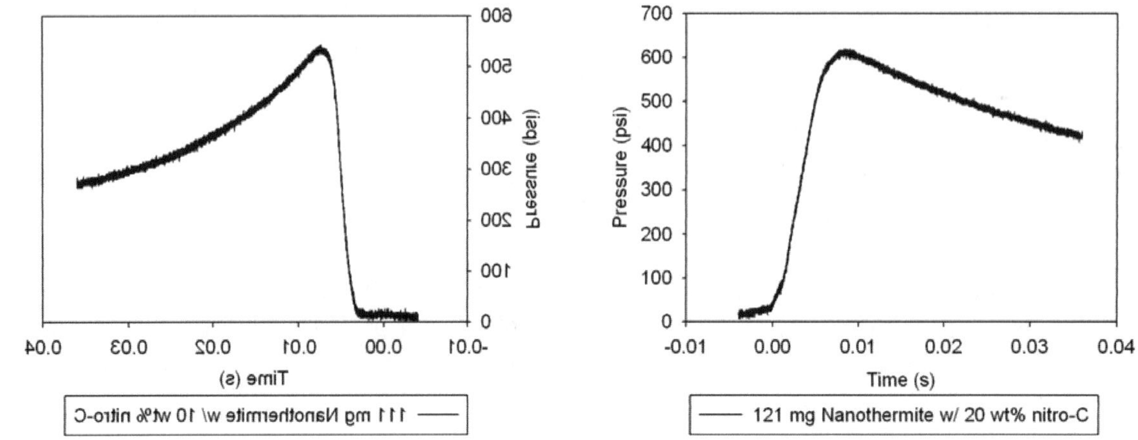

Figure 4: Pressure plots of nanothermite granules coated with 10 wt% nitrocellulose (left) and 20 wt% nitrocellulose (right).

Conclusions

It has been shown that the reactivity and ignition sensitivity of nanothermite systems can be tuned to a specific level using several different techniques. The incorporation of both iron oxide and bismuth oxide, at various ratios with respect to each other, into a nanothermite mixture containing aluminum is an easy and effect way to develop a material with specific reaction rate, pressure output, and ignition sensitivity. Other effective methods for changing reaction parameters of a material include consolidation of the material into granular form and coating of the granular material with gasifying agents.

References

[1]Puszynski, J.A., Bulian, C.J., and Swiatkiewicz, J.J., "Ignition Characteristics of Nanothermite Systems," *Energetic Material Synthesis and Combustion Characterization for Chemical Propulsion,* Begell House Inc., in press 2008.

[2]Bulian, C.J., Puszynski, J.A., and Swiatkiewicz, J.J., "The Effect of Processing Parameters on Ignition Characteristics and ESD Sensitivity of Nanothermites," *Proceedings of the 35th International Pyrotechnics Seminar,* Ft. Collins, CO, July 2008, 551-22.

[3]Puszynski, J.A., Bulian, C.J., and Swiatkiewicz, J.J., "The Effect of Nanopowder Attributes on Reaction Mechanism and Ignition Sensitivity of Nanothermites," *Materials Research Society Symposium Proceedings,* 2006, 896:147-58.

[4]Perry, W.L., Smith, B.L., Bulian, C.J., Busse, J.R., Macomber, C.S., Dye, R.C., Son, S.F., "Nano-scale tungsten oxides for metastable intermolecular composites," *Propellants, Pyrotechnics, Explosives,* 2004, 29(2):99-105.

[5]Sanders, V.E., Asay, B.W., Foley, T.J., Tappan, B.C., Pacheco, A.N., Son, S.F., "Reaction Propagation of Four Nanoscale Energetic Composites (Al-MoO_3, Al-WO_3, Al-CuO, and Al-Bi_2O_3), *Journal of Propulsion and Power,* 2007, 23(4):707-14.

[6]Puszynski, J.A., Bulian, C.J., and Swiatkiewicz, J.J., "Processing and Ignition Characteristics of Aluminum-Bismuth Trioxide Nanothermite System," *Journal of Propulsion and Power,* 2007, 23(4):698-706.

[7]Somasundaram, V., Bulian, C.J., and Puszynski, J.A., "Investigation of Al-CuO Nanocomposite Reacting Systems," *Proceedings of EUROPYRO 2003*, 2:536-41.

[8]Fischer, S.H. and Grubelich, M.C., "Theoretical Energy Release of Thermites, Intermetallics, and Combustible Metals," *Proceedings of the 24th International Pyrotechnics Seminar,* Monterey, CA, July 1998.

547e Fabrication and Characterization of Bimetallic Al-Ni Reactive Nanostructures as Nanoheaters

Qingzhou Cui[1], Julie Chen[2], and Zhiyong Gu[1]. (1) Chemical Engineering, University of Massachusetts Lowell, One University Ave, Lowell, MA 01854, (2) Mechanical Engineering, University of Massachusetts Lowell, One University Ave, Lowell, MA 01854

Heating is one of the most efficient and frequently used methods to alter material geometry, structure, and properties in both ancient and modern industries. Traditional macro- and micro-scale heating methods such as resistance and infrared heating have played a significant role in materials processing and manufacturing. However, with the continued miniaturization of electronics and devices, the expansion of multi-material systems, and the emergence of nanotechnologies, there is significant need for controlled and localized heating sources for thermal nanomanufacturing applications. There have been some methods under study to realize ultra-small heat spots with nanoscale dimensions, for example, nanoscale localized heating generated through a laser-heated AFM tip. However, most of the methods so far are serial, expensive, and difficult to scale-up.

In this presentation, we introduce a nanoscale heating source ("nanoheater") based on an exothermic reaction between aluminum (Al) and nickel (Ni). The formation process of Al and Ni alloys is a good candidate for heat-generation purposes due to its large exothermic reaction enthalpies. The fabrication and characterization of various types of nanoheater structures, e.g., bi-metallic nanowires and core-shell nanoparticles, will be presented. These Ni-Al hetero- nanostructures are synthesized by either chemical or physical methods, and are characterized with SEM and TEM. Upon ignition using different methods such as electrical, magnetic field, or infrared heating, Ni-Al alloys are formed and one-time controlled and localized heat is generated from the exothermic reaction process. The potential applications of these nanoheaters in nanomaterial processing and biomedical field will also be discussed.

Metastable Intermolecular Composites (MIC) for Small Caliber Cartridges and Cartridge Actuated Devices

Extended Abstract

This report documents demonstration testing of MIC primer compositions designed to eliminate lead from the primers used in small arms ammunition and cartridge actuated devices. This effort has shown that an Al/Bi_2O_3 composition can be successfully mixed and wet-loaded into existing US Army and US Navy primer hardware. Subsequent testing in M855 5.56 mm ammunition as well as a variety of Navy impulse and delay cartridges has shown that the primer provides performance essentially equivalent to the lead-based compositions presently in use. Thus, the new primer composition meets the objective for a drop-in replacement for the lead compounds. While the test results are positive and the MIC primer meets all performance specifications, additional work is needed to refine the primer composition to achieve faster action time in small arms cartridges and to eliminate occasional misfires.

Introduction

1.1 Background

Current percussion primers in small and medium caliber ammunition (i.e. 5.56mm-40mm) use a lead styphnate based primer formulations that pose a long term hazard to the environment and the operator of the weapon since airborne vaporized lead results from each successfully fired cartridge. Lead styphnate based primer compositions are currently specified in all of the US Army's combat small caliber ammunition and in many Cartridge Actuated Devices and Propellant Actuated Devices (CAD\PAD) used in US Navy aircraft ejection systems, countermeasure applications, and stores release systems. The CAD/PAD devices are used by all DOD components and foreign military that utilize US manufactured aircraft. Lead is a known toxic material, which pollutes test ranges and exposes the manufacturers and users of these devices to serious health hazards liabilities. Lead is regulated by the Environmental Protection Agency (EPA) and the Occupational Safety Health Administration (OSHA). Current EPA and OSHA regulations are directly impacting range and testing operations. Stricter regulations in the future will seriously impact or force closing of production, testing and range operations. With the current production rate for all small caliber ammunition (less than 20 mm), the quantity of lead to be consumed for percussion primer production alone is well over 23,686 pounds or nearly 12 tons annually.

Small caliber percussion primers generally consist of a brass cup loaded with the charge composition and a brass anvil pressed into the charge. Figure 1 shows the US Army No. 41 primer prior to inserting the anvil. When the cup is struck by a firing pin, the friction and impact sensitive charge is crushed between the bottom of the cup and anvil, causing ignition. The hot ignition products flow out of the cup around the legs of the anvil to ignite the next element in the ignition train of the weapon system. For the Army No. 41 primer, this would be the ball powder main charge in small caliber ammunition. For the Navy PVU-1/A primer, this would be either a transition charge or an output charge, depending on the particular application.

Figure 1 – No. 41 Small Caliber Percussion Primer Showing Loaded Cup and Tripod Anvil

While the No. 41 and PVU-1/A primers are similar in size, there are some important differences. The No. 41 primer is designed to provide sufficient pressure to quickly ignite the double base ball powder main charge, and thus utilizes a tripod anvil and the FA-956 primer composition, which contains PETN and aluminum powder for added brisance (Table 1(a)). The PVU-1/A is designed for "soft" ignition of delay cartridges, and uses a bipod anvil and the 5086 primer composition (Table 1(b)), which provides a lower output pressure than FA-956. Figure 2 is a photograph of a loaded PVU-1/A, illustrating the small size.

Ingredient	Weight %
Normal Lead Styphnate	37.0
Tetracene	4.0
Barium Nitrate	32.0
Antimony Sulfide	15.0
Aluminum Powder	7.0
PETN	5.0

(a) FA-956 Primer Mix (No. 41 Primer)

Ingredient	Weight %
Normal Lead Styphnate	26.0
Barium Nitrate	41.5
Tetracene	2.0
Calcium Silicide (Treated)	10.5
Antimony Sulfide	20.0

(b) 5086 Primer Mix (PVU-1/A Primer)

Table 1. Primer Mix Compositions for No. 41 and PVU-1/A Primers

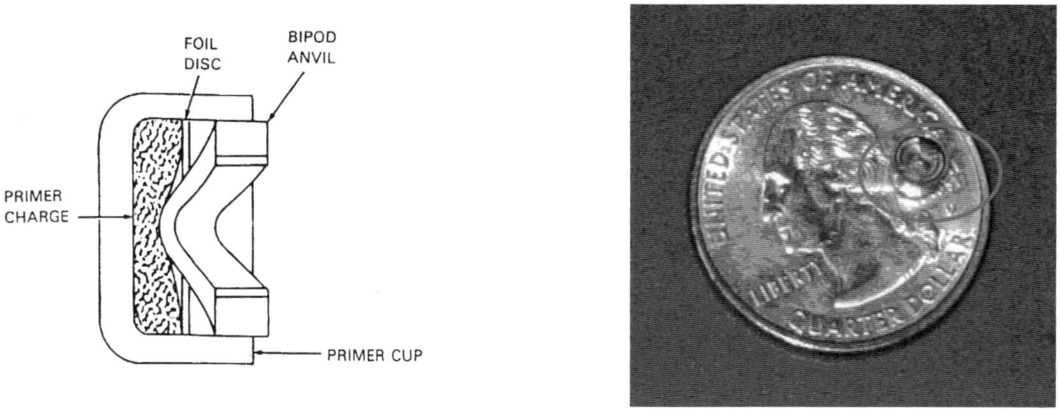

Figure 2 – PVU-1/A Percussion Primer

Technology Description

2.1 Technology Development and Application

Metastable Intermolecular Composite (MIC) material has the potential to replace the current conventional energetic composition in the initiation subcomponents of ammunition and cartridge actuated devices known as the percussion primer. The novel properties associated with nanostructure materials have resulted in the development of thermite-like formulations of energetic materials by Los Alamos National Laboratory (LANL). These materials being of nano-sized particles offer the possibility of tunable energy release and high temperatures without appreciable gas generation and attendant high pressures. There are various examples of MIC applications that attracted a great deal of interest recently for weapon enhancement. One unique feature of MIC materials is its ability to produce particles hot enough to ignite a bed of propellant. Additionally, the MIC materials are impact sensitive which makes them a good percussion primer mix candidate. MIC can be utilized as an initiation composition for replacing the current established FA-956 and 5086 primer formulations which are based on lead styphnate, barium nitrate and antimony sulfide. The MIC mixture is an environmentally friendly, lead free composition.

In general terms, the MIC material is an engineered energetic composition consisting of a metal fuel (most often nano-scale aluminum) and metallic oxidizer that are exothermically reactive with each other. By utilizing nano-sized particles, the near atomic scale proximity of the reactants minimizes distances over which the fuel and oxidizer molecules must diffuse in order to reach each other, resulting in a dramatically increased reaction rate relative to that of conventionally sized mixtures. Two of the most commonly used MIC compositions have the following chemical reactions:

$$2Al + MoO_3 \rightarrow Mo + Al_2O_3$$
$$2Al + Bi_2O_3 \rightarrow 2Bi + Al_2O_3$$

For Army small caliber ammunition applications, the MIC primer must meet #41 primer all-fire and no-fire energy requirements and also ignite the propelling charge rapidly enough to meet the action time requirement for each individual cartridge application.

For CAD/PAD applications, the MIC primer must meet the PVU-1/A primer all-fire and no-fire energy specifications and also must function such that the performance requirements for each individual application are met. Because the applications chosen for the demonstration represent a cross-section of the CAD/PAD spectrum, the performance requirements vary considerably from one application to another.

The current primer mix for all 5.56mm 7.62mm and 0.50 cal ammunition is the Frankford Arsenal composition FA 956. 20mm ammunition uses a primer mixture of slightly different amounts of the same basic components plus a carbon compound to make the mixture electrically conductive. These primer mixtures are manufactured by Lake City Army Ammunition Plant (LCAAP). The manufacturing process used in producing the FA 956 primer mix is basically a five step process that includes the manufacture of trinitroresorcinol (TNR), lead styphnate, tetracene, pentaerythritol tetranitrate (PETN), wash, and a final wet mix operation.

- Lead styphnate is formed by mixing TNR with magnesium oxide to form magnesium trinitroresorcinol. The magnesium trinitroresorcinol is in turn mixed with lead nitrate to form lead styphnate.

- Other heavy metals chemical compositions besides lead that are added during the final mixing process of producing the FA 956 primer mix are antimony sulfide and barium nitrate.

- The following flow chart summarizes the current manufacturing process for the FA 956 formulation:

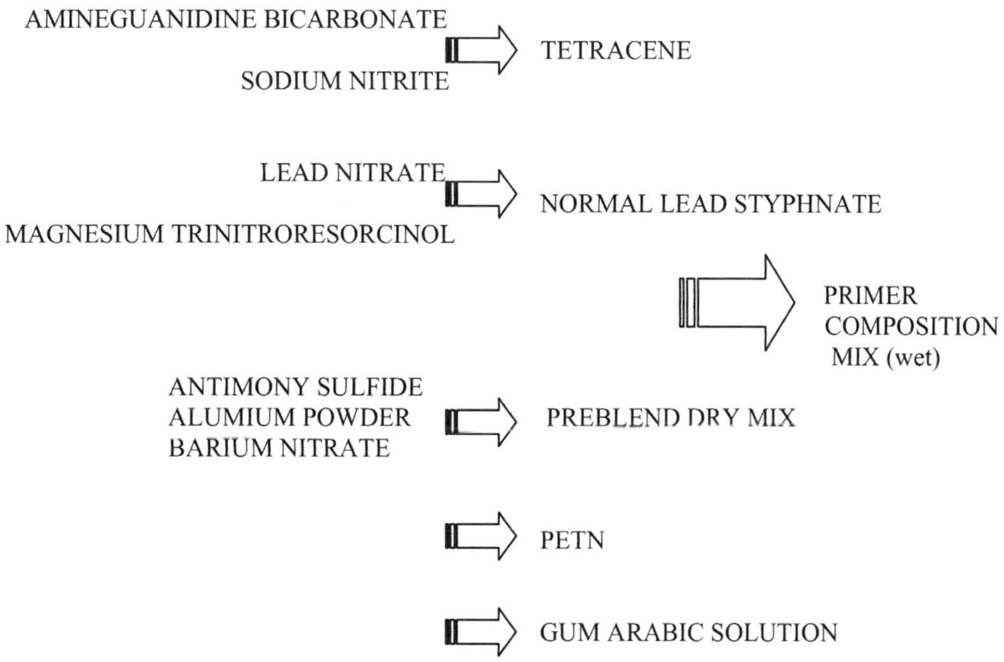

The wet-mixing procedure consists of the following steps:

Lead styphnate is placed into the mixer. Tetracene is then added followed by the gum Arabic solution and the PETN. The dry fuels (antimony sulfide and aluminum) are added to the mixture followed by the oxidizer (barium nitrate). The finished wet primer mix is transferred to a conductive container and transported to the primer pelleting. These are labor intensive processes.

Pellets are made by hand-pressing the wet primer mix onto a plate with holes that correspond to the number of primer cups that are to be filled. The pellets are then transferred to primer cups and the mix is consolidated. The anvils are then inserted into the consolidated mix and the finished primer is moved to a drying area for removal of water. After a period of time suitable to ensure the mixture is dry, the primers are tested for sensitivity and taken to the bullet assembly line to be inserted into cartridge cases

The PVU-1/A primer has been designed for use in aircrew escape systems for Army, Navy, and Air Force aircraft. Delay cartridges are used extensively in these systems, and because delay columns cannot tolerate high impact forces, it is necessary to use a primer mix that is less brisant than the FA 956 used in the #41 primer. In addition, to attain the high reliability required for man-rated systems, the PVU-1/A hardware has been designed for increased sensitivity to friction and impact. Thus, the output pressure generated by the PVU-1/A is roughly 25% of that for the #41, and the all-fire energy is considerably less (25.5 inch-ounces versus about 46 inch-ounces for the #41). These differences between the PVU-1/A and #41 primers highlight the difficulty in creating a common mixture which can be used in both.

The process for manufacturing the 5086 mix is specified in the PVU-1/A drawing package (NAVAIR 851AS110). The process is similar to that used for the FA 956, and is briefly summarized below:

- Inert ingredients (barium nitrate, calcium silicide, and antimony sulfide) are dried and sieved.
- lead styphnate and tetracene are weighed out to the correct proportions for the mix, and then re-wetted with ethanol.
- mixing speed is adjusted, the inert ingredients are added, and mixed.
- The lead styphnate and tetracene are added separately and mixed.
- The wet mix is stored in polyethylene bottles for later use.
- Prior to loading, the wet mix is dried and then sieved.

PVU-1/A primers are hand-loaded in manufacturing facilities using the dry loading procedure illustrated in Figure 3. Lot sizes are typically 10,000 primers. After a lot has been completed 900 are withdrawn for sensitivity and dud testing in a ball-drop apparatus as per WS 21535B.

The following flow chart characterizes the current MIC manufacturing process demonstrated at Armament Research, Development & Engineering Center (ARDEC), Los Alamos National Laboratory (LANL) and Naval Surface Warfare Center/Indian Head Division (NSWC/IHDIV):

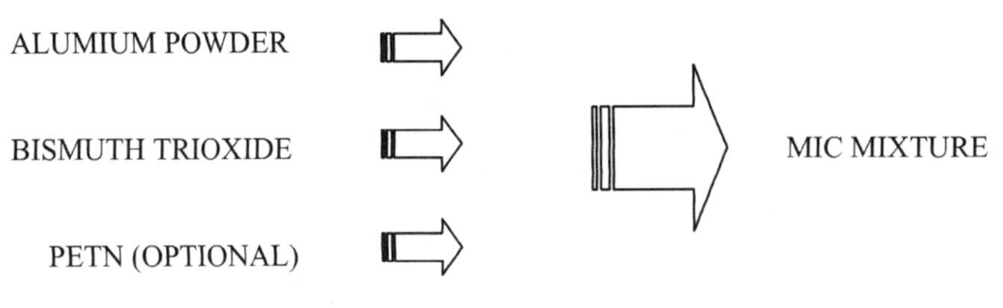

A brief description of the current procedure for manufacturing the MIC primer mix is illustrated in Figure 4. Recent technical advances in the MIC program have shown that MIC mixtures utilizing bismuth trioxide as the oxidizer can be wet mixed and loaded without significant degradation of the mix. Bismuth trioxide is currently being used as the oxidizer in all MIC development programs conducted at ARDEC and NSWC/IHDIV. Many of the details of the process have been omitted due to the sensitivity of the information. Qualified organizations with a bona fide need-to-know can contact ARDEC or NSWC/IHDIV directly for access to this information.

The correct amounts of nano aluminum, oxidizer (bismuth trioxide) and other additives are placed into a mixing vessel. The optimum amount of solvent (water, hexane, cyclohexane or isopropyl alcohol) is added, and the composition is then mixed with either an ultrasonic probe or in an ultrasonic bath. The mixing time is dependant on the method used. After mixing, the material is loaded into a primer cup, the anvil is inserted, and the loaded primer is placed into a vacuum oven for drying.

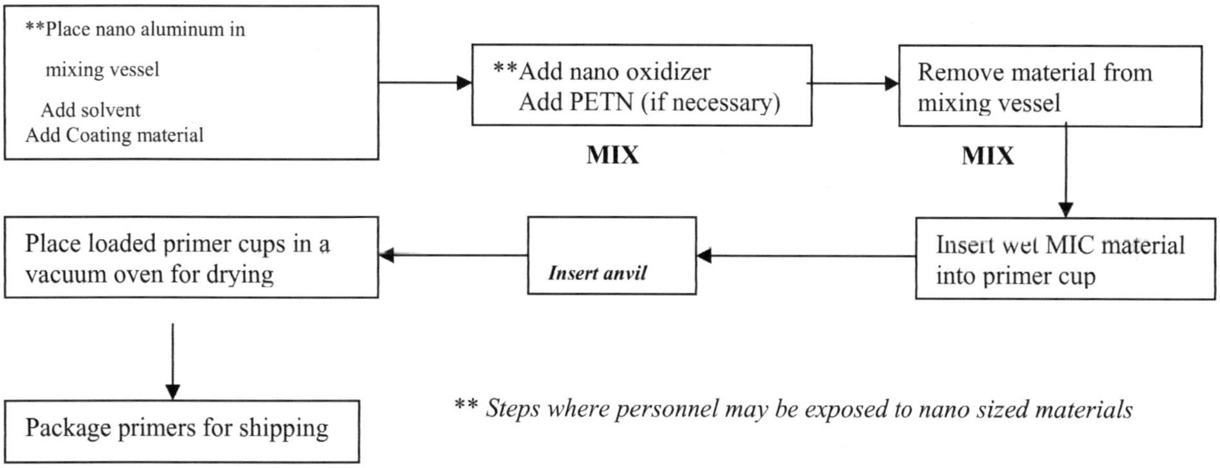

Figure 4 - Proposed MIC Primer Manufacturing Process

547g Filled Nanocomposite Materials for Biocidal Applications

Mirko Schoenitz[1], Demitrios Stamatis[1], Edward L. Dreizin[1], and Sergey A. Grinshpun[2]. (1) Chemical Engineering, New Jersey Institute of Technology, University Heights, Newark, NJ 07102, (2) Department of Environmental Health, University of Cincinnati, 3223 Eden Ave., POB 670056, Cincinnati, OH 45267

Filled nanocomposite materials were developed as a new class of energetic materials for the purpose of inactivating aerosolized biological agents. These materials, when combusted, release halogenated gaseous species in the atmosphere, neutralizing hazardous biological materials, in particular viable airborne bio-agents, such as stress-resistant viruses and bacteria.

Filled nanocomposite materials are based on previously developed fully-dense energetic nanocomposite materials prepared by Arrested Reactive Milling. By carrying out the preparative milling process at cryogenic temperatures, significant quantities of components can be embedded in a continuous metal matrix, which would be volatile under normal atmospheric conditions. In this research, elemental iodine was embedded in an aluminum matrix as biologically active component, together with molybdenum trioxide as an oxidizer. Details of the synthesis approach, the ability to contain volatile components in a metal matrix, and stability data of the filled nanocomposite materials will be presented.

An appropriate test protocol was established, and initial tests were conducted. Aerosols of two bio-agent simulants, Bacillus subtilis bacterial endospores and MS2 bacteriophage virus were exposed to an environment with burning filled nanocomposite materials. The time scale of the test required to moderate the burn velocity of the nanocomposite with paraffin wax as an endothermic binder. The biological viability of the aerosolized bio-agents after exposure was determined by cultivation.

Design for Demilitarization – Be Creative Now or Pay Me later

Synopsis: This presentation will share the progress made while serving on the Army Program Manager (PM) for Demil's Integrated Progress Team (IPT) established in fiscal year 2003 to implement "Design for Demil" (DFD) among the services. The comments on the IPT participation are from the standpoint of a Navy Activity, Indian Head Division, Naval Surface Warfare Center, which is experienced in basic and applied munitions R&D, munitions production and demil.

Abstract: The Army PM Demil Strategic Plan included DFD as a key goal. The program included six pillars deemed important for DFD implementation: establishment of official policy; development of metrics; creation of helpful tools, one of which will be a DFD handbook; publishing lesson learned from demil experience; devising a template of contractual requirements for munitions contractors; and participation on the IPT.

Former practice lead munitions design engineers to focus on performance. As a result there were problems costing time and money and possibly increasing risk when the munition was subjected to demil. These may have involved toxic constituents, hard to disassemble components or other challenges. Demil can drastically raise life cycle cost and diminish readiness. The importance of life cycle management, reaching from the genesis of the munition to the demilitarization, has refocused design.

This presentation will give examples of munitions which have designs leading to problematical demil, such as the Army's Adam Mine; discuss desirable design characteristics such as modularity, energetic material designed for easy demil; resource recovery and recycling; and minimization of toxic materials of construction. All of these can affect the management of the munition throughput its life cycle, not just at its end.

As well, it will discuss progress in educating government and contract design engineers about DFD concepts, progress in informing program managers of the importance of DFD; and success in placing stronger DFD requirement language on acquisition directives.

EXTENDED ABSTRACT

Design for Demilitarization (DFD) - what is DFD and why is my organization interested in this concept? Indian Head Naval Surface Warfare Center's mission is integrally related to the environment and life cycle management of weapons systems. Located about twenty miles south of the Washington, DC beltway, the facility is recognized as a national resource for energetics technology. The base provides research, development, test and evaluation and in-service support of energetics and energetic materials for warheads, propulsion systems, ordnance and pyrotechnic devices and fuzing for the Navy. Indian Head operates production plants for Navy ordnance such as warheads, rocket motors and cartridge and propellant actuated and cartridge actuated devices. Also located at Indian head NSWC is a permitted treatment, storage and disposal open burn and open detonation facility as well as a car - bottom furnace for closed treatment of materials and equipment contaminated with energetic materials. There are facilities for sawing weapons to access components and energetics and for performing high pressure water washout, both of which can be used to access or remove the energetics which must be done for demil or reuse. In short, Indian Head NSWC is a full service energetics facility with responsibilities from 'cradle to grave.'

The Detachment where the author works is located in Virginia and shares the same mission as the parent facility but specializes in explosives. Designated by Navy Instruction, the Yorktown facility writes documents called Naval Munitions Data, which, relying experimental loading of explosives into a weapon, provide documentation specifying safe loading procedures for conventional ammunition. In conducting studies to support writing the NMDs, a weapon my be loaded with explosives which then may be removed by water washout or melting or other means for additional tests. Other work also includes product improvement programs which can periodically introduce new energetics or designs into a weapon.

The execution of many of the mission responsibilities results in evaluation of assembly/disassembly and loading /downloading techniques. It is this direct interaction with weapons components, the actual energetics and raw batch additives leading to the safe assembly of the weapon that I think gives Indian Head a unique insight into to concepts for DFD. With this background we saw participation in the Army-sponsored Design for Demil Integrated Product Team (IPT) as a way to interact with other organizations responsible for design, testing and demil and to learn from them and to share design for demil history and lessons learned.

<u>Basic Definitions</u>: For those not familiar with the end of the life cycle of a munition, <u>DoD 4160.21-M-1,</u> "Defense Demilitarization Manual", October 21, 1991; Incorporating Change 1 - February 14, 1995, defines:

demilitarization as the act of destroying the military offensive or defensive advantages inherent in certain types of equipment or material. The term comprehends mutilation, dumping at sea, cutting, crushing, scrapping, melting, burning, or alteration designed to prevent further use of this equipment and material for its originally intended military or lethal purpose and applies equally to material in unserviceable or serviceable condition that has been screened through an Inventory Control Point (ICP) and declared surplus or foreign excess; and **disposal** as the process of redistributing, transferring, donating, abandoning, destroying or other disposition of DOD personal property.

packaging

In simpler terms which says it all, Merriam Webster online defines demilitarization as: to rid of military characteristics or uses. Of course, times change and new environmental thinking, security and safety needs are always influencing demilitarization, defining boundaries and pointing in new directions, many of which require research and development..

The Defense Logistics Agency (DLA) handles surplus government property. Field activities under the DLA comprise the Defense Reutilization and Marketing Service (DRMS) which manages disposal of hazardous property for DOD activities, getting the most use or value out of each item and minimizing environmental and safety risks and costs. DRMS manages the DOD surplus property sales program. Excess property that is not reutilized, transferred or donated may be sold to the public as surplus.

Looking at environmental impact, the Resource Recovery and Recycling Program (R3) conserves natural resources, reduces waste products and returns revenue to the military services. The Precious Metals Recovery Program significantly reduces the need for DOD to purchase metals such as gold, silver and platinum family metals by recycling excess and surplus scrap containing precious metals.

There are many components of weapons systems that can be refurbished and reused as units, removed and reprocessed for weapons use or by reclaiming raw materials for sale or reuse, the R3 above.

This then, brings us to demilitarization. "Certain property is demilitarized (i.e., rendered useless for its originally intended purpose). Surplus property with inherent military characteristics must undergo "demil." Offensive and defensive weapons and associated material are demilitarized prior to sale or as a condition of sale. (Wickepedia, Sep 2008). The next paragraphs point out some documents which create constraints on popular demil methods. For example, dumping at sea, an extreme method is not now allowed by Federal regulations

There are many DOD regulations to ensure the security and safety of demilitarized weapons to define what must be demilitarized and regarding environmental impact. Three of the several demil documents are DoD 4160.21-M-1, "Defense Demilitarization Manual" and, October 21, 1991; Incorporating Change 1 - February 14, 1995; DoD 4160.21-M, "Defense Materiel Disposition Manual", August 18, 1997; and specific guidance for property identified as Munitions List Items (MLI)/Strategic List Items (SLI) is found in DoD 4160,2 1-M-1, Defense Demilitarization and Trade Security Control Manual. These instructions are under the authority of DoD 4140.1 -R, "Department of Defense Materiel Management Regulation. There are exceptions from this authority such as nuclear weapons. The Defense Reutilization and Marketing Service (DRMS) operations are also applicable to Federal civil agency excess and surplus property turned-in to a Defense Reutilization and Marketing Office (DRMO), for disposition under applicable interagency agreements.

Because DFD is a process which can make the entire life cycle of a weapons system greener or reduce the carbon footprint (in today's vernacular), less costly to demil, easier to modernize or upgrade there are many reasons why an organization such as ours is interested in the policy. Since the Facility where the author works and wherein he author's experience lies, this presentation abstract focuses on demil of conventional ammunition containing energetic materials, and participation in the U.S. Army Design for Demil IPT.

Environmental pressures have made the use of the traditional treatment method of munitions at the end of the life cycle, open burning/open detonation less acceptable. Alternative treatment methods s known as "closed demil technologies" were implemented as research and development and evaluation fielded new methods. In closed technologies pollutant releases are contained and minimized or eliminated. In about 1996, by dollars spent, closed demil technologies overtook OB/OD as the primary overall method of demil. (Conventional Munitions Design for Demilitarization Implementation Plan, U.S. Army Product manager for Demilitarization, Jan 04.) In addition, demil began to be looked at as an asset, producing components and materials which could be recycled or reused or sold, and Resource Recovery and Recycling (R3), as it known as, includes more than 70% of demil operations today. More sophisticated demil methods began to show design decisions made very early in the ammunition life cycle would have a dramatic impact on end-of-life-cycle demil operations in cost (time to disassemble) is a large part of cost), complexity, environmental footprint and opportunity for R(3).

The Army and many others in DOD saw the value in "design for demil." No coordinating body existed in the DOD to implement a DFD program in the ammunition design community. The stand up of the Program Manager for Joint Services and the Product Manager for Demilitarization (PM Demil) within the Program Executive Office for Ammunition (PEO AMMO) changed this. Under charter the PM Demil has the responsibility for the life cycle management of recycling, demilitarization and disposal of energetic materials and munitions for the Army, Navy, Air Force, Marines and other Department of Defense and U.S. Government Activities. PM Demil is the Single Manager for Conventional Ammunition (SMCA) for demil management and intends to assure a design for demil approach early in the life cycle or during ammunition system modifications or product improvements. Almost 70% percent of the cost of a product is set by decisions made early in the engineering process. (Executive Summary, "Retooling Manufacturing: Bridging Design, Materials and Production", Committee on Bridging Design and manufacturing, National Research Council, ISBN 978-0-309-09266. (Executive Summary available online at www.nap.edu.) As a result, PM Demil directed that a "design for demil" IPT be established to determine how to develop and implement a design for demil program.

The first IPT organizational meeting occurred on 7 May 03 at Picatinny Arsenal with about 32 multiservice members. Other Navy participants included NSWC Crane, IN; Naval Air Systems Command, PMA-201; Marine Corps, Systems Command, Quantico, VA and Navy Ammunitions Logistics Center, Mechanicsburg, PA. A representative, Mr. William Houchins, on assignment from Office of Secretary of Defense to Dahlgren NSWC, although not formally joining the IPT, added additional Navy emphasis through the Joint Ordnance Commanders Group Demil/Disposal Group. He proposed an initiative of creating a "design for demil handbook" which is now at the forefront of the IPT agenda. Mr. Gary Mescavage of U.S. Army ARDEC, Picatinny Arsenal, NJ, (973) 724-3349, gary.mescavage@us.army.mil has been the IPT Leader from the stand up of the IPT. The original IPT Charter dated 7 May 2003, included 27 members and has fluctuated in numbers and sponsoring organizations over the years.

Quickly formed were three subcommittees, the names of which describe the goals: Metrics, Policy and Design for Demil Handbook. Metrics for documenting DFD have proven challenging to document and this is a continuing effort. Time to disassemble may prove a good metric. The National Academy of Sciences states in systems approach to design and engineering the cost of a product (consider weapons as a product) over its lifetime is considered. Gathering the necessary data may be expensive and arbitrary in assessments of life cycle costing: first there is acquisition including design, development and manufacturing; second there is operating cost or cost of ownership. Third, says the Academy, there "is the environmental impact of manufacturing processes and end of life recycling or disposal. The ownership cost is particularly relevant in defense systems which may last for generations, and "design decisions can have a profound impact on the adaptability of defense systems to modifications or retrofits." (Retooling Manufacturing: Bridging Design, Materials and Production," National Academy of Sciences (http://books.nap.edu/catalog/11049.html)). "Metrics for quantifying all these assessments are challenging. Accurate assessment is difficult because gathering the necessary data is expensive and also may be subjective or arbitrary. One reason is that recycling is often done by widely distributed small businesses that operate with a variety of business models making the economics of the industry opaque."

The policy is driven by documents such as executive orders such as and federal acquisition documents. These will be briefly discussed historically in the presentation but the goal of the DFD IPT was to put "teeth onto the requirement, and that has finally occurred with the with the signing of the Design for Demil Policy Letter by the Undersecretary of Defense for Acquisition, Technology, and Logistics dated 4 Aug 08. "To Implement Design for Demilitarization, Military Departments, Defense agencies and the U.S. Special Operations Command will include in their acquisition documentation for all pending (i.e. pre-milestone A) and future conventional ammunition programs how they intend to address demilitarization requirements throughout system design. Six specific requirements for Design for Demilitarization of conventional ammunition are attached to the letter. Mr. Jose Gonzalez of the OUSD (AT&L) is the Undersecretary's point of contact at 703-693-9203. Armament Research, Development, Development and Engineering Center is the Army point of contact at 703-693-9203.

The Design for Demil Handbook will be an on-line living document to explain to he design engineer what is expected in design for demil and will be located on a web site under the auspices of the Army. The handbook will also give demil lessons learned, .sources of information to assist in design and templates.

The presentation will discuss specific examples of designs that have presented demil problems. When open burning or open detonation were the primary demil techniques, engineers and designers did not need to be concerned with demil issues. Performance was the primary concern. Design had littler impact on the performance of efficient demil. Mr Mescavage discusses implementation of Design for Demil (DFD) in the Joint Services in a recent presentation at the 16th Global Demil Symposium and Exhibition, Salt Lake City, Utah, 5-9 May 2008. He gives an example of a weapon which is problematical for demil, inadvertently by design, Use of a depleted uranium salt in a polymer used in the mine body requires the use of cryogenic fracture to separate the uranium matrix from the mine. Other examples of demil problems related to Navy Weapons will be outlined.

Design for Demil is a major change in design approach, but with an investment early in the life cycle, it will improve lifecycle management and control the end of life cycle demil cost.

Lauritzen, E. K. , Egebart, C. L. and Larson, S. G. , "The Challenge of Demilitarization and Disposal of Ammunition – Environmental & Economic Considerations of Weapon Systems," presentation at the ISSC Seminar no. 560/01, Trends in Weapon Developent in the 21st Century, Brussels, 3,4 May, 2001.

Characterizing and Modeling of Particle Size Distributions from Propellant Open Burning or Open Detonation (OB/OD)

Mitchell Lindsay [1], *Jonathon Koenig* [1], *Chris Merrill* [1], *Glenn Palmer* [2]

Abstract

The US Department of Defense (DoD) demilitarization activities conducted at the Utah Test and Training Range (UTTR) include destruction of ballistic missile motors, which contain solid propellants, by open burning or open detonation (OB/OD). Detonations and burns of these solid propellants generate alumina-based particulate matter (alumina particulate) as an emission. Airborne particulate matter (PM) generated by OB/OD, such as PM2.5 (any PM with a diameter less than or equal to 2.5 microns) and PM10, is regulated as a pollutant emission. The data historically used for permitting activities at the UTTR do not provide test-based particle size information (fraction of PM2.5 and PM10), and no modeling approach is known for the prediction of particle size distributions (PSDs) or PM emission rates.

A predictive particle model was developed with two primary components: the Cheetah combustion model, and a MATLAB stochastic particle model. The predictive particle model uses an incremental time step approach to address nucleation, coagulation, and surface growth of alumina particulate in the detonation plume.

Laboratory work has been performed to validate sampling methodologies for measurement of pollutant emissions and PSDs from solid propellant OB/OD at the Utah Test and Training Range. Preliminary PSD and PM emission rate data have been acquired for small-scale burns of a surrogate propellant. Sub-scale detonation data will be used to refine and validate the particle prediction model.

These initial project phases are directed towards development of a data set to support a model of OB/OD phenomena that is acceptable to regulators and is sufficient to predict environmental emissions.

Introduction

The Utah Test and Training Range (UTTR) is the only installation on US soil permitted to perform open detonation (OD) of munitions with greater than 10,000 lbs net explosive weight (NEW) and open burn (OB) of much larger items. The OB/OD mission is one of the major emission events on the UTTR. The airborne PM released from these events is a regulated pollutant.

Historical data for PM emissions from OB/OD events is based on Bang Box emission factors. These factors do not provide PSDs. No modeling approach is known that can provide emission factors and PSDs for OB/OD events.

This paper will describe a modeling approach that provides predictive information about PSDs and emission rates from OB/OD events. The model consists of two primary

---

[1] CH2M HILL, Salt Lake City, UT

[2] OO-ALC, Hill AFB, Ogden, UT

components: the Cheetah combustion model, and a MATLAB particle model. Cheetah is a kinetic combustion and detonation dynamics model developed by Lawrence Livermore National Laboratory (LLNL). The MATLAB particle model provides particle size information by stochastically solving the general dynamics equation (GDE) of particle formation.

Laboratory work has been performed following the methodology used in previous work (Palmer, et al, 2008). A Grimm 1.107C particle monitoring system was used to measure PSDs from small scale burns of a surrogate propellant at the University of Utah Combustion Research Center.

Modeling Procedure

The two component modeling approach separates the detonation process into chemical reactions and particulate formation steps. This separation contrasts with typical soot formation models, which simultaneously simulate chemical reactions and particle processes. However, the complexity of developing a traditional soot formation model precluded this approach. Instead, the Cheetah combustion model and a MATLAB particle model were connected in series. The model components are connected as shown in Figure 1.

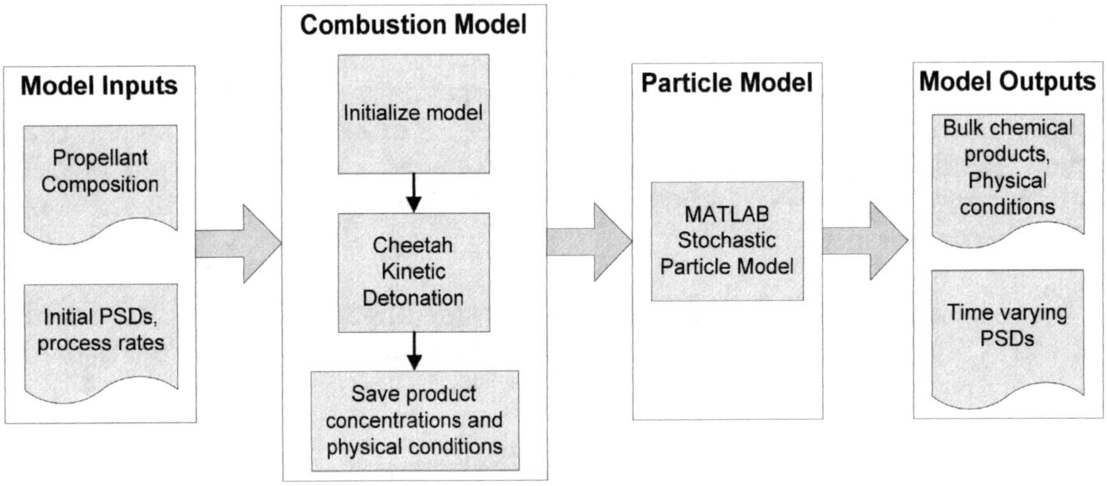

Figure 1. Model Components

The two component modeling framework unites the disparate timescales inherent in the detonation process. Cheetah provides chemical and detonation dynamics information (on the order of microseconds), while the MATLAB component models particle dynamics occurring on the order of milliseconds or seconds. By using Cheetah's output as an input to the MATLAB model, these timescales are united into a single modeling approach.

Cheetah

Cheetah 4.0 models the combustion and dynamics of the detonation process. Cheetah's combustion solver uses chemical kinetics to give time-varying product concentrations. Detonation dynamics are modeled according to the 2D Wood-Kirkwood (WK) theory, which considers a cylindrical charge of infinite length expanding radially from the centerline (Fried, et al, 1997).

A WK kinetic detonation was run in Cheetah. The reactant composition was 90% cyclotetramethylene-tetranitramine (HMX) and 10% aluminum (Al) by weight. All other parameters were set to their default values. The Cheetah reactant composition does not align well with the composition of the surrogate propellant used in the laboratory. The 90% HMX, 10% Al composition was used as an example case in this work. Future work will include the appropriate propellant composition.

MATLAB Particle Model

The MATLAB particle model stochastically simulates particle formation. Modeling begins with an initial PSD. The initial distribution for this work, shown in Figure 2, is a bimodal distribution fit to preliminary laboratory data. The model creates an initial number of simulation particles (N_{s0}) within an initial simulation volume (V_{s0}). These simulation particles are assumed to represent the entire particulate distribution.

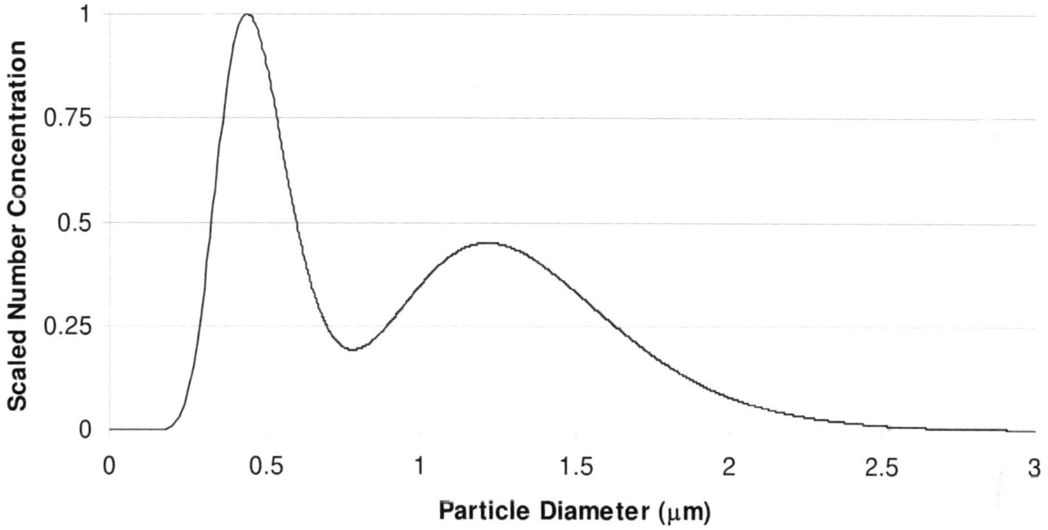

Figure 2. Initial PSD for MATLAB particle model.

The particle model simulates three particle formation processes: particle inception (nucleation), particle growth by condensation (surface growth), and particle collision (coagulation). Kernels (size dependent process rates) are used to define the rates of these three processes over the course of the simulation. Constant, size independent kernels are used throughout this work.

The rate of a process determines its likelihood of occurring during the simulation time step. Rates and probabilities for each process are calculated as shown in Table 1. For each incremental time step, a process is randomly selected based on its probability of occurring. Particles are then randomly chosen for the process, the process is performed, and time is incremented by a rate dependent time constant, $\tau = (R_n + R_s + R_c)^{-1}$.

Table 1. Particle Formation Process Rates

Process	**Process Rate** (1/s)	**Probability**
Nucleation	$R_n = \frac{V_s}{N_s} \cdot \sum_{m=1}^{Ns} J_m$	$P_n = \frac{R_n}{R_n + R_s + R_c}$
Surface Growth	$R_s = C_0 \cdot \sum_{m=1}^{Ns} Ks_m$	$P_s = \frac{R_s}{R_n + R_s + R_c}$
Coagulation	$R_c = \frac{1}{V_s} \cdot \sum_{m=1}^{Ns} \sum_{n=m+1}^{Ns} B_{m,n}$	$P_c = \frac{R_c}{R_n + R_s + R_c}$

J, Ks, and B are the kernels for nucleation ($m^{-3}s^{-1}$), surface growth (m^3/s), and coagulation (m^3/s).

N_s is the number of particles in the simulation.

V_s is the simulation volume (m^3).

Because the number of simulation particles (N_s) fluctuates over the course of the simulation, it is necessary to adjust the simulation volume (V_s), by bounding N_s so that $\frac{N_{s0}}{2} \leq N_s \leq 2N_{s0}$. When N_s is too high, V_s is halved, and simulation particles are randomly discarded until $N_s = N_{s0}$. When N_s is too low, V_s is doubled, and the simulation particles are replicated. Although these volume changes lead to small errors, they are necessary to maintain an appropriate number of simulation particles.

Test Procedure

Using the laboratory setup of Palmer et al, 2008, a Grimm 1.107C particle monitoring system was used to measure PSDs from small scale burns of a surrogate propellant. Distributions were measured every six seconds with thirty one bins, with diameters ranging from 0.265-32 microns (μm).

Results

Laboratory Data

Data were collected for multiple burns of the surrogate propellant. Figure 3 is an example of the time-varying total suspended particulate (TSP), PM10, and PM2.5 curves for a single burn. Figure 4 provides mass concentration contours (μg/m3) for the same event.

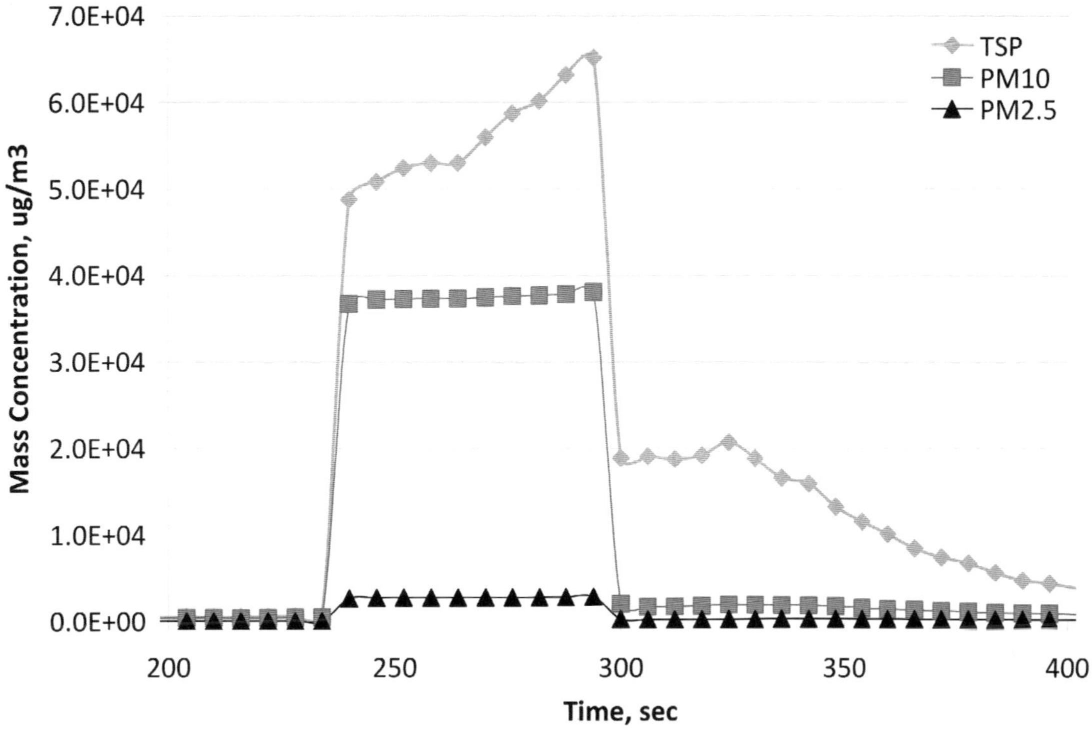

Figure 3. Total suspended particulate (TSP), PM10, and PM2.5 for run 1 on June 6, 2008.

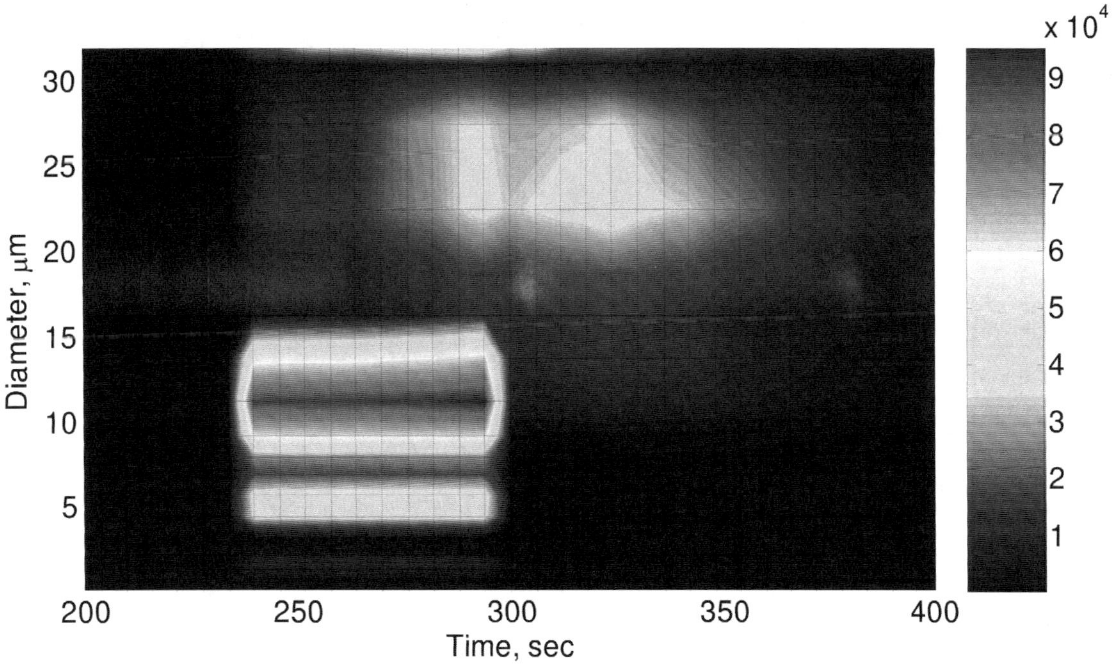

Figure 4. Mass concentration (μg/m3) of alumina over time for run 1 on June 6, 2008. The grid shows time and particle diameter sampling intervals.

Cheetah Kinetic Detonation

Cheetah models both thermodynamic and chemical aspects of the detonation. Figure 5 shows the pressure and temperature decay over the course of the detonation. Figure 6 shows the conversion of aluminum to aluminum oxide over the course of the detonation.

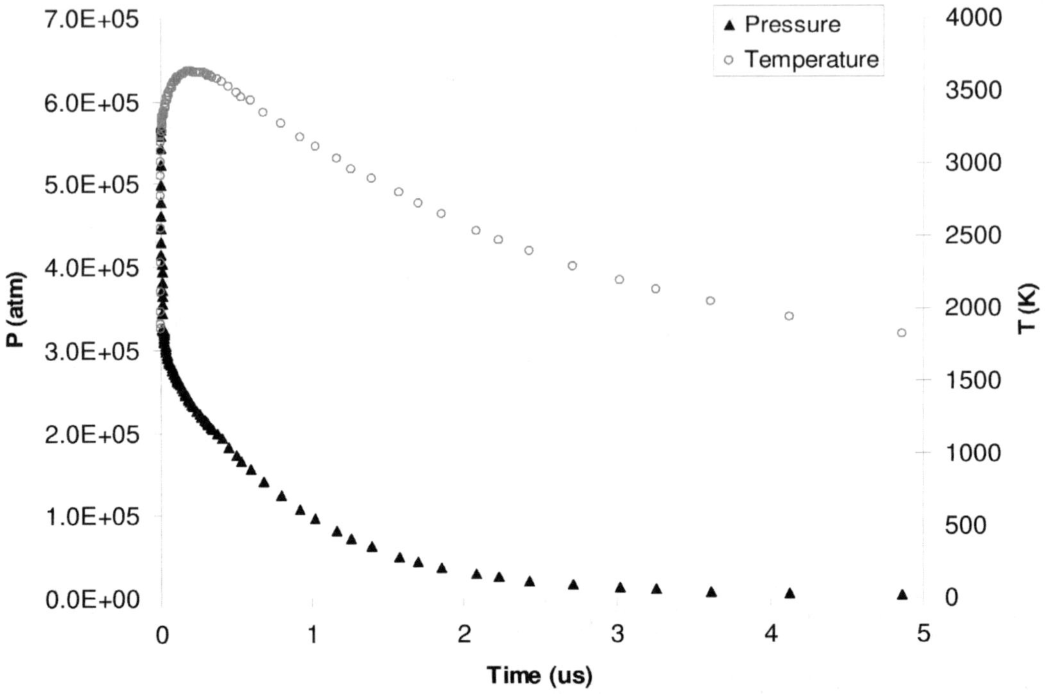

Figure 5. Pressure and temperature over time for Cheetah WK detonation of a charge composed of 90% HMX, 10% Al.

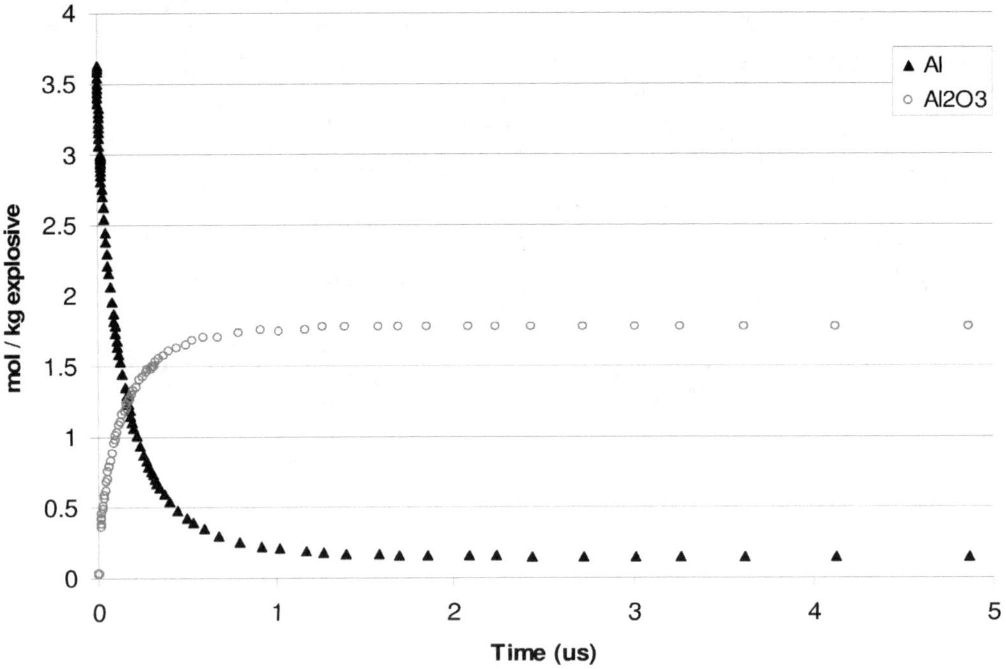

Figure 6. Al and Al_2O_3 concentrations over time for Cheetah WK detonation of a charge composed of 90% HMX, 10% Al.

MATLAB Particle Model

Particle model results are reported as mass concentration contours. As an example output, a ten-run average with constant process rates (coagulation: 10^{-10} m^3/s, nucleation: 10^{10} $m^{-3}s^{-1}$, surface growth: 0 m^3/s) is shown in Figure 7.

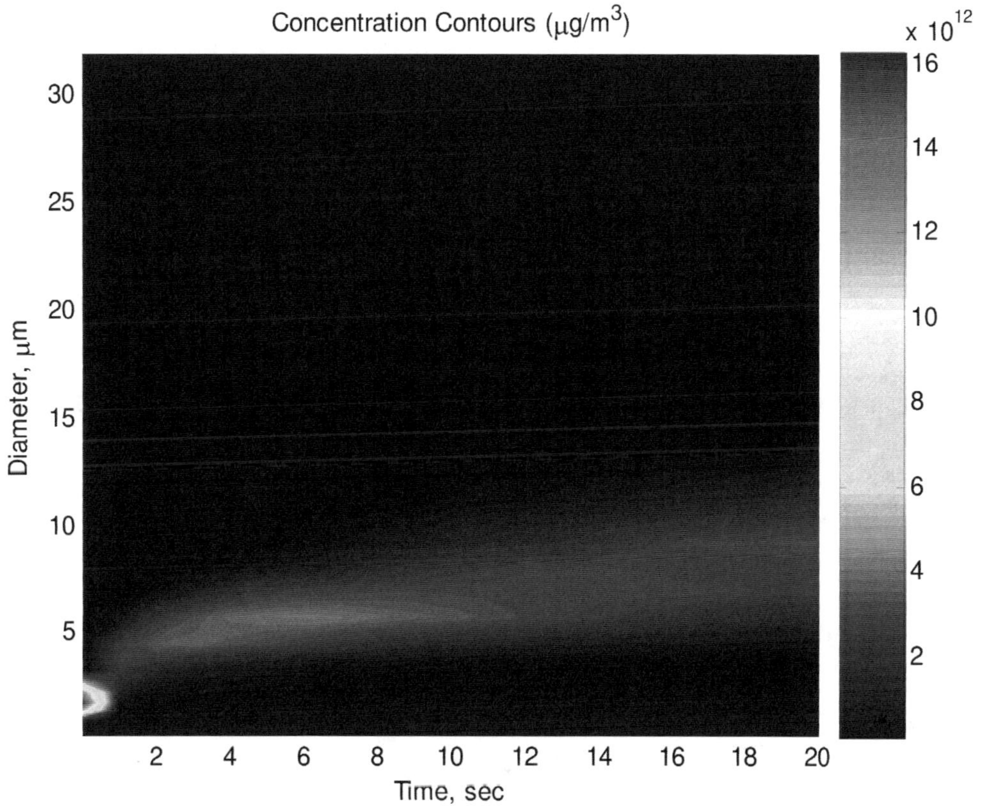

Figure 7. Average mass concentration (μg/m3) from ten particle formation model runs. (coagulation: 10^{-10} m^3/s, nucleation: 10^{10} $m^{-3}s^{-1}$, surface growth: 0 m^3/s)

Analysis

Laboratory particulate data (such as that presented in Figures 3 and 4) show high variability from run to run. Measures such as mean diameter of the distribution, maximum number and mass concentration, and total mass emitted, vary significantly between runs. Due to this variability, a more thorough experimental protocol and an improved understanding of the Grimm instrument are needed before laboratory data can be compared to model results.

Cheetah has been shown to provide accurate predictions of detonation parameters in ideal and moderately non-ideal explosives (Lu, 2001). According to the LLNL website, Cheetah "has become the Department of Defense's preferred code for designing new explosives…and, to a lesser extent, propellants" (Heller, 1997). This usage of Cheetah validates its accuracy as a combustion and detonation modeling tool.

Sub-scale detonation data are needed to confirm Cheetah's applicability to the OD events considered in this modeling approach. These data, which will include high speed pressure and temperature measurements, will be valuable for implementing and refining the Cheetah component of this model.

Stochastic methods have been effectively used to model time-varying PSDs. They have been applied to soot particles in premixed flames, aerosols in global weather prediction, and mineral dust particle size in mining operations. These applications highlight the computational advantage of stochastic models over deterministic models. The similarity of these applications to OB/OD activities supports the use of a stochastic model.

The MATLAB particle formation model component can be compared to deterministic solutions to the GDE for simple cases. For constant rate coagulation and nucleation (Maisels, et al, 2004), number concentration develops over time according to:

$$\frac{N}{N_0} = B\frac{1 + B\tanh(\tau / 2)}{B + \tanh(\tau / 2)} \qquad \text{(Equation 1)}$$

As shown in Figure 8, the 10-run average output from the MATLAB particle model agrees with the theoretical value with a maximum error of 9%. While this error value is higher than that in the Maisels, et al, report, increased simulation accuracy in the particle model can be achieved by increasing the initial number of simulation particles (N_{s0}). This increase leads in turn to an increase in computational time.

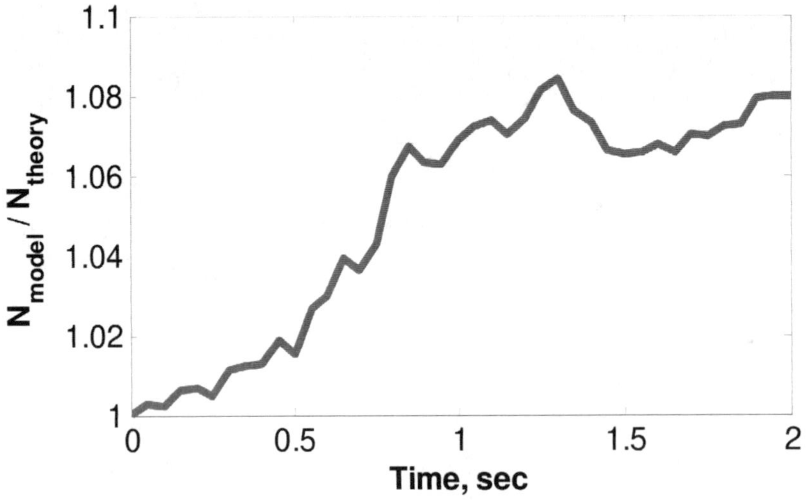

Figure 8. Comparing number concentration for theoretical and 10-run average stochastic particle model. (coagulation: 10^{-10} m^3/s, nucleation: 10^{10} $m^{-3}s^{-1}$, surface growth: 0 m^3/s)

Conclusion

A two component modeling approach has been developed for predicting particulate emissions from OD events. The Cheetah combustion model and MATLAB particle model components have been described. Cheetah predicts chemical products and detonation dynamics of a detonation. The MATLAB particle model uses a stochastic method to predict the development of PSDs over time. Preliminary laboratory work has been performed to validate sampling methodologies.

More extensive laboratory and field work is needed to better validate this modeling approach. Improved experimental protocols will be developed to reduce run to run variability in the laboratory data. Sub-scale detonation data will be acquired to refine the particle prediction model. These data are critical for the validation of this modeling approach. Refinement and validation of this modeling approach will be important initial steps to support a model of OB/OD phenomena that is acceptable to regulators and is sufficient to predict environmental emissions.

References

Fried, L. E., Howard, W. M., Souers, P. C., Haselman, L. *Adding Kinetics and Hydrodynamics to the CHEETAH Thermochemical Code*. Jan 1997. http://www.osti.gov/bridge/servlets/purl/488808-o1yxFr/webviewable/488808.PDF. Visited 9/19/08

Heller, A. *Improving Detonation Modeling with Cheetah*. Nov 1997. http://www.llnl.gov/str/Fried.html. Visited 9/19/08

Lu, Jing Ping. *Evaluation of the Thermochemical Code - CHEETAH 2.0 for Modelling Explosives Performance*. Aug 2001. http://stinet.dtic.mil/cgi-bin/GetTRDoc?AD=ADA396687&Location=U2&doc=GetTRDoc.pdf. Visited 9/19/08

Maisels, A., Kruis, F. Einar, Fissan, H. *Direct simulation Monte Carlo for simultaneous nucleation, coagulation, and surface growth in dispersed systems*. Chemical Engineering Science 59 (2004) 2231 – 2239

Palmer, G., Merrill, C., Lindsay, M. *Large Ballistic Missile Motor Detonation Support Activities*. AWMA Conference, June 2008

597a Analysis of Transport Phenomena & Modeling the Behavior of Single Particle during Drying Process

Srikanth Bellur Ramachandra, Basavaraju Agasanapur Nagendra, and Charles J. Coronella. Chemical Engineering, University of Nevada, Reno, Mailstop 170, Reno, NV 89557

The aim of the present work is the formulation of a computational model describing the transport phenomena in transient biomass drying. This work is specifically focused on simulation of the transport of momentum, heat and mass occurring between warm dry air and a single particle of moist sludge. The simulation in the gas phase was connected to the simulation in the solid phase by way of temperature continuity, and moisture equilibrium by an empirical correlation. Heat and mass transport are simulated in the solid phase, assumed to have physical properties similar to pure water. Comsol Multiphiysics was used to perform the simulations, with two spatial dimensions. With a time step of 0.1 second, we find that the solid reaches thermal equilibrium with the surrounding gas rapidly, consistent with empirical correlations. Further, we find that the rate of moisture loss is well correlated by the Chilton-Colburn j-factor analogy.

597b Gas-Solid Heat Transfer In Rotating Fluidized Beds In a Static Geometry

Axel de Broqueville, Inventor, Grez-Doiceau, 1390, Belgium and **Juray De Wilde**, Materials and Process Engineering Dept. (IMAP), Universite catholique de Louvain (UCL), Reaumur, Place Sainte Barbe, 2, Louvain-la-Neuve, B - 1348, Belgium.

Abstract:

Introduction and Concept

Gas-solid heat transfer in rotating fluidized beds in a static geometry (Figure 1(a)) is investigated theoretically and by means of computational fluid dynamics (CFD) simulations. In rotating fluidized beds in a static geometry, the fluidization gas flow rate influences both the centrifugal force and the radial gas-solid drag force in a similar way (Figure 1(b)). Hence, rotating fluidized beds in a static geometry can be operated over a very broad fluidization gas flow rate range, offering increased flexibility with respect to cooling or heating via the fluidization gas. Furthermore, dense operation at much higher fluidization gas velocities and, hence, gas-solid slip velocities than those in conventional fluidized beds is possible. The higher gas-solid slip velocities result in significantly higher gas-solid heat transfer coefficients. Intensification of the gas-solid heat transfer by one order of magnitude can be easily achieved.

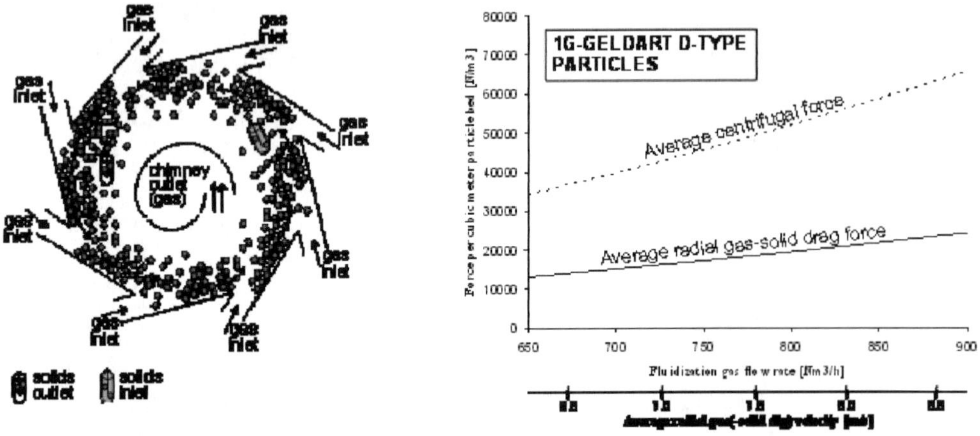

(a) (b)

Figure 1: Rotating fluidized beds in a static geometry. (a) Schematic representation;

(b) Flexibility in the fluidization gas flow rate: experimentally measured behavior with 1G-Geldart D-type particles in a 24-cm diameter, 11.5 cm long fluidization chamber.

CFD Simulations of gas-solid heat transfer

2D CFD simulations of the response of the particle bed temperature to a step change in the fluidization gas temperature (Figures 2 and 3) confirm the theoretically predicted significantly higher allowable fluidization gas velocities and resulting gas-solid heat transfer coefficients in rotating fluidized beds in a static geometry compared to conventional fluidized beds. Furthermore, as a result of the rotational motion of the particle bed and the combined radial-tangential fluidization, the particle bed temperature is shown to be much more uniform in rotating fluidized beds in a static geometry than in conventional fluidized beds (Figure 3).

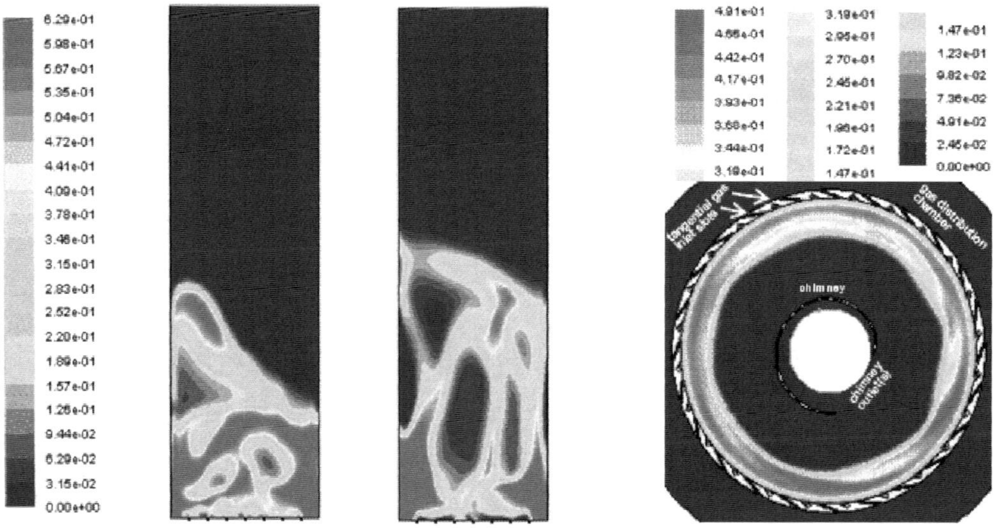

(a) (b) (c)

Figure 2: Snap-shot of the solids volume fraction. (a)-(b): Conventional fluidized bed at a fluidization gas flow rate of (a) 540 $m^3/(h \times m_{length\ fluid.\ chamber})$; (b) 1080 $m^3/(h \times m_{length\ fluid.\ chamber})$; (c) Rotating fluidized bed in a static geometry at a fluidization gas flow rate of 29800 $m^3/(h \times m_{length\ fluid.\ chamber})$.

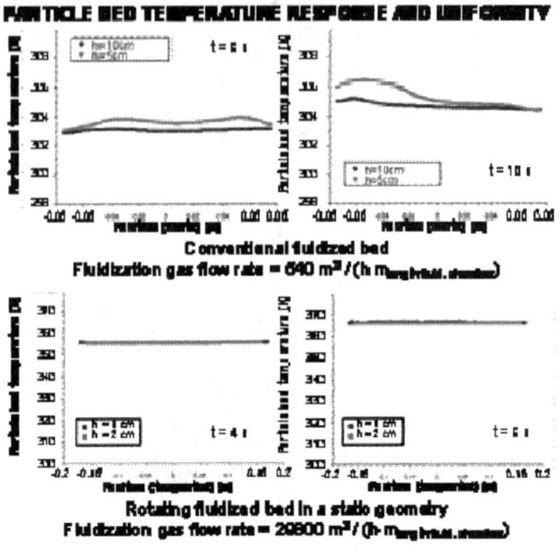

Figure 3: Simulated response of the average particle bed temperature to a step change in the fluidization gas temperature from 300 K to 400 K at time $t_0 = 0$ s. Comparison of a conventional fluidized bed and a rotating fluidized bed in a static geometry.

The combination of high fluidization gas velocities and high gas-solid heat transfer coefficients make rotating fluidized beds in a static geometry potentially interesting for carrying out fast and highly endothermic or exothermic reactions.

Particle Size Optimization for Syngas Chemical Looping Process

Fei Wang, Fanxing Li, Andrew Tong, Deepak Sridhar, and Liang-Shih Fan[*]
Department of Chemical and Biomolecular Engineering
The Ohio State University
Columbus, OH, USA

Abstract

The syngas chemical looping (SCL) process developed at the Ohio State University can efficiently convert coal derived syngas into hydrogen with in-situ CO_2 capture. In this novel process, two moving bed units are used to convert syngas into hydrogen through the assistance of an oxygen carrier particle. A third unit, an entrained bed combustor, is used to convey the oxygen carrier particles to complete the redox (reduction-oxidation) cycle. The reaction kinetics and thermodynamics of the combustor are well understood. Major challenges for the entrained bed design lie in the clear understanding of the hydrodynamics as well as the optimization of the oxygen carrier particle size. A desirable particle should be easy to convey in the third reactor while not being fluidized in the two moving bed units.

In this paper, a systematic approach is adopted to study the entrainment behavior of the oxygen carrier particles. The configurations and operating conditions for all three reactors in the SCL process are considered. A set of particle sizes are first determined based on theoretical estimation and then tested in the sub-pilot scale SCL demonstration reactor constructed at OSU. The hydrodynamics of particles conveying in the entrained bed reactor are studied. The required conveying gas velocities for the particles are then determined using the entrained bed combustor. The optimized particle size range for the operation of all three reactors is then determined. Key information for the SCL reactor demonstration such as particle attrition is also obtained.

Introduction

Energy is of great importance to the modern economy. Although renewable energy sources such as solar, wind, hydro, and biomass have been widely used, their contribution to the overall energy demands can be limited within the foreseeable future due to higher costs, geological constraints, and intermittent issues. Despite high crude oil and natural gas prices, fossil fuels is expected to provide a major portion of the overall world energy consumption in the next several decades [1, 2]. Abundantly available throughout the world, coal can play a more important role in the overall energy mix, especially when the crude and natural gas prices are volatile.

In this paper, a novel syngas chemical looping (SCL) process is studied. The SCL process converts coal derived syngas into hydrogen, which is an important feedstock for oil refining, fertilizer production, and liquid fuel synthesis. The process consists of three main units: (1) a moving bed reducer, (2) a moving bed oxidizer; and (3) an entrained bed combustor. The two moving bed units convert syngas into hydrogen while the entrained bed combustor conveys the iron based oxygen carrier particles to complete the redox (reduction-oxidation) cycle. The reaction kinetics and thermodynamics in the combustor are well understood [3]. The major challenge for the combustor design resides in the in-depth understanding to the hydrodynamic properties of the oxygen carrier particles. In this paper, the

[*] To whom correspondence should be addressed.

pneumatic conveying properties and attrition rates of the particle obtained from a sub-pilot scale SCL unit are discussed. Such information can help determine important SCL process operation parameters such as optimum particle size, shape, and fresh particle makeup rate.

Experimental Setup

Figure 1 is the schematic diagram of the sub-pilot scale SCL demonstration unit located at OSU west campus SCL demonstration lab. Figure 2 is the photography of the demonstration unit. The demonstration unit consists of three main reactors: a reducer (heating section A), an oxidizer (heating section B), and a combustor (conveying system). In the SCL reactor operation, oxygen carrier particles fed from the top of the reducer moves through the reducer and the oxidizer by gravity. Valve systems and rotary solid feeder assemblies are used to assist solid movement while avoiding gas leakage. After being discharged from the oxidizer outlet, the partially regenerated oxygen carrier will be introduced to the combustor through a slant stainless steel tube and then conveyed vertically back to the reducer inlet with air to complete the redox/solid conveying cycle. The combustor, air inlet, and slant tube are connected by a "Y" shaped tube. A programmed gate valve is mounted on the slant pipe to control the particle feed to the vertical section of the combustor. A distributor made of Hastelloy X is installed at the bottom of the combustor to ensure even distribution of the air. The main section of the combustor is a vertical pipe with a 2 in ID and a height of 20 ft. A 90^0 elbow and a horizontal pipe are connected to the top of the vertical pipe with a cyclone connected on the other end of the horizontal pipe. The cyclone allows the separation of the fines and ensures that larger particles will be re-introduced to the reducer. All the parts for the combustor are made of 304L stainless steel. An air compressor, with a capacity of ~30 l/s at 180 psig, is connected to the bottom of the distributor to provide air for both particle conveying and oxygen carrier combustion purposes.

During the reactor operation, particles are loaded into the slant pipe from the bottom of the oxidizer with the programmed ball valve. After the particles in the declined pipe are accumulated to a preset amount, all the particles are dumped into the bottom section of the vertical combustor by opening the programmed gate valve for short period of time (< 10s). After the programmed ball valve is closed, the air source from the compressor is introduced into the combustor to convey all the particles at the bottom of the vertical pipe into the reducer inlet. Exhaust air coming out of the cyclone will then be introduced to a powder separator for the separation of fine powders before ventilation. The powder separator is not shown in Figure 1.

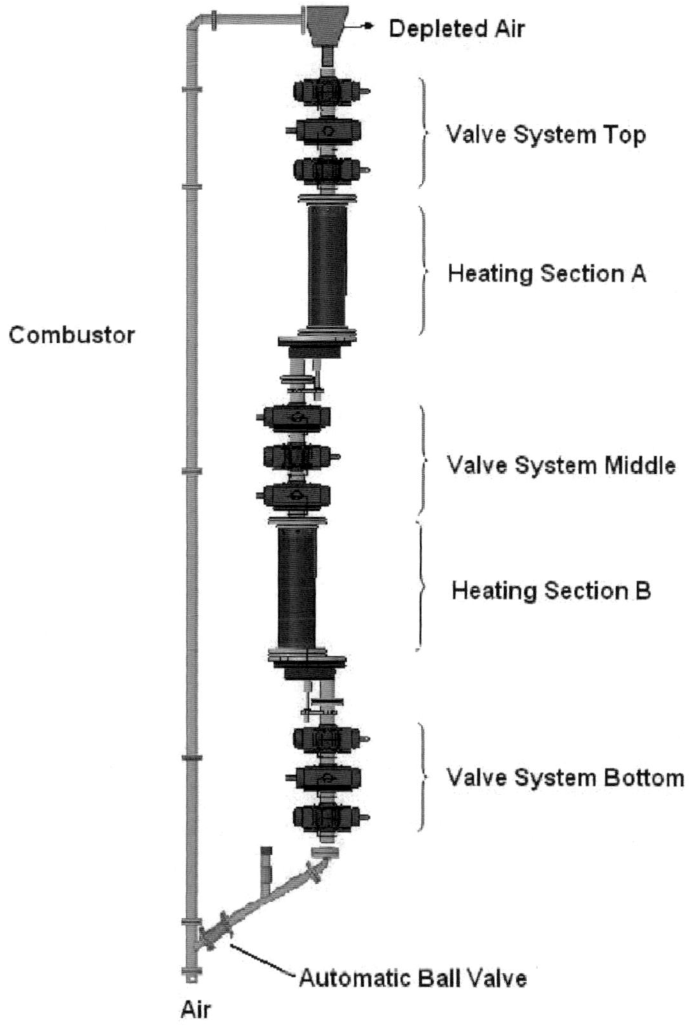

Figure 1. Schematic diagram of the units for the SCL process

Figure 2. Picture of the actual units for the SCL process in west campus SCL demonstration lab at OSU

Results and Discussion

Pneumatic Conveying of Particles

The pneumatic conveying system consists three parts: (1) vertical flow through the combustor pipe, (2) change in direction through a 90^0 elbow, and (3) horizontal flow. The flow pattern in the gas-solid pipe is a dilute suspension flow with a very low particle concentration. Several sets of particles are selected to test the combustor operation. The particles are pellets processed from iron oxide and inert support with a density of $2500\,kg/m^3$. A 3kW rotary pelletizer (ZP-35) is used to fabricate pellets with a production rate of up to 15 kg/hour. The pellets have a cylindrical shape with a diameter of 5 mm and a thickness of 1-5 mm depending on the setting of the pelletizer. Before testing, the pellets are sintered at $900\ ^0C$ for more than 20 hours to achieve better physical and chemical properties.

It was determined through experiments that the terminal velocity for the pellets with a diameter of 5 mm and a thickness of 1.5 mm, or type A particles, is around 9.3 m/s. For this type of particles, the pneumatic conveying system needs to be operated at a gas velocity ~1 m/s higher than its terminal velocity since horizontal pneumatic transport requires higher gas velocity than vertical pneumatic transport. This due mainly to the higher drag forces and the various frictions in the horizontal conveying step [4]. However, the effects of the drag forces and the frictions in the horizontal pneumatic transport are not significant since the length of the horizontal part is relatively short. It was also determined that the terminal velocity for the pellets with a diameter of 5 mm and a thickness of 4.5 mm, or type B particles, is around 15 m/s. For type B particles, the pneumatic conveying system can be operated at a gas velocity slightly higher than its terminal velocity due to the better flow properties of the particle in the horizontal pipe. The higher terminal velocity and better flow properties of type B particles can be explained by it near sphere shape which allows for a higher fluidity performance, lower effects of drag forces, and less friction than type A particles.

Particle Attrition

In order to reduce the particle purging rate, it is desirable for the oxygen carrier particles to maintain its chemical reactivity and physical integrity for multiple cycles. Since particles are in vigorous motion in the combustor, the attrition rate of the particle is an important parameter for the SCL combustor design and particle optimizations. Particle attrition affects the flow performance and economics of the SCL process. A low attrition rate of the particles is highly desirable. Due to the turbulent movements at high gas velocities, the frequent particle-particle/ particle-wall collisions during the pneumatic conveying step in the combustor and the particle separation step in the cyclone play an important role for particle attrition, which is defined as the degradation of particles due to mechanical stress [5]. Abrasion and fragmentation of the particles are the main modes of the attrition [6, 7]. The attrition effects due to the combustor and the cyclone are measured in the SCL process. For the attrition test in combustor, particles are loaded at the bottom of the vertical pipe and then conveyed out of the horizontal part of the combustor. These particles are subsequently collected using a bucket. The particles are then sieved and weighted according to different size fractions. All the particles are then well mixed and reloaded into the combustor for the next conveying cycle. To test the attrition in both the combustor and the cyclone, a similar testing procedure is adopted. The only difference is that the particles are collected after the cyclone. The attrition rates for both type A and type B particles were test with 20 runs.

Figure 3 and Figure 4 shows the combustor attrition test results for type A particles and type B particles respectively. A superficial gas velocity of 10 m/s is used to convey the type A particles. From Figure 3, the particle attrition rate is 0.56% per cycle when the cutoff is set to be 600 μm, i.e., particles smaller than 600 μm will be purged out from the system. A superficial gas velocity of 15 m/s is used to convey the type B particles. From Figure 4, the particle attrition rate is 0.23% per cycle when the cutoff is set to be 600 μm. The particle attrition rate is around 0.38% per cycle when the cyclone is used for the test at the same experimental condition, which means an additional attrition of 0.15% was introduced due to the addition of the cyclone. The type B particles with 5 mm diameter and 4.5 mm thickness perform better on attrition than type A particles due to its close-to-spherical shape. Such a shape leads to reduced effects of the drag forces and the various particle-particle and particle-wall frictions. As can be seen, both type A and type B particles possess excellent physical strength.

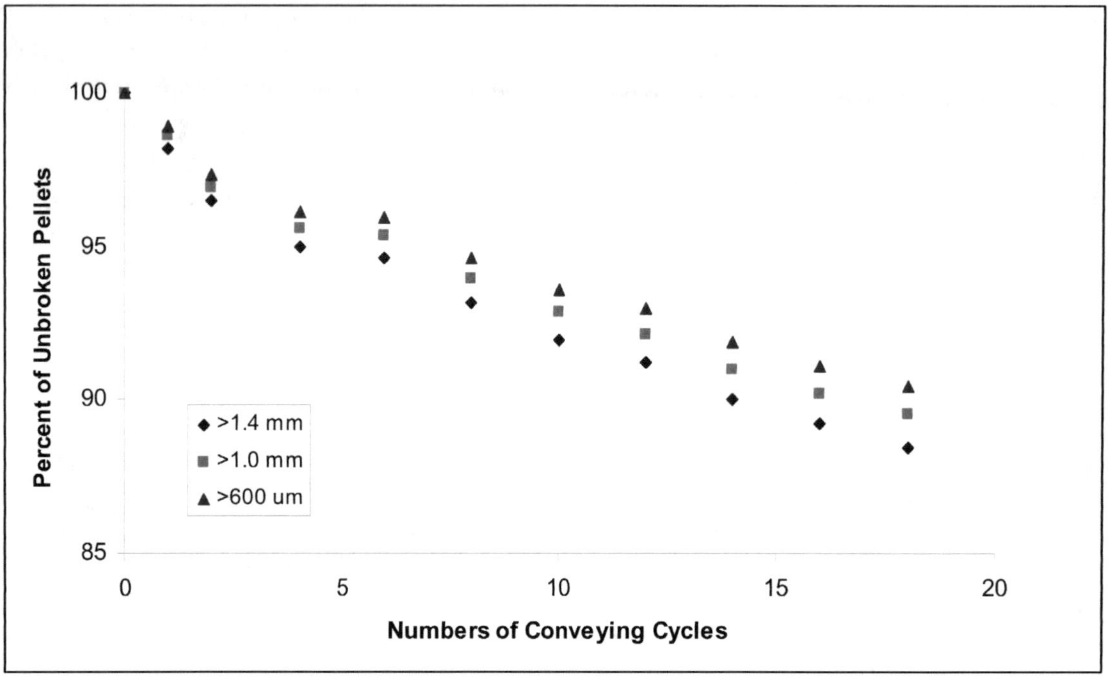

Figure 3. Attrition test results using Type A particles with 5 mm diameter and 1.5 mm thickness

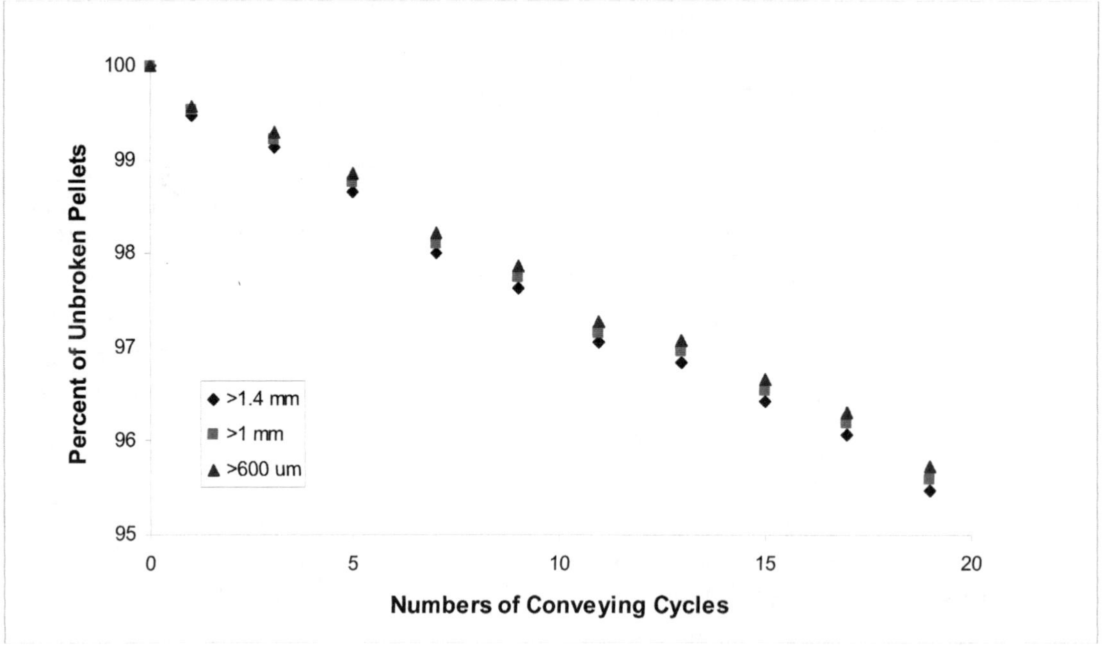

Figure 4. Attrition test results using Type B particles with 5 mm diameter and 4.5 mm thickness

Conclusion

The syngas chemical looping (SCL) process developed at the Ohio State University can convert coal derived syngas into hydrogen with in-situ CO_2 capture. Two sets of particles with different sizes are selected and tested in the entrained bed reactor constructed at OSU. The hydrodynamics of particles conveying in the entrained bed reactor are studied. It was found that the particles can be pneumatically conveyed at a linear gas velocity of 10 m/s and 15 m/s respectively. Particles with 5 mm diameter and 4.5 mm thickness show much lower attrition than particles with 5 mm diameter and 1.5 mm thickness. Both types of particles possess excellent physical strength and can potentially be recycled for multiple cycles without losing its physical integrity.

Acknowledgements

This work has been supported by the Ohio Coal Development Office of the Ohio Air Quality Development Authority, U. S. Department of Energy, Ohio State University and an industrial consortium.

References

1. Jacobsson S, Johnson A. The diffusion of renewable energy technology: an analytical framework and key issues for research. *Energy Policy*. Jul 2000;28(9):625-640.
2. McVeigh T, Burtraw D, Darmstadter J, Palmer K. Winner, loser, or innocent victim? Has renewable energy performed as expected? *Solar Energy*. Mar 2000;68(3):237-255.
3. Li FX, Fan L-S. Clean Coal Conversion Processes – Progress and Challenges. *Energy & Environmental Science*. 2008;1:248-267.
4. Fan L-S, Zhu C. *Principles of Gas-Solids Flows*. New York: Cambridge University Press; 1998.
5. Werther J, Reppenhagen J. Attrition. In: Yang W-C, Dekker M, eds. *Handbook of Fluidization and Fluid-Particle Systems*; 2003.
6. Bliniche.V.N, Streltso.V.V, Lebedeva ES. An Investigation of Size Reduction of Granular Materials during Their Processing in Fluidized Beds. *International Chemical Engineering*. 1968;8(4):615-618.
7. Pell M. Gas fluidization. *Handbook of Powder Technology*. Amsterdam: Elsevier; 1990:97-106

597d Computational Fluid Dynamics of Gas Bypassing In Fluidized Beds with Imposed Solids Flux

Ray A. Cocco, Allan Issangya, S.B. Reddy Karri, and Ted Knowlton. Particulate Solid Research, Inc., 4201 West 36th Street, Suite 200, Chicago, IL 60632

Well-designed fluidized beds are assumed to have uniform gas and temperature distributions, which in many cases is true. However, some fluidized bed processes experience poor productivity and wide temperature variations despite the fact that traditional design criteria have been met. Particulate Solid Research, Inc. (PSRI) has found that deep or dense fluidized beds are prone to gas bypassing or jet streaming where a large gas jet develops and precesses around the walls of the fluidized bed. The remainder of the bed is either defluidized or poorly fluidized. In some cases, multiple gas jets can occur in the bed. This phenomenon is not typically seen in laboratory beds or pilot plants, as bed heights in these units are usually not high enough to produce this behavior. Gas bypassing tends to be an issue for large pilot plants and commercial units. Research at PSRI has shown that gas bypassing can be mitigated by decreasing the bed height, increasing the gas velocity, adding fines, increasing the freeboard pressure, or adding internals.

Recently, PSRI has found that the addition an imposed solids flux through a fluidized bed can trigger gas bypassing in what was previously a well-behaved fluidized bed. CFD studies suggest that the gas bypassing is stabilized in the region between the existing bed and the material entering the bed. It some cases the entering material can short-circuit the existing bed to a bottom standpipe and the freeboard region. Details of possible techniques for detecting gas-bypassing are presented.

597e Equilibrium Moisture Content and Drying Analysis of Biosolids

Srikanth Bellur Ramachandra, Charles J. Coronella, and Victor R. Vasquez. Chemical Engineering, University of Nevada, Reno, Mailstop 170, Reno, NV 89557

Thermo chemical conversion of biosolids to syngas or to fuels often includes a thermal drying step as the moisture content of the biosolids must be around 30%. Equilibrium moisture content (EMC) data of biosolids are required for good design and control of the drying process. The EMC data for Municipal sludge were measured in the relative humidity range of 11-85% at the temperatures from 30°C to 60°C. The data were fitted to several well known models, including one model accounting for mixing and swelling contributions to water activity in the biosolids .Using this new model, we characterized both bonded and non bonded water. This information is used in the design of a fluidized bed dryer for handling and processing biosolids including for a preliminary cost analysis.

659a Precipitation of Green Tea Polyphenols Using a Supercritical Anti-Solvent Process

Darja Meterc[1], Maria Jose Cocero[2], Ángel Martín Martínez[2], and Eckhard Weidner[1]. (1) Chair for Process Technology, Ruhr-University Bochum, Universitaetsstr. 150, IB 6/126, Bochum, 44801, Germany, (2) Departamento de Ingeniería Química y Tecnología del Medio Ambiente, Facultad de Ciencias, Universidad de Valladolid, c/ Prado de la Magdalena s/n, Valladolid, 47011, Spain

One of the most important components present in green tea are polyphenols. There have been numerous papers published about their health benefits. Due to their antioxidant activity there are numerous fields for their application.

In this research work, supercritical anti solvent process was used as a process for precipitation of polyphenols from ethanol/water solution, starting form 100 % ethanol to 0 % of ethanol in the initial solution.

In this SAS process supercritical carbon dioxide was used as media to precipitate particles dissolved in a solvent. Schematic presentation of SAS process presented in Figure 1.

By changing different parameters such as pressure, temperature, mass flow and concentration of polyphenols in the initial solution, different powders were produced. Water content was determined with Karl Fisher and composition of powders (polyphenol and caffeine content) with HPLC. Particle morphology was determined with SEM.

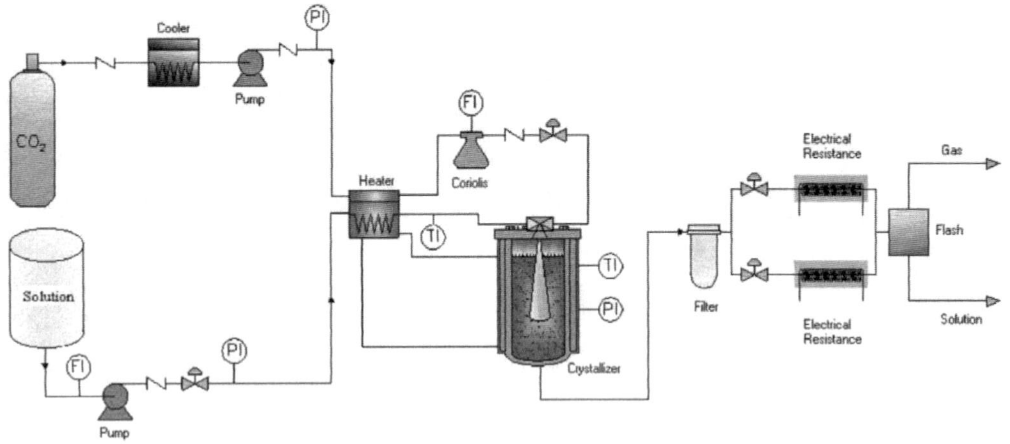

Figure 1: Schematic presentation of SAS process

659b Nanoencapusulation Using Water-in-Supercritical CO2 Microemulsion Technology

Yuichiro Otsuji, Masahiro Ikeda, Chihiro Fushimi, and Atsushi Tsutsumi. Institute of Industrial Science, The University of Tokyo, 4-6-1 Komaba, Meguroku, Tokyo, 153-8505, Japan

A novel technique for the nanoparticle production and nanoencapsulation of hydrophilic pharmaceutical are investigated using the rapid expansion of water-in-supercritical CO2 microemulsion in which coating material is dissolved in the supercritical CO2 phase. The aqueous cores of micellar droplets filled up with highly water-soluble drug are formed and dispersed in the supercritical CO2 to form the inverse microemulsion system.

In the last research, the synthetic surfactant was used as a stabilizer of the microemulsion system. However, considering the materials for drug delivery applications, it should be replaced with the natural surfactant. In this research, bio-surfactant is applied to form the microemulsion system. Bio-surfactant applications are promising due to their biodegradability, low toxicity and effectiveness in enhancing biodegradation.

The rapid expansion of the supercritical microemulsion into the atmosphere is carried out for the encapsulaton of nanoparticles. The morphology of particle samples is examined by the SEM observation with EDS analysis. The experimental results suggest that the rapid expansion of water-in-supercritical CO2 microemulsion is a promising method for encapsulation.

Title: Precipitation of Ultra-fine Particles of Fenofibrate with Controlled Size Distribution by RESOLV: Rapid Expansion Vs Stabilizer Concentrations

Session: 03A06 Particle Formation in Supercritical Fluids for Food and Pharmaceuticals

Abstract: Supercritical fluids (SCF) are being used to precipitate fine particles of variety of chemicals ranging from pharmaceuticals to explosives. While there are different SCF processes available, **R**apid **E**xpansion of **S**upercritical Solutions in **SOLV**ent (RESOLV) in particular, offers an attractive alternative to conventional precipitation processes as it avoids the use of organic solvents and aqueous suspension of fine particles is readily obtained. Literature demonstrates the use of RESS to precipitate ultra-fine particles but the problem of stabilization of particles is poorly reported. High and rapid supersaturation produced by rapid expansion of supercritical solutions results in ultra-fine particles however, growth and agglomeration of particles after precipitation also controls the particle size and its distribution. Use of water as a solvent for collection of particles allows one to add stabilizers such as surfactants and polymers to stabilize the particle size and size distribution. This work focuses on precipitation of Fenofibrate (FNB) particles using RESOLV with controlled particle size and distribution by identification and manipulation of process parameters and stabilizer concentrations. Attempts are also made to co-precipitate FNB with a poly (lactic co-glycolic acid) to control the particle size and distribution.

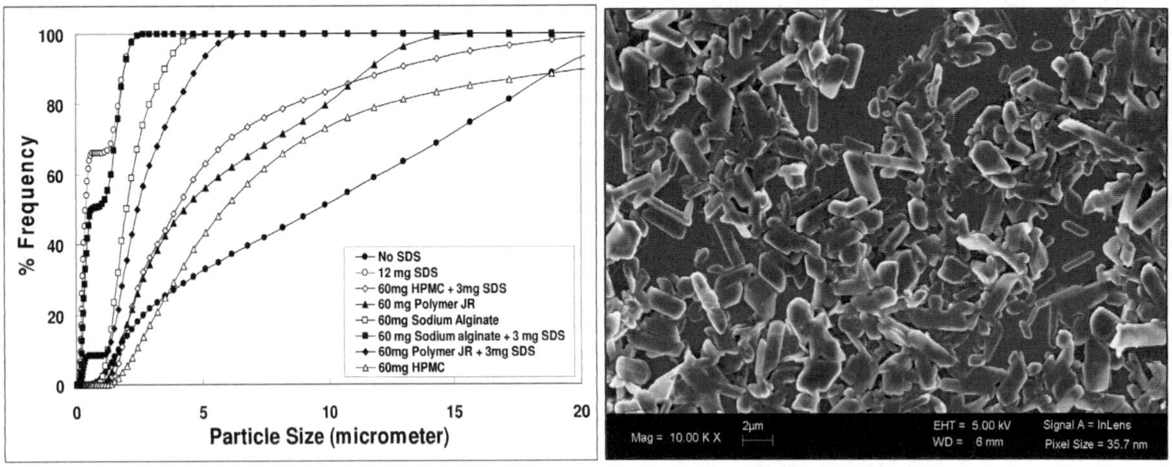

Fig. 1. (a) Cumulative particle size distribution of FNB precipitated by RESOLV with various stabilizers in 300 ml water with pre-expansion pressure of 200 bar, temperature of 60 °C, and 125 µm nozzle ID with length of 3 cm and (b) SEM micrograph of FNB particles collected in water (SDS:Drug of 1:20) at pre-expansion pressure of 200 bar, temperature of 60 °C, and nozzle ID of 125 µm and length of 3 cm

659d a Study of Production of Micronized Powders Using Supercritical Fluid Atomization Techniques

Sriram Kiran Annapragada, Department of Chemical & Petroleum Engineering, University of Wyoming, Laramie, WY 82072, **M. P. Sharma**, Chemical and Petroleum Engineering, University of Wyoming, Laramie, WY 82071, and William Lindberg, Mechanical Engineering, University of Wyoming, Laramie, WY 82071.

Fine solid materials at the nano and micro level are used in wide applications like catalysts, coatings, electronics, ceramics, superconductors, dyestuff, pigments, pharmaceuticals, food, etc. The properties of these fine scale materials are related to their particle size. A class of the current technologies to produce these fine scale materials employs many organic solvents as the process fluids which pollute the environment and could be potential contaminants. The biggest challenge in all these technologies is to obtain a narrow distribution of particle sizes. The development of a single scalable process that removes many of the constraints and limitations of the current methods is very essential. The Supercritical Fluid Assisted Atomization (SCFA) techniques are 'Green Technologies' which do not pollute the extracts, residues, and also the environment in most cases. Several of these methods are known to produce particle size in the range of nano and micro meters with narrow particle size distributions.

An experimental examination of the influence of operating conditions in production of micronized powders using the Supercritical Fluid Atomization (SCFA) technology along with an investigation of the atomization processes in these sprays is presented in this work. A theoretical model to predict the choked flow conditions across the nozzle and the size of the particles produced is also discussed. The experimental investigation involved supercritical carbon dioxide system to produce spray dried powders and concurrently examine the atomization processes using a high speed high magnification photography technique. High speed photography techniques revealed a dominant atomization mechanisms in thc decompressive supercritical sprays. Various operating conditions were used to produce spray-dried powders and the effect of each of them on particle size was investigated. Under certain conditions, bimodal particle size distribution in sprays were observed and investigated. The shape and morphology of the powders produced were also investigated. Trends comparable to the experimental results were predicted using a proposed theoretical model.

659e Effect of Co-Precipitation on Crystallinity of Pharmaceutical Compounds Using Supercritical Anti-Solvent Process

Tau Yee Lim, Wai Kiong Ng, and Reginald B. H. Tan. Crystallisation and Particle Sciences, Institute of Chemical and Engineering Sciences, 1, Pesek Road, Jurong Island, Singapore 627833, Singapore, Singapore

Over the past two decades increasing attention has been focused on solid-state chemistry of pharmaceuticals because the selection of solids with suitable physicochemical properties is a key factor within pharmaceutical development strategies (York, 1999). Recently, a new approach using supercritical fluids for particle design of pharmaceutical materials is being actively pursued due to its major advantages over conventional pharmaceutical processing such as high purity of products, ability to control particle size and narrow particle size distribution, single step process and free from residual solvent. Among these advantages, one of the major challenges is to improve bioavailability and stability of poorly soluble active pharmaceutical ingredient (API) using supercritical fluids particle design technology (Majerik, 2007). Therefore, the aim of this work is to investigate the influence of co-precipitation of API and excipient using supercritical anti-solvent (SAS) process. The API-excipient composites were generated by varying process conditions such as API-excipient ratios, relative flow rates of supercritical carbon dioxide and API solution, pressure and types of solvent. Besides that, stress test on SAS powders were carried out at 75% RH and room temperature in order to evaluate its physical stability. The untreated and SAS powders (after and before storage) were characterized using scanning electron microscopy (SEM, morphology), powder X-ray diffractometry (PXRD, crystallinity), dynamic vapor sorption (DVS, sorption isotherm) and differential scanning calorimetry, (DSC, crystallinity, Tm and ΔHf).

Our preliminary results revealed that it was technically feasible to generate a fully X-ray amorphous SAS API-excipients co-precipitated powders, which remained amorphous after 1-week storage at 75% RH and room temperature. The amorphous content of the SAS co-precipitated products were influenced by the API-excipients compositions. Besides that, different morphologies of SAS powders were obtained as compared to the untreated powders.

References:

1. York, P., Strategies for particle design using supercritical fluid technologies, PSST, 2 (1999), 430-440

2. Majerik, V., Charbit, G., Badens, E., Horvath, G., Szokonya, L., Bosc, N., and Teillaud, E., Bioavailability enhancement of an active substance by supercritical anti-solvent precipitation, J. Supercrit. Fluids, 40 (2007), 101-110

659f Re-Engineering Particle Sizes by Precipitation with Compressed Antisolvent (PCA) Process

Candy S. Lin, Department of Chemical Engineering, The Hong Kong University of Science and Technology, Clear Water Bay, Kowloon, Hong Kong, China, Ka M. Ng, Department of Chemical Engineering, Hong Kong University of Science and Technology, Clear Water Bay, Kowloon, Hong Kong, China, and Christianto Wibowo, ClearWaterBay Technology, 4000 W. Valley Blvd. Suite 100, Pomona, CA 91789.

Particle design is becoming one of the main features in the production of pharmaceuticals, food, cosmetics and other specialty chemical products. With optimized drug particle attributes (particle size, size distribution and particle morphology), improved drug targeting, controlled drug release, solubility and bioavailability can be achieved. The application of supercritical fluid processing in pharmaceutical research is increasing particularly in the field of particle formation for drug delivery systems, mainly due to advantages over the conventional particle formation in terms of improved control, flexibility, ease of operation and level of toxicity from the presence of organic solvents and contaminants.

Particle formation of salicylic acid using supercritical carbon dioxide (CO_2) was investigated using the Precipitation with Compressed Anti-solvent (PCA) technique using ethanol and acetone as solvents. The solubility of salicylic acid (SA) in supercritical CO_2 was measured in its pure form and with modifiers, at 45°C and 55°C over a pressure range of 95 to 250 bar. The output of the study gave promising conditions in particle formation by adjusting parameters based on the phase diagram obtained. Particles in the nano-range can be produced in the higher supersaturation region without changing the nozzle size of the PCA system. All the experimental conditions investigated here gave an average particle size of approximately 100 nm, with the smallest average particle size of 63.35 nm by using 1% acetone modifying condition at 45°C. The largest average particle size produced was 102.75 nm at condition with 1% ethanol operating under temperature of 55°C. A dissolution experiment was also carried out to show significant improvement of drug dissolution rate with the smaller particle size.

Hydrodynamics Modeling and Analysis Of Rapid Expansion Systems Of Supercritical Solutions (RESS)

Y. Samyudia*, F. Panau and Y.I. Liong

Department of Chemical Engineering
Curtin University of Technology, CDT 250, 98009 Miri, Sarawak, Malaysia
*Corresponding author. Ph: +60 85 443835; fax. +60 85 44 3837
Email: yudi.samyudia@curtin.edu.my

Abstract

In this paper, we study the hydrodynamics aspects of the RESS (e.g. rapid expansion of supercritical solutions) process to produce fine particles for heat-sensitive organic compounds. The Computational Fluid Dynamic (CFD) modeling and analysis of the supercritical solutions passing through a capillary nozzle were studied. The CFD simulation is focused on the pre-expansion chamber of the RESS process consisting of three successive steps that are at the stagnation chamber (i.e. the reservoir), nozzle inlet and along the nozzle itself to the outlet. The solute considered is benzoic acid and the supercritical solvents are CO_2 and CHF_3 respectively. The aim of this study is to examine the effect and sensitivity of the design parameters (i.e. temperature, pressure and density) on the formation of particulates in the chamber. The study reveals that at lower pre-expansion temperature with constant pre-expansion pressure the solute nucleation seems to start inside the nozzle. Moreover, it is also found that pre-expansion pressure has insignificant effect on the nucleation process, i.e. the nucleation rate slightly increases when the pressure is set higher. Furthermore, the study also reveals that high pre-expansion pressures and low pre-expansion temperature favors small particles. The results demonstrate that the variation of thee parameters in the ranges studied may lead to the increased super-saturation and simultaneously increase in nucleation rate.

Keywords: RESS, Hydrodynamics; CFD Modeling; Particle formation

1. Introduction

Particle formation by methods utilizing supercritical fluids (*e.g.* CO_2) is a subject of great interests in many industrial applications such as pharmaceuticals, foods, natural products and specialty chemicals[1,2,3,4]. Strict regulations on the use of organic solvents and their residual level in the end products form a major limitation to the traditional processing techniques. The particle formation using supercritical fluids (SCF) involves minimal or no use of organic solvents while the processing condition is relatively mild. Additionally, the SCF technology involves a growing of the particles in a controlled fashion to attain the desired morphology. This feature offers advantages as compared to some conventional techniques of particle size reduction (e.g. milling and grinding) so that the SCF-based particle formation is especially desirable for thermally sensitive materials.

The SCF technology for particle formation is rapidly evolving, as reflected by the number of modified processes reported since its inception [5,6,7]. These include static supercritical fluid process (SSF), rapid expansion of supercritical solutions (RESS), particle from gas-saturated solutions (PGSS), gas anti-solvent process (GAS), precipitation from compressed anti-solvent (PCA), aerosol solvent extraction system (ASES), supercritical anti-solvent process (SAS), and supercritical anti-solvent process with enhanced mass transfer (SAS-EM).

Previous studies have shown that the RESS process be one of the most promising techniques. The technique offers several advantages over conventional processes, which are either mechanical (crushing, grinding and milling), or equilibrium controlled (crystallization from solution). The RESS process can lead to very high super-saturation ratios, which leads to uniform conditions within the nucleation medium, and hence in principle to narrow particle size distributions and to form small particles. RESS can be used to precipitate mainly organic solutes upon expansion of a dilute solution from a supercritical or near critical solvent. This RESS concept can be implemented in relatively simple equipment, although particle collection from the gaseous stream is not easy.

The expansion step in RESS occurs through such a throttling device as an orifice, capillary tube or needle valve, which can support a large pressure drop without freezing. Changes in pressure, throttling device dimensions, concentrations of solute in the supercritical fluid and temperature cause wide variations in precipitate morphology (e.g. fibers to spheres). An increase in pre-expansion temperature under certain experimental conditions can lead to a transition of powder to fibrous morphology. Similarly, a decrease in pressure can also result in a powder to fiber morphology transition. This suggests that solutions with lower density are more likely to produce particles of a fibrous morphology in the RESS process. In addition to the dynamic variables of pressure, temperature and concentration, a fluid dynamic process variable – the geometry of the throttling device - has also been found to influence the morphology of the precipitate [8,9].

In this paper, we study the hydrodynamic behavior through a capillary nozzle in the RESS process using Computational Fluid Dynamics (CFD). The CFD simulation is focused on the pre-expansion chamber of the RESS process consisting of three successive steps. They are at the stagnation chamber (i.e. the reservoir), at the nozzle inlet and along the nozzle itself. The solute considered in this study is benzoic acid and the supercritical solvents are CO_2 and CHF_3 respectively. The study is aimed at identifying the effect and sensitivity of the design parameters (i.e. operating conditions such as pressure and temperature) in the pre-expansion path of the RESS process on the particulate formation. From the CFD simulation, it is found that high pre-expansion pressures and low pre-expansion temperature favor for the formation of small particles. Furthermore, for mixture of benzoic acid/carbon dioxide, the nucleation process inside the capillary nozzle could not be avoided. According to the classical nucleation theory, our analysis reveals that the lower nucleation rates should results in a delayed precipitation. Hence, it would be less time available for growth, resulting in smaller particles. The CFD modeling and analysis of RESS process provides information for understanding the basic trends that relate the particle formation with the expansion process conditions. It is therefore of important for optimization, control and design studies [10].

2. Modeling of the Capillary Nozzle in the RESS Process

In the Rapid Expansion of Supercritical Solution (RESS) process, the solutes are dissolved in a supercritical fluid through an extraction process from which the supercritical solution is then rapidly expanded into a low temperature and pressure environment in the expansion unit. The sudden change in conditions leads to a large cooling rate, resulting in high super saturations with homogeneous nucleation and particle growth.

The expansion can take place across a capillary nozzle within a very short time (see Figure 1). The formation of particles from RESS process leads to the production of thin films, crystalline or amorphous powders with narrow and controllable particle size distribution, and intimate mixtures of amorphous materials.

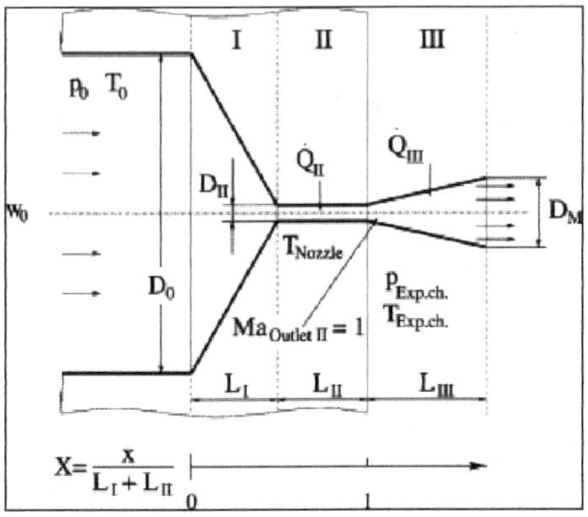

Figure 1. A schematic diagram of capillary nozzle in RESS

They are a number of process parameters that could influence the particle formation process when applying this RESS technology. They are the temperature and pressure in the pre expansion and post expansion chamber, the nozzle diameter and geometry, the solute and its concentration. Supercritical fluids are very dense gases with many properties superior to liquids or solvents. They combine properties of gases and liquids. The high compressibility of supercritical fluids leads to a fact that a small change in pressure can give a significant change in density that, in turn, affect diffusivity, viscosity, dielectric and salvation properties of these fluids, thus dramatically influencing the kinetics and mechanisms of chemical reaction.

The hydrodynamic model can be derived using the first principle of mass, momentum and energy balances, and the thermodynamic properties of the supercritical fluids were computed by means of the Bender equation of state as follows:

$$w\frac{d\rho}{dx} + \rho\frac{dw}{dx} + \frac{\rho w}{A}\frac{dA}{dx} = 0 \tag{1}$$

$$\rho w\frac{dw}{dx} + \frac{dp}{dx} = -\frac{2fw^2\rho}{D_{II}} \tag{2}$$

$$\frac{dh}{dx} + w\frac{dw}{dx} = \frac{dq}{dx} \tag{3}$$

$$P = RT\rho + B\rho^2 + C\rho^3 + D\rho^4 + E\rho^5 + F\rho^6 + (G + H\rho^2)\rho^3 e^{(-a_{20}\rho^2)} \tag{4}$$

Where w is the velocity, ρ the density, x the distance along the expansion device, A the flow area, f the Fanning friction factor, h the enthalpy and q the heat. P is the pressure (MPa), T is the temperature (K), and R is the universal constant for gases considered on a weight basis. The B, C, D, E, F, G, and H coefficients are defined by means of polynomial expressions that are temperature dependent and are correlated with the regression parameters. The proposed model of the expansion device is shown in Figure 1.

In this work, the hydrodynamic model for the RESS process is developed and implemented in CFD for a fixed geometry of the nozzle. Figure 2 and Table 1 shows a summary of the nozzle under study.

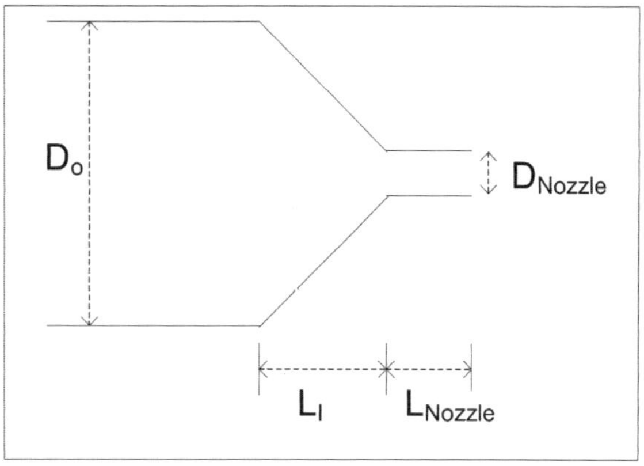

Figure 2: Schematic of capillary Nozzle for RESS Process

Table 1: Dimensions of the capillary Nozzle for RESS Process

Geometry	**Dimensions in μm**
Internal diameter D_o (μm)	1200
Diameter of capillary nozzle D_{Nozzle}(μm)	50
Length of inlet region L_I (μm)	330
Length of capillary nozzle L_{Nozzle} (μm)	50
Volume of the expansion chamber (dm^3)	22

The modeling objective is to study the effect and sensitivity of the design parameters (i.e. operating conditions) in the pre-expansion path of the RESS process on the particle formation in the chamber. Only simulation of the flow (i.e. hydrodynamic) in the capillary nozzle (capillary inlet – capillary nozzle) is carried out for the solvents of carbon dioxide (CO_2) and trifluoromethane (CHF_3) and for the organic solute of benzoic acid. The following assumptions were made on the flow inside the nozzle when developing and simulating the model:

1. The flow is steady and one-dimensional (i.e. flow along the axial direction)
2. The residence time inside the nozzle is very short

3. Adiabatic process is assumed so that no heat is transferred between the fluids inside the nozzle to the surrounding.
4. No phase change along the nozzle, one phase flow.

The boundary conditions are set as given in Table 2.

Table 2 Boundary conditions in CFD simulation

Boundary Type	Wall
Wall Influence On Flow	No Slip
Wall Roughness	Smooth Wall
Heat Transfer	Adiabatic
Boundary Type	Inlet
Flow Regime	Subsonic
Mass and Momentum	Static Pressure, 260bar
Flow direction	Zero Gradient
Turbulence	Zero Gradient
Heat Transfer	Static temperature, 383K
Boundary Type	Outlet
Flow Regime	Subsonic
Mass and Momentum	Static Pressure, 1 atm

The geometry of the nozzle is kept constant, only the pre-expansion temperature and pressure are changed. The outlet of the nozzle is kept at ambient condition. The sets of parameters are shown in Table 3.

Table 3: Sets of pre-expansion pressures and temperatures

Sets	Pre-expansion temperature (K)	Pre-expansion Pressure (MPa)
Set 1	383	26
Set 2	383	30
Set 3	418	26
Set 4	418	30

The CFX software package was used to simulate the nozzle expansion system. The heat transfer model and turbulence model are selected as the **total energy** and **k-epsilon models.** Heat transfer model are essentially used to predict the temperature throughout the flow. The types of heat transfer are via conduction, convection and also via turbulent mixing and viscous work. The **total energy** option to model the heat transfer was selected to model the transport of enthalpy, which includes kinetic energy effects. This model is used for gas flows where the Mach number exceeds 0.2, and high-speed liquid flows where vicious heating effects arise in the boundary layer, in which the kinetic energy effects become significant. The flow inside the pre-expansion nozzle is approximately at supersonic condition which is consistent with the high-speed condition of the **total energy** model. In addition, the Turbulent Wall functions were selected to deal with fine meshes (0.01).

3. Simulation Results and Discussions

3.1 Pressure and Temperature Profiles

Figure 3 shows the profiles of pressure, temperatures and velocity (in terms of Mach number) along the nozzle. The results are comparable with the experimental data gathered from literature. From the figure, it can be observed that the temperature remains relatively constant until it reaches the position-x at 75 mm. However, the pressure would experience a significant drop across the x-direction. Hence, condensation becomes the prevailing precipitation mechanism up to the capillary exit. Prior to the exit, the pressure and temperature would drop from which the nucleation would occur. From the inlet to the capillary nozzle, it can be observed that the Mach number increase exponentially until it reaches over the supersonic condition, where Ma=1.8 at the nozzle exit. At Ma=1, the speed of the fluid can approach to the speed of sound. However, from the simulation of 100mm nozzle, the Ma number obtained at the end of the nozzle is significantly larger than those portrayed at the literature. Hence, due to the high speed, rapid expansion of the supercritical solution through the nozzle can be observed.

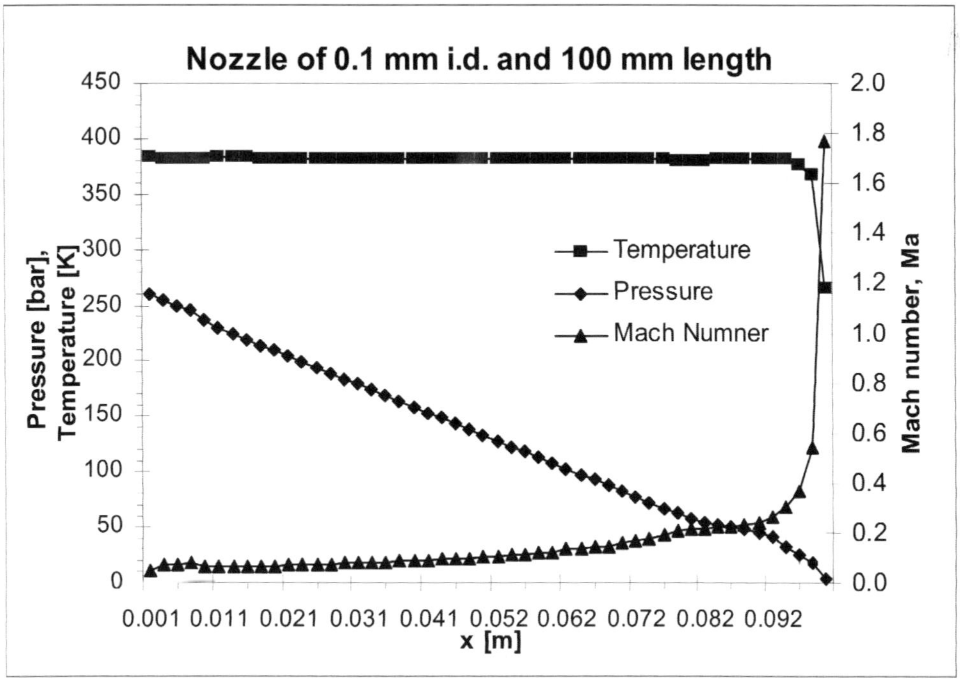

Figure 3: Profiles of pressure, temperature and Mach number along the nozzle.

3.2 Flow Profile Across the Nozzle

Figure 4 shows the profile of physical transport flow. It can be seen that the majority of bulk fluid (supercritical solutions) that enters the nozzle are from the peripheral region whereby at this region, the rate of deceleration of the flow is much lower because it is directed by the contracting wall and the friction force effect. Thus, this flow is more likely to enter the nozzle as it is more stable. Besides, there is a reduction in the probability of sudden languorous behavior as compared to the central bulk fluid where its surroundings are disordered medium itself. Hence the turbulent instability, eddies and vortices are likely to occur and slightly to shift the course of flow allowing a greater part of peripheral flow to enter.

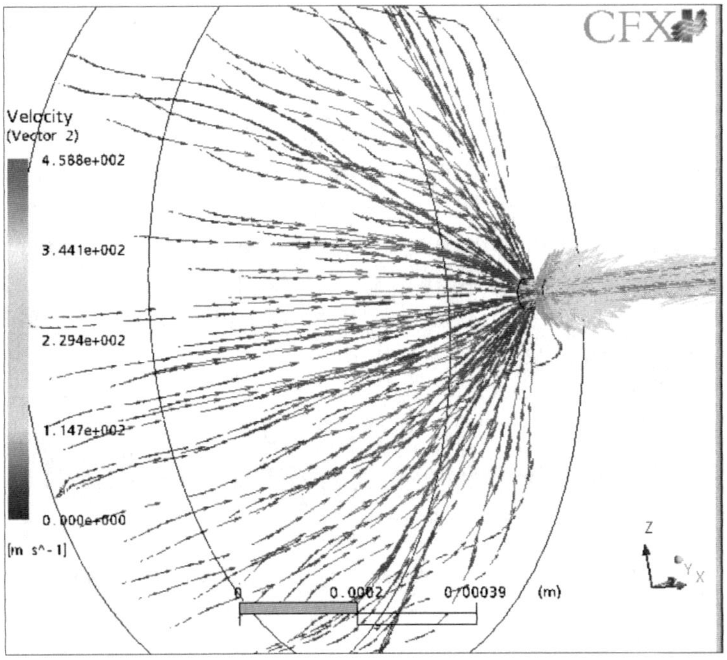

Figure 4: Flow profile across the nozzle.

3.3 Temperature Analysis

Figure 5 shows the temperature distribution of both supercritical solvents in pre-expansion. It can be seen that the temperature distribution for the reservoir region is not affected by the change in inlet pressure. Only a small deviation for both supercritical solvents because both solvents possess similarity in their properties where no effect of the pressure changes at higher temperature of 418K (SET 3 and SET 4) is observed. However, at lower inlet temperature of 383K, inlet pressure change will cause some significant effect on the temperature distribution. It is desirable to have a higher temperature distribution along the reservoir region because such high temperature increases its solubility.

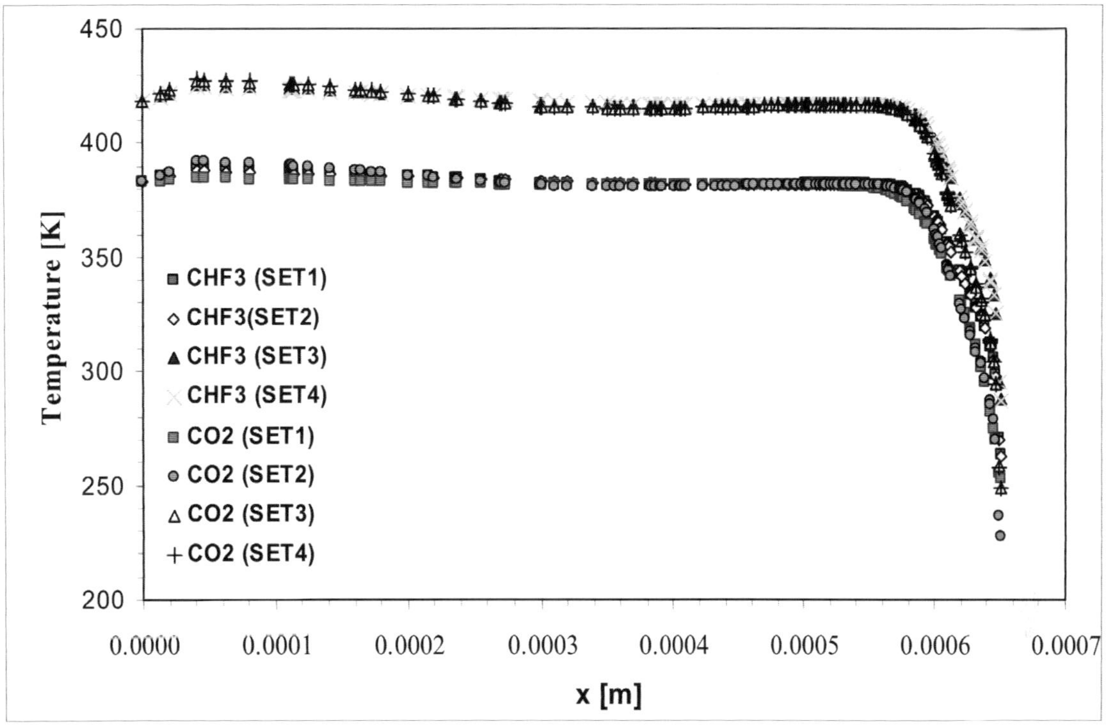

Figure 4: Central Temperature Distribution along Reservoirs and Nozzle for RESS

Figure 5 shows the temperature distribution at the nozzle region. From this figure, the higher temperature distribution is posed by CHF_3 as compared to CO_2 at similar condition. Therefore, one can infer that the rate of temperature decrement is much higher for CO_2 than CHF_3. By analyzing both Figures 4 and 5, we find that at higher constant inlet temperature or pressure, the differences in outlet temperature for CO_2 and CHF_3 increase from 18^oC to approximately 40^oC. Therefore, the discrepancy of solubility for both solvents will be more obvious when the inlet temperature increases. In addition, we observe from Figure 5 that for CO_2 the particle formation may take place earlier than that for CHF_3 because the temperature profile of CO_2 will intersect its critical boundary earlier than that of CHF_3 which implies that CO_2 solubility will decrease significantly.

3.4 Pressure Analysis

Figure 6 shows the pressure distribution for the RESS process. It is observed that the pressure gradient increases as it goes towards the nozzle with a huge degree of drop. This may be due to the rapid rise in velocity. Additionally, by looking at the trends for different supercritical solvents, increasing the inlet temperature from 383K to 418K at lower pressure of 26MPa (SET 1 & SET 3) will not affect the pressure distribution for CHF_3. In contrast, by varying the inlet temperature at higher pressure of 30MPa (SET 2 & SET 4), the pressure trends for CHF_3 will slightly diverge with the lower temperature inlet achieving a higher pressure as it approaches the nozzle. This is because at lower temperature, the viscosity of the solution will be slightly higher and therefore velocity is lower causing higher-pressure distribution. On the other hand for CO_2, the behavior is different whereby at higher pressure (30MPa), increasing the inlet temperature will not effect the pressure distribution while at lower pressure (26MPa), the temperature effect is quite significant whereby the pressure distribution will remain constant for a longer period for the case of the lower inlet temperature. Furthermore, by comparing both solvents, one will able to predict that pressure distribution for CHF_3 will always slightly higher than CO_2 due to the change in velocity.

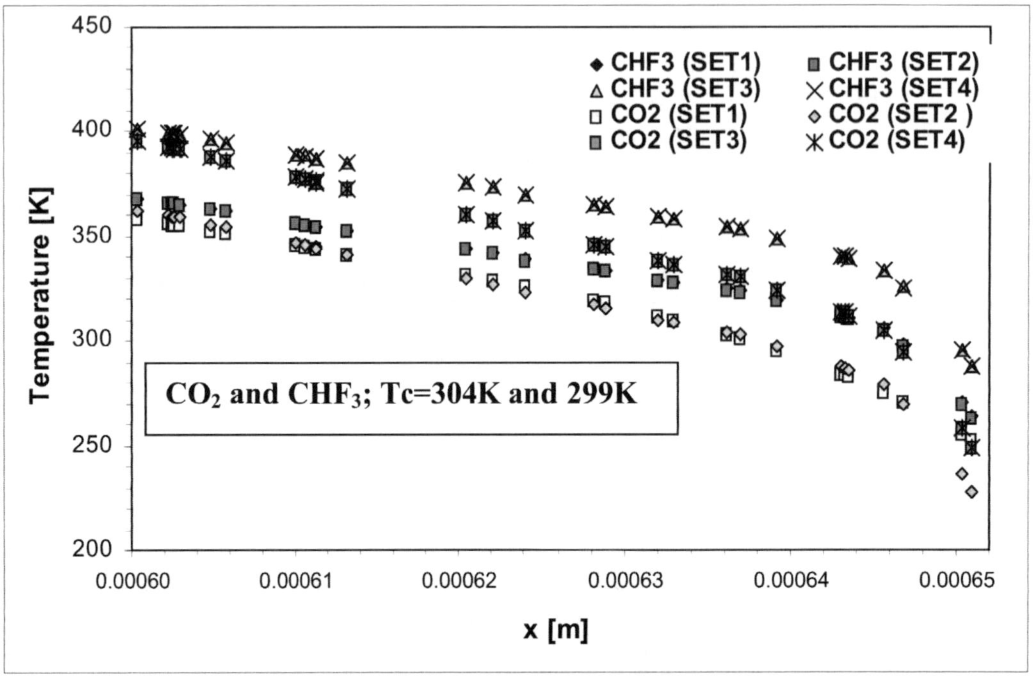

Figure 5: Central Temperature Distribution along Nozzle for RESS

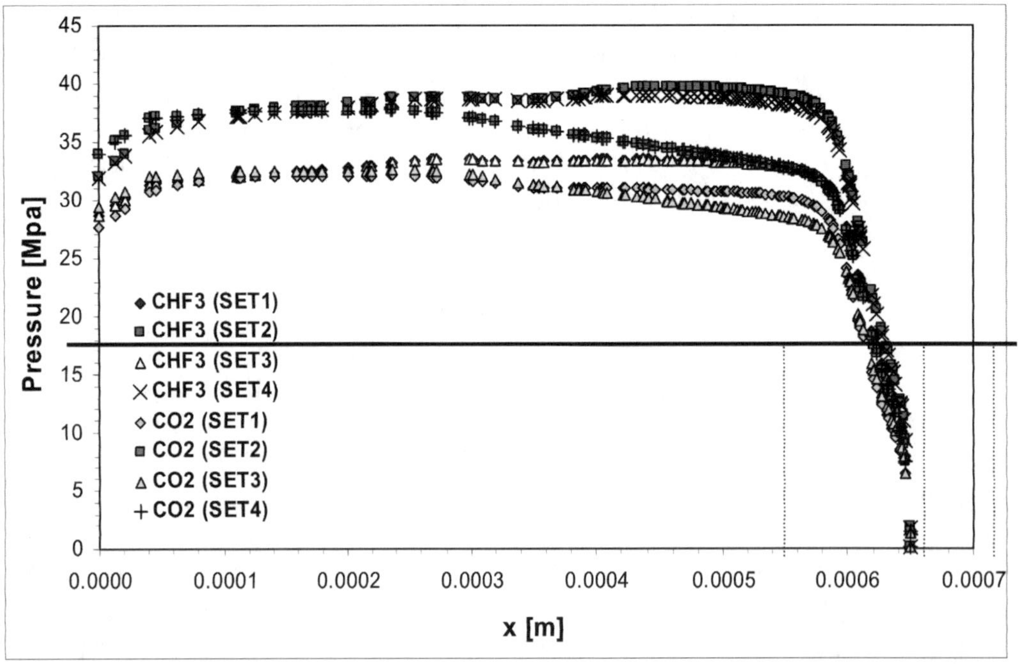

Figure 6: Central Pressure Distribution along Reservoirs and Nozzle for RESS

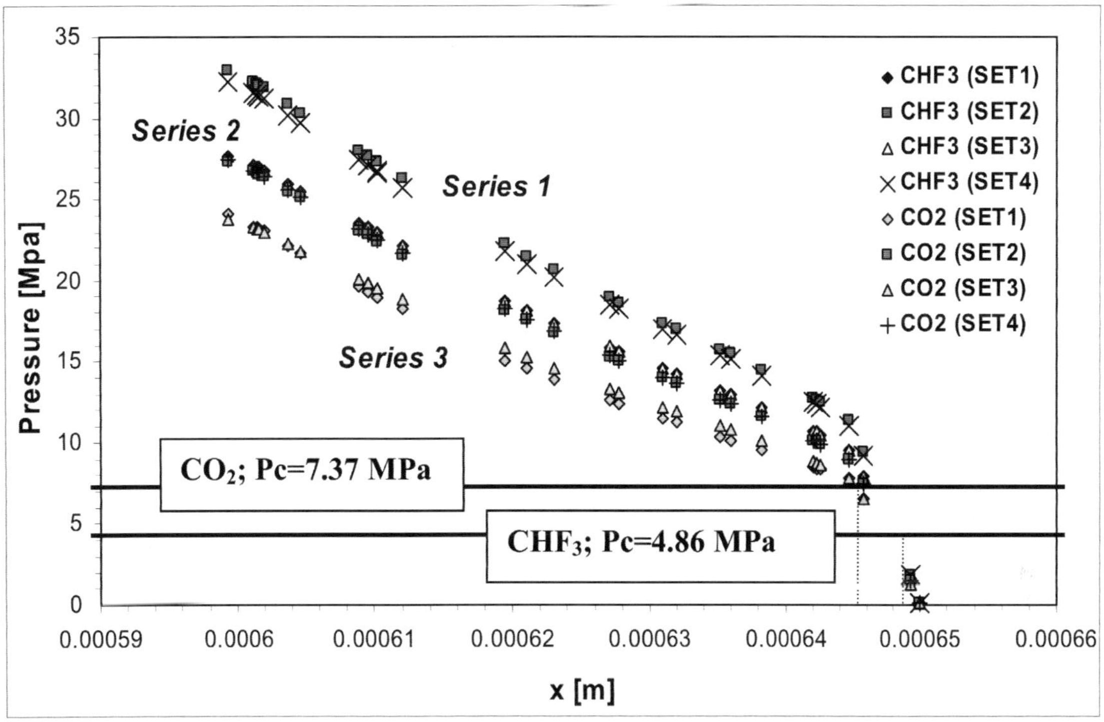

Figure 7: Central Pressure Distribution along Nozzle for RESS

Figure 7 shows the pressure distribution at the nozzle. From this Figure, it can be analyzed that the trends of pressure distribution for three series of data. Series 1 consists of SET 2 and SET 4 of CHF_3 with possessing highest-pressure distribution. As for Series 2, it composes of 4 sets and they are SET 1 and SET 3 of CHF_3 with SET 2 and SET 4 of CO_2. Hence, it can be analyzed that the effect of temperature and pressure to both supercritical solvents is contradictory to each other, i.e. the convergence of lower pressure distribution trend for CHF_3 and the higher-pressure distribution trend for CO_2. In addition, Figure 7 shows that the supercritical solvent of CO_2 reaches out of its critical region before CHF_3 does. Therefore, it can be predicted that the particle formation of benzoic acid can take place faster in CO_2 than CHF_3 due to the possibility of lost of solvent power. Likewise, from combined analysis of Figures 6 and 7, the relative pressure (P/Pc) for CO_2 is also much closer to 1 as compared to CHF_3.

3.5 Density Analysis

Figure 8 shows the density distribution along the pre expansion unit. Density is the most important parameter in the RESS process. If the density of the supercritical solvent is decreasing, the solvent power also decreases so that it leads to increase in super-saturation ratio.

It is desirable to attain the lowest density for a particular solvent at its supercritical region and at the nozzle exit because it increases the efficient use of supercritical solvent. Furthermore, the main purpose for the design parameter of pressure and temperature is to alter and control the density and salvation power of the supercritical fluid so as to increase the performance of RESS process. At higher pressure, the density of the supercritical fluid (SCF) will increase due to compression of fluid to higher solidity. Conversely, at higher temperature, the density of SCF will drop as atoms in the fluid obtain higher thermal energy to pulsate from its usual position.

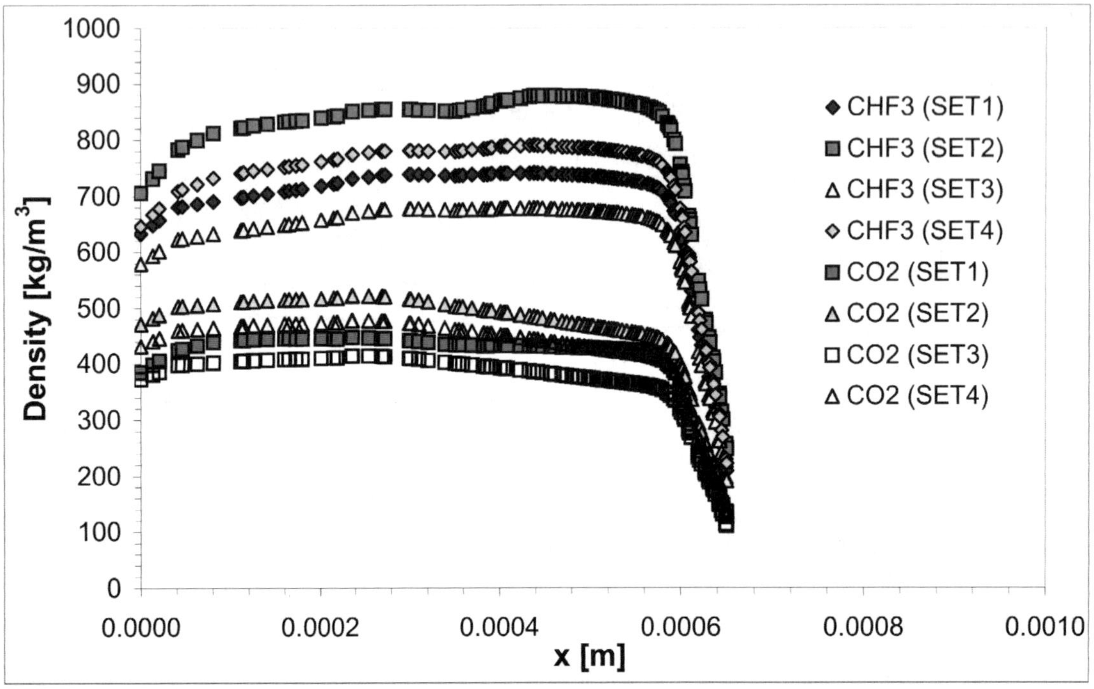

Figure 8: Central Density Distribution along Nozzle for RESS

Note that the compressibility factor should not be affected by the rate of dispersion of fluid jet or else depressurizing effect will be minimized and therefore deteriorate the RESS performance.

3.6 Discussions

Since high super saturation ratio is commonly used to calculate the critical nucleus radius, the high super saturation ratio corresponds to a low critical diameter. Moreover, a decrease in density means a decrease in solvent power. Consequently, this implies an increase in super-saturation power and the possibilities of increasing in the homogeneous nucleation rate.

From the simulation results, it is observed that at lower pre-expansion temperature with a constant pre-expansion pressure the solute nucleation may start inside the nozzle. This is due to the fact that the temperature of the fluid falls well below the critical points resulting in decreasing solvent power. In contrast, at higher pre-expansion temperature, the nucleation starts only at the nozzle outlet. This finding is true for both SCF of CO_2 and CHF_3.

It is also observed from simulation results that pre-expansion pressure slightly affects the nucleation process. Generally the nucleation rate is little higher when the pressure is higher. Thus, high pre-expansion pressures and low pre-expansion temperature favor small particle formation. High super-saturation rate tends to decrease the particle size.

4. Conclusions

Hydrodynamics modeling and analysis of RESS process has been presented in this paper. A CFD model has been developed for the specified nozzle from which the temperature, pressure and density profile analysis was conducted to condition the particle formation. The results demonstrate that the variation of thee parameters in the ranges studied may lead to the increased super-saturation and simultaneously increase in nucleation rate.

Acknowledgment

This research was supported by Malaysian Pepper Board (MPB). The authors would like to thank MPB for the research grant.

References

1. Sitompul, J.P., I.N. Widiasa and Y. Samyudia, Prediction of Selectivity and Solubility of Ethanol in the Supercritical CO_2 Solvent, *Proc. of the National Seminar on Fundamental of Chemical Engineering, Institut Teknologi 10 November Surabaya,* 5-6 November 1997
2. Sitompul, J.P., Ratnawati, A. Suwono and Y. Samyudia, Solubility prediction of cholesterol in supercritical CO_2 involving cosolvent, *Proc. Of Int'l Conf. on Fluid Thermal and Energy Conversion, 21-24 July, Yogyakarta,* 207-214, 1997
3. Sitompul, J.P., A. Prasetyaningrum, T.H. Soeriawidjaja, Ratnawati, and Y. Samyudia, Extraction of β-carotene from its carrier by supercritical CO_2: Prediction of solubility and selectivity, *Proc. Of Int'l Conf. on Fluid Thermal and Energy Conversion, 21-24 July, Yogyakarta,* 257-264, 1997
4. Sitompul, J.P., I.N. Widiasa and Y. Samyudia, Combination of Powel's Dogleg and Hybrid Algorithms: Application to the Nonlinear Equation Systems of Phase Equlibrium on a Supercritical conditions (in Indonesian), *Proc. of the Seminar on Fundamental of Chemical Engineering*, Institute of Technology 10 November, Surabaya, 25-26 November, 1998
5. Tom, J.W. and P.G. Debenedetti, Particle formation with supercritical fluids – A review, *J. Aerosol Sci.*, 22 (5), 555-584, 1991
6. Türk, M., Formation of small organic particles by RESS: experimental and theoretical investigations, *J. of Supercritical Fluids*, 15, 79-89, 1999
7. Türk, M, B. Helfgen, P. Hils, R. Lietzow and K. Schaber, Micronization of pharmaceutical substances by rapid expansion of supercritical solutions (RESS): Experiments and Modeling, *Part.Part. Syst. Charact.*, 19, 327-335, 2002

8. Vemavarapu, C., M.J. Mollan, M. Lodaya and T.E. Needham, Design and process aspects of laboratory scale SCF particle formation systems, *Int. J. of Pharmaceutics*, 292, 1-16, 2005
9. Weber, M., L.M. Russell, and P.G. Debenedetti, Mathematical modeling of nucleation and growth of particles formed by the rapid expansion of a supercritical solution under subsonic conditions, *J. of Supercritical Fluids*, 23, 65-80, 2002
10. Samyudia, Y., P.L. Lee and I.T. Cameron, Control strategies for a supercritical fluid extraction process, *Chem. Eng. Science,* 51(5), 769-787, 1996

Energetic Materials;

What are they?

Bruce Cranford, P.E., F. AIChE

Draft September 23, 2008

Presented
November 20, 2008
at the
Energetic Materials Group Annual Meeting
American Institute of Chemical Engineers Annual Meeting
Philadelphia, Pa

Abstract:

The formation of the Energetic Materials Group focused on many issues involving the science and engineering of energetic materials. One unresolved issue was the meaning of "energetic materials." This paper attempts to identify a rational for defining the category of Energetic Materials. The paper first reviews the categories of energetic materials identified in the literature, addressing the advantages and disadvantages of each. Based upon the results of the analysis, a concise definition of Energetic Materials is be proposed.

1 Introduction

What is an energetic material? If you ask 10 professionals, you will get 12 opinions with an exothermic discussion. Depending on your definition of energetic materials, the history can go back to China 1000 years ago (e.g. black power) or early Greek Civilization (Greek Fire), over 2500 years ago.

The formation of the Energetic Materials Group within the Particle Technology Forum of the American Institute of Chemical Engineers focused on many issues involving the science and engineering of energetic materials. One unresolved issue was the meaning of "energetic materials."

This paper attempts to identify a rational for defining the category of Energetic Materials. This paper draws attention to the characteristics of materials qualifying them to be considered Energetic Materials. A review of the literature and the internet reveals the following definitions of energetic materials, see Figure 1, Definition of Energetic Materials.

The most common energetic materials are listed in the Appendix 1, Table 1: Energetic Materials. A comprehensive list of all energetic materials, however is beyond the scope of this paper.

The paper first reviews the categories of energetic materials identified in the literature, and then addresses the advantages and disadvantages of each.

2 Categories

Many categorization methodologies were identified in the literature. They included categorization based upon function, composition, class of chemicals, physical properties, and forces.

2.1 Function

Attempts to date have used the definition of materials by function, e.g., explosives, propellants, pyrotechnics, batteries, bio processes, high energy mixtures. The drawback with these terms is they are vaguely defined and may vary from one author to the next. Official definitions gleaned from several sources are shown in Figure 2, Definitions, explosive, propellant, propellent, pyrotechnics, etc.

Many energetic materials can be considered

- Energetic materials are a class of materials with high amount of stored chemical energy that can be released.(Wikipedia)
- The term energetic material is used to describe any material which can react to release energy. The category of energetic material is extremely broad and includes everything from common fuels used to power automobiles such as gasoline, and diesel, all the way up to high explosives such as gun-powder, dynamite, and TNT.(University of Missouri)
- Energetic materials include but are not limited too energetic reactions, explosives, propellants, and pyrotechnics.(Energetic Materials)

Figure 1 Definition of Energetic Materials

Burn - Subsonic oxidation, thermal transfer. (Cooper)

Deflagration - Subsonic oxidation, thermal transfer. (Cooper)

Detonation - Supersonic chemical reaction, shock wave. (Cooper)

Explosive - Tending to explode. (Neufeld, Stein) (NB Not much help).
- A violent expansion or bursting with noise.(Stein)
- A release of mechanical, chemical, or nuclear energy in a sudden and often violent manner with the generation of high temperature and usually with the release of gases.(Answer.com)
- "...wide range of energetic materials that can react chemically to produce heat, light, and gas.(Cooper)
  - High explosives: Chemicals that detonate.
    - Primary, can detonate easily.
    - secondary, more difficult to detonate, do not easily fo from burning (deflagration) to detonation.
    - tertiary, most difficult to detonate, insensitive high explosive.
  - Low explosives, pyrotechnics, propellants, blasting agents, cannot be caused to detonate by meas of a common blasting cap.
- Achieved by:(Cooper)
  - bursting a vessel containing a pressurized fluid.
  - rapid heating of air and plasma by an electric arc.
  - Very fast burning reaction.
  - detonating an explosive.
- Derived from the Latin word *explōdere*, "to drive out by clapping".(Answer.com)

Fireworks - Pyrotechnics intended for amusement, public entertainment or artistic purposes (Conkling)

Propellant - A propelling agent (Stein). (NB Not much help).
- The explosive charge that propels a projectile from a gun.(Neufeld)
- Produces gases used to perform mechanical work.(Cooper)
- from "L. propellere "push forward," from pro- "forward" + pellere "to push, drive." Meaning "to drive onward."(Etymology on line)
- Mixtures of chemicals that produce large volumes of high temperature gas at controlled, predetermined rates.(Kirk-Othmer)
- Categories include:
  - Mono propellant, single compound (internal oxidation)
  - Heterogeneous mixtures
    - double base
    -elastomer-modified composite double base
    - composites
    - composite-modified double base
  - Solid
  - Liquid

Propellent - Propelling or tending to propel.(Stein, Neufeld)

Pyrotechnics - The art of making and using fireworks.(Neufeld)
- Dazzling display, as of eloquence, wit, virtuosity.(Newfeld)
- The area of technology that deals with the application of self-contained and self-sustained exothermic chemical reactions of solids to produce heat, sound, smoke, motion, combinations of these, and/or useful reaction products.(Conkling)
- Derived from the french *pyrotechnique*, which is derived from the Greek word *pyr* for fire, and *technē* for art.(Newfeld)

Figure 2 Definitions, explosive, propellant, propellent, pyrotechnics, etc.

an explosive and a propellant and a pyrotechnic. An example is black powder which is used for all three purposes. Black powders used in pyrotechnics, are also used as a propellant to propel the pyrotechnic charge into the air, where additional black powder explodes providing, sound, light, and smell. Numerous examples exist of black powders intended use being that of a propellant, malfunctions, and acts as an explosive!

2.2 Composition

The literature provides examples of energetic materials defined by composition. Energetic materials are composed of the same star dust as the human body, and Mount Everest! The historical classification of materials falls into two groupings, the Greek Elements (or variations thereof) and the Periodic Table.

2.2.1 Greek Elements

Energetic material composition can be based upon the four Greek elements, earth, wind, fire, and water. The Japanese, Chinese, Hindus and Buddhist all have similar elements. All the energetic materials can be considered fire which produces wind and comes from the earth. Not much help.

2.2.2 Periodic Table

Modern science uses the periodic table to identify and classify all the elements. All energetic and non energetic materials are composed of the same elements. Not much help in defining energetic materials.

2.3 Class of Chemicals

Chemists, chemical engineers, and other scientists have spent years developing names and naming metrologies for various class of chemicals based upon the chemical structure or function. Examples include Inorganic oxidizers, cyclic aliphatic, and azides. The chemical structure of energetic materials is not unique to any one class.

Consequently, class of chemicals is not useful in segregating energetic materials from non energetic materials.

2.4 Physical Property

Categorization by physical properties does not help either. A given chemical may be a solid, a liquid, or a gas[(1)], depending on the manufacturing processes, additives, or temperature. All matter falls into one of these categories. Many other physical properties are used to identify and classify materials, including the size of the solids, (e.g., nano particles), flow properties of the liquids and gases (rheology), physical tests (United Nations System, 27 CFR 55, numerous federal regulations, and tests to determine sensitivity and performance).

2.5 Forces

In reviewing the literature, it becomes evident that categorization by forces may add some insight and provide some assistance in defining energetic materials. The four forces found in nature are gravity, electromagnetic, strong interaction and weak interaction.[(2)] See Figure 3, Energetic Materials Flow Chart. The approach discussed in this paper defines energetic material through the four known forces.

Energetic Materials

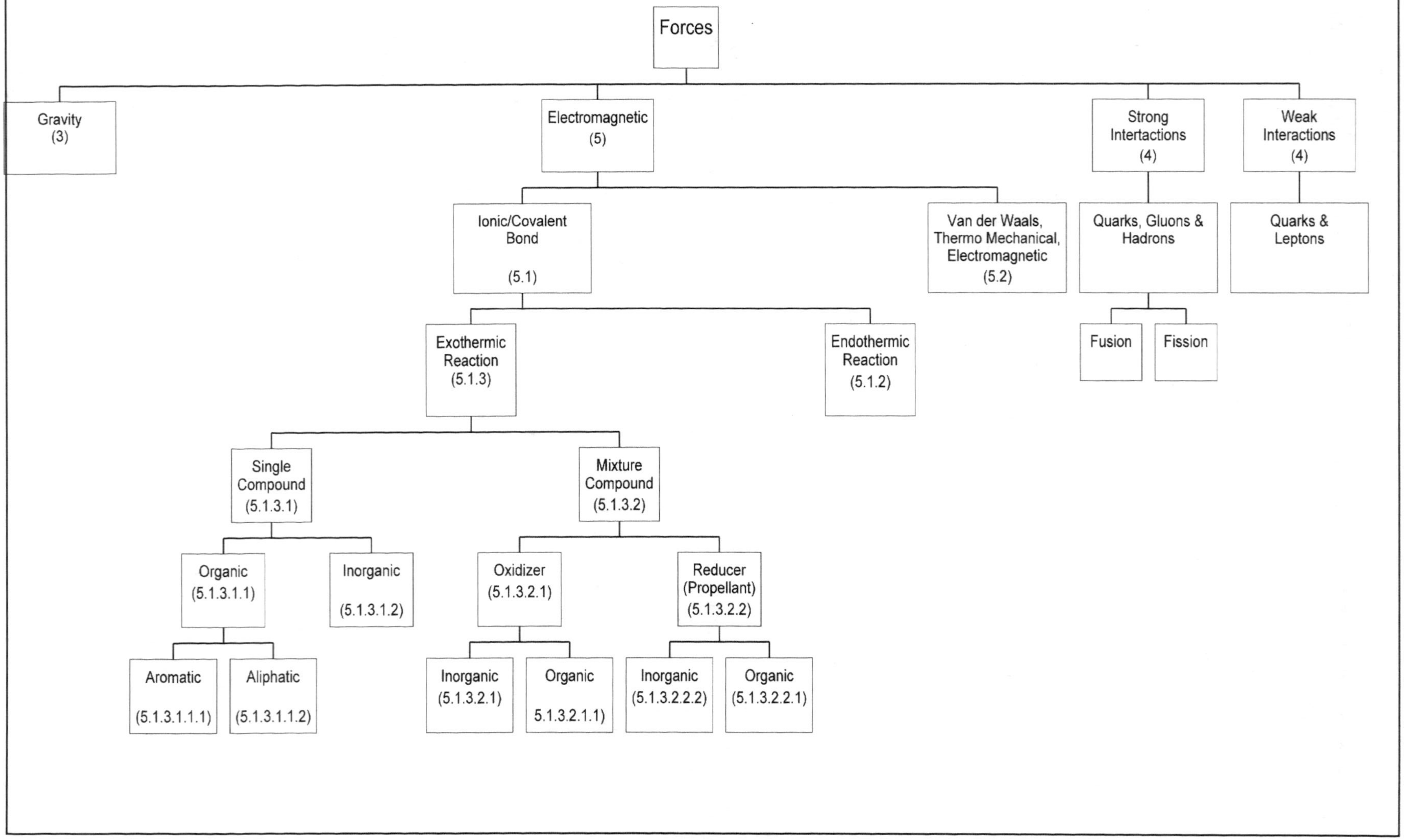

Figure 3 Energetic Materials Flow Chart

3 Gravity:

Gravity acts on all matter. Gravity acts equally on non-energetic materials as well as energetic materials. A pound of lead weights as much as a pound of HMX. The force of gravity does not help in defining energetic materials.

4 Strong and Weak interactions:

The strong and weak interaction forces act on the atomic nucleus and sub atomic particles of energetic and non energetic materials, alike. The unleashing of these forces produces very large amounts of energy, e.g., nuclear fusion, fission. Conventional chemical reactions are capable of releasing up to 10^3 J/g where nuclear fission can release up to 10^{11} J/g, eight orders of magnitude greater than chemicals. However, all atoms contain a nucleus and sub atomic particles. If the strong and weak interactions are included, than all matter can be considered energetic materials. The Strong and Weak interactions are not much help in defining energetic materials vs. non energetic materials.

5 Electromagnetic:

The electromagnetic force also acts on all the elements of the periodic table and all chemicals, but not equally. Individual atoms are attracted to other atoms by the electromagnetic forces. This is described through the concept of Ionic, Covalent, and Van Der Waals[3] bonds. Ionic and Covalent bonds are involved in chemical reactions. Chemical reactions can involve the combining and/or disassociation of chemicals. The covalent bond is typically stronger than the ionic bond which is stronger than Van der Waals forces. Van der Waals bonds are so weak, they are not typically involved in chemical reactions. The following discussion is keyed to Figure 3, Energetic Materials Flow Chart.

5.1 Electromagnetic, Ionic and covalent bond:

The ionic and covalent bond is examined first. The ionic and covalent bonds determine how atoms bond to each other, forming chemicals. The Ionic and covalent bonds are also known as electronic transfer, oxidation/reduction reaction, see Figure 4 Chemical Bonds.

Covalent bond;	Valence electrons are shared by atoms in a molecule. This is also known as electron pair bonding. The greater the attraction between atoms in a molecule, the closer the atoms are to each other and the greater the energy required to break the bond and the greater the energy release when the bonds are broken. Also, the number of electron pairs involved in a bond influence the bond energy. A single pair (single bond) contains less energy that a double bond which contains less energy then a triple bond. E.g. A single bond between two carbon atoms, represented by C-C is 80.5 kcal/mole, where the triple bond C≡C is 195 kcal/mole. Covalent bonds are typically stronger than ionic bonds but the bonds between molecules are typically weaker than ionic bonds.
Ionic bond;	Valence electrons are transferred by one atom to another atom. Now the atoms have an unbalanced charge and are attracted to each other by electromagnetic forces. The first step of the reaction, removing an electron from an atom, creates a positively charged ion and requires energy. The second step is when that electron is attracted to the other atom forming a negatively charged ion, releasing energy. The third step is when the negatively charged ion is attracted to the positively charged ion, releasing additional energy, and forming a molecule. The energy needed for step one is less that the energy released by steps tow and three. The energy in step one is known as ionization potential. Ionic bonds are typically weaker than covalent bonds, by the bonds between molecules are typically stronger than covalent.
Van der Walls bonds:	Attraction between molecules, which do not play a significant role in the formation of molecules.

Figure 4, Chemical Bonds

5.1.A Electromagnetic, Ionic and covalent bond, Energy Release

In order to understand energy release, one should understand the laws of thermodynamics. See Figure 5, Laws of Thermodynamics.

Zeroth Law ≡ Two systems in thermal equilibrium with a third are in thermal equilibrium with each other.

First Law ≡ Internal Energy of systems (ΔU) is the heat absorbed by the system (q) minus the work done by the system (w) on its surroundings,
$\Delta U \equiv q - w$

Second Law ≡ Entropy (S) of the Universe increases or in isolated systems processes for which the entropy change is negative are not spontaneous.

Third Law ≡ Entropy of a perfectly ordered system is zero.

Figure 5 Laws of Thermodynamics

From these laws we get the following relations applicable to chemical thermodynamics and Energetic Materials.
- Enthalpy (H), Figure 6
- Entropy (S), Figure 7
- State Variables, Figure 8
- Ideal Gas Law, Figure 9
- Gibbs Energy (G), Figure 10
- Helmholtz (A), Figure 11.

The word Enthalpy is derived from the Greek *Enthálpein*, to heat[Stein]. It is also known as the heat of reaction, heat of formation. Enthalpy can be expressed as joules per mole or kilo joules per mole (kJ/mole), kilo calories per gram (kcal/g) or kilo joules/kilogram (kJ/kg).

Figure **6** Enthalpy (H)

Entropy is from the German *Tropê*, turning toward.[Neufeld] The greater the disorder, the greater the Entropy Term. In general the Entropy of a solid is less than a liquid which is less than a gas.

Figure **8** Entropy (S)

<u>State Variables</u>
Density (ρ)
Energy (E)
Enthalpy (H)
Entropy (S)
Gibbs (G)
Helmholtz (A)
Internal energy (U)
Mass (m)
Pressure (p)
Temperature (T)
Volume (V)

Figure **7**, State Variables

Ideal Gas Law

$$pV = nRT$$

n = number of moles
R = Universal gas constant = 8.3145 J/mol K

Figure 9 Ideal Gas Law

Gibbs Energy:- Maximum useful work (excluding PV work associated with volume changes of the system) that a system can do on the surroundings when the process occurs reversibly at constant temperature and pressure(Thermodynamics of Chemical Equilibrium)

- Energy of a system that is free to do work at constant T, and p.(Chemical Thermodynamics)

$$G \equiv H - TS = U + pV - TS$$

G ≡ Gibbs Energy (T & p = constant)
H ≡ Enthalpy
p ≡ pressure
S ≡ Entropy
T ≡ Temperature (absolute °K or °R).
U ≡ Internal Energy
V ≡ Volume

Figure **10** Gibbs Energy (G) Equation

Helmholtz Energy:- Maximum amount of work a system can do a constant T and V.(MIT)

$$A \equiv U - TS$$

A ≡ Helmholtz Energy (T, V = constant)

Figure 11 Helmholtz (A) Energy Equation

From the analysis of these relations we get the following that apply to Energetic Materials:

- Spontaneous Reaction Definition, Figure 12
- Spontaneous Reactions, Figure 13, which represent the direction a reaction will precede to equilibrium.
- Spontaneous Reactions table, Figure 14.

Spontaneous Reactions: - Process will precede to equilibrium once it has started.(Journal of Pyrotechnics)

Figure **12** Definition of Spontaneous Reaction

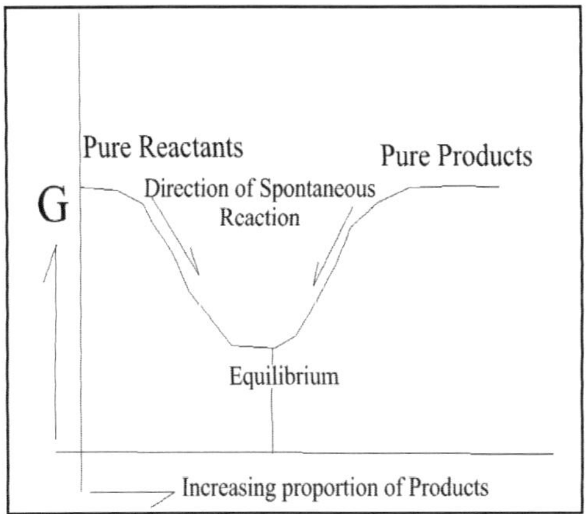

Figure 13 Spontaneous reactions (Chemical Thermodynamics)

Enthalpy Change (dH or ΔH)	Entorpy Change (dS or ΔS)	Spontaneous Reaction (dG or ΔG)
Exothermic (dH < 0)	Increase (dS > 0)	Yes dG <0
Exothermic (dH < 0)	Decrease (dS < 0)	Only at low temperatures if |TdS|< |dH|
Endothermic (dH > 0)	Increase (dS > 0)	Only at high temperatures if |TdS|> |dH|
Endothermic (dH > 0)	Decrease (dS < 0)	No dG > 0

Figure **14** Spontaneous reaction table(Chemical Thermodynamics)

When atoms form bonds, becoming molecules, the bonds either absorb energy or release energy in the process, E.g., endothermic or exothermic.

G is composed of two types of energy, Enthalpy and Entropy. Enthalpy (H) is the heat content, Figure 6 Enthalpy. If the Enthalpy is negative, then the reaction is exothermic. If the Enthalpy is positive, then the reaction is endothermic. Chemical reactions can be complex processes. In a chemical reaction where two or more chemicals react, part of the reaction may be endothermic, part may be exothermic.

The reaction must be started using an additional energy source, some type of initiation. This is defined as the "activation energy", Figure 15, Definition Activation Energy. The chemical bonds of the original compound(s) must be broken before they are free to react with other compounds to form new chemicals. See Figure 16, Activation Energy. Some compounds have low activation energies at ambient conditions and decomposes or age at these conditions.

Typical energy generation per gram of an energetic material compound is about 2 to 9 kJ/g (.... kcal/g)

Definition of Activation Energy:- Energy applied to initiate the reaction.[Journal of Pyrotechnics]

Figure **15** Definition Activation Energy

5.1.B Electromagnetic, Ionic and covalent bond, Reaction Rate

A physical property of the exothermic compound that is important is the reaction rate. The faster the reaction rate, the more energy and gases are liberated per unit of time, see Figure 17 Thermodynamic Mechanisms.

Of interest is the bulk reaction rate of large quantities of the compound(s). The reaction rates vary greatly in absolute value as well as different terminology. Slow reaction rates are fractions of a mm (or inch) per second. Fast reaction rates can be more than 10 km/sec (21,000 miles/h), about nine orders of magnitude. E.g., (Potassium Dichromate, Boron & Silicon mixture) reaction rates are as slow as 1.7 mm/sec [Journal of Pyrotechnics] (0.06 in/sec)($1.7x10^{-6}$ km/sec) where CL-20 is more than 10 km/sec[Teipel] (21,000 miles/h). Terms like aging, rotting and fermentation are used to describe very slow reaction rates. As the reaction rates increase, terms used to describe them include burning, sub sonic[4] combustion and deflagration. As reaction rates increase into the supersonic region, terms like detonation and explosion, and supersonic combustion are used to describe them. Additional descriptions are used in very narrow fields of specializations. No hard rules determine the burn rate with descriptive terms. Of course, terms like "Oh s..t" can describe many types of reaction rates.

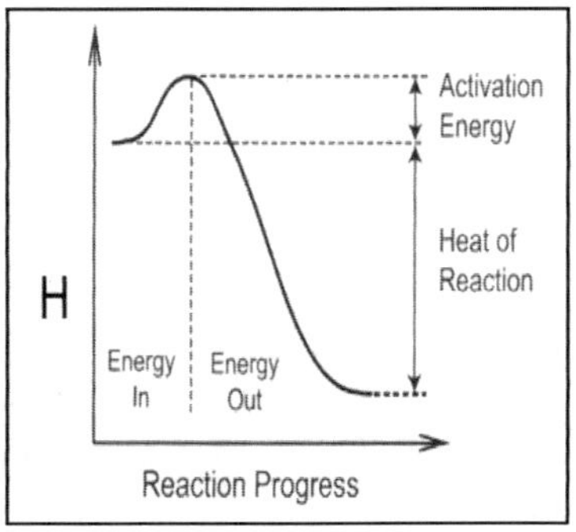

Figure 16 Activation Energy[Journal of Pyrotechnics]

Thermodynamic Mechanisms that affect reaction rates:

- Conduction
- Convection
- Radiation

Figure **17** Thermodynamic Mechanisms

Typically the reaction rate of a gas is faster than the reaction rate of a liquid which is faster than the reaction rate of a solid. As an example, in the space shuttle, there is no reaction of a solid propellant in its solid form. However, when the solid propellant is heated to a liquid/gas state, the reaction rate is very fast and is called combustion. During combustion, the components actually disassociate and recombine. Further discussion is provided in the discussion of specific categories of energy materials below.

Many properties affect the reaction/burning/decomposition rates, including chemical composition/formulation, pressure, density, diameters, particle size, mixing, additives, bulk temperature, grain geometry, ignition technique and physical state of the chemical/mixture to name a few.

In mixture compounds, one or more of the chemicals may be endothermic, but the bulk reaction must be exothermic. Also, the chemical/mixture must sustain itself until completion.

Typical operating pressures vary but rage between 500 psi (3 MPa) to about 100,000 psi (689 MPa). Gas volume generation varies from 13 cm^3/g to over 1,000 cm^3/g.

5.1.C Electromagnetic, Ionic and covalent bond, Combustion temperature:

Combustion temperatures vary from ambient temperatures to approaching 4000˚K (6800˚F, 3763˚C, 7264 ˚R) for some NC/NG formulations[(Teipel)]. HMX flame temperature is 3255 ˚K (3018˚C, 5464˚F). Fluorine and Hydrogen combustion temperatures may reach 4700˚K (4463˚C, 8065˚F). When temperatures reach more than 3550˚K (3313˚C, 6000˚F), combustion product disassociation may play a significant role.

5.1.2 Electromagnetic, Ionic and covalent bond, Endothermic:

Endothermic reactions require energy to form chemical bonds, or require energy to break the chemical bonds. See figure 18, Endothermic Reaction, decomposition of Potassium perchlorate

$$KClO_4 + energy_1\downarrow \rightarrow K + Cl + 2 \bullet O_2$$
$$energy_1 \text{ (change in Enthalpy)} = 430 \text{ kJ/mole}$$

Figure **18** Endothermic Reaction, decomposition of Potassium perchlorate

Potassium perchlorate requires 430 kJ/mole of energy in order for it to dissociate. This energy must come from another source[(5)].

5.1.3 Electromagnetic, Ionic and covalent bond, Exothermic:

Some chemicals combine or disassociate exothermically, producing heat and simpler compounds, e.g., CO_2, H_2O, N_2, C. The combustion of aluminum with potassium perchlorate liberates 6725 kJ/mole in the form of heat, increased pressure and/or increased volume of gases of potassium chloride and aluminum oxide, and any other material in contact with these reaction products, e.g., air is a good example. See Figure 19, Exothermic Reaction, potassium perchlorate plus aluminum.

$$3 \bullet KClO_4 + energy_1\downarrow + 8 \bullet Al \rightarrow 3\ KCL + 4 \bullet Al_2O_3 + energy_2\uparrow$$
$$3 \times energy_1 \text{ (430 kJ/mole)} + energy_2 \text{ (} 3 \times \text{(-437 kJ/mole)} + 4 \times \text{(-1676 kJ/mole))} = \text{-6725 kJ/mole}$$

Figure **19** Exothermic Reaction, potassium perchlorate plus aluminum [(Journal of Pyrotechnics)]

Four types of reactions are possible:
1) disassociation (endothermic) →recombination (exothermic)
2) disassociation (exothermic) →recombination (exothermic)
3) disassociation (endothermic) →recombination (endothermic)
4) disassociation (exothermic) →recombination (endothermic)

Two types of reactions occur, those involving a single molecule (single compounds) and those involving the interaction of different molecules (mixture compounds).

5.1.3.1 Electromagnetic, Ionic and covalent bond, Exothermic, Single Compound:

In the single compound category, all the components are contained within one molecule for a chemical reaction. Typically, a small amount of energy (known as activation energy) shown as $energy_1$ is applied to the molecule which starts the disassociation. The intra molecular chemical bonds are broken, releasing large amounts of energy. The reaction products now recombine to form simpler molecules, releasing more energy. Often, the product molecules are small enough and the temperatures high enough, that the products gasify, creating high pressures. The molecule disassociates exothermically. The individual atoms or small groups of atoms then recombine into very simple molecules, e.g., CO_2, H_2O, and release additional energy. Ideally, all the resulting products should be gases. E.g., See figure 20, (2, 4, 6-Trinitrotoluene). The molecule disassociates releasing more $energy_2$, then recombines releasing additional $energy_3$. In the lexicon of energetic materials, $energy_1$ is often referred to as the activation energy.

$$C_7H_5N_3O_6 + energy_1\downarrow = 7\cdot C + 5\cdot H + 3\cdot N + 6\cdot O + energy_2\uparrow = 1.5\cdot N_2 + 2.5\cdot H_2O + 3.5\cdot CO + 3.5\cdot C + energy_3\uparrow$$

Figure **20** (2, 4, 6-Trinitrotoluene)

Single Compound can be composed of organic and/or inorganic compounds.

5.1.3.1.1 Electromagnetic, Ionic and covalent bond, Exothermic, Single Compound, Organic compounds:

Organic compounds are composed of hydrogen and carbon atoms. Most energetic materials of interest are either aromatic or aliphatic compounds. The difference is the shape of the molecules

5.1.3.1.1.1 Electromagnetic, Ionic and covalent bond, Exothermic, Single Compound , Organic compounds, Aromatic

The aromatic compounds are composed of a benzene ring with various side chains attached to the benzene ring, see Figure 21. The benzene ring is composed of six carbon atoms linked by chemical bonds in such a way as to form a ring structure. Attached to the carbon atoms are other groups of atoms. In the example shown in Figure 22,Trinitrobenzine Aromatic Molecule , they are H and NO_2. Note, carbon double bond, C=C, contains more energy than the carbon single bond C-C.

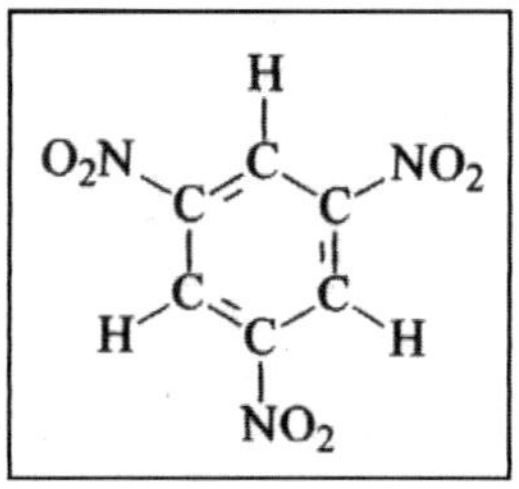

Figure 21 Trinitrobenzine Aromatic Molecule (Cooper)

5.1.3.1.1.2 Electromagnetic, Ionic and covalent bond, Exothermic, Single Compound, Organic compounds, Aliphatic

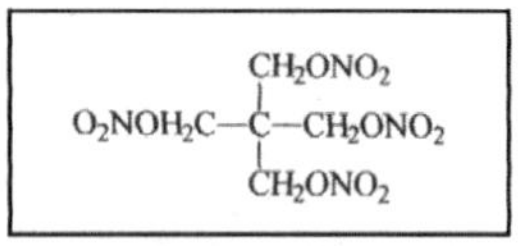

Figure 22 Nitroglycol Aliphatic Molecule (Cooper)

The aliphatic compound is a long chain molecule with various side chains attached to the chain and do not contain the benzine ring. Some of the chemical bonds are single, double, or even triple. See Figure 14, Nitroglycol Aliphatic Molecule. Now things get complicated.

Subsets of Aliphatic molecules are ring molecules and known as cyclic aliphatic compounds. They are not considered aromatic because they do not contain the benzene ring. See Figure 23, RDX Cyclic Aliphatic Molecule. This molecule contains three carbon and three nitrogen atoms in a ring structure not the six carbon atoms in the ring structure

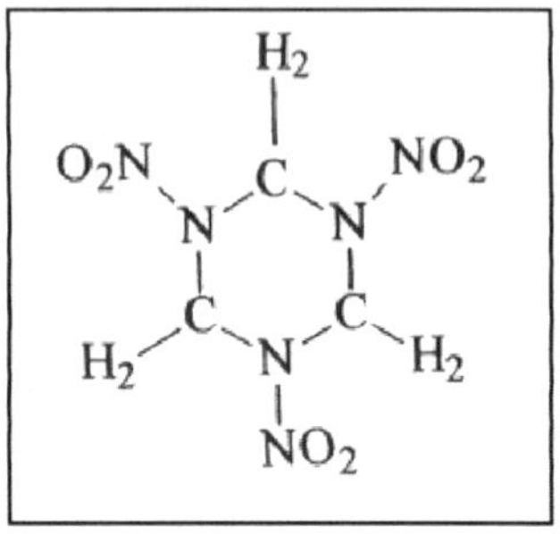

Figure 23 RDX Cyclic Aliphatic Molecule (Cooper)

5.1.3.1.2 Electromagnetic, Ionic and covalent bond, Exothermic, Single Compound, Inorganic compounds:

Inorganic compounds do not contain carbon and/or hydrogen atoms in the molecule. When these molecules dissociate they produce heat. The molecule disassociates exothermically. E.g., see Figure 24, Ammonium Nitrate.

$$NH_4NO_3 + energy_1 \downarrow \approx 2{\bullet}N + 4{\bullet}H + 3{\bullet}O + energy_2 \uparrow = N_2 + 2{\bullet}H_2O + 0.5{\bullet}O_2 + energy_3 \uparrow$$

Figure **24** Ammonium Nitrate

5.1.3.2 Electromagnetic, Ionic and covalent bond, Exothermic, Mixture compounds:

A mixture compound contains two or more different chemicals which are necessary and sufficient for the exothermic reaction. The chemicals act as an oxidizer and a reducer.

5.1.3.2.1 Electromagnetic, Ionic and covalent bond, Exothermic, Mixture compounds, Oxidizer

An oxidizer provides oxygen atoms or similar chemicals to the reaction. This is where definitions and reactions can become complicated. Some single component chemicals also function as an oxidizer in dual component compounds. An example of this type of reaction can be seen in the nitroglycerine molecule. Nitroglycerine can be considered a single compound, because it will completely disassociate producing CO_2, H_2O, and N_2, or an oxidizer because it produces excess oxygen. See Figure 25, Nitroglycerine disassociation.

$$C_3H_5N_3O_9 + energy_1 \downarrow \rightarrow 3{\bullet}C^+ + 5{\bullet}H^+ + 3{\bullet}N + 9{\bullet}O^- + energy_2 \uparrow \rightarrow \approx 3{\bullet}CO_2 + 2.5{\bullet}H_2O + 0.25{\bullet}O_2 + 1.5{\bullet}N_2 + energy_3 \uparrow$$

$$energy_1 < energy_2 < energy_3$$

Figure **25** Nitroglycerine disassociation

The oxidizer may or may not decompose. If it decomposes, it may or may not decompose exothermically. However, the oxidizer/fuel mixture must react and produce a net energy surplus.

The oxidizers can be organic chemicals (contain both hydrogen and carbon) or inorganic chemicals (does not contain hydrogen and/or carbon)

5.1.3.2.1.1 Electromagnetic, Ionic and covalent bond, Exothermic, Mixture compounds, Oxidizer, Organic

An example of an organic oxidizing compound is nitroglycerine, see Figure 25, Nitroglycerine disassociation. The nitroglycerine is an organic chemical (contains both carbon and hydrogen) and decomposes liberating oxygen for use by the fuel. Nitroglycerine is also known as nitroglycerol, glycerol trinitrate, propane-1,2,3-Tryl Trinitrate.

5.1.3.2.1.2 Electromagnetic, Ionic and covalent bond, Exothermic, Mixture compounds, Oxidizer, Inorganic

An example of an inorganic oxidizing compound is ammonium perchlorate. See Figure 26 Ammonium perchlorate. It produces excess oxygen during decomposition to be used by the fuel.

$$NH_4ClO_4 + energy_1 \downarrow \rightarrow N + 4{\bullet}H + Cl + 4{\bullet}O + energy_2 \uparrow \rightarrow \approx 0.5{\bullet}N_2 + 1.5{\bullet}H_20 + HCl + 1.25{\bullet}O_2 + energy_3 \uparrow$$

Figure **26** Ammonium perchlorate

Some oxidizers decompose. An example is liquid oxygen which decomposes, and must vaporize before it can react with the fuel. See Figure 27, Liquid Oxygen

$$\text{Liquid } O_2 + \text{energy}_1\downarrow + C \rightarrow \text{gaseous } O_2 + C \rightarrow CO_2 + \text{energy}_2\uparrow$$

Figure **27** Liquid Oxygen

5.1.3.2.2 Electromagnetic, Ionic and covalent bond, Exothermic, Mixture compounds, Reducer (fuel)

The fuel must decompose prior to reacting with an oxidizer to complete the reaction and produce a net excess of energy. The fuel decomposition may or may not decompose, and may or may not decompose exothermically. However the oxidizer/fuel mixture must react and produce a net energy surplus. These types of fuels can be categorized into organic and inorganic.

5.1.3.2.2.1 Electromagnetic, Ionic and covalent bond, Exothermic, Mixture compounds, Reducer (fuel), Organic

An example of an organic fuel is Lactose, Figure 28, Lactose decomposition .

$$C_{12}H_{22}O_{11}H_2O + \text{energy}_1\downarrow \rightarrow 12\bullet C + 12\bullet H_2O$$

Figure **28** Lactose decomposition

5.1.3.2.2.2 Electromagnetic, Ionic and covalent bond, Exothermic, Mixture compounds, Reducer (fuel), Inorganic

An example of an inorganic fuel is Aluminum, Figure 29, Aluminum decomposition.

$$\text{Solid Al} + \text{energy}_1\downarrow \rightarrow \text{Gaseous Al}$$

Figure **29** Aluminum decomposition

Another class of energetic materials requires additional energy. No chemical change takes place. The forces include momentum, pressure, temperature, and electromagnetic potential. Examples of these materials include gases stored in high pressure cylinders or an Ion Thruster powered by electricity with a Xenon working fluid. Since no chemical reaction takes place, these will not be considered.

5.2.Van der Waals bonds

Van der Waals forces are weaker than ionic or covalent bonds and do not play a significant role in most chemical reactions. Van de Waals forces will not be discussed.

6 Conclusion

Based upon the above discussion, many types of definitions are possible, and are best determined by the technical background of the audience.

The following are suggested definitions for the chemist/chemical engineer/scientist:

- $\Delta G > 0$, spontaneous reaction with sufficient energy to perform useful work.

- Single or multiple chemicals engaged in spontaneous exothermic rapid reactions, producing liquid or gaseous products sufficient to perform useful work.
- Spontaneous chemical reaction where Enthalpy is negative, Entropy positive, sufficient to perform useful work.

For the lay audience:

- Chemical reaction producing heat sufficient to perform work (propulsion, explosion, pyrotechnics).
- Universal: "Oh wow!"

Appendix 1:

Table 1: Energetic Materials

Name	**Abbreviation**	**Table**	**Formula(6)**	**Ref**
?	HTPS	?	$C_{4.765}H_{7.505}O_{2.131}N_{0.088}$	Journal of Pyrotechnics
(2-difuoro-2,2 dinitroethyl-2, 2-dinitropropoy) formal	MF-1	?	?	Teipel
1, 1-diamino-2, d-dinitroethene	Fox-7	Single compound, Aromatic	?	Teipel
1, 3, 3-trinitroazetidine	TNAZ	Single compound Inorganic, ring	?	Teipel
2, 4, 6-Trinitrotoluene	TNT		$C_7H_5N_3O_6$	Journal of Pyrotechnics, Conkling
3, 3-bis(azidomethyl) oxetane polymer	BAMO	Mixture Compounds Reducer organic (Propellant)	$C_5H_8N_6O$	Yang
3-azidomethyl-3methyloxetane polymer	AMMO	Mixture Compounds Reducer organic linear (Propellant)	$C_5H_9N_3O_3$	Yang
3-nitratomethyl-3-methyl oxetane	NMMO	Mixture Compounds Reducer organic (Propellant)		Yang
Accroides resin	Red gum	Mixture reducer organic	?	Journal of Pyrotechnics,
Aluminum	Al	Mixture Compounds Reducer Inorganic (Propellant)	Al	Conkling
Amber powder		Mixture reducer compound organic	?	Journal of Pyrotechnics
Ammonia		Mixture Compounds Reducer Inorganic (Propellant)	NH_4OH	Koelle
Ammonium dinitramide	ADN	Mixture Compounds Inorganic Oxidizer	$NH_4N(NO_2)_2$	Yang
Ammonium dinitramine	AND	Mixture Compounds Inorganic Oxidizer	$NH_4N_4O_4$	Yang
Ammonium nitrate	AN	Single compound Inorganic	NH_4NO_3	Conkling
Ammonium nitrate	AN	Mixture Compounds Inorganic Oxidizer	NH_4NO_3	
Ammonium perchlorate	AP	Mixture Compounds Inorganic Oxidizer	NH_4ClO_4	Conkling
Ammonium picrate		Single compound, Aromatic	$C_6H_2(NO_2)_3ONH_4$	Sutton
Anthracene		Mixture Compounds Reducer Organic (Propellant)	$C_{14}H_{10}$	Conkling
Antimony trisulfide		Mixture reducer compound inorganic	Sb_2S_3	Journal of Pyrotechnics
Argon	Ar	Mono Propellant Inorganic	Ar	
Arsenic sulfide		Mixture reducer compound inorganic	AS_2S_2	Journal of Pyrotechnics
Asphalt		Mixture Compounds Reducer Organic (Propellant)	?	Koelle
Barium Chlorate		Mixture Compounds Inorganic Oxidizer	$Ba(ClO_3)_2H_2O$	Journal of Pyrotechnics
Barium Chromate		Mixture Compounds Inorganic Oxidizer	$BaCrO_4$	Conkling
Barium nitrate		Mixture Compounds Inorganic Oxidizer	$Ba(NO_3)_2$	Journal of Pyrotechnics
Barium peroxide		Mixture Compounds Inorganic Oxidizer	BaO_2	Conkling
Beryllium	Be	Mixture Compounds Reducer Inorganic (Propellant)	Be	
Bis(2-fluoro-2, 2-dinitroethyl)-difluoroformal	DFF		$C_5H_4N_4O_{10}F_4$	Teipel
Bis(2-fluoro-2,3-dinitroethyl formal	FEFO		$C_5H_6N_4O_{10}F_2$	Teipel
Bis(2,2-dinitropropyl) formal	BDNPF		$C_7H_{12}N_4O_{10}$	Teipel
Bis(fluorodinitroethyl) nitramine	FDNEN		?	
Boron	B	Mixture Compounds Reducer Inorganic (Propellant)	B	Conkling
Bromine pentafluoride			BrF_5	Koelle
Carbon	C	Mixture Compounds Reducer Inorganic (Propellant)	C	
Carbon Dioxide		Mono Propellant Inorganic	CO_2	
Caroxyl-terminated polybutadiene	CTPB	Mixture Compounds Reducer Organic (Propellant)	?	
Cellulose(7)		Mixture Compounds Reducer Organic (Propellant)	$(C_6H_{10}O_5)n$	Cooper
Chlorine terifluoride		oxidizer	ClF_3	Koelle

Table 1: Energetic Materials

Name	Abbreviation	Table	Formula(6)	Ref
Copper azide		Single compound Inorganic	?	
Copper oxide		Mixture Compounds Inorganic Oxidizer	CuO	Cooper
Cyanotetrazolatopentaamine cobalt III perchlorate	CP	Single compound Inorganic	?	
Cylotetramethyllenetetranitramine	HMX	Single compound organic aromatic	$C_4H_8N_8O_8$	
Dextrine		Mixture compound, organic, fuel	$(-C_6H_{10}O_5-)nH_2O$	Journal of Pyrotechnics
Diazdinitrophenol	DINO or DDNP	Single compound, Aromatic	?	
Diborane		oxidizer	B_2H_6	Koelle
Dibutyl phthalate	DBP	Mixture Compounds Reducer Organic (Propellant)	$C_{16}H_{22}O_4$	Chemical Industry
Diethyl phthalate	DEP	Mixture Compounds Reducer Organic (Propellant)	$C_{12}H_{14}O_4$	Chemical Industry
Diethylene glycol dinitrate	DEGDN/DEGN	Mixture Compounds Organic Oxidizer alphatic	$(CH_2CH_2ONO_2)O_2$	Sutton
Diethylenetriamine		oxidizer	$(NH_2(CH_2)_2)_2NH_2$	Koelle
Dimethyl phthalate	DMP	Mixture Compounds Reducer Organic (Propellant)	$C_{10}H_{10}O_4$	Chemical Industry
Dinitramidic acid	DNA	Inorganic, ring	$HN(NO_2)_2$	Yang
Dinitrotolunene	DNT		$CH_3C_6H_3(NO_2)_2$	Sutton
Dioctile phthalate		Mixture Compounds Reducer Organic (Propellant)	?	
Ethyl alcohol		Mixture Compounds Reducer Organic (Propellant)	C_2H_5OH	
Ethyl centralite	EC	Mixture Compounds Reducer Organic (Propellant)	$C_{17}H_{20}N_2O$	Chemical Industry
Ethyl picrate		Single compound, Aromatic	?	
Ethyl Tetryl		Single compound, Aromatic	?	
Ethylene oxide		mono propellant	$(CH_2)_2O$	Koelle
Ethylenediamine		Mixture Compounds Reducer organic (Propellant)	$NH_2(CH_2)_2$	Koelle
Ethylenedinitramine	EDNA	Single Compound, Organic, Aliphatic	$C_2H_6N_4O_4$	Chemical Industry
Ferro Silicon		Mixture reducer compound inorganic	FeSi	Journal of Pyrotechnics
Fluorine	F	Mixture Compounds Inorganic Oxidizer	F_2	
Glucose		Mixture Compounds Reducer Organic (Propellant)	$C_6H_{12}O_6$	
Glycidyl azide polymer	GAP	Mixture Compounds Reducer organic (Propellant)	$C_3H_5N_3O_2$	Yang
Guanidine Nitrate			$NH{:}C(NH_2)HNO_3$	Journal of Pyrotechnics
Helium	He	Mono Propellant Inorganic	He	
hemicellulose		Mixture Compounds Reducer Organic (Propellant)	?	
Hexachloroethane		Mixture Compounds Reducer Inorganic (Propellant)	C_2Cl_6	Conkling
Hexahydrotrinitrotriazine (Cyclotetramethylene trinitramine	RDX	Single Compound, Organic aromatic	$C_3H_6N_6O_6$	Conkling
Hexanitroazobenzene	HNAB	Single compound, Aromatic	?	
Hexanitrohexazaisowurtzetane	HNIW, CL-20	Single compound, other, organic, aliphatic	$C_6H_6O_{12}N_{12}$	Yang, Teipel
Hexanitrostibene	HNS	Single compound, Aromatic	?	
Hydrazine		Single compound Inorganic	N_2H_4	
Hydrazine		Mixture Compounds Reducer Inorganic (Propellant)	N_2H_4	
Hydrazinium diperchlorate	HP2		?	Yang
Hydrazinium monoperchlorate	HP		?	Yang
Hydrazinium nitrate	HN		?	Yang
Hydrazinium nitroformate	HNF		?	Yang
Hydrazoic Acid		?	HN_3	Yang
Hydrocarbon liquid fuels[8]		Mixture Compounds Reducer Organic (Propellant)		
Hydrogen Peroxide		Single compound Inorganic	H_2O_2	
Hydrogen Peroxide		Mixture Compounds Inorganic Oxidizer	H_2O_2	
Hydrooxylammonium perchlorate	HAP		?	
Hydroxy-terminated polybutadiene	HTPB	Mixture Compounds Reducer Organic (Propellant)	?	
Hydroxyl Ammonium Nitrate	HAN	Single compound Inorganic	$NH^2OH^+NO_3$	Sutton

Table 1: Energetic Materials

Name	Abbreviation	Table	Formula(6)	Ref
hydroxyl-terminated polyethylene	HTPE	Mixture Compounds Reducer Organic (Propellant)	$C_{5.194}H_{9.84}O_{1.608}N_{0.194}$	Journal of Pyrotechnics
Hydyne		Mixture Compounds Reducer Inorganic (Propellant)	?	
Iron		Mixture reducer compound inorganic	Fe	Journal of Pyrotechnics
Iron Oxide (black)		Mixture Compounds Inorganic Oxidizer	Fe_3O_4	
Iron Oxide (red) hematite		Mixture Compounds Inorganic Oxidizer	Fe_2O_3	
Krypton	Kr		Kr	Sutton
Lactose		Mixture Compounds Reducer Organic (Propellant)	$C_{12}H_{22}O_{11}H_2O$	
Lead azide		Single compound Inorganic	$Pb(N_3)_2$	Conkling
Lead chromate		Mixture Compounds Inorganic Oxidizer	$PbCrO_4$	Conkling
Lead dioxide		Single compound inorganice	PbO_2	Conkling
Lead oxide			PbO	
Lead styphante		Single compound, Aromatic	$(NO_2)_3C_6HO_2Pb$	Brown
Lead tetroxide (red lead)		Mixture Compounds Inorganic Oxidizer	Pb_3O_4	Conkling
Led Peroxide		Mixture Compounds Inorganic Oxidizer	PbO_2	Conkling
Lignin		Mixture Compounds Reducer Organic (Propellant)		
Lithium perchlorate		Mixture Compounds Inorganic Oxidizer	$LiClO_4$	Cooper
Magnalium		Mixture reducer compound inorganic	MgAl	Journal of Pyrotechnics
Magnesium	Mg	Mixture Compounds Reducer Inorganic (Propellant)	Mg	Journal of Pyrotechnics
Mercury fulminate		Single compound Inorganic	$Hg(ONC)_2$	Conkling
Methane		Mixture Compounds Reducer Organic (Propellant)	CH_4	Sutton
Methyl nitrate		mono propellant, organic	CH_2NO_3	Koelle
Methylinitrate		Single Compound, Organic, Aliphatic	?	
Monomethyl hydrazine	MMH	Mixture Compounds Reducer Inorganic (Propellant)	CH_3NHNH_2	Sutton
n-Propyl Nitrate		mono propellant	$C_3H_7NO_3$	Koelle
Naphthalene			$C_{10}H_8$	Conkling
Nitric Acid		Mixture Compounds Inorganic Oxidizer	HNO_3	Suttton
Nitrocellulose		Mixture Compounds Reducer Organic (Propellant)	$C_6H_7O_2(ONO_2)_3$	Sutton
Nitrocelluole			$(C_6H_{10-x}O_{5-x}(ONO_2)_x)n$	
Nitrogen	N	Mono Propellant Inorganic	N	
Nitrogen Tetroxide		Mixture Compounds Inorganic Oxidizer	N_2O_4	
Nitroglycerine	NG	Single Compound, Organic, Aliphatic	$C_3H_5N_3O_9$	Conkling, Cooper
Nitroglycol		Single Compound, Organic, Aliphatic	$C_2H_4N_2O_6$	Chemical Industry
Nitroguanidine	NQ, NIGU	Single Compound, Organic, Aliphatic	$CH_4N_4O_2$	Chemical Industry, Teipel
Nitromethane		mono propellant, organic	CH_3NO_2	Koelle
Octahydrotetranitrotetrazine	HMX, octogen	Single Compound, Organic, Aliphatic	$C_4H_8N_8O_8$	Chemical Industry
Octanitrocubane	ONC	Single compound Inorganic, ring	$C_8(NO_2)_8$	Teipel
Oil		Organic fuel	C_8H_{18}	Brown
Oxygen	O, LOX	Mixture Compounds Inorganic Oxidizer	O_2	
Oxygen difluoride			OF_2	Koelle
Pentaborane		Mixture Compounds Reducer Inorganic (Propellant)	B_5H_9	Koelle
Pentaerythriloltetranitrate	PETN	Single Compound, Organic, Aliphatic	?	
Pentaerythritoltrinitrate	PETRIN	Single Compound, Organic, Aliphatic	?	
Perchloryl fluoride		oxidizer	ClO_3F	Koelle
Phosphorus	P	Mixture Compounds Reducer Inorganic (Propellant)	P	Conkling
Picric Acid, Trinitrophenol, Lyddite		Single compound, Aromatic	$C_6H_3N_3O_7$	Chemical Industry
Picryl chloride		Single compound, Aromatic	$C_6H_2ClN_3O_6$	Chemical Industry
Polobutadiene acrylonitrile acrylic acid	PBAN	Mixture Compounds Reducer Organic (Propellant)	$C_{10}H_{13}NO_2$	Chemical Industry

Table 1: Energetic Materials

Name	Abbreviation	Table	Formula(6)	Ref
polybutadiene acrylic acid	BPAA	Mixture Compounds Reducer Organlc (Propellant)	?	
Polycarprolactone glycol	PEG	Mixture Compounds Reducer Organic (Propellant)	?	
Polycarprolactone polyol	PCP	Mixture Compounds Reducer Organic (Propellant)	?	
Polyglycol adipate	PGA	Mixture Compounds Reducer Organic (Propellant)	?	
Polyisobutylene		organic fuel	?	Cooper
Polypropylene glycol	PPG	Mixture Compounds Reducer Organic (Propellant)	$C_6H_{14}O_3$	Chemical Industry
Polysulfide		organic fuel	?	Cooper
Polyurethane polyester or polyether	PU	Mixture Compounds Reducer Organic (Propellant)	$C_{25}H_{42}N_2O_6$	Chemical Industry
Polyvinl chloride	PVC	Mixture Compounds Reducer Organic (Propellant)	$(-CH_2CHCL-)n$	Conkling
Potassium Chlorate		Mixture Compounds Inorganic Oxidizer	$KClO_3$	Conkling
Potassium dichromate		Mixture Compounds Inorganic Oxidizer	$K_2Cr_2O_7$	Journal of Pyrotechnics
Potassium dinitramide	KDN	Mixture Compounds Inorganic Oxidizer	$KN(NO_2)_2$	Yang
Potassium nitrate	KN	Mixture Compounds Inorganic Oxidizer	KNO_3	Conkling
Potassium perchlorate	KP	Mixture Compounds Inorganic Oxidizer	$KClO_4$	Conkling
Propanedial-dinitrate	PDN	Mixture Compounds Organic Oxidizer	?	
Red Lead			PB_3O_4	Journal of Pyrotechnics
Rocket Propellant 1	RP-1	Mixtruc Compounds Organic Fuel	$CH_{(1.9\text{-}2.0)}$	Brown, C.
Rosin		Mixture reducer compound organic	?	Journal of Pyrotechnics
Shellac		Mixture Compounds Reducer Organic (Propellant)	$C_{16}H_{24}O_5$	Conkling
Silcon	Si	Mixture Compounds Reducer Inorganic (Propellant)	Si	Journal of Pyrotechnics
Silver acetylide		Single compound Inorganic	C_2AG_2	Cooper
Silver azide		Single compound Inorganic	AgN_3	Cooper
Silver fulminate		Single compound Inorganic	AgONC	Cooper
Sodium azide		Single compound Inorganic	NaN_3	Cooper
Sodium nitrate		Mixture Compounds Inorganic Oxidizer	$NaNO_3$	Joumal of Pyrotechnics
starch		Mixture Compounds Reducer Organic (Propellant)	$(C_6H_{10}O_5)n$	Conkling
Stearic acid			$C_{18}H_{36}O_{11}$	
Strontium nitrate		Mixture Compounds Inorganic Oxidizer	$Sr(NO_3)_2$	Journal of Pyrotechnics
Sugar, sucrose	cane sugar	Mixture Compounds Reducer Organic (Propellant)	$C_{12}HJ_{22}O_{11}$	Joumal of Pyrotechnics
Sulfur	S	Mixture Compounds Reducer Inorganic (Propellant)	S	Journal of Pyrotechnics
Tetranitroniline	TNA	Single compound, Aromatic	?	
Tetronitromethane		mono propellant	$C(NO_2)_4$	Koelle
Tetryl		Single compound, Aromatic	?	
Titanium	Ti	Mixture Compounds Reducer Inorganic (Propellant)	Ti	Journal of Pyrotechnics
Triacetin	TA	Mixture Compounds Reducer Organic (Propellant)	$C_9H_{14}O_6$	Chemical Industry
Triaminotrinitrobenzene	TATB	Single compound, Aromatic	?	Teipel
Triethylene glycol dinitrate	TEGDN	Mixture Compounds Organic Oxidizer, alphatic	$C_6H_{12}N_2O_8$	Teipel
Trimethylolethane trinitrate	TMETN	Mixture Compounds Organic Oxidizer, alphatic	$C_5H_9N_3O_9$	Teipel
Trinitroaniline, Picramide	TNA	Single compound, Aromatic	$C_6H_4N_4O_6$	Teipel, chemical industry
Trinitroanisol		Single compound, Aromatic	?	
Trinitrobenzene	TNB	Single compound, Aromatic	$C_6H_3N_3O_6$	Chemical Industry
Trinitrobenzoic acid	TNBA	Single compound, Aromatic	$C_7H_3N_3O_8$	Chemical Industry
Trinitrocresol	TNC	Single compound, Aromatic	?	

Table 1: Energetic Materials

Name	Abbreviation	Table	Formula(6)	Ref
Trinitroresorcinol (styphnic acid)		Single compound, Aromatic	?	
Tungsten	W	Mixture Compounds Reducer Inorganic (Propellant)	W	
Unsymmetrical Dimethylhydrazine	UDMH	Mixture Compounds Reducer Inorganic (Propellant)	$(CH_3)_2NNH_2$	
Zinc	Zn	Mixture Compounds Reducer Inorganic (Propellant)	Zn	Journal of Pyrotechnics
Zirconium	Zr	Mixture Compounds Reducer Inorganic (Propellant)	Zr	

Appendix 2:

Table 2: List of Abbreviations

Abbreviation	Name
ADN	Ammonium dinitramide
Al	Aluminum
AMMO	3-azidomethyl-3methyloxetane polymer
AN	Ammonium nitrate
AN	Ammonium nitrate
AND	Ammonium dinitramine
AP	Ammonium perchlorate
BAMO	3, 3-bis(azidomethyl) oxetane polymer
BDNPF	Bis(2,2-dinitropropyl) formal
Be	Beryllium
BPAA	polybutadiene acrylic acid
cane sugar	Sugar, sucrose
CP	Cyanotetrazolatopentaamine cobalt III perchlorate
CTPB	Caroxyl-terminated polybutadiene
DBP	Dibutyl phthalate
DEGDN/DEGN	Diethylene glycol dinitrate
DEP	Diethyl phthalate
DFF	Bis(2-fluoro-2, 2-dinitroethyl)-difluoroformal
DINO or DDNP	Diazdinitrophenol
DMP	Dimethyl phthalate
DNA	Dinitramidic acid
DNT	Dinitrotolunene
EC	Ethyl centralite
EDNA	Ethylenedinitramine
F	Fluorine
FDNEN	Bis(fluorodinitroethyl) nitramine
FEFO	Bis(2-fluoro-2,3-dinitroethyl formal
Fox-7	1, 1-diamino-2, d-dinitroethene
GAP	Glycidyl azide polymer
HAN	Hydroxyl Ammonium Nitrate
HAP	Hydrooxylammonium perchlorate
HMX	Cylotetramethyllenetetranitramine
HMX, octogen	Octahydrotetranitrotetrazine
HN	Hydrazinium nitrate
HNAB	Hexanitroazobenzene
HNF	Hydrazinium nitroformate
HNIW, CL-20	Hexanitrohexazaisowurtzetane
HNS	Hexanitrostibene
HP	Hydrazinium monoperchlorate
HP2	Hydrazinium diperchlorate
HTPB	Hydroxy-terminated polybutadiene
HTPE	hydroxyl-terminated polyethylene
HTPS	?
KDN	Potassium dinitramide
KN	Potassium nitrate
KP	Potassium perchlorate
MF-1	(2-difuoro-2,2 dinitroethyl-2, 2-dinitropropoy) formal
MMH	Monomethyl hydrazine
N	Nitrogen
NG	Nitroglycerine
NMMO	3-nitratomethyl-3-methyl oxetane
NQ, NIGU	Nitroguanidine
O, LOX	Oxygen
ONC	Oxtanitrocubane
PBAN	Polobutadiene acrylonitrile acrylic acid
PCP	Polycarprolactone polyol
PDN	Propanedial-dinitrate
PEG	Polycarprolactone glycol
PETN	Pentaerythriloltetranitrate
PETRIN	Pentaerythritoltrinitrate
PGA	Polyglycol adipate
PPG	Polypropylene glycol
PU	Polyurethane polyester or polyether
PVC	Polyvinl chloride
RDX	Hexahydrotrinitrotriazine (Cyclotetramethylene trinitramine
Red gum	Accroides resin
RP-1	Rocket Propellant 1
TA	Triacetin
TATB	Triaminotrinitrobenzene
TEGDN	Triethylene glycol dinitrate
TMETN	Trimethylolethane trinitrate
TNA	Tetranitroniline
TNA	Trinitroanilinc, Picramidc
TNAZ	1, 3, 3-trinitroazetidine
TNB	Trinitrobenzene
TNBA	Trinitrobenzoic acid
TNC	Trinitrocresol
TNT	2, 4, 6-Trinitrotoluene
UDMH	Unsymmetrical Dimethylhydrazine
Zn	Zinc
Zr	Zirconium

Nomenclature

C - degrees Calceus
F - degrees Fahrenheit
g - gram
h - hour
in - inches
J - Jouls
K - degrees Kelvin

Kcal - kilocalories
KJ - kilojoules
km - kilometer
mm - milimieter
R - degrees Rankin
sec - second

↓ - added to the reaction
↑ - given off by the reaction
→ - next reaction
= - equals
≡ - defind as
≈ - approximatlye equals

Throughout this paper, examples of chemical reactions are used to clarify the descriptions of the concepts. The nomenclature for chemical formulas is the following: The chemical equation is highlighted in a figure 30, Combustion of Methane

$$CH_4 + 2 \bullet O_2 + energy_1 \downarrow \rightarrow C + 4 \bullet H + 4 \bullet O \rightarrow CO_2 + 2 \bullet H_2O + energy_2 \uparrow$$

Figure **30** Combustion of Methane

The standard chemical formula is give for the reactions. As an example when methane (NH_4) is burned with oxygen (O_2), it produces carbon dioxide (CO_2), water (H_2O), and heat (energy↑). The equation is balanced. Energy↓ is endothermic, energy is added to the reaction. Energy↑ is exothermic, energy is given off by the reaction.

References:

Answer.com http://www.answers.com/topic/explosion, 2008

Ball, P., "Life's Matrix", University of California Press, 2001

Brown, C., "Spacecraft Propulsion", AIAA Education Series, 1996

Brown, G.I., "The Big Bang, A History or Explosives", Sutton, 2000

Hyperphysics, 2008, http://hyperphysics.phy-astr.gsu.edu/hbase/chemical/bond.html

Chemical Industry, 2008, http://www.chemindustry.com/apps/chemicals

Chemical Thermodynamics, http://www.shodor.org/unchem/advanced/thermo/index.html, 2008

Conkling, J., "Chemistry of Pyrotechnics" CRC Press, 1985

Cooper, P., et al, "Introduction to the Technology of Explosives", Wiley-VCH, 1996

Contemporary Physics Education Project, CPEC, http://www.cpepweb.org/ 2008, Fundamental Particles and Interactions

Cranford, B., "Spacecraft & Satellite Dictionary 2007 Unabridged Edition", Cranford, Potomac, MD, 2007

Energetic Materials, http://www.ematerials.org/, 2008

Etymology on line, http://www.etymonline.com/, 2008

Gibbs, "Gibbs Free Energy". http://en.wikipedia.org/wiki/Gibbs_free_energy, 2008

Hill, P., et al, "Mechanics and thermodynamics of Propulsion, Addison Wesley, 1965.

Kirk-Othmer "Encyclopedia of Chemical Technology", Fourth Edition, 1991, Fifth Edition, 2004, John Wiley & Sons.

Journal of Pyrotechnics, "Pyrotechnic Chemistry", Journal of Pyrotechnics, 2004

Koelle, "Handbook of Astronautical Engineering", McGraw Hill, 1961

Krauss, L., "Atom",Back Bay Books, Boston 2002

Marshall, A., "Explosives, Their History, Manufacture, Properties and Tests", Vol. III, Gorden Press, Reprinted, 1932

MIT, http://pruffle.mit.edu/3.00/lecture_02_web/node16.html, 2008

National Research Council "Advanced Energetic Materials", National Academy Press, Washington, D.C., 2004

Neufeld, V., et al, "Webesters New World Dictionary", Webester's New World, 1988

Sienko, M., et al, "Chemicstry", McGraw-Hill, 1961

Snyder, C., "The Extraordinary Chemistry of Ordinary Things", Wiley, 2003

Sorum, C., "Fundamentsl of General Chemistyr", Prentice-Hall, 1963

Stein, J., "The Random House Dictionary of the English Language", Random House, 1971

Sutton, G., et al, "Rocket Propulsion Elements", Wiley Interscience, New York, New York,. 2001, 1956

Teipel, U., "Energetic Materials, Particle Processing and Characterization", Wiley-VCH, Germany, 2005

Thermodynamics of Chemical Equilibrium, http://www.chem1.com/acad/webtext/thermeq/index.html, 2008

Universithy of Missouri, http://undergradresearch.missouri.edu/forums-conferences/abstracts/abstract-detail.php?abstractid=843, 2008

University of Delaware, http://www.udel.edu/pchem/C443/Lectures/Lecture16.pdf, **2008**

Van Wylen G, et al, "Fundamentals of Classical Thermodynamics", John Wile and Sons, 1978

Yang, V., et al, "Solid Propellant Chemistry, Combustion and Motor Interior Ballistics", AIAA, 2000

Wikipedia, http://en.wikipedia.org/wiki/, 2008

Zemansky, M., "Heat and Thermodynamic", McGraw-Hill, 1957

Endnotes

1. Of the five states of matter, solid, liquid, gas, plasma, and Bose Einstein condensate, only the first three will be considered.

2. Discussion of dark matter and dark energy were not included, since this is an emerging area of study and little is known compared to the forces of gravity, electromagnetic, strong and weak interactions.

3. Dutch scientist Johannes Diderik van der Waals, (1837-1923) Nobel Laureate

4. Sonic, sub sonic and super sonic refer to the speed of sound in the energetic materials, not the speed of sound in air.

5. Note the change in enthalpy is positive (+) when energy is added, and negative (-) when energy is liberated.

6. The formulation varied between some references.

7. Cellulose is a general category of products (solids, liquids, gases) derived from plant material (cellulose, hemicellulose, lignin).

8. Hydrocarbon Fuels is a general category which includes liquid fuels derived from petroleum, coal, natural gas, and cellulosic feedstocks.

666b History of Energetic Materials

Boyd Hill, Hazards Analysis, ATK Launch Systems, 5000 S 8400 W, Magna, UT 84084

In honor of AIChE's 100th Anniversary, this presentation discusses the history and contribution of Energetic Materials and discusses major accidents with Energetic Materials. The history and contribution begins with the pre-energetic material Greek fire in 424 BC to the first energetic material black power in 850 AD to the rocket motor propellants of today. Story of Energetic Materials in the United States will be told through the heritage of ATK which includes 1802 DuPont founding, 1912 DuPont anti-trust founding of Hercules Powder, 1929 founding of Thiokol and 2001 ATK purchase of Thiokol. A few major historical energetic material accidents will also be discussed.

666c Incorporation of Pressure Dependent Wall Slip to the Simulation of the Continuous Processing of Energetic Suspensions

Dilhan Kalyon, Stevens Institute of Technology, Castle Point St., Hoboken, NJ 07030 and **Moinuddin Malik**, HfMI, Stevens Institute of Technology, Castle Point St., Hoboken, NJ 07030.

Recent studies have indicated that the wall slip behavior of polymeric melts and polymer based suspensions are affected by pressure with the slip velocity increasing with decreasing pressure in the die and processor [Tang and Kalyon, J. Rheology, 2008]. The ramifications of the pressure dependence of wall slip will be elucidated in conjunction with flow of energetic suspensions through dies and their processing in single and twin screw extruders.

666d Prediction of Conditions Under Which Flow Instabilities Are Onset to Give Rise to Extrudate Distortions during the Extrusion of Highly Filled Energetic Suspensions

Suzanne Prickett[1], Hansong Tang[2], Halil Gevgilili[3], and Dilhan Kalyon[3]. (1) Code 910 M, Naval Surface Warfare Center, Indian Head Division, Indian Head, MD 20640-5035, (2) Civil Engineering, City University of New York, New York, NY 10021, (3) Stevens Institute of Technology, Castle Point St., Hoboken, NJ 07030

Extrudate distortions upon exit from the die in batch or continuous processing are commonly encountered and very difficult to correct on a trial and error basis. Here a time dependent source code is introduced which allows the conditions under which flow instabilities are onset in extrusion processes. Comparisons of theory versus experiments will be presented for various well-defined suspensions.

"Twin Screw Extrusion Processing of a Nanoalumina Based Formulation in Conjunction with a Molding Powder Preparation Methodology"

R. Patel[1], P. Redner[2], D. Kapoor, M. Fair, J. Bolognini, D. Stec

(1) Energetics and Warheads Division, RDECOM-ARDEC, Bldg 3028, AMSRD-AAR-MEE-W, Picatinny Arsenal, NJ 07806-5000, (2) Energetics and Warheads Division, US Army ARDEC, Bldg 3022, AMSRD-AAR-MEE-W, Picatinny, NJ 07806,

and

H. Gevgilili, A. Ergun, S. Vural, M. Malik, D. M. Kalyon

Stevens Institute of Technology, Hoboken, NJ 07030

Abstract

Various energetic formulations that rely on the mixing together of plasticized cellulosic polymers compounded together with various solvents and myriad energetic particles exhibit gel-like characteristics. Gelation during processing retards or prevents the distributive and dispersive mixing of the ingredients and the shaping of the formulation. Incorporation of various types of nanoparticles into such formulations is expected to accelerate the formation of such gels during processing and increase their moduli. Here it is demonstrated that the effective distributive and dispersive mixing of the ingredients of an energetic simulant based on a nanoaluminized cellulose acetate butyrate formulation, which cannot be achieved with conventional technologies including roll-milling, can be achieved upon the use of the twin screw extrusion technology starting from a molding powder, in conjunction with a judicious selection of operating conditions and extrusion geometries.

Introduction

Rapid and reliable processing of energetic formulations is an important challenge that becomes more complicated upon the use of nanoparticles as part of the formulation. Under such conditions, the abilities to use scaleable and industrially relevant processing methods are valuable assets [1]. Other challenges associated with the manufacturing of energetic grains with nanometallic powders (5-200nm range) involve the control of the particle sizes of nanoparticles by preventing the formation of particle clusters, the selection of proper wetting agents and binders, the tailoring of the process to the rheological behavior of the dispersion, and the need to work at a relatively very small processing rates.

Twin screw extrusion process allows the melting and mixing together of multiple ingredients, the devolatilization of the air or other gases from the system as well as the shaping of the energetic grain [2-4]. Here, a novel twin-screw extrusion processor i.e., the smallest industrially relevant twin-screw extruder, ever designed and constructed [5, 6] is used for the compounding of a molding powder (energetic simulant) with nanoparticles of alumina in conjunction with an elastomeric binder, which is notorious in giving rise to strong gelation in the process. Two different methods were used for the preparation of a molding powder, followed by its compounding with nanoparticles of alumina in the extruder. The procedures used are discussed along with the rheological behavior of the resulting compounds. Some of the challenges associated with the rheological characterization of solvated systems are also elucidated.

Experimental

Materials

The goal of this project was to process an inert formulation that would be representative of a likely future formulation containing nanomaterials. The following inert formulation was chosen: 75 % Dechlorane Plus (inert simulant of HMX), 5% 100-200 nm nano-aluminum oxide (inert simulant of nano-aluminum), 8% cellulose acetate butyrate (CAB) and 12% acetyl triethyl citrate (ATEC). The final formulation was processed upon the compounding of an apriori prepared molding powder together with a nano-aluminum slurry using the twin screw extrusion technology.

Molding Powder Preparation.

The raw materials, ATEC, CAB, Dechlorane plus 515, and 50% ATEC-50% nano alumina paste, were received from Picatinny Arsenal. The details of all the formulations are given in Figure 1, including the final formulation, the molding powder, and Al-slurry formulation. The molding powder was prepared using a solvent concentration of 14% by weight. The mixing of the molding powder, which consisted of Dechlorane Plus, CAB (elastomer), and ATEC (plasticizer) was carried out using two separate procedures. In the first procedure, a Haake torque rheometer was utilized for 10 min. at room temperature.

The Haake Torque Rheometer is an intensive mixer with two counter-rotating and intermeshing roller type agitator blades. The rotational speed of the counter-rotating rollers was 32 rpm. The volume of the mixing head was 60 cc. Such intensive mixers are used partially full and a degree of fill of 70% was employed during the batch mixing. The roller blades were operated at 32 rpm speed and at ambient conditions. The ingredients of the molding powder were introduced as a premix to the mixing chamber. The solvent mixture (80% ethyl acetate and 20% ethanol) was added in steps up to the point at which the solvents constituted 14% by weight of the mixing powder. The molding powder was mixed for 10 minute after the last drop of solvent was added.

The torque, T, and hence the specific energy, E_s, input generated during the mixing process could be monitored during mixing as:

$$E_s = \frac{\Omega \int^{t_m} T dt}{M_t} \qquad (1)$$

where Ω is the rotational speed of the blades of the mixer, t is time, M_t is the total weight of the mixture in the mixer and t_m is the duration of the batch mixing.

In the second method of the preparation of the molding powder, the ingredients of the molding powder were mixed in a Baker Perkins batch mixer for 15 minutes. Here, the ingredients were placed in the mixer with a solvent mixture (80% ethyl acetate and 20% ethanol) of 14% by weight of the total material. The material was mixed at 60 rpm at room temperature. Afterwards, the material was dried in an oven overnight at 70° C and then broken up into powder using an attritor ball mill. The ball mill used 2000 g of steel balls (3/16") to process 175 grams of molder powder mixed with 500 g of distilled water. The mill was run for 45 minutes at 400 rpm. Afterwards, the pulverized molding powder was collected. Upon the inspection of the powder, it became evident that there were chunks of relatively harder material that could be distinguished from the rest of the powder. These chunks were further broken down upon an additional 30 minutes of milling at 600 rpm.

Nano aluminum slurry

The Nano-Aluminum slurry was prepared by first manual mixing of 10 g of nano-Al, 10 g of ATEC, and 22 g of acetone. This mixture was then sonicated under heat to drive off the solvent. After 6 hours of sonication, the mixture was placed into a drying oven at 70° C for overnight drying.

Twin screw extrusion set up:

The MPR 7.5 mm twin screw extruder [5, 6] shown in Figure 2 was used. Some of the critical dimensions of the extruder are: the bore diameters are 7.5mm and the screws are 7.38mm in diameter, the clearance between the screw and the barrel is D/128 (+/- 60 microns), the centerline distance between screws is 6.27 mm, the screw root diameter of the screw is 5.04 mm and the tolerance on the screw diameter is +/- .5"/1000". The fully-intermeshing co-rotating twin screw extruder is designed with a slit die as an integral part of the assembly (Figure 2). The dimensions of the die are: width of 10 mm, gap of 1 mm and length is 50.8 mm including the converging flow section and 36.8 mm long for the straight land section.

The assembly included two temperature control zones, one at the die, and the second for the extruder barrels. The machine is fitted with three thermocouples for controlling the metal temperature of the top and bottom barrel halves and the die assembly. The screw speed is controllable up to 200 rpm. The barrels and screws are machined from a heat-treated alloy of stainless steel and a ferritic nitro carburized to reduce the potential of adhesive wear between the screws and the barrel bores. The drive is a ¼ HP motor, 100% explosion proof.

The screw configuration was designed through mathematical modeling of the mixer/extruder and die area. The screws are monolithic (Figure 2), that is, they are constructed from a solid piece of bar stock with no welds, joints, cracks or crevices for material to migrate into. The screw design consists of three sealed mixing and vacuum zones, as shown in Figure 2. The fully-flighted conveying/plasticization section (Section I) is sealed with a reverse fully-flighted element. As shown with mathematical modeling, the screw elements need to be completely full to be able to generate the pressure necessary to force the binder of the suspension to go through the reversely configured section, which follows the forwarding elements. The mixing of the nanoparticles into the molding powder (Section II) is sealed with four pairs of 90 degree kneading discs. The screw is also expected to be completely full at the second kneading disc section and immediately preceding the set of kneading blocks and at the fully-flighted section preceding the die. The imposition of vacuum and consequent devolatilization takes place in the section between the kneading disks and the die (Section III).

Data Acquisition and Control:

A PC based and field point capable data acquisition and process control system were used (Figure 2). The system is a state-of-the-art open architecture PC based system, and includes the instruments to monitor and control zone temperatures, product temperature, process pressure and screw speed. The software allows the remote running of the equipment. The field point technology also allows the data to be accessible remotely and the extruder to be run remotely (turn on, off, set the screw speed, set the temperatures of the oil circulating, using the remote Internet connections and also wireless.)

There is one explosion proof pressure transducer and thermocouple combination at the converging section of the die inlet. There is one temperature control thermocouple in the die assembly. Two temperature control thermocouples are placed in the barrel, one in the upper, and one in the lower half. Temperature control is through the circulation of heat transfer medium at desired temperature. An intrinsically safe tachometer pickup is located at the coupling to the parallel shaft reducer to monitor screw speed.

Twin Screw Extrusion runs:

The wet molding powder was fed to the extruder using a Harvard apparatus infusion pump with an aluminum syringe with a diameter of 23 mm. The calibration of the infusion pump was performed to enable accurate control of the feeding rate of the molding powder (Figure 2). The n-Al-slurry (n-Al and solvents) was fed into the extruder using a Biochem micro-diaphragm pump. The micro-pump was calibrated using the time-dependent weight change of the vial holding the n-Al slurry at differing pulse rates, for example, at 20 pulses/min. At this pulse rate, the pump generated a 10.4 mg/pulse feed rate, which was equal to 9.650 μl per pulse.

The MPR 7.5 mm, L/D=15 co-rotating fully intermeshing twin-screw extruder was run at the targeted flow rate of 20 g/h at ambient temperature. The molding powder was fed from an infusion pump at a 16.24 g/h flow rate. The n-Al-slurry was fed from a liquid vial feeder by micro-diaphragm pump at a flow rate of 3.8 g/h. Thermal images were taken at every 20 s interval during extrusion. A twin-flighted set of complex screws containing reversely configured kneading discs and full-flighted elements with 7.5 mm diameter and screw lead length were used. At the targeted flow rate of 20 g/h, the actual flow rate was determined to be 18.8 g/h.

A dual pressure, temperature transducer was placed along the rectangular slit section at a distance of 32.8 mm away from discharge. The molding powder was introduced first into the extruder, followed by the n-Al-slurry down stream in the extruder. During the twin screw extrusion processing experiments, the rotational speed of the screws was 40 rpm. The minimum residence time in the twin screw extruder was about 15 min.

The system took an additional 10 minutes to reach steady-state upon the first exit of the extrudate from the die (25 minutes upon the initiation of the experiment). The samples were collected at 15 s intervals to validate the flow rate and the solvent concentration upon thermogravimetric analysis. The typical flow rate was determined to be 18.8 g/h and the actual solvent content was determined from thermo-gravimetric analysis as 22.2%. During the twin screw extrusion process, the torque value was around 19% of the maximum available and the pressure drop through the die was negligible. The temperature was 19°C. After the achievement of the steady-state was assured upon the flow rate and solvent concentration measurements, the extruded samples were collected into a vial for subsequent characterization of their dynamic properties.

Roll Mill runs

The conventional roll milling process has been suggested as an alternative processing method for the compounding of the molding powder with the nano-Al and was investigated to provide a benchmark for the efficacy of the mixing in the twin screw extrusion process. The process of roll milling began with 20 g of molding powder being wet with 50 g of butyl carbitol. Into this slurry, 2.2 g of the ATEC-nano-Al paste was added. This mixture was premixed manually and placed into an Exakt 3 Roll Mill set. The roll speed was set to 70 rpm. After each pass, the material was collected and put back through the mill, starting a new pass. The simulant underwent six consecutive passes through the roll mill. During the first 3 passes, the material became visibly more viscous. This process was repeated 3 times to create three separate trials. The roll milled suspension was dried in an oven at 60° C under a steady flow of air to remove the butyl carbitol.

Dynamic properties:

A number of rheological material functions can be characterized to follow the progress of the mixing and compounding processes [7-10]. Here the small-amplitude oscillatory shear was used to characterize the dynamic properties (linear viscoelastic material functions) the storage modulus, G", the loss modulus, G" and the magnitude of complex viscosity. Dynamic properties are very sensitive to the changes in the microstructure of the suspension, i.e., provide an accurate fingerprint indication of the distributive and dispersive mixing efficacies of the ingredients of the formulations and changes in their physical properties [9]. The ARES rotational rheometer was used to characterize the linear viscoelastic properties of the processed suspension samples. The experiments were conducted at room temperature without the impingement of air/N_2 stream on the fixtures holding the specimen to minimize the solvent loss during testing. Disks with diameters of 8 mm and 25 mm were used in the small-amplitude oscillatory shear experiments. The dynamic properties of the samples of the molding powder and the extruded complete simulant formulation were characterized.

Results and Discussion

The mixing that could be achieved with the conventional process of roll milling [11-12] was very poor in comparison to what could be achieved with the twin screw extrusion process. A number of quantitative methods are available for the quantitative characterization of the degree of mixedness at various scales of examination [13-19]. However, the product of the roll milling process exhibited gross segregations, an example of which is the occurrence of color inhomogeneities that were encountered with the conventional roll milling process, as shown in Figure 3. Such gross inhomogeneities were not evident in the product of the twin screw extrusion process. Furthermore, as will be indicated in the discussion of the dynamic properties, the confidence of the data emanating from the compounds obtained with the twin screw extrusion process were relatively narrow, indicative of the relatively higher degree of mixedness achieved with the twin screw extrusion process. The rest of the program was carried out using the twin screw extrusion process for the compounding of the nano-Al with the molding powder.

a. Dynamic properties of the molding powder:

A visual inspection of the molding powder prepared by the Baker Perkins Mixer indicated that the degree of mix of this material was not acceptable. Thus, an additional test involving the thermo gravimetric analysis of the second molding powder samples was carried out. The TGA results (results not shown here) indicated that there was indeed significant scatter in the degree of mixing of the ingredients when the second method of molding powder preparation was utilized. All of the results to be discussed next were obtained with the Haake intensive mixer (mini- Banbury mixer).

The first dynamic oscillatory shear experiment conducted on the molding powder prepared with the Haake intensive mixer was the strain sweep (at 5 rps frequency) to determine the typical linear viscoelastic range. The dynamic properties were sensitive to strain magnitude over a wide range of strains indicating that the molding powder suspension was highly nonlinear and susceptible to wall slip. Based on these findings the rest of the dynamic oscillatory experiments were conducted at the smallest possible strain (0.03%) below which the material properties were relatively insensitive to the strain magnitude. The solvent content of the sample was determined upon the completion of the strain sweep experiment to be 15 %.

Time sweep experiments were conducted under constant frequency and strain to follow up on the rapid solvent loss issue. The change in material properties were monitored and related to solvent concentrations, which were measured at the beginning and end of the experiments. These time sweep experiments were conducted at 0.03% strain amplitude and 5 rps frequency. The sample data are shown in Figure 4. The magnitude of complex viscosity values increased from approximately 100 kPa.s at the beginning of the test, to over 1 MPa.s within 5 minutes of testing and similarly significant increases were also observed in the time dependent values of the storage and loss moduli (Figure 3). These results clearly suggest the sensitivity of the material properties to the solvent concentration and the difficulties associated with the determination of the rheological properties of such solvent incorporated formulations [7-10]. All of the data was subsequently collected using time scans and the mean values of the solvent concentrations determined during the course of the small-amplitude oscillatory shear experiment. The results are reported as a function of the solvent concentrations.

b. Dynamic properties of the twin screw extruded simulant formulation:

Similar to the experiments carried out on the molding powder, the strain sweep experiment was first carried out on the extruded simulant sample at 5 rps frequency. As expected from the behavior of the molding powder, the dynamic properties of the extruded sample were also sensitive to strain magnitude over a wide range of strains, indicating that the simulant suspension is highly non-linear. Similar to the molding powder characterization experiments, the rest of the dynamic oscillatory experiments were conducted at the smallest possible strain (0.03%) below which the material properties were relatively insensitive to strain magnitude. The typical solvent concentration of the sample was determined to be 22 % upon the completion of the strain sweep test. The significant increases of the dynamic properties (Figure 5) were thus a result of the solvent loss mainly through the free surfaces of the suspension during small-amplitude oscillatory shear characterization. Time sweep experiments were also performed at 0.03% strain and 5 rps frequency for the twin screw extruded sample. The changes in material properties were monitored and related to solvent concentrations, which were measured at the beginning and end of the experiments.

c. The comparison of the dynamic properties of the molding powder and the n-aluminized simulant formulation:

The dynamic properties of the molding powder and the extruded simulant samples are compared as shown in Figures 6 and 7. All the dynamic properties of the extruded final formulation are lower than those of the molding powder. The lower dynamic properties of the final formulation are expected primarily due to the higher solvent (22%) and the increased amount of plasticizer (ATEC) in the final extruded formulation. These two factors compensate for the presence of the nano particles. Thus, although the final formulation has n-Al particles, which would typically increase the viscosity of the paste, its binder phase is also significantly richer in the solvent and plasticizer, which gives rise to a reverse, i.e., competing effect. In the case of the final extruded formulation, the increase in the solvent and plasticizer content predominates over the addition of the nano particles.

The narrowness of the confidence intervals observed in the dynamic properties of the compounded nanosuspensions is indicative of the relative homogeneity of the compounded formulation (Figures 6 and 7). There are a number of methods that can be utilized to determine quantitatively the degree of mixedness of energetic formulations relying on wide-angle x-ray diffraction, energy dispersive x-ray, thermo gravimetric analysis and rheological characterization of various material functions [11-19]. If the degree of mix is relatively poor, one would observe a significant variation of the dynamic properties since they are accurate fingerprints of the structure and hence the degree of mix of the ingredients. Generally, a relatively poor degree of mix would generate a relatively higher yield stress and shear viscosity and elasticity that would be immediately evident in the spread of the dependence of the dynamic properties on frequency.

Acknowledgements: This project was supported by RDECOM-ARDEC of Picatinny Arsenal, NJ for which we are grateful.

VI. References

1. S. Ozkan, H. Gevgilili, D.M. Kalyon, J. Kowalczyk, and Mark Mezger, "Twin screw extrusion of nano-alumina based simulants of energetic formulations involving gel based binders", Journal of Energetics Materials, 25, 3, 173-201 (2007).
2. D. M. Kalyon and M. Malik, "An integrated approach for numerical analysis of coupled flow and heat transfer in co-rotating twin screw extruders", International Polymer Processing, **22**, 293-302 (2007).
3. E. Demirkol and D. M. Kalyon, "Batch and continuous processing of polymer layered organoclay nanocomposites", Journal of Applied Polymer Science, 104, 1391-1398 (2007).
4. M. Erol and D. Kalyon, "Assessment of the Degree of Mixedness of Filled Polymers: Effects of Processing Histories in Batch Mixer and Co-Rotating and Counter-rotating Twin Screw Extruders", International Polymer Processing, 20, 228-237 (2005).
5. J. E. Kowalczyk, M. Malik, D. M. Kalyon, H. Gevgilili, D. F. Fair, M. Mezger and M. Fair, "Safety in Design and Manufacturing of Extruders used for the Continuous Processing of Energetic Formulations", Journal of Energetics Materials, 25, 247-271 (2007).
6. http://www.mprus.com
7. D. Kalyon, "An overview of the rheological behavior and characterization of energetic formulations: Ramifications on safety and product quality", Journal of Energetic Materials, 24, 213-245 (2006).
8. D. Kalyon, H. Gevgilili, J. Kowalczyk, S. Prickett and C. Murphy, "Use of adjustable-gap on-line and off-line slit rheometers for the characterization of the wall slip and shear viscosity behavior of energetic formulations", Journal of Energetic Materials, 24, 175-193 (2006).
9. E. Birinci, H. Gevgilili, D. Kalyon, B. Greenberg, D. Fair and A. Perich, "Rheological characterization of nitrocellulose gels", Journal of Energetic Materials, 24, 247-269 (2006).
10. D. Kalyon, "Apparent Slip and Viscoplasticity of Concentrated Suspensions", J. Rheology, 49, 3, 621-640 (2005).
11. J. M. McKelvey, "Polymer Processing", John Wiley, New York (1962).

12. Z. Tadmor, and C. G. Gogos, "Principles of Polymer Processing" John Wiley, New York (1979).
13. R. Yazici and D. M. Kalyon, "Degree of Mixing Analyses of Concentrated Suspensions by Electron Probe and X-Ray Diffraction," Rubber Chem. and Techn., 66 (4), 527-537 (1993).
14. R. Yazici and D. M. Kalyon, "Quantitative Measurement of Mixing Quality", Proceedings of 1998 Joint Army, Navy, NASA, Air Force Propellant Development and Characterization Meeting, Houston, Texas, April 20-24 (1998).
15. R Yazici, D. Kalyon and D. Fair, "Microstructure and Mixing Distribution Analysis in M30 Triple-Base Propellants", ARDEC INFORMATION RESEARCH CENTER, ARWEC CR 98015, September (1998).
16. R. Yazici and D. M. Kalyon, "Quantitative Characterization of Degree of Mixedness of LOVA Grains," Journal of Energetic Materials, 14 (1), 57-73 (1996).
17. D. Kalyon, E. Birinci, R. Yazici, B. Karuv, and S. Walsh, " Electrical Properties of Composites as Affected by the Degree of Mixedness of the Conductive Filler in the Polymer Matrix", Polym. Eng. Sci., 42, 7, 1609-1617 (2002).
18. D. M. Kalyon, D. Dalwadi, M. Erol, E. Birinci and C. Tsenoglu "Rheological Behavior of Concentrated Suspensions as affected by the Dynamics of the Mixing Process", Rheologica Acta, 45, 641-658 (2006).
19. B. Greenberg, D. Kalyon, M. Erol, M. Mezger, P. Redner, K. Lee and S. Lusk, "Structural analysis of slurry coated CL-20 (PAX 12 granules) using a novel grazing incidence x-ray diffraction method", Proceedings of Joint Army, Navy, NASA, Air Force Propellant Development and Characterization Meeting, Charlottesville, VA, March 26 (2003).

Final Formulation containing 22% Solvent
Corresponding Binder Phase

	Volume% Binder	Wt % Binder
ATEC	21.26%	24.9%
CAB	13.63%	16.6%
Ethyl acetate 80%	50.41%	46.8%
Ethanol 20%	14.69%	11.7%
Total Solvent=	65.1%	58.5%

Molding Powder Formulation containing 14% Solvent
Corresponding Binder Phase

	Wt %	Volume%
ATEC	24.05%	21%
CAB	27.49%	23%
Ethyl acetate 80%	38.77%	43%
Ethanol 20%	9.69%	13%
Total Solvent=	48.46%	55.56%

Figure 1

Experimental set-up

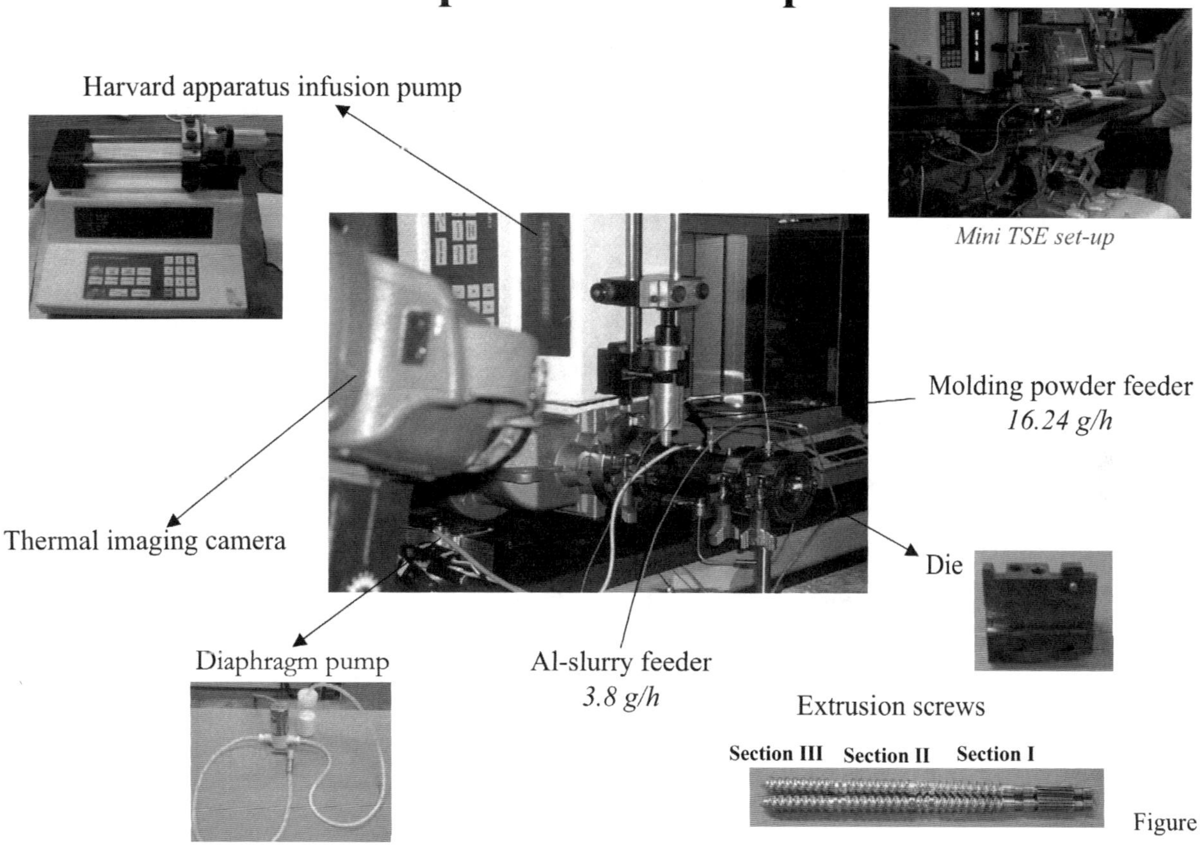

Figure 2

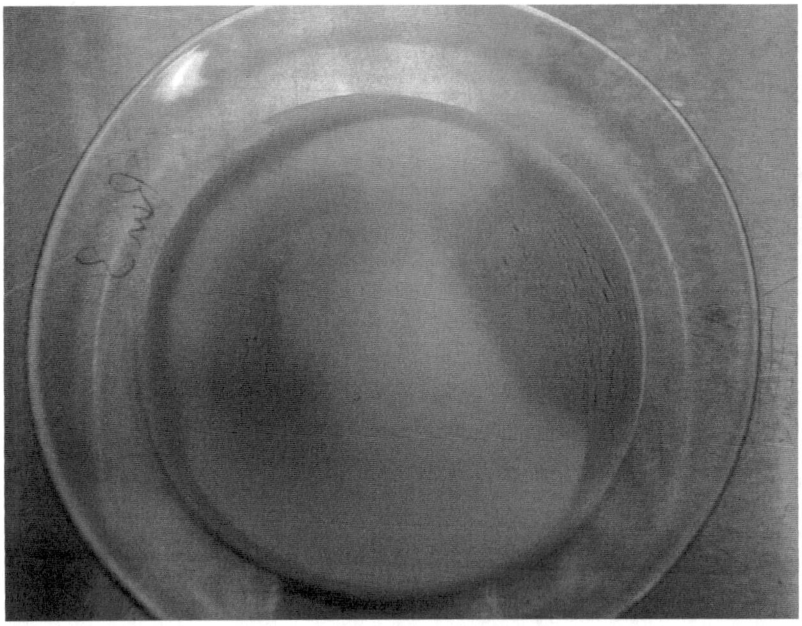

Typical color inhomogeneities of roll milled formulation

Figure 3

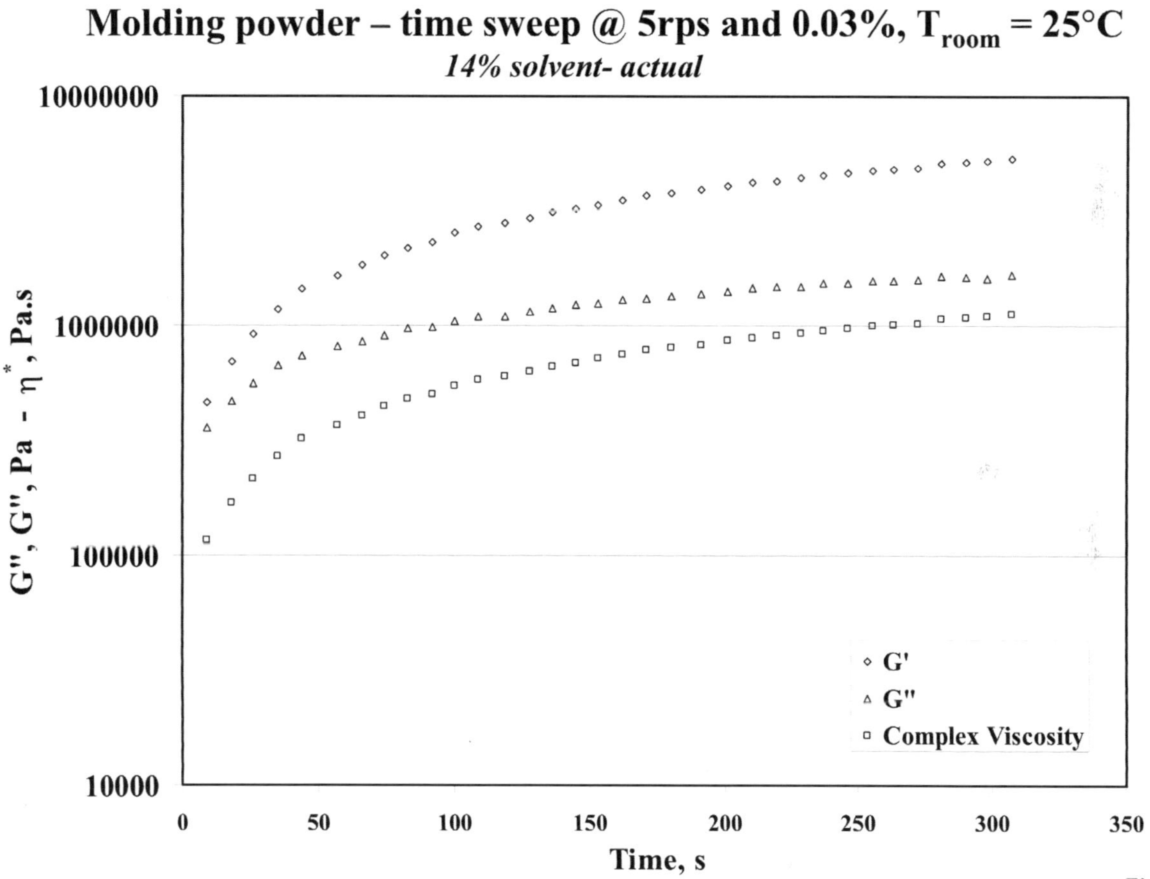
Molding powder – time sweep @ 5rps and 0.03%, T room = 25°C
14% solvent- actual
10000000
1000000
100000
10000
G", G", Pa - η*, Pa.s
0
50
100
150
200
250
300
350
Time, s
G'
G"
Complex Viscosity

Figure 4

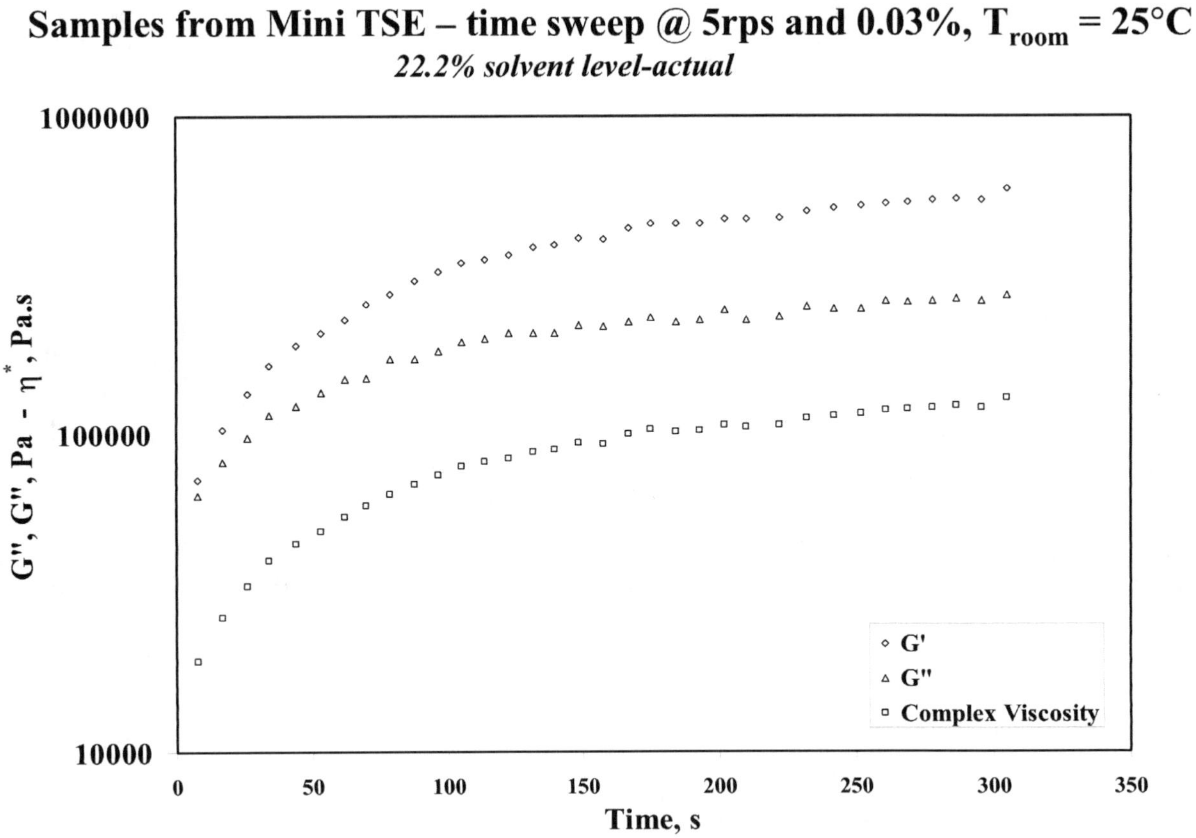

Samples from Mini TSE – time sweep @ 5rps and 0.03%, T room = 25°C
22.2% solvent level-actual
1000000
100000
10000
G", G", Pa - η*, Pa.s
0
50
100
150
200
250
300
350
Time, s
G'
G"
Complex Viscosity

Figure 5

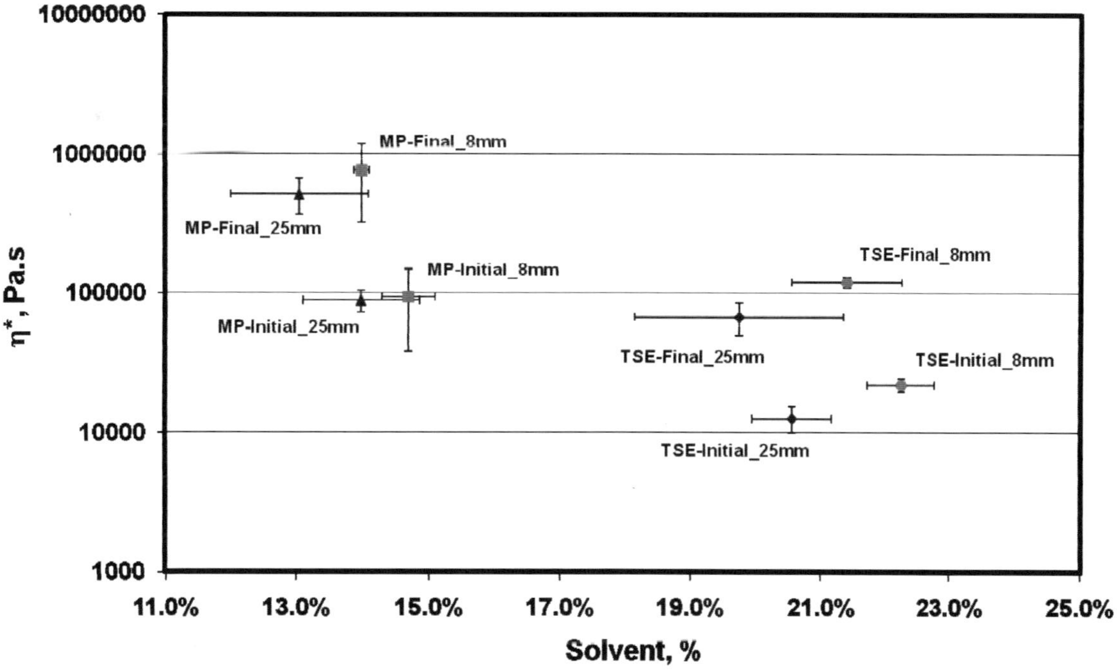

Comparison of Molding Powder and Final Extruded Formulation
Tested in the 8 and 25 mm Fixtures

Figure 6

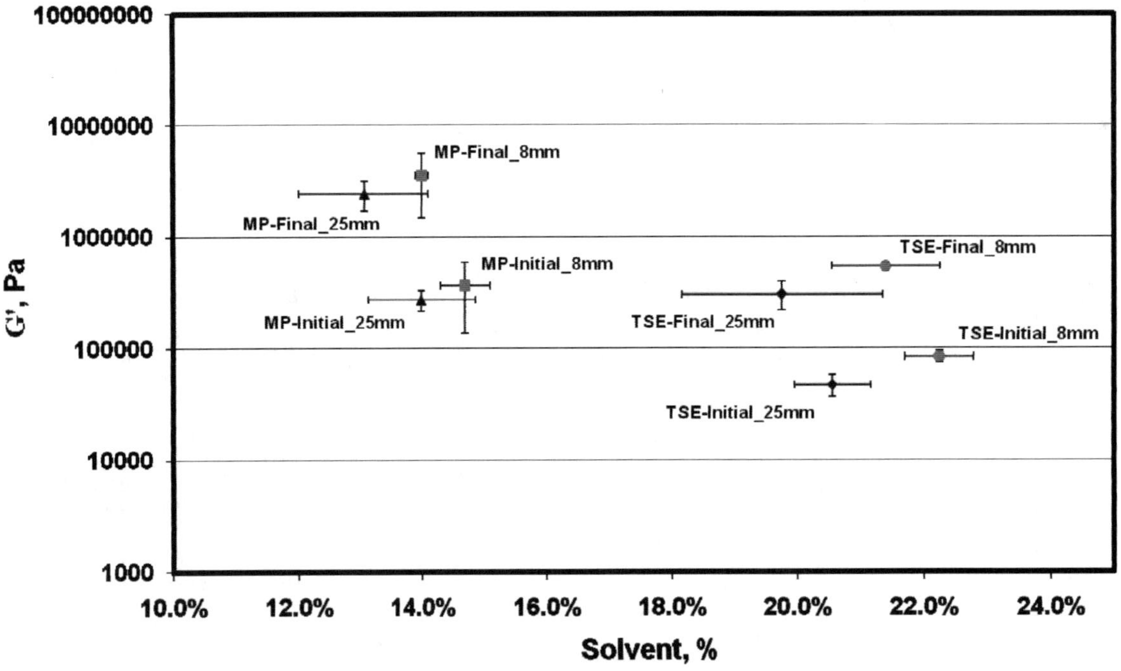

Comparison of Molding Powder and Final Extruded Formulation
Tested in the 8 and 25 mm Fixtures

Figure 7

666f Inverse Problem Solution of Squeeze, Rectangular Slit, Steady Torsional and Capillary Flows to Enable the Determination of Parameters of Wall Slip and Shear Viscosity for Energetic Suspensions

Hansong Tang[1], Halil Gevgilili[2], and Dilhan Kalyon[2]. (1) Civil Engineering, City University of New York, New York, NY 10021, (2) Stevens Institute of Technology, Castle Point St., Hoboken, NJ 07030

The accurate determination of the parameters of wall slip and the shear viscosity material function of energetic suspensions is a must for realistic simulation of their processing, especially when continuous processing methodologies are utilized. However, due to the difficulty of handling energetic materials at various relevant shear rate ranges novel methodologies are necessary for the characterization of parameters of slip and shear viscosity. Here inverse problem solution methodologies involving squeeze, rectangular slit, steady torsional and capillary flows to enable the determination of parameters of wall slip and shear viscosity for energetic suspensions will be discussed.

716a Mesoscale Modeling of Energy Release Processes: Dissipative Particle Dynamics Simulations of Shear Initiated Behavior

John K. Brennan, Weapons and Materials Research Directorate, U.S. Army Research Laboratory, Building 4600, Aberdeen Proving Ground, MD 21005, Martin Lisal, Institut for Chemical Process Fundamentals, Academy of Sciences of the Czech Republic, Rozvojova 135, Prague, CZ-165 02, Czech Republic, and Josep Bonet Avalos, Departament d'Enginyeria Química, Universitat Rovira i Virgili, Tarragona, 43007, Spain.

Mechanically-stimulated phenomena such as shock or shear of energetic materials can occur over a wide range of spatial and temporal scales. For example, localized regions of elevated thermal energy, or *hot spots*, can occur in the material due to shock heating. Typically in energetic materials, these hot spots occur near microscale defects in the material such as voids and grain boundaries but the energy transfer processes can be atomistically governed. While modeling and simulation of such phenomena can offer fundamental insight, it is computationally challenging. Computing power and resources continue to grow, but there still exists much behavior that cannot be studied using molecular simulation techniques. Moreover, field-based mesoscale modeling techniques have fidelity limitations with respect to gaining a fundamental understanding.

In this study, particle-based mesoscale modeling is used to help characterize energy release processes that occur when an energetic material is sheared. Effects on the energy transfer are characterized by considering several factors, including, the shear rate and the orientation of grain boundaries relative to the shear. We implement the energy-conserving version of the Dissipative Particle Dynamics method (DPDE) [1-4]. In contrast to the standard Dissipative Particle Dynamics method, the DPDE method simulates the hydrodynamic behavior of materials while conserving both momentum *and* energy. The DPDE method assigns an internal energy to each particle which allows particles to exchange both momentum and thermal energy. This particle energy is included as a separate equation of motion along with the equations of motion for the particle's position and momentum. The atomic degrees of freedom which are explicitly resolved into the mesoparticle internal energy account for the atomic vibrations in an averaged way. We demonstrate that the DPDE method is a viable tool for simulating energy release processes of energetic crystalline materials.

[1]Bonet Avalos, J., and A.D. Mackie, Europhysics Letters, 40, 141 (1997).

[2]Español, P., Europhysics Letters, 40, 631 (1997).

[3]Mackie, A.D., J. Bonet Avalos, and V. Navas, Phys. Chem. Chem. Phys., 1, 2039 (1999).

[4]Bonet Avalos, J., and A.D. Mackie, J. Chem. Phys., 111, 5267 (1999).

Molecular Dynamics Simulation of the Kinetic Reaction of Ni and Al Nanoparticles

Brian J. Henz[1], Takumi Hawa[2,3], and Michael Zachariah[2,3]

[1]*U.S. Army Research Laboratory, Aberdeen Proving Ground, MD*
[2]*National Institute of Standards and Technology, Gaithersburg, MD*
[3]*Department of Mechanical Engineering and the Department of Chemistry and Biochemistry, University of Maryland, College Park, MD*

Abstract

Molecular dynamics simulations are used to simulate the kinetic reaction of Ni and Al particles at the nanometer scale. The affect of particle size on reaction time and temperature for separate nanoparticles has been considered as a model system for a powder metallurgy system. Coated nanoparticles in the form of Ni-coated Al nanoparticles and Al-coated Ni nanoparticles are also analyzed as a model for nanoparticles embedded within a matrix. Simulation results show that the sintering time for separate and coated nanoparticles is nearly linearly dependent upon the number of atoms or volume of the sintering nanoparticles. We have also found that nanoparticle size and surface energy is an important factor in determining the adiabatic reaction temperature for both systems at nanoparticle sizes of less than 10nm in diameter.

1. Introduction

Nanoparticles have interesting physical properties that often vary from the bulk material. Some of these properties, including increased reactivity [1], are due to the high surface area to volume ratio of nanoparticles. With that in mind nanoparticles may provide enhanced energy release rates for explosive and propellant reactions [2].

There is considerable interest in the self-propagating high-temperature synthesis (SHS) reactions of intermetallic compounds because of the associated energy release that takes place [3] during the alloying reaction. In addition to the energetic reaction observed in these materials it is possible to produce structural materials that contain this energy release property. Once ignited, the SHS reaction releases a large amount of energy in a short period of time. The SHS process has many potential applications where heat generation is required and oxygen is not available or gaseous products are not desirable. These include alloy formation, net-shape processing, propellants, and as initiators. One of the compounds formed from the SHS reaction, and studied here, is NiAl or nickel aluminide. NiAl is an important alloy because of its desirable high temperature strength and oxidation resistance [4] and the high energy of formation [5].

The focus of this paper is to use atomistic simulation to model the reactive behaviour of Ni-Al nanoparticles in various configurations. Fortunately, there have been numerous efforts to determine accurate empirical potentials for simulating the Ni-Al material system [6]. Prior simulations using these potentials have investigated the diffusion of Ni and Al atoms [6], point-defect concentrations in NiAl [7], and plasticity [8] in addition to many other mechanical and chemical properties. These efforts have primarily focused on bulk materials rather than nanoparticle systems [9], even though there are many manufacturing processes that produce nanometer sized powders for SHS reactions [10]. For this simulation effort we have chosen a set of embedded atom method (EAM) parameters that reproduce reasonably well the properties of Ni, Al, and NiAl in the temperature range of interest.

2. Simulation Approach

In this work we employ classical molecular dynamics (MD) with an EAM interatomic potential to study the SHS reaction. The EAM is used because of its accuracy and capability to scale up to material systems with over 10^6 atoms. The MD simulations are compared with thermodynamic analyses in order to provide validation of the simulation results and assess the expected energy release.

The MD simulation was conducted using the LAMMPS software package [11]. For the Ni-Al interactions the Finnis-Sinclair EAM potential [12] from Angelo et al. [13] was used. The Finnis-Sinclair EAM potential allows for non-symmetric embedding potential terms, potentially providing improved accuracy for metallic alloys [14]. In addition to the parameters for NiAl from Angelo et al. other authors have also developed parameters for the Ni-Al system [7] that may also be described by using the Finnis-Sinclair EAM.

Three primary nanoparticle sizes were considered in this work from smallest to largest are nanoparticles with 1289, 5635, and 36523 atoms each, which correspond approximately to 3nm, 5nm, and 10nm, respectively. The range of sizes was chosen because it represents nanoparticles that may be produced in the laboratory, and which offers reasonable computational time to conduct parametric studies. For the largest system studied, the 10nm diameter nanoparticle kinetic reaction simulation requires approximately two days and 64 processor cores to complete a few nanoseconds of simulated time on 3.0 GHz Intel Woodcrest processors.

3. Thermodynamic Analysis of Separate Nanoparticles

The separate nanoparticle system is used as a model for powder metallurgy systems where Ni and Al particles are compressed into a structural component. In addition to mechanical properties, the structural component will contain stored energy for future release through a SHS reaction. A thermodynamic analysis of the SHS reaction for the separate Ni and Al nanoparticle system is used here to determine the expected trends and data points for simulation validation. In the thermodynamic analysis we are interested in determining the system parameters of the Ni-Al nanoparticle system that contribute to the combustion temperature and reaction time. Here we have assumed an adiabatic process so that energy released to the surroundings can be ignored. This is a good approximation since the reaction occurs on relatively short time scales and the nanoparticles are expected to be included in a much larger system where the overall surface to volume ratio is small, limiting convective and radiative heat loss. The validity of this assumption is explored in a later section. The SHS reaction of an equimolar Ni and Al mixture is written as

$$0.5Ni + 0.5Al \rightarrow Ni_{0.5}Al_{0.5} \tag{1}$$

In order to compute the adiabatic temperature for the synthesis reaction the enthalpy of the products and reactants must be equal.

$$H_{prod}(T_{ad}) = H_{reac}(T_0) \tag{2}$$

Assuming that the reaction begins with the reactants at 600K, above the simulated melting temperature of the Al nanoparticles, the enthalpy of the reactants is computed as,

$$H_{reac} = (0.5)(H_{Al,fusion}) + (0.5)(H_{Al,600K} + H_{Ni,600K}) = 11.85\,{}^{kJ}\!/_{mol} \tag{3a}$$

This enthalpy result includes the enthalpy of solid Ni and liquid Al [5]. The Al nanoparticle is assumed to be liquid because for small nanoparticles the melting temperature is known to be appreciably below the bulk melting temperature [15]. Additionally, for the EAM potential used here [13] the aluminum is liquid for these nanoparticle sizes at 600K. The choice of initial temperature will have a nearly linear affect on the adiabatic temperature as long as the temperature is between the melting temperature of the Al and Ni nanoparticles. This linear affect has been observed in experiments [16], and is a reasonable assumption so long as the heat capacities of the solid phases of Ni and NiAl are relatively insensitive to temperature in the ranges studied.

For the products of the SHS process the enthalpy calculation must take into account contributions from the melting of the nickel and the NiAl nanoparticle, enthalpy of formation for the NiAl alloy, and changes in surface energy. The first of these, the enthalpies of melting for Ni and NiAl is experimentally determined to be $17.2\,{}^{kJ}\!/_{mol}$ and $31.4\,{}^{kJ}\!/_{mol}$, respectively. The enthalpy of mixing for Ni and Al has garnered close scrutiny in the experimental community with a wide range of reported values. The enthalpy of formation that is used here is approximately in the middle of the reported values at about $-65\,{}^{kJ}\!/_{mol}$ [5,17,18].

The last contribution to the enthalpy of the products, results from the change in surface energy, due to the reduced total surface area of the combined nanoparticle [19]. The contribution to the change in system energy from the change in surface area is given as equation 4.

$$\Delta E_{surf} = \sigma_{NiAl} \cdot a_{NiAl} - (\sigma_{Ni} \cdot a_{Ni} + \sigma_{Al} \cdot a_{Al}) \tag{4}$$

In equation 4, a_{NiAl}, a_{Ni}, and a_{Al} are the surface area of the NiAl, Ni, and Al nanoparticles, respectively. For the 3nm, 5nm, and 10nm Al nanoparticles the reactant surface area is computed from the Gibbs surface [20] as 36.32 nm^2, 98.17 nm^2, and 343.7 nm^2, respectively. For the associated Ni nanoparticles the surface area is 27.15 nm^2, 73.59 nm^2, and 257.87 nm^2, respectively. The surface energy is approximately $1115\,{}^{mJ}\!/_{m^2}$ for Al and $2573\,{}^{mJ}\!/_{m^2}$ for Ni at 600K [21]. The surface area of the sintered NiAl nanoparticles is 50.77 nm^2, 137.18 nm^2, and 480.25 nm^2 for the 3nm, 5nm, and 10nm nanoparticle case, respectively. In experimental analysis of the free surface energy of NiAl near its melting point, the free surface energy has been reported as $1400\,{}^{mJ}\!/_{m^2}$ [22]. The approximate change in energy versus nanoparticle size is tabulated in table 1.

Table 1. change in surface energy versus nanoparticle size.

Nanoparticle Radius (nm)	ΔE_{surf} (kJ/mol)
3	-18.35
5	-11.41
10	-6.17

In table 1 the trend is for a lower surface energy contribution to the reaction as the nanoparticle size increases. Intuitively, one may expect this because the surface area to volume ratio is also decreasing with increasing particle size, and therefore has less influence on the sintering process. With the enthalpy of formation for NiAl around $-65\,{}^{kJ}\!/_{mol}$, the surface energy contribution to the change in enthalpy for coalescence of 10nm diameter nanoparticles is less than 10% of the total enthalpy change. This means that even at relatively small nanoparticle sizes, e.g. 10nm, the affect of nanoparticle size on energy release is minimal.

With the preceding discussion it is possible to take into account many of the sources of enthalpy change in the reaction products including phase and surface area changes. The enthalpy of the products is now estimated as

$$H_{prod} = H_{form,NiAl} + \Delta H_{surf} + \int_{298K}^{T_{ad}} C_{p,NiAl}(T)dT + H_{melt,Ni} \tag{5}$$

The heat capacity for solid and liquid NiAl is given in Kubaschewski et al [12]. For the 3nm case, assuming the NiAl nanoparticle melting temperature to be about 1350K, or the melting point of a similarly sized Ni nanoparticle it is possible to compute the adiabatic reaction temperature, table 2.

Table 2. Computed adiabatic temperature versus nanoparticle radius, including contact of flat surfaces or infinitely sized spheres.

Nanoparticle Radius (nm)	T_{ad} (K)
3	2115
5	1920
10	1772
∞	1599

Notice in table 2 that if no surface energy contribution is considered, i.e. infinitely large spheres, the final adiabatic temperature is computed to be 1599K. In the simulation section we will observe that these results are reasonable and accurately predict the simulated increase in temperature attributable to the contribution from the surface energy.

4. The Coalescence Processes

For Ni and Al nanoparticles the SHS reaction consists of two processes, namely coalescence and alloying. In this work we have considered the coalescence of a two nanoparticle system with an Al and a Ni nanoparticle with an atomic ratio of unity. A complete SHS reaction of this system will result in a single NiAl nanoparticle. The MD simulations used to work model adiabatic conditions with constant number of atoms and total system energy. The purpose of these simulations is to analyze the affect of nanoparticle size on sintering time, adiabatic combustion temperature, and to visualize the process. The assumed process is illustrated in figure 1.

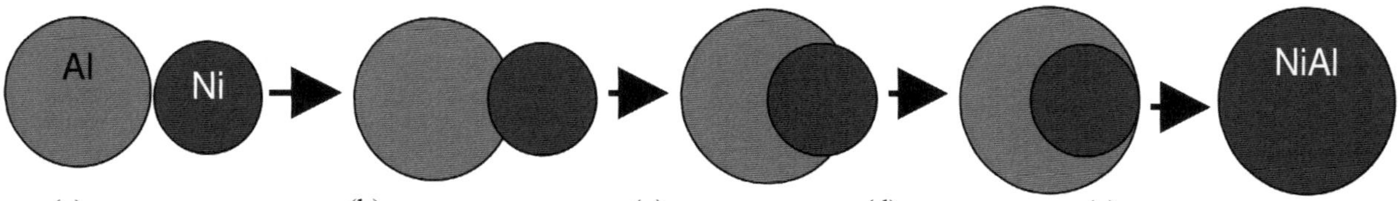

(a) (b) (c) (d) (e)

Figure 1. Illustration of sintering process showing liquid Al nanoparticle first coating the solid Ni nanoparticle and then complete alloying after the Ni nanoparticle has melted.

In figure 1 the nanoparticles are initially in contact at a point (a) and the Al nanoparticle is larger than the Ni nanoparticle because of the longer Al-Al bond length. The simulations are initialized at 600K so that the Al nanoparticle is liquid and the Ni nanoparticle is solid. In figure 1 the sintering process proceeds with the liquid Al nanoparticle initially coating the solid Ni nanoparticle while forming some Ni-Al bonds on the surface (b-d). Next, the alloying process proceeds with the Ni nanoparticle being heated above its melting point and becoming liquid so that mixing may occur (e). The formation of Ni-Al bonds beyond the interfacial surface requires diffusion of Al into the Ni nanoparticle or Ni into the liquid Al. Either of these processes is possible but since diffusion is a relatively slow process in solid materials it is expected that the Ni nanoparticle must melt before the coalescence process proceeds appreciably.

The nanoparticle sintering process is driven by two sources of energy as previously discussed. The first of these is a decrease in surface area that lowers the total surface energy of the system. This energy release mechanism is also observed in the sintering of homogeneous material systems such as silicon nanoparticles [19,23]. The second source of energy is from the reactive synthesis that occurs initially at the interface between the nanoparticles and later throughout the entire system. The energy release from the surface sintering is proportional to the surface area of the Ni nanoparticle that is coated by Al and in the whole system to the total number of Ni and Al atoms. Additionally, with the temperature increase there is a decrease in the viscosity of the liquid aluminum that will affect the predicted coalescence time.

The coalescence of nanoparticles in the liquid and solid phases has been examined extensively [24-26]. These studies are primarily concerned with the coalescence of two liquid or two solid nanoparticles. The analysis for the Ni-Al system requires considering the coalescence of a liquid Al nanoparticle and a solid Ni nanoparticle. Lewis et al [24] considered the coalescence of a liquid and a solid gold nanoparticle, this is similar to the situation here except that the material system considered was homogeneous.

In Lewis et al [24] the author is able to simulate two phases occurring simultaneously for a single material by choosing the size of each nanoparticle such that at a specific temperature the phase of the nanoparticles is different. Lewis found that coalescence proceeded in two stages, first the contact area was maximized and secondly "sphericization" took place driven by surface diffusion. The first stage is much faster than the second and is very similar to the process observed here where the Al nanoparticle maximizes the contact area and partially coats the Ni nanoparticle. In this case there is an added driving force in addition to the surface energy, specifically the energy release on forming of Ni-Al bonds as compared to the Al-Al and Ni-Ni bonds. During the second stage the atoms in the two nanoparticles diffuse and rearrange until the system becomes a single spherical nanoparticle. This stage is driven strongly by the formation of Ni-Al bonds and is expected to occur on a much shorter time scale than for two nanoparticles of the same material. The analytical model and MD simulation results shown in the following sections will explore this assumption.

5. Phenomenological Model of Nanoparticle Reactive Sintering

To gain further insight we have developed a phenomenological model for the reactive sintering of Ni and Al nanoparticles. The model includes energy release from surface energy, bond formation, and viscous dissipation through deformation. Frenkel [27] has developed a model for the coalescence of two homogeneous nanoparticles, however his model did not account for any phase change, kinetic sintering, or heterogeneous materials. Here we extended Frenkel's model to consider the coalescence of two liquid drops to consider the coalescence of a liquid and a solid drop with reactive synthesis.

By conservation of energy the rate of energy dissipated by viscous forces must be equal to the rate of energy generated by the surface and bond formation terms.

$$\frac{dE_{viscous}}{dt} = \frac{dE_{surf}}{dt} + \frac{dE_{reactive}}{dt} \tag{6a}$$

$$\tfrac{8}{3}\pi r_{Al,0}^{3}\eta\left(\frac{dD}{dt}\right)^{2} = \frac{d}{dt}\left[\sigma_{Ni}^{s}\cdot S_{Ni,\exp osed} + \sigma_{Al}^{l}\cdot S_{Al,\exp osed}\right] + \frac{d}{dt}\left[2\beta\pi v_{Ni} r_{Ni}\right] \tag{6b}$$

After writing equation 6b in terms of dθ/dt and simplifying the right and left hand sides we find that equation 6b is only linearly dependent on dθ/dt. Even with this simplification, equation 6, is most easily solved numerically using an iterative solver. In order to solve equation 6b we need some physical properties of Al, Ni, and NiAl. The dynamic viscosity of bulk molten Al at the melting temperature is about $\eta = 1.3\cdot 10^{-3}\,Pa\cdot s$ [28]. Based upon a comparison of the configurational energy in MD simulations of separate nanoparticles and Al-coated Ni nanoparticles the energy release per unit area, $\beta_{density}$, is estimated to be $20.7\,{}^{eV}\!/_{nm^2}$. This number is computed by subtracting the system energy of an Al coated Ni nanoparticle system from the energy of a system with separate nanoparticles and dividing by the interfacial surface area. This method results in the net change in energy during coating of the Ni surface with Al since some Al-Al bonds are lost during the coating process while some Ni-Al bonds are formed at the interface. By numerically solving equation 6b we are able to compute the contact angle, θ, as a function of time and relate this to total exposed surface area of the coalescing nanoparticles. This result is presented in figure 2 along with a comparison to the MD simulation results.

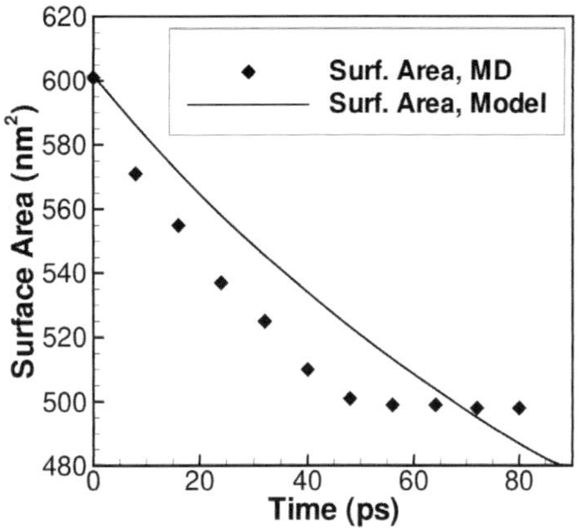

Figure 2. Total system surface area versus time from mathematical model and MD simulations for the sintering of 10nm diameter nanoparticles, where the final surface area of the NiAl nanoparticle is approximately 480 nm^2.

Although qualitatively the results in figure 2 show similar trends the absolute rate of coalescence is slightly under predicted by the model. This difference can be attributed to the obvious simplicity of the model and more specifically to the difficulty in obtaining accurate material parameters. For instance, it is difficult to compare the viscosity of a nanoparticle to the bulk material [29] and since the coalescence time is linearly dependent upon the viscosity a change in viscosity is directly proportional to a change in modeled coalescence time. The deviation of the model time from the simulation results at about 50 ps is due to the switch from stage 1 to stage 2 in the kinetic coalescence process as described by Lewis et al [24]. As described by Lewis, during stage 2, surface diffusion is the predominant factor in continued coalescence and is a much slower process than contact area maximization. The actual simulation results, as compared with the illustration in figure 1, of the observed coalescence process are given in figure 3.

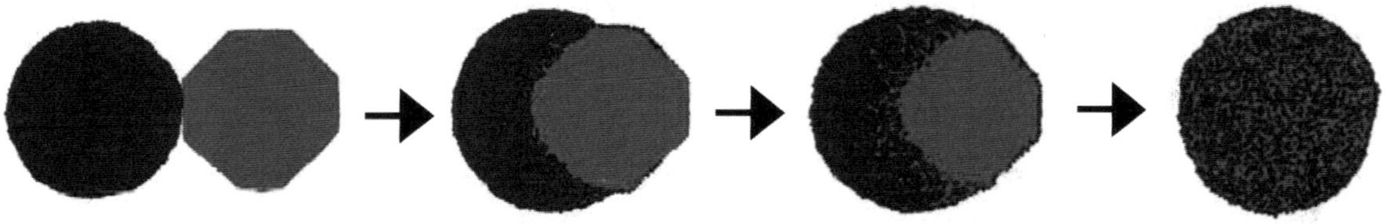

Initial	Stage 1 Fast (50 ps)	Stage 2 Slow (450 ps)	Completed Slow (400+ ps)

Figure 3. Cross sectional view from MD simulations of Ni/Al nanoparticle sintering process showing the start of the second stage of coalescence where diffusion is the driving force as opposed to contact area maximization. Aluminum atoms are blue and nickel atoms are red.

In figure 3 each of the steps in the coalescence process are shown with plots from an MD simulation of the coalescence of 10nm diameter Al and Ni nanoparticles. The correlation of the sintering stages to the reaction temperature and time is illustrated in figure 4 for the sintering of separate 10nm diameter nanoparticles. In the initial step the liquid Al nanoparticle, blue atoms in figure 3, has melted and is spherical in shape. The solid Ni nanoparticle, red atoms, has large faceted sides and is a single crystal, a typical configuration for a crystalline nanoparticle at low temperatures. During stage 1 the Al nanoparticle is attracted to the Ni surface because of the dual driving forces of surface energy minimization and Ni-Al bond formation. This period lasts about 50 ps in this simulation as noted in figures 3 and 4. Between stages 1 and 2 the driving forces associated with the surface energy are counteracted by a resistance to flow in the Al nanoparticle, causing the coalescence process to slow down dramatically. During stage 2, lasting about 450 ps, the surface area is not changing so that energy release from the surface energy terms has ceased to contribute to the change in system potential energy. The subsequent energy release is entirely attributable to the formation of Ni-Al bonds. This stage lasts a much longer time than the initial nanoparticle coalescence stage and is governed by the material diffusion coefficients. Initially at stage 2 the Ni nanoparticle is still solid and the formation of Ni-Al bonds is only possible by Al diffusing into the Ni core or Ni on the surface of the core melting and diffusing away from the interface. This process proceeds until the Ni core has reached its melting point and mixing of the remaining Ni and Al atoms occurs more rapidly, driven by the enthalpy of formation of NiAl. From stage 2 until complete alloying has occurred, taking approximately 400 ps, diffusion and mixing of Ni and Al atoms is the primary driving force.

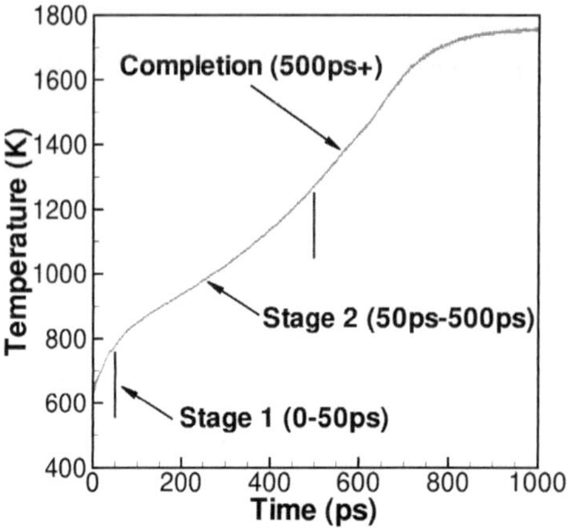

Figure 4. Time versus temperature plot for sintering of separate 10nm diameter Al and Ni nanoparticles. The various stages if the coalescence processes are denoted on the curve, including the final completion stage that occurs after the Ni nanoparticle has melted.

6. MD Simulation Results of Separate Nanoparticle Reactivity

We have previously predicted the adiabatic temperature and sintering time for the reactive sintering process of separate equimolar nanoparticles of Al and Ni. In figure 5, the MD simulation results for the equimolar nanoparticles are plotted along with the computed adiabatic temperature for each considered particle size.

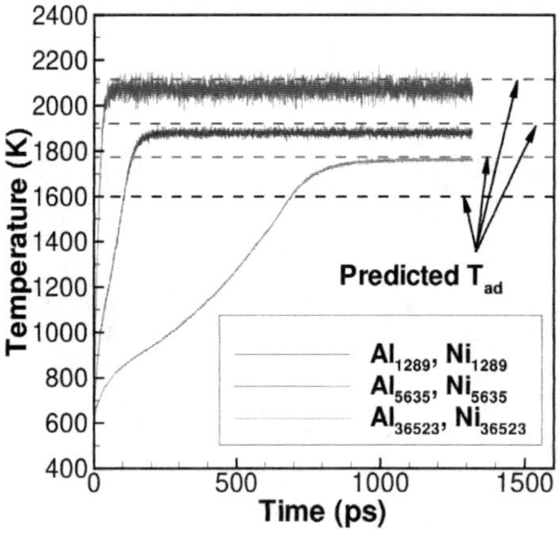

Figure 5. Temperature versus time in the sintering of nanoparticles with an Ni:Al ratio of 1:1. The subscripts in the legend refer to the number of atoms of each material and correspond to nanoparticles of diameter approximately 3nm, 5nm, and 10nm. The color coded dashed lines are the computed adiabatic temperature from the thermodynamic analysis. The black dashed line is the predicted temperature for coalescence of bulk Al and Ni.

From figure 5 it is apparent that the predicted adiabatic temperature is in close agreement with the simulated temperature. Variability of the computed temperature arises from the wide range of experimental results for the surface tension for liquid Al and solid Ni, the reported enthalpy of formation for NiAl, and the assumed melting temperature for the Ni and NiAl materials at this scale. Each of these experimental data points are used in the thermodynamic analysis and contribute to the small inaccuracies in the predicted temperature.

The characteristic time for reactive synthesis that we use here is defined by Zhao et al [30] as t when

$$T(t) = T_0 + 0.8(T_1 - T_0) \tag{7}$$

where T_0 is the initial temperature, T_1 is the maximum size dependent temperature reached, and $T(t)$ is the transient temperature.

7. Reactive Sintering of Core-Shell Nanoparticles

7.1. Aluminum Coated Nickel

In this section we will discuss the sintering process for an Al-coated Ni nanoparticle followed by a discussion of a Ni-coated Al nanoparticle. Both of these systems can be used as a model for highly compacted Ni and Al nanoparticles or one material serving as a matrix for nanoaprticles of the other. In the first model system we assume that a Ni nanoparticle has been coated with Al and equilibrated without the Ni melting, or any further reaction occurring. Results for the reaction time and temperature will be presented and a comparison with the separate nanoparticle case will be given. Here again we have considered three system sizes with 1289, 5635, and 36523 atoms each of Al and Ni.

An initial estimate is that the coalescence process for the fully coated nanoparticle system will be a truncated version of the separate nanoparticle case. In the coated nanoparticle system we do not have the first stage of coalescence occurring and only observe the second stage, namely diffusion of Ni and Al atoms to form Ni-Al bonds. The sintering temperature versus time plot is given in figure 6 and shows an interesting result. Whereas the maximum temperature reached increases with decreasing nanoparticle size for coalescence of separate nanoparticles, the opposite is true here, the temperature decreases with decreasing nanoparticle size.

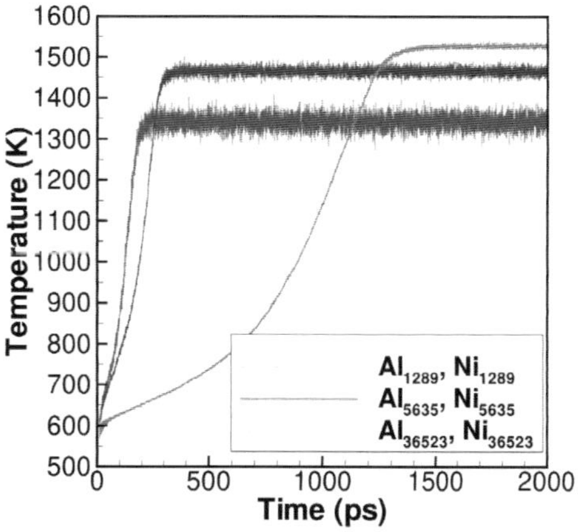

Figure 6. Temperature versus time in the sintering of Al-coated Ni nanoparticles with an Ni:Al ratio of 1:1.

In figure 6 the observed decrease in adiabatic temperature is due to the fact that the ratio of atoms near the interfacial region to the atoms in the bulk nanoparticle decreases as the nanoparticle size increases. Atoms in the interfacial region have already formed Ni-Al bonds and are therefore already at a lower configurational energy than if they were contained in a homogeneous nanoparticle of either pure Al or Ni. If we extend the adiabatic temperature relationship to infinitely large particles we would approach the result obtained from the analysis of separate nanoparticle as they increase in size. The thermodynamic analysis is similar except that the surface energy term is zero and the enthalpy of formation is lowered by a factor proportional to the ratio of surface area to volume. The enthalpy of the products, equation 6, modified for coated nanoparticles becomes

$$H_{prod} = \left(1 - \frac{t \cdot A_{surface}}{V}\right) H_{form,NiAl} + \int_{298K}^{T_{ad}} C_{p,NiAl}(T)dT + H_{melt,Ni} \tag{8}$$

where t is a computed thickness value for the interfacial layer, $A_{surface}$ is the area of the interfacial region, and V is the volume of the Ni core. In order to determine the correct empirical thickness value, t, for equation 8 we have used the adiabatic temperature computed in the MD simulation results for the Al-coated Ni nanoparticle. These results indicate that an interface thickness of 0.07nm is able to accurately predict the adiabatic temperature observed in the MD simulations, figure 7.

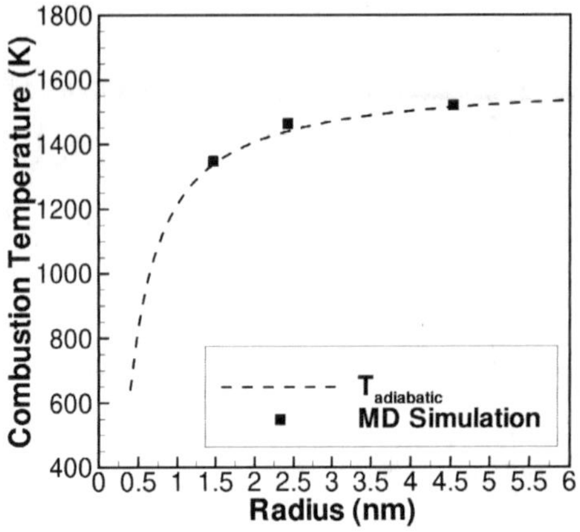

Figure 7. Comparison of thermodynamically determined adiabatic temperature for Al-coated Ni nanoparticle and results from MD simulation.

In figure 7 it is apparent that the adiabatic combustion temperature is highly size dependent for nanoparticles of less than 10nm in diameter. For very small nanoparticles, less than 1nm diameter, there is little predicted change in temperature from the initial temperature of 600K since most of the potential Ni-Al bonds have already been formed.

From the sintering of separate nanoparticles it is expected that the reaction time will be linearly related to the radius of the nanoparticle to a power of about 2.5. In figure 8 this appears to be the case for this range of nanoparticle sizes. A slight deviation from the separate nanoparticle result is probably related to the fact that the coalescence process, stage 1, is not included in this model system and diffusion takes longer to initialize the kinetic reaction process.

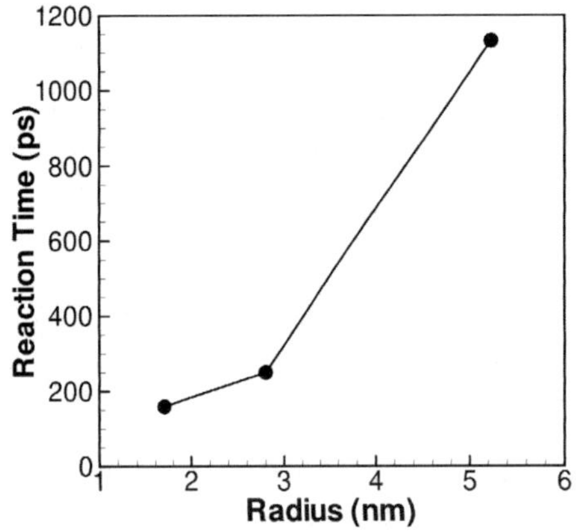

Figure 8. Reaction time versus number of Al atoms in the Al-coated Ni nanoparticle system.

The results for the Al-coated Ni nanoparticle indicate the trends that one might expect from a material system that included an Al matrix with embedded Ni nanoparticles. From the results in figures 7 and 8 there are two competing reaction results, namely reaction time and maximum temperature. In figure 8 we see that as the Ni nanoparticle size decreases the reaction time decreases, causing the energy release rate to increase. A second observation that can be made from figure 7 is that the reaction temperature decreases with decreasing Ni nanoparticle size, potentially minimizing the effect of the rapid energy release.

Looking more closely at the reaction time versus number of atoms for the separate nanoparticle and Al-coated Ni nanoparticle cases we observe a similar relationship of reaction time to nanoparticle size as that found in separate nanoparticles. In both cases the reaction time appears to have a power law relationship with radius, with an exponent of 2.5. The accelerated temperature increase in figure 6 after about 900ps for the Al_{36523} curve is the convergence rate discussed previously.

Conclusions

We have analyzed two model systems for the energetic reaction of Ni and Al. In the first case we considered the coalescing and sintering of separate nanoparticles and found that the energy release from the change in surface area is only significant at small, less that 10 nm diameter, nanoparticles. These separated nanoparticle reaction simulations and thermodynamic analyses show that the reaction time will decrease and the adiabatic reaction temperature will increase with decreasing nanoparticle sizes. This may be important for applications where high energy release rates are desired. The simulation data closely match a classical thermodynamic analysis.

In the second part of this work we considered the sintering of Al-coated Ni nanoparticles and Ni-coated Al nanoparticles as a model material system for nanoparticles embedded in a matrix of the other metal. This work revealed that the reaction time is again inversely related to nanoparticle size but the adiabatic temperature decreases with decreasing nanoparticle size. Mechanically the Al-coated Ni nanoparticle system is a model system for a light weight Al matrix with embedded Ni nanoparticles, a system with relatively high strength compared to a loosely bonded powder of Al and Ni nanoparticles. This Al matrix system could be used in systems where mechanical strength is important in addition to energy release from kinetic sintering of the Ni and Al atoms. In the Ni-coated Al nanoparticle system we investigated possible rupture and fragmentation of the Ni shell but were unable to observe any fragmentation.

Acknowledgements

The authors would like to acknowledge the support received by the U.S. Army Major Shared Resource Center (MSRC) at the Aberdeen Proving Ground, MD. Additional support was provided by the National Institute for Standards Technology (NIST) and the U.S. Army Research Office (ARO).

References

1. X. Phung, J. Groza, E. A. Stach, L. N. Williams, and S. B. Ritchey. "Surface characterization of metal nanoparticles", *Materials Science and Engineering A*, **359**, 261–268, 2003.
2. A. Rai, D. Lee, K. Park, and M. R. Zachariah. "Importance of Phase Change of Aluminum in Oxidation of Aluminum Nanoparticles", *Journal of Physical Chemistry B*, **108**, 14793–14795, 2004.
3. H. P. Li. "Influence of ignition parameters on micropyretic synthesis of NiAl compound", *Materials Science & Engineering A*, **404**, 146–152, 2005.
4. S. Gennari, U. A. Tamburini, F. Maglia, G. Spinolo, and Z. A. Munir. "A new approach to the modeling of SHS reactions: Combustion synthesis of transition metal aluminides", *Acta Materialia*, **54**, 2343–2351, (2006).
5. P. Nash, and O. Kleppa. "Composition dependence of the enthalpies of formation of NiAl", *Journal of Alloys and Compounds*, **321**, 228–231, 2001.
6. R. Hu, and P. Nash. "The enthalpy of formation of NiAl", *Journal of Materials Science*, **40**, 1067–1069, 2005.
7. J. C. Trenkle, T. P. Weihs, and T. C. Hufnagel. "Fracture toughness of bulk metallic glass welds made using nanostructured reactive multilayer foils", *Scripta Materialia*, **58**, 315–318, 2008.
8. S. Dong, P. Hou, H. Cheng, H. Yang, and G. Zou. "Fabrication of intermetallic NiAl by self-propogating high-temperature sysnthesis reaction using aluminum nanopowder under high pressure", *Journal of Physics: Condensed Matter*, **14**, 11023–11030, 2002.
9. L. J. Lewis, P. Jensen, and J.-L. Barrat. "Melting, freezing, and coalescence of gold nanoparticles", *Physical Review B*, **56**(4), 2248–2257, 1997.
10. T. Hawa, and M. R. Zachariah. "Coalescence kinetics of bare and hydrogen-coated silicon nanoparticles: A molecular dynamics study", *Physical Review B*, **71**, 165434, 2005.
11. T. Hawa, and M. R. Zachariah. "Coalescence kinetics of unequal sized nanoparticles", *Aerosol Science*, **37**, 1–15, 2006.
12. M. R. Zachariah, and M. J. Carrier. "Molecular Dynamics Computation of Gas-Phase Nanoparticle Sintering: A Comparison with Phenomenological Models", *Journal of Aerosol Science*, **30**(9), 1139–1151, 1999.
13. S. H. Ehrman. "Effect of Particle Size on Rate of Coalescence of Silica Nanoparticles", *Journal of Colloid and Interface Science*, **213**, 258–261, 1999.
14. S. Arcidiacono, N. R. Bieri, D. Poulikakos, and C. P. Grigoropoulos. "On the coalescence of gold nanoparticles", *International Journal of Multiphase Flow*, **30**, 979–994, 2004.
15. S. Yu, C.-Y. Wang, T. Yu, and J. Cai. "Self-diffusion in the intermetallic compounds NiAl and Ni_3Al: An embedded atom method study", *Physica B*, **396**, 138–144, 2007.
16. Y. Mishin, M. J. Mehl, and D. A. Papaconstantopoulos. "Embedded-atom potential for B2-NiAl", *Physical Review B*, **65**, 224114, (2002).
17. J. Mei, B. R. Cooper, and S. P. Lim. "Many-body atomistic model potential for intermetallic compounds and alloys and its application to NiAl", *Physical Review B*, **54**(1), 178–183, 1996.
18. F. Delogu. "Numerical simulation of the thermal response of Al core/Ni shell nanometer-sized particles", *Nanotechnology*, **18**, 505702, 2007.
19. H. X. Zhu, and R. Abbaschian. "Reactive processing of nickel-aluminide intermetallic compounds", *Journal of Materials Science*, **38**, 3861–3870, 2003.
20. S. J. Plimpton. "Fast Parallel Algorithms for Short-Range Molecular Dynamics", *Journal of Computational Physics*, **117**, 1–19, 1995.

21. M. W. Finnis and J. E. Sinclair. "A simple empirical N-body potential for transition metals", *Philosophical Magazine A*, **50**, 45–55 (1984).
22. J. E. Angelo, N. R. Moody, and M. I. Baskes. "Trapping of hydrogen to lattice defects in nickel", *Modelling Simul. Mater. Sci. Eng.*, **3**, 289–307, (1995).
23. G. J. Ackland, and V. Vitek. "Many-body potentials and atomic-scale relaxations in noble-metal alloys", *Physical Review B*, **41**(15), 10324–10333, 1990.
24. P. Pawlow. *Z Phys. Chem.*, **65**, 545, 1909.
25. P. Zhu, J. C. M. Li, and C. T. Liu. "Adiabatic temperature of combustion synthesis of Al-Ni systems", *Materials Science and Engineering A*, **357**, 248–257, 2003.
26. R. Arroyave, D. Shin, and Z.-K. Liu. "Ab initio thermodynamic properties of stoichiometric phases in the Ni–Al system", *Acta Materialia*, **53**, 1809–1819, 2005.
27. F. Z. Chrifi-Alaoui, M. Nassik, K. Mahdouk, and J. C. Gachon. "Enthalpies of formation of the Ni–Al intermetallic compounds", *Journal of Alloys and Compounds*, **364**, 121–126, 2004.
28. M. R. Zachariah, M. J. Carrier, and E. Blasiten-Barojas, *J. Phys. Chem.* **100**, 14856 (1996).
29. M. P. Allen and D. J. Tildesley. Computer Simulation of Liquids. New York: Oxford, 1996.
30. V. I. Nizhenko. "Free Surface Energy as a Criterion for the Sequence of Intermetallic Layer Formation in Reaction Couples", *Powder Metallurgy and Metal Ceramics*, **43**(5–6), 273–279 (2004).
31. A. Y. Lozovoi, A. Alavi, and M. W. Finnis. "Surface Stoichiometry and the Initial Oxidation of NiAl(110)", *Physical Review Letters*, **85**(3), 610–613, 2000.
32. D. Mukherjee, C. G. Sonwane, and M. R. Zachariah. "Kinetic Monte-Carlo Simulation of the Effect of Coalescence Energy Release on the Size and Shape Evolution of NanoParticles Grown as an Aerosol", *Journal of Chemical Physics*, **119**, 3391, 2003.
33. J. Frenkel. *J. Phys.*, **9**, 385, (1945).
34. M. A. Assael, K. Kakosimos, R. M. Banish, J. Brillo, I. Egry, R. Brooks, P. N. Quested, K. C. Mills, A. Nagashima, Y. Sato, and W. A. Wakeham. "Reference Data for the Density and Viscosity of Liquid Aluminum and Liquid Iron", *J. Phys. Chem. Ref. Data*, **35**(1), 285–300, (2006).
35. S. Zhao, T. C. Germann, and A. Strachan. "Atomistic simulations of shock-induced alloying reactions in Ni/Al nanolaminates", *Journal of Chemical Physics*, **125**, 164707, (2006).

EXPERIMENTAL STUDY OF IGNITION OF MAGNESIUM POWDER BY ELECTRO-STATIC DISCHARGE

Authors:

Ervin Beloni and Edward L. Dreizin

ABSTRACT

Ignition sensitivity of powders to electro-static discharge (ESD) stimulation is commonly tested to assess the safety of powder handling. However, the mechanisms of powder ignition by electric spark remain unclear. This paper represents an experimental study in which the spark parameters are quantified and the ESD ignition of a metal powder is investigated. A spherical powder of Mg, for which thermal ignition kinetics was described in the literature, was used in experiments. The experimental setup was built based on a commercially available apparatus for ESD ignition sensitivity testing. Additional diagnostics enabled measurements of electrical current, voltage, and spark and ignited powder emission in real time. The spark duration was of the order of a few µs. The spark current and voltage were always observed to have significant AC components. The electrical impedance of the spark discharge was determined experimentally using the recorded current traces and assuming that the spark could be represented as a series LRC circuit. The optical emission was filtered to separate the signals produced by the spark plasma and by the heated and igniting powder. The radiation signal produced by the igniting powder was always delayed after the spark. The delay time decreased from about 3.5 to 0.5 ms as the spark energy increased from 10 to 60 mJ; the delay remained nearly constant when the spark energy continued to increase to over 100 mJ. For experiments where the powder volume decreased or where binder was introduced delay times were reduced, particularly above 60 mJ the delay time went down to 0 ms.

INTRODUCTION

Electro-static spark is a common ignition stimulus and a variety of tests, standards, and evaluation methodologies are developed to investigate ignition behavior of different substances struck by a spark. Electro-static discharge (ESD) ignition for flammable gases was studied in great detail and a number of important trends were established. Effects of electrode materials, electrode shape, discharge duration, and other similar parameters have been quantified and reported. Reviews by Mellor et al., [1-2] published in 1990's described these trends and suggested that a similar systematic study would be necessary to establish a scientifically sound test of ESD sensitivity of powders. However, such a study has never been undertaken. Note that ignition of various aerosolized powders has been studied by many investigators, e.g., [3-5]. However, the mechanisms of flame propagation in aerosols are different from those of non-aerosolized powders routinely tested in industry. In fact, an experimental test of ESD ignition sensitivity for non-aerosolized powders is one of the most commonly used to evaluate a variety of powdered materials, including agricultural [6], food [7], textile [8], pharmaceutical [9], metallic [10-12], and of course, energetic components [13-14]. Typically, a spark is generated by a high-voltage capacitor discharging over a gap between a sharpened electrode and a powder bed. The minimum capacitor energy at which the powder ignites is specified as the minimum ignition energy (MIE), a parameter defining the sensitivity of a powder to ESD ignition stimulation. It has been suggested in ref. [14] that the spark represents primarily a thermal source capable of raising the temperature of a flammable powder above the point at which thermal runaway occurs. However, it remains unclear how the spark heats the powder particles, what portion of the spark energy is being transferred to the powder and by which mechanism. For example, the spark's plasma can heat the powder surface directly while the current of the spark discharge can result in Joule heating distributed in the powder volume, along the current path. It remains unclear whether the polarity of the spark discharge is a factor affecting the ignition energy. The transport properties of powders including their thermal and electrical conductivities are initially governed by the respective contact resistances between the particles. While a powder is heated, gasified, and/or being melted, its transport properties are expected to change dramatically [15-16] affecting its interaction with the spark. The spark itself does not behave as a steady heat source and the energy distribution is expected to change as a function of both time and location. However, it is unclear whether the ignition governed by a runaway chemical reaction occurs within the same time frame as the relatively short-lived spark discharge, and thus whether the temporal changes in the spark energy distribution are significant in affecting the ignition sensitivity of various materials.

Understanding the mechanisms of the ESD ignition of powders becomes increasingly important with dramatic acceleration in recent research aimed to develop new powdered materials, including nanomaterials for a variety of applications. Current protocols used for ESD ignition sensitivity testing, e.g., described in various standards, e.g., [17-18] are not based on such understanding and the results are often inconsistent between different testers and inconclusive in nature. Furthermore, the current protocols are not suitable for a number of newly developed powders, which are not available in the quantities required for the standardized testing. At the same time, often the development of the new material manufacturing processes necessary to produce sufficient material quantities may hinge upon its positive safety evaluation, of which a conclusive ESD sensitivity evaluation is a necessary part.

The goal of this work is to initiate a systematic study of the mechanisms of ESD powder ignition. The overall study is viewed as a combination of systematic experimental and modeling efforts ultimately resulting in quantitative mechanisms describing ignition of various powder-like materials by electric sparks. This paper represents the first part of this study and reports an experimental investigation of ESD ignition of a spherical metal powder. The paper describes measurements of optical emission produced by the spark and ignited powder and the spark current and voltage measurements. Magnesium was selected for this study because there are relatively reliable descriptions for its thermal ignition available in the literature [19-20]. Thus, it is expected that the results of the measurements reported in this paper would be useful for quantitative modeling of the ESD stimulated powder ignition in which the thermal runaway reaction in the powder is described with reasonable accuracy.

EXPERIMENTAL

Powder

The powder used in the ignition experiments was 1-11 μm spherical Mg by Hart Metals, Inc. The particle size distribution was measured using a Beckman-Coulter LS230 Enhanced Particle Analyzer. Based on the volumetric size distribution, the mean particle size was 10.3 μm. This same powder was earlier used in experiments on thermal ignition of Mg using an electrically heated filament [20] so that direct comparisons between some of the results presented in ref. [20] and in this paper are possible.

Electrostatic discharge apparatus

A schematic diagram of the experimental setup is shown in Fig. 1. A Model 931 Firing Test System (FTS) by Electro-Tech Systems, Inc. was used to generate spark discharges. FTS includes a capacitor bank with capacitance varied in the range of 100 – 10000 pF. The capacitors can be discharged through the spark gap directly or through an additional resistor varied from 500 to 5000 Ω. Additional resistors were not used in this project. The capacitors can be charged to a voltage varied from 100 V to 26 kV.

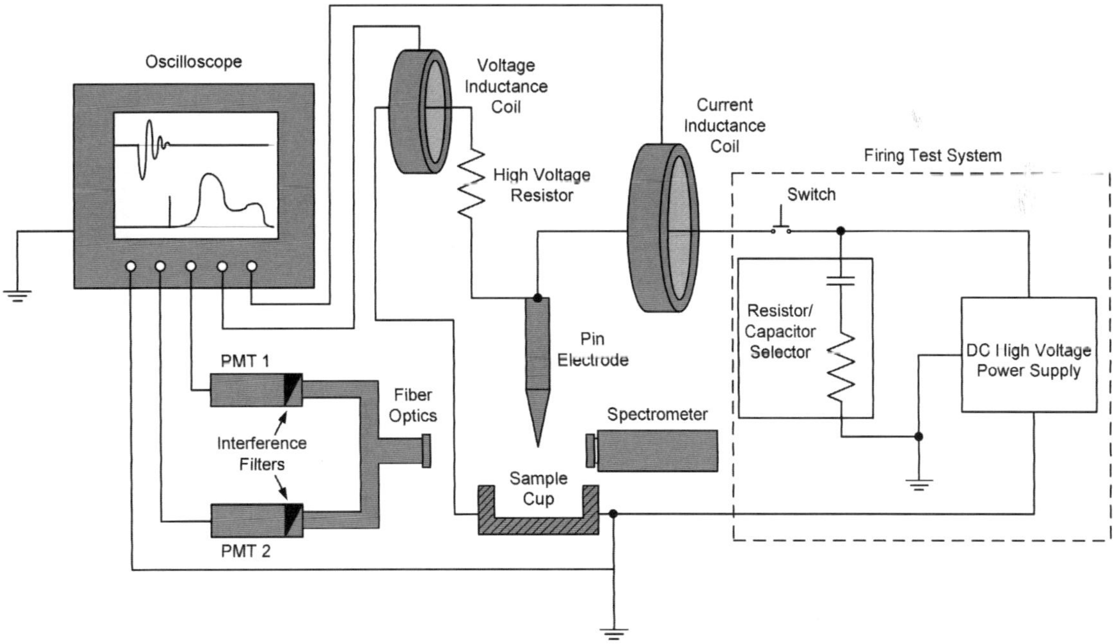

Figure 1. Schematic diagram of the experimental setup.

The polarity of the electric output can be changed so that the high-voltage electrode can be either positive or negative. For a standard test [17], the powder is placed into a stainless steel cup (15 mm diameter, 3 mm deep) affixed to a grounded base. The high voltage pin-electrode (a stainless steel needle) is placed 0.2 mm above the surface of the powder being tested. If sufficient energy is delivered to the powder, ignition occurs. In a standard test [17], ignition is determined to have occurred when the ignited powder produces a bright white flame.

Sample holders

Three sample holders were used in this study. A standard sample holder is a stainless steel cup described above. In addition, two custom-made sample holders were made and used to assess the effect of the powder amount on its ignition. For both custom-made sample holders, it was desired to avoid the possibility for the spark to strike the sample holder surface rather than the powder directly. Therefore, the only conductive surfaces of each of the two customized sample holders were their bottom surfaces on which the powder was placed. The sample holders comprised aluminum plates with dielectric washers affixed by an epoxy resin. A 6 mm internal diameter, 0.5 mm-thick nylon washer was used to make a larger customized sample holder. Similarly, a 2.1 mm internal diameter, 0.6 mm-thick polycarbonate washer was used to make a smaller sample holder.

Binder

Several experiments were conducted with a small amount of binder added to Mg powder in order to investigate the effect of powder ejection by electric spark (see below) on its ignition. The binder chosen for these experiments was Star Brite liquid electrical tape which contains a vinyl acetate co-polymer, hydrocarbons, and ketones [21]. 0.1 g of binder was diluted in 10 mL of acetone. 1 mL of the diluted binder solution was added to 0.4 g of powder, thoroughly mixed, added to the sample holder, and air dried prior to testing. The binder amount was estimated to be only 0.3 wt. % of the binder/powder mixture. The binder was only used in selected tests utilizing a larger (6 mm diameter) customized sample holder described above.

Diagnostics

Inductance coils by Pearson Electronics were used to measure spark current and voltage across the pin electrode and sample gap. The current was measured using a model 110A coil with a 1 V / 10 A ratio. To measure the voltage drop, a 1 kΩ high voltage resistor was connected in parallel between the discharge pin and sample cup substrate. The current through the resistor proportional to the spark voltage was measured using a model 4100 coil with a 1 V / 1 A ratio. Both current and voltage traces were visualized and recorded by a LeCroy WaveSurfer 64Xs Series oscilloscope.

Optical emission produced by the spark and by the igniting powder was monitored in real time. In preliminary measurements, optical spectra produced respectively by sparks between the pin electrode and an empty sample cup and between the pin electrode and igniting Mg powder were recorded using an EPP2000 spectrometer by Stellarnet Inc. The recorded spectra are shown in Fig. 2 without correction for sensitivity of the spectrometer. The lack of correction results in a distortion of the black-body emission envelope, while the uncorrected spectra are still helpful for identifying the spectral regions most suitable for monitoring the emissions produced by the spark and by the heated and ignited powder. The emission of the spark itself (generated over an empty sample cup) was dominated by an ultraviolet peak (around 280 nm).

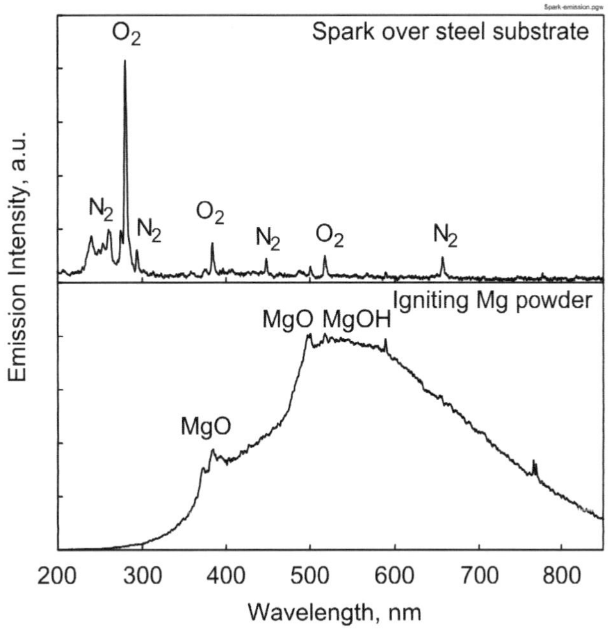

Figure 2. Emission spectra of the sparks between the pin electrode and an empty steel sample cup (top) and between the pin electrode and igniting Mg powder (bottom).

Based on refs. [22- 23], it was assigned to molecular oxygen. The emission of the igniting Mg powder was heavily dominated by a broad black-body spectrum with one of the strong peaks observed around 500 nm, assigned to MgO. Respectively, interference filters at 280 nm and 500 nm were selected to separate between the radiation signatures of the spark and igniting powder. A bifurcated fiber optics cable with a single input window was used to split the optical signal between two outputs. Each output was connected to a respective interference filter and a photomultiplier.

Preliminary characterization of the spark

In initial experiments, emission and electrical current traces were acquired for sparks striking solid metal substrates. The distance between the pin electrode and the substrate surface was maintained at 0.2 mm. The spark polarity, the value of the discharging capacitor, and the material of the substrate varied while the initial voltage was fixed at V_0=5 kV. Figure 3 shows typical examples of the current trace and radiation traces recorded using ultra-violet (280 nm) and green (500 nm) interference filters. Both the current and 280-nm filtered traces showed a strong AC component. Note that change in the pre-set polarity of the spark discharge using the FTS polarity selection did not appreciably change the shapes of the recorded current traces. The oscillations in current corresponding to the repeated re-charging of the capacitor correlated with the oscillations in the filtered spark emission signal.

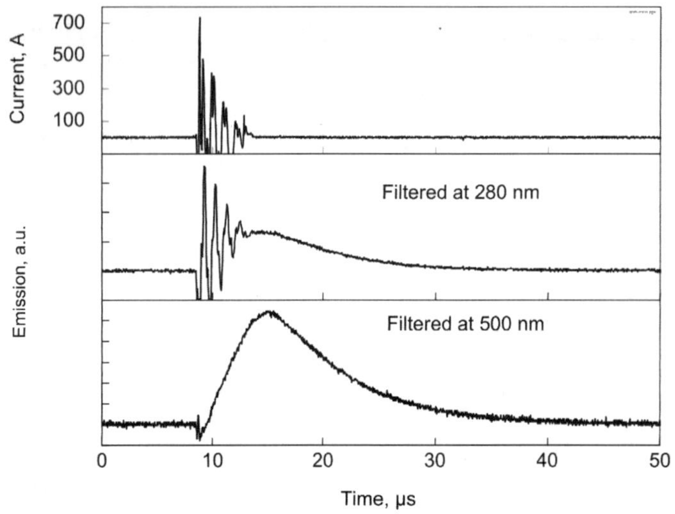

Figure 3. Examples of the spark current and emission traces (traces shown were obtained from a spark striking a stainless steel substrate, pin electrode was negative, $V_0 = 5$ kV, $C = 5000$ pF, and no additional resistor).

The green filtered trace showed no oscillations and the signal continuously increased during the time the spark current was measured. This clearly indicates the thermal radiation produced by the surface heated by the spark. Both radiation traces were observed to decay after the current trace was reduced to zero.

Varied parameters

Spark energy was changed by changes in both applied voltage and selected capacitor. Experiments were performed at both spark polarities. The spark gap was varied between 0.2 mm, recommended by the ESD ignition testing standard [17] and 1.5 mm. The other experimental conditions and parameters included use of different size sample holders and use of binder for the Mg powder.

RESULTS

Powder Ejection

Initial ignition tests showed that a portion of the powder placed in a sample cup is ejected from the cup for both ignition and non-ignition cases. A small crater, of about 1 – 3 mm diameter was always formed at the location where the spark struck the powder. Examples of the produced craters for the powder placed in the standard sample holder are shown in Fig. 4. The electrode cup on the left shows powder that has not been struck by a spark. The middle cup shows Mg powder that has been struck by a spark but did not ignite. A crater of about 3 mm diameter is clearly seen in the center of the cup. The cup on the right shows Mg powder that has been struck by a spark and the spark energy was sufficient to ignite the powder. A crater of about 3 mm diameter is again observed in the center of the cup. There is also a white oxide layer covering almost the entire powder surface. It should be noted that the oxide layer formed after the spark was over while the powder away from the crater was burning for several seconds. Note that when the 6 mm diameter, custom sample holder was used the powder ejection and crater formation were observed, similar to the case illustrated in Fig. 4. For the smaller, 2.1 mm diameter sample holder, the ejection resulted in effective removal of the entire powder charge from the sample holder.

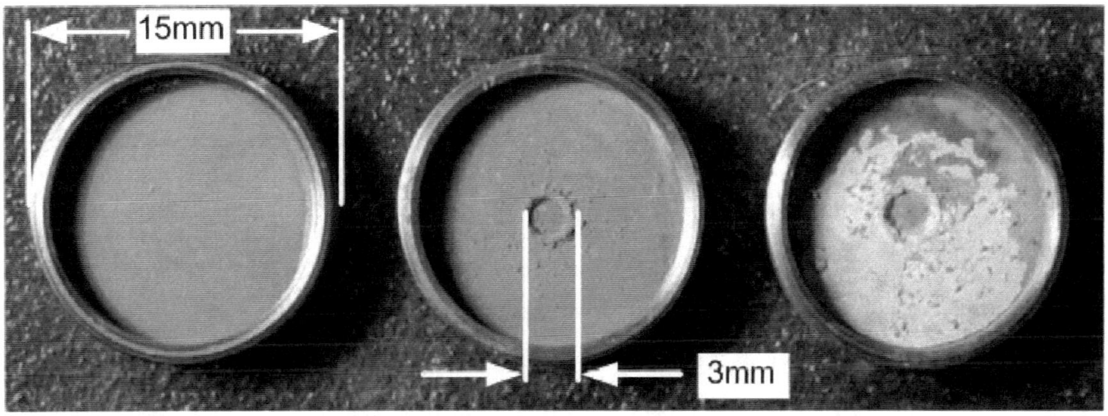

Figure 4. Electrode cups (15 mm diameter) containing Mg powder; left cup contains powder that has not been struck by a spark, middle cup contains powder that has been struck by a spark and has not ignited, and the right cup contains powder that has been struck by a spark and has ignited.

To determine the velocity of the ejected particles, the sample was illuminated using a modulated green laser sheet placed perpendicular to the powder sample holder. The spark energy was kept lower than the powder ignition threshold. Ejected, but not ignited particles scattered the laser light and were photographed using a digital camera with an open shutter. It was observed that the initial velocities at which the particles moved (measured about 1 mm from the powder surface) varied from 30 to 70 cm/s. The velocities were not affected by the spark voltage, systematically varied in these experiments.

The ejection of powder by the spark resulting in the formation of the central crater can be attributed to the effect of the spark-produced shock wave [24]. This shock wave passes through the powder layer, reflects from the bottom of the sample holder and the powder becomes accelerated and is lifted by the pressure produced in the reflected wave. The role of the powder ejection in the ignition mechanism has not been previously discussed in the literature.

Circuit impedance and measured spark energy

Current and voltage measurements were used to determine the total spark energy. In addition, the spark/powder impedance was roughly evaluated from processing the recorded current traces. The spark/powder system was assumed to be represented by an LRC series circuit, with constant values of resistance, R, inductance, L, and capacitance, C. Figure 5 shows typical experimental traces for the current and voltage. In addition, the best fit trace for the current signal is shown. The fit was obtained using the equation of current for an LRC series circuit with no voltage source applied after the capacitor is charged to a specified voltage [25]:

$$I(t) = -\frac{2CV_A}{\sqrt{4LC - R^2C^2}} \cdot \sin\left(\sqrt{\frac{1}{LC} - \frac{R^2}{4L^2}}t + \theta\right) \cdot \exp\left(-\frac{R}{2L}t\right) \tag{1}$$

where $I(t)$ is the spark current as a function of time, t; V_A is the voltage applied to charge the capacitor, and θ is the phase angle. The values of L, R, and C were adjusted to obtain the fit of the current trace predicted by Eq. (1) and the experimental trace. Note that in order to produce estimates meaningful for description of spark ignition events, it should be recognized that the assumption of the constant impedance made above to find the L, R, and C values by matching the experimental current trace and prediction using Eq. (1) cannot be entirely accurate.

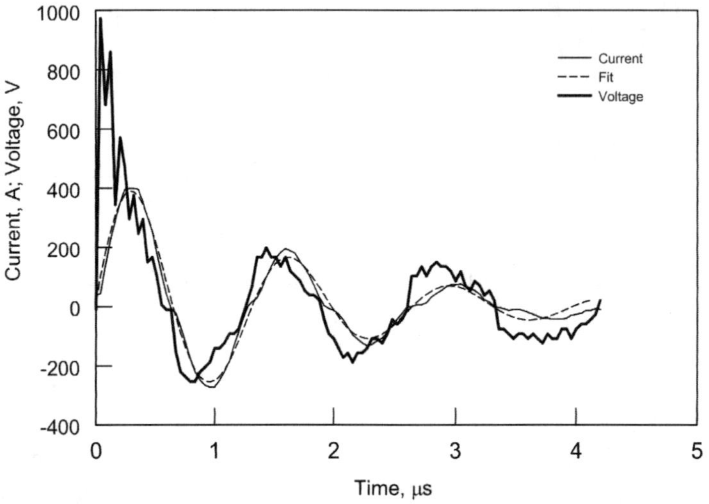

Figure 5. Experimental current and voltage traces for a spark and a current trace obtained fitting the experimental data and assuming the spark as a series LRC circuit.

The impedance of the spark/powder system is likely to change in time, both due to substantial changes in the conductivity of non-equilibrium plasma existing in the spark gap [26-27], and due to changes in the properties of Mg powder while it is being heated and melted. In initial analyses it was noticed that the quality of match between the experimental and calculated curves (as shown in Fig. 5) was consistently different for the earlier and later spark times. The earlier times play a greater role for the energy transfer from the spark to the powder, both because of the initially higher conductivity of the spark kernel [26-27] and because of the rapid decrease in the amplitude of the discharge current with time. Therefore, to obtain the current fitting curves and identify the values of R, L, and C, most useful for description of the energy transfer from the spark to powder, the experimental current trace was truncated as shown in Fig. 5 to remove two of the final and weakest oscillation periods observed. This resulted in a description matching better with the experimental current at the earlier times.

The spark energy was determined as:

$$E = \sum IV\Delta t \tag{2}$$

where the summation was performed over the entire spark duration with the time step Δt equal to the sampling time of the digital oscilloscope and using the recorded values of I and V as a function of time. For all current and voltage measurements the time step was Δt = 40 ns. Equation (2) estimates the true or active portion of the total released energy available for heating the conductive elements of the circuit. It does not include the reactive energy, which might result in some losses through electromagnetic radiation produced by the phase-shifted components of the current and voltage existing due to the inductive nature of the spark circuit impedance. Similar measurements were performed and processed for empty sample holders and for the sample holders with Mg powder. The results of this signal analysis and obtained values of the spark energy are shown in Figs. 6, 7. Each measurement here and below was repeated 10 times. The error bars show one standard deviation of the 10 repetitions.

The spark energy measured for empty sample holders under various experimental conditions is shown as a function of the energy stored in the capacitor in Fig. 6. Only about 1/6th of the capacitor's energy is recovered as determined from the recorded current and voltage traces and using Eq. (2).

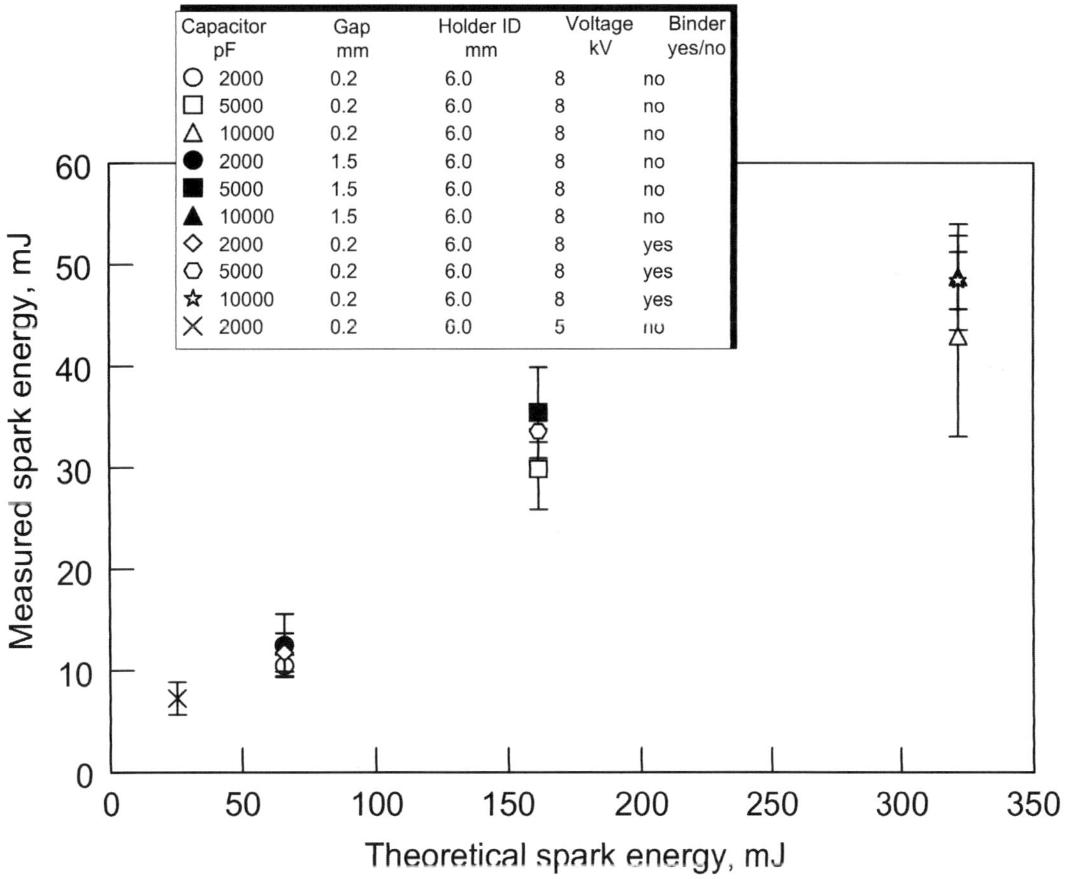

Figure 6. Measured active spark energy as a function of the energy stored in the capacitor for empty sample holders with different diameters and for different spark gaps and capacitor values.

A small part of the capacitor's energy is unaccounted for because of the truncated current trace. It is also possible that the capacitor retains some small charge after the spark discharge is over. Furthermore, it is hypothesized that a substantial part of the capacitor's energy is unaccounted for by Eq. (2) due to the inductive nature of the spark's impedance. Slightly higher measured spark energies correspond to the longer spark gap, while the effect is weak. No effect of the sample holder size can be detected. A similar plot for the spark energy measured for the sample holders filled with Mg powder as a function of the energy stored in the capacitor is shown in Fig. 7.

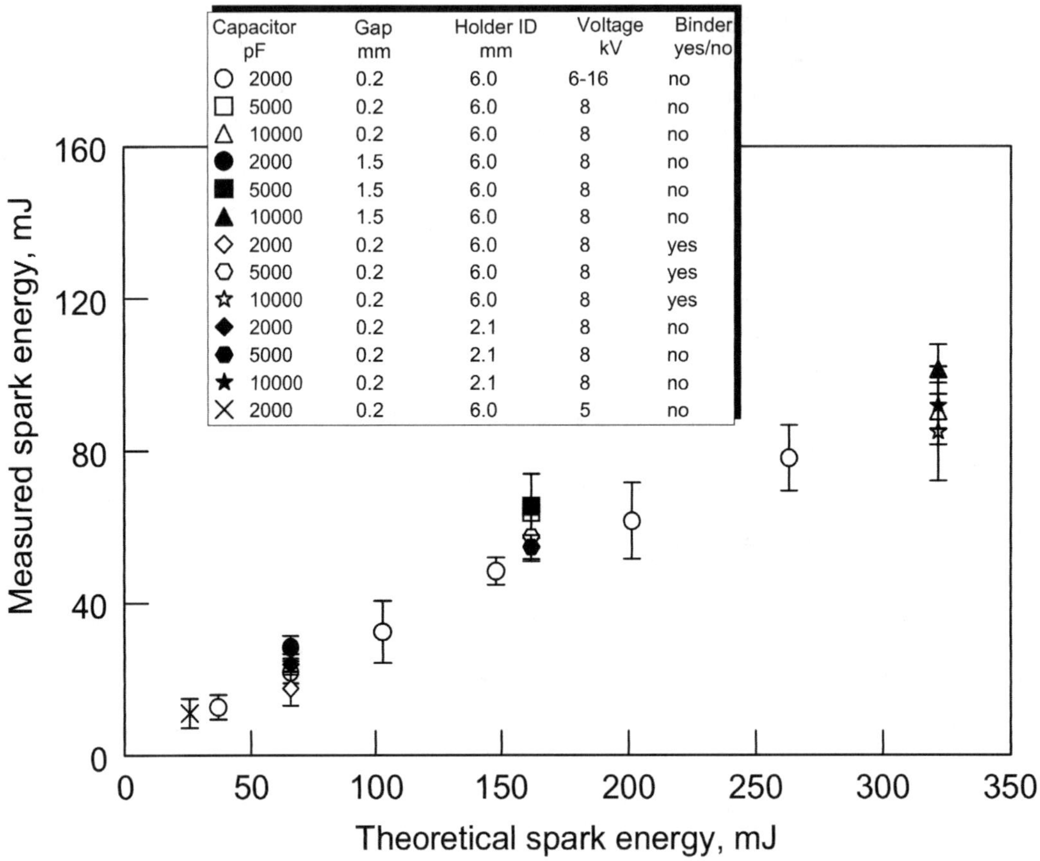

Figure 7. Measured active spark energy as a function of the energy stored in the capacitor for sample holders with different diameters filled with Mg powder with and without binder and for different spark gaps and capacitor values.

It is interesting that in the latter case, the measured spark energy is about 1/3rd of the capacitor's energy, or about two times greater than it is for the same experimental configuration but without Mg powder. This can be attributed to the added active resistance caused by the Mg powder. With this additional resistance, the overall spark impedance becomes less inductive, so that a greater portion of the capacitor's energy is recovered using the measured traces of I and V and Eq. (2).

The spark impedance values obtained from matching the current traces using the current description given by Eq. (1) are shown in Figs. 8 and 9. The L, R, C values corresponding to the empty sample holder are shown in Fig. 8. The capacitance shown is a correction to the pre-set capacitance of the FTS required for the best fit for the measured current trace using Eq. (1). Clear trends of reduced resistance and inductance and increased capacitance at the higher spark energies are observed. The effect of the spark gap on the determined spark impedance is negligible.

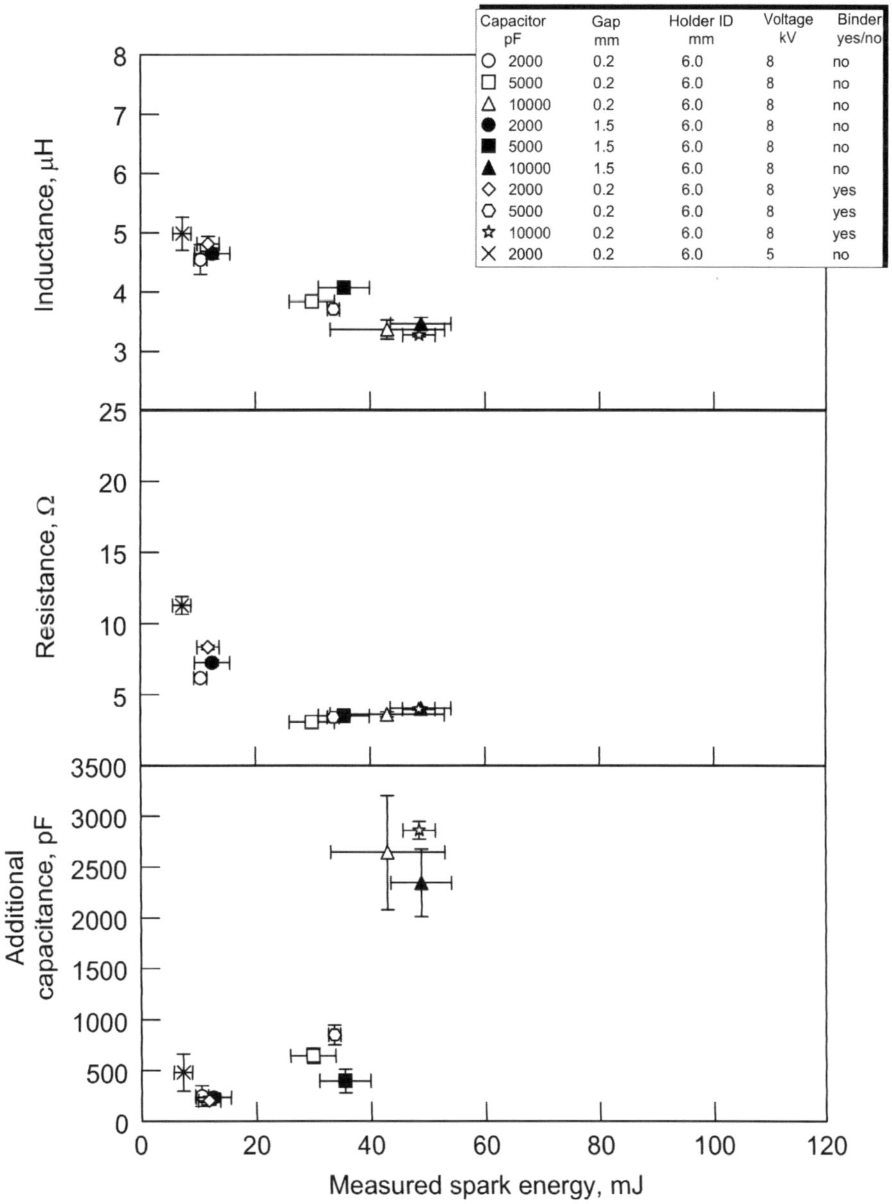

Figure 8. Circuit impedance as a function of the measured spark energy for the experiments with both the 6 mm diameter and 2.1 mm diameter empty sample holders (spark only).

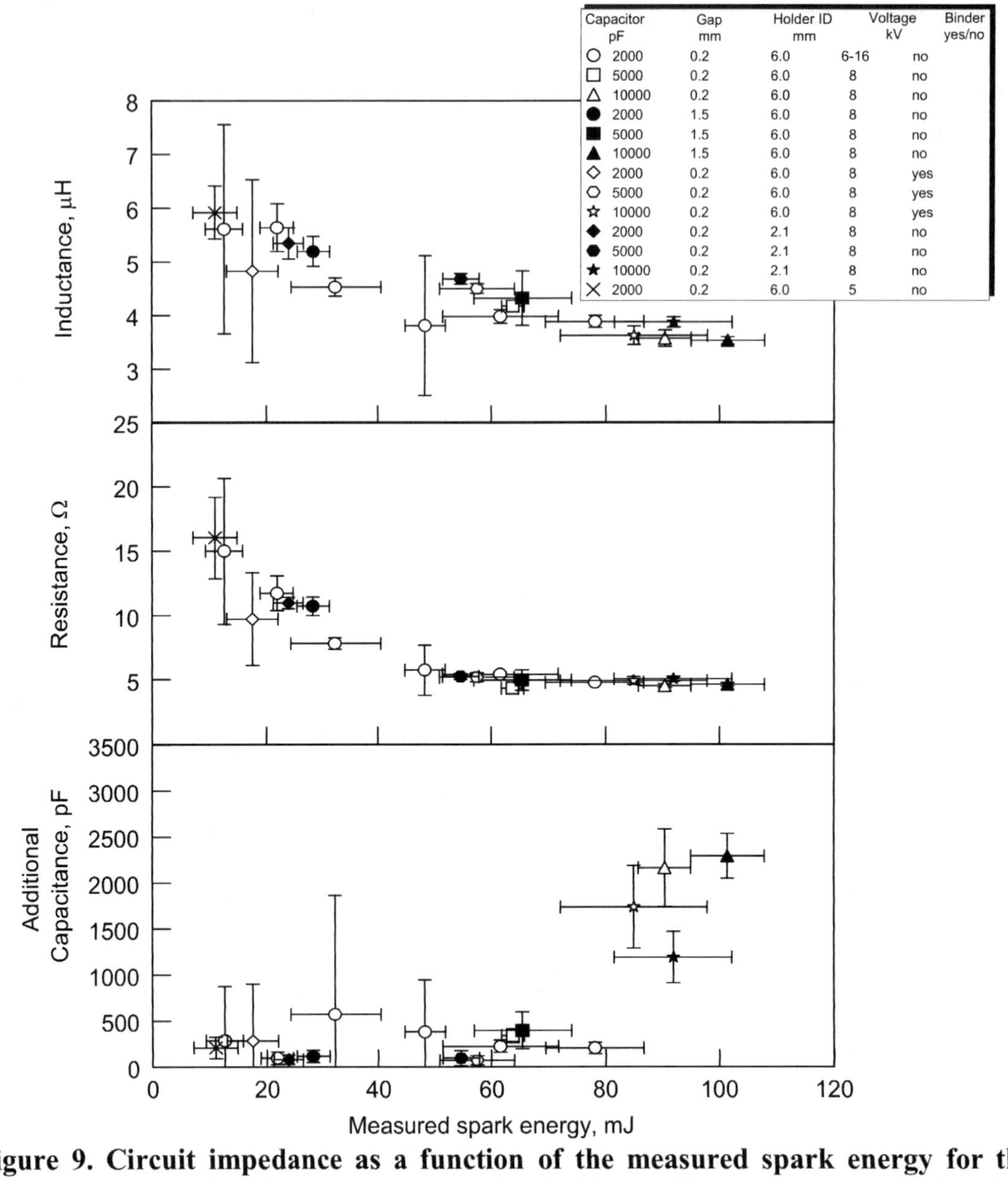

Figure 9. Circuit impedance as a function of the measured spark energy for the experiments with igniting loose Mg powder in 6 mm diameter sample holder, Mg with binder in a 6 mm diameter sample holder, and loose Mg in a 2.1 mm diameter sample holder.

For experiments with Mg powder, impedance values are shown in Fig 9. Results are shown for ignited Mg placed in both 2.1 and 6 mm diameter custom-made sample holders with the latter used with and without binder added to the powder. In addition, results obtained for measurements with different spark gap lengths are shown. It is observed that adding powder slightly increases the values of resistance and inductance necessary to fit the experimental data with the shape predicted by Eq. (1) for all spark energies. This is consistent with the observed greater portion of the capacitor's energy recovered as calculated by Eq. (2). There is no detectable effect of the spark gap or binder on the resistance and inductance values. The trends of reduced inductance and resistance and increased capacitance with increased spark energies are clearly visible and similar to those observed from Fig. 8.

The correction for the capacitance value is relatively small for all, except for the largest used capacitor (10,000 pF), for which it reaches about 20% of the capacitor value. In most cases, the additional capacitance decreases with powder included. Binder addition and sample holder size do not have a clear affect on the additional capacitance.

Minimum Ignition Energy

The experiments using the standard sample holder and spark gap of 0.2 mm suggested by the standard procedure [17] established the minimum ignition energy for Mg powder to be 25 mJ. This value is the energy stored in the capacitor. The measured spark energy available to the powder is 11 mJ and the energy delivered from Joule heating is 5 mJ.

Ignition delays

It was observed that the optical signal produced by the ignited powder was noticeably delayed following the spark emission and current traces. The ignition delay is illustrated in Fig. 10. The top plot shows the initial (shaded) portion of the recorded signal with the logarithmic time scale, so that it is clearly observed that the emission decreases to its baseline level after the spark and the ignition peak is only detected after a substantial delay. The delay times were measured using the recorded emission traces. Two different signal processing techniques were used and respectively two different values of ignition delays were obtained for each of the recorded traces. In both cases, the baseline signal level was determined as the signal recorded prior to the spark ignition.

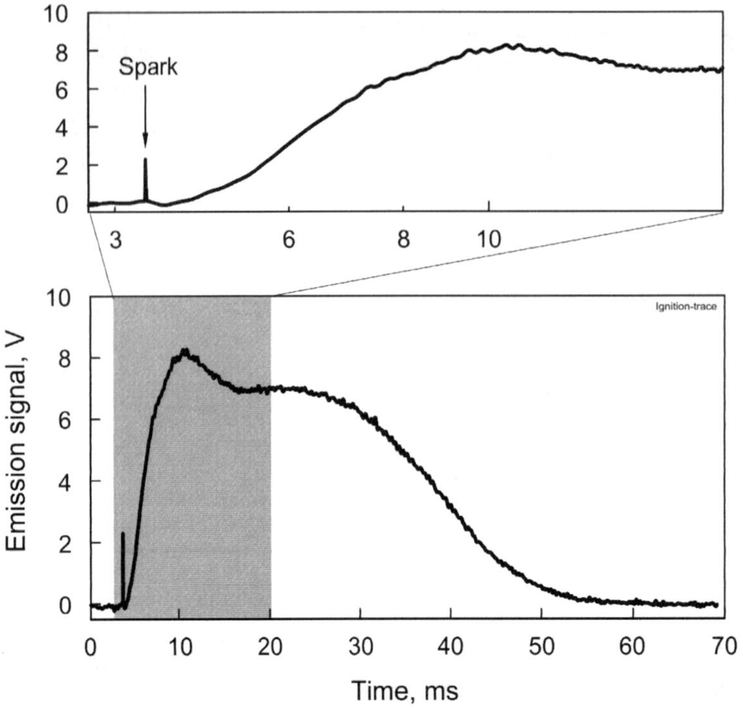

Figure 10. An emission signal produced by a spark-ignited Mg powder. A shaded portion of the signal is expanded with a logarithmic horizontal scale to clearly illustrate the short spark pulse soon after 3 ms preceding the Mg ignition that produces the main broad peak.

In the first moment considered to determine the ignition delay, the ignition was assumed to occur when the emission signal increased abruptly. Thus, the rate of ensuing combustion affecting the slope of the emission signal influenced the identification of the ignition moment. To find the delay by this method, a time derivative of the emission signal was obtained and the instant and signal value of the emission trace corresponding to the maximum in its derivative were found. The signal slope at this point was projected as a straight line to cross the baseline signal level, as illustrated in Fig. 11, so that the ignition delay could be determined.

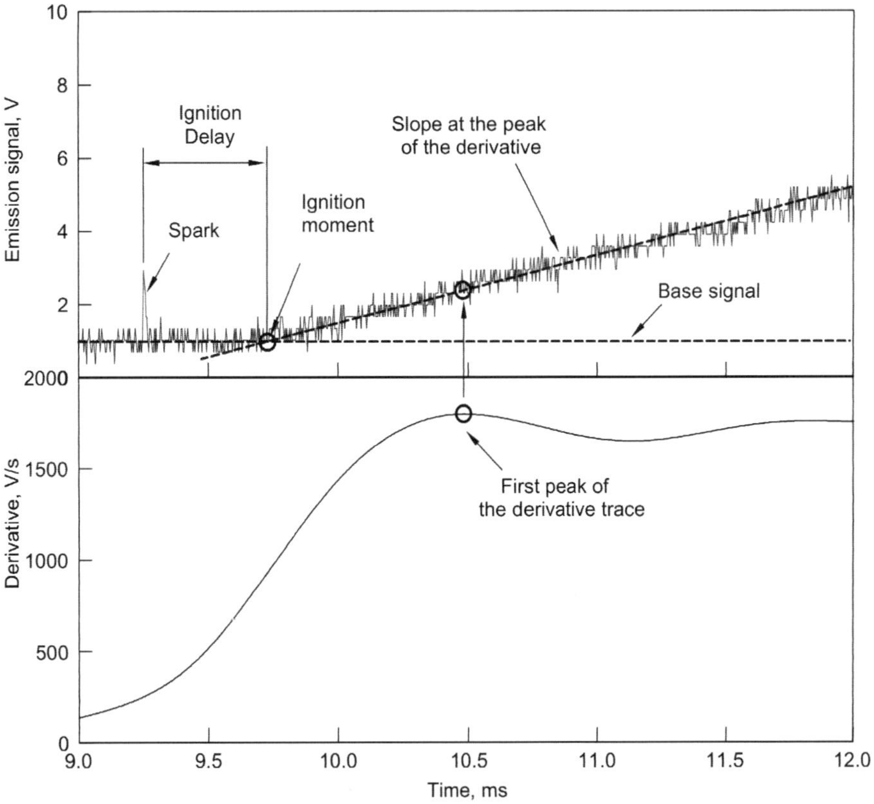

Figure 11. Illustration of the first method used to determine ignition delay. The bottom plot shows a derivative of the emission signal (the spark is not visible due to the signal smoothing). The top plot shows the respective portion of the recorded unprocessed signal and the identified slope corresponding to the first peak in its derivative. The time between the spark and the point where the slope line crosses the zero-level signal line is the measured delay.

In the second method, ignition was assumed to occur when the emission signal produced by the ignited powder exceeded the baseline signal. Specifically, the ignition instant chosen was when the emission signal following the spark increased 3 standard deviations above the baseline level, as illustrated in Fig. 12.

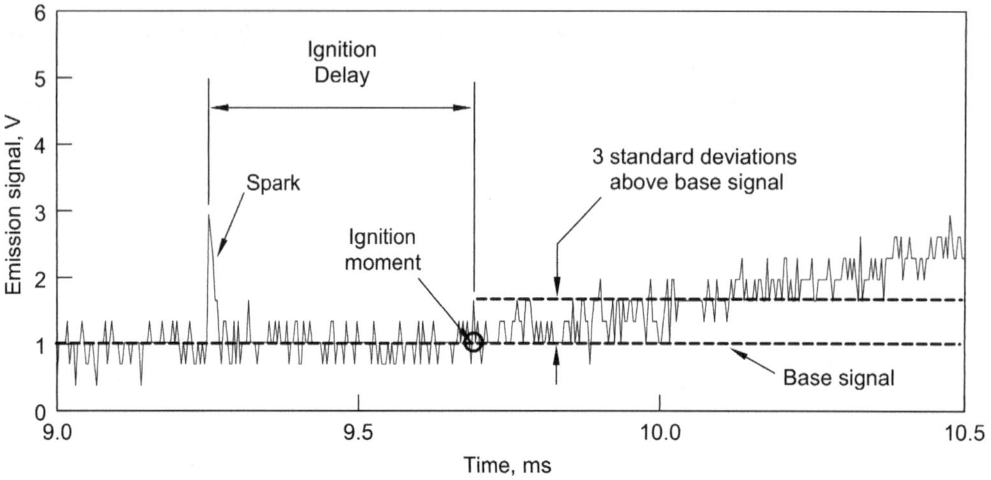

Figure 12. Illustration of the second method used to determine ignition delay. The ignition moment is chosen when the emission signal increases 3 standard deviations above the base signal noise after the spark.

Figures 13 – 14 show ignition delays versus measured spark energies for loose Mg powder in the 6 mm diameter sample holder where delays are determined by methods 1 and 2, respectively. Ignition delays vary in the range of 0.5 – 3 ms and correlate with the measured spark energy, with shorter delays corresponding to greater spark energies. This trend is also observed in experiments with Mg/binder in the 6 mm diameter sample holder and loose powder in the 2.1 mm diameter sample holder, as shown in Figs. 15 and 16, respectively. Results presented in Figs. 13 and 14 show no effect of the spark gap on ignition delay. Ignition delays are shorter in Figs. 15 – 16 as compared to Figs. 13 – 14 (loose powder on a 6 mm diameter sample holder). Interestingly, the effect of the spark energy on the ignition delay for experiments with binder is different from that for the loose powder. For the powder with binder, the ignition delay continues to decrease when the spark energy increases above 60 mJ (Fig. 16); at the same time the ignition delay becomes roughly constant above 60 mJ for the loose powder (Figs. 13 – 15). The specific value assigned to the ignition delay is affected by the method used for its quantification. According to Figs 15 – 16, method 2 yields shorter delays than method 1. However, according to Figs. 13 – 14, below 30 mJ method 2 yields shorter delays but above 30 mJ method 2 yields longer delays than method 1.

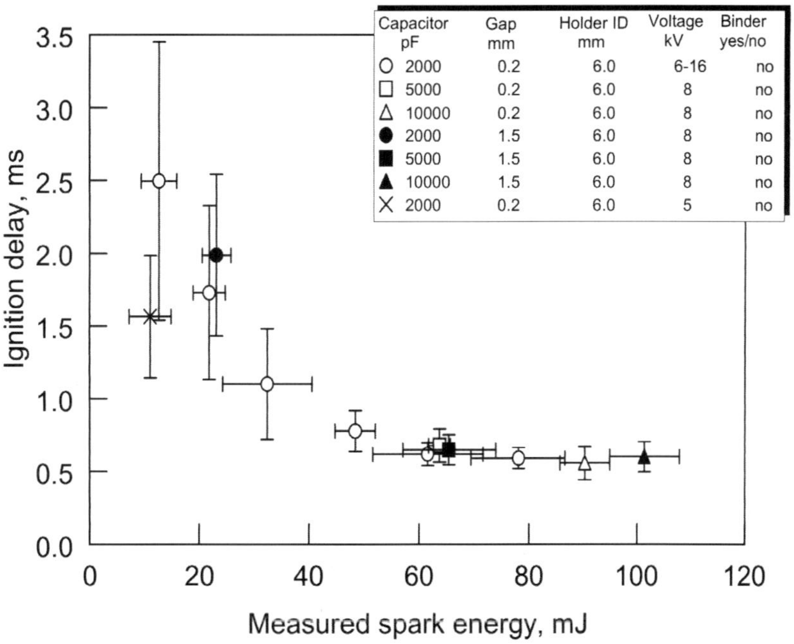

Figure 13. Ignition delay (method 1) v measured spark energy for experiments conducted with the 6 mm diameter sample holder and no binder.

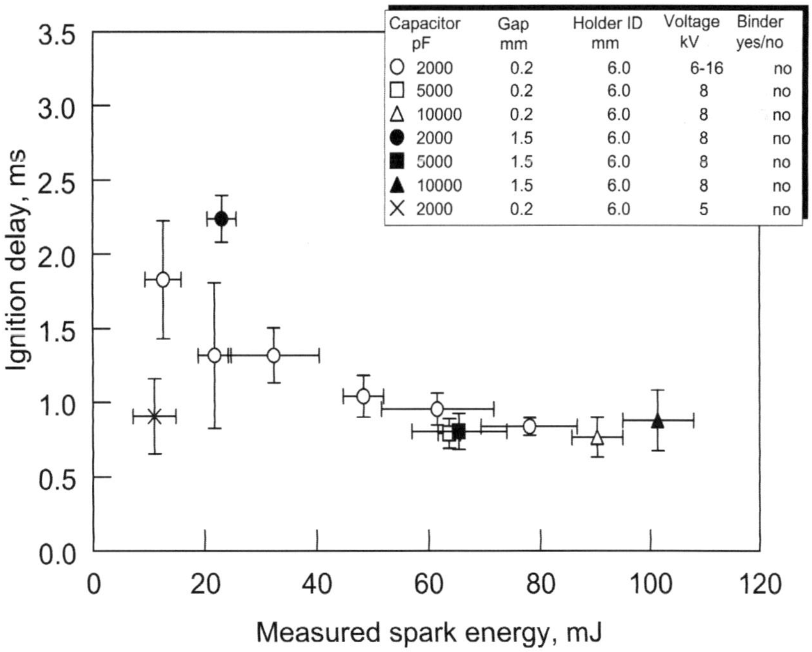

Figure 14. Ignition delay (method 2) v measured spark energy for experiments conducted with the 6 mm diameter sample holder and no binder.

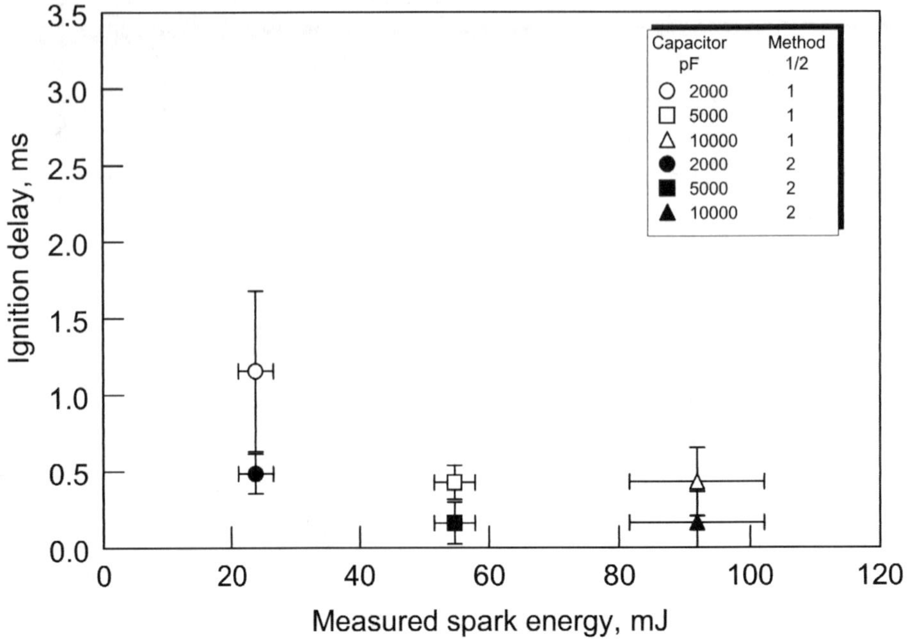

Figure 15. Ignition delay (both methods) v measured spark energy for experiments conducted with the 2.1 mm diameter sample holder and no binder at 8 kV and a gap of 0.2 mm.

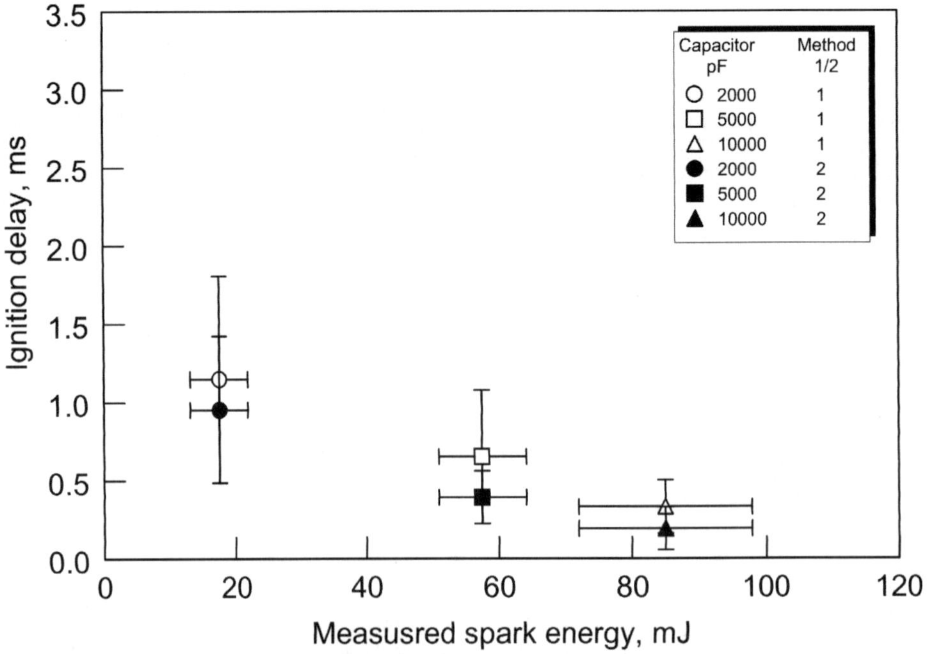

Figure 16. Ignition delay (both methods) v measured spark energy for experiments conducted with the 6 mm diameter sample holder and binder at 8 kV and a gap of 0.2 mm.

DISCUSSION

Spark impedance and measured spark energy

The values of inductance on the order of a few μH are consistent with the measurements reported in [28] for a similar spark configuration. The relatively large additional capacitance values in Figs. 6-8 are clearly not affected by powder or powder/binder mixtures. The additional capacitance values are mostly likely associated with deviations of the capacitor characteristics from their nominal values, stray capacitances produced by the contact between the aluminum substrate and sample holder support, and parasitic capacitance of the output circuit of the FTS. The effect of stray capacitance is further supported considering that the two custom sample holders used different sized aluminum substrates with different contact areas were characterized with consistently different capacitance corrections in experiments.

The resistance values recovered from the current versus time curve fitting for the experiments performed with and without Mg powder can be used to assess the powder electrical resistance. Knowing the current and resistance of the powder enables us to quantify the Joule heating of the powder. Thus, the role of Joule heating in the powder spark ignition mechanism can be clarified. The powder resistance, R_p, was estimated as:

$$R_p = R_w - R_{w/o} \tag{3}$$

where R_w is the circuit resistance obtained for the experiment with powder in the sample holder (cf. Fig. 9), and $R_{w/o}$ is the circuit resistance obtained for the identical capacitance, voltage, and spark gap and using an empty sample holder (cf. Fig. 8). The energy transferred to the powder as a result of its Joule heating is then determined as:

$$E_{Joule} = \sum I^2 R_p \Delta t \tag{4}$$

This processing was performed for each ignition experiment and Fig. 17 shows the result. In addition, the spark total energy for the same conditions obtained using Eq. (2) is also shown for reference. For the experiments with no binder used, the Joule heating energy varies from 50 to 90 % of the total measured spark energy. For the experiments with binder, the values of Joule heating energy and the total spark energy are approximately the same. This discrepancy likely points out at an inaccuracy in quantifying the powder resistance for cases when the binder was used.

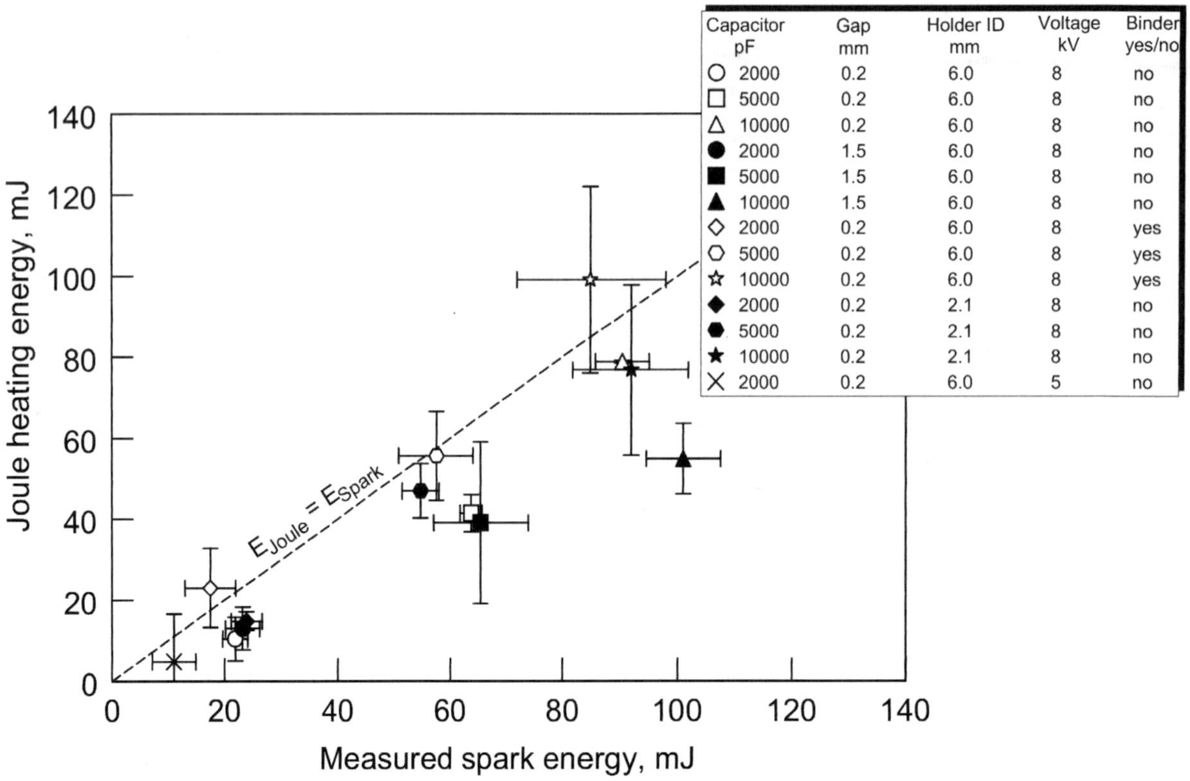

Figure 17. Joule heating v measured spark energy.

The energy transferred to the powder as a result of its Joule heating is substantial and must result in its significant heating. The role of this heating in triggering powder ignition is further discussed below.

Powder heating and ignition by the spark

Spark discharge breaks down the gas gap producing a relatively narrow (~ 100 µm diameter [27, 29]) plasma channel with a small resistance. When an empty sample holder is used, this channel is reaching the metal surface with electrical resistance negligible compared to that of the plasma. However, when metal powder is placed in the sample holder, its electrical resistance is much greater than that of a bulk metal because of the substantial contact resistance between adjacent particles. Thus, in order to achieve comparable currents in both cases, the spark plasma must be extended through the powder layer. In other words, local micro-discharges occur between particles placed under the plasma channel, effectively reducing the contact resistance between these particles. The local discharges between the particles serve as hot spots heating the powder during the spark discharge, while the resistance of the individual particles is much smaller so that their Joule heating by the current existing inside the particles is negligible.

Assuming that the spark discharge directly heats only the powder layer directly under the plasma channel, both diameter, D, and initial temperature, T_{in} of the directly heated powder cylinder can be estimated. This estimate will use the experimentally determined minimum ignition energy for the spark ignition of Mg powder.

First, note that the distance the heat can travel within the powder during the spark duration is negligible. Indeed, using the thermal diffusivity for a Mg powder layer measured in ref. [20], $\alpha = 2.29\times10^{-7}\, m^2/s$, it can be readily estimated that in τ= 1 μs (characteristic spark duration) the heat can only travel within the powder for about $\sqrt{\alpha\tau} \approx 0.5$ μm. Thus, for a crude estimate it can be assumed that the powder cylinder of a fixed diameter, *D*, is heated adiabatically and uniformly by the current passing through it. The height of the cylinder, *H*, is taken to be 0.5 mm, which is close to the height of the powder sample placed in the sample holder. For such a cylinder, the temperature achieved as a result of its Joule heating will be a function of the cylinder radius, *R*, as expressed by Eq. (5):

$$
\begin{aligned}
&if\ T_{in} < T_m;\quad T_{in} = T_0 + \frac{E_{Joule}}{\pi\rho\eta C_{PS} R^2 H} \\
&if\ T_{in} > T_m;\quad T_{in} = T_m + \frac{E_{Joule}}{\pi\rho\eta C_{PL} R^2 H} - \frac{C_{PS}}{C_{PL}}\left(T_m - T_0\right) - \frac{L}{C_{PL}}
\end{aligned}
\qquad (5)
$$

where T_0 is the initial powder temperature assumed to be equal to the room temperature, 298 K, T_m=923 K is the magnesium melting temperature, η is packing density, and for densely packed spheres η=0.75, C_{PS} and C_{PL} are the values of specific heat for solid and liquid magnesium respectively, *L*=357.9 kJ/kg, is the latent heat of melting for magnesium, E_{Joule} is the spark energy released in the powder as a result of its Joule heating, and ρ is the powder density. Assuming that E_{Joule}= 5 mJ corresponding to the experimentally determined minimum ignition energy, the temperature, T_{in}, predicted to be achieved in the heated volume is plotted as a function of the cylinder radius (the horizontal axis shown at the top of the plot) in Fig. 18.

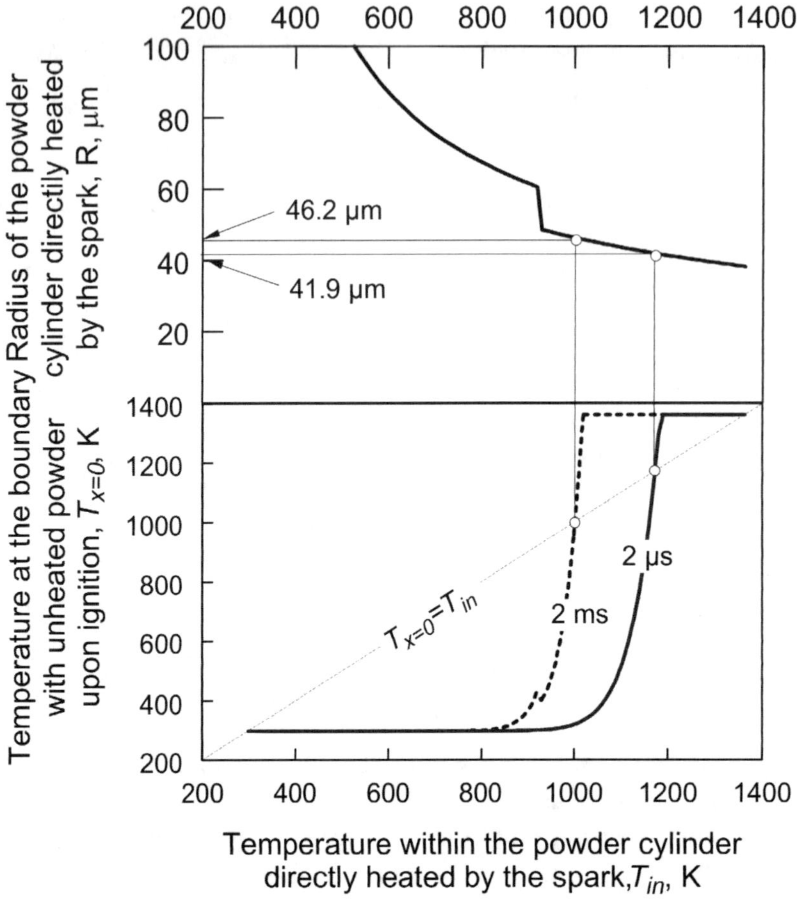

Figure 18. Illustration for evaluation of the radius of the powder cylinder directly heated by the spark. The top plot shows how this radius changes depending on the temperature achieved in the cylinder and assuming that the spark energy is equal to the minimum ignition energy for Mg. The bottom curves show temperatures at the boundary with the cold portion of the powder required to remove the heat generated by chemical reaction within the directly heated cylinder for the time delays of 2 μs and 2 ms.

In order to determine which diameter corresponds to the experimental situation, two separate estimates can be made. In the first case, consider the heat balance between the powder heated by the discharge directly and the rest of the sample, assuming that the powder remains in the sample holder (i.e., neglecting powder ejection). The powder heated by the spark directly will start oxidizing and heat the rest of the sample. Assuming that the oxygen is available to the powder directly heated by the spark (after some ignition delay), the respective heat flux can be described as

$$q_s'' = \frac{A}{A_v} Z\Delta H \exp\left(-\frac{E_A}{R_u T_{in}}\right) = \frac{3R\eta}{2r} Z\Delta H \exp\left(-\frac{E_A}{R_u T_{in}}\right) \qquad (6)$$

where A is the area over which oxidation occurs equal to the total surface area of all the particles inside the directly heated cylinder and A_v is the surface area of the heated cylinder in contact with the rest of the surrounding powder; it is assumed that all particles have the same radius, r, equal to the mean radius of the powder used in experiments (5.15

μm), Z is the pre-exponent, ΔH is the enthalpy of oxidation, E_A is the activation energy, R_u is the universal gas constant. The values of Z=10^{10} kg/m^2 s and E_A=215 kJ/mol are taken from ref. [20] where the same Mg powder was studied. For powder ignition to occur, this heat flux needs to be balanced by the heat flux produced in the conductive colder powder outside of the directly heated volume. For a simple estimate, the transient temperature profile in that portion of the sample, T_x as a function of, x, the distance from the interface with the directly heated volume, can be approximately described by an analytical expression available for a semi-infinite solid heated at its boundary by a constant heat flux [30]. For x=0, the heat flux can be expressed through temperature at the interface, $T_{x=0}$ as:

$$q_s^{"} = \frac{2}{k} \cdot \frac{\sqrt{\alpha t / \pi}}{T_{x=0} - T_0} \tag{7}$$

where α is thermal diffusivity, k is thermal conductivity, and t is elapsed time. The value of $q_s^{"}$ can be taken as a function of temperature from Eq. (6), so the radius R of the directly heated cylinder can be found, for which the condition (8) below is satisfied:

$$T_{x=0} = T_{in}(R) \tag{8}$$

The elapsed time is assumed to be equal to the characteristic ignition delay, t=2 ms. For this time, the condition (8) is fulfilled for R=46.2 μm. This calculation is illustrated in Fig. 18. Another borderline case, with the elapsed time close to the spark duration, t=2 μs is also considered.

The simple estimates above can be used to assess the radius of the powder cylinder directly heated by the spark. Based on results presented in Fig. 18, this radius is limited by 41.9 and 46.2 μm. This size compares well with the reported dimension of the plasma kernel produced in similar spark discharges [27, 29].

The second approach for estimating the size of the powder cylinder directly heated by the spark can be considered assuming that the powder is ejected by the discharge, so that observed ignition is effectively equivalent to combustion of the individual ejected particles. To determine to what temperature the particle needs to be heated prior to its ejection so that it will ignite when exposed to cold air, a heat transfer balance between chemical reaction heat release and convective and radiative heat losses can be considered for an individual particle:

$$Z\Delta H \exp\left(-\frac{E_A}{RT_{Particle}}\right) = \frac{Nu}{D} k\left(T_{Particle} - T_\infty\right) + \varepsilon\sigma\left(T_{Particle}^4 - T_\infty^4\right) \tag{9}$$

where $T_{Particle}$ is the temperature of the particle that needs to be reached in order for ignition to occur, Nu is the Nusselt number, D is the particle diameter, k is the thermal conductivity of air taken at the film temperature, T_∞ is the temperature of the surrounding air and surfaces, ε is the emissivity of the particle surface, and σ is the Stefan-Boltzmann constant. The particle is assumed to have a low Reynolds number and therefore the value of $Nu=2$ is used corresponding to a stationary sphere. The temperature at which Eq. (9) is satisfied is assumed to be the ignition temperature for a single Mg particle in cold air. Using the heat balance (9) it is estimated that for a particle of 10.3 μm diameter (the mean diameter of the Mg powder used in experiment) the ignition temperature is 1067 K. From the top curve shown in Fig. 18, this temperature corresponds to the radius of the heated powder cylinder of approximately 45 μm, which is very close to the previous estimate.

Assuming that this dimension does not change significantly as a function of the spark energy, the ignition delays for different spark energies can be evaluated. It is generally clear that shorter delays will be predicted to correspond to the greater spark energies, as observed experimentally. However, the simple estimates are limited because once the powder in the volume directly heated by the spark starts boiling, the regime of heat transfer between this boiling powder and the colder external portion of the sample change substantially. It also is clear that the description of the oxygen transfer to the heated powder is of critical importance since this transport will determine the rate of reaction in the powder cylinder directly heated by the spark, which, in turn, will likely determine the observed ignition delay. By using Eq. (6) it was assumed that oxygen is available to the powder, while this may not be the case immediately after the spark is over. Indeed, the oxygen available for reaction is contained inside the porous powder layer. An estimate shows that if all this oxygen is consumed, the powder temperature can only be increased by about 2 K. Additional oxygen is therefore required and must be supplied by diffusion and/or convection flux which is likely established during the observed ignition delay.

Effect of powder ejection

Based on the observed velocities of the ejected particles, it can be concluded that the powder effectively remains in place during the spark. The particle velocities do not exceed 1 m/s, so the particles do not move more than 1 μm during the 1 μs long spark discharge. On the other hand, the particles do get displaced during the time comparable to the typical ignition delay. The displacement is of the order of 0.5 – 1 mm, which is close to the height of the sample. The powder ejection during the ignition delay affects the heat transfer between the powder heated by the spark directly and the rest of the sample. It can be generally expected that the heat transfer becomes less efficient when part of the directly heated powder is removed, resulting in longer ignition delays. This effect is qualitatively consistent with the results presented in Figs. 14 and 16: shorter delays are observed for the powder with binder. Also, the delays at higher spark energies and no binder do not decrease with further increase in the spark energy, indicating that the powder removal due to ejection impedes the heat transfer within the powder more significantly.

CONCLUSIONS

It was found that for the characteristic ESD ignition testing configurations, the spark duration is of the order of a few μs. The current in the discharge has substantial AC component so that the polarity of the electrodes was observed to be insignificant for the spark energy transfer to the powder. It was also found that only about 1/3 of the energy stored in the capacitor is supplied to the igniting powder as a result of its Joule heating. For Mg powder used in the experiments, the powder ignition was observed to be delayed by 0.5 – 3.5 ms after the spark discharge is over. Shorter ignition delays were observed for greater spark energies. Portion of the powder was observed to be ejected by the spark independently whether ignition was or was not observed. It was hypothesized that the ejection is due to a shock wave produced by the spark and reflected from the bottom of the sample holder. Limiting the powder ejection by adding a small amount of binder to the powder affects the dependency of the ignition delay on the spark energy. For the powder with the binder, the decrease in the ignition delays is almost linear as a function of the spark energy. With no binder, the ignition delays level out for the spark energy of about 60 mJ and do not decrease at greater spark energies. A simplified estimate shows that the ignition can be described considering that the spark first adiabatically heats a powder cylinder located directly under the spark plasma channel. Considering the experimentally determined minimum ignition energy for magnesium powder, the radius of this cylinder directly heated by the spark is estimated to be between 42 and 46 μm. This directly heated powder cylinder then starts oxidizing and the released enthalpy of oxidation heats the rest of the powder causing its subsequent ignition.

ACKNOWLEDGEMENT

This work was supported by Dr. Ralph A. Anthenien of the U.S. Army Research Office.

REFERENCES

[1]Mellor, A. M., Stoops, D. R., "Optimization of spark and ESD propellant sensitivity tests, "*Propellants, Explosives, Pyrotechnics,* 15, 1990, pp. 1-7

[2]Mellor, A. M., Baker, P. J., "Propellant properties conducive to electrostatic discharge ignition," *Journal of Energetic Materials,* 12:1, 1994, pp. 1-62

[3]Ryzhik, A. B., "Critical conditions of the spark ignition of metal power suspensions in gas," *Fiz. Goreniya Vzryva,* 14, 1978, pp. 53-57

[4]Ryzhik, A. B., Makhin, V. S., Kititsa, V. N., Il'in, V. V., "Experimental study of the spark ignition of aerosol suspensions of lightweight metal powder, *Tr. Vses. N.-i.i Proekt.In-t Alyum., Magn.i Elektrod.Prom-sti.,* 1977, pp. 106-112

[5]Kim, S. W., Colver, G. M., "Spark ignition of aluminum powder in mixtures of oxygen, nitrogen, and carbon dioxide - a new testing method," *American Society of Mechanical Engineers, Heat Transfer Division,* 148, 1990, pp. 97-104

[6]Kao, C. S., Duh, Y. S., "Accident investigation of an ABS plant," *Journal of Loss Prevention in the Process Industries,* 15 (3), 2002, pp. 223-232

[7]Glor, M., "Hazards due to electrostatic charging of powders," *Journal of Electrostatics,* 16 (2-3), 1985, pp. 175-191

[8]Wu, Z., Chen, Y., Hu, X., Liu, S., "Research on ESD hazards of textiles," *Journal of Electrostatics,* 57, 2003, pp. 203-207

[9]Tunnicliffe, G., Thomson, M., "Explosion protection," *Cleanroom* Technology, 9 (9), 2003, pp. 29-32

[10]Senecal, J. A., "Manganese mill dust explosion," *Journal of Energetic Materials,* 4, 1991, pp. 332-337

[11]Matsuda, T., Yamaguma, M., "Tantalum dust deflagration in a bag filter dust-collecting device," *Journal of Hazardous Materials,* 77 (1-3), 2000, pp. 33-42

[12]Ebadat, V., Pilkington, G., "Dust explosion hazards," *Chemical Processes* 58 (9), 1995, pp. 5

[13]Zeman, S., Kočí, J., "Electric spark sensitivity of polynitro compounds: Part IV. A relation to thermal decomposition parameters," *Journal of Energetic Materials*, 8 (1), 2000, pp. 23-26

[14]Skinner, D., Olson, D., Block-Bolton, A., "Electrostatic discharge ignition of energetic materials," *Propellants, Explosives, Pyrotechnics,* 23, 1997, pp. 23-42

[15]Matsugi, K., Hatayama, T., Yanagisawa, O., "Effect of direct current pulse discharge on specific resistivity of copper and iron powder compacts," *Journal of the Japan Institute of Metals,* 59 (7), 1995, pp. 740-745

[16]Belyaev, A. A., Gutso, D. E., Kiseev, V. M., "Formation of interparticle contact resistance in powder materials," *Poroshkovaya Metallurgiya,* 10, 1992, pp. 83-86

[17]MIL-STD-1751A, Manual of Data Requirements and Tests for the Qualification of Explosive Materials for Military Use, AOP-7, 2003

[18]Electromagnetic compatibility (EMC) - Part 4-2: Testing and measurement techniques - Electrostatic discharge immunity tests IEC 61000-4-2 Edition: 1.2 International Electrotechnical Commission, 2001

[19]Roberts, T. A., Burton, R. L., Krier, H., "Ignition and combustion of aluminum/magnesium alloy particles in O2 at high pressures," *Combustion and Flame,* 92 (1-2), 1993, pp. 125-143

[20]Ward, T. S., Trunov, M. A., Schoenitz, M., Dreizin, E. L., "Experimental methodology and heat transfer model for identification of ignition kinetics of powdered fuels," *International Journal of Heat and Mass Transfer,* 49 (25-26), 2006, pp. 4943-4954

[21]Star Brite liquid electrical tape MSDS, Star Brite product catalog, available at: http://www.starbrite.com/msdssheets/84104%20msds%202-29-08.pdf.

[22]Pearse, R. W. B., Gaydon, A. G., The Identification of Molecular Spectra 4th ed., Chapman and Hall, London, UK, 1976, pp. 209-211

[23]Atomic Spectra Peaks, NIST Atomic Spectra Database v3, available at: http://physics.nist.gov/PhysRefData/ASD/index.html.

[24]Borghese, A., D'Alessio, A., Diana, M., Veintozzi, C., "Development of hot nitrogen kernel produced by a very fast spark discharge," *Twenty-Second Symposium on Combustion/The Combustion Institute*, 1998, pp. 1651-1659

[25]Kerchner, R. M., Corcoran, G. F., Alternating Current Circuits 4th ed., John Wiley & Sons, New York, 1960, pp. 571-572

[26]Sher, E., Ben-Ya'ish, J., Kravichik, T., "On the birth of spark channels," *Combustion and Flame*, 89, 1992, pp. 186-194

[27]Howatson, A. M., An introduction to gas discharges, Pergamon Press, Oxford, 1965, pp. 47-100

[28]McCahill, M. J., Lee, R. J., Remmers, D. L., "Electrostatic Discharge Sensitivity Testing Revisited," JANNAF

[29]Ono, R., Masahuru, N., Fujiwara, S., Horiguchi, S., Oda, T., "Gas temperature of capacitance spark discharge in air," *Journal of Applied Physics*, 97, 2005, pp. 123307, 1-7

[30]Incropera, F. P., Dewitt, D. P., Bergman, T. L., Lavine, A. S., Introduction of Heat Transfer 5th ed., John Wiley & Sons, New York, 2007, pp. 286-292.

T-JUMP/TIME-OF-FLIGHT MASS SPECTROMETRY FOR TIME RESOLVED ANALYSIS OF ULTRA-FAST SOLID STATE REACTIONS

Lei Zhou, Nicholas Piekiel, Snehaunshu Chowdhury and Michael R. Zachariah*

Department of Mechanical Engineering and Department of Chemistry and Biochemistry
University of Maryland, College Park, 20742, USA

*Corresponding author.
Email:mrz@umd.edu. Phone: 301-405-4311. Fax: 301-314-9477

ABSTRACT

We describe a new T-jump/time-of-flight mass spectrometer for time-resolved analysis of solid state reactions, with a focus on energetic materials. The instrument employs a thin wire substrate which can be coated with the material of interest, and can be rapidly heated (10^5 K/s). The T-Jump probe is inserted with the extraction region of a linear-TOF, which enables multiple spectra to be obtained during a single reaction event. By monitoring the electrical characteristics of the heated wire, the temperature could also be obtained and correlated to the mass-spectra. As example, we present time-resolved spectra for the ignition of nitrocellulose and RDX. The fidelity of the instrument is demonstrated in the spectra presented which show the temporal formation and decay of several species in both systems. A simultaneous measurement of temperature enables us to extract the ignition temperature and the characteristic reaction time. The time resolved mass spectra obtained show that these solid energetic material reactions, under a rapid heating rate, can occur on a time scale of milliseconds or less. While the data sampling rate of 10000 Hz were used in the present experiments, the instrument is capable of a maximum scanning rate of up to ~30 kHz. The capability of high speed time resolved measurements offers an additional analytical tool for characterization of the decomposition, ignition, and combustion of solid energetic materials

KEYWORDS

Temperature-jump, time of flight mass spectrometry, energetic material

INTRODUCTION

Here we report on a new Time-of-Flight mass spectrometer (TOFMS) electron impact (EI) ionization source that can obtain time resolved mass-spectra during the ignition of energetic materials. The unique feature of this apparatus is a) implementation of TOFMS/EI with a Temperature Jump (T-Jump) technique to monitor highly reactive solid state-samples at high heating and decomposition rates, and 2) measurement of the chemistry in a bi-molecular gas-phase-free kinetic environment. Due to its low detection limits and fast time response, the instrument developed here allows for a time resolved characterization of the decomposition, ignition, and combustion of solid energetic materials

Quantitative measurement of the condensed phase reaction kinetics are usually performed using conventional thermal analysis techniques[1] such as TGA and DSC. However, those methods fail in the measurement of fast chemistry processes such as rapid thermal decomposition, ignition and combustion of energetic materials where high heating rates are involved. It is well established that the high heating rates in those processes are critical and must be attained in order to study rapid condensed phase reactions[2-4]. In recent years, many experimental diagnostic methods have been developed to characterize rapid reaction processes[1, 5-12]. In particular, T-Jump/FT-IR spectroscopy was developed for studying reaction kinetics of condensed-phase propellants[2, 13]. In the T-Jump/FT-IR the sample is placed on a Pt filament and rapidly heated to a chosen temperature and the gaseous species are detected and quantified using FTIR spectroscopy. The thermal decomposition behavior of numerous energetic materials under isothermal conditions have been studied using this technique[14-16]. However, for rapid solid state reactions especially those associated with an ignition event, the relevant time scale can be on the order of milliseconds or less. The nominally low IR spectra scanning rate greatly limits the application of the T-Jump/FT-IR spectroscopy in characterizing ignition, and combustion.

Our objective in developing the T-Jump-MS system was first be able to characterize chemistry under high heating rate conditions (i.e. fast chemistry), and second to conduct the experiments under conditions where the secondary gas phase chemistry can be minimized. In the former case high heating rates correspond more closely to the environment usually encountered by energetic materials but more profoundly one should expect reaction channels to increasing favor the higher activation channels possessing the lowest entropy constraints. The later emphasis is that minimizing gas-phase chemistry allows a more direct probe of what is happening in the condensed state. The essence of the experiment is that the T-Jump probe is directly inserted into the Electron Impact (EI) ionization chamber of the mass spectrometer, and the species from T-Jump excitation are monitored by the TOF mass spectrometer continuously. The time-resolved mass spectrometric capabilities of the instrument enable the characterization of rapid solid state reactions, which should provide an insightful complement to conventional thermal analysis. The purpose of this initial paper is to describe the operation and capabilities of this new instrument.

EXPERIMENTAL SECTION

a. EI/TOF Mass Spectrometer.

The EI/TOF mass spectrometer is comprised of a linear Time-of-Flight chamber, adapted from a previously developed Single Particle Mass Spectrometer (SPMS)[11, 17] and includes an electron gun for ionization, and the T-Jump probe with an electrical feedthrough, as shown in figure 1. The sample loading chamber is separated from the ionization chamber by a gate valve, which enables the T-Jump probe to be rapidly changed without the need to break vacuum in the TOF chamber. An electron gun (R. M. Jordan Company) is mounted between the extraction plates of the TOF, and perpendicular to the orientation of the T-jump probe. The electron beam is nominally operated at 70 eV, and 1 mA, with the background pressure in the TOF chamber at ~10^{-7} Torr.

b. T-Jump Sample Probe.

For the T-Jump we have primarily used a 76 μm diameter platinum wire, with a total heated length of ~ 1 cm, which is replaced after each heating event. In each experiment, the wire is coated with a thin layer of either sample powder as in the case of particulates, or solution dipped to prepare organic coatings. Using an in-house built power source, the heating rate of the T-Jump probe can be varied by changing the pulse voltage or pulse width, at a rate of up to ~ 5 x10^5 K/s for the present filament configuration.

c. Control and Data Acquisition System.

The present design is based on a previously developed Single Particle Mass Spectrometer (SPMS) which is configured for a standard laser ionization source[11, 18]. To ensure a field-free region for EI ionization, one DC high voltage power supply is used with a "T" splitter to bias both the repeller plate, and the extraction plate. In the presence of a field-free region, electrons are injected between the plates and ionization takes place. After a predetermined ionization period the voltage on the extraction plate is changed by a high voltage pulser, to create the field for ion extraction region between the plates. The extracted ions drift in the linear TOF tube, and are counted at the MCP detector. Following the ion extraction period, the voltage on the extraction plate is pulsed back, and a new ionization period begins. Serial pulses generated from a TTL pulse generator are used to trigger the high voltage pulser so that the ionization and extraction processes occur continuously. The pulse timing sequence of the high voltage pulse is also traced from the monitor signal output of the high voltage pulser. Both the detector signal and the monitor signal are recorded with a 500 MHZ digital oscilloscope and transferred to a PC for further analysis. The heating of the T-Jump probe is also synchronized with the time-of-flight measurement system by triggering the probe power supply from the TTL pulse generator. The temporal voltage and current of the T-Jump probe during the heating event is recorded, so that a resistivity measurement can be obtained, and related to the instantaneous temperature, which can be mapped against the mass spectra.

RESULT AND DISCUSSION

Before testing the T-Jump/TOF mass spectrometer, T-Jump probe heating experiments were conducted by heating an un-coated wire to evaluate the performance of T-Jump probe. The heating rate of the probe can be varied by changing the heating pulse width and the output pulse intensity. The pulse width can be varied from ~1 ms to ~100 ms, with a maximum output voltage of ~50 V. Figure 2 shows the resulting temporal temperature of the platinum wire from the voltage and current trace. Since the rise time of the heating pulse is in the range of 10 to ~100 us depending on the output pulse voltage, the resistance and the corresponding temperature is calculated after the rise time of the heating pulse. Thus the filament temperature is estimated to be ~400 K initially, and reaches ~1800 K after the 2.5 ms; i.e. a heating rate ~640,000 K/s.

Another important factor to consider in the design of the T Jump/TOF mass spectrometer is the nature of the ion extraction in the presence of the T-jump probe. The nominal configuration of the ion extraction electrode assembly ensures a uniform extraction field between the plates[18]. However the presence of the probe and in particular its location was found experimentally to be a sensitive parameter to both signal sensitivity and resolution. One might reasonably expect that placing the probe too close to the extraction plates would distort the electric field, and result in a decrease in the mass spectrometer's resolution. Placing the probe too far away from the ionization region would lower the concentration of reaction product species in the ionization region, and consequently decrease the sensitivity of the measurement. The effect of the T-Jump probe position was examined experimentally by inserting the probe at different distances from the plates, and monitoring the ion signal from the background gas. The relative water (H_2O^+) peak intensity are plotted in figure 3 as a function of probe position. Each experimental data point is an average of 40 mass spectrum measurements, and normalized by the peak intensity measured without the presence of the T-Jump probe. As a comparison to the experimental data, the effect of the T-Jump probe on the electric field and ion detection was also evaluated by conducting ion-trajectory simulations using Simion[19]. In the simulation, water ions with +1 charge were placed in the center plane of the ion extraction region with a uniform distribution, and their flight trajectories were calculated for the voltages used in the experiment. The relative ion abundance calculated from trajectory simulation is also plotted as the function of probe position in figure 3. Both simulation and experimental data show that the ion signal is significantly decreased when the probe is placed close to the extraction plates. As the probe moves away from the plates, the ion signal increases, and reaches a plateau at a distance of 1.3 cm, suggesting that the presence of the grounded probe significantly perturbs the electric potential in the ion source region. Extraction of the probe further out results in a slight decrease of the ion signal in the experimental data, even though the effect of the probe on the electric field is minimized. This implies that at larger distances sensitivity will be lost for material originating from the probe, and that a distance 1.3 cm would seem to be a near optimal for this system.

Nitrocellulose and hexahydro-1,3,5-trinitro-1,3,5-triazine (RDX) samples were used to test the performance of the T-Jump/TOF mass spectrometer as examples of a slow and fast "burners". In these experiments, Nitrocellulose sample (Mallinckrodt Baker Inc.) or RDX was mixed with diethyl ether or acetone, and a small amount of solution is coated on the T-Jump filament surface using a dropper. While the eventual goal of this instrument is to use the temporal mass spectra, and temperature, to extract mechanistic information, the purpose of the present paper is to illustrate the capabilities of the instrument, and we defer any detailed analysis to the future.

The mass spectra obtained for rapid pyrolysis of nitrocellulose are shown in figure 4. The heating duration is about 9 ms with a heating rate of ~1.3 $\times 10^5$ K/s, with a total of 95 spectra sampled with a temporal resolution of 100 μs per spectrum (10000Hz). Out of the 95 spectra obtained in the experiment, we plot 17 of them in figure 4, along with a more detailed view of a spectrum at t = 2.5 ms. Since the heating pulse is synchronized with the first EI ionization duration, the mass spectrum at t = 0 ms is actually the background in the ion source region, and is primarily water (m/z +18), N_2 (m/z +28) and O_2 (m/z +32). We sampled up to m/z ~ 300 for each spectrum, but no heavy ions were observed, and major peaks are only seen for m/z <100. At t =1.7 ms, the estimated temperature of the probe is ~575 K, and a new peak of m/z 31 (HNO^+) appears which suggest the start of the reaction. At t = 1.8 ms which corresponds to a probe temperature ~590K, the ion signal intensity of m/z 31 increases along with peaks at m/z +27 ($C_2H_3^+$), m/z +29 (COH^+), m/z +45 (HCO_2^+) and m/z +59 (2-oxoethanolate, $C_2H_3O_2^+$).. As the reaction time advances to t = 1.9 ms (T~ 600K), peak m/z 31 achieves its maximum intensity, and now peaks at m/z +30 (CH_2O^+/NO^+) and m/z +46 (NO_2^+) appear, along with m/z +16 (O^+), m/z +43 ($C_2H_3O^+$), m/z +44 (CO_2^+). There is significant increase for m/z +28 during the heating event, which can be attributed to the release of CO from nitrocellulose combustion. The species identified in this experiment are similar to those reported previously at slower heating[20, 21]. These species last for the whole duration of the heating pulse (~9ms), and some species are still present well after the end of the wire heating due to the self-burning of nitrocellulose. The time-resolved feature of the spectra allows us to extract the characteristic time of the reaction. As the highest peak intensity for most of major peaks were achieved at t = 2.3 ms, following which the ion signals gradually decreases with no noticeable changes after 4 ms, it is suggested that the most aggressive reaction (ignition) happens within ~4 ms, and the whole reaction lasts ~9 ms.

For a second example we present results for the ignition and combustion of RDX. Similar to the nitrocellulose experiment, we use a sampling rate of 100 μs per spectrum (10,000Hz) to capture the progression of the reaction. The heating pulse is about 8 ms at a heating rate of ~1.5×10^5 K/s, and a total of 95 spectra obtained. Figure 5 shows that species, other than background (water/N_2/O_2), only appear from 0.7 ms - 2.6 ms, which corresponds to a wire temperature of 370K to 670K. These results clearly show as expected that RDX is more facile than Nitrocellulose, and occurs over an interval of only ~2 ms. Although a m/z range up to 400 was recorded for each spectrum, no heavy ions were observed above m/z 150. The major ions from RDX decomposition observed are m/z +15, +28, +29, +30, +42, +46, +56, +75 and +127. Small peaks of m/z +14, +16, +41, +43, +81, +120 are also found in some spectra.

Based on the experimental results presented above, it can be clearly see that the characteristic reaction time for energetic materials decomposition/combustion is in the order of milliseconds or even less. We note that while a time resolution of 100 μs was used to test and demonstrate the instrument, the T-Jump/TOF mass spectrometer can be operated with a scanning rate up to ~ 30 kHz. The experimental results suggest that the time-resolved spectra obtained using T-jump/TOF mass spectrometer shod have sufficient sensitivity, and time resolution to probe the reaction dynamics of extremely fast condensed state reactions at high heating rates.

CONCLUSION

A new Time-of-Flight mass spectrometer (TOFMS) combined with a temperature jump technique is described. The instrument allows for the time resolved characterization of the decomposition, ignition, and combustion of solid energetic materials or other highly reactive condensed state reactions. Using heating rates of up to 10^5 K/sec, samples of nitrocellulose and RDX were ignited, and time resolved mass spectra were obtained. By monitoring the electrical characteristics of the heated wire, the temperature could also be obtained and correlated to the mass-spectra. When combined with the time dependent temperature information, the results indicate that the instrument can capture the signature of rapid solid-state reactions in a time resolved manner.

REFERENCES

(1) Vyazovkin, S. *Analytical Chemistry* **2006**, *78*, 3875-3886.
(2) Brill, T. B.; Brush, P. J.; James, K. J.; Shepherd, J. E.; Pfeiffer, K. J. *Applied Spectroscopy* **1992**, *46*, 900-911.
(3) Brill, T. B.; Arisawa, H.; Brush, P. J.; Gongwer, P. E.; Williams, G. K. *J. Phys. Chem.* **1995**, *99*, 1384-1392.
(4) Thynell, S. T.; Gongwer, P. E.; Brill, T. B. *Journal of Propulsion and Power* **1996**, *12*, 933-939.
(5) Furutani, H.; Fukumura, H.; Masuhara, H.; Kambara, S.; Kitaguchi, T.; Tsukada, H.; Ozawa, T. *J. Phys. Chem. B* **1998**, *102*, 3395-3401.
(6) Roberts, T. A.; Burton, R. L.; Krier, H. *Combustion and Flame* **1993**, *92*, 125-143.
(7) Ward, T. S.; Trunov, M. A.; Schoenitz, M.; Dreizin, E. L. *International Journal of Heat and Mass Transfer* **2006**, *49*, 4943-4954.
(8) Dauerman, L.; Salser, G. E.; Tajima, Y. A. *Aiaa Journal* **1967**, *5*, 1501-&.
(9) Blais, N. C.; Fry, H. A.; Greiner, N. R. *Review of Scientific Instruments* **1993**, *64*, 174-183.
(10) Makashir, P. S.; Mahajan, R. R.; Agrawal, J. P. *Journal of Thermal Analysis* **1995**, *45*, 501-509.
(11) Mahadevan, R.; Lee, D.; Sakurai, H.; Zachariah, M. R. *J. Phys. Chem. A* **2002**, *106*, 11083-11092.
(12) Trunov, M. A.; Schoenitz, M.; Dreizin, E. L. *Propellants Explosives Pyrotechnics* **2005**, *30*, 36-43.
(13) Kim, E. S.; Lee, H. S.; Mallery, C. F.; Thynell, S. T. *Combustion and Flame* **1997**, *110*, 239-255.
(14) Brill, T. B.; Gongwer, P. E. *Propellants Explosives Pyrotechnics* **1997**, *22*, 38-44.
(15) Brill, T. B.; Ramanathan, H. *Combustion and Flame* **2000**, *122*, 165-171.
(16) Hiyoshi, R. I.; Brill, T. B. *Propellants Explosives Pyrotechnics* **2002**, *27*, 23-30.
(17) Lee, D.; Park, K.; Zachariah, M. R. *Aerosol Science and Technology* **2005**, *39*, 162-169.
(18) Wiley, W. C.; McLaren, I. H. *Review of Scientific Instruments* **1955**, *26*, 1150-1157.
(19) Dahl, D. A., V.8.0 ed.; Scientific Instrument Services, Inc. www.simion.com: Ringoes, NJ.
(20) Fowler, A. H. K.; Munro, H. S. *Polymer Degradation and Stability* **1985**, *13*, 21-29.
(21) Dauerman, L.; Tajima, Y. A. *Aiaa Journal* **1968**, *6*, 1468-&.

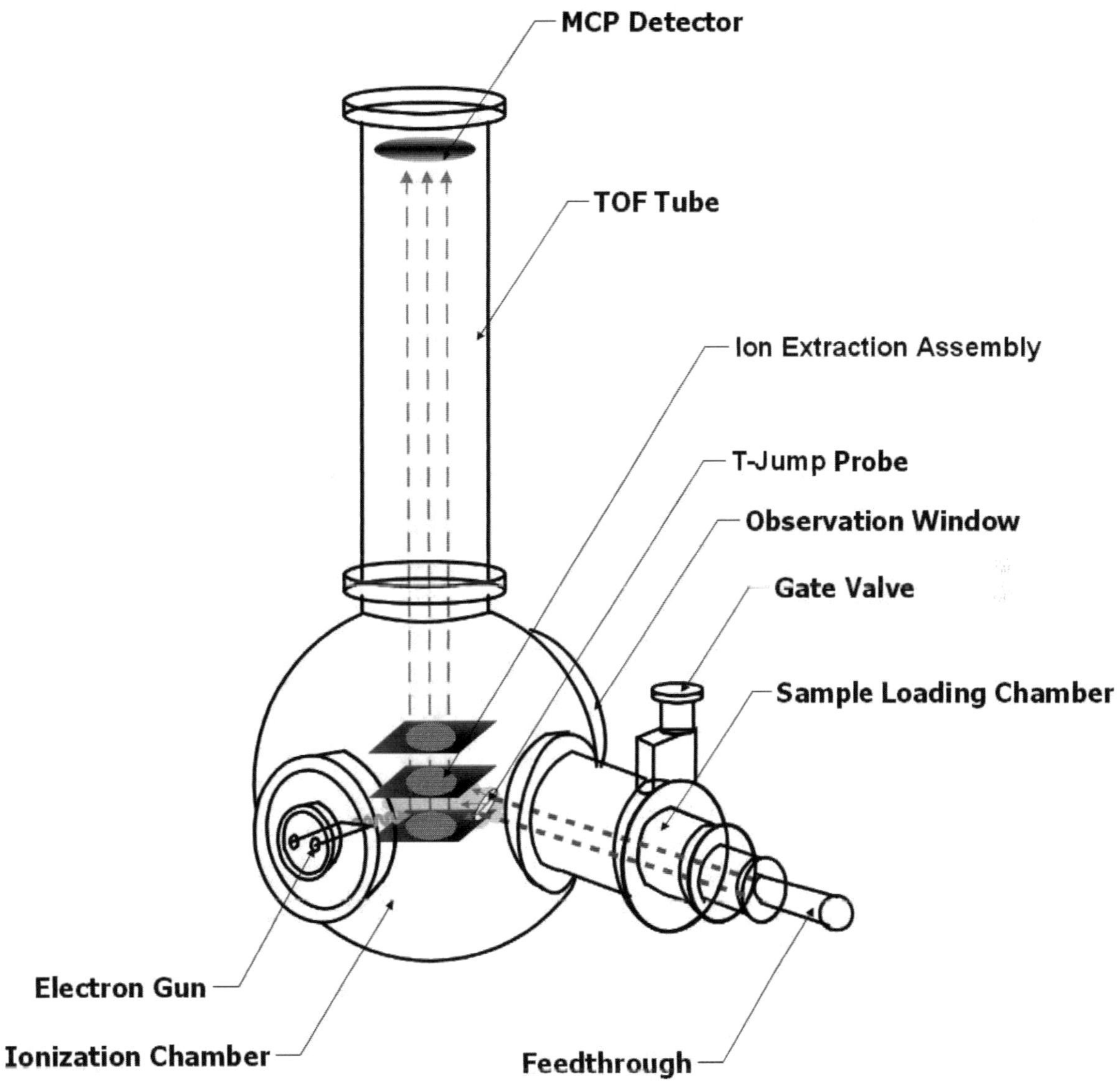

Figure 1. Diagram of T-Jump/TOF mass spectrometer.

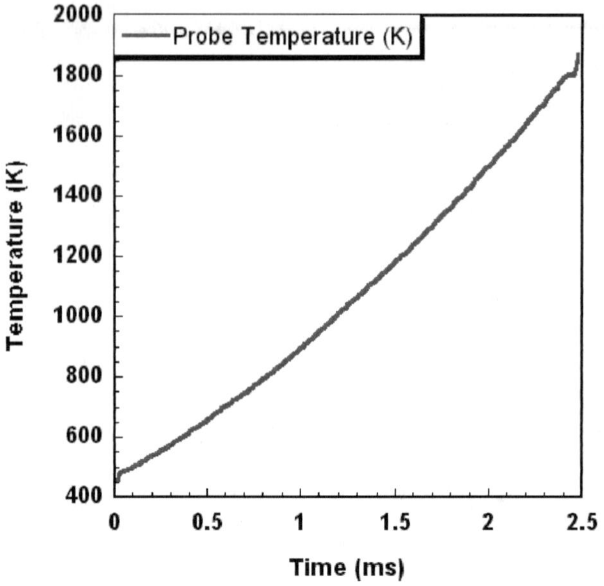

Figure 2 Estimated probe temperature from electrical resistance

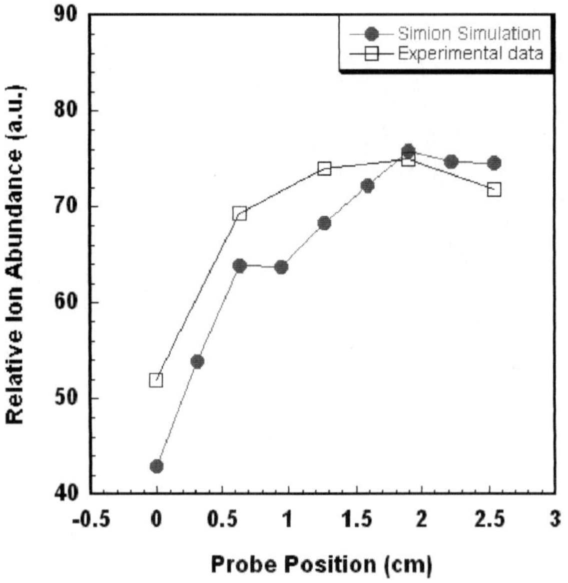

Figure 3 Relative water ion abundance as a function of the probe filament position.

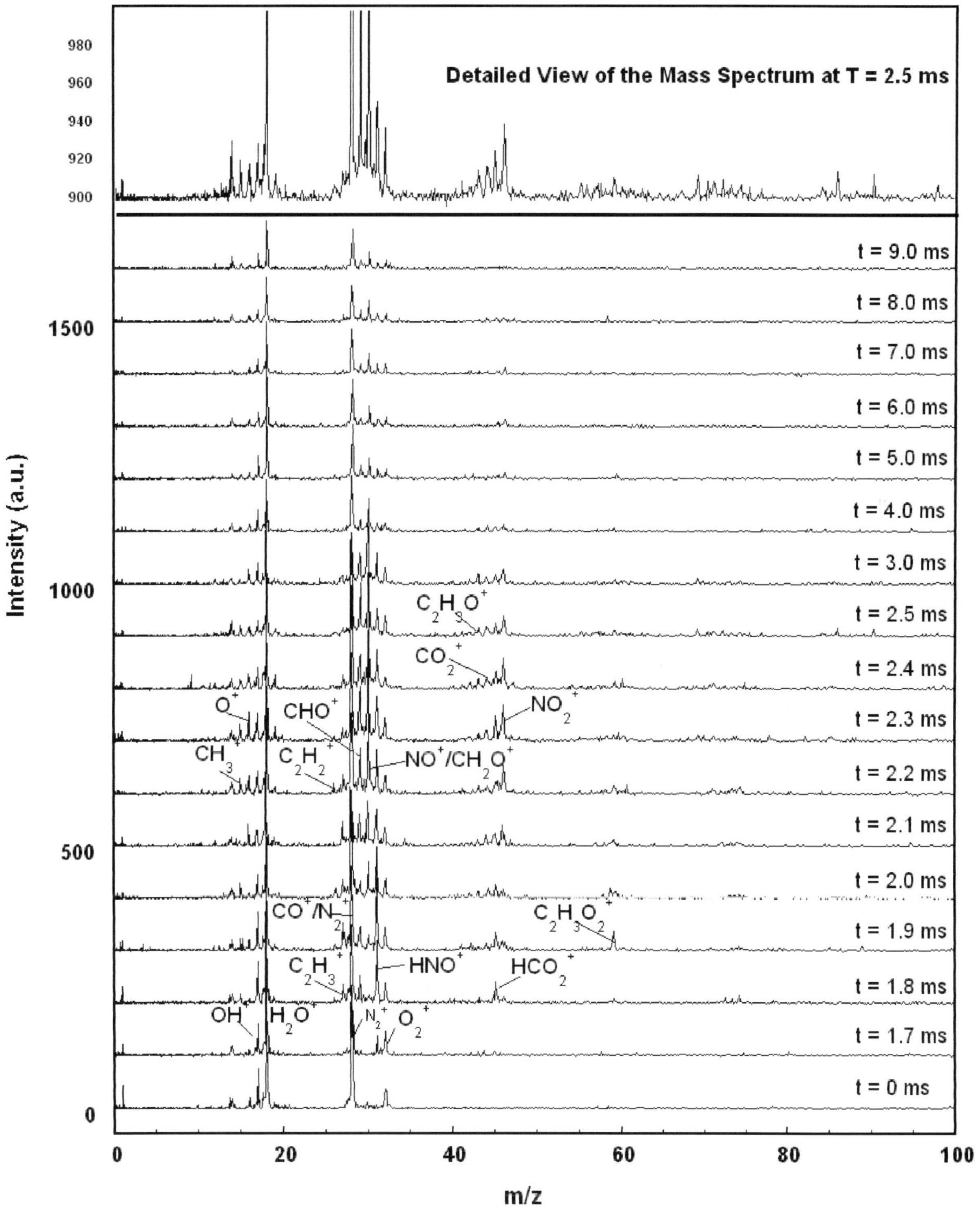

Figure 4. Time resolved mass spectrum from rapid heating of nitrocellulose. Heating rate ~ 1.3×10^5 K/s

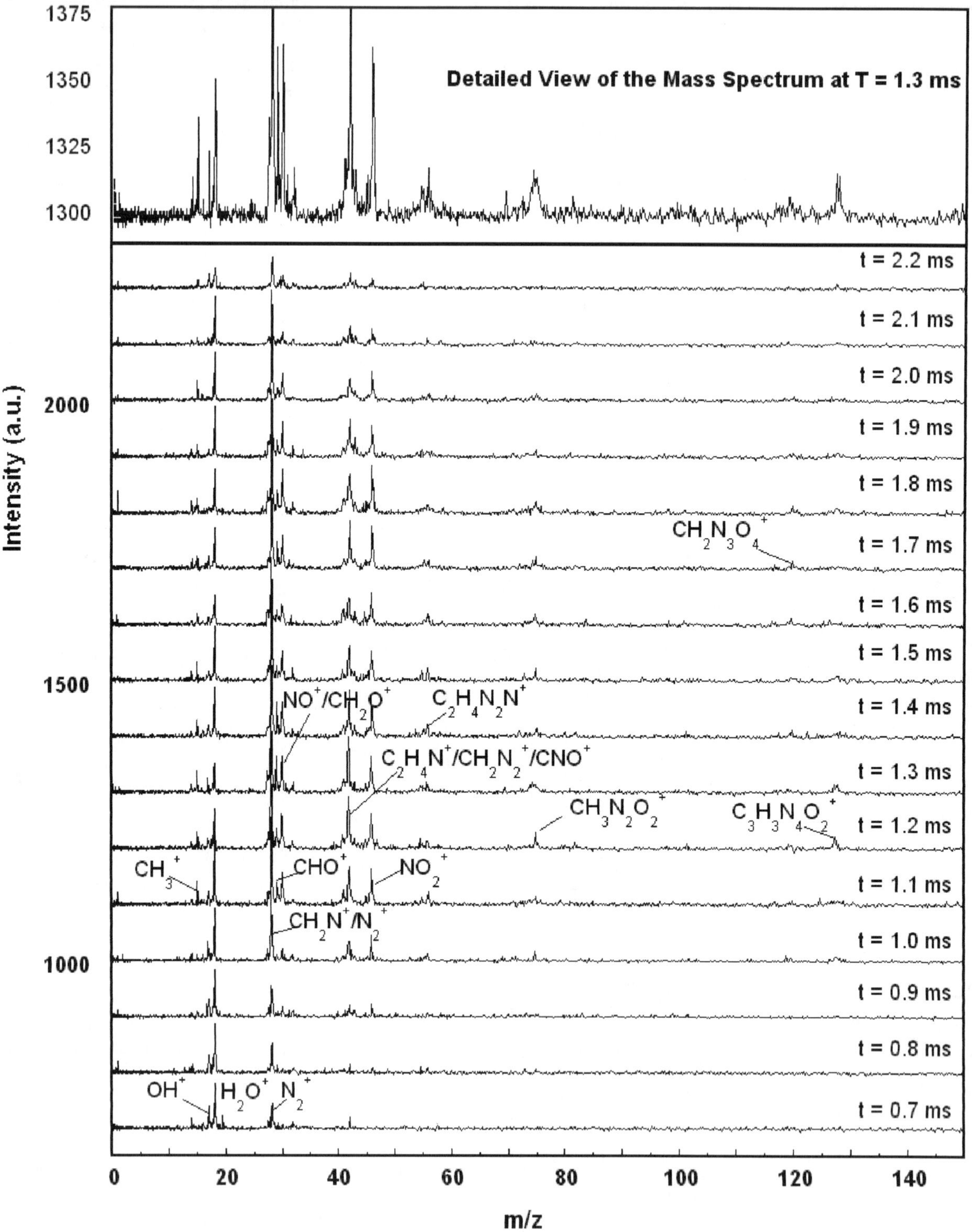

Figure 5. Time resolved mass spectrum from rapid heating of RDX. Heating rate ~1.5 $\times 10^5$ K/s

716e Thermophysical Property Prediction of Energetic Materials with Atomistic Computer Simulations

Nandhini Sokkalingam, Marybeth H. Ketko, and Jeffrey J. Potoff. Department of Chemical Engineering, Wayne State University, 5050 Anthony Wayne Dr, Detroit, MI 48202

There has been considerable interest in the development of new Insensitive Munitions (IM) compounds due to their thermal stability and low shock sensitivity over traditionally used explosive compounds. Six such compounds are 2,4-dinitroanisole (DNAN), N-methyl-p-nitroaniline (MNA), Dinitropyrazole (DNP), Nitrotriazolone (NTO), 1-methyl-2,4,5-trinitroimidazole (MTNI) and 1,3,5-triamino-2,4,6-trinitrobenzene (TATB). Due to the increased environmental and safety concerns, it is necessary to determine how these compounds behave in the environment in terms of bioaccumulation potential and aqueous solubility. A pre-biological screen [1] based on the knowledge of octanol-water partition coefficients (log K_{ow}) and Henry's law constants (log H) is useful in this aspect. Atomistic simulation is attractive as a predictive tool for assessing the environmental impact of these compounds due to their hazardous nature. An accurate atomic force field with an appropriate methodology is the key to predicting accurate thermophysical properties.

In this work, we develop molecular models, or "force fields" for the explosives to predict their octanol-water partition coefficients, Henry's law constants and also vapor-liquid equilibria, vapor pressure and critical points. For the non-bonded interactions, Lennard-Jones parameters for each interaction site are obtained from the TraPPE-UA [2,3,4] and TraPPE-EH force fields [5] from similar functional groups. The partial charges are fitted to reproduce partition coefficients and Henry's law constants from the experiment. NPT molecular dynamics simulations coupled with the free energy perturbation technique is used to compute the physicochemical properties and Gibbs-Duhem integration is used to determine the vapor-liquid equilibria, vapor pressure and critical parameters. The log K_{ow} values predicted for DNAN and MNA are 1.648 and 1.934, respectively and are within 3 % of experiment.

[1] J.W.Gillett, Environmental Toxicology and Chemistry. 2, 463 (1983).

[2] C.D.Wick, J.M.Stubbs, N.Rai, J.I.Siepmann, J.Phys.Chem.B, 109 (18974).

[3] M.G.Martin, J.I.Siepmann, J.Phys.Chem.B, 104 (8008).

[4] J.M.Stubbs, J.J.Potoff, J.I.Siepmann, J.Phys.Chem.B, 108 (17596).

[5] N.Rai, J.I.Siepmann, J.Phys.Chem.B, 111 (2007).

716f Monte Carlo Simulations of 1,3,5-Triamino-2,4,6-Trinitrobenzene (TATB): Pvt Data, Phase Equilibria, and Solubility

J. Ilja Siepmann[1], Neeraj Rai[1], John L. Lewin[2], Divesh Bhatt[2], and Laurence E. Fried[3]. (1) University of Minnesota, Department of Chemistry, Minneapolis, MN 55455-0431, (2) Department of Chemistry, University of Minnesota, 207 Pleasant St. SE, Minneapolis, MN 55455-0431, (3) Chemical Sciences Division, Lawrence Livermore National Laboratory, Livermore, CA 94550

The transferable force field for phase equilibria (TraPPE) force field is extended to nitro and amino substituents for aromatic rings via parameterization to the vapor-liquid coexistence curves of nitrobenzene and aniline, respectively. These groups were then transferred to model 1,3,5-triamino-2,4,6-trinitrobenzene (TATB). Without any further parameterization to solid state data, the TraPPE force field is able to predict TATB's unit cell lengths and angles at 295 K with mean unsigned percentage error (MUPE) of 0.3 and 1.1% and the specific density within 1%. These predictions are comparable in accuracy to the GRBF model [Gee et al., J. Chem. Phys. 120, 7059 (2004)] that was parameterized directly to TATB's solid state properties. Both force fields are able to reproduce the pressure dependence of TATB's unit cell volume, but they underestimate its thermal expansion. Due to its energetic nature and unusually large cohesive energy, TATB is not chemically stable at temperature in its fluid range and its solubility in common solvents is very low. Gibbs ensemble simulations are used to predict TATB's vapor-liquid coexistence curve at elevated temperatures, melting point, and solubility in common solvents and two ionic liquids.

Web Page: www.chem.umn.edu/groups/siepmann/

Prediction of physicochemical properties of energetic materials via EPI Suite

Siri Chakka (Speaker), Veera M. Boddu*, Stephen W. Maloney
Environmental Processes Branch, U.S. Army Engineer Research and Development Center
Construction Engineering Research Laboratory (ERDC-CERL)
P.O. Box 9005, Champaign, IL 61826-9005
and
Reddy Damavarapu
Energetics, Warheads & Technology Directorate; AMSRD-ARR-AEE-E
U.S. Army Armament Research, Development and Engineering Center (ARDEC)
Picatinny Arsenal, NJ 07806-5000

Abstract

In order to maintain environmental compliance and mission readiness for the Army industrial base, it is necessary to assess the environmental impacts of munitions compounds. Informed decisions on development of waste treatment technologies can be made by close examination of physical properties of the munitions compounds. Information on the physical properties will help predict the environmental fate of these materials and make it possible to select appropriate methods of prevention and treatment of hazardous effects on the environment.

Here, we present the estimations of physicochemical properties such as solubility, octanol-water partition coefficient (K_{ow}), Henry's Law Constant (K_H), organic carbon adsorption (K_{oc}), bioconcentration factor (BCF), aquatic toxicity (LC50), dermal permeability coefficient (K_p), the hydrolysis rate constant, biodegradation probability, boiling point (BP), vapor pressure (VP), and melting point (MP), using the Estimation Programs Interface (EPI) Suite.(1) Predictions at room temperature are carried out for the energetic materials dinitroanisole (DNAN), n-methyl-p-nitroaniline (MNA), nitro-triazolene (NTO), triaminotrinitrobenzene (TATB), cyclotetramethylene-tetranitramine (HMX), cyclotrimethylenetrinitramine (RDX) and trinitrotoluene (TNT). The EPI Suite requires only the chemical structure or the Chemical Abstracts Service (CAS) number to estimate the properties. Results of the EPI Suite predictions of the physical properties of the above munitions compounds and comparisons with the available literature values are presented.

Introduction

EPI Suite predicts physicochemical properties and is a relatively convenient means of studying organic materials. When experimental data are not available to assess environmental risk, a possible way to estimate the necessary values is the use of estimation models. The EPI Suite was developed to help environmental scientists to prepare profiles for a wide array of chemical profiles.(1) The fact that the program simply requires the chemical structure or Chemical Abstract Service (CAS) number to generate all the predicted and experimental values has simplified its use.

Aqueous solubility of a chemical compound is useful to predict the aquatic distribution of the material and to determine the most efficient method of treating aqueous wastes containing the compound. Solubility is also important to determine the many other parameters being

* Veera.Boddu@usace.army.mil, phone: 001-217-398-5511

estimated by the EPI Suite. The octanol-water partition coefficient, K_{ow}, represents the differential solubility of the compound between two solvents, one being water and the other being 1-octanol. K_{ow} predicts potential migration of the dissolved hydrophobic organic compounds to soil and groundwater matrices. This value is important in estimating the environmental distribution of a compound in groundwater which could result in pollution of waterways and biota. Prediction of the Henry's Law Constant, K_H, value is important in order to calculate the vapor pressure, which helps estimate the condition of the aerial surroundings of wastestreams. The K_H value also estimates the partition between water and air, providing an indication of which of the two phases a chemical will lean towards at equilibrium. The organic carbon adsorption, K_{oc}, value is important in predicting the amount of a compound that is able to be absorbed by soil, which may affect plant growth and groundwater supplies. The K_{oc} value also depends upon temperature, pH, particle size distribution, concentration, ratio between solids and solution, volatility of the compound, degradation of the compound, and time. The bioconcentration factor, BCF, is important in predicting a compound's tendency to build up in marine organisms. The Lethal Concentration, LC50 is important to predict toxicity of a compound in the aquatic organisms. Dermal permeability is important to assess the extent a toxic munitions compound can penetrate into the skin of animals and its effects upon organisms' organs. Predicted hydrolysis rate constants are important in order to estimate how fast a munitions compound will undergo hydrolysis at various pHs of a solution. This information would aid in developing more efficient methods in cleaning up wastewater from industrial facilities. Biodegradation probability is important to predict the fate of a compound within the environment if left untreated. The boiling point (BP), melting point (MP), and the vapor pressure (VP) are important properties useful to predict the relationship between the munition and temperature and their distribution in the atmosphere. The BP also indicates the volatility of a compound. The VP is important in evaluating the distribution of a compound into the environmental compartments of soil, air, and water.

* Veera.Boddu@usace.army.mil, phone: 001-217-398-5511

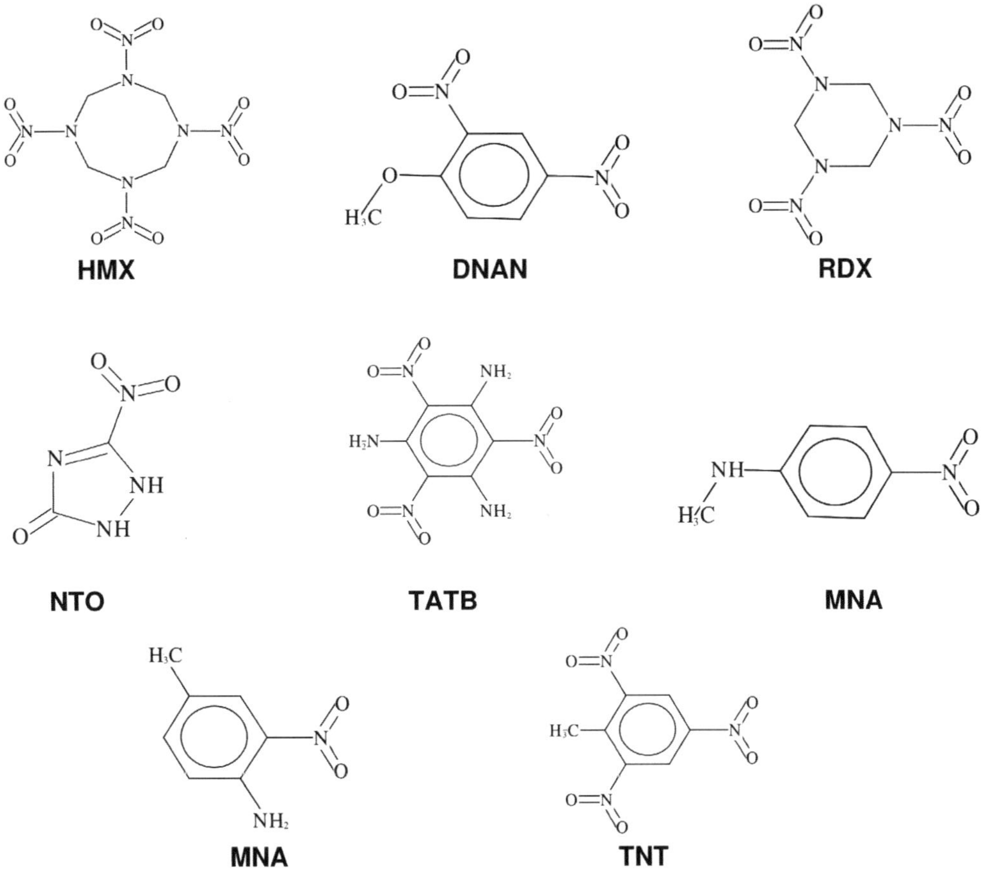

Figure 1. Chemical structures of the munitions compounds studied in this work

Methods

The water solubility of an organic compound is estimated by utilizing the K_{ow} and the chemical structure.[(2)] The EPI Suite presents two correlations for the solubility. The equation (1) is based on a validation set of 85 substances and an experimental $\log(K_{ow})$ but no melting point. The equation (2) is based upon 817 compounds with measured water solubilities and melting points. The two solubility correlations are:

$$\log S\ (mol/l) = 0.796 - 0.854 \log K_{ow} - 0.00728MW + c_f \qquad (1)$$

$$\log S\ (mol/l) = 0.693 - 0.96 \log K_{ow} - 0.0092(MP\text{-}25) - 0.00314MW + c_f \qquad (2)$$

where MW is the molecular weight, MP is the melting point, and "cf" is the correction factor.

The K_{ow} value is calculated by taking the ratio of a chemical's concentration in the octanol-phase and the concentration in the aqueous phase of a two-phase system at equilibrium. The $\sum(f_i n_i)$ is the summation of f_i (the coefficient for each atom or fragment) and times n_i (the number of times, the atom/fragment occurs in the structure). The $\sum(c_j n_j)$ is the summation of c_j (the coefficient for each correction factor) and times n_j (the number of times, the correction factor occurs or is applied in the structure). The atom/fragment contribution method was developed by multiple linear regressions of experimental log (K_{ow}) values in two sets. The first set corresponded to the log (K_{ow}) values and the second set corresponded to the

* Veera.Boddu@usace.army.mil, phone: 001-217-398-5511

correlated correction factors. The log (K_{ow}) was then estimated by summing up the values from the chemical structure of the chemical compound as shown in equation (3).[3]

$$\log K_{ow} = \Sigma(f_i n_i) + \Sigma(c_j n_j) + 0.229 \quad (n=2351,\ r^2=0.982,\ s_d=0.216) \qquad (3)$$

The K_H value is calculated by using both the group contribution and the bond contribution methods of Hine and Mookerjee over the temperature range of 0 to 50 °C.[4] The EPI Suite modifies this calculation by developing new fragment constants from experimental data.[5] The K_H estimations are based upon group and bond contributions which are derived from experimentally measured log ($K_{air\text{-}water}$) values.

$$K_H = \text{vapor pressure x molecular weight / water solubility [Pa m}^3\text{/mol]} \qquad (4)$$

The K_{oc} value is the ratio of the chemical adsorbed per unit weight of organic carbon in the soil to the concentration of the chemical in solution at equilibrium. Chemical structure of the compound is input following the Simplified Molecular Input Line Entry System (SMILES) notation in the EPI Suite. A value from a correlation to the molecular connectivity indices and correction factors depending on the organics' chemical class is incorporated into the final value of the K_{oc}.[6]

The BCF is estimated by the program by retrieving the BCF data in a file that contains information on measured BCF and other key experimental details. The log (BCF) was regressed against log (K_{ow}), and chemicals with significant deviations from the line of best fit were analyzed according to chemical structure. The results in the algorithm classifies substances as either ionic or nonionic. The ionic substances were further divided into carboxylic acids, sulfonic acids and their salts, and quaternary N compounds. Log (BCF) for nonionics is estimated from log (K_{ow}) and a series of correction factors specific to each chemical.[7]

LC50 values are extracted from a collection of experimental data submitted by industry based on measured test data, or they are developed by other sources for chemicals with similar structures. By combining experimental aquatic toxicity values and estimated K_{ow} values, regression equations are developed for each class of chemicals. The toxicity values for new chemicals are calculated (eq. 5) using the estimated K_{ow} value into the regression equation and adjusting the result based on the molecular weight of the compound. [8]

$$\log LC50\ (96h) = -0.73 \log(K_{ow}) - 2.16 \qquad (5)$$

The melting point is estimated by both the Joback group contribution method [10] and by a correlation drawn between the MP and BP values.[11] The two MP values derived by both methods are then compared to the chemical structure of the organic compound. The magnitude of the difference of both values and a corrected MP is estimated. The BP is estimated by the Stein method (eq. 6).[9] The group increment value is "g_i," and "n_i" is the number of times the group occurs in the compound. The resulting BP is then corrected with either equation 7 or 8 based on the temperature.

* Veera.Boddu@usace.army.mil, phone: 001-217-398-5511

$$BP = 198.2 + \sum n_i\, g_i \qquad (6)$$

$$BP(corr.) = BP - 94.84 + 0.5577\, BP - 0.0007705\, (BP)\ [BP<700^{o}K] \qquad (7)$$

$$BP(corr.) = BP + 282.7 - 0.5209\, BP \qquad [BP>700^{o}K] \qquad (8)$$

The vapor pressure is estimated by the Antoine equation [12], the modified Grain method, and the Mackay method.[11] The program then calculates a suggested VP based on which method is recommended for the chemical's class. is The isothermal retention index is I(x), and A, B, and C are constants that depend on the substance.

$$I(x) = A + \frac{B}{Tc + C} \qquad (9)$$

The dermal permeability K_p is estimated using one general estimation equation and three specific class estimation equations. The dermally absorbed dose per event is predicted by both the input of the duration of the event and the concentration of the chemical in water.[13] The hydrolysis rate constant was predicted by calculating a second-order acid or base-catalyzed hydrolysis rate constant at 25°C, and the acid and base catalyzed half-lives are calculated within pH values of 7 and 8. [14]

$$K_p = \text{acid-catalyzed} + \text{base catalyzed} + \text{neutral hydrolysis rate constant} \qquad (10)$$

The biodegradation probability is based on multiple linear and non-linear regressions and biodegradation experimental data. It calculates the probability of rapid biodegradation from the experimental data and then estimates the primary and holistic biodegradation times for complete degradation using evaluations of 200 chemicals by 17 biodegradation experts.[16] The updated model works upon this already standing model along with two new predictive models, based on the Japanese Ministry of International Trade and Industry (MITI) biodegradation test.[15] The predictions for the hydrolysis rate and biodegradation probability were unable to be extracted from the EPI Suite for the munitions compounds.

The hydrolysis rate is only available for esters, carbamates, epoxides, halomethanes and specific alkyl halides.

Summary:

The EPI Suite predictions of the physical properties of the munitions compounds and comparisons with the available literature values are summarized in Table 1. The comparison with literature values shows that the predicted values differ substantially with experimental values. The table shows that there are very large variations in the predicted values depending on the model chosen for solubility and Henry's Law Constant (K_H). In the case of solubility, none of the predicted values developed so far are close to the values measured in the laboratory at ERDC-CERL. The variation in octanol-water coefficient (K_{ow}) is predicted in a

* Veera.Boddu@usace.army.mil, phone: 001-217-398-5511

much narrower range (although it is on a logarithmic scale), and comes closer to the measured values. However, the EPI Suite provides an estimate where no experimental data are available. The work is in progress, and a detailed analysis of the results will be reported later.

* Veera.Boddu@usace.army.mil, phone: 001-217-398-5511

Table 1. Summary of physical properties of the munitions compounds

Munitions component	S_w (mg/L)	Log(K_{ow})	K_H (atm-m^3/mole)	K_{oc}	BCF	LC50 (mg/L)	K_p (cm/hr)	BP (°C)	VP (mmHg)	MP (°C)
DNAN	632.1 1741 632 (Henry's LC) .14 (Method of Klopman)[17] .159 (COSMOtherm)[17] 3706.5 (Ideal Solubility Eq.)[17] 155 @ 15° C[18] 276.2 @ 25° C[24] 399.2 @ 30° C[24] 560.0 @ 35° C[24]	1.71 .537 (1C Model [32]) 1.471(4C Model [32]) 1.648(6C Model [32]) 1.948(10C Model [32])[17] 1.70(Ghose, group)[17] 1.38 (Broto, group)[17] 1.92 (COSMOtherm)[17] 1.612 @ 25° C[24] 1.549 @ 30° C[24] 1.472 @ 35° C[24]	3.01E-7 [4] 4.96E-9 [4] 5.91E-8 1.16E-6 (1C Model[32]) 1.12E-4 (6C Model[32]) 4.15E-4 (10C Model[32]) 1.366 @ 25° C[24] 1.397 @ 30° C[24] 1.442 @ 35° C[24]	159.9	4.14	475.9	1.93E-3	319.62 (Stein and Brown)[9] 206 @ 12mmHG (Exp.)	.000145 (Modified Grain)	96.56 (Weighted) 94.5 (Exp.)
MNA	1344 645.5 1.34E3 (Henry's LC) 980.91 [33] 1.17 (Method of Klopman)[17] .347 (COSMOtherm)[17] 971.3 (Ideal Solubility Eq.)[17] 85.42 @ 25° C[21] 111.51 @ 30° C[21] 141.93 @ 35° C[21]	2.02 .902 (1C Model [32]) 1.724 (4C Model [32]) 1.734 (6C Model [32]) 2.523 (10C Model [32]) 1.574 (Ghose, group)[17] 1.47 (Broto, group)[17] .807 (COSMOtherm)[17] 2.04 (Hansch, 1995) 2.1028 @ 25° C[21] 1.9846 @ 30° C[21] 1.9377 @ 35° C[21]	1.65E-8 [4] 1.25E-7 2.19E-4 (6C Model[32]) .6100 25° C[21] .5566 @ 30° C[21] .5160 @ 35° C[21]	74.97	3.16	196.3	6.31E-3	259.24 (Stein and Brown)[9]	.000839 (Modified Grain)	60.96 (Weighted) 152 (Exp.)
NTO	3.859E5 3.86E5 (Henry's LC) 7.56 (Method of Klopman)[17] 9.97 (COSMOtherm)[17] 55.8 (Ideal Solubility Eq.)[17]	-2.72 -.82 (Broto, group)[17] -1.19 (COSMOtherm)[17]	4.07E-13 [4] 2.58E-13	125.9	3.16	2.3E6	3.59E-6	389.94 (Stein and Brown)[9]	5.83E-7 (Modified Grain)	161.19 (Weighted)
TATB	2.626E5 663.95 [33] 2.544E6 .16 (Method of Klopman)[17] .197 (COSMOtherm)[17] 2.5 (Ideal Solubility Eq.)[17]	-1.28 -.44 (Ghose, group)[17] -1.5 (Broto, group)[17] 4.74 (COSMOtherm)[17]	8.6E-17 [4] 2.08E-19 [4] 2.044E-17	873.2	3.16	2.5E5	6.29E-6	481.26 (Stein and Brown)[9]	1.58E-11 (Modified Grain) 3.00E-3 mmHG @ 175 C (Exp.)	203.85 (Weighted) 350 (Exp.)
HMX	9438 5[25]	.82 .16[20]	8.67E-10 [4] 9.951E-13	1853	3.16 11[23]	4239.4	3.86E-5	436.41 (Stein and Brown)[9]	2.41E-8 (Modified	182.89 (Weighted)

* Veera.Boddu@usace.army.mil, phone: 001-217-398-5511

	1E6 [33]								Grain) 3.3E-14 (Exp.)	
RDX	6062 59.7[18] 59.9[25] 44.32 @ 22° C[26] 40.0[27] 36.7 @ 20° C[28] 50.0 @ 20° C[29] 44.42 @ 20° C[26] 40.0 @ 20° C[30] 40.2[31] 1E6 [33]	.68 .87 (Exp.)	6.32E-8 [4] 6.46E-11	195.4	3.16 15[23]	4210.2 22.98[2] 2 25.64[2] 2	3.49E-4	353.43 (Stein and Brown)[9]	1.34E-6 (Modified Grain) 4.1E-8mmHG (Exper.)	205.5 (Exp.)
TNT	552.5 75.537 [33] 130[19] 100.5[25]	1.99 1.60[20]	3.65E-10 [4] 3.30E-9 [4] 4.57E-7 (Exp.)	1834	3.40	311.15	.000801 (Mackay) 5.24E-7 (Koa)	364.14 (Stein and Brown)[9]	1.72E-5 (Grain) 8.02E-6mmHG (Exper.)	124.36 (Weighted) 80.1 (Exper.)

*Unless specified, all values are at 25° C.
Exp. – Taken from EPI's experimental database

* Veera.Boddu@usace.army.mil, phone: 001-217-398-5511

References

1. "Estimation Software." Syrres.com. 1999. Syracuse Research Corporation. 15 February 2008. <http://www.syrres.com/esc/est_soft.htm>.
2. Meylan, W.M. P.H. Howard and R.S. Boethling. 1996. Improved method for estimating water solubility from octanol/water partition coefficient. Environ. Toxicol. Chem. 15: 100-106.
3. Meylan, W.M. and P.H. Howard. 1995. Atom/fragment contribution method for estimating octanol-water partition coefficients. J. Pharm. Sci. 84: 83-92.
4. Hine and Mookerjee. Journal of Organic Chemistry, volume 40, pp 292-8 (1975).
5. Meylan, WM & Howard, PH. 1991. Bond contribution method for estimating Henry's law constants. Environ. Toxicol. Chem. 10: 1283-1293.
6. Meylan, W., P.H. Howard and R.S. Boethling, "Molecular Topology/Fragment Contribution Method for Predicting Soil Sorption Coefficients", Environ. Sci. Technol. 26: 1560-7 (1992).
7. Meylan, WM, Howard,PH, Boethling,RS et al. "Improved Method for Estimating Bioconcentration / Bioaccumulation Factor from Octanol/Water Partition Coefficient", Environ. Toxicol. Chem. 18(4): 664-672 (1996).
8. "New Chemicals Program." EPA.gov. 2008. U.S Environmental Protection Agency. 20 March 2008. <http://www.epa.gov/oppt/newchems/>.
9. Stein and Brown method Journal of Chemical Information and Computer Science, Volume 34, pp. 581-7 (1994)
10. Reid, RC. The Properties of Gases and Liquids. NY: McGraw Hill, 1987
11. Neely, WB, Blau, GE. Environmental Exposure From Chemicals. CRC Press, 1985
12. Lyman, WJ. Handbook of Chemical Property Estimation Methods. Washington, DC: American Chemical Society, 1990
13. EPA document Dermal Exposure Assessment: Principles and Applications (EPA/600/8-91-011B, January 1992).
14. Mill,T., Haag,W., Penwell,P.,Pettit,T. and Johnson, H. "Environmental Fate and Exposure Studies Development of a PC-SAR for Hydrolysis: Esters, Alkyl Halides and Epoxides". EPA Contract No. 68-02-4254. Menlo Park, CA: SRI International (1987).
15. Tunkel, J., Howard,P.H., Boethling,R.S., Stiteler,W. and Loonen,H. 2000. Predicting Ready Biodegradability in the MITI Test. Environ. Toxicol. Chem. accepted for publication.
16. Boethling, R.S., Howard, P.H., Meylan, W., Stiteler, W., Beaumann, J. and N. Tirado. 1994. Group contribution method for predicting probability and rate of aerobic biodegradation. Environ. Sci. Technol. 28: 459-65.
17. Rebecca K. Toghiani, Hossein Toghiani, Stephen W, Maloney, Veera M. Boddu, Prediction of physicochemical properties of energetic materials. Fluid phase Equilibria 264 (2008) 86-92
18. Myrdal, Paul, Ward, Gary H, Dannenfelser, Rose-Marie, Mishra, Dinesh, Yalkowsky, S.H. AQUAFAC 1. Aqueous functional group activity coefficients; application to hydrocarbons. Chemosphere. 24:8 (1992) 1047-1061
19. Seidell, Atherton. The solubility of acetanilide, phenacetine, caffeine and salol in several solvents. J. Am. Chem. Soc., 29, 7 (1907) 1088 – 1091
20. Monteil-Rivera, Fanny, Groom, Carl, Hawari, Jalal. Sorption and Degradation of Octahydro-1,3,5,7-tetranitro-1,3,5,7-tetrazocine. Soil. Environ. Sci. Technol., 37 (17), 3878 -3884, 2003.
21. Boddu, V.M., Abburi, K., Maloney, S.W., Damavarapu, R. Physicochemical properties of an insensitive munitions compound, N-methyl-4-nitroaniline (MNA). Journal of Hazardous Materials, 155 (1-2), 288-294. 2008.
22. Mukhi, Sandeep, Pan, Xiaoping, Cobb, George P, Patino, Reynaldo. Toxicity of hexahydro-1,3,5-trinitro-1,3,5-triazine to larval zebrafish (*Danio rerio).* Chemosphere 61 (2005) 178-185.
23. Rochelau, Sylvie, Lachance, Bernard, Kuperman, Roman G, Hawari, Jalal, Thiboutot, Sonia, Ampleman, Guy, Sunahara, Geoffrey I. Toxicity and uptake of cyclic nitramine explosives in ryegrass *Lolium perenne.* Environmental Pollution xx (2008) 1-8.
24. Boddu, Veera M, Abburi, Krishnaiah, Maloney, Stephen W, Damavarapu, Reddy. Thermophysical Properties of an Insensitive Munitions Compound, 2,4-Dinitroanisole. American Chemical Society. 2008.
25. Brannon, J.M., Price, C.B., Yost, S.L., Hayes, C., Porter, B.. Comparison of environmental fate and transport process descriptors of explosives in saline and freshwater systems. *Marine Pollution Bulletin* 50 (3) 247-251.
26. S. Larson, A. Martin, B. Escalon, M. Thompson, Dissolution, Sorption, and Kinetics involved in Systems Containing Explosives, Water, and Soil, Environ. Sci. Technol. 42(2008) 786-792.
27. C. Groom, A. Halasz, L. Paquet, P. D'Cruz, J. Hawari, Cyclodextrin-assisted capillary electrophoresis for determination of the cyclic nitramine explosives RDX, HMX, and CL-20, Comparison with high-performance liquid chromatography, J. Chromatography. Data 2003, *999*, 17-22.
28. D. Burton, S. Turley, G. Peters, Water, Air and Soil Pollution. Data 1994, *76*, 449-457.
29. U.S. Army: 1984, Military Explosives, Tech. Manual TM 9-1300-214, Headquarters, Department of the Army, Washington DC.

* Veera.Boddu@usace.army.mil, phone: 001-217-398-5511

30. Jenkins, T.T., J.C. Pennington, T.A. Ranney, T.E. Berry, P.H. Miyares, M.E. Walsh, A.D. Hewitt, N.M. Perron, L.V. Parker, C.A. Hayes, E.G. Wahlgren, 2001, Characterization of explosives contamination at military firing ranges. Tech. Rep. ERDC TR-01-5. USAVE Engineer Research Development Center, Vicksburg, MS.
31. J. Hawari, A. Halasz, in: G.Buton (Ed.), Encyclopedia of Environmental Microbiology, Wiley, New York, 2000, p.1.
32. Potoff, Jeffrey J. Prediction of Environmental Impact of High-Energy Materials with Atomistic Computer Simulations. Wayne State University. Research project under contract (Personal communication with Dr. Boddu).
33. Boethling RS, D Mackay, eds. 1999. *Handbook of Property Estimation Methods for Chemicals: Environmental and Health Sciences.* CRC, Boca Raton, FL, USA..

* Veera.Boddu@usace.army.mil, phone: 001-217-398-5511

Mechanisms and Kinetics of Thermal Decomposition of CL-20

Olexandr Isayev,[1] Leonid Gorb[1,2], Mo Qasim[1,2], and Jerzy Leszczynski[1,2]

[1]Computational Center of Molecular Structure and Interactions,
Jackson State University, Jackson, MS 39217, USA;
[2]US Army ERDC, Vicksburg, MS, 39180, USA

This summary is abstracted from the paper, "An ab initio Molecular Dynamics Study on the Initial Chemical Events in Nitramines: Thermal Decomposition of CL-20," which is accepted for publication in the *Journal of Physical* Chemistry, ID: jp-2008-04765m and provides the basis for our AICHE presentation.

In contrast to past studies of thermal decomposition mechanisms of explosives designed to determine the detailed chemical pathways by which these large organic molecules are transformed into small reaction products (N_2, NO, N_2O, HCN, CO, CO_2, H_2O, etc.)—crucial to predicting long-term stability for storage; sensitivity to heat and mechanical impact; and risk assessment associated with their usage, the purposes of this study were to investigate the initial chain of chemical events during CL-20 decomposition of isolated and condensed phases and, additionally, to analyze the nature, energetic, and kinetic patterns of the primary and secondary products. To improve atomistic understanding of the thermal decomposition of solid phase CL-20, (a high-energy nitramine explosive possessing $R_2N\text{-}NO_2$ functional groups) we performed a series of *ab initio* molecular dynamics simulations, finding that during unimolecular decomposition, unlike other nitramines, CL-20 has only one distinct initial reaction channel—homolysis of the $N\text{–}NO_2$ bond. No HONO elimination reaction was observed during unimolecular decomposition; the ring breaking reaction was followed by NO_2 fission.

Therefore, in spite of limited sampling, we propose a scheme of unimolecular decomposition of CL-20. The averaged product population over all trajectories was estimated at 4 HCN, 2-4 NO_2, 2-4 NO, 1 CO, and 1 OH molecule per one CL-20 molecule. Simulations provided a detailed description of the chemical processes in the initial stages of thermal decomposition of condensed CL-20, allowing elucidation of key features of such processes as composition of primary reaction products, reaction timing, and Arrhenius behavior of the system. The primary reactions leading to NO_2, NO, N_2O and N_2 occur at very early stages. We also estimated potential activation barriers for the formation of NO_2, which essentially determines overall decomposition kinetics and effective rate constants for NO_2 and N_2. The calculated solid phase decomposition pathways correlate with available condensed phase experimental data.

Approach: The elementary reactions involved in the complicated decomposition process were obtained directly from the MD trajectories. First, a set of trajectories was collected by simulating an isolated CL-20 molecule at various temperatures. In addition to unimolecular decomposition, significant effort was also spent on analysis of condensed phase reactions. Consequently, the unimolecular decomposition process was analyzed only *qualitatively*. In the second part of the research, we produced an overview of atomistic mechanisms of decomposition of crystalline CL-20 and estimated the rate constants of the primary and secondary reactions under constant high temperature conditions.

For the first time, a series of *ab initio* molecular dynamics simulations were performed to shed light on the chain of chemical events of thermal decomposition of isolated and crystalline CL-20, revealing a qualitative picture of reaction pathways of unimolecular decomposition of CL-20. High temperature simulations perfectly agree in prediction of key reactions and leading products over a broad temperature range on the ten picoseconds scale. This process was found to proceed via formation of cyclic intermediates. The averaged product population was also estimated.

The activation barrier for the formation of NO_2, which essentially determines the overall decomposition kinetics, was estimated, and the effective rate constants for NO_2 and N_2 was evaluated. Unfortunately, the short time scale accessible to the first principle simulations did not allow estimation of characteristic reaction time and product lifetime.

Several important differences were found from the earlier empirical (FeaxFF) and tight-binding simulations of other cyclic nitramines. These include dramatic differences in concentration and reaction rates for water, carbon monoxide and carbon dioxide. Therefore, it is not clear whether differences are due to the distinct chemistry of monocyclic (RDX, HMX) and polycyclic (CL-20) nitramines, to the nature of underlying electronic-structure methods, or even to the simulation setup, e.g., type of dynamics, choice of thermostat, etc.

Overall, doubling the system size of the simulation box did not introduce significant differences as to facilitating exploration of the reaction processes. Therefore, a one unit cell was enough to observe the most important features of the process and to make semi-quantitative estimations of thermodynamic and kinetic parameters. However, due to the larger temperature fluctuations in the smaller system, it appears to have slightly faster kinetics and larger fluctuations in product concentrations. Thus, a larger system size is desired for obtaining a more precise product time evolution series and higher accuracy.

Isolated System: For calculation of the gas-phase molecule, a 18 Å cubic supercell was used with a cell length chosen to be large enough to prevent interactions between periodic images. The effect of the unit cell length was checked by expanding the box size to approximately 20 Å. Only a minor variation in the total energy was found. The geometry of the system was minimized and subsequently equilibrated at the NVT ensemble. After equilibration, five independent production runs with target temperatures of 1000, 1500, 2000, 2500 and 3000K were initiated. To improve observed statistics, each simulation was repeated at least one more time, starting from a different configuration. The length of simulations varied from 5 to 10ps and actual temperature fluctuations were within 50-150 K from those desired.

Crystal Phase: In order to obtain better statistical sampling and test system size dependence, simulations were performed with larger supercells (1×1×2 unit cells, eight CL-20 molecules, 288 atoms). The system was constructed by merging two equilibrated unit cells along the c axis (smallest axis). After a 1ps quick re-equilibration, production runs were initiated at 2000 and 3000K. The lengths of simulations were 5.1 (aborted due to hardware failure) and 15.4ps respectively; the actual temperatures were 1947 and 2882 K, respectively.

It was observed that homolysis of the $N–NO_2$ bond occurs on the 0.1-0.5 ps scale, depending on temperature. As expected, the higher the temperature, the faster the reaction proceeds. No clear temperature dependence of the number of broken $N–NO_2$ bonds was found. Seven of ten trajectories (including the most and least extreme) resulted in fission of all six nitrogroups, which is in perfect agreement with the mass spectral pattern, where peaks corresponding to elimination of 3-6 NO_2 groups were observed. No distinct differences between axial and equatorial nitrogroups wee observed, probably due to the fact that, under such extreme conditions, the activation barrier difference of several kcal/mol[44] becomes negligible. It is quite probable that, due to the limited statistical sampling, this list is not exhaustive of all possible geometrical variations. However, seven out of ten trajectories resulted in 12-atomic intermediates. This species oscillates around planar geometry and is stabilized by adoption of an aromatic structure with typical C-N bond lengths, ranging from 1.2 to 1.45Å, and having a lifetime of approximately 0.5ps at 3000 K and quickly increasing to 9ps at 1500 K.

Three major distinctive channels were found for subsequent decomposition of CL-20 intermediates. These secondary reactions include: a) fragmentation, b) oxidation by a NO_2 radical, and c) hydrogen abstraction by a NO_2 radical. Fragmentation of the intermediate results in stepwise breaking into HCN molecules. Oxidation by NO_2* yields N-methylenformamide derivatives and NO (Figure 1), leading to further decomposition. Methylenformamide derivatives undergo fast decomposition, eliminating 1-2 HCN molecules (depending on molecular length) and yielding simplest methylenformamide carbenes, and leading to final the CN-COH molecule degradation. The third pathway represents hydrogen abstraction by NO_2, resulting in HONO molecules. This is an interesting variation of HONO molecular formation, formed during the later stages of the reaction ($\tau > 2$ ps) by interaction of the NO_2 free radical and the intermediate. Based on our knowledge, this reaction has not before been described in nitramines, neither experimentally nor theoretically. This finding is in contrast to the suggested concerted *intramolecular* mechanism of HONO formation in RDX/HMX, [20,21,45] where departing NO_2 groups capture hydrogen radical. Our results strongly favor the stepwise red-ox *intermolecular* process. Evidence was also found relating to the possibility of formation of an NO_3. From ten simulations, NO_3 occurred only once, at T = 1500K. Two NO_2 radicals combined form NO and NO_3. For a short time (< 1ps) they formed a complex very similar to the ion pair, nitrozonium nitrate ($NO^+NO_3^-$).[49] After that step the system recombined to the original NO_2 radicals.

Overall, total decomposition (starting from CL-20) into two- or tri-atomic molecules occurred within 4.5-5ps at 3000K. A similar process took approximately twice as long at 2000-2500K. The present timescale was not sufficient to observe total decomposition at 1500 and 1000K. For the latter, a much longer simulation (perhaps in the order of 0.1ns – 1.0ns) is needed to reveal full decomposition of CL-20.

Therefore, in spite of the fact that CPMD simulations of isolated CL-20 presented in this section provide a mostly *qualitative* picture, high temperature simulations on a ten picoseconds scale perfectly agree as to primary reactions and leading products over a broad temperature range. The averaged product population over all trajectories was estimated to yield 4 HCN, 2-4 NO_2, 2-4 NO, and approximately one CO, and one OH molecule per one CL-20 molecule. Due to the lack of CL-20 gas phase experiments, those population numbers cannot be compared with observed ones.

Decomposition of CL-20 in the condensed-phase: There is clear enhancement of the decomposition process with temperature increase. Therefore, starting with the fastest reaction, the kinetics of the predominant intermediates and reaction products can be analyzed in detail, despite available limited computational facilities allowing simulation of only the initial steps of decomposition.

The eight picosecond time scale was almost sufficient to reach product equilibrium concentration at T=3000 K. All nitrogroups quickly decayed within the first 800fs. At the same time, the number of NO_2 radicals proportionally increased, reaching a maximum of 13 molecules at t=400fs. By the time ~2.5 ps was reached, practically the entire concentration of NO_2* was consumed. Similarly, nitric oxide accumulated to ~1.2 ps, with a maximum concentration of 16 molecules per unit cell, after which it slowly reacted into N_2, N_2O, and NOH, reaching equilibrium at a concentration of 5-6 molecules per unit cell by the end of the simulation. It should be noted, that the equilibrium established here (and for all species later on) is dynamic, i.e. previously formed molecules react and undergo rearrangement, and new species are formed—maintaining approximately constant concentration.

The other experimentally determined ratios are 3.3, 1.2 and 0.82 for CO_2, CO and N_2O, respectively, with calculated values amounting to 1.25, 1.0 and 1.0 respectively. Significant deviation in the concentration of CO_2 can be attributed to the short time scale and small system size accessible to the AIMD simulation. Indeed, carbon dioxide is a product of the latter stage of the reaction. The length of simulation needed to be at least an order of magnitude longer to observe the steady state formation of CO_2, as well as to detect the CO conversion reaction ($CO + H_2O = CO_2 + H_2$). The concentration of CO_2 and a few other important products did not establish a well behaved tendency (especially for low temperature simulations) and were found in significantly smaller amounts than the species described above. By the end of the simulation, the reaction mixture included three molecules of H_2O and CO; two of OH radical; and one of HCN, CO_2, NH_3, HCNO, CN, NOH, and CNC(O)H.

Decrease of simulation temperature to 2500 K slowed the whole reaction process. On one hand, the decaying of nitrogroups was only slightly affected; just one picosecond was needed to complete this process. On the other hand, as much as twice the time (about 4 ps) was needed for formed NO_2 to react, starting from a peak concentration of 14 molecules per unit cell to equilibrium. Further, the decomposition reaction did not reach equilibrium concentrations of both NO and N_2. The observed values were 8 and 15 molecules, respectively. Assuming similar kinetics and C_∞ for molecular nitrogen, the estimated time needed to reach steady stage concentration was found to be ~20 ps, and $k=0.3\ ps^{-1}$. Concentrations of other molecules were found as follows: H_2O, OH radical, and CO_2 – 3; HONO, HCN, and CO – 2.

Further decrease of temperature leads to more dramatic differences in the reaction process. Significant prolongation of the lifetimes for both NO_2 and NO were immediately observed. Concentration of NO_2 at approximately 10 molecules per unit cell was maintained for an additional 3ps. Decay of the NO concentration was barely noticeable, but was still higher than that of N_2. Formation of molecular nitrogen was expected to be delayed for 4.5 ps. The resulting concentrations at t=10.5 ps were NO – 12, N – 11, N_2O, CO_2 – 4, H_2O, CO, OH radical – 2, very far from the values expected at the equilibrium stage. Finally, our least extreme simulation, at T = 1500 K was barely long enough to observe full fission of the nitrogroups. Consequently, the concentration of NO does not reach its maxima. Decrease of temperature also stabilized the concentration of N_2O. Overall, all mentioned approaches provide *qualitatively the same* picture of thermal decomposition for the whole class of cyclic nitramine high-energetic materials. Numbers from larger system size are in better agreement with experimental data than ones derived from the smaller system, which means that the larger simulation system is needed to obtain trustworthy concentration values.

Limited computational resources forced the abortion of the other large simulation at T=2000 at not longer than 5 ps. Compared to the one unit cell run, the large simulation exhibited similar behavior, but slightly slower kinetics. For example, NO_2 fission required 3.5-4 ps to proceed. Produced NO_2 reached its maximum concentration of 16 molecules per unit cell at about 2 ps. Both NO and NO_2 temporary stabilized at the same concentration, ~10 molecules/unit cell, but only after 4 ps of simulation.

Overall, our simulations revealed the systematic behavior of the entire decomposition process in a wide temperature range of 1500-3000 K. Correspondence between the low temperature experiments and high temperature simulations suggest a unified reaction mechanism for both temperature regimes. A careful analysis of the trajectories indicates complex multimolecular and multichannel processes.

Despite such complexity there are several important differences between the studied phenomena and unimolecular reactions that can be identified. Due to higher density and crystal packing, departing nitrogroups "bombard" neighboring molecules, causing a higher oxidizing and conversion rate, in turn, leading to numerous branching reactions where no distinct intermediate was formed. Another difference relates to the concentrations of N_2O and N_2 forming during the decomposition process. In contrast to the unimolecular process, N_2 makes up the prevalent reaction product in solid phase CL-20.

This difference originates in the fact that rapid branching produces a large number of chain molecules containing the NNO pattern. These molecules formed either directly from original nitrogroups of CL-20 or from simpler compounds (e.g. R—[N=C=N]$^{\bullet}$ and NO). Resulting species with NNO motif are not stable and dissociate yielding N_2O. Finally, the nitrous oxide oxidizes various compounds, producing molecular nitrogen. Unfortunately, it was not possible to identify several distinct reaction channels due to the branching and great number of secondary and other types of reactions.

Finally, all condensed phase simulations have been discussed in terms of the thermal decomposition process. However, similar conditions were also achieved during the steady detonation (Chaman-Jouget) state or Zeldovich-von Neumann detonation. Therefore, the above discussion is useful in describing the detonation process. For detonation, formation of condensed carbon (carbon clustering) is a very important process. No carbon clustering was observed in the present work.

Reaction kinetics: Several experimental studies suggest the predominant role of homolysis of $N–NO_2$ in the first stages of decomposition. It was shown that NO_2 fission reaction is rate limiting and determines kinetics of crystal decomposition. Our calculated time evolution profiles for this reaction mostly follow exponential decay—fitting first order kinetics. However, several curves have a sigmoid shape, implying first order kinetics with autocatalysis. Experimental data suggest that either case could be possible, depending on the decomposition stage. It is not clear whether the origin of the sigmoid shape originates from the simulation artifacts. However, no evidence of autocatalysis has been found at this stage.

This process exhibits typical Arrhenius behavior, allowing estimation of the activation barrier of CL-20 crystal decomposition. According to the linear fit, estimated values of Ea and A are 137±24 kJ/mole and 2.6×10^{9}, respectively. Corresponding experimentally determined values varied from 150 to 200 kJ/mol and $\sim10^{13}$ – $\sim10^{17}$, depending on experimental setup. Our simulations supported the lower values of the activation barrier. However, statistical uncertainty is quite high. The calculated value of the exponential prefactor A, is at least four orders of magnitude lower than the experimental ones. This discrepancy originates from the number of approximations on the simulation setup. All calculations were performed at NVT ensemble, keeping the volume and number of particles of the system constant. In reality, however, the system expands and undergoes phase transitions during heating. Evolving gaseous products contribute to the mass transport between the sample and environment. Obviously, confined gases like NO_2, NO, and N_2 significantly affect the entropy of the system and, consequently, the value of A.

Our simulations provide a detailed description of the chemical processes in the initial stages of thermal decomposition of condensed CL-20, allowing elucidation of key features of such processes such as: composition of primary reaction products, reaction timing, and Arrhenius behavior of the system. They clearly indicate that the primary reactions leading to NO_2, NO, N_2O and N_2 occur at very early stages.

742a Identifying Emulsion Drop Breakage Mechanisms for Population Balance Equation Models of High Pressure Homogenization

Neha B. Raikar[1], Surita R. Bhatia[2], Michael F. Malone[3], Cristhian P. Almeida-Rivera[4], Peter Bongers[5], and Michael A. Henson[1]. (1) Department of Chemical Engineering, University of Massachusetts, 686 North Pleasant Street, Amherst, MA 01003, (2) Chemical Engineering, University of Massachusetts, Amherst, 686 North Pleasant Street, Amherst, MA 01003-9303, (3) College of Engineering, University of Massachusetts, Dean of Engineering, 125 Marston Hall, Amherst, MA 01003-9303, (4) Process Science, Unilever R&D Vlaardingen, Olivier van Noortlaan 120, POBox 114, Vlaardingen, 3130AC, Netherlands, (5) Process Science Deparment, Unilever R&D Vlaardingen, Olivier van Noortlaan 120, POBox 114, Vlaardingen, 3130AC, Netherlands

Previously we have used an inverse method to develop a population balance equation (PBE) model for drop breakage in a simulated oil-in-water emulsion (Raikar et al., 2007). A potential advantage of inverse methods is that the drop breakage functions are constructed non-parametrically without the need for mechanistic understanding of the breakage mechanisms. From an emulsion design perspective, the major shortcoming of inverse methods and many other PBE modeling techniques is that the drop breakage functions do not depend explicitly on formulation and processing variables. Such drop breakage kernels have limited predictive capabilities as the PBE model must be refit to each new set of experimental data in which formulation and/or processing variables were changed.

In this contribution, we demonstrate that drop distribution data from a high-pressure homogenizer can be used to identify relevant drop breakage mechanisms. Two breakage rate functions that depend explicitly on the emulsion properties (disperse phase volume fraction, density, and viscosity, interfacial tension, critical capillary number) and the homogenization pressure were considered. The first function was derived by other researchers under the assumption that drop breakage results from collisions with turbulent eddies, while we derived the second function assuming that breakage is due to turbulent shear. We show that neither breakage rate function nor the combination of the two functions provide satisfactory predictions when binary breakage is assumed and the daughter drop distribution function is a truncated normal distribution. Our experiments reveal that substantial breakage of the pre-emulsion occurs during the first homogenization pass even under zero pressure operation, suggesting a pressure independent breakage mechanism. The use of pressure dependent and pressure independent breakage rate functions along with judicious choice of the other breakage functions are shown to provide superior predictions than can be achieved with a single breakage rate function.

742b Population Balance Equation Modeling of Emulsion Drop Coalescence in High Pressure Homogenization

Jason D. Rosenberg, Neha B. Raikar, Michael F. Malone, and Michael A. Henson. Chemical Engineering, University of Massachusetts, Amherst, MA 01002

Previously we have developed population balance equation (PBE) models of drop breakage in oil-in-water emulsions prepared with high pressure homogenization. These models were based on the assumption of negligible drop coalescence, which is reasonable for emulsions with low dispersed phase volume fractions and large surfactant loadings. However, many industrial emulsions do not satisfy these conditions and drop coalescence must be understood for rational emulsion design and processing. Other researchers have proposed inverse PBE methods for the extraction of the coalescence kernel from transient drop distribution measurements. The major shortcoming of inverse methods and many other PBE modeling techniques is that the coalescence functions do not depend explicitly on formulation and processing variables. Such coalescence kernels have limited predictive capabilities as the PBE model must be refit to each new set of experimental data in which formulation and/or processing variables were changed.

In this contribution, we demonstrate that drop distribution data from a high-pressure homogenizer can be used to identify drop coalescence mechanisms. Collision frequency and coalescence efficiency functions that depend explicitly on the emulsion properties (disperse phase volume fraction, density, and viscosity, interfacial tension) and the homogenization pressure were considered. To minimize the impact of drop breakage and isolate the coalescence phenomenon, we performed a series of homogenization experiments in which the emulsion was processed at very high pressure until a constant drop volume distribution was obtained and the resulting emulsion was further processed at a lower pressure. Drop volume distribution measurements at the lower pressure were used to estimate adjustable parameters in the collision frequency and coalescence efficiency functions using nonlinear parameter estimation. We evaluated the ability of the PBE model to predict experimental results not used for parameter estimation. By combining the coalescence kernel with the breakage functions determined in our previous studies, we developed a single model capable of predicting the combined effects of drop breakage and coalescence.

742c A Generalized Geometric Population Balance Model for Asphaltene Aggregation

Tabish Maqbool, Sasanka Raha, and H. Scott Fogler. The University of Michigan, 2300 Hayward St., Ann Arbor, MI 48109-2136

The precipitation of asphaltenes from crude oils is a major concern for oil production. A generalized geometric population balance model for asphaltene aggregation along with the thermodynamics of asphaltene destabilization are brought into a unified framework. This study focuses on the nano-aggregation of asphaltenes destabilized by the addition of heptane. The destabilization of asphaltenes is dictated by equilibrium thermodynamics and is relatively rapid while the kinetic phenomena associated with the reaction limited flocculation process is slow. The model is in agreement with the experimental data of the the evolution of centrifuged aggregates of varying heptane composition. The flocculation process accelerates with increasing amounts of heptane added to the asphaltenic crude because of the larger quantity of destabilized asphaltene.

742d A Combined Mechanistic Model for Nucleation, Aggregation and Breakage In a Population Balance Model of Granulation

Rohit Ramachandran, Department of Chemical Engineering and Chemical Technology, Imperial College London, Exhibition Road, London SW7 2AZ, United Kingdom, Charles D. Immanuel, Centre for Process Systems Engineering, Department of Chemical Engineering, Imperial College London, South Kensington Campus, London, SW7 2AZ, United Kingdom, James D. Litster, Chemical Engineering and Industrial Pharmacy, Purdue University, Forney Hall of Chemical Engineering, Room 2154, 480 Stadium Mall Drive, West Lafayette, IN 47907-2100, Francis J. Doyle III, Department of Chemical Engineering, University of California, Santa Barbara, Santa Barbara, CA 93106-5080, and Frantisek Stepanek, Department of Chemical Engineering, Institute of Chemical Technology, Prague, Technicka 5, Prague 6, 166 28, Czech Republic.

Granulation is the generic term for particle agglomeration processes, whereby fine powdery solids are agglomerated together with a liquid/melt binder to form larger aggregates. It finds application in many industries such as pharmaceuticals, fertilisers, detergents and food processing. Despite its widespread use, granulation processes are highly inefficient and current industrial plants operate with high recycle ratios [1]. Thus, there is a clear need for better process design, operation and control of these processes. However, the complexity of the process, the limited number of potential manipulated variables and the large time delay associated with the downstream units make control and operation of granulation processes a challenge. Therefore, the development of a first-principle mathematical model for granulation is vital for the application of model-based strategies for better design, optimisation and control of granulation processes.

The application of population balances, which accounts for the individual sub-processes (e.g. nucleation, aggregation, breakage and consolidation), provides a convenient framework from which the granulation process is modelled and these sub-processes will be the emphasis of this study [2-5]. One major challenge in developing these population balance models is the identification of appropriate kernels for the sub-processes. Due to the limited knowledge of the mechanisms that underlie the granulation process, many of the proposed kernels are empirical or semi-empirical, which may not be valid over a wider operating range. In order to further improve the predictive capabilities of the kernels and to extend their region of validity, one has to directly incorporate the mechanistic features of the process and develop kernels strongly based on first-principles.

In this study a three-dimensional volume-based population balance model for the granulation process is presented, incorporating combined mechanistic formulations for the nucleation, aggregation and breakage phenomena, in combination with an empirical relation for consolidation. Immanuel and Doyle III have presented a methodology for developing the aggregation mechanistic kernel reported in this work [6]. Similarly, the mechanistic formulation of the nucleation kernel has been developed by Poon et al. [7] and both these mechanistic kernels have been quantitatively validated against experimental data for different operating conditions [8]. Therefore, the major focus of this work is the development and validation of the mechanistic breakage kernel.

The breakage kernel is modelled as a quotient of applied external stress on a granule over intrinsic strength of a granule. The external stress in turn is calculated from the derivation of external forces over their contact area. These external forces are a result of (i) fluid forces in the granulator, (ii) particle-particle collisions and (iii) particle-wall collisions. The intrinsic strength of the granule is derived from the capillary, viscous and frictional forces that are responsible for the liquid bridge strength of the granule. Subsequently the individual breakage kernels for the three types (i-iii) of breakage are transformed into an effective breakage kernel that is based on the probability of an ith size particle colliding with another particle of jth size, or a wall. Fluid forces act in addition to forces (ii) and (iii). Qualitative validation of breakage kernel/model was performed and trends of lumped properties (i.e., total particles, average size, binder content and porosity) and distributed properties (i.e., granule size, fractional binder content and porosity distributions) show good agreement with the expected phenomenological behaviour. Successful high-shear granulation experiments were then carried out to mimic breakage only behaviour whereby the rate of breakage is significantly greater than that of nucleation and aggregation. Quantitative validation of the breakage kernel/model against experimental data is currently underway. For the solution of the population balance model, a robust and efficient numerical solution technique was employed [6, 9].

References

1. S.M. Iveson, J.D. Litster, K. Hapgood, and B.J. Ennis. Nucleation, growth and breakage phenomena in agitated wet granulation processes: a review. Powder Technology, 117: 3-39, 2001.

2. L.X. Liu and J.D. Litster., Population balance modelling of granulation with a physically based coalescence kernel. Chemical Engineering Science, 57: 2183-2191, 2002.

3. H.S. Tan, A.D. Salman, and M.J. Hounslow., Kinetics of fluidized bed melt granulation – II: Modelling the net rate of growth. Chemical Engineering Science, 61: 3930-3941, 2006.

4. Biggs, C.A., Sanders, C., Scott, A.C., Willemse, A.W., Hoffman, A.C., Instone, T., Salman, A.D. and Hounslow, M.J., Coupling granule properties and granule rates in high shear granulation. Powder Technology, 130: 162-168, 2003.

5. I.T. Cameron, F.Y. Wang, C.D. Immanuel, and F. Stepanek., Process systems modelling an applications in granulation: A review. Chemical Engineering Science, 60: 3723-3750, 2005.

6. C.D. Immanuel and F.J. Doyle III. Solution technique for a multi dimensional population balance model describing granulation processes. Powder Technology, 156: 213-225: 2005.

7. J. Poon, C. D. Immanuel, F.J. Doyle III and J.D. Litster. A three-dimensional population balance model of granulation with a mechanistic representation of the nucleation and aggregation phenomena.

Chemical Engineering Science, 63: 1315-1329: 2008.

8. J. Poon, R. Ramachandran, C.F.W. Sanders, T. Glaser, C.D. Immanuel and F.J. Doyle III and J.D. Litster, F. Stepanek, F.Y. Wang and I.T. Cameron. Experimental Validation Studies on a Multi-Dimensional and Multi-Scale Population Balance Model of Batch Granulation. Submitted to Chemical Engineering Science.

9. M. A. Pinto and C. D. Immanuel and F. J. Doyle III. A feasible solution technique for higher-dimensional population balance models. Computers and Chemical Engineering, 31: 1242-1256: 2007.

742e The Elasticity and Breakage of Colloidal Aggregates In Shear and Turbulent Flows

Alessio Zaccone, Institute for Chemical and Bioengineering, Department of Chemistry and Applied Biosciences, ETH Zurich, Hoenggerberg, HCI, Zurich, 8093, Switzerland, Hua Wu, Institute for Chemical and Bioengineering, Department of Chemistry and Applied Biosciences, ETH Zurich, Wolfgang-Pauli-Str. 10, HCI F 138, Zurich, 8093, Switzerland, Marco Lattuada, Chemistry and applied biosciences, Institute for chemical and bioengineering, ETH Zurich, Wolfgang-Paulistr. 10, HCI-F133, Zurich, Switzerland, and Massimo Morbidelli, ETH Zurich, Hoenggerberg, HCI, Zurich, 8093, Switzerland.

Recently, we proposed a free energy expansion-based model for the shear modulus of random colloidal aggregates as they are formed in shear and turbulent flows, resorting to the Cauchy-Born theory of solids. Considering only harmonic terms, application of the fundamental relations of linear elasticity yields an expression for the shear modulus of the aggregates which is proportional to the packing fraction, to the bond rigidity and the coordination number n. The latter term is modelled based on statistical mechanical theories of the liquid state. The main assumption of the model is that interparticle bonds are rigid and the particles are multiply connected. n accounts for the short-range liquid-like order which is a good approximation for dense aggregates that have a high fractal dimension (>=2.5) and introduces a further dependence on the packing fraction. However, also empirical scattering data for the structure can be employed. The model has already been validated against data for cohesive packings of beads. We will show how this approach, combined with the most detailed description of breakage rate in turbulent flows and accounting for turbulence intermittency via a multifractal description, is able to predict the experimentally observed asymptotic scaling between size and shear rate in turbulent breakage of aggregates, without any free parameters.

Zaccone et al., Journal of Chemical Physics, 2007

Zaccone et al., Europhysics Letters, submitted

752a Study on Aggregation of Gold Nanoparticles by Tuning the Surface Properties

Varun Kumar[1], Chungyin Cheng[2], R. O. Fox[3], and Robert K. Prud'homme[1]. (1) Chemical Engineering, Princeton University, A-217; Engineering Quadrangle, Olden Street, Princeton, NJ 08544, (2) Chemical and Biological Engineering, Iowa State University, Ames, IA 50011, (3) Department of Chemical & Biological Engineering, Iowa State University, 2114 Sweeney Hall, Ames, IA 50011-2230

The precipitation technique to formulate drug nanoparticles has recently been used to encapsulate a broad range of hydrophobic drug molecules (Horn & Rieger, 2001; Hu et al., 2004; Johnson & Prud'homme, 2003). The governing principle behind this kinetically driven process is the stabilization of the aggregates formed by the solutes by the block copolymer. Though particles of the desired size can be obtained by the addition of inert filler, a more detailed study needs to be performed to fully understand the factors controlling the nanoparticles size. The estimation of particle size for gold colloid stabilized by PEG-b-PCL was recently studied (Gindy et al., 2008) based on the assumption that the sticking probability between two colliding gold particles is unity. However a more unified approach needs to be developed to estimate the nanoparticle size for solutes with weaker interparticle interaction. The aggregation kinetics for colloids have already been theoretically analyzed for both the diffusion and reaction limited regime (Smoluchowski, 1916; Weitz et al., 1984; Ball et al., 1987).

In this study, the particle size is first tuned by the addition of inert filler during the formulation. By varying the amount of PCL homopolymer, the size of the gold nanoparticle changed from 60 to above 200 nm. However, estimation of the particle size for any given solute and need for the formulation of even smaller particles make it imperative to fully understand the aggregation kinetics of the solutes. The study provides a starting point for the development of a correlation between the aggregation kinetics of the colloids and their surface properties. We tuned the sticking probability of the gold colloids by using both ionic and non-ionic surfactants. Ionic surfactants including SDS and TDTAB have been used to characterize the surface properties of the gold colloid. The anionic surfactant, SDS, was found to be better stabilizing agent compared to the cationic surfactant. The nanoparticle size is then tuned by varying the concentration of the surfactant and thereafter the stabilizing effect of the block copolymer, PEO-b-PS (3k-b-1k) on the assembly has been analyzed.

The experiments are analyzed by comparing them with an aggregation model that solves the population balance equations (PBEs) as functions of mixing time and physical properties of the inlet streams. In this work, a multivariate population balance model is developed to account for the copolymer arresting and stabilizing particle growth mechanisms in the NanoPrecipitation process. In order to achieve more economical computation, the quadrature method of moments (QMOM) is applied to solve PBEs in this model. The QMOM transforms the PBE into a set of lower moment equations, which contain useful information regarding the particle size, surface area, solid volume, and mean particle size. As a result, instead of solving the particle size distribution (PSD), QMOM requires less computation and has proven a good numerical approach in terms of its efficiency and accuracy.

752b Effect of Particle Surface Charge Groups on the Aggregation of Elastomer Particles Under Static and Flow Conditions

Miroslav Soos, Cornelius Gauer, Hua Wu, and Massimo Morbidelli. Institute for Chemical and Bioengineering, Department of Chemistry and Applied Biosciences, ETH Zurich, Wolfgang-Pauli-Str. 10, Zurich, 8093, Switzerland

Aggregation (coagulation or floculation) is a commonly used process to separate colloidal particles from liquid, where the particles after addition of appropriate amount of coagulant are aggregated to form large clusters. Besides Brownian motion, which is the controlling mechanism for sub-micron particles or clusters, effect of shear, induced by stirring of the suspension, contribute additionally to the cluster growth as well as controls final size of formed aggregates. The final size of produced aggregates is strongly connected to the material and surface properties of the primary particles, internal structure of aggregates and inter-particles forces between primary particles composing an aggregate. When primary particles undergo coalescence the situation becomes even more complicated.

In this work we investigate aggregation behavior of elastomer particles with sub-micron size and different surface charge groups under both, static and flow conditions. In all experiments the amount of added coagulant (acid and various types of salts) was well above the critical coagulation concentration. Depending on the size of produced aggregates either light scattering (dynamic as well as static light scattering) or image analysis were used for characterization. Moreover, for selected conditions cryogenic scanning electron microscopy and confocal scanning electron microscopy were used to analyze the internal structure of the produced aggregates.

It was found that depending on the nature of the surface charge groups of the elastomer particles, coalescence between primary particles can occur, resulting in significantly different steady state aggregate sizes. Moreover, depending on the experimental conditions one may obtain fractal aggregates under static conditions while under flow conditions particles exhibit coalescence. This will be demonstrated by the series of experiments carried out at different pH values.

752c Distinguishing Between TiO2 Aggregates and Agglomerates by High Pressure Dispersion

Alexandra Teleki, Department of Mechanical and Process Engineering, Swiss Federal Institute of Technology Zurich, Particle Technology Laboratory, Sonneggstrasse 3, Zurich, CH-8092, Switzerland, Robert Wengeler, Institut für Mechanische Verfahrenstechnik und Mechanik, Universität Karlsruhe (TH), Karlsruhe, 76128, Germany, Hermann Nirschl, Chemical and Process Engineering, University of Karlsruhe, Institute for Mechanical Process Engineering and Mechanics, Strasse am Forum 8, Karlsruhe 76131, Germany, and **Sotiris E. Pratsinis**, Particle Technology Laboratory, Institute of Process Engineering, Department of Mechanical and Process Engineering, ETH Zurich, Sonneggstrasse 3, Zurich, CH-8092, Switzerland.

The potential of high pressure dispersion (HPD) and dynamic light scattering (DLS) is explored [1] for rapid and quantitative estimation of the extent of particle aggregation and agglomeration by analyzing the entire particle size distribution [2]. Commercially available and tailor-made TiO_2 particles by flame spray pyrolysis (FSP) were characterized by X-ray diffraction, nitrogen adsorption and transmission electron microscopy (TEM). Volume distributions of these particles were obtained by DLS of their electrostatically stabilized (with $Na_4P_2O_7$) aqueous suspensions. Dispersing these suspensions through a nozzle at 200 to 1400 bar reduced the size of agglomerates (particles bonded by weak physical forces) resulting in bimodal size distributions composed of their constituent primary particles and aggregates (particles bonded by strong chemical or sinter forces). Sintering FSP-made titania particles from 200 to 800 °C for four hours progressively increased the minimum primary particle size (by grain growth) and aggregate size (by neck growth and phase transformation).

[1] Wengeler, R., A. Teleki, M. Vetter, S.E. Pratsinis, and H. Nirschl, "High pressure liquid dispersion and fragmentation of flame-made silica agglomerates," Langmuir 22, 4928-4935 (2006).

[2] Teleki, A., R. Wengeler, L. Wengeler, H. Nirschl, and S.E. Pratsinis, "Distinguishing between aggregates and agglomerates of flame-made TiO_2 by high pressure dispersion," Powder Technol. 181, 292-300 (2008).

752d Use of Asymmetric Flow-Field Flow Fractionation to Characterize Aggregating Colloidal Dispersions

Marco Lattuada[1], Carlos Olivo[1], Cornelius Gauer[2], Giuseppe Storti[1], and Massimo Morbidelli[1]. (1) Institute for Chemical and Bioengineering, Department of Chemistry and Applied Biosciences, ETH Zurich, Wolfgang-Pauli-Str. 10, HCI F 133, Zurich, 8093, Switzerland, (2) Institute for Chemical and Bioengineering, Department of Chemistry and Applied Biosciences, ETH Zurich, Wolfgang-Pauli-Str. 10, HCI F 138, Zurich, 8093, Switzerland

The estimation of the size distribution of complex colloidal dispersions, such as dispersions of clusters of nanoparticles is a notoriously challenging task. Common characterization techniques include electron microscopy, light scattering, centrifugation, ultrasound spectroscopy etc.. Most of the above mentioned techniques, as for light scattering, provide some average size of the dispersion, but the reconstruction of the entire size distribution is often an ill-posed problem. Other techniques, like electron microscopy, are capable of analyzing individual particles and clusters, and can provide detailed information about the distribution, but require long times and require to dry the sample, a harsh process which can alter the structure of delicate clusters. Flow Field Flow Fractionation, on the other hand, is a well established chromatographic technique able to fractionate mixtures of colloidal nanoparticles having different sizes under gentle conditions. Usually, the extraction of a size distribution is carried out by analyzing the elution times of the samples and by estimating their hydrodynamic radii using a calibration curve. In this work, we make use of Asymmetric Flow Field Flow Fractionation coupled to a two angle static light scattering detector (SLS) to characterize suspensions of different polymeric colloidal nanoparticles undergoing stagnant aggregation under both diffusion and reaction-limited conditions. Two types of particles have been used: rubbery fluorinated colloids undergoing coalescence and hard polystyrene latexes forming fractal clusters. The cluster size distribution is quantitatively reconstructed not from the elution times, but from the analysis of the SLS data, which requires the knowledge of the scattering properties of both clusters and particles. The estimated cluster size distributions favorably compare with the calculated ones from Population-Balance-Equations. These results prove the effectiveness of AF4 in combination with multi-angle light scattering as a tool to quantitatively characterize the size distribution of complex colloidal dispersions.

752e A Structural Model for Aggregation and Its Application to Small-Angle Scattering

Gregory Beaucage, Department of Chemical and Materials Engineering, University of Cincinnati, Cincinnati, OH 45221-0012

Aggregation is widely seen in nanoparticles especially under conditions of high concentration and in temperature and annealing time regimes below those allowing for coalescence and sintering. A quantifiable description of aggregates has proven difficult since they are by nature disordered and show a wide dispersion in structure. The scaling of the density of aggregates with size follows mass-fractal laws and can be used to quantify one aspect of aggregates but such a description fails to quantify features such as asymmetry and branch content. In this presentation a fundamental, statistical description of aggregates is outlined based on their tortuosity and connectivity. Tortuosity is measured by the mass fractal dimension of a minimum path through the aggregate; while connectivity is related to the mass fractal dimension of a stick structure connecting branch and end points in the aggregate. These two measures can be used to calculate the mole fraction branches, the asymmetry and other aspects of the aggregate. This approach can predict the diffusion coefficient as well as the spring constant for aggregates and yields hope in predicting the overall dynamic properties of aggregates. Small angle scattering can directly measure the connectivity and tortuosity in a simple static measurement. Observed variation in aggregate structure using this description during nanoparticle growth in flames will be presented based on in situ x-ray scattering studies.

Partially supported by the US NSF through CBET 0626063 as well as by LyondellBasell Corporation and the Swiss NSF.

Web Page: www.eng.uc.edu/~gbeaucag/BeaucageResearchGroup.html

Size and shape engineering of ZnO nanoparticles

Renata Marczak, Robin Klupp Taylor, Michael Voigt, Doris Segets, Wolfgang Peukert
Institute of Particle Technology, Friedrich-Alexander-University Erlangen-Nuremberg, Cauerstr. 4, 91058 Erlangen, Germany

Introduction

Zinc oxide is an attractive material for a broad range of electronic, optical, and piezoelectric applications due to its direct band gap and excellent thermal, chemical, and structural properties. For example, ZnO has been suggested for use in application such as light emitting diodes,[1] transparent electrodes,[2] solar cells,[3] sensors,[4] and many other devices.[5]

Different synthesis routes have been developed, including solid-vapor phase thermal sublimation,[6] spray pyrolysis,[7] RF plasma synthesis,[8] sonochemical or microwave-assisted synthesis[9] and hydrothermal processing.[10] However, wet-chemical synthesis of ZnO is an area of particular interest, since it provides a low-temperature, economical way to produce various ZnO nanostructures.[11, 12] Growth from solution typically starts with nucleation and is followed by growth of the nuclei by addition of dissolved species until the metal cation concentration reaches the solubility of the oxide. Then particle ageing proceeds in the mother liquors. Synthesis conditions such as temperature,[13] the overall concentration of the precursors,[14] water concentration,[15] ions present in solution[16], and solvent[17] have been shown to impact particle nucleation and growth. Therefore, the investigation of ZnO nanoparticles is of great interest due to the wide range of potential applications for this material. Despite this, a thorough understanding of the optical, electronic and structural properties of ZnO nanoparticles with respect to their synthesis is still lacking. We tackled this problem by synthesizing ZnO nanoparticles with nanocrystal and nanorod form. Dynamic light scattering (DLS), absorption and emission spectroscopies, thermogravimetric analysis (TGA), X-ray powder diffraction (XRD), Fourier infrared spectroscopy (FTIR) as well as transmission and scanning electron microscopies (HRTEM and SEM) were employed to characterize and compare the nanoparticles composition, their shape, size and crystallinity. By varying different parameters of the synthesis processes like precursors and their concentrations, solvent used, surfactant, temperature, and reaction time we were able to control the size and shape of the particles. In particular we noted the tendency of ZnO nanocrystals to self-assemble into dense structures which preserve the anisotropic crystal structure.

Experimental session

All chemicals were analytical grade reagents used without further purification. ZnO nanoparticles were prepared from zinc acetate dihydrate in alcoholic solution under basic conditions. To learn more about the growth mechanism, we have performed the experiments in different ways.

Quasi-spherical ZnO nanoparticles were obtained using a preparation procedure adapted from Spanhel, Anderson[11] and Meulenkamp[12]. A 2.19 g (0.01 mol) of zinc acetate dihydrate was dissolved in 100 mL of boiling ethanol at atmospheric pressure and cooled down to the synthesis temperature. A white powder of anhydrous zinc acetate precipitated close to room temperature.[12] A 0.34 g (0.014 mol) sample of lithium hydroxide was dissolved in 100 mL of boiling ethanol and cooled to the synthesis temperature. Then, the lithium hydroxide solution was added dropwise to the zinc acetate solution under vigorous stirring. The reaction mixture became transparent after addition of about 1/3 volume of the lithium hydroxide solution. The ZnO colloid was stored at ≤ 10 °C to prevent rapid particle growth. Next, the ZnO suspension was washed by repeated flocculation of ZnO affected by addition of n-heptane. The supernatant was removed by centrifugation and decantation. For colloidal characterization the ZnO flocculates were redispersed in ethanol. In order to obtain the powder, the ZnO flocculates were dried under nitrogen for about 5 minutes.

ZnO nanorods were prepared according to a method described by Harnack.[18] Briefly, 29.5 g (0.13 mol) of zinc acetate dihydrate was dissolved in 125 mL of methanol at 65 °C. Then, a solution of 14.8 g (0.23 mol) of potassium hydroxide in 65 mL of methanol was added. The reaction mixture was stirred for several days at 65 °C. During this time, flocculation of the nanoparticles was observed. The length and the width of the resulting nanorods depended on the reaction time. The length of the nanorods increased considerably with longer reaction times, but the width of the nanorods grew only slightly. The ZnO floccs were washed with methanol. The supernatant was removed by decantation and the sample was dispersed in chloroform.

Particle size distribution, zeta potential and conductivity of ZnO nanoparticle suspensions were determined via dynamic light scattering (DLS) using a Malvern Nano ZS Instrument with a 633 nm "red" laser. For each sample ten measurements were performed and the average value per sample was calculated. The optical properties of the nanoparticles were determined from Uv-vis absorption spectra recorded using a Cary 100 Scan Spectrometer (Varian) with a 10mm path length cuvette. Structural analysis of the ZnO nanoparticles was performed using a D8 Advance (Bruker AXS) X-Ray diffractometer (XRD) using Cu-Kα radiation (0.154 nm). HRTEM images were obtained using a Philips CM 300 UltraTwin microscope with the particles deposited on a standard copper grid supported carbon film. The SEM images were obtained by using a Zeiss Gemini Ultra 55 SEM operating at an acceleration voltage of 20kV.

Results and discussion

Quasi-spherical ZnO particles were formed by precipitation in ethanolic solution. The zinc acetate and lithium hydroxide solutions in ethanol were mixed at different temperature under vigorous stirring. In order to follow the progress of the oxide formation, the suspensions of the nanoparticles were aged in their mother liquors and the crystal growth was monitored by the photospectroscopy and DLS.

The influence of nanocrystal size on the electronic structure of semiconducting material is represented by the band gap increasing with decreasing of the particle sizes, which is attributed to the so-called quantum confinement effect. ZnO shows this effect for particles smaller than 8 nm.[19] Hence, the measurement of absorption spectra provides a convenient way to investigate particle growth. Figure 1 shows the absorption spectra of the ZnO suspension recorded immediately after the preparation (about 3 minutes) and performed in 20 minutes intervals for about 14 hours. Just after mixing the zinc acetate with the lithium hydroxide the absorbance spectrum shows a well-defined exciton peak, which suggests a very fast nucleation. In addition, a marked red shift in the absorption edge was observed during the early formation stage, an indication of fast crystal growth. These absorption measurements could also provide a qualitative indication of the crystal size distribution. The sharp excitonic peak in the absorption spectra in the case of small nanocrystals (at shorter times) is indeed indicative of the narrow size distribution of the nanoparticles in the sample. For the larger particles, sharp excitionic features are not present in the absorption spectra (at longer times), but only a broad and featureless absorption edge. The latter is due to the fact that a number of exciton peaks appears at different energies corresponding to different sized nanocrystals which overlap with each other. Therefore, the nanocrystals are expected to have a broadened size distribution as the ageing time increased. This was confirmed by the number-weighted size distributions recorded by using DLS at different times during the aging of the particles (Figure 2).

The influence of the ageing temperature on the particle growth was also examined (data not shown). It was found that the particle size continued to evolve, even when stored at lower temperatures. The ability to obtain various particle sizes is based on the phenomena that the growth of the particles is governed by ageing temperature and time but also by the presence of the reaction byproduct lithium acetate during the ageing process. The reaction could be almost completely stopped by removing this byproduct by reversible flocculation of the ZnO nanoparticles by addition of n-heptane[12] and redispersing in ethanol or drying under nitrogen for about 5 minutes to get white powder. To verify this restriction on crystal growth, DLS, Uv-vis (not shown) and HRTEM (Figure 3) measurements were carried out on redispersed crystals flocculated at different temperatures and times of ageing.

Herein, we present also that the shape of the ZnO particles was sensitive to the overall concentration of the precursors and the solvent used. As we showed above at a zinc acetate dihydrate concentration in ethanol of below 0.01 mol quasi-sperical particles were formed, whereas mainly nanorods were formed at 10 times higher concentrations. The ZnO nanorods were prepared from zinc acetate dihydrate in methanolic solution hydrolyzed by potassium hydroxide. This colloidal solution was refluxed at 65 °C for different length of time. The formation of ZnO nanorods is based on oriented attachment of preformed ZnO quasi-spherical nanoparticles. The growth from individual particles by crystallographically oriented and partially fused dimers and oligomers to almost perfect rods was monitored by SEM (Figure 4). During the early stage of growth (up to 1.5 h), aggregated quasi-spherical particles are recognized. Longer refluxing led to a coexistence of spherical and already formed short rod-like aggregates (Figure 4c and 4d). HRTEM studies of particles at intermediate growth stages revealed aggregated dimers (Figure 5) where it is seen that the lattice planes of the constituent particles are almost perfectly aligned. Moreover, the lattice planes go straight through the contact area, that is, the particles are epitaxially fused together, the neck between the particles still being visible. This oriented aggregation was observed along the ZnO *c* axis (0002). In the final formed nanorods (Figure 4e) the necks between particles have been filled up since the faces parallel to the *c* axis are seen to be smooth. A possible explanation is that this occurs by conventional mechanism of dissolution and growth of primary particles, which might also result in an overall growth of the rods, especially after a long reflux time. Thus, Ostwald ripening assists the growth process by oriented attachment in order to yield finally high-quality nanorods.

Structural analysis of both of the ZnO samples, i.e. the nanospheres and nanorods, was performed by using X-Ray diffraction. In the XRD pattern of the ZnO nanoparticles presented in Figure 6 no impurity peaks are observed. All the diffraction peaks are well assigned to standard hexagonal phase ZnO with a wurtzite structure reported in JCPDS card (No. 36-1451, a = 3.249 Å, c = 5.206 Å). The broadening of the XRD peaks observed for the ZnO spheres is due to the small particle size.

Conclusions

We have shown that by varying different parameters of the wet synthesis processes like precursors and their concentrations, solvent used, temperature, and reaction time we were able to control the size and shape of the ZnO nanoparticles. The presented growth mechanisms of the spheres and rods offer excellent tools to design advanced materials with anisotropic properties.

References

1. Colvin, V. L.; Schlamp, M. C.; Alivisatos, A. P. *Nature* **1994**, *370*, 354.
2. Duran, P.; Capel, F.; Tartaj, J.; Moure, C. *Adv. Mater.* **2002**, *14*, 137.
3. (a) O'Regan, B.; Gra¨tzel, M. *Nature* **1991**, *353*, 737. (b) Quintana, M.; Edvinsson, T.; Hagfeldt, A.; Boschloo, G. *J. Phys. Chem. C* **2007**, *111*, 1035. (c) Zeng, L.; Dai, S.; Xu, W.; Wang, K. *Plasma Science & Technology* **2006**, *8*, 172. (d) Redmond, G.; Fitzmaurice, D.; Grätzel, M. *Chem. Mater.,* **1994**, *6*, 686.
4. (a) Yang, M.; Wang, D.; Peng, L.; Zhao, Q.; Lin, Y.; Wei, X. *Sens. Actuators B* **2006**, *117*, 80. (b) Rout, C. S.; Raju, A. R.; Govindaraj, A; Rao, C. N. R. *Solid State Commun.* **2006**, 136.
5. (a) Johnson, J. C.; Yan, H.; Schaller, R. D.; Haber, L. H.; Saykally, R. J.; Yang, P. *J. Phys. Chem. B* **2001**, *105*, 11387. (b) Zhang, X. T.; Liu, Y. C.; Zhang, L. G.; Zhang, J. Y.; Lu, Y. M.; Shen, D. Z., Xu, W.; Zhong, G. Z.; Fan, X. W.; Kong, X. G. *J. Appl. Phys.* **2002**, *92*, 3293.
6. Wang, Z. L. *J. Phys. Condens. Matter* **2004**, *16*, R829.
7. (a) Park, S. B.; Kang, Y. C. *J. Aeroslo Sci.* **1997**, *28*, S473. (b) Okuyama, K.; Lenggoro, I. W. *Chem. Eng. Sci.* **2003**, *58*, 537.
8. Sato, T.; Tanigaki, T.; Suzuki, H.; Saito, Y.; Kido, O.; Kimura, Y.; Kaito, C.; Takeda, A.; Keneko, S. *J. Cryst. Growth* **2003**, *255*, 313.
9. (a) Hu, X. L.; Zhu, Y. J.; Wang, S. W. *Mater. Chem. Phys.* **2004**, *88*, 421. (b) Hong, R. Y.; Shen, Z. H.; Li, H. Z. *J. Process Eng.* **2005**, *5*, 693.
10. (a) Zhang, H.; Yang, D.; Ji, Y.; Ma, X. Y.; Xu, J.; Que, D. L. *J. Phys. Chem. B* **2004**, *108*, 3955., (b) Liu, B.; Zeng, H. C. *J. Am. Chem. Soc.* **2003**, *125*, 4430.
11. Spanhel, L; Anderson, M. A. *J. Am. Chem. Soc.* **1991**, *113*, 2826.
12. Meulenkamp, E. A. *J. Phys. Chem. B* **1998**, *102*, 5566.
13. Wong, E. M.; Bonevich, J. E.; Searson, P. C. *J. Phys. Chem. B* **1998**, *102*, 7770.
14. (a) Pacholski, C.; Kornowski, A.; Weller, H. *Angew. Chem. Int. Ed.* **2002**, *41*, 1188. (b) Verges, M. A. Mifsud., A.; Serna, C. J. *J. Chem. Soc. Faraday Trans.* **1990**, *86*, 959.
15. Hu, Z.; Escamilla Ramirez, D. J.; Heredia Cervera, B. E.; Oskam, G.; Searson, P. C. *J. Phys. Chem. B* **2005**, *109*, 11209.
16. (a) Hu, Z.; Oskam, G.; Penn, R. L.; Pesika, N.; Searson, P. C.; *J. Phys. Chem. B.* **2003**, *107*, 3124. (b) Hu, Z.; Herrera Santos, J. F.; Oskam, G.; Searson, P. C.; *J. Colloid. Interface Sci.* **2005**, *288*, 313.
17. Hu, Z.; Oskam, G.; Searson, P. C. *J. Colloid Interface Sci.* **2003**, *263*, 454.
18. Harnack,O.; Pacholski, C.; Weller, H.; Yasuda, A.; Jurina M.; Wessels, J. M. *Nano Letters* **2003**, *3*, 1097.
19. (a) Koch, U.; Fojtik, A.; Weller, H.; Henglein, A. *Chem. Phys. Lett.* **1985**, *122*, 507., (b) Wong, E. M.; Bonevich, J. E.; Searson, P. C. *J. Phys. Chem. B* **1998**, *102*, 7770., (c) Brus, L. E. *J. Phys. Chem. B* **1986**, *90*, 2555.

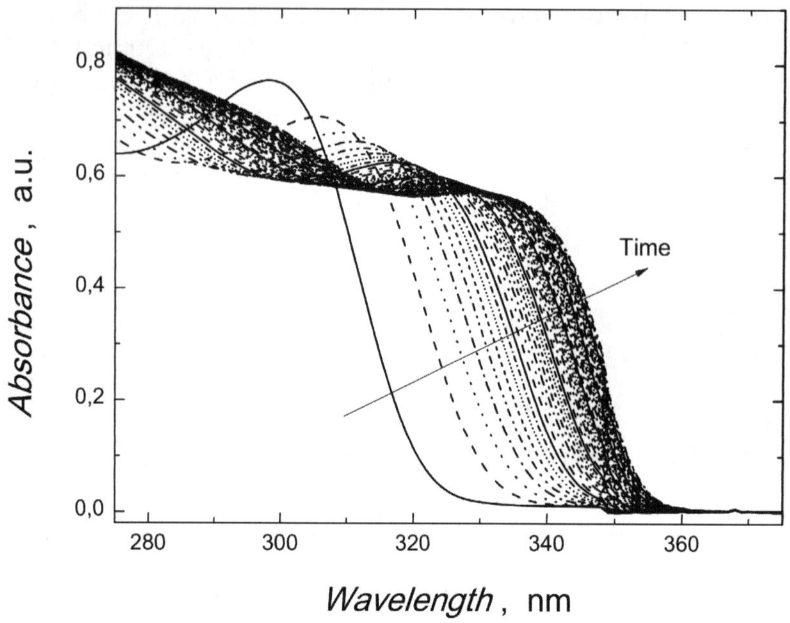

Figure 1. Room temperature absorption spectra recorded at different aging time of ZnO nanocrystals in ethanolic suspension. Red-shift in the absorption onset was observed between interval time 0 and 20 minutes.

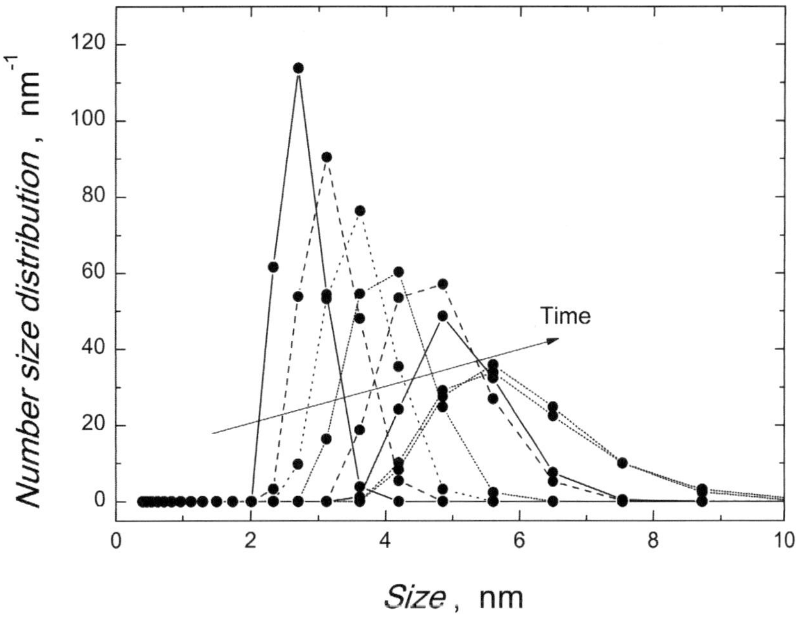

Figure 2. Number size distributions recorded by DLS at different aging time of the ZnO crystals in ethanolic suspension at 25 °C.

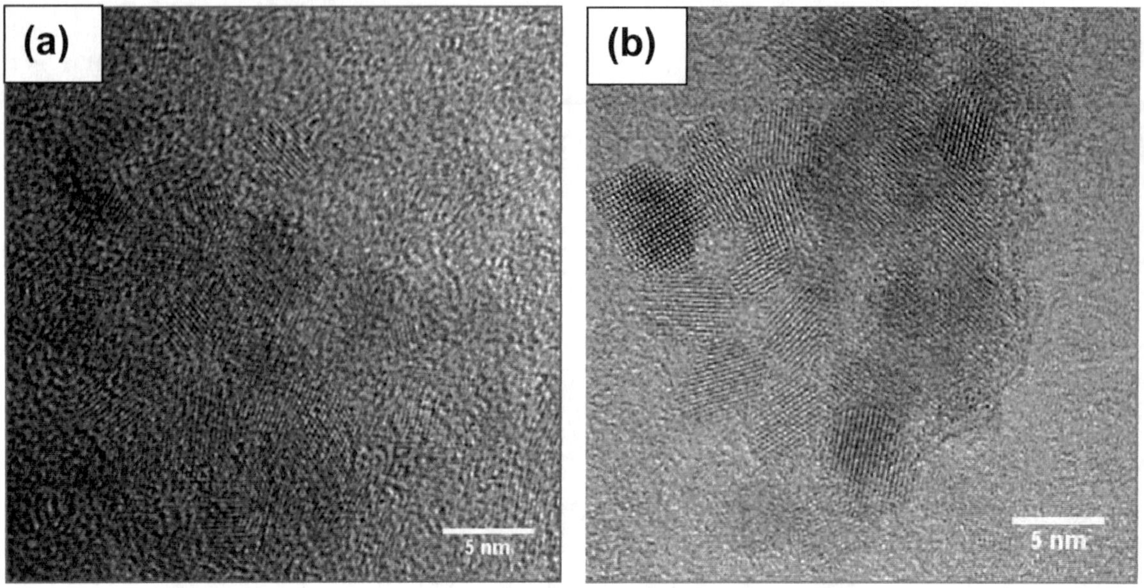

Figure 3. High resolution transmission electron microscopy images of ZnO nanospheres aged (a) for 4 h at 25 °C and (b) for 4 h at 35 °C. The crystalline nature of the ZnO nanoparticles is evidenced by the clearly observed lattice features.

Figure 4. Scanning electron microscopy images of ZnO nanoparticles taken at different refluxing times: (a) 5, (b) 90, (c) 150, (d) 300 and (e) 1860 minutes.

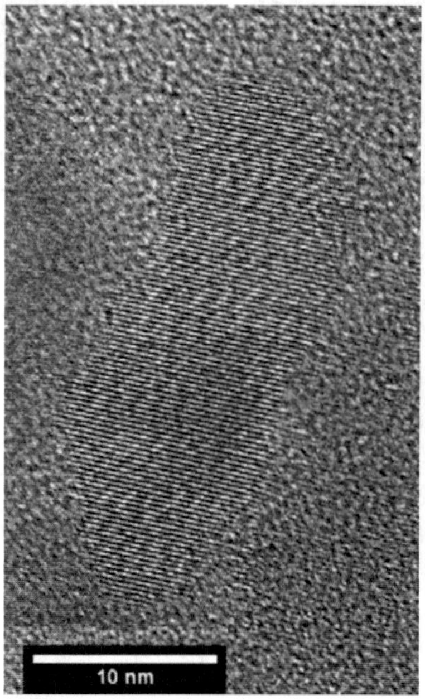

Figure 5. High resolution transmission electron microscopy image of the ZnO concentrated solution after reflux for 4 h showing dimers formation by oriented attachment. The crystalline nature of the ZnO nanoparticles is evidenced by the clearly observed lattice features.

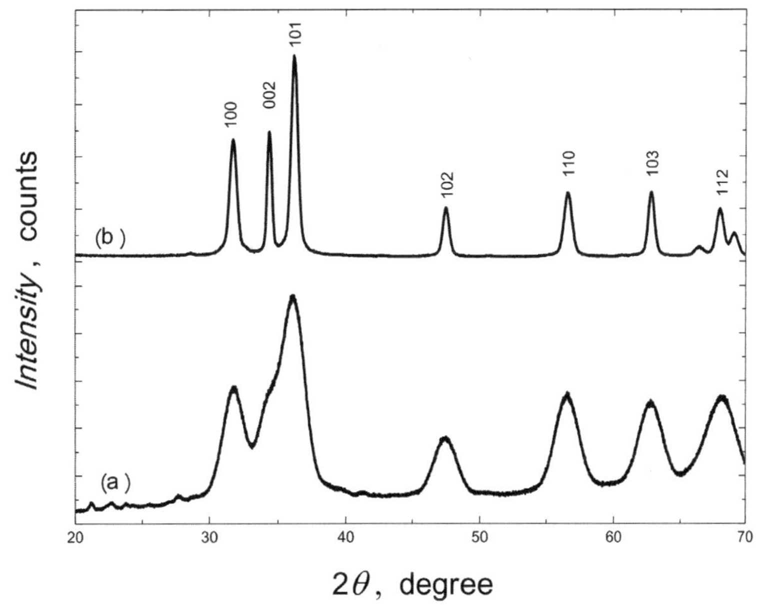

Figure 6. X-ray pattern of (a) ZnO nanospheres and (b) ZnO nanorods. All peaks correspond to ZnO wurtzite structure. The broadness of the XRD peaks reveals the nanocrystalline nature of the ZnO powders.

763b Multifunctional Particles: Magnetic Nanocrystals and Gold Nanorods Coated with Fluorescent Dye-Doped Silica Shells

Danielle K. Smith, Andrew T. Heitsch, and Brian Korgel. Department of Chemical Engineering, The University of Texas at Austin, 1 University Station C0400, Austin, TX 78712

Gold nanorods and magnetic nanocrystals are two materials that are being studied for their suitability as imaging contrast agents in biological systems. Magnetic nanocrystals can be used to enhance magnetic resonance imaging (MRI) contrast, for labeling of cancer cells, and also for intracellular labeling. The magnetic response of a nanocrystal can be tuned by varying the size of the nanocrystal as well as the composition.

Gold nanorods are being explored for biological and medical use as optical contrast agents for dark field and two-photon luminescence diagnostic imaging and photothermal therapy of cancer cells. They are attractive candidates for medical imaging because their optical response can be tuned to near-infrared wavelengths, which penetrate deep into cells and tissue; furthermore, they do not photobleach or blink, and are chemically inert and biologically compatible.

Gold nanorods have been produced in many laboratories by a colloidal seed-mediated, surfactant-assisted approach that has been widely studied and improved. Nonetheless, the reproducibility of the synthesis (i.e., the nanorod size, shape, and yield) has been a persistent challenge facing the technique. Many studies have addressed this issue, attributing differences in reproducibility to a wide variety of factors including seed aging time, the method of mixing the seed and growth solutions, variations in salt concentration and temperature of the growth solution, as well as nanorod growth time. We discovered that one significant factor in nanorod synthetic reproducibility is the supplier of the surfactant, hexadecylcetyltrimethylammonium bromide (CTAB), and that CTAB obtained from some suppliers does not yield gold nanorods and produces only spherical gold nanoparticles instead. It turns out that an impurity in the CTAB surfactant greatly affects nanorod formation.

Though as-synthesized gold nanorods are soluble in water, the bilayer of CTAB on their surface may render these particles harmful to living systems. As-synthesized magnetic nanocrystals are not soluble in water because they are capped with organic ligands. By coating both of these materials with silica, they are made water soluble and can thus be introduced into biological systems. Furthermore, when a fluorescent dye is embedded into the silica coating of the nanocrystals, an additional imaging modality is added. Thus, the final fluorescent dye doped silica coated nanocrystal is a heterostructure with dual imaging modality.

This presentation will reveal the importance of CTAB in the synthesis of gold nanorods and describe their subsequent coating with fluorescent dye doped silica. A similar coating strategy which has been applied to magnetic nanocrystals will also be discussed.

Phospholipid-assisted formation and dispersion of aqueous nano-C_{60}

*Yanjing Chen and Geoffrey D. Bothun**
Department of Chemical Engineering, University of Rhode Island, Kingston, RI, USA

Introduction

Water-soluble carbon fullerenes have been widely explored as candidates in the areas of medicine, materials and cosmetics [1, 2]. Despite of their potential application, their instability in aqueous solution has limited their application [3, 4]. Recent efforts have attempted to overcome this. For example, Deguchi stabilized nano-C_{60} by protein adsorption [5] and Govindaraj used cetyltrimethylammonium bromide (CTAB) to prepare nano-C_{60} dispersion Water-soluble nano-C_{60} is commonly prepared based on the method of Deguchi et al. [6]. This method is laborious, solvent intensive, and time consuming. In our study, DPPC vesicle assemblies were used to aid in the rapid formation of aqueous nano-C_{60}.

Dipalmityolphosphatidylcholine (DPPC, 734 MW) lipid dispersions form bilayer vesicles, or liposomes, spontaneously in aqueous solution because of their hydrophilic choline headgroups and their hydrophobic dual hexadecane (C_{16}) tails. DPPC is a main component of mammalian cell membranes and a primary lung surfactant. In addition to their use as drug delivery vehicles, DPPC liposomes have been used as nanoscale reactors and templating agents for synthesizing inorganic materials.

In the area of lipid-C_{60} complexes, Ikeda et al. [3] used a C_{60} fullerene exchange method to prepare lipid membrane-incorporated C_{60} fullerenes and Szymanska [7] utilized fullerene-modified supported lipid membranes as sensors. These reports suggest that C_{60} fullerenes are able to partition into liposome bilayers and enhance electron flow across the bilayer [8, 9]. However, the effect of C_{60} and nano-C_{60} aggregate on the stability of the bilayer, and how bilayer stability relates to the degree of dispersion, is unknown.

In this study, we examine the extent as which C_{60} nanoparticles can be adsorbed or entrapped in the DPPC liposome bilayers, and if DPPC liposomes can assist in the formation and dispersion of aqueous nano-C_{60}. In this study, DPPC/C_{60} vesicle assemblies were prepared by the reverse phase evaporation (REV) method [10]. We attempted to elucidate (i) the preference of molecular C_{60} to partition into the vesicle bilayer or form nano-C_{60} in solution, (ii) the formation of small nano-C_{60} in the bilayer due to aggregation, and (iii) the ability to back extract the lipids and isolate nano-C_{60} as a function of the DPPC/C_{60} molar ratio and presence of magnesium perchlorate ($Mg(ClO_4)_2$). Effects of nanoparticle-lipid interactions on bilayer phase behavior investigated were by differential scanning calorimetry (DSC). Structure and morphology of the liposomes and nano-C_{60} were imaged by cryogenic transmission electron microscopy (cryo-TEM). In addition to aiding the formation of small, potentially water-soluble nano-C_{60}, this work also provides insight into the ability of C_{60} to partition into biological membranes and its effect on membrane structure.

Experimental

A modified reverse phase evaporation (REV) method for liposome formation was used to prepare the DPPC/C_{60} assemblies. A set of samples were prepared at DPPC concentrations of 20 mM with DPPC/C_{60} molar ratios of 500:1, 200:1, and 100:1. DPPC dissolved in chloroform and C_{60} dissolved in carbon disulfide where placed in round-bottom flask. Deionized water was then added to form an emulsion. The solvents were removed rotary evaporation at 50°C (above the DPPC melting temperature of ~42°C) and 300 mbar for 15 minutes, at which point the pressure was reduced to 100 mbar for 10 minutes. The procedure is described in Figure 1, from A to C.

To observe the morphologies and sizes of DPPC/C_{60} assemblies, artifact-free images were taken by cryo-TEM. Samples were vitrified from room temperature in liquid ethane. Lipid bilayer phase behavior in the DPPC/C_{60} assemblies was analyzed by differential scanning calorimetry (DSC) where the samples were equilibrated at 25 and sequential heat/cool cycles were performed from 25-50°C at scan rate of 1°C min^{-1}.

The Folch Method was used to extract DPPC from the DPPC/C_{60} assemblies and recover the nano-C_{60}. We added a chloroform-methanol (2:1 by volume) mixture into C to extract the lipids. The bottom phase was composed of chloroform-methanol- water at 86:14:1 (by volume) and contained virtually all of the lipids, while the top phase was composed of chloroform-methanol-water at 3:48:47. Nano-C_{60} in the top phase (aqueous) was dried and imaged by TEM.

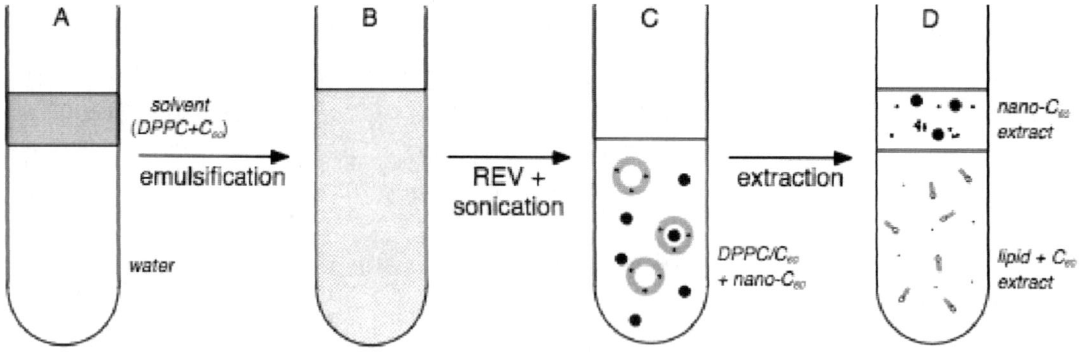

Figure 1. Schematic illustration for preparation of C_{60}/DPPC assembly and extraction of lipids from DPPC/C_{60} assemblies.

Results and Discussion

The size, morphology, and location of nano-C_{60}, relative to DPPC vesicles, that were formed during the REV process were observed by cryo-TEM. In Figures 2A-C, the C_{60} nanoparticles are observed in DPPC/C_{60} assemblies. From the observation, the size of C_{60} nanoparticles ranges from approximately 20 to 80 nm, which is similar to the average size of C_{60} nanoparticles reported previously (50 nm) in aqueous solution [8]. Moreover, the images show the location of nano-C_{60} that some C_{60} nanoparticles existed within liposomes, and others existed outside liposomes.

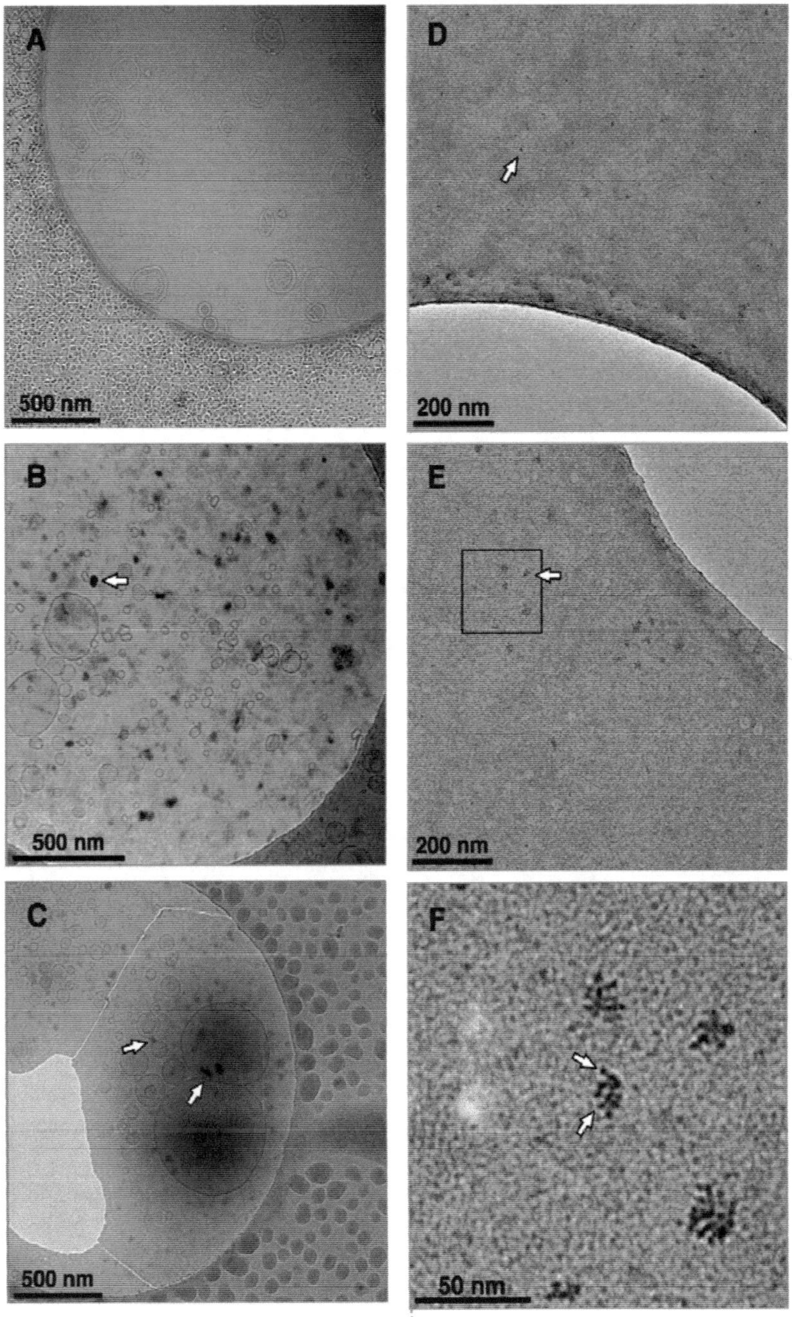

Figure 2. Cryo-TEM images of the DPPC vesicle control (A,no C_{60}) and samples at a DPPC/C_{60} ratio of 200:1 containing DPPC vesicles and nano-C_{60} (B and C). TEM images of the aqueous phase after extracting lipids from DPPC/C_{60} at a ratio of 200:1 (D-F; F is a magnification of the selected region in E).

Figure 2 D-F shows TEM images of C_{60} nanoparticles and nano-C_{60} aggregates that remained in the top aqueous phase after extracting the lipids. In Figure 2F, we observe large nano-C_{60} aggregates (ca. 20-40 nm) composed of small individual C_{60} nanoparticles (ca. 5 nm). The small C_{60} nanoparticles have diameters similar to that of the thickness of a lipid bilayer (4-5 nm), which suggest that these small nanoparticles may have formed within the bilayer.

Based on cryo-TEM, the presence of lipid assisted in the rapid formation of aqueous nano-C_{60} during liposome formation; however, C_{60} is hydrophobic and has also been shown to partition into the acyl region of lipid bilayers. This may occur as molecular C_{60} or, as suggested by back-extraction studies, small nano-C_{60}. To detect the presence of C_{60} within the bilayers, we have examined the effects of C_{60} on the lipid bilayer phase behavior using DSC. Figure 3 shows the DSC transition curves of the samples at different DPPC/C_{60} molar ratios. The presence of C_{60} broadened the temperature range of main phase transition (i.e. bilayer melting) and lowered the melting (T_m). Meanwhile, the pre-transition of DPPC/C_{60} liposome assemblies is eliminated. This suggests that nano-C_{60} can be incorporated in lipid bilayers.

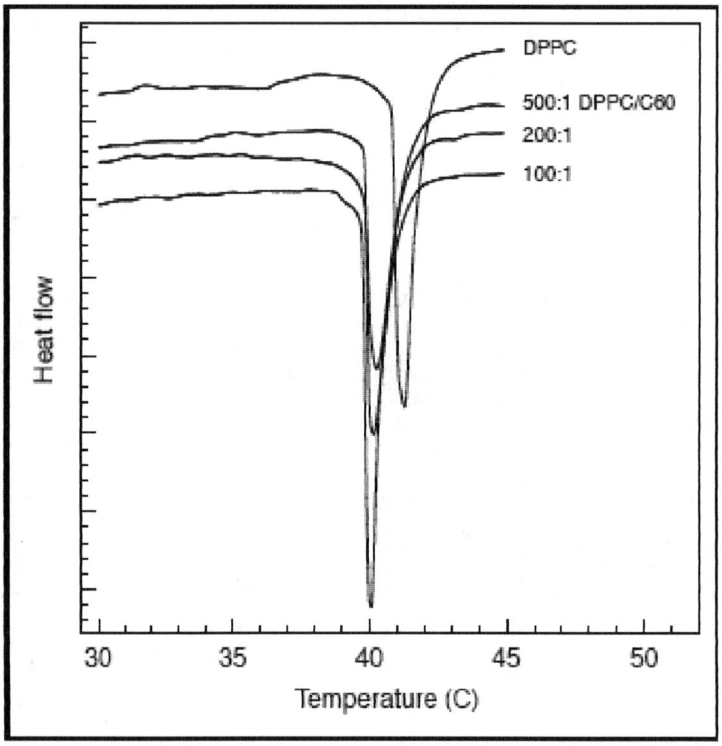

Figure 3. DSC transition curves of DPPC liposome and DPPC/C_{60} assemblies.

Conclusion

In this study, C_{60} nanoparticles were form in aqueous solution assisted with DPPC liposomes, which improved the stability of C_{60} and increased the rate of formation. From the cryo-TEM results, two kinds of C_{60} nanoparticles were observed: (i) large aqueous nano-C_{60} existing outside or encapsulated within the vesicles and (ii) small nano-C_{60} trapped within the bilayer. DSC results suggest that small nano-C_{60}, and likely molecular C_{60}, trapped within the bilayer disrupted the bilayer, decreasing the melting temperature and broadening the phase transition range. We propose that the DPPC/C_{60} assemblies provide a convenient means of rapidly forming nano-C_{60} in aqueous solution.

References

1. Innocenzi, P.; Brusatin, G., Fullerene-Based Organic-Inorganic Nanocomposites and Their Applications *Chemical Materials* **2001,** *13*, (10), 3126-3139.
2. Colvin, V. L., The potential environmental impact of engineered nanomaterials. *Nat Biotechnol* **2003,** *21*, (10), 1166-70.
3. Ikeda, A.; Sato, T.; Kitamura, K.; Nishiguchi, K.; Sasaki, Y.; Kikuchi, J.; Ogawa, T.; Yogo, K.; Takeya, T., Efficient photocleavage of DNA utilising water-soluble lipid membrane-incorporated [60]fullerenes prepared using a [60]fullerene exchange method. *Org Biomol Chem* **2005,** *3*, (16), 2907-2909.
4. Chen, K. L.; Elimelech, M., Aggregation and deposition kinetics of fullerene (C60) nanoparticles. *Langmuir* **2006,** *22*, (26), 10994-11001.
5. Deguchi, S.; Yamazaki, T.; Mukai, S.-a.; Usami, R.; Horikoshi, K., Stabilization of C60 Nanoparticles by Protein Adsorption and Its Implications for Toxicity Studies *Chem. Res. Toxicol* **2006,** *20*, (6), 854 -858.
6. Fortner, J. D.; Lyon, D. Y.; Sayes, C. M.; Boyd, A. M.; Falkner, J. C.; Hotze, E. M.; Alemany, L. B.; Tao, Y. J.; Guo, W.; Ausman, K. D.; Colvin, V. L.; Hughes, J. B., C60 in water: nanocrystal formation and microbial response. *Environ Sci Technol* **2005,** *39*, (11), 4307-16.
7. Szymanska, I.; Radecka, H.; Radecki, J.; Kikut-Ligaj, D., Fullerene modified supported lipid membrane as sensitive element of sensor for odorants. *Biosens Bioelectron* **2001,** *16*, (9-12), 911-5.
8. Sayes, C. M.; Gobin, A. M.; Ausman, K. D.; Mendez, J.; West, J. L.; Colvin, V. L., Nano-C60 cytotoxicity is due to lipid peroxidation. *Biomaterials* **2005,** *26*, (36), 7587-95.
9. Braun, M.; Hirsch, A., Fullerene derivatives in bilayer membranes: an overview *Carbon* **2000,** *38*, (11-12), 8.
10. Gregoriadis, G., Liposome Technology. *New York : Informa Healthcare* **2007,** *1*, 71-79.

763d Synthesis of Doped Zinc Selenide Quantum Dots Using Microemulsions as Templates

Tracy Heckler, Qi (Grace) Qiu, Jun Wang, and T. J. Mountziaris. Chemical Engineering, University of Massachusetts, Amherst, MA 01003

The doping of II-VI Quantum Dots (QDs) with transition metals is attracting significant attention because of the possibility of tuning their emission wavelength to longer wavelengths than the corresponding bulk gap emission and for the development of magneto-optical QDs for spintronic applications.

In this presentation we will discuss the synthesis of ZnSe:Mn and ZnSe:Cu QDs in a microemulsion template, their extraction from the template, surface modification, and stabilization in water. The synthesis route is based on the templated growth of ZnSe nanocrystals reported by Karanikolos et al. [1-3]. The template consists of a microemulsion having p-xylene as the continuous phase, water as the dispersed phase, and a poly(ethylene oxide)-poly(propylene oxide)-poly(ethylene oxide) (PEO-PPO-PEO) block copolymer as the surfactant. This template is very stable and has very slow droplet-droplet coalescence kinetics, thus eliminating particle aggregation and allowing precise control of particle size. The precursors for Zn, Mn, and Cu were acetate salts of these metals that were dissolved in the aqueous dispersed phase. The concentration of Zn acetate in the dispersed phase determines the final size of the QDs and the concentration of Mn or Cu acetate determines the dopant concentration in the nanocrystal. The Se precursor was hydrogen selenide gas diluted in hydrogen and was bubbled through the microemulsion.

The fluorescence spectra of the resulting QDs exhibit a blue-violet peak attributed to ZnSe gap emission and either a yellow-orange peak or a green peak attributed to Mn^{2+} or Cu^{2+}, respectively. The evolution of the intensity and ratio of the ZnSe:Mn peaks was studied as a function of dopant salt concentration in the precursor solution and time. The results indicate that ZnSe QDs are formed first and Mn slowly diffuses into the ZnSe lattice.

To make the doped ZnSe QDs useful as multi-color tags for biological imaging and sensing applications, a procedure was developed that allows QD removal from the microemulsion template, capping with functional hydrophilic ligands, and stabilization in an aqueous solution. The microemulsion-based synthesis technique is easy to scale up and employs less expensive chemicals compared to the common QD synthesis technique that involves hot injection of organometallic precursors in a coordinating solvent.

References

1. G. N. Karanikolos, P. Alexandridis, G. Itskos, A. Petrou, and T. J. Mountziaris, Langmuir 20, 550 (2004).

2. G.N. Karanikolos, P. Alexandridis, R. Mallory, A. Petrou, and T. J. Mountziaris, Nanotechnology, 16, 2372-2380 (2005)

3. G.N. Karanikolos, N.-L. Law, R. Mallory, A. Petrou, P. Alexandridis, and T. J. Mountziaris, Nanotechnology, 17, 3121-3128 (2006).

763e Zeolite Crystal Growth: A High-Resolution Cryo-TEM Study

Sandeep Kumar[1], R. Lee Penn[2], and Michael Tsapatsis[1]. (1) Chemical Engineering and Materials Science, University of Minnesota, 421 Washington Ave SE, Mail Box 111, Minneapolis, MN 55455-0132, (2) Chemistry, University of Minnesota, Smith Hall, 207 Pleasant St SE, Minneapolis, MN 55455

Investigating early stages of crystal growth in solution is of considerable fundamental and practical significance to various materials including zeolites. TPA-silicalite-1 (hereafter refer to as silicalite-1) is very often used as a model system. Here, we present a study of silicalite-1 crystallization by cryogenic transmission electron microscopy (cryo-TEM). Our results show silicalite-1 crystal growth by aggregation of structurally diverse nanoparticles. The present study also demonstrates that zeolites can be imaged by cryo-TEM with structural resolution.

Synthesis of Size Distribution Controllable Zeolite Nanocrystals *via* a Novel Confined Space Strategy

Yi Huang, Monash University, Melbourne, VIC, Australia
Dr. Anita J. Hill, Commonwealth Scientific and Industrial Research Organization Materials Science & Engineering, Melbourne, VIC, Australia
Dr. Huanting Wang (corresponding author), Monash University, Melbourne, VIC, Australia

Abstract

A novel confined-space synthesis method has been developed to synthesize zeolite nanocrystals. Colloidal silica nanoparticles serving as the silica source as well as the hard template were embedded in the mesoporous carbon *via* in situ polymerization of furfuryl alcohol (FA) in the presence of tri-block copolymer (Pluronic P123), and then reacted with an alkaline aqueous solution (Na_2O-Al_2O_3-H_2O) infiltrated through the mesoporous channels of the carbon matrix. The synthetic zeolite nanocrystals possess similar size distribution corresponding to that of colloidal silica nanoparticles. The synthesis of zeolite NaA nanocrystals is demonstrated. The amount of FA, aging time and hydrothermal time showed significant effects on the morphology and crystal size of zeolite NaA. N_2 sorption, X-ray Diffraction (XRD) and Scanning Electron Microscopy (SEM) were used to characterize the mesoporous carbon-silica composite and zeolite NaA nanocrystals. Dynamic Light Scattering (DSL) was used to measure the particle size and size distribution of zeolite nanocrystals as well as colloidal silica nanoparticles. Zeolite nanocrystals synthesized in this study are useful to fabricate zeolite-polymer nanocomposite membranes and the as-synthesized carbon-zeolite nanocomposite has high potential to serve as hierarchical nano-structured zeolite adsorbent for wastewater treatment.
Keywords: confined space synthesis; carbon-silica composite; NaA zeolite; nanocrystals

Introduction

Zeolite nanocrystals with high crystallinity have been attracted great interests because it is very important for various studies and applications, such as the fundamental understanding of the crystal growth process at the zeolite external surface, especially nucleation and growth processes zeolite [1-9] and the emerging nontraditional applications [10-11]. Moreover, small zeolite crystals are favorable in many catalytic reactions because of the reduced transport path and a substantial increase of the specific external area [12]. Therefore, to synthesize high-quality zeolite nanocrystals is of great interest. Different types of zeolite nanocrystals such as NaA [1], MFI [3] and FAU [8] have been already reported. For instance, Zeolite NaA nanocrystal, a highly versatile member of the LTA family, is one of the most important zeolites which have been extensively used in ion exchange, adsorption, catalysis and other aspects [12-15].

Many efforts have been made to synthesize zeolite with nanometer size and narrow size distribution. In most syntheses, organic templates, structure directing agent, seeding crystals and many other additives were often added to synthesize zeolite crystals with reduced sizes. Recently, confined-space synthesis has been developed for preparation of zeolite nanocrystals within an inert matrix. Many types of zeolite have been prepared using this procedure within carbon black matrixes, such as ZSM-5, Zeolite Beta, X and Zeolite A [16-17]. In this method, the hard media such as carbon black, and carbon nano-tubes were often used as space-confining porous media. The carbon black matrix provides not only limited space which is effective in confining zeolite growth but also preventing intergrowth of zeolite crystal during crystallization and aggregation of zeolite nanocrystals during calcination. However, careful investigation should be always drawn to identify the suitable porous system of inert support materials for the purpose of yielding zeolite nanocrystals with desirable particle size and size distribution.

In this work, a novel confined-space synthesis method that takes advantages of both space confined synthesis and a templating approach was developed to prepare zeolite NaA nanocrystals with tunable crystals size and controllable crystal size distribution. The synthesis of zeolite NaA nanocrystals is demonstrated by using colloidal silica particles (SNOWTEX ST-ZL: 70-100 nm) as the starting silica source as well as templates. Firstly, silica-carbon composite was prepared by carbonization of a gel prepared by mixing colloidal silica particle, FA and Pluronic P123. Secondly, colloidal silica particles wrapped silica-carbon composite were completely converted into zeolite NaA nanocrystals in a template-free alkaline alumina solution under hydrothermal treatment. The major advantages of this method are: (i) the absent of structure directing agent, seeding crystals and many other additives in model system and (ii) tunable crystals size and controllable crystal size distribution characteristics of as-synthesized zeolite.

Experimental Details

1. Confined space synthesis of NaA zeolite

The detailed procedures of our novel confined space synthesis can be found in the paper we have reported [18]. Typically, 2g of P123 $[(EO)_{20}(PO)_{70}(EO)_{20}]$ (*M*n ca. 5 800, Aldrich) was mixed with 5g Furfuryl alcohol (FA, 98%, Aldrich) until a uniform and homogeneous solution was obtained. Then 2.5 g of aqueous colloidal silica (SNOWTEX ST-ZL: 70-100 nm) was added. 1g of moderate concentrated acid solution (2 M, HCl) was added to complete the starting gel for the purpose of achieving an acidic circumstance and controlling the polymerization rate of FA [19]. The solvent evaporation and acid-catalyzed polymerization of FA were carried out in an uncapped polypropylene bottle at room temperature for 2 days and then at 70 °C in an oven for 2-3 days, leading to a black SiO_2-P123-PFA monolith with a mass ratio of 1 SiO_2: 2 P123: 5 PFA. After polymerization, a black solid was obtained. To prevent silica from dissolving out of mesoporous silica-carbon composite particles during the penetration of the alkaline solution, the SiO_2-P123-PFA monoliths were crushed into particles with a size range of 3-6 mm for surface coating. A silica free P123-PFA solution with the same mass ratio of the monolith was employed to produce the mesoporous carbon coating. After re-polymerization, the coated SiO_2-P123-PFA monoliths were calcined under flowing nitrogen gas in the furnace up to 500 °C at the heating rate of 1 °C /min and then kept for 5 h to develop the mesoporous silica-carbon composite. After calcination, 31.17 g of alkaline solution was ultimately mixed with the given amount of mesoporous silica-carbon composite to form the molar ratio of 5.85 Na_2O: 1.00 Al_2O_3: 182 H_2O: 2 SiO_2. The final mixture was sealed and then left it in the fume cupboard aging for 2-7 days after which the mixture was transferred into an 80 °C electrical oven for zeolite transformation. After hydrothermal treatment, carbon-zeolite composite was rinsed thoroughly with doubly deionized (DDI) water and dried at room temperature for around 2 days. Zeolite nanocrystals were retrieved by burning off mesoporous carbon in air at 550 °C for 6 h, a temperature that was reached at a heating rate of 1 °C /min.

2. *Characterization*

Powder X-ray diffraction (XRD, Philips PW1140/90 diffractometer using Cu Kα radiation) was used to determine sample phase and crystallinity. Data was collected at a scan rate of 1° /min from 5° to 45° with a step size of 0.02°. Nitrogen adsorption-desorption experiments were carried out at 77 K on a Micrometrics ASAP 2020 to determine the Brunauer-Emmett-Teller (BET) surface area. The micropore volume and pore size distribution were determined by t-pot method and BJH (Barrett-Joyner-Halenda) method, respectively. The total pore volume, V_{total}, was evaluated from the desorption branch of the isotherm at P/P_o =0.98, assuming complete pore saturation. SEM images were taken with a JEOL JSM-6300F scanning electron microscope operating at a voltage of 15 kV. Elemental analysis of the samples was conducted by an energy dispersive X-ray spectrometer (EDXS) attached to the JEOL JSM-6300F scanning electron microscope. Suspended particle size distributions of colloidal silica and synthesized Zeolite NaA nanocrystals were determined using a Malvern Instruments (Malvern, UK) Zetasizer Nano ZS series (ZEN 3600) at pH around 9 [18].

Results and discussion

To successfully synthesize zeolite nanocrystals with designed particle size and particle size distribution, it is very important to prepare SiO_2-P123-PFA monoliths with preferable mass ratio which would subsequently supply sufficient carbon confinement and expected mesoporous system in silica-carbon matrix. Therefore, four mass ratios including 1 SiO_2: 2 P123: 3 FA, 2 SiO_2: 2 P123: 3 FA, 1 SiO_2: 2 P123: 5 PFA, and 2 SiO_2: 2 P123: 5 FA were examined. However, uniform silica-P123-PFA composite was only prepared by adopting the mass ratio of 1 SiO_2: 2 P123: 5 FA [18]. The SiO_2-P123-PFA monolith and its calcined sample were denoted as SiO_2-5F and SiO_2-5C, respectively. SEM images show that SiO_2-5F and SiO_2-5C monoliths are quite uniform (Figure 1a, b). Silica particles are well separated in both samples by respective P123-PFA copolymers and mesoporous carbon walls.

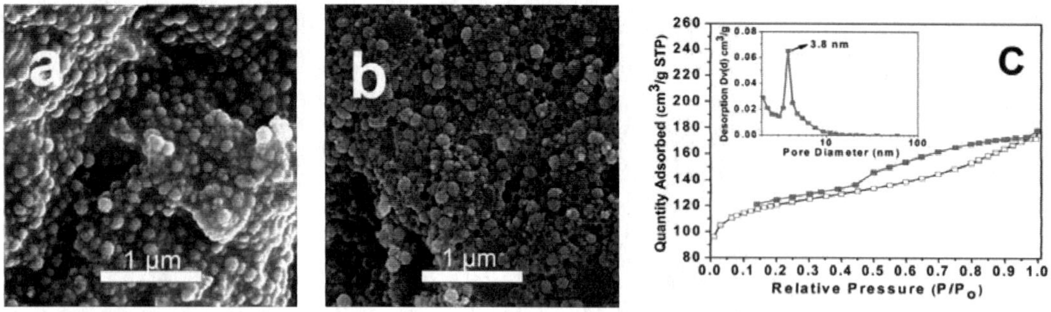

Figure 1. SEM images for (a) SiO_2-5F, (b) SiO_2-5C and (c) Nitrogen adsorption and desorption isotherm and pore size distribution (inset) of SiO_2-5C.

Figure 1 (c) shows the Nitrogen analysis for SiO_2-5C sample. The nitrogen sorption isotherm shows a hysteresis at relative pressures starting from approximately 0.45, indicating a mesopore structure. Despite a wide pore size distribution, the SiO_2-5C composite possessed relatively more uniform mesopores and a smaller mean mesopore size (3.8 nm) than other compositions [18]. SiO_2-5C exhibits a BET surface area of 225 m^2/g, and a pore volume of 0.30 cm^3/g.

SiO_2-5C monoliths were subsequently soaked in a designed amount of alkaline solution under decreased pressure for certain period. The lost weight was compensated by the addition of DDI water, after which the mixture was sealed for 2-7 days before subjection of hydrothermal treatment at 80 °C for a designed period. It is well known that longer aging time would lead to more pre-nuclei, and then favor formation of small particles. However, in our mesoporous carbon confined synthesis, crystal sizes become less sensitive to the aging and crystallization conditions, making nanocrystal growth more controllable. According to experimental investigation, when the aging time and crystallization time was fixed in the range of 4-7 days and 6-8 h respectively, highly crystallized zeolite nanocrystals with expected particle size and size distribution were finally synthesized [18].

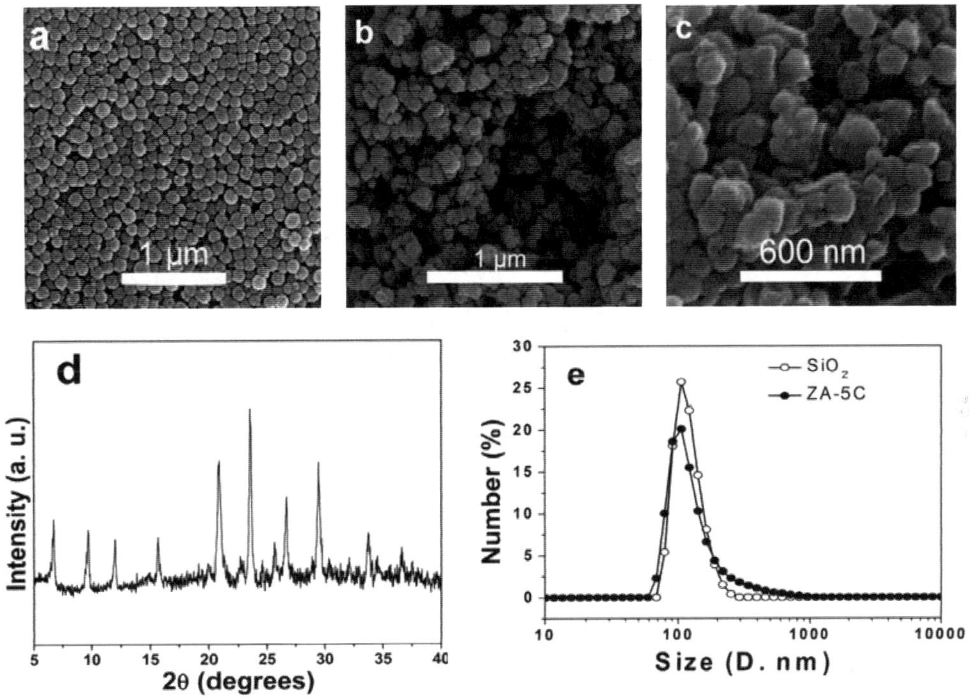

Figure 2. SEM images for (a) silica particles (SNOWTEX ST-ZL), (b) zeolite NaA (ZA-5C) and (c) high magnification of ZA-5C; (d) XRD pattern for ZA-5C and (e) DLS results for silica and ZA-5C. Note: ZA-5C was synthesized from SiO_2-5C under identical hydrothermal conditions (80 °C for 7h) after 5-day aging.

Figure 2a shows the spherical shape of commercial SNOTEX ST-ZL silica samples processing the particle size in the range of 70-240 nm. By comparison, the as-synthesized zeolite NaA nanocrystals (denoted as ZA-5C) exhibit similar morphology and particle size to those of starting silica particles (Fig. 2b). The high magnification SEM image (Fig. 2c) clearly reveals the round cube morphology and uniform particle size of ZA-5C. Their crystal sizes manually measured on the SEM image range from ca. 60 to ca. 250 nm. XRD pattern indicates that only pure zeolite NaA zeolite phase was crystallized (Fig. 2d). Energy-dispersive X-ray spectroscopy (EDXS) detected that the Si/Al ratio of the nano-zeolite bodies rapidly increased in the first six hours, from 1.00:0.58 at 2 h to 1.00:0.71 after 1 h to 1.00:1.01 after 2 h, indicating the rapid incorporation of aluminum ions to form the crystalline zeolite frameworks. The Si/Al molar ratio of final zeolite nanocrystals prepared via 7-hour hydrothermal treatment was detected to be maintained around 1.00:1.01 which is in good agreement with the standard composition of zeolite A. Although the limitations of EDXS in quantifying the aluminum content accurately, the trend of the change in aluminum content should be reliable. Moreover, the chemical analysis results revealing the fast crystallization of zeolite are consistent with the XRD studies.

Dynamic Light Scattering (DLS) studies of the size and size distribution of ZA-5C and starting silica (DDI H_2O diluted) were conducted using a Malvern Instruments (Malvern, UK) Zetasizer Nano

ZS series (ZEN 3600). The particle sizes of starting silica were detected in the range from 70-250 nm, and their mean particle size is around 106 nm (Fig. 2 open circle), which is larger than of 70 - 100 nm in the SNOWTEX ST-ZL specifications. This may be due to the slight agglomeration of silica particles. Zeolite NaA nanocrystals exhibit the mean particle size around 103 nm (Fig. 2 solid circle), which is slightly smaller than that of starting silica [18]. It also shows the size distributions of starting silica and most Zeolite NaA nanocrystals match perfectly. 94.3% quantities of Zeolite NaA nanocrystals possess similar size to that of starting silica (70-250 nm), and 80.1% quantities of which are within 150 nm. 5.7% of relative large particles are still present. However, from SEM investigation, the large particle should arise from zeolite nanocrystals aggregates.

In N_2 analysis, ZA-5C shows that the BET surface area is 24.9 m^2/g, which is much higher than normal micron-sized NaA prepared without porous carbon (~ 6 m^2/g). It is noted that the micropore volume is extremely small (<0.001 cm^3/g), and thus their BET surface area should arise from the external surfaces of zeolite crystals. This indicates that the zeolite NaA nanocrystals synthesized in the present work are well crystallized. Furthermore, the BET surface area of ZA-5C is fairly comparable to that of the starting colloidal silica nanoparticles (24.0 m^2/g), which is consistent with the particle size results obtained with SEM and DSL studies. As a result, the mass ratio of 1 SiO_2: 2 P123: 5 FA appears to be suitable for the formation of continuous mesoporous carbon networks for successful confined conversion of silica nanoparticles into zeolite nanocrystals [18].

Zeolite NaA nanocrystals were successfully crystallized in mesoporous carbon confinement using our novel method. The crystal growth is prohibited within the expected range of confinements. In our system, the synthetic conditions such as aging time and crystallization time show a less influence on the crystal size; while the crystallinity of zeolite nanocrystals can be enhanced by prolongating either of them. These results further support the space-confined growth mechanism and silica templating effect.

Conclusion

Zeolite nanocrystals with duplicated particle size distribution from starting colloidal silica were successfully synthesized without structure-directing agent, seeding crystals and other additives *via* our novel confined space strategy. XRD investigation indicated the capability of this method for the synthesis of pure highly crystalline zeolite nanocrystals. SEM coupled with DSL results doubly confirmed the similar particle size and size distribution of starting colloidal silica and as-synthesized zeolite nanocrystals. It should also be possible to synthesize other types of silica-containing zeolite nanocrystals, such as zeolites X/Y, ZSM-5 and sodalite by altering the composition of the alkaline solution (with or without SDA) and the reaction conditions. The crystal size and size distribution of as-synthesized zeolites could be readily controlled by simply varying those of starting silica particles.

Acknowledgement

This work was supported by the Australian National Research Flagships through Water for a Healthy Country Research Cluster and Australian Research Council. H.W. thanks the Australia Research Council for the QEII Fellowship. We thank Nissan Chemical Industries, Japan, for kindly providing colloidal silica.

References

[1] Mintova S, Olson NH, Valtchev V, Bein T. Mechanism of zeolite a nanocrystal growth from colloids at room temperature. Science. 1999 Feb;283(5404):958-60.

[2] Zhou Q, Qiu SL, Pang WQ. Studies on the crystallization mechanism of beta zeolite synthesized with nucleation gel. Chemical Journal of Chinese Universities-Chinese. 2000 Jan;21(1):1-4.

[3] Mintova S, Olson NH, Senker J, Bein T. Mechanism of the transformation of silica precursor solutions into Si-MFI zeolite. Angew Chem-Int Edit. 2002;41(14):2558-+.

[4] Murayama N, Yamamoto H, Shibata J. Mechanism of zeolite synthesis from coal fly ash by alkali hydrothermal reaction. International Journal of Mineral Processing. 2002 Feb;64(1):1-17.

[5] Gonthier S, Thompson RW. EFFECTS OF SEEDING ON ZEOLITE CRYSTALLIZATION, AND THE GROWTH-BEHAVIOR OF SEEDS. Advanced Zeolite Science and Applications. Amsterdam: Elsevier Science Publ B V 1994:43-73.

[6] Boudreau LC, Kuck JA, Tsapatsis M. Deposition of oriented zeolite A films: in situ and secondary growth. Journal of Membrane Science. 1999 Jan;152(1):41-59.

[7] Xu XC, Yang WS, Liu J, Lin LW. Synthesis and perfection evaluation of NaA zeolite membranc. Scparation and Purification Technology. 2001 Oct;25(1-3):475-85.

[8] Wang HT, Holmberg BA, Yan YS. Homogeneous polymer-zeolite nanocomposite membranes by incorporating dispersible template-removed zeolite nanocrystals. Journal of Materials Chemistry. 2002;12(12):3640-3.

[9] Huang AS, Lin YS, Yang WS. Synthesis and properties of A-type zeolite membranes by secondary growth method with vacuum seeding. Journal of Membrane Science. 2004 Dec;245(1-2):41-51.

[10] Mintova S, Mo SY, Bein T. Humidity sensing with ultrathin LTA-type molecular sieve films grown on piezoelectric devices. Chemistry of Materials. 2001 Mar;13(3):901-5.

[11] Wang ZB, Wang HT, Mitra A, Huang LM, Yan YS. Pure-silica zeolite low-k dielectric thin films. Adv Mater. 2001 May;13(10):746-9.

[12] Landau MV, Vradman L, Valtchev V, Lezervant J, Liubich E, Talianker M. Hydrocracking of heavy vacuum gas oil with a Pt/H-beta-Al2O3 catalyst: Effect of zeolite crystal size in the nanoscale range. Ind Eng Chem Res. 2003 Jun;42(12):2773-82.

[13] Robinson SM, Arnold WD, Byers CH. MASS-TRANSFER MECHANISMS FOR ZEOLITE ION-EXCHANGE IN WASTE-WATER TREATMENT. Aiche J. 1994 Dec;40(12):2045-54.

[14] Yahiro H, Lund A, Aasa R, Benetis NP, Shiotani M. Association forms of NO in sodium ion-exchanged A-type zeolite: Temperature-dependent Q-band EPR spectra. J Phys Chem A. 2000 Aug;104(34):7950-6.

[15] Chen XB, Yang WS, Liu J, Lin LW. Synthesis of zeolite NaA membranes with high permeance under microwave radiation on mesoporous-layer-modified macroporous substrates for gas separation. Journal of Membrane Science. 2005 Jun;255(1-2):201-11.

[16] Madsen C, Jacobsen CJH. Nanosized zeolite crystals - convenient control of crystal size distribution by confined space synthesis. Chemical Communications. 1999 Apr(8):673-4.

[17] Pham-Huu C, Wine G, Tessonnier JP, Ledoux MJ, Rigolet S, Marichal C. BETA zeolite nanowire synthesis under non-hydro thermal conditions using carbon nanotubes as template. Carbon. 2004;42(10):1941-6.

[18] Y. Huang et al., Micropor. Mesopor. Mater. (2008), doi:10.1016/j.micromeso.2008.07.036.

[19] Yao JF, Wang HT, Chan KY, Zhang LX, Xu NP. Incorporating organic polymer into silica walls: A novel strategy for synthesis of templated mesoporous silica with tunable pore structure. Microporous and Mesoporous Materials. 2005 Jul;82(1-2):183-9.

AUTHOR INDEX

Aarons, Lee R. 14
Abbas, Ali 47
Abu-Nahar, A. 157
Adschiri, Tadafumi 250
Akhtar, Kamal M. 97, 173, 267
Akhuemonkhan, Osifo 90
Akkisetty, Pavan Kumar 448
Aksomaitytė, Gabrielė 202
Allen, Todd Michael 587
Allendorf, Mark D. 232
Almeida-Rivera, Cristhian P. 752
Alonso, Esther 202
Amundsen, P. A. 492
Anand, Anshu 467
Anderson, Kelly 145
Angelov, Ivan 5
Annapragada, Sriram Kiran 627
Antonini, Alessandro 387
Antunes, Elisabete 483
Arastoopour, Hamid 479
Armenante, Piero M. 465
Aromaa, Mikko 89
Arrington, Yeook 33
Atashfaraz, Mehrnoosh 250
Athanassiou, Evagelos K. 539
Avalos, Josep Bonet 688
Ayala, Vanessa 349
Azurdia, Jose 88
Badiola, Carlo 517
Balani, Prashant N. 438
Bang, Youkyung 556
Barghi, Shahzad 430
Barrera, Carola 349
Bartel, Melissa 352
Bastos, Jaci C. S. C. 204
Bastos, Jaci Carlo Schramm Camara 191
Beach, Lauren 98, 99
Beaucage, Gregory 92, 248, 763
Belforte, Luca 387
Beloni, Ervin 569, 699
Benyahia, Sofiane 235
Bharadwaj, Rahul 188, 382
Bhatia, Surita R. 752
Bhatt, Divesh 736
Bhattacharya, Tathagata 134
Bhowmick, Tridib 388
Bichay, Magdy 587
Birkmire, Andrew 36
Boddu, Veera M. 737
Bongers, Peter 752
Bonnett, Peter 252
Borchert, Christian 3
Bothun, Geoffrey D. 775
Breault, Ronald W. 58
Brennan, John K. 688
Bucalá, Verónica 159
Buesser, Beat 170, 176, 254
Bulian, Chris J. 579
Cabrejos, Francisco J. 297
Cameron, I. T. 6
Caram, Hugo S. 470, 471
Caramellino, Micaela 465
Carpen, Ileana C. 233
Carroll, Nick 555
Castellanos, Antonio 229
Castner, David G. 96
Castro, Miguel 150
Chakka, Siri 737
Chalermsinsuwan, Benjapon 58
Chan, Sui Yung 438
Chandross, Michael E. 145
Chang, Jane P. 152

AUTHOR INDEX

Chaudhuri, Bodhisattwa 136
Chen, Alvin U. 269
Chen, Jiaqi 559
Chen, Julie 586
Chen, Yanjing 775
Chen, Yuhua 98, 99
Cheng, Chungyin 759
Cheng, Yi 559
Chester, Arthur W. 25
Cho, Dong-Lyun 171
Cho, Sung-Yong 172
Choi, Wonjoon 537
Chopra, Sachit 92, 248
Chowdhury, Snehaunshu 725
Chun, Byung Soon 100
Chung, Moon 567
Chung, Sang-Ho 175
Churchman, Charles 275
Cocco, Ray A. 33, 622
Cocero, Maria José 202, 624
Conway, Stephen L. 35
Cooke, Michael 124
Coronella, Charles J. 610, 623
Cotabarren, Ivana 159
Couzis, Alexander 268
Cranford, Bruce 643
Cui, Jingqin 353
Cui, Qingzhou 586
Cuitino, Alberto 55
Curtis, Jennifer S. 188, 316, 381, 382, 467
Custers, Johan P. A. 340
Da Rocha, Sandro R. P. 343
Da Rosa, Leonardo Machado 191
Dalvi, Sameer 626
Damavarapu, Reddy 737
Dang, Hongyi 148
Data, Biplab 492
Dave, Rajesh 98, 138, 229, 528, 562, 626
Davies, Clive E. 40, 306
Davis, Jeffrey M. 351
Davis, Zenda 230
De Broqueville, Axel 563, 611
De Wilde, Juray 83, 611
Decker, Rodrigo K. 204
Deng, Shuang 350
Dhodapkar, Shrikant 315
Dias, Daniel De Brito 218
Diemer, R. Bertrum 449
Ding, S. 26, 244, 492
Ding, Yulong 449
Domingues, Maria Cristina 443
Donahue, Carly M. 44
Doorenbos, Zac 529
Doyle, Francis J. 755
Drake, Joshua 270
Dreizin, Edward L. 516, 517, 528, 569, 594, 699
Du, Bing 56, 57
Du, Juan 560
Dubey, Atul 137
Duffadar, Ranojoy D. 351
Duvvuri, Subbarao 411, 499
Dycus, Eric 36
Dyrøy, Are 244
Eggers, Jan 1
Eggersdorfer, Max L. 176
Ehrl, Lyonel 7, 156
Ehrman, Sheryl 90, 388
Ekatpure, Rahul P. 563
Elsner, Martin Peter 5
Eom, Hyoung-Choon 247
Epstein, Norman 100
Ermoline, Alexandre 516

AUTHOR INDEX

Fan, L. -S. 76, 391, 615

Favier, John...379

Ferreira, Paulo 355, 483

Figueroa, Isabel...16

Fogler, H. Scott...46, 754

Follansbee, David M. 409

Ford, Kenneth J. 470, 471

Fox, R. O.................... 270, 354, 473, 474, 478, 481, 759

Francis, Todd M..561

Franka, Nathan P...75

Freireich, Ben..189, 383

Fricke, Michael..146

Fried, Laurence E. 736

Fritsching, Udo..204

Furlan, Marco 348

Fushimi, Chihiro 625

Gamble, Lara J. 96

Gandhi, Devkant S. 442

Garcia, Fernando A. P......................................355, 483

Garg, Rahul 474

Gauer, Cornelius 190, 558, 760, 762

Gellman, Andrew J. 352

Gentzler, Michael...35

George, Steven M. 253

Gevgilili, Halil...666, 667, 687

Ghadiri, Mojtaba 449

Gidaspow, Dimitri..58

Gilchrist, James F. 470, 471

Glasser, Benjamin J. 10, 135, 187, 234, 468

Glasser, David 187, 468

Glotzer, Sharon C. 384

Gnichwitz, Jan-Frederik 541

Gogos, Costas G. 251, 252

Gollamandala, Deepika R...233

Gorb, Leonid 747

Gorrepati, Elizabeth A. 46

Grace, John R. 100

Graeve, Olivia A. 344

Graf, Markus 345

Grass, Robert N. 539

Grest, Gary S. 145

Grinshpun, Sergey A. 594

Gu, Zhiyong 586

Guenther, Chris 285

Guldi, Dirk M. 541

Guney-Altay, Ozge 385

Guo, Bing 93

Gupta, Ram B. 144

Hadinoto-Ong, Kunn 316

Haines, Chris 529, 567

Hajra, Suman K. 24, 134

Hall, Talesha T. 482

Halvorsen, Britt M. 66

Han, Ting(Tim).................................... 56, 57, 153, 498

Hancock, Bruno C............................ 188, 381, 382, 467

Harris, Michael T. 482

Harshe, Yogesh M. 45

Hausberger, Brendon 187, 468

Hawa, Takumi 255, 689

Heckler, Tracy 780

Heindel, Theodore 75, 270

Heine, Martin C. 173, 267

Heitsch, Andrew T. 774

Hennart, Stephen.. 443

Henson, Michael A. 752, 753

Henthorn, Kimberly H. 469

Henz, Brian J. 255, 689

Herrera, Adriana P. 349

Heynderickx, Geraldine J. 563

Hildebrandt, Diane 187, 468

Hill, Anita J. 782

Hill, Boyd 664

AUTHOR INDEX

Hill, Priscilla J. 442
Hirlinger, John M. 587
Hirsch, Andreas 541
Hodapp, Maximilian J 178, 286
Horio, Masayuki 11
Howes, Tony 6
Hrenya, Christine M. 44, 473, 478
Huang, Bingyuan 88
Huang, Dale 174, 538
Huang, Yi 782
Igci, Yesim 476
Ikeda, Masahiro 625
Immanuel, Charles D. 755
Irizarry, Roberto 150, 151
Isayev, Olexandr 747
Ismail, Ahmed E. 145
Issangya, Allan 111, 566, 622
Jacob, Karl 315
Jallo, Laila 98, 99
Jamaleddine, Tarek 378
Jang, Hee Dong 246
Janka, Kauko 89
Jayarathna, Chameera K. 66
Jayarathna, Sanoja A. 66
Jiang, Xianjin 528, 569
Johnson, Samantha 266
Johnston, Jack L. 595
Jones, E. Nicholas 269
Jonnavittula, Divya 39
Kalasin, Surachate 351
Kalyon, Dilhan 390, 665, 666, 667, 687
Kang, Choon-Hyoung 171, 172
Kang, Dong-Yuk 171
Kapoor, Deepak 529, 567
Karimipour, Shayan 87
Karlsen, Morten 244
Karri, Reddy S. B. 111, 566
Karri, S. B. Reddy 33, 622
Kashyap, Mayank 58
Kasper, Gerhard 463
Kawajiri, Yoshiaki 5
Kemmere, Maartje F. 346
Kempkes, Michel 1
Keskinen, Jorma 89
Ketko, Marybeth H. 735
Ketterhagen, William R. 381, 467
Keurentjes, J. T. F. 340
Keurentjes, Jos T. F. 346
Khanna, Rajesh 499
Khinast, Johannes G. 10, 135, 234
Khumalo, Ngangezwe 187, 468
Kim, Bok-Min 172
Kim, Chul-Joo 245, 246, 247
Kim, Huiyong 556
Kim, Hyung Rae 391
Kim, Jong-Ho 155
Kim, Min 88
Kim, Sang Done 100
Kim, Seonmin 388
Kim, Seung-Jai 172
Kim, Sung-Don 245, 246
Kim, Young-Chul 155
King, David M. 253, 266
Kirchhof, Martin J. 463
Kitchens, Christopher L. 94
Klinzing, George E. 317, 326
Knieke, Catharina 450, 454
Knowlton, Ted 33, 111, 566, 622
Kodam, Madhusudhan 188, 382
Koenig, Jonathon 601
Kong, Jin-Hwa 155
Korgel, Brian 774

AUTHOR INDEX

Koynov, Athanas A. 55
Kreider, Peter...561
Kretzschmar, Ilona 353
Krinke, Thomas J. 89
Kruis, Einar...475
Krumeich, Frank 173, 267
Kumar, Sandeep..781
Kumar, Varun...759
Laine, Richard M. 88
Lall, Anshuman.............................95, 174, 228, 538
Lane, J. Matthew D...145
Langroudi, Mehrdad Kheiripour 466
Latorre, Magda 349
Lattuada, Marco 8, 45, 156, 348, 758, 762
Lee, Dae-Won...175
Lee, Dong Hyun..100
Lee, Kwang Soon...556
Lee, Kwan-Young 175
Lee, Yoong..100
Leon, Madeline...150
Lepek, Daniel..229
Leszczynski, Jerzy...747
Leu, Lii-Ping...249
Lewin, John L...736
Li, Fanxing...391, 615
Liang, Xinhua..253, 266
Lichty, Paul 561
Lim, Tau Yee...628
Lin, Candy S. 629
Lindberg, William...627
Lindsay, Mitchell 601
Liong, Yuh Inn 630
Lisal, Martin..688
Litster, James D. 6, 189, 383, 755
Littman, Howard 409
Liu, Sha..91
Lohrmann, Maike..342
Lorenz, Chris 145
Luss, Dan...568
Ma, X. 95, 174, 228, 511, 538
Maetani, Toshio 11
Mäkelä, Jyrki M. 89
Malik, Moinuddin 665
Malone, Michael F. 752, 753
Maloney, Stephen W. 737
Mann, Reginald 124
Mao, Yuanbing 152
Maqbool, Tabish..754
Marashdeh, Qussai..76
Marczak, Renata 541, 764
Marin, Guy B. 563
Marini, Fabio 286
Marjamäki, Marko...89
Martignoni, Waldir Pedro 191, 218
Martin, Darold..529, 567
Martin, Lealon L. 409
Martínez, Ángel Martín 624
Martirosyan, Karen S. 568
Mazumder, Jahirul I. 419
Mazzotti, Marco 1
McCann, Ryan...448
McCarthy, J. J. 16, 24, 134
Meesters, G. M. H..6, 443
Merrill, Chris..601
Meterc, Darja..624
Metzger, Matthew J. 187, 468
Meuldijk, Jan..346
Milani, Paolo..387
Min, Byoung-Jun..172
Min, Jian 270
Monastiriotis, Spyros 268
Moon, Dong-Ju 155

AUTHOR INDEX

Morbidelli, Massimo 7, 8, 45, 156, 190, 348, 558, 758, 760, 762
Mori, Milton 178, 191, 204, 218, 286
Morris, Kenneth 448
Mountziaris, T. J. 780
Mudde, Robert F. 125
Muhle, Michael E. 354
Mujumdar, Ajit 11, 98, 99, 138
Mujumdar, Kaustubh S. 9
Mukasyan, Alexander S. 527
Mulholland, G. M. 95, 228
Muzzio, Fernando J. 34, 137
Nagendra, Basavaraju Agasanapur 610
Nair, Sachin 230
Nakagawa, Kenshiro 44
Nakamura, Naomichi 11
Nere, Nandkishor 3, 448
Ng, Ka M. 629
Ng, Wai Kiong 438, 628
Nirschl, Hermann 761
Ocone, R. 472
Oetama, Ratna 388
Ogunsola, Oluwatosin 388
Olivo, Carlos 762
Olson, Jon H. 449
Olusanmi, Dolapo 449
Ostrowski, Peter P. 587
Otsuji, Yuichiro 625
Paap, Scott M. 462
Paccione, John 409
Pahk, Jae Bum 317
Palmer, Glenn 601
Panau, Freddie 630
Park, Ah-Hyung Alissa 100
Park, Hyungkyu 245, 247
Park, In-Seon 175
Park, Nam-Cook 155
Park, Young-Moo 175
Passalacqua, Alberto 473, 474
Patel, Rajen 567, 667
Pati, Ranjan 90
Penn, R. Lee 781
Pestov, Dmitry 385
Petsev, Dimiter N. 555
Peukert, Wolfgang 450, 454, 541, 764
Pfeffer, Robert 110, 229, 389
Piekiel, N. 511, 725
Piña, Juliana 159
Pingali, Kalyana 34
Pira, Nello Li 387
Piseri, Paolo 387
Piumsomboon, Pornpote 58
Plawsky, Joel L. 409
Pomfret, Michael B. 90
Potoff, Jeffrey J. 735
Prasad, D. V. N 380
Pratsinis, Sotiris E. 97, 170, 173, 176, 254, 267, 345, 761
Prickett, Suzanne 666
Procopio, Adam 35
Prud'Homme, Robert K. 759
Prulliere, Frederic 36
Pugsley, Todd S. 87
Puszynski, Jan 529, 579, 587
Qasim, Mo 747
Qiu, Qi (Grace) 780
Quevedo, Jose A. 110
Quintanilla, Miguel Angel Sanchez 229
Radeke, Charles 25, 39
Radl, Stefan 135
Raha, Sasanka 754
Rai, A. 511
Rai, Neeraj 736
Raikar, Neha B. 752, 753

AUTHOR INDEX

Rajesh, Dave ... 99
Rajniak, Pavol ... 453
Ramachandra, Srikanth Bellur ... 610, 623
Ramachandran, Rohit ... 755
Raman, Venkatramanan ... 481
Ramkrishna, Doraiswami ... 3, 448
Rasteiro, Maria Da Graca ... 355, 483
Ray, Ajay K. ... 419
Ray, Madhumita ... 378
Razzak, Shaikh ... 430
Redner, Paul ... 567, 667
Reklaitis, Gintaras V ... 448
Remy, Brenda ... 10, 234
Rho, Seon-Gyun ... 172
Rinaldi, Carlos ... 349
Roberts, Christopher ... 149
Rodgers, Thomas L. ... 124
Rokkam, Ram G. ... 354
Romagnoli, Jose A. ... 47
Romanski, Frank ... 557
Rosenberg, Jason D. ... 753
Rossit, José ... 159
Rostedt, Antti ... 89
Rothenbacher, Sonja ... 463
Roy, Papiya ... 499
Saasen, A. ... 492
Sadoff, Hillary ... 90
Sahni, Ekneet ... 136
Samavedham, L ... 37
Samokhvalov, Alexander ... 230
Samyudia, Yudi ... 630
Sanganwar, Ganesh ... 144
Santore, Maria M. ... 351
Sarkar, Avik ... 143, 169
Saunders, Steven R. ... 149
Scheffe, Jonathan R. ... 232
Schmid, Hans-Joachim ... 463
Schoenitz, Mirko ... 516, 517, 528, 594
Scicolone, James V. ... 138, 562
Segets, Doris ... 764
Seidel-Morgenstern, Andreas ... 5
Seipenbusch, Martin ... 463
Shariaty-Niassar, Mojtaba ... 250
Sharma, M. P. ... 627
Shen, Zheng ... 101
Shende, Rajesh ... 529
Shi, Deliang ... 24
Shin, Jae-Soon ... 155
Shukla, Nisha ... 352
Siegel, Nathan P. ... 232
Siepmann, J. Ilja ... 736
Silva, Maria Das Graças E. ... 286
Silva, Maria G. E. ... 178
Simons, Jo M. M. ... 346
Singh, Chetana ... 384
Singh, Dhananjay ... 626
Singh, Jagdeep ... 92, 248
Sinha, Kaustav ... 344
Smith, Benjamin ... 36
Smith, Danielle K. ... 774
Smith, Marty J. ... 35
Snider, Dale ... 366
Sohn, Jung-Soo ... 245
Sokkalingam, Nandhini ... 735
Solt, Paul ... 294
Soos, Miroslav ... 7, 156, 190, 558, 760
Spicka, Peter ... 453
Sridhar, Deepak ... 391, 615
Srinivasan, Rajagopalan ... 37
Stamatis, Demitrios ... 569, 594
Stark, Wendelin J. ... 539
Stepanek, Frantisek ... 755

AUTHOR INDEX

Stephenson, David 124
Storti, Giuseppe 762
Strano, Michael S. 537
Strumendo, Matteo 479
Subramaniam, Shankar 474
Sun, Jin 14
Sundaresan, Sankaran 14, 476
Sundmacher, K. 146
Sundmacher, Kai 3
Sung, Yonduck "james" 481
Swiatkiewicz, Jacek J 579
Swihart, Mark T. 91, 148
Tallon, Steve J. 306
Tan, Reginald B. H. 438, 628
Tang, Hansong 666, 687
Tardos, Gabriel I. 466
Tatarchuk, Bruce 230
Taylor, Robin Klupp 764
Techane, Sirnegeda D. 96
Teleki, Alexandra 97, 173, 267, 761
Tepper, Gary 385
Tester, Jefferson W. 462
Thakur, Ranjit 342
Theuerkauf, Joerg 9, 56, 57, 153, 315, 380, 498
Threlfall-Holmes, P. 472
Tomassone, M. Silvina 39, 557
Tomassone, M. Silvlna 25
Tomassone, Maria. S. 137
Tong, Andrew 615
Torres-Lugo, Madeline 349
Tricoli, Antonio 345
Troxel, Tom G. 297
Trunov, Mikhaylo A. 516, 528
Tsapatsis, Michael 781
Tsutsumi, Atsushi 625
Tungapindi, Navina D. 469
Tüzün, U. 157
Valverde, Jose Manuel 229
Van Bokhoven, Jeroen 92, 248
Van Den Broeke, Leo J. P. 340
Van Der Hoeven, M. J. 6
Van Nispen, Sjoerd F. G. M. 340
Vargas, Watson L. 24
Vasilenko, Alisa 34
Vasquez, Nestor A. 56, 57, 315
Vasquez, Victor R. 623
Vats, Alok 529
Vedula, Prakash 473
Venkatasubramanian, Venkat 448
Ventura, Carla A. F. 355
Vetter, Thomas 1
Vicent, Arol 568
Vinati, Simone 387
Viswanathan, Harish 478
Voigt, Andreas 3, 146
Voigt, Michael 764
Von White, Gregory 94
Walker, Robert A. 90
Walter, Kim T. 36
Wang, Dawei 101, 401
Wang, Fei 76, 391, 615
Wang, H. G. 112
Wang, Huanting 782
Wang, Jun 780
Wang, Leizheng 568
Wasito, W. 76
Wassgren, Carl R. 143, 169, 188, 189, 381, 382, 383, 467
Weber, Alfred P. 463
Webster, Emile S. 40, 306
Wegner, Karsten 387
Wei, J. 475
Weidner, Eckhard 624

AUTHOR INDEX

Weimer, Alan W. 232, 253, 266, 561

Weng, Peng 251, 252

Wengeler, Robert 761

Werner, Fabian 541

White, Jeremiah D. E. 527

Wibowo, Christianto 629

Widenski, David 47

Wildeboer, W. J. 6, 443

Winter, Jessica O. 350

Wolfe, Michael S. 462

Wong, Vivian S. Y. 438

Wu, Hua 190, 558, 758, 760

Wu, Libo 343

Wu, Wenjun 66

Xiao, Guoxian 26

Xu, Delong 26

Yang, W. Q. 112

Yau, Robert 136

Ydstie, B. Erik 560

Yim, Hoon 93

Yoon, Ho-Sung 245, 246, 247

York, Trevor A. 124

You, Jun 401

Young, Ming-Wan 251, 252

Yu, Yung-Sz 249

Zaccone, Alessio 758

Zachariah, Michael 95, 174, 228, 255, 511, 538, 689, 725

Zelinskaya, Alexandra 44

Zeng, Liang 391

Zhang, Hongwang 91

Zhou, Lei 95, 228, 511, 725

Zhou, Ying 37

Zhu, Chao 101, 401

Zhu, Jesse 419, 430

Zhu, Linjie 251, 252

Zhu, Wusheng 557

Zhu, Xiaoying 517

Ziomek, Grzegorz 5

American Institute of Chemical Engineers
3 Park Avenue
New York, New York 10016-5991

CURRAN ASSOCIATES INC
proceedings.com